Encyclopedia of
HUMAN
GEOGRAPHY

Encyclopedia of
HUMAN GEOGRAPHY

Edited by
Barney Warf

Florida State University

A SAGE Reference Publication

SAGE Publications
Thousand Oaks ■ London ■ New Delhi

For information:

 SAGE Publications, Inc.
2455 Teller Road
Thousand Oaks, California 91320
E-mail: order@sagepub.com

SAGE Publications Ltd.
1 Oliver's Yard
55 City Road
London EC1Y 1SP
United Kingdom

SAGE Publications India Pvt. Ltd.
B-42, Panchsheel Enclave
Post Box 4109
New Delhi 110 017 India

Printed in the United States of America.

Library of Congress Cataloging-in-Publication Data

Encyclopedia of human geography / editor, Barney Warf.
    p. cm.
"A SAGE reference publication."
Includes bibliographical references and index.
ISBN 0-7619-8858-0 (alk. paper : cloth)
    1. Human geography—Encyclopedias. I. Warf, Barney, 1956-
GF4.E54 2006
304.203—dc22   2005036239

This book is printed on acid-free paper.

06   07   08   09   10   11   9   8   7   6   5   4   3   2   1

Publisher:	Rolf Janke
Acquisitions Editor:	Robert Rojek
Developmental Editor:	Yvette Pollastrini
Reference Systems Coordinator:	Leticia Gutierrez
Project Editor:	Tracy Alpern
Copy Editor:	D. J. Peck
Typesetter:	C&M Digitals (P) Ltd.
Indexer:	David Luljak

Contents

List of Entries

Reader's Guide

Human geography is such a broad field of study that it is nearly impossible to categorize the multitude of topics it covers. This list is designed to assist readers in finding articles on related topics. Headwords are organized into six major categories: Economic Geography, Urban Geography, Political Geography, Social/Cultural Geography, Geographic Theory and History, and Cartography/Geographic Information Systems. Note, however, that many topics defy easy categorization and belong to more than one grouping.

CARTOGRAPHY/GEOGRAPHIC INFORMATION SYSTEMS

Agent-Based Modeling
Automated Geography
Cartogram
Cartography
Cellular Automata
Computational Models of Space
Digital Earth
Ecological Fallacy
Fractal
Geodemographics
Geoslavery
GIS
GPS
Humanistic GIScience
Information Ecology
Limits of Computation
Location-Based Services
Multicriteria Analysis
Neural Computing
Ontology
Overlay
Social Informatics
Spatial Autocorrelation
Spatial Dependence
Spatial Heterogeneity
Spatially Integrated Social Science
Tessellation
Time, Representation of
Uncertainty

ECONOMIC GEOGRAPHY

Agglomeration Economies
Agriculture, Industrialized
Agriculture, Preindustrial
Agro-Food System
Applied Geography
Capital
Carrying Capacity
Cartels
Census
Census Tracts
Circuits of Capital
Class
Class War
Colonialism
Commodity
Comparative Advantage
Competitive Advantage
Conservation
Consumption, Geography and
Core–Periphery Models
Crisis
Debt and Debt Crisis

GEOGRAPHIC THEORY AND HISTORY

Subject and Subjectivity
Theory
Tobler's First Law of Geography

POLITICAL GEOGRAPHY

Anticolonialism
Boundaries
Bureaucracy
Civil Society
Communism
Critical Geopolitics
Decolonization
Democracy
Electoral Geography
Environmental Determinism
Environmental Justice
Geopolitics
Gerrymandering
Hegemony
Imperialism
Institutions
Justice, Geography of
Law, Geography of
Local State
Nationalism
Nation-State
Political Ecology
Political Geography
Power
Redistricting
Resistance
Socialism
Social Movement
Sovereignty
State
World Systems Theory

SOCIAL/CULTURAL GEOGRAPHY

AIDS
Animals
Art, Geography and
Behavioral Geography
Body, Geography of
Children, Geography of
Communications, Geography of
Crime, Geography of
Critical Human Geography
Cultural Ecology

Cultural Geography
Cultural Landscape
Cultural Turn
Culture
Culture Hearth
Cyberspace
Demographic Transition
Diaspora
Diffusion
Disability, Geography of
Domestic Sphere
Emotions, Geography and
Empiricism
Enlightenment, The
Environmental Perception
Epistemology
Ethics, Geography and
Ethnicity
Femininity
Feminisms
Feminist Geographies
Feminist Methodologies
Fertility Rates
Fieldwork
Film, Geography and
Food, Geography of
Gays, Geography and/of
Gender and Geography
Geography Education
Health and Healthcare, Geography of
Heterosexism
Historic Preservation
Historical Geography
Home
Homophobia
Hunger and Famine, Geography of
Identity, Geography and
Languages, Geography of
Lesbians, Geography of/and
Literature, Geography and
Malthusianism
Masculinities
Medical Geography
Mental Maps
Migration
Mobility
Modernity
Mortality Rates
Music and Sound, Geography of
Natural Growth Rate

URBAN GEOGRAPHY

Editorial Board

Contributors

Paul Adams
University of Texas

Stuart Aitken
San Diego State University

Derek Alderman
East Carolina University

Luc Anselin
University of Illinois

Trevor Barnes
University of British Columbia

Rob Bartram
University of Sheffield

Dean Beck
University of Illinois, Urbana–Champaign

Robert Bednarz
Texas A&M University

F. L. Bein
Indiana University/Purdue University at Indianapolis

Lawrence Berg
University of British Columbia

William Beyers
University of Washington

Nicholas Blomley
Simon Fraser University

Barry Boots
Wilfrid Laurier University

Scott Bridwell
University of Utah

Kath Browne
University of Brighton

Brian Ceh
Indiana State University

Thomas Chapman
Florida State University

Jianer Chen
Texas A&M University

Paul Cloke
University of Bristol

Michael Conzen
University of Chicago

Meghan Cope
State University of New York at Buffalo

Susan Craddock
University of Minnesota

Jeremy Crampton
Georgia State University

Altha Cravey
University of North Carolina

Tim Cresswell
University of Wales, Aberystwyth

Jeff Crump
University of Minnesota

Nicholas Dahmann
University of Chicago

Christopher Dalbom
Louisiana State University

Bruce D'Arcus
Miami University

Vincent Del Casino
California State University, Long Beach

Dydia DeLyser
Louisiana State University

Michaela Denny
Florida State University

Lary Dilsaver
University of South Alabama

Teresa Dirsuweit
University of the Witwatersrand

Deborah Dixon
University of Wales, Aberystwyth

Jerome Dobson
University of Kansas

Rebecca Dolhinow
California State University, Fullerton

James Duncan
University of Cambridge

James Eflin
Ball State University

Glen Elder
University of Vermont

Colin Flint
University of Illinois

Jay Gatrell
Indiana State University

Wil Gesler
University of North Carolina at Chapel Hill

Mary Gilmartin
University College Dublin

Jim Glassman
University of British Columbia

Pat Gober
Arizona State University

Michael Goodchild
University of California, Santa Barbara

Jon Goss
University of Hawaii

William Graves
University of North Carolina at Charlotte

Richard Greene
Northern Illinois University

John Grimes
Eastern Kentucky University

Jeanne Guelke
University of Waterloo

Daniel Hammel
Illinois State University

Stephen Hanna
University of Mary Washington

Holly Hapke
East Carolina University

Francis Harvey
University of Minnesota

Maureen Hays-Mitchell
Colgate University

Michael Heiman
Dickinson College

Andrew Herod
University of Georgia

Ken Hillis
University of North Carolina at Chapel Hill

Steve Hoelscher
University of Texas

Briavel Holcomb
Rutgers University

Gail Hollander
Florida International University

Louise Holt
University of Brighton

Mark Horner
Florida State University

Ed Jackiewicz
California State University, Northridge

Daniel Jacobson
University of Calgary

Donald Janelle
University of California, Santa Barbara

Nuala Johnson
Queen's University, Belfast

Lynda Johnston
University of Waikato

Ronald Kalafsky
University of North Carolina at Charlotte

David Kaplan
Kent State University

Paul Kingsbury
Simon Fraser University

Andrew Klein
Texas A&M University

Dan Klooster
Florida State University

Larry Knopp
University of Minnesota–Duluth

Olaf Kuhlke
University of Minnesota–Duluth

Richard Kujawa
St. Michael's College

Jonathan Leib
Florida State University

Jonathan Lepofsky
University of North Carolina at Chapel Hill

Paul Longley
University College London

Susan Mains
University of the West Indies–Mona

Jo Margaret Mano
State University of New York at New Paltz

Sallie Marston
University of Arizona

Tom Martinson
Auburn University

Robert Mason
Temple University

Kent Mathewson
Louisiana State University

Jon May
University of London

Kendra McSweeney
The Ohio State University

Christopher Merrett
Western Illinois University

Peter Meserve
Fresno City College

Don Mitchell
Syracuse University

Karen Morin
Bucknell University

Alison Mountz
Syracuse University

Thomas Mueller
California University of Pennsylvania

Peter Muller
University of Miami

Beverley Mullings
Syracuse University

Garth Myers
University of Kansas

David Nemeth
University of Toledo

Elizabeth Oglesby
University of Arizona

Shannon O'Lear
University of Kansas

Kathleen O'Reilly
University of Illinois

David O'Sullivan
University of Auckland

Ruth Panelli
University of Otago

Thomas Paradis
Northern Arizona University

Hester Parr
University of Dundee

Robert Pennock
Florida State University

Donna Peuquet
Pennsylvania State University

Scott Pike
Texas A&M University

Gabriel Popescu
Florida State University

Jeff Popke
East Carolina University

Deborah Popper
College of Staten Island/City University of New York

Marcus Power
University of Durham

Valerie Preston
York University

Darren Purcell
University of Oklahoma

Neil Reid
University of Toledo

George Rengert
Temple University

Susan Roberts
University of Kentucky

Clayton Rosati
Syracuse University

Robert Ross
Syracuse University

Grant Saff
Hofstra University

Joseph Scarpaci
Virginia Tech

Andrew Schoolmaster
Eastern Kentucky University

Joan Schwartz
Queen's University

Anna Secor
University of Kentucky

Joanne Sharp
University of Glasgow

Fred Shelley
University of Oklahoma

Betty Smith
Eastern Illinois University

Jonathan Smith
Texas A&M University

Michael Solem
Association of American Geographers

Kristin Stewart
Florida State University

Roger Stough
George Mason University

Christa Stutz
Mesa College

Daniel Sui
Texas A&M University

Emily Talen
University of Illinois, Urbana–Champaign

Jean-Claude Thill
University at Buffalo, The State University of New York

Mary Thomas
Ohio State University

Waldo Tobler
University of California, Santa Barbara

Paul Torrens
University of Utah

Carlos Tovares
California State University, Northridge

James Tyner
Kent State University

Michael Urban
University of Missouri

Robert Vanderbeck
University of Leeds

Richard Van Deusen
Syracuse University

Peter Vincent
Lancaster University

Andy Walter
State University of West Georgia

Barney Warf
Florida State University

Gerald Webster
University of Alabama

Elizabeth Wentz
Arizona State University

David Wilson
University of Illinois

Charles Withers
University of Edinburgh

David Wong
George Mason University

John Wylie
University of Sheffield

Emily Yeh
University of Colorado

Junbo Yu
Tsinghua University

May Yuan
University of Oklahoma

Jingxiong Zhang
Wuhan University

About the Editor

Barney Warf is Professor and Chair of Geography at Florida State University in Tallahassee. His research and teaching interests lie within the broad domain of human geography, particularly social, economic, and urban issues. He has studied New York as a global city, telecommunications and electronic capital markets, offshore banking in Panama, information networks in the Dominican Republic, international networks of legal and engineering services, mergers in the telecommunications industry, the geographies of cyberspace, military spending, the lumber industry, the political economy of ports, Indonesia, and Cleveland, among other things. He has coauthored or coedited 5 books, 25 book chapters, and 80 articles in journals, and he has won teaching and research awards.

Introduction

Human geography over the past decade has undergone a conceptual and methodological renaissance that has transformed it into the most dynamic and innovative of the social sciences. Geography, especially human geography, long suffered from a negative popular reputation (particularly in the United States) as a trivial discipline with little analytical substance, a view that centers on the "capes and bays" approach. That misconception has been decisively annihilated by the intellectual advances of the past four decades. The *Encyclopedia of Human Geography* offers a comprehensive overview of the major ideas, concepts, terms, and approaches that characterize a notoriously diverse field. No single volume can hope to capture the breadth and variety to be found in a discipline, but this book aspires to encapsulate at least the most important highlights of human geography at this moment in time. The reader will find a variety of themes characterizing different schools of thought and subject areas in this volume. The emphasis throughout has been on topics and ideas, and this focus has required the omission of other possible entries. For example, there are no biographical summaries of well-known geographers.

Human geography—the study of how societies construct places, how humans use the surface of the earth, how social phenomena are distributed spatially, and how we bring space into consciousness—has matured along multiple fronts. Starting as early as the 1950s, many geographers turned to mathematical models of spatial phenomena, developing increasingly complex understandings of, for example, the spatial structures of urban areas, transportation systems, and public services. These approaches, although now less prominent, made great contributions to the study of spatial diffusion, networks, and industrial location. Philosophically, this approach elevated the abstract over the concrete—the general over the specific—and reduced geography to geometry. Its rigorous methodology reduced the role for armchair speculation and was useful in uncovering regularities in the landscape. The so-called positivist school of geography has been challenged and supplemented by various other philosophies and approaches, but the growth over the past two decades of geographic information systems (GIS) has given this way of looking at space new popular appeal. The explosion of GIS has had wide-reaching and generally highly beneficial consequences for human geography, providing new means to model and simulate spatial phenomena with an unprecedented degree of analytical sophistication. The presence of GIS, both as a tool and as a language, has energized human geography in ways that were unthinkable just a generation ago. Although this encyclopedia addresses several topics of significance to positivism (e.g., the gravity model, location theory), its focus leans more heavily toward more contemporary approaches.

Several postpositivist perspectives have contributed significantly to the diversity of human geography today. Marxists injected into the field a concern with class and power along with a far richer understanding of production and the spatial division of labor, uneven development, and the need to historicize our understanding of space (i.e., embed geographies in their temporal contexts). Marxism illustrated that geography cannot be understood independently of social structures—of how resources are organized and opportunities and constraints are produced differentially for and challenged by different groups—and raised the ethical obligation to confront inequality and injustice. Similarly, feminists brought to the field the notion that social and spatial life always is gendered and that gender permeates social relations, crosscutting class and ethnicity in complex ways and shaping the daily lives and access to resources of men and women in a manner that often perpetuates, but occasionally challenges, patriarchy. An emerging line of thought concerns the spatiality of sexuality, introducing views drawn from queer theory to study sexual minorities. More recently, many geographers have turned to the spatial analysis of race and ethnicity, revealing that race and racial inequality are far from biologically given "natural" categories; rather, they are social products of domination and subordination that play out unevenly over space and time. Humanistic geographers, drawing on the rich tradition of phenomenology and existentialism, emphasized what it means to be human, the constitutive role of language in shaping human consciousness, the intangible dimensions of place as repositories of meaning, and the ways in which landscape and identity constitute one another, in the process "humanizing" social and spatial structures and processes by revealing the active role played by people in everyday life.

Moving beyond the usual definitions of culture as the sum total of learned behavior or a "way of life," many human geographers have effectively overcome the long-standing "micro–macro" division in the social sciences. Because culture is acquired through the process of socialization, individuals never live in a social vacuum; rather, they are socially produced from cradle to grave. Social and spatial structures consist of the rules and resources that people draw on in their daily lives and that in turn structure their actions. Thus, time and space are reproduced through the very same structures that enable people to carry out their daily existence. The socialization of the individual and the reproduction of society and place are two sides of the same coin. People reproduce the world, largely unintentionally, in their everyday lives, and in turn the world reproduces them through socialization. In forming their biographies every day, people recreate and transform their social worlds primarily without meaning to do so; individuals are both produced by and producers of history and geography. Hence, everyday thought and behavior do not simply mirror the world; they constitute it. This way of looking at human geography emphasized the contingent, open-ended nature of landscapes and the active role of people as agents, and it softened the blunt edges of earlier structuralist theories.

Recently, many of the dualities that long characterized social science–nature versus society, the individual versus the social, the historical versus the geographic, and consumption versus production have broken down in the face of postmodern and poststructuralist approaches. Postmodernism, a term that has suffered from its popularity, emphasized the complexity of the world, the difficulty or impossibility of finding absolute truth, the deep linkages between knowledge and power, and the ways in which some ways of truth making cover up, ignore, or annihilate other perspectives. This trend forced a reevaluation, among other things, of the nature of the human subject; whereas classical theories portrayed human identities as stable and consistent, postmodernism holds them to be constantly in flux as individuals move among different categories of meaning. Geographically, identities are both space forming and space formed, that is, inextricably intertwined with geographies in complex and contingent ways. Space affects not only what we see in the world but also how we see it. Likewise, the human body has become an inspirational topic for human geographers, particularly the multiple ways in which identity, subjectivity, the body, and place are sutured together. Although bodies appear as "natural," they are in fact social constructions deeply inscribed with multiple meanings and "embodiments" of class, gender, ethnic, and other relations.

Human geographers often are fascinated by the question of how space is encoded and brought into consciousness through language. In a poststructuralist light, every representation is a simplification filled with silences, for the world is inherently more complex than our language allows us to admit. Representations of space—whether maps, stories, diagrams, or narratives—always are social products with social origins, even if they become taken for granted as "natural" or "objective." Moreover, it is widely accepted that spatial representations always are linked to power; that is, they serve someone's interest and never are neutral or value free. Representations of space inevitably have social consequences (albeit not always intended ones), and geographic knowledge is less an objective mirror of the world than a contested battleground of views linked to different social interests. Discourses are socially produced sets of representations that simultaneously enable and constrain our understanding of the world. In short, geographic representations are part of the reality they help to construct; word making is also world making. That is, discourses do not simply mirror the world; they constitute it. This line of thought led to a "cultural turn" in economic geography, demonstrating that culture as a complex contingent set of relations is every bit as important as "economic" factors in the structuring of economic landscapes.

The growth of culturalist explanations and the concern for the social nature of representation also infiltrated into the study of GIS. An earlier literature denaturalized maps, revealing them to be far from objective views of space but rather partial, inevitably biased discourses that represent the world in some ways and not others, naturalizing what they portray by obscuring social origins and processes. Geographic information systems, for all of their technological sophistication, long labored under the assumption that they too were, or at least could be, atheoretical, objective representations of the world. Human geography, however, has engaged in a mutually beneficial dialogue with practitioners and theoreticians of GIS, a dialogue that has pointed to GIS as a culturally laden discourse that selectively filters the ways in which the world is portrayed and analyzed. Thus, the process of pixelizing the social has been complemented by a parallel process of "socializing the pixel."

The explosion of the Internet has unleashed, perhaps predictably, analyses of the geography of cyberspace. Electronic communications have contributed to a massive worldwide round of time–space compression that reconfigured social relations and the rhythms of everyday life. Human geographers have charted the multiple impacts of this universe, including the growth of cybercommunities and their associated virtual selves, the "digital divide" that separates information haves and have-nots globally and locally, the growth of e-commerce, digital pedagogy, and the political uses of the Internet. In studying cyberspace, most human geographers jettisoned the technological determinism that holds that telecommunications simply affects space in favor of views that emphasize the *coevolution* of communications and space. The Internet is a social product that is interwoven with relations of class, race, and gender and inescapably subject to the uses and misuses of power. In an age when ever broader domains of everyday life are increasingly mediated electronically, this literature has moved beyond simplistic dichotomies such as on-line and off-line to suggest the ways in which the real and the virtual are shot through with one another. Moreover, far from signaling the "death of distance," cyberspace itself is deeply structured geographically, with multiple topologies at different spatial scales.

Globalization, the latest chapter in the expansion of capitalism, has rapidly increased the scope, volume, and velocity of international linkages, and as a result geographers have produced an ocean of literature on topics such as transnational capital, international trade, global commodity chains, global cities, international financial and telecommunications systems, and how the global economy is reshaping geopolitics and governance. By revealing how the global and the local are shot through with one another, or "glocalized," this literature has contributed mightily to more nuanced understandings of how globalization is manifested differently in different places, challenging simplistic views that globalization inevitably leads to homogeneous landscapes and the eradication of local differences.

In several disciplines, including human geography, postcolonialism has turned the study of globalization back in time, noting that the European colonial conquest of the world was as much a cultural and ideological project as an economic and political one. Thus, colonialism took many forms, including the pervasive Eurocentrism of Western social science that portrayed the West as the dynamic active motor of history and the rest of the world as passive recipients. This view has been increasingly challenged, in part by human geographers. Geography as a way of knowing space—the active "geo-graphing" of various parts of the globe—was part and parcel of the Western control of colonized regions, naturalizing Western dominance and non-Western inferiority. Postcolonial geographers confront the discursive and ideological presumption that non-Western societies were not every bit as much intellectually vibrant and original as the West and that non-Western ways of knowing have been marginalized through the power relations of colonialism. Indeed, the very dichotomy between the West and the "Rest" has itself been undermined in favor of an emphasis on hybridity.

One of the healthiest products of human geography's sustained intercourse with social theory includes a widespread "denaturalization" of many phenomena once assumed to lie outside the domain of human control. As topic after topic has fallen sway to social constructivism, including gender, time and space, poverty, and the body, it is not surprising that the discipline recently has exhibited a renewed appreciation of how social relations are intertwined with the physical environment. Some human geographers have argued for the social construction of nature, a perspective that refutes long-standing assumptions that nature lies "outside" of human affairs. By enfolding nature within social relations and discourse, the biophysical environment is depicted as shaped, molded, and even created through human action. In jettisoning the artificial dualism between "humans" and "nature," the discipline has worked to overcome the long-standing schism between human geography and physical geography through the use of perspectives such as political ecology and the social production of nature.

All of these changes, schools, and perspectives have made human geography both considerably more complicated and much richer. Long a borrower of ideas from other disciplines, geography has become a contributor in its own right, and a "spatial turn" is evident in disciplines as diverse as sociology, anthropology, and literary criticism. The editors hope that users of this encyclopedia will appreciate the diversity and sophistication of contemporary human geography and in turn use its themes and concepts for their own purposes. Those who would like to pursue these topics further will find "Suggested Reading" samples at the end of each entry. For broad overviews of the topic, see the entries at the end of this Introduction.

Finally, I thank the numerous people who were so generous with their time in this project. The associate editors—Dydia DeLyser, Dan Sui, Larry Knopp, David Wilson, and Altha Cravey—worked long and hard to secure great entries from good authors. This project and I owe them an enormous debt of gratitude. The authors and contributors themselves—all 157 of them—contributed a wonderful series of entries on a bewildering array of topics; I have learned more from them than they will ever know. Sage's Leticia Gutierrez, Tracy Alpern, Yvette Pollastrini, and D. J. Peck were enormously helpful throughout the editorial and production process. Any mistakes are my own. And of course, I am constantly thankful for my wife Santa Arias and my son Derek for their love, energy, humor, and support.

—Barney Warf

Suggested Reading

Anderson, K., Domosh, M., Pile, S., & Thrift, N. (Eds.). (2003). *Handbook of cultural geography.* London: Sage.

Cloke, P., Philo, C., & Sadler, D. (1991). *Approaching human geography.* New York: Guilford.

Holt-Jensen, A. (2003). *Geography: History and concepts* (3rd ed.). London: Sage.

Hubbard, P., Kitchin, R., Bartley, B., & Fuller, D. (2002). *Thinking geographically: Space, theory, and contemporary human geography.* London: Continuum.

Hubbard, P., Kitchin, R., & Valentine, G. (Eds.). (2004). *Key thinkers on space and place.* London: Sage.

Johnston, R., Gregory, D., Pratt, G., & Watts, M. (Eds.). (2001). *Dictionary of human geography* (4th ed.). Oxford, UK: Blackwell.

Johnston, R., & Sidaway, J. (2004). *Geography and geographers: Anglo-American human geography since 1945* (6th ed.). London: Edward Arnold.

Low, M., Cox, K., & Robinson, J. (Eds.). (2003). *Handbook of political geography.* London: Sage.

Peet, R. (1998). *Modern geographical thought.* Oxford, UK: Blackwell.

Sheppard, E., & Barnes, T. (Eds.). (2000). *A companion to economic geography.* Oxford, UK: Blackwell.

ACQUIRED IMMUNODEFICIENCY SYNDROME

SEE AIDS

AGENT-BASED MODELING

Agent-based modeling (ABM) is a technique used to build computer simulations. ABM allows for the creation of synthetic, but ultimately realistic, artificial geographic worlds in which events, phenomena, processes, and scenarios can be created and studied flexibly. ABM is an important tool in human geography employed in evaluating hypotheses and ideas that might not be easily experimented with, evaluating "what if" scenarios that cannot be tested otherwise, or relating to future conditions that cannot be sampled.

ABM is a part of a growing geographic methodology based on geocomputation and geosimulation. Both approaches mark a departure from traditional models focused on exchange of human geographic units between coarsely represented divisions of space. Newer models based on ABM are more likely to be built as simulations with massive amounts of intelligent geographic entities, each represented at the atomic scale, connected and interacting dynamically in space as complex adaptive systems.

Agent-based models belong to a family of models called automata. Automata have distinguished origins in pioneering work on digital computing during the 1930s and 1940s. Automata tools were first employed in geography as cellular models during the early 1970s, with the methodology evolving toward ABM during the 1990s. An automaton is a simple information processor just like the processors in digital watches and computers. Automata have some key properties that render them useful for model building. They have states that allow attributes to be encoded to them, changed, and stored. Automata have some representation of time that catches state conditions at discrete temporal points. They also contain transition rules that govern changes between states as time progresses. Rules are formulated as (computational or mathematical) functions that accept state information input from other automata, and this can be derived from neighboring automata within a specified local neighborhood of influence, as is characteristic with cellular automata.

Agent automata extend this basic framework, adding attributes borrowed from research on behavior and artificial intelligence. These attributes are very relevant for work in human geography. Agents are heterogeneous, contrasting with more traditional models that treat entities as "average individuals." Agents are also proactive and may act to realize a goal or set of goals. They may have perception—the ability to sense other agents and environments—often based on an internal cognitive model. Importantly, agent interaction may take many forms: communication, active and intentional querying of other agents, human–environment effects, and so on. Agents are also adaptive and may change their rules of behavior based on experience within a simulation.

Agent tools are used in a variety of applications in human geography: pedestrian and crowd motion, vehicular traffic, residential mobility, gentrification, land use and land cover change, urban growth and sprawl, spatial epidemiology, civil violence, sociospatial segregation, and economic geography.

During recent years, research in this area has focused on applying agent-based models to new phenomena in human geography, and a growing integration between ABM and geographic information systems (GIS) and geographic information science (GIScience). In particular, authors have begun to develop geography-specific methodologies and toolkits based on ABM but with geography as a central building block.

—*Paul Torrens*

See also GIS; Social Informatics

Suggested Reading

Batty, M. (2005). *Cities and complexity: Understanding cities with cellular automata, agent-based models, and fractals.* Cambridge: MIT Press.

Benenson, I., & Torrens, P. (2004). *Geosimulation: Automata-based modeling of urban phenomena.* London: Wiley.

Torrens, P., & Benenson, I. (2005). Geographic automata systems. *International Journal of Geographic Information Science, 19,* 385–412.

AGGLOMERATION ECONOMIES

By clustering in close proximity to one another, firms can lower their production costs. This fact forms the basis of agglomeration economies, or the benefits derived from clustering together, one of the most important forces shaping the economic geography of different types of production. By forming dense webs of production and embedding themselves within them, firms usually can produce more efficiently and profitably.

Agglomeration economies take several forms. *Production linkages* accrue to firms locating near other producers that manufacture their basic raw materials. By clustering, distribution and assembly costs are reduced. *Service linkages* occur when enough firms locate in one area to support specialized support services. For example, the advertising industry in New York is concentrated within a short distance of Madison Avenue. By locating near one another in dense networks, firms can monitor up-to-date information and gossip on the latest trends, markets, clients, hires, and products. *Marketing linkages* occur when a cluster is large enough to attract specialized distribution services. For example, the firms of the garment industry in New York City have collectively attracted advertising agencies, showrooms, buyer listings, and other aspects of finished product distribution that deal exclusively with the garment trade. Firms within the cluster have a cost advantage over isolated firms that must provide these specialized services for themselves.

Agglomeration economies may be temporary, are found to different extents in different industries, and may offset through various forms of economic, technological, and geographic change. Typically, agglomeration economies reflect some kinds of firms' need for close interaction with clients and suppliers. Thus, they are most pronounced in vertically disintegrated types of production in which firms have many linkages "upstream" and "downstream" in the production process. (In contrast, vertically integrated firms, with relatively few external linkages, are less dependent on agglomeration.) Firms in markets with low degrees of uncertainty (usually due to slow rates of technical change, market structure, or the regulatory environment), in contrast, are less reliant on agglomeration to minimize costs and maximize profits. As firms grow, they often become more vertically integrated and more capital intensive, have fewer external linkages, and come to substitute economies of scale for agglomeration economies.

Because agglomeration economies provide powerful incentives for firms to locate in close proximity to one another, they are most heavily manifested in large metropolitan areas. The prime motivation behind the agglomeration of firms in metropolitan regions is the ready access they offer to clients, suppliers, and ancillary services, most of which is accomplished through face-to-face interaction. Often personal relationships of trust and reputation are of paramount significance. Thus, agglomeration maximizes access to information, much of which is irregular and unstandardized, and helps firms to minimize uncertainty. Firms in these locations have an advantage, within limits, over similar firms in more rural areas. Cities provide markets, specialized labor forces and services, utilities, and transportation connections required by manufacturing. *Urbanization economies,* therefore, are a combination of production, service, and marketing linkages concentrated at a particular location. Agglomeration forms the basis for the comparative advantage of cities in forms of production that typically consists of relatively small, vertically disintegrated firms in highly competitive markets with high degrees of uncertainty and change.

Agglomeration economies have been manifested in different industries throughout the historical geography of industrial capitalism. They were critical during the early Industrial Revolution, when many small firms in industries such as watch making and gun manufacturing clustered in the cores of British cities. Since the emergence of post-Fordist "flexible production" during the late 20th century, the competitiveness of regions such as California's Silicon Valley, Italy's Emilia-Romagna, and Germany's Baden-Württemberg has relied heavily on agglomeration. Finally, producer services (business and financial services that cater primarily to other firms) rely heavily on agglomeration economies, often in "global cities," forming complexes of service firms comparable to other types of highly concentrated production.

—Barney Warf

Suggested Reading

Stutz, F., & Warf, B. (2005). *The world economy: Resources, location, trade, and development* (4th ed.). Upper Saddle River, NJ: Pearson/Prentice Hall.

AGRICULTURE, INDUSTRIALIZED

Geographers have tended to study industrialization from an urban perspective, largely overlooking its relationship to rural landscapes. This urban bias limits our ability to see that urbanization could not have occurred without technological change in agriculture that allowed fewer farms to produce more food. This freed other farmers to become part of the urban working class. This entry describes the origins and impacts of industrialization on agricultural production and rural landscapes.

INDUSTRIALIZATION OF AGRICULTURE

Industrialization includes the mechanization of processes previously done by human hands. It also involves the reorganization of labor practices and the application of new energy and transportation technologies to increase the rate at which humans transform nature into goods. Increased output also requires new markets. Hence, the industrialization of agriculture involves widespread change in four areas: (1) supply of farmland, (2) public policy, (3) technological change, and (4) agribusiness consolidation along the value chain.

Prior to the 20th century, American farmers practiced an extensive form of agriculture. As demand for food increased, farmers expanded into new territory. New plows invented by John Deere made it easier to till fertile but heavy prairie soils. However, agricultural production increased because more acres were planted, not because yields per acre increased during this time. Conditions changed when the frontier closed at the end of the 19th century. With no new land to cultivate, output could grow only through increasing yields. This marked the beginning of intensive agriculture.

Public policy decisions created a foundation for industrial agriculture. The U.S. Department of Agriculture (USDA) was established in 1862. In the same year, the Morrill Act stipulated that each state should have one "land grant" college where agricultural sciences could be taught. During the early 20th century, political leaders noticed that American agricultural productivity lagged behind that of England and Germany. President Theodore Roosevelt launched the Country Life Commission, which concluded that productivity could increase only if the infrastructure of rural America was modernized. Recommendations included reforming rural schools to teach students agronomy and improving the road system to better transport produce to markets. The USDA continues to promote industrial agriculture through subsidies, research, and supply management, for example, by redistributing surplus commodities through the Food Stamp program.

Productivity increased dramatically through the mechanization of farms. In 1910, there were an estimated 1,000 tractors in use. By 1940, that number had risen to 1.6 million—a number that tripled to 4.8 million by 1965. Increasing horsepower and tractor versatility also contributed to productivity increases. Wheat and corn yields were 15.4 and 30.0 bushels per acre, respectively, in 1940. By 1970, the corresponding numbers had more than doubled to 31.8 and 80.8 bushels per acre. Productivity increases also occurred because of advances in genetic engineering, pesticides, and fertilizers, among other farm inputs.

Industrialization has transformed the agricultural sector beyond the farm as well. As a raw commodity such as corn leaves the farm, it follows a so-called value chain that includes processing, distribution, and retail on its way to consumers. At each step, value is added to the commodity as it is transformed into products and moved closer to consumers. The problem for farmers is that large agribusinesses are working to control more and more of the value chain. Multinational firms (e.g., ConAgra, Cargill) sell inputs such as fertilizer, but they also process commodities. Under these increasingly monopolistic conditions, farmers have less bargaining power to affect the price of inputs they must buy or the crops they must sell. In the end, farmers earn a smaller portion of each dollar spent by consumers because the prices that farmers receive for their crops have stagnated, even though the prices that consumers pay continue to increase.

GEOGRAPHIC IMPLICATIONS

Commodity prices have stagnated because the industrialization of agriculture has increased the supply of farm commodities. Laws of supply and demand suggest that as farm productivity increases, the unit price for the commodity is going to drop. To remain profitable in the face of dropping prices and profit margins, some farmers will increase productivity by investing in new equipment, crop hybrids, or other inputs. To pay for these improvements, farmers are forced to amortize their costs over a larger farm area. This drives a tendency toward farm consolidation as larger, more successful farms take over smaller marginal operations. Consolidation is also driven by changes along the value chain as market pressures force marginal farmers out of business. The result is that farms are getting bigger while the total number of farms is decreasing. In 1940, there were more than 6 million farms with an average size of 180 acres. By 1970, that number had dropped to 2.9 million with an average size of 400 acres. The decline in farm numbers continues but has slowed. There are currently 2.2 million farms with an average size of 440 acres.

Farm consolidation contributes to rural depopulation and out-migration. There are rural counties in the Great Plains and Midwest with populations that peaked during the early 20th century and have slowly declined due in large part to farm consolidation. At the beginning of the 20th century, roughly 50% of Americans were directly involved in agriculture. That figure is now less than 2%. We live in an urban culture because the industrialization of agriculture contributed to a larger, more affordable food supply. However, these benefits have come at a price. Many farm communities struggle demographically and economically. There are also questions about the environment and the sustainability of the current system because of its heavy reliance on petrochemicals.

—*Christopher D. Merrett*

Suggested Reading

Cochrane, W. (1993). *The development of American agriculture* (2nd ed.). Minneapolis: University of Minnesota Press.

Hart, J. (2003). *The changing scale of American agriculture.* Charlottesville: University of Virginia Press.

Hudson, J. (1994). *Making the Corn Belt: A geographical history of middle-western agriculture.* Bloomington: Indiana University Press.

Reynolds, D. (1999). *There goes the neighborhood: Rural school consolidation at the grass roots in early twentieth-century Iowa.* Iowa City: University of Iowa Press.

U.S. Department of Agriculture. (2004). *Quick Stats: Agricultural statistics data base* [computer database]. Available: www.nass.usda.gov/QuickStats/

AGRICULTURE, PREINDUSTRIAL

Throughout much of the world today, and throughout the bulk of human history (indeed, dating back to the Neolithic Revolution 8,000–10,000 years ago), societies fed themselves through an assortment of preindustrial agricultural systems. Preindustrial or nonindustrial agricultural systems differ from industrialized ones in a variety of respects. Perhaps most important, preindustrial systems do not use the inanimate sources of energy that are vital to industrialized agricultural systems (e.g., fossil fuels) and, therefore, are markedly less energy intensive in nature. Rather, work in preindustrial farming systems is accomplished entirely through human or animal labor power. Thus, these types of farming are much more labor intensive. In societies fed predominantly through preindustrial agriculture, the vast bulk of people are engaged as farmers or peasants. Second, because many preindustrial societies are not fully commodified (i.e., capitalist social relations have not come to dominate every aspect of production), preindustrial agricultural systems are generally organized around production for subsistence rather than production for profit. In other words, food is grown mostly for local consumption rather than for sale on a market.

Preindustrial agricultural systems played an enormous role in history, including the variety of slave-based and feudal social systems that unfolded across much of the world. For example, Roman *latifundia*—large estates worked by slaves—formed the backbone of agricultural production during the empire. The expansion of medieval agriculture into the dense soils of Northern Europe was made possible by the introduction of the heavy plow and, later, the three-field system. The manorial system that formed the social and economic basis of feudal Europe involved peasants and serfs who rented land from large land owners, paying rent with a fraction of their output. Variations of peasant-based production continue to be important in many contexts.

Today, there is a great diversity of types of preindustrial agricultural systems throughout the world, with large variations in the types of crops grown, the methods used, their productivity, and their relative vulnerabilities to drought or other hazards. Although Nomadic herding is not technically a form of agriculture, many observers classify it in this category; however, it involves only the domestication of animals, not crops. Typically, nomadic herders measure their wealth in terms of livestock (generally cattle, goats, or reindeer) and often follow their herds in annual migratory cycles such as transhumance, the movement between summer pastures in higher elevations and winter pastures in lower ones. Nomadic herding has been slowly vanishing throughout the world over the past two centuries, but contemporary examples include the Masai of East Africa, the Mongols of Mongolia and northern China, the Tuareg of northern Africa, and the Lapps of northern Finland.

The best-known example of preindustrial agriculture is slash-and-burn, also known as swidden or shifting cultivation. This form is found only in tropical areas such as parts of Central America, the Amazon rain forest, West and Central Africa, and Southeast Asia, to which it is ideally suited. Roughly 50 million people continue to be fed this way in these regions. Due to heavy rainfall and the leaching of nutrients, tropical soils are generally quite poor and most nutrients are stored in the biomass. The first step in slash-and-burn, therefore, is to cut down existing trees and bushes in a given plot of land and to burn them, releasing nutrients into the soil through the ash. Crops are then planted for several years. However, because the rate of nutrient extraction exceeds the rate of replenishment, the site can be used for only a brief period—generally 2 to 6 years—before the farmers must move on to a new site. Abandoned sites may gradually recover with a sufficient fallow period. If rapid population growth occurs and fallow periods are reduced, the soil may permanently decline in fertility. This form of farming was widely practiced in the Mayan kingdoms prior to the Spanish conquest, and declining soil fertility may have played a role in the collapse of the Mayan states.

A third form of preindustrial agriculture is that of Asian rice paddy cultivation, which is widely practiced throughout a region stretching from Japan, Korea, and southern China throughout Southeast Asia into eastern India. This form may be partially or even completely commodified. Rice is the staple crop for billions of people in Asia, and its cultivation in this form goes back millennia. Young rice plants require standing pools of water, and to create spaces in which this occurs, Asian societies carved terraces out of hillsides, controlling the flow of water with vast networks of dikes and small levees. Furrows may be dug using water buffalos. Often small fish may be grown in these pools of water as a source of protein. The planting of rice is extremely laborious and is often associated with stereotypes of Asian peasants engaged in arduous labor in their fields. The supply of water may rely on monsoon rainfalls.

Preindustrial agricultural systems have functioned effectively for thousands of years and continue to do so in many parts of the developing world. In contrast to common stereotypes that such systems are stagnant or unchanging, Ester Boserup showed that rising populations in such places often stimulate productivity growth. In most places, preindustrial systems are marginalized or threatened by the expansion of globalized, capitalist, industrialized farming systems, including imports of subsidized grains from Europe or North America. However, preindustrial systems enjoy advantages of their own, including a diversity of crops (in contrast to industrialized monocultures) and freedom from a dependence on pesticides and petroleum. Thus, it may be helpful to view these systems not as backward remnant forms of food production but rather as historical adaptations to particular social and environmental contexts, that is, as nonindustrial rather than preindustrial.

—*Barney Warf*

See also Agriculture, Industrialized; Food, Geography of; Peasants

Suggested Reading

Boserup, E. (1965). *The conditions of agricultural growth: The economics of agrarian change under population pressure.* Chicago: Aldine.

Food and Agriculture Organization of the United Nations. (1984). *Changes in shifting cultivation in Africa.* Rome: Author.

Peters, W. (1988). *Slash and burn: Farming in the Third World forest.* Moscow: University of Idaho Press.

AGRO-FOOD SYSTEM

The term *agro-food system,* sometimes called *agrifoods,* captures the increasingly long and complicated path that food takes to get to our table. Although we may like to think that the food we eat comes from a farm, that is only one place among many involved in the system that produces our food. Most farming is possible only with industrial inputs such as tractors, combines, and chemical inputs (e.g., fertilizers, pesticides). Farmers often require loans of money (called "capital") each season to buy what is needed to produce a crop. Farming is also dependent on energy to run the machines, pump water, produce fertilizer, and transport the finished product because most of the places where food is produced are not where consumers are located. Farmers need expert information on what and when to plant, how to diagnose and treat blights and pests, how to obtain and use weather information, and how to decide when and at what price to sell their crop. When we think about what goes into farming, we realize that farms are linked to and dependent on many other places such as places of industrial production, petrochemical and fuel production, banking centers, and universities and government where research and policy are created. Where and what is done with the outputs of farms is equally complicated.

Farm output can remain in its original form and simply be graded, washed, and shipped to consumers. But most food we consume in the developed world is not in an unprocessed or "raw" form. Most of the food we consume has been modified and transformed substantially by processing and been made durable through canning, freezing, or other methods. This is important because only with durable foods is long-distance trade possible. In fact, the distinction between agriculture and industry has become so blurred that many farm products transformed by an industrial process have become known by that industrial process, including *homogenized* milk, *pasteurized* cheese, and *refined* sugar. Agricultural products can be further industrialized by processing that breaks them down into their constituent parts. For example, a starch, a sweetener, oil, and protein can be extracted from grain. Processors attempt to break the product of the farm into as many parts as possible and then find profitable uses for them. These different "fractions" of whole farm products are then often used as generic inputs for manufactured foods or used in other industrial processes.

The producers of manufactured foods capture a greater part of the dollars spent on food and increasingly have an advantage over farmers. Manufactured food producers have flexibility in where they get their ingredients. For example, the manufactured food requires a sweetener, but not necessarily sugar from the sugarcane plant. It requires oil, but not necessarily oil from corn. It requires a starch, but that could be derived from a potato, wheat, or a number of other grains. The production of potato chips provides a good example of this substitution effect; producers can fry the chips in whatever oil is cheapest at the moment of production. This illustrates how producers of manufactured foods have flexibility in where they source their ingredients and how they can make places compete against one another and reduce farming into ingredient production for complexly constructed industrial foods. These characteristics of the agro-food system illustrate why farmers are at a disadvantage.

More toward the consumer end of the agro-food system is food distribution. Food reaches consumers via food wholesalers, food retailers, and the restaurant and catering industry. Powerful economic entities in food distribution can shape the agro-food system by their purchasing power such as when fast-food restaurant chains decide to fry their french fries in healthier oil or to add salads to their menus. Large grocery chains have a similar power when they decide to carry some items and not others.

At the end of the agro-food system are the final consumers. Food is unlike other commodities because we must eat daily to survive. Food is taken into our bodies and metabolized (used by our cells to provide energy). Our food choices affect our own bodies but also reverberate back and reshape the agro-food system. What we eat reflects demographic characteristics such as the size and growth of the population, purchasing power, and social relations (e.g., the structure of the family). Consumers choices shape, but are also shaped by, the agro-food system. Obviously, advertising influences our food choices. But more subtly, the ever quickening pace of the economy and its demands have led to the proliferation of "fast" foods (those that can be consumed without utensils) and other convenience foods meant to be consumed "on the go," in the car, or at the desk.

The recent dietary trend of avoiding foods high in carbohydrates has reduced the consumption of potatoes, rice, bread, donuts, and orange juice (to name just a few) and has affected their places of production and sale. But these changing attitudes toward food also provide opportunity. For example, a food that is high in fat (e.g., fried chicken), criticized during the time when a healthy diet was thought to be a low-fat diet, can present itself as a healthier food choice now that the food trends have changed and carbohydrates are to be avoided.

The geographies of the agro-food system are continuing to change as food technologists attempt to bypass the farm altogether by creating "nonfood foods" or foods that are consumed but not metabolized by the body. These substances are made in the laboratory—not grown on the farm—and allow food producers to avoid the risks inherent in farming, such as unreliable weather, pests, and blights, while providing greater control over the production process. The most recent and visible nonfood foods are fat and sugar substitutes. More common in the agro-food system and growing in number are "functional foods" (also called "nutraceuticals") that attempt to marry foods and pharmaceuticals to create a substance consumed to create a desired effect in the body. Examples include oat-based breakfast cereals promoting themselves as "heart healthy," orange juice with added calcium for "strong bones," grape juice with added antioxidants to fight cancer, and "smart drinks" with added ginseng, caffeine, and vitamins. Functional foods blur the line between drugs and foods, and their producers know that foods that make health claims often have an advantage over their competitors in a competitive marketplace.

Not only does the changing agro-food system have impacts on our bodies, but also its changing technology and consumer choices have significant impacts in reshaping our geography.

—*John Grimes*

Suggested Reading

Bonanno, A., Busch, L., Friedland, W., Gouveia, L., & Mingione, E. (Eds.). (1994). *From Columbus to ConAgra: The globalization of agriculture and food.* Lawrence: University Press of Kansas.

Friedmann, H. (1993). The political economy of food. *New Left Review, 197,* 29–57.

Goodman, D., & Redclift, M. (1990). *Refashioning nature: Food, ecology, and culture.* London: Routledge.

Schlosser, E. (2001). *Fast food nation: The dark side of the all-American meal.* Boston: Houghton Mifflin.

AIDS

The geography of acquired immunodeficiency syndrome (AIDS) encompasses a number of spatial approaches to understanding the epidemic. More recent geographic studies of AIDS have focused less on the virus and macro diffusion patterns and more on the human geographies of risk and experience of AIDS. One category of investigation focuses on regionally specific contexts of human immunodeficiency virus (HIV) vulnerability. In these studies, social, economic, political, and cultural practices at multiple spatial scales are examined for their impact on individuals' vulnerability to HIV in particular regional locations. These place-specific investigations are critical to understanding micro patterns of transmission given the substantial evidence that factors driving transmission of HIV in one place do not necessarily explain transmission patterns and levels in another place. Clearer understandings of what makes people engage in risky behaviors and become vulnerable to HIV is, in turn, pivotal in implementing more effective prevention and treatment strategies.

Examining geographies of everyday life with HIV/AIDS constitutes another important part of a geography of AIDS. How and whether persons living with HIV and AIDS (PLWHAs) are able to access healthcare and other services are critical to providing the best treatment possible. Earlier geographic studies focused on mapping residence patterns with location of clinics and other services, but more recent studies have recognized that access is more complicated and includes, among other things, individuals' social networks, the degree of flexibility in the workplace, how much stigma individuals face in their lives, income levels, child care responsibilities, and quality of care available. Other work has looked at the ways in which PLWHAs cope with reduced spaces and places in which they live their lives. This can be because stigma works to block access to particular places such as housing, jobs, countries, and individuals' homes or because deteriorating physical status reduces mobility. The ways in which people experience space and place when coping with AIDS are vital to implementing better outreach programs and services.

Earlier geographic studies of AIDS focused on the virus itself, investigating theories of HIV's origins and transmission patterns. Many scientists and social scientists thought that determining sites of the first HIV cases would assist in understanding where, when, and how HIV subsequently spread to the rest of the world. Much geography of AIDS during the 1980s consequently focused on mapping spatial routes of transmission over time, tracing likely patterns of HIV diffusion across continents using data of first known cases in each region together with travel and migration routes. Although none of these patterns was conclusive, they provided models for illuminating continued transmission of HIV as well as likely points of intervention. Critics of origin theories, however, contended that finding origins does little to understand current patterns of HIV transmission and instead generates negative consequences such as blame for causing a deadly epidemic. Focusing on large-scale geographic patterns also did little to further understanding about the complex network of behaviors and practices underlying transmission of HIV.

—*Susan Craddock*

See also Health and Healthcare, Geography of; Medical Geography

Suggested Reading

Brown, M. (1995). Ironies of distance: An ongoing critique of the geographies of AIDS. *Environment and Planning D: Society and Space, 13,* 1391–1396.

Kalipeni, E., Craddock, S., Oppong, J., & Ghosh, J. (Eds.). (2004). *HIV and AIDS in Africa: Beyond epidemiology.* Boston: Blackwell.

Shannon, G., Pyle, G., & Bashshur, R. (1991). *The geography of AIDS: Origins and course of an epidemic.* New York: Guilford.

Takahashi, L., Wiebe, D., & Rodriguez, R. (2001). Navigating the time–space context of HIV and AIDS: Daily routines and access to care. *Social Science and Medicine, 53,* 845–863.

ANIMALS

Animal geographers study the interplay among animals, culture, and society, exploring a broad range of human–animal concerns such as habitat loss and species endangerment, domestication, animal entertainment and display, and wildlife restoration. Animal geographies are essentially about nonhuman animals and their place in society, with *place* meaning both material borders (societal practices that shape the spaces where some animals are welcomed and others are not) and conceptual boundaries that call up questions of human identity and animal subjectivity. We can think in terms of three basic themes in contemporary animal geographies: (1) animals and the making of place, (2) human identity and animal subjectivity, and (3) the role of ethics and how humans ought to treat animals. These organizational themes are not independent of one another, and they frequently overlap and dovetail with concepts such as animal instrumentalism, anthropocentrism, and the human–animal continuum. Moreover, animal geographers recognize the fluidity of boundaries, emphasizing not only the distinctions but also the connections, overlaps, and similitudes between human and animal worlds.

MATERIAL BOUNDARIES: ANIMALS AND THE MAKING OF PLACE

Discussions in human geography about the social construction of landscapes have led to the exploration of how animals and their networks leave their imprint on places, regions, and landscapes over time. Animal geographers consider tangible places such as zoos, farms, experimental laboratories, and wildlife reserves as well as economic, social, and political spaces such as the worldwide trade of captive wild animals. Even a relatively new space through which animals are woven into human culture, the "electronic zoo," has been explored as an emerging form of animal display trading in digital images rather than animal bodies like traditional zoos and aquariums.

Animal geographers also study places characterized by the presence or absence of animals and how human–animal interactions create distinctive landscapes. Researchers have considered the impact of land use practices on wildlife survival in the Peruvian Amazon, the boundary-making policy conflicts between urban and rural New Yorkers over the proper place of wolves, and the changing relationships between people and mountain lions in California. In addition, some animal geographers foreground the links between humans and other animals—those used for meat, medicine, clothing, and beauty products, for example—that go largely unseen in contemporary society given the distance engendered by modern commodity chains. Other researchers focus on domesticated animals that share the most intimate spaces with humans, including beloved family pets and service animals. Borderland communities, where humans and animals share public and/or private space and where some animals are loved, others are despised, and so many are unconsciously consumed, reveal the contingent and often contradictory ways in which humans and animals interact with one another.

Borderland communities can span various places and spaces. Investigating human–dolphin encounter spaces, for example, requires a look at the well-defined boundaries of zoos and aquariums, where dolphins are confined and cared for by humans, as well as natural habitats, where a growing number of tourism operators seek out dolphins to sell a "magical experience" to customers who wish to closely interact with, or even touch and swim with, wild dolphins. On the other hand, U.S. government officials strive for just the opposite, calling such activities illegal "harassment" and working to keep people a defined distance apart from all wild dolphins. And how do the dolphins encourage or defy the human ordering of these border waters? Each of these material places, from the zoo and the open ocean to the economic and policy arenas considered by investigating human–dolphin encounter spaces, helps illuminate the complex relationships between human and nonhuman worlds.

CONCEPTUAL BOUNDARIES: HUMAN IDENTITY AND ANIMAL SUBJECTIVITY

Breaking from the traditional geographic approach to animals, contemporary animal geographers think about nonhuman animals as more than simple biotic elements of ecological systems. Not only are animals appreciated as foundational to countless cultural norms and practices, they also are valued as individuals with mental and emotional lives. Thus, animal geographers call for a more theoretically inclusive approach to thinking about humans and animals; both are considered to be embedded in social relations and networks with others on whom their social welfare

depends. Such thinking suggests a reconceptualization of the human–animal divide that portrays humans as vastly different from (and superior to) animals and points instead toward a continuum that allows for a "kinship" with animals while still acknowledging the differences between humans and other animals.

Animal geographies also encourage thinking about animal agency and subjectivity, recognizing that animals have intentions and are communicative subjects with potential viewpoints, desires, and projects of their own. For example, some animal geographers suggest that nonhuman animals are best seen as "strange persons" or as marginalized, socially excluded people. But because animals cannot organize and challenge human activities for themselves, animal geographers recognize that animals require human representatives to speak and act in their interests.

ETHICS, HUMANS, AND OTHER ANIMALS

Human relationships with animals have been and remain multifarious and deeply complex, ranging from magnificent to malignant. In every case, humans remain the regulators of whether animals are conceived of as either "in place" or "out of place," and it is moral sensibility that defines such orderings with significant ethical implications. In many cases, animal geographers attribute instances of instrumentalism, exclusion, and exploitation of the nonhuman world to a history of anthropocentric, or human-centered, thinking. Critical of such activities, much of the animal geographies literature is concerned with the ethical task of advancing the well-being of animals.

One way of advancing this unmistakably normative project is to explicate societal values, which certainly determine human treatment of animals. For example, some animal geographers have considered how an animal's position in the scientific community's hierarchy of value (as determined by the rarity of the species) can have a significant influence on its fate. A crocodile that belongs to a species that is included in a global conservation policy, for instance, is "protected" and therefore privileged over animals that are not included in such a policy. With a change in conservation policy, or the "downlisting" of a particular species from the rank of endangered species, the same crocodile once protected and perhaps flourishing in its natural habitat could very well be removed for human use to an impoverished (and shortened) life as a factory farm animal.

In striving to advance the well-being of both humans and animals, some animal geographers explicitly locate animals in the moral landscape, recognizing that ethical questions are present in all human and animal geographies. In these instances, animal geographers argue for the inclusion of animals in the moral community, valuing animals as ends in themselves (rather than as means to human ends). The practical consequences of such inclusion are considerable. For example, how are we to decide what is most important in environmental policymaking? And who, exactly, gets to decide? Especially when human–animal needs clash in a world of finite space, a framework of normative principles suggested by animal geographies—principles inclusive of animal interests and desires—can guide human–animal relations and resolve the moral dilemmas that relate to conflicting wants and needs of both humans and animals. This is where animal geography largely departs from the theoretical positioning in the contemporary nature–culture debates in geography that remain largely anthropocentric. Granting subjectivity and moral inclusion to animals requires an emphasis on the well-being of both humans and other animals. As such, animal geographies implicitly call for a social and environmental justice that is widened to include animal justice.

—*Kristin L. Stewart*

Suggested Reading

Lynn, W. (1998). Contested moralities: Animals and moral value in the Dear/Symanski debate. *Ethics, Place, and Environment, 1,* 223–242.

Midgley, M. (1983). *Animals and why they matter.* Athens: University of Georgia Press.

Philo, C., & Wilbert, C. (Eds.). (2000). *Animal spaces, beastly places: Critical geographies.* London: Routledge.

Wolch, J., & Emel, J. (Eds.). (1998). *Animal geographies: Place, politics, and identity in the nature–culture borderlands.* London: Verso.

ANTHROPOGEOGRAPHY

The term *anthropogeography* refers to a perspective and program in human geography with both major and minor traditions, expressions, and manifestations. Friedrich Ratzel (1844–1904) is credited with coining the term. His two-volume work, *Anthropogeographie* (published in 1882 and 1891), is usually cited as the founding document. The first volume, in which he offered an overview of human history as adaptation to physical environment, often has been misrepresented as an environmental determinist tract. It is true that many subsequent environmentalists, perhaps most famously Ellen Churchill Semple (1863–1932), interpreted Ratzel's anthropogeography in this light. These misreadings of Ratzel led to anthropogeography's major tradition—that of the study of the effects of the biophysical environment on human culture and history. By the time of Ratzel's death, an increasing number of geographers were producing studies that superficially could be attributed to Ratzel's example. This remained the case through the 1920s, but thereafter their industry and influence waned.

In North America, this eclipse was due in no small part to Carl O. Sauer's (1889–1975) attacks on environmental determinism in geography and Franz Boas's (1858–1942) condemnations from his base in anthropology. Sauer's critiques included alternative views of human–environment relations, ones that incorporated much of what Ratzel proposed for cultural geographic studies in the second volume of *Anthropogeographie*. According to Sauer, Ratzel pioneered the study of the distribution of culture traits, first stated the case for cultural diffusion as the prime process, and anticipated the culture area concept. This is all second-volume Ratzel. By the 1940s, when environmental determinism had been largely discredited and the term anthropogeography had fallen into disuse, Sauer began his rehabilitation of the term. He and some of his students, such as Fred Kniffen (1900–1993) and George Carter (1912–2004), used it to identify their approach to a cultural geography centered on locating cultural cores or hearths, tracing diffusions of culture traits, and more generally reconstructing the making and breaking of cultural landscapes through "all human time." By the 1950s, Sauer had begun to self-identify with anthropogeography explicitly. Although few have applied this appellation to Sauer's collective enterprise, the Berkeley School, it is perhaps the most apt way to encompass the problems, perspectives, and practices associated with this school. This, then, can be considered anthropogeography's minor, if explicitly antithetical, tradition.

One maxim of this minor tradition is that where cultural historical questions are concerned, "it is always earlier than you think." Accordingly, the origins of the term anthropogeography probably antedate Ratzel's deployment. The earliest detectable English use seems to be from the 1650s, when it appeared in alchemical discourse pertaining to the symmetries and correspondences between the human body and the earth. Its most common current use resides in bibliothecal categories. The U.S. Library of Congress indexing system equates anthropogeography and human ecology and puts this major heading (GF) between environmental science (GE) and anthropology (GN). Future cross-fertilizations between disciplinary sectors of geography and anthropology may be expected to bring about new meanings of this adaptable term and concept.

—Kent Mathewson

See also Berkeley School; Cultural Geography; Culture Hearth

Suggested Reading

Mathewson, K., & Kenzer, M. (Eds.). (2003). *Culture, land, and legacy: Perspectives on Carl O. Sauer and Berkeley School geography* (Geoscience and Man, Vol. 37). Baton Rouge, LA: Geoscience Publications.

Speth, W. (1999). *How it came to be: Carl O. Sauer, Franz Boas, and the meanings of anthropogeography.* Ellensburg, WA: Ephemera.

ANTICOLONIALISM

Anticolonialism is a broad term used to describe the various resistance movements directed against colonial and imperial powers. The ideas associated with anticolonialism—namely justice, equality, and self-determination—commingled with other ideologies such as nationalism and antiracism.

Colonial rule assumed many different forms. Consequently, anticolonial movements likewise varied, influenced in part by the particularities of foreign rule. Whether the colony was ruled directly, through force, or indirectly would significantly determine how anticolonial movements originated and progressed. In Vietnam, for example, the anticolonial and communist organization known as the *Viet Nam Doc Lap Dong Minh Hoi* (League for the Independence of Vietnam [or Vietminh]) waged a lengthy anticolonial war against French colonial rule. Led by Ho Chi Minh, the Vietminh resorted to guerrilla warfare during the 1940s when France attempted to reassert its colonial rule following World War II. Likewise, in the former British and French colonies of Kenya and Algeria, respectively, anticolonial resistance movements used force to restore indigenous rule. For example, the Mau Mau in Kenya conducted a violent campaign to remove British colonists, and the *Front de Libération Nationale* (National Liberation Front [or FLN]) waged an 8-year war against French forces in Algeria.

Some colonies were spared the violence and destruction of the decolonization process. The former British colony of Ceylon (present-day Sri Lanka) achieved independence relatively smoothly in 1948. The British had acquired the colony from the Dutch in 1815 following the Napoleonic Wars and granted the colony its independence following World War II.

It was not uncommon for simultaneous anticolonial movements to emerge in a single colony. For example, during the late 19th century, the Philippines, long a colony of Spain, was the site of two anticolonial movements. During the late 1800s, there first emerged a reform movement known as the *ilustrados*. Composed mostly of highly educated and wealthy Filipinos, these individuals, embodied in the Propaganda Movement, demanded moderate administrative and religious reforms such as greater political representation and the curtailment of the excessive power of the friars. Many of the ilustrados were Chinese mestizos who were schooled in Barcelona and Madrid, Spain. Concurrently, there emerged a more radical revolutionary movement that advocated the complete overthrow of the Spanish colonial government. Founded in 1892, the *Kataastaasan Kagalang-galang na Katipunan ng mga Anak ng Bayan* (Highest and Most Honorable Society of the Sons of the Country [or Katipunan]) was a secret society committed to overthrowing Spanish rule and replacing it with a Filipino nationalist government. The founder of the Katipunan was Andres Bonifacio. Unlike the ilustrados, Bonifacio grew up in poverty and was self-taught. The contrast between Bonifacio and the ilustrados conveys the importance of class and ethnic differences in anticolonial movements.

Anticolonial movements should not be viewed as isolated events; indeed, many anticolonial leaders and organizations learned from other movements. Ania Loomba, an English professor, noted that there were important political and intellectual exchanges between different anticolonial movements and individuals and that even the most rooted and traditional of these was shaped by a syncretic history.

Many of the classic writings associated with anticolonial movements continue to hold salience in contemporary society. For example, the works of Aimé Césaire, Frantz Fanon, and Kwame Nkrumah resonate strongly in current antiglobalization movements. This continuity is testimony to the powerful ideas that embraced anticolonial movements, namely concerns with sovereignty, equality, and social justice.

—*James Tyner*

See also Colonialism; Imperialism; Postcolonialism

Suggested Reading

Césaire, A. (1972). *Discourse on colonialism* (J. Pinkham, Trans.). New York: Monthly Review. (Original work published 1955)

Chamberlain, M. (1999). *Decolonization: The fall of European empires* (2nd ed.). Malden, MA: Blackwell.

Fanon, F. (1963). *The wretched of the earth* (C. Farrington, Trans.). New York: Grove. (Original work published 1961)

Loomba, A. (1998). *Colonialism/Postcolonialism*. New York: Routledge.

APPLIED GEOGRAPHY

Many public policy problems facing society today have geographic components or dimensions. For example, redrawing boundaries during political redistricting, locating a new public housing project, identifying a suitable site for a sanitary landfill, and mapping coastal area vulnerability to flooding all could be conceptualized as geographic problems. Applied geography focuses on the use of geospatial information and research techniques to build perspective and knowledge that can be used to identify, understand, and solve human and environmental problems from local to global scales. Another characteristic of applied geography is that it extends the scientific method often used in academic geography to include the implementation and evaluation of geospatial information in addressing problems of social relevance in nonacademic settings. This extension often requires applied geographers to work as part of an interdisciplinary team and to collaborate with a variety of public- and private-sector decision makers.

The problem-solving approach of applied geography is further illustrated using the example of finding the best location to build a new municipal fire station. Here the applied geographer would use geospatial information and research techniques to answer the following four interrelated questions. Where are the existing fire stations located? What has been the spatial pattern for the type, number, and frequency of emergency calls received from across the service area? How and where is land use change taking place in the city that could influence future demands for emergency services? What is the current and planned municipal infrastructure, including transportation networks and utility availability? Answers to these questions could be presented visually through a series of maps and supporting information, enabling city officials, fire department representatives, and the general public to view different scenarios as part of the decision-making process.

Applied geography has a long and rich tradition as a subdiscipline or specialty area within American human geography. Some of the earliest work can be traced back to the land surveys of the American West during the middle 1800s. The writings of John Wesley Powell on the arid West and the need to develop reliable water sources for agricultural development contributed much to the passage of the Reclamation Act of 1902, which ushered in the involvement of federal agencies such as the Bureau of Reclamation in developing western water resources.

During the 1920s, cultural geographer Carl Sauer played a leading role in the Michigan Land Economic Survey, which emphasized the need for improved land management planning to offset environmental degradation caused by soil erosion and deforestation. During the Great Depression and New Deal period, geographers such as Harlan Barrows and Gilbert White were involved with multiple-purpose resource management agencies such as the Public Works Administration and the National Resource Planning Board. The contributions of geography and geographers to logistics and transportation planning, military intelligence, area and regional studies, and cartography during World War II are well documented.

The practice of applied geography in the private sector, particularly in business and marketing, began in earnest during the 1930s with the work of William Applebaum in the retail food distribution industry. Since then, applied geographers have made contributions in market area analysis, retail site selection, and shopping center development for a number of companies, including J. C. Penney, Kroger, and Stop and Shop.

Prior to the late 1970s, most applied geographers were employed by federal land management and environmental planning agencies, by city and regional planning organizations at the local and state levels of government, and in the private sector. Although applied geography was recognized and practiced by some geographers working in universities and colleges, their efforts were not well recognized or coordinated. In 1978, those conditions began to change with the inception of the Applied Geography Conference. The purpose of this conference was to provide a forum for applied geographic research and curriculum issues and to serve as a venue for bringing together geographers from a variety of professional backgrounds. The Applied Geography Conference continues to bring together academic and nonacademic geographers to discuss mutual interests, share strategies and research agendas, and demonstrate the utility of applied geography in human and environmental problem solving.

A key factor in the development of the applied geography subdiscipline has been the increased capability to collect, analyze, and display geospatial information through the use of geographic information systems (GIS). These systems have evolved from an initial combination of computer cartography and database management to include remote sensing, global positioning systems, spatial statistics, visualization and simulation, and Web-based information access and sharing. As hardware and software capabilities continue to improve, the opportunity to better use geospatial information will also improve, making the potential for applied geographic research even greater. This potential is further enhanced by the GIS software becoming more user-friendly and therefore easier to implement in a wide range of user environments where the emphasis is on application.

The future of applied geography is promising and limited only by a lack of imagination as to how geospatial information can be used to better understand our world. Although some people would maintain that geography itself has become less relevant due to advances in telecommunications and computer technology, recent world events would argue otherwise and reinforce the idea that a better understanding of geography is critical to our well-being as individuals and as a nation. Natural hazards such as hurricanes continue to demonstrate how vulnerable coastal areas are to flooding, the destruction of property, and the loss of life. Environmental hazards caused by the misapplication of pesticides and herbicides and the disposal of nuclear waste remain public health concerns. There is a geography of terrorism, and understanding its historical roots, along with the temporal and spatial patterns of recent events, is important in developing our homeland security policy. Although each of these topics is complex and multifaceted, an applied geographic perspective is an important first step in determining how best to respond to these threats.

—*Andrew Schoolmaster*

See also Geodemographics; GIS; Gravity Model; Population, Geography of; Urban and Regional Planning

Suggested Reading

Golledge, R. (2002). The nature of geographic knowledge. *Annals of the Association of American Geographers, 92,* 1–14.

Pacione, M. (2002). *Applied geography.* London: Routledge.

Torrieri, N., & Ratcliffe, M. (2003). Applied geography. In G. L. Gaile & C. J. Willmott (Eds.), *Geography in America at the dawn of the 21st century* (pp. 543–551). New York: Oxford University Press.

ART, GEOGRAPHY AND

Geography has always been highly reliant on visual imagery—and not least art—to explain the patterns and processes that lie at the heart of the discipline. Although this often has meant that art has been used as nothing more than the straightforward representation of place or landscape, during the past 40 years historical and cultural geographers have cultivated the critical interpretation of art as a specialist interest in geography. This has brought with it distinctive methods and approaches that have followed the broader contours of human geography. Before examining these in more detail, it is important to grasp two important ideas. First, it is misleading to refer to art as a homogeneous entity; art embraces numerous practices and outputs and includes sketching, etching, lithography, painting, sculpting, printing, montage work, installation work, and performance art. Sometimes the distinctions among these practices are difficult to discern, and artists invariably combine more than one technique in the production of a piece of art. Second, when we differentiate among different forms of art, we tend to refer to the genre and aesthetic styles that have been defined by the discipline of art history. Again, some of the distinctions that are made here can be misleading, although they remain important in the interpretation of art because they allow us to refer to key influences, primary practitioners, and broader cultural histories.

Although these ideas have been influential in shaping geography's interest in art, the discipline has also fashioned its own interpretive methods. These can be explained with reference to two significant developments in the study of art.

LANDSCAPE PAINTING AND REPRESENTATION

Geography's early attempt to interpret art was inspired by an overarching pursuit of generalizable rules about landscape taste and national identity. For David Lowenthal and Hugh Prince, John Constable's *The Haywain,* painted in 1821, exemplified an English devotion to rustic life and landscape. *The Haywain* was, and arguably still is, a depiction of quintessential England. However, Lowenthal and Prince argued that for every typical English landscape, there was always the aesthetic antithesis—the imposing demonic chimneyscapes of industry, as represented in L. S. Lowry's landscape art. Although Lowenthal and Prince's work on landscape art created important openings for geography, art was deemed to be not much more than a visual archive, a painted record of landscape artifacts. There was little consideration given to artistic style, technique, and genre.

The cultural turn in geography during the mid-1980s addressed this shortcoming in many ways and brought with it fresh insights to the interpretation of art. Inspired by the work of John Berger and Raymond Williams, among others, Denis Cosgrove and Stephen Daniels developed an intellectual history of the landscape concept in European art. They argued for an interpretive method they called "iconography" that allowed students and researchers of landscape to delve into the symbolical meanings represented in art. For Cosgrove and Daniels, it was not just the content of landscape art that was intriguing but also artists' use of color, texture, technique, perspective, and scale that allowed the links to be made between art and broader cultural histories. So, for example, in the interpretation of J. M. W. Turner's 1844 painting *Rain, Steam, and Speed,* Daniels argued that the artist was not interested in painting a factual local scene but instead was intent on endowing this landscape with ideas of a historical destiny shaped by the Industrial Revolution. Perhaps the most challenging aspect of iconography as an interpretive method is that it does not attempt to reveal a single truth about art; instead, it advocates multiple deconstructions of meaning. Art then becomes best understood as yielding a duplicity of meaning.

The work of Cosgrove and Daniels during the 1980s and 1990s has been inspirational to most geographers who have interpreted art during the past 15 years or so. Importantly, Cosgrove and Daniels broadened the horizons for geography by demonstrating that geographers could make valuable contributions to debates on art. Indeed, geographers have established some important collaborations with artists and galleries. But iconography has been taken forward and adapted as a methodology during recent years as an interest in visual culture has emerged.

GEOGRAPHY, ART, AND VISUAL CULTURE

Geographers are beginning to consider art in relation to visual culture. For Gillian Rose, there is an implicit set of power relations in the production and reception of visual imagery. That is, power relations are forged in the representation of an object *and* in its interpretation. For example, the female nude in Western art represents women as unclothed, passive, and a spectacle for the male gaze. This tells us much about the representation of subordinated women in Western art. It also tells us much about how masculine identities are constructed in the viewing of this art. Formulating an interpretation of art that addresses these two concerns allows us to think about the social conditions and effects of art. As a clear extension of these interests, geographers have begun to explore the spatialities of artistic practice where artistic practice, and not just the artwork, is deemed to be meaningful in its own right. In this sense, artistic practice not only is a means by which art is produced but also constitutes particular sociospatial networks.

There have been clear limitations to geography's well-established interest in art. There has been a reluctance to engage with abstract art, "non-Western" art, or art in a medium other than paint, and the art gallery as a social space remains a relatively unexplored subject matter. These areas of untapped interest suggest a potentially rich and diverse future for geography and art.

—*Rob Bartram*

See also Cultural Geography; Cultural Turn; Photography, Geography and; Spaces of Representation; Vision

Suggested Reading

Cosgrove, D., & Daniels, S. (Eds.). (1988). *The iconography of landscape: Essays on the symbolic representation, design, and use of past environments.* Cambridge, UK: Cambridge University Press.

Daniels, S. (1993). *Fields of vision: Landscape imagery and national identity in England and the United States.* Cambridge, UK: Verso.

Lowenthal, D., & Prince, H. (1965). English landscape tastes. *Geographical Review, 55,* 186–222.

Rose, G. (2001). *Visual methodologies: An introduction to the interpretations of visual materials.* London: Sage.

AUTOMATED GEOGRAPHY

Geography is the science and humanity of knowing about people and places. Automated geography is the modern, computer-assisted version of that quest. Formally, it is defined as the eclectic application of geographic information systems (GIS), digital remote sensing, the global positioning system, quantitative spatial modeling, spatial statistics, and related information technologies to understand spatial properties, explain geographic phenomena, solve geographic problems, and formulate theory. Its relationship to geographic information science (GISci) is analogous to the relationship that geography maintained with cartography for centuries and with remote sensing for decades, long before the advent of computers and satellite sensors.

Geographers have practiced their craft for at least 2,500 years, but their brand of analysis has always been extremely difficult due to the enormous volumes of data required to represent three-dimensional places and features, both physical and cultural. Thus, automated geography represents a historic leap forward for geographers and for society at large. During ancient times, one person could know and process a significant portion of all knowledge. The explosion of information generated by specialized disciplines during and after the Renaissance left geographers with three disappointing options. Those who studied large areas were limited to such coarse data that they often were dismissed as generalists. Those who insisted on detailed understanding were limited to such small areas that hardly anyone cared about their results. And those who limited themselves to a topical specialty sacrificed much of the holism that distinguishes geography from other disciplines. Today, automated geography restores geographers' ability to know and process a greater portion of all that is known. It enables them to study complex phenomena over large areas with sufficient spatial, temporal, and topical detail to reveal deep insights and generate new theories.

Collectively, GIS, remote sensing, and related geographic information technologies constitute a macroscope. Just as the microscope enabled people to see smaller things and the telescope enabled them to see farther, the macroscope enables them to see large phenomena in fine detail. Will this new scientific instrument turn out to be as powerful as those earlier ones? Will it generate revolutionary new theories in rapid succession as they did? Many conventional theories, developed in isolation by specialized disciplines with little thought for geographic relationships, spatial logic, or integration, have stood unchallenged for decades. The time is right for geographers and geographic information scientists to enter the fray. Automated geography ensures that they have much to offer.

—*Jerome E. Dobson*

See also GIS; Social Informatics

Suggested Reading

Dobson, J. (1993). A conceptual framework for integrating remote sensing, GIS, and geography. *Photogrammetric Engineering and Remote Sensing, 59,* 1491–1496.

Dobson, J. (1993). The geographic revolution: A retrospective on the age of automated geography. *The Professional Geographer, 45,* 431–439.

Longley, P., Goodchild, M., Maguire, D., & Rhind, D. (2001). *Geographic information systems and science.* New York: John Wiley.

B

BEHAVIORAL GEOGRAPHY

Behavioral geography investigates human action in geographic space as mediated through the cognitive processing of environmental information. Its emphasis is on spatial behavior and the psychology that lies beneath it at an individual level. Behavioral geography deals with the environment defined by human behavior, with people central and integral to every problem. Its major focus has been on the relations between a multidimensional environment and the multifaceted process of human action, mediated through perception and cognition as active processes of learning about places, with the mind mediating between the environment and behavior in it.

Behavioral geography grew as a reaction to the absence of individual action in the models of spatial science that arose from the quantitative revolution in geography during the late 1950s and early 1960s. Researchers became dissatisfied with the mechanistic and deterministic nature of quantitative models of human behavior that focused on so-called "rational economic man." Some of the early assumptions of spatial analysis, that individuals were both entirely rational and optimizers in their spatial actions, were too simplistic. Randomness was introduced to empirical studies, soon to be followed by a set of cognitive variables that led to common ground with psychology. Behavioral geography seeks to understand the geographic world through the windows of individuals—their thoughts, knowledge, and decisions—aiming to provide an insight into human spatial processes by studying the processes themselves. In this manner, behavioral geography attempts to comprehend reasons for overt spatial behavior by incorporating behavioral variables and through understanding the ways in which humans come to know the geographic world in which they live.

The environment in which spatial behavior takes place—the myriad of decision-making processes that are undertaken each day to travel, to work, to shop, and so on—is far too complex to be incorporated into a computational model using individual normative rational beings. This would involve mapping at a scale of nearly one to one. As a result, aggregate models of spatial behavior were developed. Alternative models of aggregate behavior developed within behavioral geography, built on the concepts of satisfier rather than optimizer. Within behavioral geography, there is interest in the environment beyond the physical, economic, social, political, and legal, and this is expanded to include the cognitive, perceptual, ideological, philosophical, and sociological. The focus is at a more micro level and is process based, and generalizations are based on behavioral responses rather than arbitrary criteria such as location, demographics, and socioeconomic indexes.

THEMES WITHIN BEHAVIORAL GEOGRAPHY

Research during the peak of behavioral geography's popularity advanced around several themes. Locational analysis was reshaped to incorporate the more grounded ideas of decision making and the awareness that decisions were based not only on economic and other quantifiable variables but also on values, cultural biases, and habit. The concerns and actions of the decision-making actors in the geography of environmental hazards were clearly at odds with mathematical rational decision making. One example is the study of people relocating to and investing in property located in hurricane- and storm surge–prone areas. Here behavioral geography is critically used to study individuals' spatial actions, in choosing whether or not to evacuate, along with their perceptions of extreme weather events. Behavioral geography continued to expand into areas of environmental perception, the evaluation of the meaning of places, the study of mental and cognitive maps, environmental learning, spatial search behavior, wayfinding, and spatial reasoning.

CRITIQUES OF BEHAVIORAL GEOGRAPHY

During the early 1980s, behavioral geography came under attack for retaining methodologies that appeared to be aligned and predicated on positivist philosophies that shaped the quantitative revolution. Further criticisms were made about the intrusive nature of its methodology, interrupting the flow of natural human action. Furthermore, by using semiformal methods of evaluation, the social context from which spatial behavior and actions originate was normalized or removed. With other theories gaining ground in human geography, further questions were leveled against behavioral geography. How could realities that are not directly observable be explored? How does the behavior of individuals relate to the contextual forces of ideology and social structure?

In its search for the cognitive component in spatial behavior—how individuals acquire, code, store, recall, and ultimately implement the information they have acquired—behavioral geography has attracted criticisms from researchers concerned with social issues. Later, behavioral geography was attacked for understanding the world rather than trying to change it. It was criticized for being passive to social problems of the geographic world. This dissatisfaction caused a split within behavioral geography into two branches: the analytical branch, which was concerned with incorporating behavioral information in spatial models, and the phenomenological branch, which rejected spatial models being concerned with a sense of place, values, and morals.

BEHAVIORAL GEOGRAPHY TODAY

More recently, behavioral geography is becoming more socially aware, focusing on the individual and acknowledging the importance of social and cultural contexts within which we live. In recent work by Reginald Golledge and Robert Stimson, the range and depth of behavioral geography can be seen as vast and relevant. If the term *behavioral geography* is replaced by *the geographic study of spatial behavior,* the utility of the research area can be reconsidered. Geographic inquiry involves exploring spatial behavior across aggregate and disaggregate situations at a variety of scales from micro to macro, over varying time spans, and in different settings. In this state-of-the-art review, many of the earlier criticisms of behavioral geography are addressed, including issues of objectivity, validity, and reliability. Today behavioral geography remains relevant even if the term is not widely used. Its concepts are central to research in the following areas: decision making and choice behaviors; technological and social change; urban patterns and trends; spatial knowledge, perceptions, attitudes, and risk; spatial cognition and cognitive mapping; activity analysis in travel and transportation modeling; consumer behavior and retail center location; causes and nature of migration; residential mobility and location decisions; geography; and disabled populations. Behavioral geography continues its search to clarify the decision-making processes that influence spatial behavior.

—*Daniel Jacobson*

See also Body, Geography of; Cognitive Models of Space; Disability, Geography of; Environmental Perception; Existentialism; Humanistic Geography; Identity, Geography and; Mental Maps; Phenomenology; Social Geography; Spaces of Representation; Subject and Subjectivity; Symbols and Symbolism; Time, Representation of; Vision

Suggested Reading

Cox, K., & Golledge, R. (Eds.). (1981). *Behavioral problems in geography revisited.* London: Methuen.

Golledge, R., & Stimson, R. (1997). *Spatial behavior: A geographic perspective.* New York: Guilford.

Spencer, C., & Blades, M. (1986). Pattern and process: A review essay on the relationship between behavioral geography and environmental psychology. *Progress in Human Geography, 10,* 230–248.

BERKELEY SCHOOL

The Berkeley School refers to the loose association of like-minded geographers associated with Carl O. Sauer (1889–1975) and his perspectives and predilections. During his long career (1923–1975) in the Department of Geography at the University of California, Berkeley, Sauer fostered an "invisible college" of geographers and a distinctive school of geography grounded in biophysical, cultural, and historical approaches. Initial members were mostly his graduate students, but subsequent affiliates included visiting faculty and lineal descendants now into the fifth academic generation. Field study conducted in Latin America is one hallmark of the Berkeley School. Some 200 geographers can be included in these ranks. Perhaps an equal number have pursued Berkeley-style studies elsewhere in the world. First-generation adherents include some of geography's major figures of the 20th century: John Leighly, Fred Kniffen, Donald Brand, Joseph Spencer, Leslie Hewes, George Carter, Dan Stanislawski, Andrew Clark, Robert West, James Parsons, Wilbur Zelinsky, Philip Wagner, David Sopher, Homer Aschmann, Fred Simoons, and Marvin Mikesell. In turn, they and their students have spawned an ongoing collectivity that has carried the enterprise forward—with modifications, of course—into the present. Some of the notables of the succeeding generations of Latin Americanists include William Denevan, Daniel Gade, Bernard Nietschmann, B. L. Turner, II, David Harris, Daniel Arreola, Thomas Veblen, and Karl Zimmerer. Others less directly in the lineage include Yi-Fu Tuan and David Lowenthal. Geographers with informal ties to the Berkeley department could also be included. J. B. Jackson, Peirce Lewis, and Robin Donkin stand out here, but the list ultimately includes all of those geographers and kindred scholars who self-identify with, and draw inspiration from, Sauerian historical–cultural landscape studies in their various modes. That cohort, past and present, numbers in the hundreds and consequently remains perhaps the largest single such grouping in geography.

Although Sauer himself on various occasions disavowed promotion of a school or issuing programmatic statements, both their outlines and output were evident within Sauer's first decade at Berkeley. Sauer's 1925 philosophical/methodological tract, "Morphology of Landscape," issued an incisive broadside against environmental determinist tendencies within human geography and the Davisian physiographic cycle as a model for physical geography. It also served to put historical chorology and cultural landscape studies at the center of a postenvironmentalist geography. Sauer's program was periodically reinforced by additional statements, most notably his entry on "Recent Developments in Cultural Geography" in the 1927 volume *Recent Developments in the Social Sciences* and his 1940 presidential address "Foreword to Historical Geography" to the Association of American Geographers. More important than his philosophical writings, however, were his substantive research interests. In this regard, his career trajectory went from regional studies in graduate school (his Ozark dissertation), to land use inventory and field methods in Michigan, to geomorphology at the outset of his California move, to historical studies of colonial California, to prehistoric investigations in northern Mexico (especially questions of plant and animal domestication), to cultural diffusions more broadly, to Pleistocene human migrations and adaptations, to tropical cultural biogeography, to anthropogenic environmental impacts globally, and finally (after retirement in 1957) to a suite of historical geographic studies of North America, the North Atlantic, and the Caribbean. Although this set of concerns scarcely encompasses the Berkeley School's bounds, it invited collaboration, elaboration, and imitation. Several of his students (e.g., Kniffen, Clark) have been credited with establishing their own distinctive schools, with multiple students producing studies that are recognizably part of the larger Berkeley tradition.

Despite a far-reaching eclecticism that embraces cultural and historical topics from the ancient and arcane to the contemporary and quotidian, the Berkeley School's overarching and unifying concern, as Sauer said many times, is for the appropriation of habitat by habit and the resultant impact of culture(s) on the earth's landscapes through all of human time.

—*Kent Mathewson*

See also Anthropogeography; Cultural Geography; Culture Hearth; Regional Geography

Suggested Reading

Kenzer, M. (Ed.). (1987). *Carl O. Sauer: A tribute.* Corvallis: Oregon State University Press.

Mathewson, K., & Kenzer, M. (Eds.). (2003). *Culture, land, and legacy: Perspectives on Carl O. Sauer and Berkeley School geography* (Geoscience and Man, Vol. 37). Baton Rouge, LA: Geoscience Publications.

Speth, W. (1999). *How it came to be: Carl O. Sauer, Franz Boas, and the meanings of anthropogeography.* Ellensburg, WA: Ephemera.

BODY, GEOGRAPHY OF

Many critical human geographers, such as feminist, socialist, antiracist, postcolonial, and queer geographers, focus on the body as one possible route to changing social, cultural, and economic relations for the better. These geographers increasingly recognize that bodies—those that have a particular skin type and color, shape, genitalia, and impairments, are a specific age, and so on—are always placed in particular temporal and spatial contexts. Questions of the body—its materiality, discursive construction, regulation, and representation—are crucial to understanding spatial relations at every spatial scale.

In some ways, attempting to define the body seems nonsensical. We all are bodies, and bodies are more than just possessions. Although there has been a long-standing preoccupation with the body, there is little agreement about the meaning of the body or even what the body is. Philosophers from the ancient Greeks to the postmodernists have attempted to understand and define the body. During the Enlightenment, philosopher Descartes argued that the mind was separate from—and superior to—the body. This dichotomy became known as the Cartesian division, or dualistic thinking, which laid the foundations for the development of modern scientific rationalization. This distinction between mind and body has been gendered, racialized, sexualized, and so on. The mind has been associated with positive terms such as rationality, consciousness, reason, whiteness, heterosexuality, and masculinity, whereas the body has been associated with negative terms such as emotionality, nature, irrationality, blackness, homosexuality, and femininity.

Claims about allegedly natural biological differences between men and women, or between whites and blacks, are known as essentialist arguments. They assume that bodies have fixed or stable essences. This has been challenged by social constructionists who argue that differences are produced through social and material practices and systems of representation rather than by biology. Dualisms have shaped geographers' understandings of society and space and the production of geographic knowledge to the point that, for example, the public has been privileged to the exclusion of the private. This has been challenged in recent feminist work that shows how bodies are constructed through a variety of public and private spaces.

Bodies are surfaces of social and cultural inscription, house people's identity, are sites of pleasure and pain, are public and private, have permeable boundaries, and are material, discursive, and psychical. Although our bodies make a difference to the experience of places, we might also think of bodies and spaces as mutually constituted. Instead of thinking about space and place as preexisting sites where bodily performances occur, some studies have argued that bodily performances themselves constitute or reproduce space and place. Geographers have looked at the way in which bodies are gendered, sexualized, racialized, aged, and so on by, for example, workplaces, schools, leisure spaces, homes, suburbs, cities, and nations.

—*Lynda Johnston*

See also Behavioral Geography; Children, Geography of; Cognitive Models of Space; Emotions, Geography and; Existentialism; Home; Human Agency; Humanistic Geography; Identity, Geography and; Mental Maps; Sense of Place; Situated Knowledge; Social Geography; Time Geography

Suggested Reading

Longhurst, R. (2001). *Bodies: Exploring fluid boundaries.* London: Routledge.

Nast, H., & Pile, S. (Eds.). (1998). *Places through the body.* London: Routledge.

BOUNDARIES

Boundaries are the edges of regions. This term often is reserved for political boundaries that mark the change from a region administered by one governing authority to that administered by a different authority. Although cities, counties, and provinces all are political entities and all have boundaries, international boundaries (those between states) are of special concern because states remain the highest level of political authority in the world. Current changes in the status and functions of states (e.g., loss of decision making over trade decisions to the World Trade Organization) are reflected in the status and functions of their boundaries.

Although boundaries limit state sovereignty and therefore the enforcement of regulations to state territory, there have been and will continue to be incidents where states attempt to or actually enforce their laws extraterritorially. Kidnapping suspected criminals residing in other states, for example, has been organized by both Israeli and U.S. government agencies. The Israelis eventually executed former Nazi leader Adolph Eichmann in 1962 after capturing him in Argentina, and Humberto Alvarez was released to Mexico in 1992 after being kidnapped to the United States and later acquitted of murder charges in federal court. International law prohibits extraterritorial actions, but the international community is unable to prevent them.

Nearly all existing international boundaries are defined by treaty and are demarcated on maps. Positional disputes over the locations of boundaries certainly exist but rarely lead to war. The United States and Canada, for example, still disagree over offshore boundaries in the Beaufort Sea and the Dixon Entrance. A few frontiers remain where areas—rather than lines—separate states, but these are impractical for regulating passage into and out of states or for developing resources and so have been progressively replaced by boundaries. Most of the few remaining frontiers are located on the Arabian Peninsula.

Although some boundaries are marked on the ground by walls or other structures (e.g., the U.S.–Mexico boundary between San Diego and Tijuana), the costs of construction are prohibitive; checkpoints along official crossing points are far more common. Aerial and electronic surveillance can be used for patrolling boundaries, but enforcement of state sovereignty along a state's boundaries is rarely absolute. Smuggling of illegal merchandise—whether it is drugs, people, weapons, or bootleg DVDs—is too lucrative for operators to cease their operations. The discovery of tunnels under the San Diego–Tijuana wall and elsewhere on the United States–Mexico boundary is evidence of the profits involved in smuggling.

State boundaries extend upward, downward, and offshore, increasing the resources and strategic locations under state authority as well as the possibility of conflict with other states. All mineral resources below a state's territory are under state authority, but states encounter difficulties when fluid resources (e.g.,

petroleum, ground water) flow into or from neighboring jurisdictions. For example, Kuwait was accused by Iraq of pumping oil from the shared Rumaila oilfield prior to Iraq's 1992 invasion. Airspace above sovereign territory—to the height of powered flight—is also under state authority. Invasion of airspace has resulted in arrests (e.g., of Matthias Rust, a German teenager who illegally landed in Moscow in 1987) as well as in destruction (e.g., the downing of Korean Airlines Flight 007 over the Soviet Union in 1983).

Offshore boundaries include a 12-nautical-mile territorial sea and a 200-nautical-mile exclusive economic zone (EEZ), with the latter extending state regulatory authority over offshore resources such as fisheries and petroleum. This specific offshore distance has changed over time. Centuries ago, offshore authority extended only 3 miles (the distance that cannon fire could control), and the current 200-mile EEZ was first claimed by Peru and Chile in 1947. In 1995, the Canadian navy boarded and seized a Spanish fishing vessel operating in international waters off the Canadian coast. The United Nations Conferences on the Law of the Sea, especially the Third Conference, have endorsed the EEZ, and the supporting convention has been signed and/or ratified by more than 140 states.

Traditionally, boundaries were perceived as separating internal domestic concerns from external international ones. This distinction, never completely true, is becoming increasingly blurred. Increased trade and trade agreements, migrations, and growing amounts of foreign investments all are components of the *globalization* process that creates increasing financial, cultural, and political linkages among states. As a result, state policy decisions unavoidably affect both domestic and international entities, and state decision makers are, in turn, subject to both domestic and international pressures to modify their policies.

State governments regulate access across their boundaries to create a geographically distinct regulatory climate that benefits constituencies. External economic threats have been met by limiting access to national markets through tariffs and other methods. For example, during recent years the United States has limited imports of beer and lumber from Canada and imports of avocados from Mexico. Security threats have been dealt with by excluding individuals fitting profiles of terrorists as well as those suspected of supporting terrorism. Immigration policies and communication policies (e.g., Iranian laws prohibiting MTV within Iran's borders) are also designed to promote national interests.

In addition, state governments regulate egress. Security concerns are the basis of regulations restricting technology or technical specialists from leaving state territory. Stringent controls on emigration remain the policy of North Korea, Cuba, and other states. Extradition is a politically sensitive decision, especially when famous or infamous individuals (e.g., Alberto Fujimoro) are involved. Expulsion, especially of diplomats, is practiced by all states to remove threats to national interests.

Boundary policies (passage regulation) generate countless disputes between states despite the processes of globalization. Disputants need not be contiguous, although contiguity ordinarily increases the flows—and therefore the potential for dispute—between states. Mexico and Canada both contest American regulations on access to the United States, but so do Japan, the United Kingdom, and Brazil. Future disputes over passage into and out of states may well increase as exports of water, toxic waste disposal, genetically modified organisms, and other controversial items become more common.

—*Peter Meserve*

See also Geopolitics; Globalization; Nation-State; Political Geography; State

Suggested Reading

Brenner, N. (Ed.). (2003). *State/Space: A reader.* Boston: Blackwell.

Newman, D. (Ed.). (1999). *Boundaries, territory, and post-modernity.* London: Frank Cass.

BUILT ENVIRONMENT

Humans are builders, and people surround themselves with the built environment—the landscapes, structures, and other artifacts that reflect their culture. Geographers studying the built environment often use it as a marker, a means of tracing the diffusion of the values, attitudes, beliefs, and traditions of the builders. Geographers such as Carl Sauer contributed early to understanding the built environment, although he used the term *cultural landscape*. His thesis was that if one knew the landscape, one would know the builder and

then could proceed to an understanding of human–environment relationships.

Many contend that philosopher Michel Foucault set the stage for contemporary urban postmodern critical inquiry by opening a dialogue on space, power, and knowledge. Others point to Jean-François Lyotard's incredulity toward metanarratives and his rejection of metatheory while advocating multiplicity. Jean Baudrillard continued the conversation by proposing that image or style (*simulacra*) has supplanted reality in our highly commodified world and that everything may best be understood as a complex of self-referential signs; it is the map that engenders the territory. These theorists accentuated new relationships between the city observer and the city observed, igniting a reexamination of the social aspects of spatiality. Cities such as Los Angeles, and even individual buildings such as the Bonaventure Hotel in that city, took privileged positions in this inquiry. Fredric Jameson's comments on the Bonaventure Hotel are especially pertinent because they challenged geographers to think of this structure as a marker for theoretical critical discourse rather than as a referent to the past.

Jameson's challenge was accepted by geographers such as David Harvey, who asserted that the era of postmodernism is characterized by massive space–time compression. In the postmodern period, space and time have virtually disappeared, losing their meaning and the structure of control they represented. The loss of the previous referents, particularly time, meant that geography was poised to contribute materially to the discourse on the built environment in the context of critical social theory.

Edward Soja responded directly to this challenge by drawing, in part, on the works of Michel Foucault to expose the connections between knowledge and power in the context of Los Angeles, the quintessential postmodern place, and the works of Henri Lefebvre, who claimed that social power derives from and is expressed in space. Soja employed Los Angeles as a test case of his contention that geography not only contributes to but also shapes the discourse on critical social theory. Thus, the built environment emerges as the marker for a new critical geography.

—*Tom L. Martinson*

See also Cultural Geography; Infrastructure; Postmodernism; Urban Geography

Suggested Reading

Ellin, N. (Ed.). (1996). *Postmodern urbanism.* Cambridge, MA: Blackwell.

Harvey, D. (1989). *The condition of postmodernity: An enquiry into the origins of cultural change.* Cambridge, MA: Blackwell.

Soja, E. (1989). *Postmodern geographies: The reassertion of space in critical social theory.* London: Verso.

BUREAUCRACY

The term *bureaucracy* refers to the growing tendency in modern societies to have power and influence embedded in institutions in the political, administrative, and economic realms. The term, academically popularized by Max Weber, points to the pervasive influence of diverse institutions in determining social and spatial outcomes in urban, suburban, and rural settings. Weber, concerned with the roots and sources of power in Western societies, argued that an ascendant bureaucratization of activities is a dominant characteristic of the modern era. Bureaucratic organization, characterized by layers of rules and regulations, casts of managers and gatekeepers, and the articulated prioritization of efficiency, is identified as a privileged instrument that governs local life.

Weber's notion of bureaucracy has been widely applied by human geographers to explain the control of resources in contemporary Western society and their social and spatial outcomes. For example, urban geographers have used this frame to understand patterns of residential differentiation and ghettoization in cities. Key gatekeepers, seen as controlling and managing scarce urban resources (e.g., mortgage loans, information on housing vacancies, government subsidies), are posited as powerful city operatives. Similarly, political geographers have used the bureaucracy notion to comprehend how political regimes in regions rely on a bureaucratic–organizational instrumentality to acquire legitimacy and assert political power. Bureaucracy here is simultaneously a method of resource allocation, a mechanism of political control, and a structure to organize the operation of institutions.

Most of the applications of bureaucracy by geographers have relied on an early rearticulation of the notion offered by sociologist Ray Pahl. This notion identifies the power and influence of local organizations whose members operate essentially autonomously and unconstrained by broader scale processes. A later version, also offered by Pahl, presents local government as the central source of organizational power in cities and society but whose members become increasingly interested in their own growth and perpetuation. This second notion, situating local organizations in a complex web of propelling forces, has been used less often by human geographers.

—*David Wilson*

See also Ghetto; Power; Segregation; Urban Geography; Urban Social Movements

Suggested Reading

Weber, M. (1978). *Economy and society.* Berkeley: University of California Press.

CAPITAL

Typically viewed as an accumulation of anything of value, interpretations of capital have multiplied into a broad range of meanings. In its simplest form, capital refers to the value of accumulated goods, although some would suggest that this definition be limited to the value of accumulated goods that will be used to generate profits. The valuation of capital may be based on its use value (the value of the capital in the production of other goods), its exchange value (the value of the capital in trade for other goods), or its labor value (the labor cost of reproducing the goods).

Although capital often is viewed as synonymous with money, capital has the additional property of convertibility. Capital can be easily converted from money into labor, into commodities, and then back into money. This conversion (or circulation or trade) can cause capital to increase in value. For example, when capital is converted into labor (via the payment of $10 in wages to a coal miner), the labor produces a commodity (coal). If the coal is later sold for $12, then $10 of money was converted into $12 of commodity via circulation. Money is a special case of capital because it can be deployed either as capital (e.g., to increase the value of existing capital stocks) or in nonproductive activities (e.g., entertainment). Capital's meaning was derived from royal capital grants of land during the 15th century. These capital grants formed the basis of estates that were intended to accumulate additional capital. The value of capital is tied to its future potential for productivity.

Some social scientists have moved beyond viewing capital as a thing and instead conceive of capital using the Marxist perspective of capital as a social relation. This definition of capital is used to refer to the class of people who possess capital. This elite group of people use their capital to employ (and control) the working class (or proletariat). Membership in the working class is defined by the lack of capital; the only commodity of value possessed by the working class is their labor power, which is traded for wages. The exchange of capital (money) for labor (and the rules that govern this exchange) comprises this social relation.

Geographic research on capital has focused largely on the flow and availability of capital for investment and on the role of the social relations created by capital in creating patterns of uneven economic development and environmental impacts at the individual, local, regional, and global scales.

—*William Graves*

See also Economic Geography; Factors of Production; Marxism, Geography and

Suggested Reading

Basgen, B., & Blunden, A. (Eds.). (2004). *Encyclopedia of Marxism: Glossary of terms.* Available: www.marxists.org

Wolff, R., & Resnick, S. (1987). *Economics: Marxian vs. neoclassical.* Baltimore, MD: Johns Hopkins University Press.

CARRYING CAPACITY

The concept of carrying capacity is borrowed from ecology, where it is defined as the maximum number of a given species that can be supported indefinitely in a given habitat without permanently reducing its productivity. This definition needs to be modified for humans because they can eliminate competitive species, import resources, and adopt technologies to sustain numbers.

Human carrying capacity is particularly difficult to predict because it depends not only on poorly understood natural constraints, such as sustainable soil fertility and climatic uncertainties, but also on a whole gamut of socioeconomic factors, such as migration, demography, values and fashions, individual versus collective choice, and even religion. And in a world where trade is global and commons such as the seas and atmosphere are shared, the notion of carrying capacity at anything but the global level is not helpful.

It is useful to distinguish between the actual population that can be sustained under a possible technological fix (biophysical carrying capacity) and the number that might be sustained under a pattern of resource consumption associated with a particular social system (social carrying capacity). At any level of economic development, the social carrying capacity will always be less than the biophysical one. After all, no one wants to live like factory-farmed animals or to live on a diet of bread; choice and freedom of action are part and parcel of development, and not all states will want to eat from the same menu. It is also worth noting that technological fixes cannot make the biophysical carrying capacity infinite because ultimately there are technical limits to photosynthetic efficiency and the production of carbohydrates.

Models of global carrying capacity have varied dramatically in their predictions. The *Limits to Growth* study published by the Club of Rome in 1972 predicted, rather pessimistically, that within 100 years of that time society would run out of renewable resources, leading to a precipitous collapse in the world economic system and food production and ultimately resulting in a soaring increase in the death rate and disastrous population decline.

In contrast, optimistic models, such as that put forward in *The Next 200 Years: A Scenario for America and the World,* predict a densely populated world with no poverty and humans in control of the forces of nature. Optimistic scenarios are contingent on continually improving technologies developed as and when needed. This is a somewhat utopian vision given the repeated catastrophic impacts of hurricanes on the U.S. Gulf Coast and the prolonged droughts in the Sahel.

Maintaining any carrying capacity requires sustainability, and there are reasons to believe that many global resources are becoming severely degraded. Irreversible land degradation is widespread, atmospheric pollution is a feature of many industrial regions, and even the oceans are not without damage as measured by the catastrophic decline of many of the oceans' fish stocks. All of these impacts indicate beyond any doubt that global social carrying capacity has already been exceeded. And carrying capacity models have not even begun to include variables such as global warming. Even if people can be persuaded to change their lifestyles, maximizing carrying capacity requires better social, political, and economic global governance.

—*Peter Vincent*

Suggested Reading

Kahn, H. (1976). *The next 200 years: A scenario for America and the world.* New York: William Morrow.

Meadows, D., Randers, J., & Meadows, D. (2004). *Limits to growth: The 30-year update.* White River Junction, VT: Chelsea Green.

CARTELS

Cartels are associations of independent business firms or nations usually involved in the same industry. Their purpose is to regulate the production, pricing, and marketing of goods by their members. Cartels aim to increase market share. They have been particularly common in the mineral sector. In market economies such as the United States, cartels are carefully monitored because of possible collusion and price fixing. As a result, noncartel nations fear unfair comparative advantages. Some market analysts consider cartels as monopolistic and guilty of triggering price distortions in commodity trading. By forcing prices up collectively, members of cartels avoid direct competition with each other yet maintain high market share and profits.

Cartel agreements identify how, when, where, and at what price a given commodity will be exploited. Perhaps the most renowned cartel is the Organization of Petroleum Exporting Countries (OPEC). It was established in 1960 to manage the network of oil production and distribution as well as to ensure tighter control over the price of crude oil. The tripling of oil prices by OPEC in 1973 demonstrated the full force that cartels can impose on the global economy and cast that particular organization as a powerful political and economic force.

Although much attention has focused on OPEC, it is a relatively new cartel. Older mineral cartels such as the one for tin were designed by suppliers to keep prices high. However, the high prices of tin during the first half of the 20th century led to tin recycling that, in turn, curtailed demand for this nonferrous metal. Competition from other materials such as aluminum and plastics also kept tin prices relatively low.

If the aim of these consortia is to control the supplies and prices of their minerals, their effectiveness has been uneven. Cartels for mercury and bauxite, for instance, have had mixed results. The record shows that when minerals are more geographically concentrated (e.g., oil in the Middle East), cartels tend to be more effective at establishing global prices and supplies than when mineral operations are more dispersed.

Cartels can afford member nations a way in which to countervail the forces imposed by transnational corporations such as Exxon, Shell, and Alcoa. Multinational corporations' first allegiance is to investors and not the needs of the exporting nations. Nonetheless, it would be inaccurate to characterize all national members of cartels as benevolent market economies; recent controversies in Saudi Arabia, Russia, and Venezuela suggest otherwise. So long as labor costs for mineral extraction help keep mineral extraction in developing nations lower than comparable sites in more developed nations, cartels will exercise considerable power, especially in the hydrocarbon sectors (oil and gas).

—*Joseph Scarpaci*

See also Transnational Corporations

Suggested Reading

Dicken, P. (2003). *Global shift: Transforming the world economy* (4th ed.). New York: Guilford.

Jumper, S., Bell, T., & Ralston, B. (1980). *Economic growth and disparities: A world view.* Englewood Cliffs, NJ: Prentice Hall.

CARTOGRAM

A cartogram is a map projection that uses purposeful distortion to represent some terrestrial phenomena on a geographic map. In this sense, Mercator's projection is a cartogram because it distorts distances to represent loxodromic (rhumb line) directions, and all maps have some distortion of the two-dimensional surface of the earth. In more conventional parlance, there are three main types. One of these uses a metric scale other than kilometers to represent distance on the map, generally from one or two places. The map scale may be in minutes of travel time, dollar costs, or other units of inconvenience. The most common type is centered at a specific place. On a normal map, the same relation often is shown by isochrones or isotims. To show simultaneous relationships from more than two places on one graphic requires an approximation, usually calculated using a trilateration, multidimensional scaling, or least squares.

The second common type of cartogram stretches space according to the density of some distribution on the earth, often population or resources by country. Within this class, there are three subtypes. One of these is the rectangular cartogram of Erwin Raisz, also called a value-by-area map. A second subtype is the noncontiguous cartograms introduced by J. Olson. The more common variant is represented by a continuous map, also called a contiguous cartogram. This latter group, a generalization of the equal area class of map projection, has been the subject of numerous construction algorithms, including many recent ones using computers. This is in part because the single defining equation is not sufficient to render a unique solution. In some cases, the value-by-area property is relaxed to better preserve recognizable shapes.

The most common use of contiguous cartograms is as a graphic display to contrast the geographic distribution of some phenomenon in comparison with the conventional map. On occasion, a second geographic variable is shown (often by distinct colors) on the cartogrammatic base map, for example, per capita income on a world population cartogram. Another use, most common in epidemiology, is to present the geographic arrangement of some distribution of concern to examine whether or not clusters are related or dependent on the distribution of people. Cartograms may also be used as an anamorphose designed to solve a specific theoretical problem.

The third map subtype maintains topological relations but not metric distances. The classic example is the London subway diagram, where the order of stations is correct but the distances between them are not. Early railroad advertising maps were similar.

—*Waldo Tobler*

See also Cartography

Suggested Reading

Dorling, D. (1995). *A new social atlas of Britain.* Chichester, UK: Wiley.

Levison, M., & Haddon, W. (1965). The area adjusted map: An epidemiological device. *Public Health Reports, 80*(1), 55–59.

Muller, J.-C. (1984). Canada's elastic space: A portrayal of route and cost distances. *The Canadian Geographer, 18,* 46–62.

Olson, J. (1976). Noncontiguous area cartograms. *The Professional Geographer, 28,* 371–380.

Raisz, E. (1934). The rectangular statistical cartogram. *Geographical Review, 24,* 292–296.

Spiekermann, M., & Wegener, M. (1994). The shrinking continent. *Environment and Planning B, 21,* 651–673.

Tikunov, V. (1988). Anamorphated cartographic images: Historical outline and construction techniques. *Cartography, 17*(1), 1–8.

Tobler, W. (2004). Thirty-five years of computer cartograms. *Annals of the Association of American Geographers, 94*(1), 58–73.

CARTOGRAPHY

Cartography can be concisely and classically defined as the art, science, and technology of making maps. The popular associations of the word with techniques of map making are a reflection of its lexical routes in *cart* (French for *map*) and *graffiti* (Greek for *writing*).

More specifically, cartography is a unique set of transformations for the creation and manipulation of visual or virtual representations of spatial information, most commonly maps, to facilitate the exploration, analysis, understanding, and communication of information about that space. Maps are a symbolized representation of a spatial reality designed for use when spatial relationships are of primary interest. This sweeping definition would encompass all types of maps, plans, charts and sections, three-dimensional models, and globes representing spatial information or any celestial body at any scale. Cartography, therefore, has many variables of meaning but can be broadly considered as the process and study of map making. It is more than an art/craft or a technology for producing artifacts (maps); it is a science seeking to abstract general truths and principles about this process.

The nature of cartography reflects the human need to have a spatial awareness and knowledge of the environment. This has been expressed from times of prehistory in cave drawings to the present day in complex computer models and virtual worlds. In this sense, maps historically have acted, and continue to act, as external aids for spatial communication and to facilitate the investigation, analysis, and discussion of spatial problems.

DEFINING MAPS

Put simply, a map is a model of spatial information. Traditionally, maps often were classified according to their subject or purpose—navigation charts, cadastral maps showing land ownership, topographic maps, general reference maps, thematic or statistical maps, maps illustrating a particular theme, and so on. It is now preferable to think of maps along different dimensions. A map can be permanent and hard copy (on paper) or virtual (existing in digital or cognitive [mental map] form). Maps can be visible (able to be seen) or invisible (stored in a computer database). Maps can be readily manipulated among these forms: paper (permanent: visible and tangible), on a computer screen (virtual: visible but not tangible), stored on a disk (virtual: invisible but tangible), and accessible over a network from a database such as the World Wide Web (virtual: invisible and intangible). Maps now have the capacity for additional functionalities; they can be dynamic, animated in real time, designed with new variables such as sound, and interactive (containing hyperlinks to connect with additional information within the related database), thereby offering sources well beyond their visible content. Maps help users to navigate through geospace via associated network-linked databases of geospatially related information. Maps can be used as single virtual images or as collections of such images accessible on CDs or over a network, they can be used as part of an interactive system in which the user/decision maker is able to select and interact with previously assembled maps, and they can be used to access databases (via an interface map) to search and customize what is needed. This facilitates a novel dynamic two-way process of interacting with spatial information.

CARTOGRAPHIC TRANSFORMATIONS

These map types have been developed due to recent transformations in cartography. Since the 1960s, cartography has become increasingly computer assisted with (a) the development of software and hardware to facilitate map production, (b) the flexibility and user-friendliness of the graphical user interface and widespread development of desktop publishing software, and (c) the rise of the use of geographic information systems (GIS), which has led to a renewed interest in cartography and the power of maps as the critical end point in the public display of complex and systematic geographic analysis. A GIS is a specialist information system that processes geographic/geospatial information combining software, hardware, data, data transfer systems, procedures, and humans, facilitating the analysis and display of geographic and related information. The advent of the Internet, and in particular the Web, has led to a proliferation of maps and mapping services. This has increased the amount of geospatial information available to nonexperts and the authoring of maps by many nontraditional cartographers. Parallel to this, new issues around ownership, access, and the security of information have developed. Further insights into cartographic products have been gained through cartographic visualization. Here a generalized, symbolized, and measurable visual image is explored in a cartographic manner to reveal previously unknown relationships or patterns within the data. Thus, an animated interactive digital terrain model is a form of cartographic visualization. With recent technological developments in positioning systems and mobile computing, a new realm of cartography is emerging in the form of portable digital mapping delivered to personal data assistants or mobile telephones with personalized content and geographically contextual relevant information. Cartography is now used on a wide range of scales, from displaying the minute DNA in medical imaging to illustrating the vast displays of interstellar systems.

CARTOGRAPHIC RESEARCH AREAS

The technical advances of cartography have been paralleled by the recognition of the social origins and consequences of maps, including the power of maps as used for colonial, navigation, war, propaganda, ownership, and territorial agendas and their role in framing and shaping the power and knowledge that led to the role they played in the understanding of geographies of the modern world. Cartography is a complex, culturally embedded process situated within historically specific contexts.

There has been a rise of technical and computational approaches that have led to an increase in analytical tools, symbolic codes, comprehension of data

values, spatial patterns, and geographic relationships derived from developments in computer science and other disciplines. At the same time, cartography has been challenged as an objective rational science; its ability to create an accurate and objective scaled representation of reality has been challenged due to inherent problems with representation and those of cartographic generalization, selection, and classification with the need to suppress, smooth, and displace features. These problems have been known for a long time but have been explored more systematically in a technical manner attempting to quantify uncertainty and imprecision.

The creativity of the artistic process involved in cartography has long been acknowledged. Art is most apparent by the use of emotive symbols, the choice of colors for graphical representation, and the use of decoration; hence, maps are prized as works of art. In addition, there is an awareness of the role of the imagination and artistic processes involved during the classic cartographic methodological problems of framing selection, classification, and composition.

Cartography is a vibrant field, combining research and ideas from many disciplines that are relevant to social and scientific inquiry. Cartography's broad reach and impact on our lives continue to evolve with new developments in visualization and on the Web, such as cybercartography, taking cartography into areas of augmented and virtual reality.

—*Daniel Jacobson*

See also GIS; Spaces of Representation; Vision

Suggested Reading

MacEachren, A. (1995). *How maps work: Representation, visualization, and design.* New York: Guilford.

Robinson, A., Morrison, J., Muehrcke, P., Kimerling, A., & Guptill, S. (1995). *Elements of cartography.* New York: John Wiley.

Slocum, T., McMaster, R., Kessler, F., & Howard, H. (2005). *Thematic cartography and geographic visualization* (2nd ed.). Upper Saddle River, NJ: Prentice Hall.

Wood, D. (1992). *The power of maps.* New York: Guilford.

CELLULAR AUTOMATA

Cellular automata are a class of abstract models that exhibit complex spatial dynamics. Cellular automata are attractive as relatively simple representations of apparently complex processes. In human geography, cellular automata have been used to model urban development and sprawl, land use and land cover change, and the spatial interactions of social groups.

In computer science, an *automaton* is a machine whose internal state changes in response to its current state and the state of its inputs. A *cellular automaton* (CA) is a collection of identical automata interconnected in a *lattice* or an *array,* so that the inputs to each automaton are the states of neighboring automata. Each automaton is a cell whose evolution is governed by its current state and by the changing states of neighboring cells. In geographic applications, two-dimensional grid arrays with each cell connected to its four or eight immediate neighbors are most common, although other array configurations are possible. The mapping that defines how each combination of the current and neighboring states of a cell leads to the next state is termed a *rule.* A rule may be deterministic or stochastic, and cell state changes may occur simultaneously for all cells or sequentially.

This description gives little sense of the variety of dynamic behavior exhibited by cellular automata. John Conway's "Game of Life" generates patterns reminiscent of the development of cell cultures on a microscope slide from a very simple rule and is the best-known example, with many free implementations available on-line. In general, there is no way to predict the global behavior of a cellular automaton from the rule governing its behavior at the local cellular level. This characteristic resonates strongly with the issue of understanding how processes scale up and down in geography.

In human geography and planning, CAs have served both as simple abstract models and as the basis for more complicated models of urban development and sprawl, land use and land cover change, and the spatial interaction of social groups. Consider, for example, how land use and land cover change can be represented in a CA. Using remote-sensed imagery, land cover classification may be assigned to every cell on a map grid. Possible and likely transitions in land cover classes can be described as a rule, so that land cover dynamics are represented by the evolution of a CA. For example, land classified as "industrial" might change to "derelict" but not immediately to "parkland." A simpler example might use only two cell states, developed and not developed, to explore urban growth and sprawl.

Models along these lines have been presented by Michael Batty, Keith Clarke, and Roger White, among others. Many departures from the standard CA architecture are typical in geographic applications. In particular, geographers have been concerned with accommodating nonlocal interaction between cells and have experimented with cell update sequences that do not require all cells to consider changes at every time step in model evolution. The implications of departing from regular grid arrays have also been explored.

Opinions differ as to the usefulness of CA-based models in geography. Although the potential for developing models that intrinsically capture how local interactions scale up to create global patterns is welcome, for some the framework is too restrictive for the development of truly useful simulation models. However, the pedagogic value of simple CA models in showing how local effects can combine to produce unexpected outcomes is widely acknowledged.

—David O'Sullivan

See also Agent-Based Modeling; GIS; Humanistic GIScience

Suggested Reading

Batty, M., Couclelis, H., & Eichen, M. (1997). Urban systems as cellular automata. *Environment and Planning B, 24,* 159–305.

Couclelis, H. (1988). Of mice and men: What rodent populations can teach us about complex spatial dynamics. *Environment and Planning A, 20,* 99–109.

CENSUS

A census is a periodic enumeration of people, the value of their property, and other general characteristics of a country. Historically, it was a way for leaders to assess how many men could be mobilized for war and how much property could be taxed. The first U.S. census was taken in 1790, under the direction of Secretary of State Thomas Jefferson, for the purpose of apportioning seats in the House of Representatives to the original 13 states. The U.S. Constitution mandated that Congress conduct a census every 10 years to collect information needed to reapportion Congress and to gauge the state of the nation. The practice of taking a census spread across Europe during the

19th century and then spread to the rest of the world after World War II.

The census provides a wealth of spatially referenced data. The provision of universal coverage—that everyone in a specified territory is counted and described—enables geographers to map the population and its characteristics comprehensively. The census's periodic quality allows analysis over both time and space. Census data are used widely to monitor income and poverty levels of a population, to locate medical services, to design transportation systems, and to track the changing skill levels of the labor force. Despite their usefulness in public policymaking, decennial census data become increasingly outdated as the decade progresses and spatial detail is sometimes sacrificed to meet census confidentiality provisions.

Although censuses often are depicted as "objective" sources of information, it is increasingly clear that there are limits to this objectivity. It is nearly impossible to count every member of a population, especially in a large, diverse, and constantly moving population. It was just such a challenge that led in 1980 to the idea of a postenumeration survey by the U.S. census to adjust the head count based on the known undercount of urban minorities and the known overcount of suburban whites. However, partisan politics interfered. In 1998, the U.S. Supreme Court prohibited the use of sampling for apportionment, and the U.S. Bureau of the Census decided not to adjust the 2000 census for redistricting purposes.

The taking of a national census is an expression of national identity—this is who we are as a people. The organization of people into categories reflects a resolve among elites to set boundaries and develop cultural identities within the larger group—to distinguish among peoples, religions, languages, and regions. Racial and ethnic classification systems derive from and reinforce race and ethnicity as sources of group identity. Groups that advocated for the opportunity to choose two or more races in the 2000 census justified it with the language of social identity. Although civil rights enforcement favors a small number of categories, a growing multiracial society requires a larger number for choice, self-expression, and cultural identity.

—*Pat Gober*

See also Census Tracts; Population, Geography of

Suggested Reading

Kertzer, D., & Arel, D. (Eds.). (2002). *Census and identity: The politics of race, ethnicity, and language in national censuses.* Cambridge, UK: Cambridge University Press.

Prewitt, K. (2003). *The American people: Census 2000.* New York: Russell Sage.

CENSUS TRACTS

Census tracts are small, relatively stable statistical areas that generally contain between 1,500 and 8,000 people, with an optimal population of 4,000. They belong to a hierarchical system for organizing the territory of the United States for census-taking purposes. The country is divided first into regions and subdivisions, then into states and counties, and finally into census tracts, block groups, and blocks. Census tracts are delineated by local census statistical area committees following U.S. Bureau of the Census guidelines. Their spatial sizes vary widely depending on the density of settlement. Census tract boundaries are defined with the idea that they will be maintained over many decades so that comparisons can be made from one census to the next, but physical changes in street patterns and new development may result in boundary revisions. In areas of rapid growth, census tracts often are split; in areas of substantial population decline, they are sometimes combined.

In 1906, the idea of collecting census information for small areas was first put forth by Walter Laidlaw, who studied neighborhoods in New York City. In response to his request, the Bureau of the Census tabulated information from the 1910 census by census tracts for eight cities with populations larger than 500,000: New York, Baltimore, Boston, Cleveland, Chicago, Philadelphia, Pittsburgh, and St. Louis. Data for the same eight cities were again tabulated in the 1920 census. In 1930, the number of cities was increased to 18. Data were not published but were made available by the Bureau of the Census for purchase. Beginning with the 1940 census, the Bureau of the Census established the census tract as an official geographic unit and published the tabulations for large cities. By 1990, census tracts were delineated for most metropolitan areas and other densely populated counties, and six states (California, Connecticut, Delaware, Hawaii, New Jersey, and Rhode Island) paid a fee to have complete tract coverage. The 2000 census was the first in which the entire United States was covered by census tracts.

Census tract boundaries are available from the Bureau of the Census's Topologically Integrated Encoding and Referencing (TIGER) files. Files from the TIGER database can be downloaded directly into geographic information systems (GIS) software and used to map a variety of census tract features, including demographic (e.g., age, race, ethnicity, gender), social (e.g., education, place of birth, ancestry), economic (e.g., income, occupation, employment, poverty status), and housing (e.g., size, age, type, value, presence of plumbing and heating facilities) characteristics. Census tract maps commonly are used to represent intraurban variation because tracts generally are small enough to be somewhat similar in income, housing, and other characteristics but large enough to avoid a visually unmanageable number of spatial units. In addition, the use of tracts usually avoids missing data problems stemming from census disclosure rules. Census participants are promised that the information collected about them as individuals and households will remain confidential.

—*Pat Gober*

See also Census; Population, Geography of

Suggested Reading

Kertzer, D., & Arel, D. (Eds.). (2002). *Census and identity: The politics of race, ethnicity, and language in national censuses.* Cambridge, UK: Cambridge University Press.

CENTRAL BUSINESS DISTRICT

The central business district (CBD), a term coined by the Chicago School of urbanists during the 1920s, refers to the downtown of urban areas. Because of its maximum proximity to all parts of the metropolis, this location is geographically advantaged, allowing firms located there the greatest access to urban labor supplies, one another, clients and customers, the infrastructure, and specialized pools of information. Thus, the CBD offers a comparative advantage in vertically disintegrated types of production where firms have many linkages to one another, and locating there allows those firms with high transportation costs (i.e., multiple inputs and outputs) to minimize costs by taking advantages of the agglomeration economies readily available there. Because accessibility is the major determinant of land values and land use, locations in the CBD typically command very high rents (including the peak value intersection) and are marked by high degrees of vertical real estate development.

The form and function of the CBD have changed significantly over time, reflecting broad structural changes in the local, regional, national, and global economies. During the 19th century, when city sizes in the United States were relatively small, CBDs were comparatively less well developed. In larger metropolitan regions, they often were characterized by webs of small manufacturing firms and smokestacks. Well into the 1920s, when Ernest Burgess and others first theorized about the CBD, the location was surrounded by blue-collar, working-class neighborhoods (the "zone of workingmen's homes" in classical social ecology). Many CBDs also contained important retail functions.

However, during the 1880s and 1890s, a period marked by the emergence of producer services and the transition from local to national economies, CBDs became larger and more complex. As multiestablishment corporations began to dominate national economies, CBDs became the command-and-control centers of cities, including the headquarters of many firms. Thus, their landscapes were increasingly given over to skyscrapers, an innovation made possible by technological developments such as structural steel, the elevator, the telephone, and mass transit. By the 1960s, this process was more or less complete in the United States, and CBDs were marked by dense complexes of steel and glass towers occupied by workers in producer services such as finance, law, accounting, and insurance.

However, suburbanization, "white flight," and industrial decentralization took their toll. In many American cities plagued by deindustrialization and capital disinvestment, neighborhoods near CBDs during the 1970s and 1980s experienced sharp economic declines, including rising levels of poverty, unemployment, and homelessness. Many downtowns, particularly in the Northeast and the Midwest, exhibited closed factories and warehouses. The suburbanization of retailing and the evacuation of middle-class purchasing power led many downtown stores to close.

By the 1990s, as globalization and the explosion of producer services ushered in a new round of growth and investments, many CBDs were reclaimed by corporations, a process accompanied by widespread gentrification and the associated influx of professional workers. Today, CBDs are typically the primary points of entry for the forces of globalization in the American city; large parts of the downtowns of many U.S. cities are owned by foreign firms, including large real estate interests. Despite the telecommunications revolution, CBDs continue to facilitate face-to-face interactions, another indication of their long-standing importance to the creation of urban agglomeration effects.

—*Barney Warf*

See also Chicago School; Urban Geography

Suggested Reading

Knox, P., & McCarthy, L. (2005). *Urbanization: An introduction to urban geography* (2nd ed.). Upper Saddle River, NJ: Prentice Hall.

CHICAGO SCHOOL

Between the two world wars, Chicago emerged as the epicenter of American social science, particularly with regards to urban analysis. As the prototype of the rapidly growing, industrialized city populated by streams of immigrants, Chicago became the prototypical example of American urbanization. The University of Chicago played a major role in disciplines such as economics, sociology, and geography. Within this context, the Chicago School of urban studies arose and was enormously influential in sociology and geography for the next several decades.

The origins and success of the Chicago School lay largely with its nominal leader, Robert E. Park, a former journalist turned teacher. The Chicago School is credited with the first systematic attempt to understand the dynamics of urban areas, including social change, urban planning, and territoriality. In 1925, Park, Ernest Burgess, and Roderick McKenzie published *The City*, a series of interpretive essays about the cultural patterns of urban life, a volume that both summarized and inspired a long tradition of urban ethnography. Chicago School practitioners, who inaugurated the creation of the tradition of detailed case studies, ranged far and wide over the city, studying the wealthy, immigrants, hobos, the destitute, dance halls, criminals, prostitutes, and anyone else they could in an attempt to draw as rich and detailed a portrait of the city as possible. In the process, they irrevocably fused the study of space and the study of society.

The first paradigm of urban structure offered by Chicago School theorists, particularly McKenzie, centered on a biological metaphor of the city as an urban jungle, a view derived in large part from the social Darwinism prevalent during the early 20th century. Thus, for example, the displacement of one ethnic group by another in a given neighborhood was framed as a process of invasion and succession, a model that drew directly from studies of how one plant species displaced another through successive stages in the evolution of ecosystems. Later, this biological metaphor would be dropped in the face of stinging criticisms that it lacked a coherent account of social relations and naturalized the inequality of urban areas. Throughout the Chicago School's worldview, competition appears repeatedly as a driving force behind ethnic and class segregation.

Chicago School theorists also drew on the urban sociology of Ferdinand Tönnies and notions such as *Gemeinschaft* and *Gesellschaft* to examine the phenomenology of urbanization in light of the massive rural-to-urban migration that was then characteristic of most U.S. cities. In this reading, urbanization represented the annihilation of mythologized rural communities in which everyone knew everyone else. In contrast to small towns in which everyone ostensibly was intimately connected to everyone else and presented the same sense of self under all contexts, urbanization was held to decompose these traditional bonds and erode the foundations of mutual trust. Cities, it was held, were not conducive to the formation of a sense of community. Louis Wirth, in particular, advocated a desolate but compelling view of city life as structured around three major axes: size, density, and heterogeneity. Size or total population, he held, created a climate that was inherently predatory, utilitarian, uncaring, and commodified; strangers were rare in small towns but were the norm in large cities. Density, he argued, led people to be close physically but not emotionally; indeed, alienation was the norm. Finally, social and cultural heterogeneity, manifested in the diverse lifestyles found in large cities, generated few of the common values necessary to the success of healthy communities. The result was allegedly the widespread presence of crime and other social pathologies ranging from suicide to psychoses. (Subsequent work, it should be noted, has rectified this stereotype

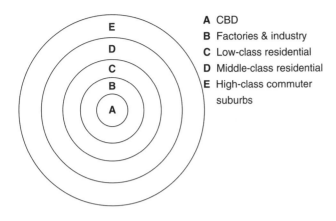

A   CBD
B   Factories & industry
C   Low-class residential
D   Middle-class residential
E   High-class commuter
    suburbs

Figure 1   Burgess's Concentric Ring Model

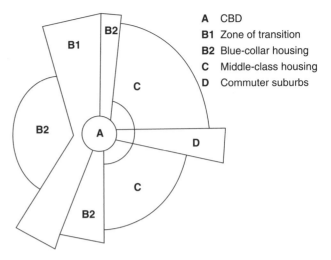

A   CBD
B1   Zone of transition
B2   Blue-collar housing
C   Middle-class housing
D   Commuter suburbs

Figure 2   Hoyt's Sector Model

by pointing to the high crime rates in many small cities and the presence of healthy, vibrant urban neighborhoods.)

Perhaps the most famous products of the Chicago School are three models of urban social structure repeated endlessly in introductory sociology and geography textbooks. The first of these, proposed by Burgess in 1927, was the *concentric ring model* (Figure 1), which, extrapolating from the specific instance of Chicago, viewed the city as a series of rings of varying size centered on the central business district (CBD), a term coined by the Chicago School. Adjacent to the CBD was a zone of factories and warehouses, sometimes called a "zone of transition." Moving outward, this was followed by the "zone of workingmen's homes" (i.e., working-class, blue-collar communities). Yet farther out were the medium-income and then high-income belts of suburbia. Burgess observed that cities tend to expand horizontally as the wealthy had new homes constructed on the urban periphery. As they came to occupy these, the relocation of families outward set off a chain of vacancies that reverberated across the urban landscape as less-well-off families, in turn, occupied the cast-off mansions of the rich, a process known as filtering or the trickle-down theory of housing supply. Moreover, Burgess observed a paradox: Low-income residents in cities lived on expensive accessible land near the urban core, whereas more-well-to-do inhabitants of wealthier rings occupied less expensive land. This paradox, he noted, was easily explained by the population density curves characteristic of cities that decline exponentially with distance from the CBD. The poor, crowded in dense

communities, collectively generate high-aggregate rents that create relatively high rates of profit in inner-city areas, whereas the low-density environments of the wealthier classes reduce the profitability of the less accessible periphery.

In contrast to the rigid geometry of the Burgess model, Homer Hoyt, an economist and another influential Chicago School theorist, proposed the *sector model* of urban growth in 1939 based on an empirical analysis of 142 cities (Figure 2). In this view, rather than concentric rings, urban growth occurred along transportation lines centered on the CBD. Once parts of the central city acquired distinctive uses, they radiated outward. High-income land uses played a determining role in shaping the rest of the city, growing along waterfronts, along high-altitude areas, or toward other high-income neighborhoods, and other uses filled in the spaces between them. Rather than belts, socioeconomic groups occupy sectors or wedges. The wealthy, with the greatest ability to pay, outcompeted less advantaged groups for the most desirable locales, generally far from the disamenities of factories and railroad lines. Low-income groups found themselves confined in zones relatively far from the wealthy.

Finally, the third of the Chicago School trilogy, proposed by Chauncy Harris and Edward Ullman in 1945, was called the *multiple nuclei model* (Figure 3). Essentially, this view attempted to rectify the perceived simplistic shortcomings of the previous two models. It maintained that American cities did not have a single city center but rather had become polycentric, with

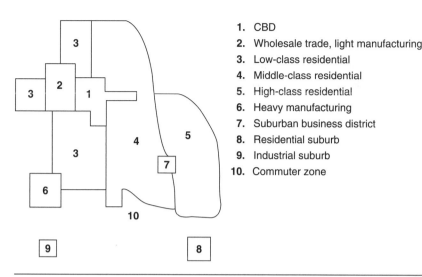

1. CBD
2. Wholesale trade, light manufacturing
3. Low-class residential
4. Middle-class residential
5. High-class residential
6. Heavy manufacturing
7. Suburban business district
8. Residential suburb
9. Industrial suburb
10. Commuter zone

Figure 3   Harris and Ullman's Multiple Nuclei Model

Suggested Reading

Bulmer, M. (1984). *The Chicago School of sociology: Institutionalization, diversity, and the rise of sociological research.* Chicago: University of Chicago Press.

Dear, M. (Ed.). (2002). *From Chicago to L.A.: Making sense of urban theory.* Thousand Oaks, CA: Sage.

Park, R. (1926). The urban community as a spatial pattern and a moral order. In C. Peach (Ed.), *Urban social segregation* (pp. 21–31). London: Longman.

Park, R., Burgess, E., & McKenzie, R. (Eds.). (1925). *The city.* Chicago: University of Chicago Press.

Wirth, L. (1938). Urbanism as a way of life. *American Journal of Sociology, 44,* 1–24.

many nuclei around which land uses were organized in a complex quilt. Rather than a single overarching logic, this perspective maintained that certain land uses would repel one another while others might be mutually attractive.

Although the Chicago School began to diminish in importance shortly after World War II, its ideas were carried into urban sociology and geography for many years afterward. For example, social ecologists during the 1960s, armed with multivariate statistical methods and census data, argued that each of the three classic models effectively captured a different aspect of urban social space. Thus, family status was distributed in rings, per Burgess's concentric ring model, reflecting the dynamics of the family life cycle. Economic class was held to occur in sectors conforming to Hoyt's theory of land use. Finally, ethnicity was theorized to reflect the dense nucleations of different immigrant groups, as proposed by the multiple nuclei model.

The Chicago School essentially defined American urban analysis throughout the 20th century. More recently, however, attempts by urban political economists to reveal the complex dynamics of globalization and immigration in a postindustrial context have yielded the so-called Los Angeles School, which takes the southern California metropolis, rather than Chicago, as its point of departure.

—*Barney Warf*

See also Central Business District; Invasion–Succession; Neighborhood; Urban Geography

CHILDREN, GEOGRAPHY OF

Geographers began researching the worlds of children during the mid-20th century, but it was not until recently that the notion of children's geographies developed as a coherent aspect of the discipline. Earlier work documented regional variations in child welfare or, spurred by the growth of behavioral and perceptual geography, focused on children's mapping abilities and environmental competences. Although this kind of work continues today, geographic research since the 1990s is perhaps most influenced by feminist and poststructural theories. For the most part, these new perspectives form a critical and reflexive engagement with the lives of young people, focusing on positionalities, playfulness, and prescriptions for spatial justice and the celebration of difference. This new research embraces the places and scales from which young people inform and are informed by their world. It elaborates the nuances and complexities of children's so-called development in a way that belies older linear and decontextualized ways of knowing. It positions children more forcefully in their local environments and at the heart of larger globalization processes.

BUILDING A PLACE FOR CHILDREN IN GEOGRAPHY

The contemporary origins of geographers' interests in children may be traced to William Bunge's 1960s expeditions in Detroit and Toronto. Focusing specifically on the spatial oppression of children, Bunge argued that young people are the ultimate victims of the political, economic, and social forces that contrive the geographies of the built environment. Starting with observations of working-class children at play in inner-city neighborhoods, Bunge's expeditions employed a myriad of quantitative and qualitative, as well as aggregate and individualistic, approaches to the study of spatial structure and interaction without losing sight of the central theme of children's oppression.

Around the same time, geographers and environmental psychologists adopted experimental science and humanistic approaches to explore children's cognitive development and wayfinding as well as their imaginative play and sense of place. Although much of this work lacked the political edge of Bunge's expeditions, it opened up for geographers some of the developmental theories elaborated by Jean Piaget, Erik Erikson, and others.

During the 1980s, a number of researchers built on this groundbreaking work, making significant contributions that focused, for example, on the effects on children of spatial inequalities in the distribution of health, housing, and educational resources or on how children were positioned in relation to environmental hazards and poverty. Other researchers were drawn to new social studies of childhood that critiqued the notion of childhood as a developmental phase that leaves children as less than adult and suggested instead that children are competent actors in and of themselves. This research joined with feminist and poststructural thinking to rekindle geographic interest in Bunge's commitment to give children a voice in an adult-oriented world.

CHILDREN AS COMPETENT GEOGRAPHIC ACTORS

Feminist sensitivity to difference, diversity, and political activism focused discussion on children as competent social and spatial actors rather than as a marginalized social group. Children are considered able to actively resist and subvert adult definitions of their lives. The concepts of competence and agency form the basis of a body of work on the ways in which young people appropriate adult public space and develop ingenious ways of adapting everyday environments to their own uses. Other studies note young people's independence within the virtual geographies

of video gaming, e-mail, and the Internet. Still other studies focus on the autonomous spaces of children, their labor, and their contributions to productive activities. Some of this work has raised the issue of children's rights, spatial justice, and the problematic relations between childhood and citizenship.

Contemporary poststructural geographic perspectives contest traditional notions of children and space with nonlinear notions of development, nomadic spaces of play, and nonmechanistic ideas of rights and discipline. Drawing from the work of Michel Foucault, geographers study the ways in which children's activities are scrutinized through panoptic surveillance and how children are disciplined through exclusion from adult spaces and placement in special, often commodified, and seemingly child-friendly places. Other poststructuralists focused on the problematic linearity of child development. Piagetian theory, for example, formalizes stages of childhood in a series of hierarchical stages of intellectual development to the extent that children's completeness is determined by biological age. Alternatively, it is possible to think of children as being rather than becoming, and so-called development need not be gauged against some normative standard that ultimately culminates in adulthood. Arguments here suggest that child development does not take place in two-dimensional spaces, with children's horizons expanding from crib to home to neighborhood and so forth. The work of Gilles Deleuze and Felix Guattari invites perspectives in which adults create striated spaces for young people through rules, routines, and structures, whereas the more chaotic spaces created by children are rendered smooth. These smooth and striated spaces coexist and subvert and taint each other; they are intertwined and entangled together. Space exists only as relations of smoothing and striation, and neither children nor adults have a monopoly on the processes that elaborate these geographies. Thus, the entangled worlds of adults and children are understood as continually maneuvering around and modifying each other, casting doubts on the certainty with which spaces of adulthood and childhood are contrived.

PLACING CHILDREN AT THE HEART OF GLOBALIZATION

New wisdom about what constitutes childhood and adulthood places children closer to the center of our understanding of consumption, production, and reproduction and at the heart of inequities generated by globalization. In a connected world of flexible capital and instantaneous market adjustments, local places are increasingly important for understanding the children. And geographers see young people as something more that a simple tabla rasa on which the will of capital is etched. Children not only become or develop through the influences of these changing objects, they also bring the totality of themselves into cultural life as they actively participate in the day-to-day workings of places. In the same sense that the processes of globalization are neither unidirectional nor even, it is impossible to characterize or position a uniform context for childhood because the local conditions of global children are so varied. In short, childhood not only is constructed in different ways at different times but also varies depending on where it is constructed.

—*Stuart Aitken*

See also Body, Geography of; Critical Human Geography; Home; Humanistic Geography; Population, Geography of; Poststructuralism; Social Geography

Suggested Reading

Aitken, S. (2001). *Geographies of young people: The morally contested spaces of identity.* New York: Routledge.

Holloway, S., & Valentine, G. (2000). *Children's geographies: Playing, living, learning.* New York: Routledge.

Katz, C. (2004). *Growing up global: Economic restructuring and children's everyday lives.* Minneapolis: University of Minnesota Press.

CHOROLOGY

Also known as aerial differentiation, chorology comes from the Greek words for the science of place, in contrast to chronology. Thus, it has a long history in geography. Strabo (64 BC–24 AD), a Greek geographer working for the Romans, advocated a form of chorology in his 17-volume *Geography*, which was essentially a handbook for administrators. In contrast, Ptolemy (87–150 AD), a Roman geographer and astronomer working in the famous museum at Alexandria, maintained that the task of geography is the description of the earth as a whole. In his eight-volume *Guide to Geography*, Ptolemy ridiculed Strabo's emphasis on regions, arguing instead for a holistic view of the earth

and that the regional emphasis was like painting a person by showing only one of their eyes or ears. Ptolemy differentiated among geography as the study of universals, topography as the study of localities, and chorography as integrating the two.

The great 17th-century geographer Varens (Varenius) (1622–1650), who wrote the highly influential *Geographia Generalis* in 1650, distinguished between what he called *specific geography* (concerned with the unique character of places) and *general geography* (concerned with universal laws). Immanuel Kant (1724–1804), a geographer as well as a philosopher, played an important role in the historical evolution of chorology by arguing that, unlike the theoretical sciences such as chemistry, geography and history were essentially concerned only with the empirical and the unique. His views were hugely influential in subsequent philosophies of space.

During the 19th century, geographers such as Carl Ritter likewise practiced a form of chorology. Perhaps its first explicit advocate was Paul Vidal de la Blache (1845–1918), considered the father of French geography, who studied small French rural areas called *pays* and their associated styles of life (or *genres de vies*). Because the climate of France did not vary much but lifestyles did, Vidal de la Blache was also crucial to the introduction of possibilism to the discipline. His German counterpart, Alfred Hettner (1859–1941), argued in the Kantian tradition that geography was the art of regional synthesis (i.e., the pursuit of interrelations in given areas), an aspect that other disciplines ignored. Thus, chorology became the basis of geography's disciplinary identity.

During the 1920s, American geographers adopted chorology or aerial differentiation in the aftermath of the catastrophe of environmental determinism. American chorology was personified by Richard Hartshorne (1899–1992), who studied under Hettner and graduated from the University of Chicago in 1924. In the tradition of Kant, Hartshorne and his fellow chorologists argued that the essence of geography was the regional description of regions, including cultural and physical phenomena. Chorologists advocated getting to know places in great depth with a healthy regard for cartography and fieldwork. Because large regions are diverse and complex, he argued that chorology should focus on small, relatively homogeneous regions. Hartshorne maintained that regions are essentially mental concepts, that is, subjective tools to find meaning and create order in the landscape. Thus, regions were necessarily simplifications and were useful only inasmuch as the gain in understanding they provided exceeded the loss of detail. Implicit in Hartshornian chorology was the view that location served as a form of explanation (i.e., proximity was synonymous with causality), leading to a crude form of spatial determinism reminiscent of Tobler's first law. Finally, Hartshorne argued that because landscapes are essentially stable from a human perspective (i.e., exhibiting relatively little change in the course of one lifetime), there was no urgent need to study the process of change. In arguing that only by sticking to the facts could we remain objective, Hartshorne's line of thought drew on the philosophical tradition of empiricism in which facts are simply true without regard for theory. Later, more theory-conscious geographers acknowledged that all data are theory laden.

Chorology collapsed during the 1950s as positivism arose to take its place, beginning with a famous attack on Hartshorne by Fred Schaefer in 1953. Schaefer claimed that Hartshorne's view of geography as an integrative science concerned only with the unique was simplistic. By refusing to search for explanatory laws, geography condemned itself to what Schaefer called an immature science. Rather than idiographic regions, geographers should seek nomothetic regularities across regions. This critique helped to open the door to the rise of positivism and the quantitative revolution.

Although traditional chorology died under the positivist onslaught, it did experience something of a resurrection during the 1980s. Some Marxists, beginning with Doreen Massey, argued that broad social processes always play out in different ways in different places. This perspective led to a renewed respect for the idiographic. What became known as the localities school approached regions in terms of their historical development as they acquired unique combinations of imprints of different divisions of labor (e.g., investments, labor market practices, cultural forms). In this view, general laws of explanation are manifested only in unique contexts and localities are transformed into objects of scientific understanding. Unlike the earlier tradition of chorology, therefore, this approach eschews empiricism and maintains a central role for theory.

—*Barney Warf*

See also Empiricism; History of Geography; Idiographic; Nomothetic; Regional Geography; Tobler's First Law of Geography

Suggested Reading

Berry, B. (1964). Approaches to regional analysis: A synthesis. *Annals of the Association of American Geographers, 54,* 2–11.

Hart, J. (1982). The highest form of the geographer's art. *Annals of the Association of American Geographers, 72,* 1–29.

Hartshorne, R. (1939). *The nature of geography.* Washington, DC: Association of American Geographers.

Massey, D. (1995). *Spatial divisions of labor* (2nd ed.). New York: Routledge.

Sack, R. (1974). Chorology and spatial analysis. *Annals of the Association of American Geographers, 64,* 439–452.

Schaefer, F. (1953). Exceptionalism in geography: A methodological examination. *Annals of the Association of American Geographers, 43,* 226–249.

Warf, B. (1993). Post-modernism and the localities debate: Ontological questions and epistemological implications. *Tijdschrift voor Economische en Sociale Geografie, 84,* 162–168.

CIRCUITS OF CAPITAL

When you stand on a city corner, in the parking lot of a factory, in a farmer's field, in the aisle of a supermarket, or even in your own living room, everything else you can see is a node in the circuit of capital. That capital circulates—that capital must circulate—seems obvious enough when seen from the viewers' gallery at a stock exchange or when looking at exchange rate and balance of payment statistics in the business pages of a newspaper, but must that be the case in these other places? In fact, everything you see and experience—the geography of the world—is influenced by the rhythms of capital circulation. Geography is built through circulating capital, and even in the most natural of landscapes your very ability to view it (or not) is defined by whether and how capital has circulated.

Circuit of capital refers, at its most basic level, to the movement of capital:

$$M + \text{MP, LP} \rightarrow C \rightarrow (M + \Delta M) = M',$$

where M is the money capital used to purchase the means of production (MP) and labor power (LP), C is the resulting commodity, and M' is the money received when the commodity is sold. ΔM indicates the surplus value produced in the production process.

There are several things to notice here. First, labor power in capitalism is itself a commodity and must be purchased. Therefore, some capital, in the form of wages, circulates in the hands of workers, who then (among other things) form a market for goods produced, returning capital back to the production process or diverting it to other capitalists who build their homes, finance their loans, repair their cars, and so on. The circuit of capital must take account of this form of circulation with all of the risks and diversions (e.g., savings) associated with it. Second, the means of production include a range of commodities from the buildings and machines to the raw materials that go into making the finished product. In each of these commodities, some portion of circulating capital is "frozen" for a period of time (relatively short for raw materials, potentially very long for buildings and machinery). Capital frozen in buildings, parking lots, and so on is critical to the circuit of capital. But such frozen capital is at risk for constant devaluation by innovation and obsolescence, economic crisis, and so on.

Third, some necessary "fixing" of capital that makes the circuit of capital possible—in roads, rails, power grids, dams, and so on, together with institutions necessary for the reproduction of labor power—is too massive for a single capitalist to undertake. Some surplus capital is diverted to the state or various consortia of capitalists to undertake such massive projects. Fourth, because of the need for massive public works to spread the risk of long-term investments and to "rationally" allocate other surplus capital not reinvested in the production that gives rise to it, financial institutions and capital markets arise. These, of course, are crucial and are the most obvious nodes for the circuit of capital, but they are also a function of the circuit through the production process itself (even if they also determine the *where* and *how* of much production through interest rates, loan approval algorithms, etc.).

Fifth, and crucially, the point of circulating capital through the production process is to create more capital—to accumulate. As opportunities for profit decline in one place, capital switches—at least ideally—to locations with lower labor or fixed costs (or more competitive factories), to products that have a better chance of being sold, to frontiers of suburban development or gentrification, or to new or distant capital markets. In reality, such shifts in investment (in the circuit of capital) are rarely smooth in their own terms (e.g., mistakes are made, investments are lost) and almost always are disruptive to those who stand in a relation to capital circulation different from

that of financial barons or captains of industry—workers put out of their jobs when a factory is shuttered, local shopkeepers who lose their markets, homeowners who cannot meet their mortgages, farmers who can neither pay their workers nor afford their machinery, and so on.

Circuits of capital are complex, crisis prone, and contradictory. Different owners of capital may have different aims and may deploy their capital in such a way as to thwart the smooth circulation of capital as a whole. Contradictions such as these are endemic within capitalism. Circuits of capital may come to a halt—be thrown into crisis—because labor is too expensive, means of production are outmoded, or commodities cannot be sold at their full value due to an overaccumulation. The highly complex geography of capital circulation is defined through the ongoing development and resolution of such crises and contradictions. The complex relations, crises, and contradictions of circulating capital, in other words, are key determinants of both the built landscape and of the commodities with which we may populate it at any time. Stand on a street corner, in your living room, in a factory, in a store, or in a farmer's field, and no matter what else you see, you certainly can see the (result of) circuits of capital.

—Don Mitchell

See also Capital; Economic Geography; Marxism, Geography and; Uneven Development

Suggested Reading

Harvey, D. (1999). *The limits to capital* (2nd ed.). London: Verso.

Harvey, D. (2001). *Spaces of capital.* New York: Routledge.

Henderson, G. (2002). *California and the fictions of capital.* Philadelphia: Temple University Press.

Marx, K. (1987–1992). *Capital* (3 vols.). New York: International Publishers.

Smith, N. (1990). *Uneven development: Nature, capital, and the production of space* (2nd ed.). Oxford, UK: Blackwell.

CITY GOVERNMENT

City governments are central and powerful institutions and actors across urban areas in the United States. They are the core administrative unit for more than 174 million people in America (76 million of whom live in cities with populations of at least 100,000 people). These 76 million people make up 62% of the country's population. States specify the administrative nature of city governments in America, with four forms of governance dominating: the mayor–council, council–manager, commission, and town meeting mechanisms. Each outlines proper procedures for cities to solicit citizen input into the constituting of programs and rules for local administrating. The five largest cities—New York, Los Angeles, Chicago, Houston, and Philadelphia—have mayor–council institutional arrangements.

Models to understand the operation of city government now increasingly depart from the "benevolent–passive" perspective. One prominent model is the Marxist–instrumentalist one. Here city governments are apparatuses to serve the interests of local economic and political elites. Such elites—amalgams of prominent builders, developers, realtors, speculators, utility companies, banks, and so on—push to drive profit accumulation via city growth that also enhances tax ratables for cities. But the accumulation process, ripe with conflicts and contradictions, requires the use of city governments ("the local state") to adjudicate these dilemmas on behalf of these elites. Local governments, situating themselves in a veneer of neutrality and objectivity, ultimately toil with this one constituency favored.

City governments are also widely understood through a Weberian–bureaucratic perspective. Here city governments historically have operated to help assist local urban elites but increasingly move away from this to advance their own interests and ambitions. It is contended that a critical moment is reached when city governments get sufficiently large and powerful to shift their "logic of operations" to benefit themselves. This shift means that they increasingly strike out (offer new programs, policies, regulatory procedures, etc.) to advance their sphere of power and influence as dominant social, political, and economic institutions that desire bolstering. Similar to the Marxist–instrumentalist perspective, this drive is performed under the cover of benevolent and constituent-neutral rhetoric.

City governments are widely seen to have changed their operations with the 1980s rise of the "neoliberal era" (beginning with Ronald Reagan's election in 1980). This new era, characterized by a deepened emphasis on private markets to determine social welfare and a reduced welfare state orientation, affects and is affected by local governments. Since 1980, local governments have more fervently privileged private markets to determine patterns of land use, amounts and types of subsidies allocated, and recipients of government largesse. In this context, the private sector (businesses and corporations) is widely identified as the positive engine of change in cities, the group that can progressively restructure cities socially and spatially. The result has been that people have needed to rely more thoroughly on private resources to live, travel to work, and physically upgrade homes and neighborhoods after public services, public funds, and access to public decision making have been slashed. City governments across America, it follows, are propelled by this neoliberal thrust that is simultaneously a vision of progressive politics, a policy experiment, and a new reality of landscape change.

In this context, city governments now operate under more severe constraints than they did before. A steady erosion of federal aid has meant fewer resources to tackle the continuance of urban blight, housing abandonment, poverty, homelessness, and other entrenched problems. In fiscal year 2006, community development block grant funds (the principal source of federal aid to cities) have been cut by $1 billion (to $4.355 billion). At the same time, President George W. Bush proposed to make the program more efficient by moving it out of the U.S. Department of Housing and Urban Development to the Economic Development Administration in the U.S. Department of Commerce. Yet his professed goal, to unleash the latent innovative capacities of local urban economies by entrepreneurializing them, has also exacerbated spatial and economic disparities in urban populations. The poorest people in cities, recently growing in number and intensity of deprivation, have been those hurt most severely by city governments' recent change.

—David Wilson

See also Neoliberalism; Political Geography; State; Urban Geography

Suggested Reading

Brenner, N., & Theodore, N. (2002). *Spaces of neoliberalism: Urban restructuring in North America and Western Europe.* Oxford, UK: Blackwell.

CIVIL SOCIETY

Civil society is a concept with varied meanings. Although somewhat simplistic, it is useful to see the concept as having one set of meanings that derive largely from liberal social theory and another set of meanings that derive largely from Marxist social theory.

LIBERAL APPROACHES TO CIVIL SOCIETY

Liberal social theorists have usually presented civil society as a space of human activity distinct from the activities of the state or government. Although early liberals did not always use the term in ways that are congruent with its later use, one can see the rudiments of this view of civil society in their work. For 17th-century British philosopher John Locke, the social contract that ends the state of nature ideally creates a space where naturally given property rights and economic freedoms are to be protected by the state. In the work of later liberal social theorists, such as 19th-century British political economist John Stuart Mill, this emphasis on society as properly being a space of economic freedom is supplemented by a more developed sense of society as a space of varied human liberties, including rights to free speech and political organization.

Throughout the 20th century, in a wide range of liberal writings, civil society was idealized as a space of both economic and political/ideological freedom. Although many modern liberals have seen these freedoms as two faces of the same coin, others have also noted that the economic freedoms associated with capitalism might not be entirely compatible with the development of other forms of freedom that are central to liberal concepts of civil society. As a consequence, even within a broadly liberal framework, there are sometimes competing emphases in the discussion of civil society. For example, organizations such as the World Bank and the International Monetary Fund (IMF) have emphasized economic liberalization as a key to the broader development of human freedoms. In this approach, civil society is presented as a space of economic and political opposition to strong and interventionist states. Yet some critics of World Bank and IMF economic liberalization policies have also presented civil society as a space of activity that is potentially at odds with both a strong state and a capitalist economy. In this approach, civil society is construed as one of three distinctive arenas of human activity: the state, the market, and civil society.

MARXIST APPROACHES TO CIVIL SOCIETY

Marxist theories start from different assumptions than do liberal theories. For Karl Marx himself, writing during the 19th century, society was fundamentally structured in its patterns of development by conflicts between different social classes. Although these conflicts centered on control of the economic surplus produced by society, they were always carried out simultaneously in various realms, including sites of production (the economic realm), sites of state power (the political realm), and sites of cultural and ideological struggle (the ideological realm). Thus, early-20th-century Italian Marxist theorist Antonio Gramsci, whose work has had significant influence on contemporary understandings of civil society, contended that the term did not refer to a realm of social activities that could be seen as autonomous from the state or the market. Rather, for Gramsci, the state, the market, and civil society were integrally interconnected with one another and were essentially different faces of the same social structure.

CIVIL SOCIETY IN CONTEMPORARY DEVELOPMENT DEBATES

Such differences between liberal and Marxist approaches help to explain contemporary debates over development and the appropriate role of states. Organizations such as the World Bank and the IMF have argued for forms of development that minimize the interventionist roles of states and involve a more active role for civil society organizations, with these often being represented as economic actors who can benefit from more decentralization of economic and political power at the local level. Such neoliberal approaches often have sanctioned the growth of activity by nongovernmental organizations (NGOs) as a way of devolving state power to nonstate actors more embedded in local contexts of development.

At the same time, some NGOs and liberal social groups arguing for a stronger civil society have favored greater political and economic decentralization but have also opposed the sort of liberalization, privatization, and free trade agenda promoted by the World Bank and the IMF. Such organizations have argued not only for devolution of political and economic decision making to local levels but also for more active participation by local groups in making those decisions—an outcome that is not ensured by the devolution of formal decision-making authority. Some such NGOs have favored forms of alternative development that bypass the state and generate development through locally embedded social, cultural, political, and economic processes.

Although this last type of liberal populist agenda has gained sympathy from many advocates of socialist forms of development, it has also been criticized from the political left for its assumption that civil society can be seen as a realm apart from the state or the market. Thus, for some leftist critics of NGOs, the NGOs are not organizations separate from the state and thus are not capable of bypassing it. Rather than being representatives of an autonomous civil society, NGOs are seen as part of the larger social structure, intertwined with the state and the market in a variety of ways. For example, NGOs can be seen as organizations that have been promoted and allowed to flourish by powerful classes active within both the state and the market as a way of trying to undo forms of state "intervention" that historically were promoted through working-class struggles. From this perspective, NGO collaboration in devolution of decision-making power to the local level might potentially pit many NGO projects against the interests of most nonelite groups.

—*Jim Glassman*

See also Development Theory; Marxism, Geography and; Neoliberalism; State

Suggested Reading

Cowen, M., & Shenton, R. (1996). *Doctrines of development.* London: Routledge.
Gramsci, A. (1971). *Selections from the prison notebooks of Antonio Gramsci* (Q. Hoare & G. Smith, Eds. & Trans.). New York: International Publishers.
Locke, J. (1980). *Second treatise of government.* Indianapolis, IN: Hackett.

CLASS

In general, the term *class* refers to a group of people who have the same social or economic status such as the working class or a professional class. Within the social sciences, two views of class have dominated: those drawing on Karl Marx (1818–1883) and those drawing on Max Weber (1864–1920). Given that he wrote after Marx, Weber often is said to be engaged in a dialogue with the ghost of Marx on the matter of class. Although he was interested in many of the same questions as Marx, Weber came to quite different conclusions. For example, Marx believed that workers' alienation (by which he meant how workers gradually lost control of the product of their labor such that the shift from feudalism to industrial capitalism in Europe transformed workers from relatively self-sufficient peasant farmers into wage laborers who did not own what they produced) would eventually be eliminated when workers finally owned the products of their labor in some future workers' state. Weber, however, believed that alienation had little to do with who owned the means of production (e.g., factories, mines, financial institutions) but was rather a consequence of bureaucracy.

For Marx, class is primarily an economic category derived from the differential ownership of the means of production, with the structure of such ownership subsequently determining how the product of society's labor is divided. According to Marx, the fundamental division in any society is between those who own the means of production and those who do not. Under capitalism, this means that workers (the proletariat), who do not own the means of production, must sell their labor power for a wage to the capitalist class, which does. Marx maintained that such a class system is not inevitable but has its origins in the historical development of human societies. Thus, he argued that class systems began to develop once humans moved beyond hunter–gatherer societies and beyond a level of social development in which everyone produced for themselves and their immediate family members. Once agriculture had been invented some 10,000 years ago, there developed a social relationship in which some people came to control the means of production (e.g., land, tools) and so, ultimately, to control the product of others' labor. In formulating a general analysis of this development of class relations over time, Marx relied on the historical materialist approach to understanding the forces of history. Finally, Marx made a distinction between a *class-in-itself,* by which he meant a group of individuals who objectively share a similar social and economic situation, and a *class-for-itself* (by which he meant a group that has developed a consciousness about its own existence, i.e., that workers recognize their class position as workers relative to capitalists). In the Marxist schema, class is defined by social relationships—the relationships between those who own the means of production and those who do not—rather than by statistical categories (which might see, say, all of those earning more than $100,000 a year as being in one class and all of those earning less than that in another). Although Marx argued that there could be movement of individuals between classes, such individualized movement should not be seen to negate the existence of classes themselves.

Weber, like Marx, viewed class as having important economic dimensions, but he did not seek to develop a theory of the class forces that drove history over the long term. Instead, he was more interested in how societies are organized into hierarchical systems of domination and subordination (on both an individual basis and a collective basis) and the significance of power in shaping social relationships. He did this by examining how what he called the three dimensions of stratification (class, party, and status) intersected. For Weber, an individual's or a group's ability to possess power relates to the control of various social resources such as capital, land, knowledge, and prestige. Within this framework, class power results from unequal access to economic resources, social power (status) results from one group or individual being seen as the social superior or inferior to another, and political power (party) relates to how the state is organized (if a particular group can influence how laws are made or public policy is implemented, it is seen as having political power). In Weber's configuration, the ability to shape a decision-making process in any of these three realms means that one holds power, whether that ability is based on economic class (one can threaten to fire workers if he or she is their employer), social status (a celebrity may command great respect from the public), or political power (one can influence whether a particular law is implemented through investing time or money). These three axes of power do not necessarily coincide—someone with social status might not be wealthy, for instance—but generally power in one realm will suggest power in another; wealthy people usually have higher status and greater ability to shape the political process than do poorer people. Whereas Marx argued that it was the economic class within which individuals were situated that shaped the possibilities of their having social and political power, Weber saw the three dimensions of power as, in theory, potentially independent of one another, even if in reality that were rarely the case. Also, Marx tended to focus on the social, economic, and political system as a whole, whereas Weber was more inclined to examine individuals and particular social groups and their different levels of power within the system.

Within geography, it is the Marxist view of class that has tended to dominate. In particular, much effort has gone into theorizing the spatial aspects of processes of class formation and class dynamics. Specifically, whereas early Marxist work saw classes largely in aspatial terms as economic groupings that were shaped essentially by the social relations of capitalism, later work recognized that classes develop within particular spatial contexts such that processes of class formation and patterns of class structure vary geographically. Likewise, this geographic variation results in the landscapes of capitalism being made in different ways in different places.

—Andrew Herod

See also Class War; Economic Geography; Justice, Geography of; Labor Theory of Value; Marxism, Geography and; Mode of Production; Uneven Development

Suggested Reading

Herod, A. (2001). *Labor geographies: Workers and the landscapes of capitalism.* New York: Guilford.

CLASS WAR

The term *class war* is used by people on both the right and left of the political spectrum. In general, the term relates to the political and economic conflicts between different socioeconomic classes over things such as the distribution of wealth and whether or not government policy should be implemented to reduce inequalities of wealth. Typically, the term is used to describe conflicts between the "haves" and the "have-nots" that work themselves out in some regulated judicial manner such as in elections to government of various political parties. Sometimes, however, actual violent conflict between different socioeconomic classes may break out. Such is the case when revolutionary situations bring about significant transformations in a society's socioeconomic structure, particularly with regard to the distribution of its wealth. Adopting the language of military conflict, political scientist James Scott, in his 1985 book *Weapons of the Weak*, distinguished between what he called "the small arms fire" and the "big guns" of class conflict. For Scott, examples of small arms fire include workers deliberately being late for work, stealing from their employers, and intentionally ruining the products of their labor (e.g., sewing the wrong-color buttons on shirts in the case of garment workers). Rather than simply being examples of antisocial behavior, Scott saw these activities as ways for workers to come to terms with their alienation in the workplace and to wrest some control of the labor process away from their employers or landlords (in the case of peasant farmers, the subject of Scott's book). The big guns of class conflict, Scott suggested, are activities such as striking and fostering political revolution.

Rhetorically, political parties on the left often have used the term *class war* to describe how the powerful in society exploit the less powerful and how, in turn, the less powerful should organize themselves to improve their position. In such a discourse, it is argued, the less powerful are victims of a class war waged against them by those in positions of economic and political power; therefore, their actions are defensive, designed to limit their own exploitation. Frequently, however, those on the political right argue that any efforts to bring about wealth redistribution are simply examples of "class envy" and are attempts by the poor or leftist politicians to wage "class war" against the wealthy. For such commentators, unequal distributions of wealth are seen either as natural or as the reward for individual sacrifice and hard work; that is, for many on the political right, the causes of poverty are seen as the result of the personal failings of the poor rather than the operation of structural forces such as institutionalized racism or the ways in which unregulated markets operate in a capitalist society. Many leftists counter that, in decrying the class war rhetoric of the left, those on the political right are themselves, in fact, engaging precisely in class war by seeking to defend the social status quo.

—Andrew Herod

See also Class; Marxism, Geography and

Suggested Reading

Scott, J. (1985). *Weapons of the weak: Everyday forms of peasant resistance.* New Haven, CT: Yale University Press.

COGNITIVE MODELS OF SPACE

Cognitive models of space refer to the different mechanisms by which humans perceive and understand the components of geographic space. Cognition is the range of intellectual activities spanning from awareness, through perception and reasoning, and finally to judgment. Spatial cognition refers to the mental process of knowing that events and processes occur in, are influenced by, and influence other events and processes in geographic space.

SPATIAL AWARENESS AND THINKING

Spatial awareness is based on a simple principle of being cognizant that human and natural events and activities occur in geographic space. Human systems are composed of human activities such as land conversion, movement of people, road construction, and energy consumption. As a complement to human systems, physical systems are based on the natural environment such as storm events, volcanoes, plant and animal life, and river systems. We use location as the organizing principle to identify where on the surface of the earth these activities occur.

The process of spatial thinking involves a continuum from spatial awareness, through spatial perception and spatial reasoning, and finally to spatial judgment. Spatial awareness is based on a simple principle of being cognizant that human and natural events and activities occur in geographic space. Spatial perception implies a personal capacity to recognize and interpret the interactions of spatial events and processes. Spatial reasoning involves logical and analytical thought to make a decision concerning spatial events, processes, and their interaction. Finally, spatial judgment is the mental ability to perceive and distinguish spatial relationships and the ability to assess alternative situations.

A well-documented geographic example that had unintended environmental consequences is the location of industrial activities in the Ohio and Pennsylvania region. The decision to locate industries that emit nitric oxide and nitrogen dioxide gases (NOx), as well as a group of chemical compounds of sulfur and oxygen (mostly sulfur dioxide) gases (SOx), into the atmosphere in the Ohio and Pennsylvania region has degraded water quality in New England lakes via acid rain deposition. The decision makers at the time these industries were built did not understand the spatial processes associated with emissions and atmospheric processes.

MAPS AS MODELS

The paper-based map (and now the digital version of it) has been the primary mechanism for conceptualizing, defining, and understanding geographic space. A map is a graphical representation of geographic space where location and attribute (where something is located and what it is) are combined into a single visual product. The relative location (where it is located in conjunction with other elements of the map) and the absolute location (the precise coordinate information) are provided along with the attribute information (a solid blue region conveying that it is a lake) as a graphic image. Maps are useful for conveying specific messages about a topic (e.g., that the distribution of chemical waste sites is concentrated in one geographic region) and useful for analyzing geographic phenomena (e.g., visualizing stream erosion).

Mental maps are internalized images of geographic space. Although many people are very talented at viewing geographic space in a manner similar to viewing paper maps, other people are not. Mental maps do not have the elements and metrics (e.g., ability to measure distances) associated with them as do their physical counterparts.

SPATIAL LANGUAGE AS MODELS

Spatial information can be conveyed through oral or written language. People frequently provide instructions on how to find a location through language rather than drawing a map (e.g., wayfinding, navigation). For example, directions to a house or other location may instruct the user to turn left at a stop sign or head north on a freeway. Spatial language can be as vague or as precise as needed and can provide as little or as much of the surrounding context as the author provides. Most often, spatial language uses relative locations (e.g., north of another known entity) rather than absolute locations (e.g., the latitude and longitude). In face-to-face interactions, some of the spatial language can also be conveyed through gestures (e.g., pointing in the direction of a school).

More formally, spatial language and ideas can be organized in a systematic fashion. These are spatial ontologies, which provide the basic elements to formally and explicitly organize objects, concepts, and the relationships among them. Spatial ontologies are descriptions of geographic "things"—categories of geographic objects, their behavior, and their relationships as they exist in space. These elements can be concrete or abstract, divisible or indivisible. They can be a simple taxonomy, a lexicon, a thesaurus, or even a fully axiomatized theory. Two basic types of ontologies exist: descriptive and formal. Descriptive ontologies are built around concepts and categories that,

taken together, form the basis for a particular view of the world. A formal ontology endeavors to define elements based on a set of concepts and then further defines the relationships between those elements. The process of developing spatial ontologies involves creating order of ideas, objects, and processes that interact in space. Creating order and describing processes require a spatial language so that ideas and information can be communicated effectively. The language and accompanying vocabulary describe explicitly ideas, objects, properties, and behaviors.

Despite attempts for explicit spatial ontologies of spatial characteristics, concepts and objects often remain fuzzy and inexact. Fuzzy set theory and other strategies are being used to represent, both conceptually and in computer data models, information that by nature cannot be defined with sharp boundaries. The fuzzy boundaries could be on the surface of the earth (e.g., the boundary between different vegetation types) or in the categorization criteria (e.g., land use designations). Spatial concepts and data may also be fuzzy when there are uncertainties at the needed level of detail—location, time, or attribute.

CULTURAL MODELS

For many human geographers, space and locations are not always the defining principles for assessment, evaluation, and analysis. The geographic context is viewed not as absolute but rather as relative with regard to human experience and is understood only by the objects and processes that constitute it. As such, cognitive models of space are based on visual perception, personal experience, and nonvisible structures of space (e.g., social class, globalization).

Some geographers question whether political boundaries (and how space is conceptualized by some) remain significant in assessing the impact of issues that have worldwide consequences, for example, when studying globalization. Globalization is defined as issues that are globally connected or worldwide in scope or application. Geographers doing research on globalization examine the politics, economics, and social issues associated with climate, poverty, terrorism, pollution, land degradation, and other issues related to human and natural activities. Proponents consider globalization as the answer to social, political, and economic problems that plague developing countries because it provides them with opportunities to advance. Globalization is also considered a problem—primarily in developing countries—because of inequalities, loss of jobs, and environmental degradation ("winners" and "losers"). Some geographers argue that the network (social, political, or economic) represents the defining linkages and could be independent of the spatial location or configuration. For example, terrorist networks, existing in cells worldwide, are linked via social connection rather than geographic location.

The recent trend in geography has been to analyze human and physical landscapes as a union. The environmental deterministic ideas created a division between human and physical features. This approach used the physical characteristics of the land to manipulate viewpoints concerning the characteristics of humans based on environmental conditions. The unifying views of the landscape can be seen within cultural geography and recent trends in geography. These ideas incorporate a more holistic approach that includes the examination of political and social conflicts in gender, race, and class differences. The cause-and-effect relations in early cultural geography (e.g., associated with environmental determinism) assume that the physical attributes are the driving agent to culture. In the recent approaches to cultural geography, these contributions are acknowledged, yet there is a difference between considering the past and regarding it as the only accepted methodology. The contemporary trends in geography are searching for broader cause-and-effect relations in space and time. Although no definitive conclusion can be drawn regarding whether a reductionist or a holistic view of the landscape is more appropriate, the evolution of geographic thought has modified how space is conceptualized.

—*Elizabeth Wentz*

See also Behavioral Geography; Computational Models of Space; Ontology

Suggested Reading

Guarino, N. (1997). Understanding, building, and using ontologies. *International Journal of Human–Computer Studies, 46,* 293–310.

Longley, P., Goodchild, M., Maguire, D., & Rhind, D. (2005). *Geographic information systems and science* (2nd ed.). Chichester, UK: Wiley.

Mark, D., Freksa, C., Hirtle, S., Lloyd, R., & Tversky, B. (1999). Cognitive models of geographic space. *International Journal of Geographical Information Science, 13,* 747–774.

COLONIALISM

Colonialism, as distinguished from imperialism, is generally defined as the appropriation, occupation, and control of one territory by another. This simple definition, however, masks a longer and more complex genealogy of the term and concept. The term *colonial*, derived from the Roman concept of *colonia*, originally referred to settlement. Roman colonies were viewed as the physical extension of the Roman Empire. This initial use was focused on Roman citizens. These settlements were places where Romans retained their citizenship, a practice reminiscent of extraterritoriality. Colonies were self-sufficient. This definition did not consider the position of the indigenous populations.

The modern use of *colonialism* includes elements and characteristics that extend far beyond the initial sense of "settlement." And whereas colonialism always entails the settlement of people from the colonial state to a colonized territory, the practice of colonialism is characterized by more than simply immigration flows. Colonialism has come to refer to the conquest and control of other peoples and other territories. This distinguishes colonialism from another equally complex term, *imperialism.* This latter term is generally defined as the ideological underpinning of colonial practices.

There is no essential colonialism. The meanings and interpretations of colonialism are contingent on different eras, different places, and different territorial relationships. These have been shaped by particular contexts of politics, economics, culture, and geography. There is, however, general agreement that our contemporary world geography is a result of European (and American) colonial practices that have occurred over the past five centuries. In effect, these powers constructed the current political world. Modern state boundaries are largely a reflection of colonial histories and rivalries. It is instructive, therefore, to consider how colonies, particularly within the past five centuries, were established, administered, and maintained.

The establishment of colonies is a reflection of geography and a reflection of political intent. Although there is no set pattern, colonies may be established initially through the use or threat of military force. Economic, cultural, and political institutions are introduced subsequently. Colonies may also be established through the imposition of (unequal) treaties. This may likewise be imposed by the threat or actual use of violence. Treaties may also establish a protectorate in which a dependent territory surrenders all or part of its sovereignty to the colonial power.

Colonial rule may be established suddenly, or it may be extended over a period of years. The imposition of French rule in Indochina, for example, was completed during a period of 25 years, from the late 1850s to the mid-1880s. The British conquest of Burma was completed over six decades and included two substantial wars. Colonized states, moreover, may witness considerable variation or sequencing of colonialism. The Caribbean island state of Grenada, for example, was initially colonized by the French in 1650; over the next three centuries, it alternated from French and British colonial control until it achieved independence in 1974. The Caribbean island of St. Eustatius, colonized by the Dutch in 1636, likewise changed in colonial affiliation among the Dutch, French, and British for more than four centuries. It remains a Dutch possession.

A number of different motivations led to the establishment of colonies. Most accounts of colonialism stress economic motives. Colonies serve as sources of labor, raw materials, and markets. Often colonial powers form monopolistic arrangements. Colonies may also serve as sources of investment. A colonial power can also increase its wealth through the appropriation of other societies' wealth. During the 16th century, for example, Spain plundered the riches of existing civilizations in the Americas and later augmented this wealth through the control and exploitation of mines and plantations.

Although it is generally agreed that colonies are established for economic reasons, there are other motivations as well. For example, colonies may be founded for religious purposes. Many Western European states attempted to spread their religious beliefs and to convert nonbelievers throughout their areas of influence. During the 16th and 17th centuries, Spain spread Roman Catholicism throughout its colonies in the Americas as well as in the Philippines. The Dutch, beginning largely in the 17th century, likewise spread Protestant beliefs throughout their colonies in the islands of present-day Indonesia.

Other cultural explanations dovetail with religious motivations. It was not uncommon, for example, for colonial powers to justify their practices on the presumption that colonial subjects were not capable of self-government. Such beliefs were used to legitimate

periods of tutelage and "benevolent" assimilation. Often racist and paternalistic attitudes were apparent, as in the United States' reference to Filipinos as America's "little brown brothers."

Strategic reasons also lead to the establishment of colonies. During the late 19th century, for example, the United States required a system of coaling stations. Navies and maritime commerce activities, such as whaling, were powered by coal. This required a network of maritime base coaling stations. Thus, colonial practices were conducted in line with the doctrines of maritime power that existed during the 19th century. Other strategic reasons for the establishment of colonies include the protection of trade routes such as the control of the Cape of Good Hope at the southern tip of Africa and the British control of Egypt to ensure continued access to, and use of, the Suez Canal.

Once established, the nature of the colonial regime and its administrative form vary greatly. In general, a distinction is made regarding the extent to which indigenous populations play a role in the administration of a colony. On the one hand, colonies may be administered by direct rule. In this case, the administrative functioning of the colony is exercised without any influence by indigenous people. The Portuguese, for example, adopted a policy of direct rule in their African colonies of Angola and Mozambique. On the other hand, colonies may be administered through a system of indirect rule. Under this system, an element of power is given to a small, carefully selected indigenous group of people. These are figureheads and do not represent the local population. In British-controlled Nigeria, for example, local kings and chiefs functioned as intermediaries, acting as links between their people and the British colonial authorities. In French Indochina, Vietnamese landlords likewise became extensions of the French colonial government. Under systems of indirect rule, it is not uncommon for the local rulers to be responsible for the collection of taxes and the enforcement of local ordinances. Whether direct or indirect, however, the ultimate administrative control of a colony is found within the colonial power.

There exists tremendous variation of administrative control both between and within colonial powers. The British, for example, had no preconceived model of a colony. British authorities did not follow a set pattern or model of colonialism and instead preferred a policy of devolution whereby different parts of the British empire were granted varying degrees of autonomy.

The type of colony was based on a combination of factors, including preexisting political and economic structures, the proximity of potential colonial rivals, and physical geography. Consequently, Britain retained a network of crown colonies, condominiums (territories ruled jointly by two or more states), trusteeship territories, commonwealth territories, and high commission territories (those administrated by a high commissioner).

Different forms of colonies would consequently entail different forms of administration. In Africa, for example, the British distinguished between colonies and protectorates. Colonies were generally coastal in location, small in scale, and ruled intensively and directly. Kenya and Lagos were considered colonies. Protectorates, conversely, often were remotely located in inland areas; these tended to be ruled indirectly through local rulers. Very few British settlers lived in protectorates. Nigeria, Uganda, and Swaziland, among others, were ruled as protectorates. The British possession of Sudan, conversely, was ruled indirectly through the Anglo-Egyptian Condominium Government. Sudan, in principle, was administered jointly by British and Egyptian officials; however, British rule remained dominant.

It is not uncommon for different ethnic groups to be treated differently within colonies. In Sierra Leone, for example, the British colony consisted of Freetown Peninsula and Shebro Island, and the rest of the country was administered as a protectorate. Ethnically, the colony was dominated by the Creoles, who were descendants of the freed slaves from Nova Scotia. However, there were also 17 other ethnic groups, with the largest groups being the Mende and Temne. These groups, whose members resided mostly in the outlying areas, were marginalized by both the British and the Creoles.

The administration of colonies is also a function of the underlying motivations of colonization. Given the prevalence of economic motives, colonial administrators generally exploited resources and markets to the detriment of the colonies. Cultural motivations, however, would also significantly influence the administration of colonies. In Africa and Asia, both France and Portugal fostered policies of acculturation that sought to encourage their colonial subjects to adopt the culture, language, and customs of France and Portugal, respectively. In French colonies, this was referred to as the *mission civilisatrice* (or civilizing mission). The French operated on the ideals of the French Revolution, including an unbinding belief in the superiority of French culture and civilization. In the Portuguese colonies of Angola, Mozambique, and Guinea-Bisseau, conversely, Africans who aspired to Portuguese citizenship were granted *assimilado* status, and these individuals were placed at the top of the social hierarchy instituted by the Portuguese. In a process of divide-and-rule, assimilados would be given preference in civil service positions in the colonies.

The maintenance of colonial rule was conditioned by a multitude of factors, including preexisting historical circumstance, local geographies, the amount of raw materials and markets, and the nature of economic and political administration. Some colonial powers maintained rule through strict policies and violence, whereas others were more "benevolent" in their approach. In general, however, colonies suffered at the expense of the colonial powers. Indigenous populations, and especially those of the peasant class, witnessed a curtailment of civil rights and a lack of political representation.

Economic arrangements were decidedly unequal. In 1830, the Dutch introduced a colonial practice known as the *culture system* on the colony of Java in present-day Indonesia. This work scheme required compulsory labor of the indigenous peoples and mandated the intensive cultivation of cash crops such as coffee, sugar, indigo, tea, tobacco, pepper, and cinnamon. Crops were subsequently sold at low prices to Dutch authorities. The system was a government-run economic monopoly. Dutch authorities determined which crops would be planted. The culture system produced substantial profits for the Dutch but contributed to the impoverishment of many Javanese.

The extent, and consequently the intensity, of colonialism was spatially uneven both between and within colonies. Spain confined its activities in the Philippines largely to the main island of Luzon and particularly the capital city of Manila. In Vietnam, the French affected a more radical change in the southern region than in the northern one. In the southern region, known by the French as Cochin China, colonial authorities attempted to make the region self-supporting. Officials introduced the concept of private ownership and subsequently encouraged the conversion of former communal lands into private estates and plantations for the cultivation of cash crops for export. As a result, social relations were ruptured and transformed for the majority of Vietnamese peasants.

Historically, European colonialism has been divided into two broad eras. The first period, which occurred approximately between the 15th and 18th centuries, began with the Portuguese taking of Ceuta off the coast of Northwest Africa in 1415, the Spanish conquest of the Americas in 1492, and the Portuguese capture of Malacca in 1511. This era was dominated by Spain, Portugal, and England. Geographically, most colonies were established throughout the Americas. These included the colonies of British North America and the Spanish colonies of present-day Central and South America. Colonial practices during the first period were underpinned by an economic system termed *mercantilism*. This system bridged feudalism and capitalism and was premised on the attempt to garner a favorable balance of trade. Wealth was measured by the accumulation of gold and silver.

The second period witnessed the colonization of much of Africa and Asia. The Berlin Conference and the subsequent scramble for Africa illustrate vividly the practices of this second era of European colonialism. In 1884–1885, the major European powers, as well as the United States, met in Berlin to establish rules for the partitioning of Africa. European powers agreed to the rules for the partitioning of Africa. A free trade zone was declared across Africa and recognized European spheres of influence, which provided rules for the occupation of colonies throughout the continent. In 1870, more than 80% of the African continent was controlled by indigenous rulers. Within a decade, however, this situation was reversed as European powers colonized all of sub-Saharan Africa with the exception of Ethiopia.

Colonialism as a concept evolves continuously. During the 1960s, for example, radical groups such as the Black Panther Party (cofounded by Huey Newton and Bobby Seale) developed the idea of *domestic colonialism*. They argued that African American communities within the United States functioned as internal colonies, with the labor of blacks being appropriated and exploited by white capitalists.

It has been argued that colonialism, as a practice, was abolished decades ago. However, the work of Derek Gregory testified quite clearly that the practices that are used to define colonialism remain as potent as ever politically, economically, and culturally.

—James Tyner

See also Anticolonialism; Dependency Theory; Globalization; Imperialism

Suggested Reading

Cohen, B. (1973). *The question of imperialism: The political economy of dominance and dependence.* New York: Basic Books.

Morgenthau, H. (1949). *Politics among nations: The struggle for power and peace.* New York: Knopf.

Smith, B. (1996). *Understanding Third World politics: Theories of political change and development.* Bloomington: Indiana University Press.

COMMODITY

A commodity is an economic good or, more specifically, a good that is produced for the purpose of exchange. So long as the commodity is exchanged, it can be tangible or intangible. Traditional definitions of commodities further specify that they are goods for which variations in quality are insignificant (i.e., all items are considered to have a similar value regardless of their sources). This homogeneity of commodities is significant because they are considered to form the basis of economic exchange—a process that may lead to the creation of additional capital. The critical role of commodities in trade and capital accumulation caused Karl Marx to refer to commodities as the "cell" of capitalist society.

The value assigned to a commodity is thought by some to be dependent on several factors: its use value (the gain that consumers will receive from the consumption of the good), its exchange value (the value of the commodities that will be received in trade for the commodity in question), and its labor value (the value of man-hours involved in the extraction or production of the commodity). However, the necessity of commodity trading requires the value of commodities to be set quickly and easily; therefore, traders commonly rely on values set by market mechanisms. This valuation shortcut may result in commodity prices that are divorced from local use, exchange, or labor values.

Because commodity values often are set by forces beyond producers' control, it is desirable for commodity producers to "de-commodify" their products to make them appear more attractive (useful) and to increase the prices received for the goods. For example, commodities such as gasoline have been perceived as homogeneous and therefore will generate minimal profits for producers. Advertising can be used to suggest that specific brands of gasoline have greater use values than do others. Such de-commodification allows producers to charge premium prices for common products. In its extreme, this de-commodification may transcend consumers' needs for a product, and marketing can result in the consumption of commodities even when consumers receive no value from such consumption.

Geographic research on commodities historically has focused on their availability as a source of comparative advantage. Current research has shifted to analyze the geography of commodity transformation. The emphasis of these studies has been on isolating the network (or chain) through which a raw material moves as it is transformed into a more sophisticated commodity. By isolating the forces that connect the network and a localities position in the commodity chain, geographers can identify the relative importance of a place within the global economy.

—*William Graves*

See also Marxism, Geography and

Suggested Reading

Amin, A., & Thrift, N. (Eds.). (2004). *The Blackwell cultural economy reader.* Oxford, UK: Blackwell.

COMMUNICATIONS, GEOGRAPHY OF

There is no process in human geography, whether economic, political, or cultural, that does not depend on communication in extensive and important ways. Communications are flows of ideas and information through space and time. Communications often also contain images of places, either generic or specific. The geographic interest in communication, accordingly, is composed of two related concerns: the spatial organization of communication flows, on the one hand, and the ways in which places are represented and socially contested, on the other hand.

SPATIAL FLOWS AND STRUCTURES

Interest in the spatial organization of communication flows and infrastructure arose during the 1960s as geographers turned to physics for the keys to understanding geographic patterns. Spatial analysts used distance decay models, analogous to models of gravitic attraction, to predict interaction between places. Rather than distance, the more sophisticated models were based on accessibility. These models generally neglected the ability of people to act at a distance through communication media, but a handful of geographers, including Ronald Abler, Donald Janelle, and Peter Gould, gave special attention to communications. An important idea to emerge from their scholarship was the dynamic changeable quality of space when viewed in terms of accessibility. The best-known aspect of this research is the concept of time–space convergence, the progressive reduction in the time required to access one location from another location. Janelle showed that time–space compresses or converges due to a combination of technological innovation and economic competition, with each encouraging the other. The end point of such convergence—absolute or complete time–space convergence—was of particular interest. In such a situation, distance presumably would no longer affect the interaction between two points. Although complete time–space convergence was only a theoretical limit in transportation studies, it already existed in practical terms for communications by the 1970s due to technological innovations such as radio, television, and the telephone. Spatial analysts who studied communication needed to develop theories in which distance was replaced by other space-shaping factors such as perceptions and policies. Janelle also introduced the idea of personal extensibility, that is, the ability of an individual to access distant points.

These ideas of accessibility, extensibility, and spatial metamorphosis were subsequently revisited during the 1990s in light of the diffusion of networked computers and other information technologies. Cultural geographers, urban geographers, political geographers, and economic geographers all contributed to this emerging topic of interest, arguing that the space created by instantaneous communication was not as simple as early observers had expected. Late-20th-century communications had given rise to both centralization and decentralization at the same time. Economic, political, and administrative power became centralized in a small number of technological growth poles, including world cities and cyberstates, even as many jobs were decentralizing to sprawling suburbs with back offices and farther afield to *maquiladora* factories and overseas sweatshops. Both centralization and decentralization resulted from a new integration of production, sales, administration, and distribution over long distances with virtually no delays involved in access via communication technologies.

This process brought the world under increasing control by world cities or, more precisely, by well-connected elites in world cities. Manuel Castells described the situation in terms of a "space of flows"—a digital context of interaction absorbing and suppressing the older "space of places" where people live, work, and struggle to achieve security. Castells's model was subsequently modified to acknowledge local variations in the understanding and appropriation of communication technologies. Technologies are understood to be socially constructed; their ubiquity does not mean that their adoption occurs in the same way in every place. First, places vary tremendously in their degrees of access to the new communication devices. Second, even where the access to communication devices is equally high, the particular mix of media uses varies due to differences in pricing, regulation, and taxation. Third, the uses of a communication device vary from place to place, so the same device may be understood in one community as an adjunct to marketing and sales even while it is understood in another community as a forum for interpersonal communication. Therefore, the absolute time–space convergence produced by new media is far from rendering space and place irrelevant.

REPRESENTATIONS OF PLACES AND SPACES

The second array of concerns in the geography of communications arises from the fact that place images are an essential part of communication content. Some communications are explicitly about place, and representations of people and things are symbolically linked to places—the cowboy in the American West, the sky-scraper in Manhattan, and so on. Geographers have asked various questions about place representations. Who are they made by and for? What purposes are they intended to serve? What are their historical social origins? How is their production funded? What specific symbols and signs do they contain? These questions are addressed through various approaches ranging from humanist geography and phenomenology to Marxist structuralism, poststructuralism, postmodernism, postcolonial theory, gender theory, and queer theory.

The 1970s was the formative decade for this approach as a group of geographers, including Yi-Fu

Tuan, Anne Buttimer, and David Seamon, began to explore literary and artistic communications for insight into experiences of space and place. In exploring personal experience, these geographers drew eclectically on anthropology, sociology, philosophy (particularly phenomenology), and history. In a more analytical vein, Robert Sack analyzed the historical transformation of social representations of territory and space. Geographers such as Allan Pred and Gunnar Olsson not only studied communication but also reconfigured communication through radically experimental language.

By the 1990s, the focus of humanistic geography had shifted from subjective place experiences toward social power relations and the instrumental uses of symbolism. Systematic distortions in maps and texts were understood in terms of how they maintained social domination, oppression, and exploitation. Representations of places are now seen not only as means of sharing experiences of the world but also as tools used by elites to mask social conflicts and maintain dominance. Key works include a critique of the practices of naming and mapping geographic space by Peter Jackson; deconstructions of mapmaking practices by J. B. Harley, Mark Monmonier, and John Pickles; and interpretations of power relations in the urban landscape by James Duncan, Denis Cosgrove, and Edward Soja. The objective of such work was to reveal, excavate, deconstruct, or destabilize taken-for-granted representations of spaces and places and often to support silenced, subaltern, or resistant place meanings. This shift of humanist geography toward political concerns was matched by a reciprocal "cultural turn" among political and economic geographers. British geographers, in particular, drew on the writings of Raymond Williams, E. P. Thompson, and European semiotic theory to justify the study of both dominant/authoritative and popular forms of communication such as romances, popular songs, and soap operas. In the United States, the situation was complex because of the competition of the critical approach with a broad-based humanistic approach influenced by Tuan and an empirical approach promoted by George Carney and others in the American Midwest; however, the "critical social theory" contingent became increasingly predominant during the 1990s.

Recently, inconsistencies in the critical approach have been revealed. Geographers on both sides of the Atlantic drawing on social constructivism have suggested that to critique representations (whether popular or elite) is to presume that one has access to a truth or reality that is ontologically prior to and outside of representational practice. Although the critic claims to be unsettling, excavating, or destabilizing authority, he or she in fact takes on an authoritative position. The contradictions inherent in this approach suggest that geographers should *engage* with representations as equal participants in meaning making rather than trying to discipline understanding through critique.

—*Paul C. Adams*

See also Spaces of Representation; Telecommunications, Geography and

Suggested Reading

Adams, P., Hoelscher, S., & Till, K. (Eds.). (2001). *Textures of place: Exploring humanist geographies.* Minneapolis: University of Minnesota Press.

Brunn, S., & Leinbach, T. (Eds.). (1991). *Collapsing space and time: Geographic aspects of communication and information.* London: HarperCollins Academic.

Cresswell, T., & Dixon, D. (Eds.). (2002). *Engaging film: Geographies of mobility and identity.* Lanham, MD: Rowman & Littlefield.

Duncan, J., & Ley, D. (Eds.). (1993). *Place/culture/representation.* London: Routledge.

Gould, P. (1991). Dynamic structures of geographic space. In S. Brunn & T. Leinbach (Eds.), *Collapsing space and time: Geographic aspects of communication and information* (pp. 3–30). London: HarperCollins Academic.

COMMUNISM

Communism, as a theory and as a social movement, centers on the lack of private property and the organization of society such that all members have equal status both economically and socially. Under a communist system, labor would be divided among citizens according to their interests and abilities, and resources would be distributed corresponding to need. By such a vision, government itself would be replaced by communism, that is, by the communal ownership of all property. Communism is also thought to be the abolition of all forms of oppression, whether in the form of oppression of people by people, of countries by countries, of classes by classes, or any other form of oppression.

Although early views of communism, promoted by Plato during the 4th century BC and by later perspectives during the 1600s, advocated communal ownership of property on a small scale, Karl Marx conceived of communism as a revolutionary movement that had potential at the global scale. He envisioned that society would move through successive phases—feudalism, capitalism, and then socialism.

In 1848, Marx and Friedrich Engels wrote *The Communist Manifesto,* in which they described communism as an inevitable outcome of the fact that in most societies the wealth and means of production (i.e., natural resources and infrastructure) are controlled by a small elite. Under capitalist systems, this group of people, whom Marx referred to as the *bourgeoisie,* purchased from the majority of the population their labor and sold the results of their work for a profit. This imbalance of power and economic wealth, Marx argued, created different classes of citizens and established an unbalanced and unsustainable distribution of wealth. At some point, Marx argued, members of the working class, which he referred to as the *proletariat,* would organize themselves to overthrow the bourgeoisie elite and to redistribute wealth more equitably.

Unlike social democrats (e.g., the Social Democratic Party in Germany, the British Labour Party) who have believed that communism could be brought about by democratic means, Vladimir Lenin maintained that revolution, initially in less economically developed states such as Russia, was necessary to transform society to communism. Lenin's work inspired Leon Trotsky and Joseph Stalin, both of whom contributed much to building and strengthening the Communist party in Russia and, shortly thereafter, in the Soviet Union. According to Lenin, the establishment of a Communist party was a political necessity within a communist system, but critics have argued that the Communist party in the Soviet Union served the interests of a politically powerful elite rather than benefiting the populace as a whole.

From a Marxist perspective, Soviet-style communism failed because it attempted to move society directly from feudalism to socialism without the intermediary phase of capitalism. Other factors recognized as contributing to the collapse of the Soviet Union, and thus the end of the cold war, include the Soviet Union's inability to afford the arms race against the United States and the plummeting world oil prices during the early 1970s that denied the oil-exporting

Soviet Union much-needed income for the purchase of food and other basic goods. Communism continues to shape political practice in China, Cuba, Laos, North Korea, and Vietnam, but Communist parties in these contexts differ greatly from each other.

—*Shannon O'Lear*

See also Marxism, Geography and

Suggested Reading

Marx, K., & Engels, F. (1998). *The communist manifesto: A modern edition.* London: Verso. Available: www.anu.edu.au/polsci/marx/classics/manifesto.html

This godless communism. (1961). *Treasure Chest* (Vol. 17, Nos. 2–20). Available: www.authentichistory.com/images/1960s/treasure_chest/godless_communism.html

COMPARATIVE ADVANTAGE

Under capitalism, different regions have long specialized in the production of different types of goods and services. In Europe during the Industrial Revolution, for example, Britain became a major producer of textiles, ships, and iron; France produced silks and wine; Spain, Portugal, and Greece generated citrus, wine, and olive oil; Germany, by the end of the 19th century, was a major exporter of heavy manufactured goods and chemicals; Czechs sold glass and linens; Scandinavia produced furs and timber; and Iceland exported cod to the growing middle classes. Within the United States, similarly, different places acquired advantages in some goods and not others. The Northeast was dominated by light industry, particularly textiles; the Manufacturing Belt became the center of heavy industry; Appalachia developed a large coal industry to feed the furnaces of the industrial core; the South grew crops such as cotton and tobacco; the Midwest became the agricultural products behemoth of the world; and the Pacific Northwest was incorporated into the national division of labor based on the expanding timber and lumber industry.

When regions or countries specialize in the production and export of some goods or services, they enjoy a comparative advantage. This notion was first introduced by 19th-century economist David Ricardo (1772–1823). Like all classical political economists,

he assumed the labor theory of value (i.e., the value of goods reflects the amount of socially necessary labor time that goes into their production) and thus ignored demand. Ricardo concluded that nations will specialize in the production of commodities that they can produce using the least labor compared with other nations.

Ricardo's classic example of this process is demonstrated in Table 1, which illustrates the allocation of labor time in England and Portugal, two longtime trading partners, before and after they specialized. In the first part, which depicts the labor hours per unit of wine or cloth that England and Portugal must each dedicate to the production of one unit of each good, it is evident that Portugal has an absolute advantage in both goods; that is, it can produce both of them with fewer labor hours than can England. If Portugal is more efficient, does it make sense for Portugal to trade? The answer is yes, implying that even the most efficient producer benefits from trade. Ricardo's analysis examined what happens when each country allocates its resources to the good it can produce most efficiently compared with its trading partners, that is, when it acquires a comparative advantage. Thus, in the second part of the table, England produces only cloth (two units at 100 hours each) and Portugal produces only wine (two units at 80 hours each). In the process of specializing (i.e., of producing for a market that consists of both economies together rather than either economy alone), each country frees up some resources that would otherwise have been dedicated to the inefficient production of a good in which it did not have a comparative advantage. England saves 20 labor hours and Portugal saves 10 labor hours; thus, the combined trading system saves 30 labor hours that can be reallocated toward investment (although the original model is static and says nothing about change over time).

The Ricardian model—the simplest of many complex notions of comparative advantage—has important implications for economic geography. First, it shows how powerfully trade and exchange shape local production systems. It demonstrates that trade allocates resources to the most efficient (i.e., profitable) ends. The costs of free trade are borne by inefficient producers, in this case English wine makers and Portuguese textile producers. Second, Ricardian notions of comparative advantage reveal that specialization reduces the total costs of production; thus,

Table 1   Ricardian Example of Comparative
            Advantage

Before specialization (labor hours/unit):

	Wine	Cloth	Total
England	120	100	220
Portugal	80	90	170
Units produced	2	2	390

After specialization (labor hours/unit):

	Wine	Cloth	Total	Savings
England	0	200	200	20
Portugal	160	0	160	10
Units produced	2	2	360	30

SOURCE: Ricardo, D. (1817). *Principles of political economy and taxation.* London: John Murray. Reprinted 1996 by Prometheus Books.

trade improves efficiency even without reallocating resources. For this reason, the vast majority of economists favor free trade as beneficial to all parties concerned. Third, this approach points out that large markets allow more specialization than do small ones. Adam Smith noted the same thing when he stated that the division of labor is governed by the size of the market. In this case, when the market expanded from one country to two countries, it allowed firms to specialize and become more efficient in the process.

Just as there is no specialization without trade, there can be no trade without transportation. Goods must be moved across space from producer to consumer, and these transport costs must ultimately be borne by those who consume the goods. To the degree that transport costs affect the delivered price of commodities, they also influence consumers' willingness to buy them and thus the competitiveness of the regions that export them. If transport costs are low, their impacts on the division of labor will be minimal. However, particularly for heavy and bulky goods, transportation costs sometimes may increase the market prices of exports/imports prohibitively; that is, transport costs may make the exports too expensive to ship across regions. Throughout the history of capitalism, declines in transport costs have made it progressively easier for regions to realize their comparative advantages; thus, lower transport costs have contributed to lower production costs. For example, New Zealand became a major producer of lamb following the introduction of refrigerated shipping during the

late 19th century. Similarly, the Pacific Northwest began to export vast quantities of wood and paper to the cities of the Midwest and East Coast following the completion of the transcontinental rail lines during the 1890s.

Ricardo's two-country, two-product theory of comparative advantage can be expanded by allowing several production factors. The multifactor approach to trade theory derives from work by two Swedish economists, Eli Heckscher and Bertil Ohlin. The Heckscher–Ohlin theory holds that a country should specialize in producing those goods that demand the least from its scarce production factors. Unlike the original Ricardian model, it includes demand and allows for the production of more than one good. In this formulation, specialization of production will be incomplete; that is, countries may continue to produce some of a good even if they do not enjoy complete superiority in the costs of production. The Heckscher–Ohlin theory argues not only that trade results in gains but also that wage rates will tend to equalize. The reasoning behind this factor–price equalization, as it came to be called, is as follows. If a country specializes in a labor-intensive good, its abundance of labor diminishes, the marginal productivity of labor rises, and wages increase. Conversely, if a different country specializes in capital-intensive goods, labor becomes less scarce, the marginal productivity of labor falls, and wages fall.

The traditional theory of comparative advantage is simplistic and unrealistic. Ricardo never gave an adequate account of why regions specialize in some goods and not others, instead offering a picture that is static with respect to time, overemphasizes labor and climate, ignores consumption as well as the role of economies of scale and agglomeration, says nothing about the nature of competition, and is silent concerning the impacts of public policy. These shortcomings were addressed in the theory of competitive advantage.

—*Barney Warf*

See also Competitive Advantage; Economic Geography; Factors of Production

Suggested Reading

Stutz, F., & Warf, B. (2005). *The world economy: Resources, location, trade, and development* (4th ed.). Upper Saddle River, NJ: Pearson/Prentice Hall.

COMPETITIVE ADVANTAGE

An alternative to the traditional theory of comparative advantage is called the theory of competitive advantage. Unlike the Ricardian model, which was useful for understanding the simpler economies of the early Industrial Revolution, this approach focuses on the social creation of innovation in a knowledge-based economy. The key to competitiveness in this view is productivity growth; over the long run, rising productivity creates wealth for everyone, if not equally. Productivity growth in turn reflects many factors, including the education and skills of the labor force, available capital and technology, government policies and infrastructure, and the presence of scale economies. In the context of global markets, all firms can maximize scale economies.

Competitive advantage is dynamic and changes over time. The goal of national development strategies is to move into high-value-added, high-profit, high-wage industries as rapidly as possible. Such goods have high multiplier effects and do the most to trigger rounds of growth. To accomplish this goal, firms and countries should seek to sell high-quality goods at premium prices in differentiated markets. Quality is a key variable here; countries often acquire reputations for producing high- or low-quality goods, earning (or not earning) brand loyalty as a result. By moving into high-value-added goods, nations should seek to automate low-wage, low-skill functions and retain knowledge-intensive ones.

Although the global economy is increasingly seamless, competitive advantage is created in highly localized contexts, that is, within individual metropolitan areas. Globalization does not eliminate the importance of a home base. Thus, countries that succeed internationally do so because a few regions within them move into "cutting-edge" products and processes. Within the United States, propulsive regions include Silicon Valley, Boston's Route 128, and New York's position in finance and producer services; in Europe, they include Italy's Emilia–Romagna, that continent's largest high-technology region, as well as Germany's Baden–Württemberg, Denmark's Jutland peninsula, and the Cambridge region of the United Kingdom; and in Japan, the government has actively constructed a series of technopolises toward this end.

The overall determinants of competitive advantage include skilled labor, good educational systems, and technical training; agglomeration economies, including pools of expertise, webs of formal and informal interactions, trust, linkages, strategic alliances, trade associations, and integrated networks of suppliers and ancillary services; and a culture that rewards innovation, adaptation, experimentation, risk tolerance, and entrepreneurship, including heavy levels of corporate and public research and development and the continual upgrading of capital and skills. Corporations must engage in ongoing and organizational learning, anticipating changes in markets and demand. Rigid corporate bureaucracies lead to complacency and short planning horizons, and uncompetitive markets (i.e., private or public monopolies) exhibit little innovation. In the world economy today, increasingly sophisticated buyers spur a constant upgrading in the quality of output, adequate financing and venture capital, and public policies that encourage productivity growth, including subsidized research, export promotion, educational systems, and an up-to-date infrastructure (e.g., airports, telecommunications).

The theory of competitive advantage maintains that four attributes of a nation combine to increase or decrease its global competitive advantage and world trade: (1) factor conditions, (2) demand conditions, (3) supporting industries, and (4) firm strategy, structure, and competition. Factor conditions (or production factors) include human resources (quantity of labor, skill, educational level, productivity, and cost of labor), physical resources (raw materials and their costs, location, access, and transport costs), capital resources (funds to finance the industry and trade, including the amount of capital available; savings rate; health of money markets and banking in the host country; government policies that affect interest rates, savings rates, and the money supply; levels of indebtedness; trade deficits; and public and international debt), knowledge-based resources (research, development, scientific and technical community within the country, its achievements and levels of understanding, and the likelihood of future technological support and innovation), and infrastructure (all public services available to develop the conditions necessary for producing the goods and services that provide a country with a competitive advantage, including transportation systems, communications and information systems, housing, cultural and social institutions, education, welfare, retirement, pensions, and national policies on healthcare and child care). These five factors are identified in current international and economic circles as the keys to the competitive advantage of a nation in the foreseeable future.

Demand conditions are the market conditions in a country that aid the production processes in achieving better products, cheaper products, scale economies, and higher standards in terms of quality, service, and durability. Demand conditions cause firms to become innovative and thus to produce products that will sell not only in the domestic market but also in the world market.

To be competitive internationally, firms require access to networks of other firms that specialize in different tasks in the economy. For example, large financial institutions require law firms, marketers, and advertisers. Often large companies use management consultants or similar business services, subcontracting tasks that require heavy investments in human capital. Access to these industries that generally provide expertise often is done through face-to-face contact.

Firm strategy, structure, and competition relate to the conditions under which firms originate, grow, and mature. For example, because stockholders demand that U.S. companies show short-term profits, U.S. corporate performance may be less successful in the long run than it would be if it were judged over a much longer time period, as is Japanese and German corporate performance.

State support of corporate strategy and performance is important. For example, a country can regulate taxes and incentives so that investment by a firm is high or low. In addition, competition within a country can impose demands on company performance; new business formations often pressure existing firms to improve products and lower prices and thus to increase competitiveness.

—*Barney Warf*

See also Comparative Advantage; Economic Geography; Factors of Production

Suggested Reading

Porter, M. (1998). *The competitive advantage of nations* (2nd ed.). New York: Free Press.

COMPUTATION, LIMITS OF

SEE LIMITS OF COMPUTATION

COMPUTATIONAL MODELS OF SPACE

Computational models of space are ways in which space is represented to solve spatial problems from given inputs by means of algorithms. The development of computational models of space relates closely to how space is conceptualized: as discrete objects or as continuous fields. Both object- and field-based conceptualizations of space have been represented in various forms to facilitate geographic computation. The nature of a geographic problem determines the suitability and effectiveness of computational models as to how the models represent space, ingest input data, and support algorithm development to derive solutions for the given problem.

Computational models of objects define space by identifiable entities of interest. In traffic analysis, for example, computational space is defined by the transportation network of interest; and trips outside the network are excluded from consideration. Similarly, computational models for power grids may include transmission lines and transformers, and those for census demographics may include areas of enumeration. Object space often is implemented by vector models of points, lines, and polygons with object identifiers, dimensions, coordinates, and attributes. These geometric objects can be further combined to form complex objects to represent geographic entities of complex shape and structure such as rings to represent lakes with islands and aggregates of line segments to represent delivery routes. Computational geometry serves as the foundation for the development of vector algorithms to quantify individual objects and their spatial distributions, topological relationships, and spatial interactions.

A field describes the distribution of a geographic variable for which value is determined by location; that is, value is a function of location such as a temperature field. A field space is said to be planar and spatially exhaustive because every location has one and only one value for a given variable. In field-based computational models, space is partitioned into regular or irregular units, each of which has a fixed location and, therefore, defines a field value. The most commonly used field model is a matrix of squares (i.e., rasters or grids). Remote sensing technologies provide rich sources for raster data. Other possible partitions of space include triangles, hexagons, and irregular polygons. Among all types of spatial partitions, fields of regularly spaced squares are the most computationally efficient because of geometric simplicity and regular tessellation of space. There are two basic approaches to the development of raster algorithms. One is based on cellular automata that consider how a particular cell value (e.g., fire cells) propagates in a raster layer (e.g., to examine how a fire spreads in space). The emerging technique of agent-based modeling takes a similar approach to examine the evolution of spatial patterns aggregated from individual behaviors when discrete cells of a certain value (e.g., individual pedestrians) animate on a raster over time under a specified set of rules and assumptions (e.g., allow moving only to adjacent cells). The other approach is map algebra in which each raster serves as a spatial variable to formulate algebraic expressions. All input and output variables in map algebra are rasters. Computation may be performed on a cell-by-cell basis or on a group of cells.

Both object- and field-based models are essential to meet the computational needs of diverse geographic problems. In some cases, conversion between vector and raster data is deemed necessary to support geographic problem solving.

—*May Yuan*

See also GIS; Humanistic GIScience; Ontology

CONSERVATION

Conservation, the principle or practice of managing the use of natural resources, is fundamental to successful human societies but became part of widespread political and economic discussion in 19th-century America. It enters into human geography through its mediation of human–environment interactions.

Although practiced—or abused—by all societies, conservation was chiefly an American ideology until recently. Today, its three strands are central to debates within environmental ethics. One strand—more aptly called nature preservation—emerged from the Romantic movement with its spiritual reverence for creation and an intrinsic value of nature. Naturalist John Audubon, an early advocate, called for protection of natural habitat against human abuse. Later preservationists John Muir, Henry David Thoreau, and Ralph Waldo Emerson championed this distinctly biocentric ethic. In contrast, a second strand borrowed from the Enlightenment principle of rationalism, prompting scientific studies of land and water resources to provide understanding of the extent that nature could yield to American society. Conservation in this form holds the anthropocentric notion that nature is instrumental to the human purpose of resource development and became the dominant view of conservation by the end of the 19th century. A third strand emerged midway through the 20th century as the study of ecology provided a science-based but nonanthropocentric understanding of human–environment interlinkages, best attributed to the work of Aldo Leopold. Holistic in its approach, this view gravitates toward an ecocentric ethic and evolved into ecosystem management. The three strands created a dynamic tension that continues in 21st-century American resource management.

CONSERVATION AS NATURE PRESERVATION

"In wildness is the preservation of the world," wrote Thoreau in 1851. This sentiment is the essence of the preservation movement today and is most closely associated with its greatest proponent, Muir. After walking from Indiana to Florida in 1867, Muir set out to explore the Sierra Nevada. With Thoreau, he advocated for conservation borne of human transcendence over nature. For Muir, living in the wilderness was the greatest spiritual experience in which to be "born again in the spirit." Connecting conservation with Romantic transcendentalism first appeared in Emerson's 1836 essay "Nature": "In the woods, we return to reason and faith."

Nature preservation gained scientific credibility in 1864 with George Perkins Marsh's *Man and Nature,* a scientific perspective on the fragility of the North American environment. This helped to legitimize the creation of the first national parks—Yellowstone (1872) and Yosemite (1891)—and wildlife preserves, eventually extended to hundreds of protected areas both domestically and internationally. In 1964, a century after *Man and Nature* appeared, the Wilderness Act became law, an effort culminating from decades of dedicated work by leaders of new nationwide organizations such as the Sierra Club and the Wilderness Society. By the end of the 20th century, preservation of wild nature remained a leading ideal for many environmentalists.

CONSERVATION AS EFFICIENT RESOURCE DEVELOPMENT

An expanding America with a frontier rich in natural resources began to see a need for efficient use of those resources by the end of the 19th century when the American frontier was declared closed and New England was largely deforested. Marsh precipitated the conservation movement in *Man and Nature*. Its chief resource focus, the role of forests in maintaining soil and water quality, remained the early emphasis of conservation in America and led to establishment of the Division of Forestry in 1881 and of New York's Adirondack Forest Preserve in 1885. The leading conservation proponent at this time was Gifford Pinchot, first chief of the U.S. Forest Service. Pinchot held that conservation is founded on three principles: (1) development of resources to benefit people who are alive "here and now," (2) prevention of waste and destruction of natural resources, and (3) resource management for the benefit of the many, not of the few—the utilitarian ethic that drove early conservation efforts. Not surprisingly, the Forest Service was created under the Department of Agriculture rather than the Department of the Interior.

With westward expansion into the so-called Great American Desert following John Wesley Powell's exploration of the Colorado River in 1869, conservation of soil and water resources became key concerns for conservation. Powell recognized that aridity made 160-acre homesteads impractical in the West, and he advocated a new policy of 2,560-acre homesteads. Congress rejected this notion, and over the next several decades a great migration to the arid lands was promoted under the myth that "the rain follows the plow." The outcome would later lead to establishment of the Bureau of Land Management, the Bureau of Reclamation, and the Soil Conservation Service (now the Natural Resource Conservation Service) and precipitated a shift of America's population and political power toward the Sunbelt.

CONSERVATION AS ECOSYSTEM MANAGEMENT

Ecology, traced by some to Marsh, was championed most emphatically by wildlife biologist Aldo Leopold and popularized in his *Sand County Almanac*, published in 1949. As ecology matured, it provided a dispassionate scientific approach to understanding nature. It did not advocate for either preservation or development of natural resources, but it gave resource management more insightful and holistic methods that were eventually adopted by the Nature Conservancy in its establishment of a network of privately protected and managed ecosystems and by the Forest Service and other federal agencies that began taking an ecosystem approach to conservation. In this late-20th-century manifestation, humans are included as coequals with the ecosystem, and polarized champions of development versus preservation are included in decision making with other stakeholders. Less developed than either of the earlier strands of conservation, ecosystem management represents the newest wave—and might not be the last.

CONSERVATION IN THE 21ST CENTURY

The Nature Conservancy, the World Conservation Union, and a growing number of other nongovernmental organizations have extended the practice of resource conservation globally, so it is no longer a uniquely American practice. Conservation has been extended beyond resource commodities to protect threatened and endangered species and now extends to include whole ecosystems. The principle remains one that is contested by differing ideologies; recent momentum toward more holistic and sustainable views of conservation contrasts sharply with continued political wrangling over resource exploitation. Conservation in the 21st century remains a principle focused on resource management, but that concept has been extended to include managing the global environment.

—*James Eflin*

See also Enlightenment, The; Nature and Culture; Resource; Wilderness

Suggested Reading

Meffe, G., Nielsen, L., Knight, R., & Schenborn, D. (2002). *Ecosystem management: Adaptive, community-based conservation*. Washington, DC: Island Press.

Nash, R. (1982). *Wilderness and the American mind* (3rd ed.). New Haven, CT: Yale University Press.

Petulla, J. (1977). *American environmental history: The exploitation and conservation of natural resources*. San Francisco: Boyd & Fraser.

Weeks, W. (1997). *Beyond the ark: Tools for an ecosystem approach to conservation*. Washington, DC: Island Press.

CONSUMPTION, GEOGRAPHY AND

In contrast to production, which has been studied in exhaustive detail in geography, consumption has long been ignored or taken as unproblematic. The reasons for this silence are not clear but may reflect, among other things, Marxism's emphasis on production and labor as the central acts of social life and, conversely, neoclassical economics' sterile and asocial view of consumption. Consumption and production cannot be neatly separated and are closely intertwined; most people work in order to consume and consume in order to live. Historically, the growth of mass production was accompanied by mass consumption and advertising during the 19th century and by Keynesian demand management during the 1930s. During the late 20th century, changes in the world economy, including deindustrialization and the explosive growth of producer services, induced concomitant changes in consumption, including increasingly specialized niche markets and sophisticated consumers. By any measure, consumption is enormously important as an economic act (constituting the bulk of gross national products of most countries), environmentally (e.g., energy use, the act of turning products into trash), and in terms of the lifestyles and self-images of much of the population. The geography of consumption is critical to understanding related issues such as travel and transportation, tourism, standards of living, and uneven development.

THEORETICAL PERSPECTIVES ON CONSUMPTION

The historically dominant view of consumption came from neoclassical economics, which analytically privileges demand. In this perspective, individual consumers, personified by the desolate, self-centered, asocial character *Homo economicus*, maximize their utility or happiness by allocating incomes among different goods. This topic has been examined in exhaustive detail, including topics such as the impacts of changing incomes and prices, consumer surplus, elasticities of supply and demand, and imperfect information. Inevitably, the conclusion of such views is that markets are optimally efficient (and hence morally optimal as well). Although the neoclassical view is internally consistent within its own terms of reference, it is ultimately sterile and ahistorical, failing to do justice to the rich semiotics and social

dimensions of consumption. In part, this failure arises because neoclassical economics does not represent consumers, or consumption, as a social act, that is, embedded within broader relations of class, gender, ethnicity, and power. For example, it offers no account of the origins of utility curves or why they assume their particular form. Social categories, if they arise at all, are defined largely by their relations to consumption; for example, class in conventional Weberian social analysis refers to income and socioeconomic status.

A second interpretation of consumption comes from Marxism, which argues that social science must penetrate the veneer of outer appearances to reveal the social relations that lie beneath them. In this vein, Marx argued that commodities are not only *things* but also embodiments of social relations. To view commodities separately from their social origins is to commit the error of commodity fetishism; the opaqueness by which market relations obscure relations among producers is functional for capitalism. Rather, Marxism draws on classical economics to differentiate the use value of commodities—the qualitative subjective dimensions—from their exchange value, that is, the quantitative price they command on the market. For example, the use value of an apple is its taste and the relief from hunger it offers, whereas its exchange value is the price at which it sells. Critically, for Marxists, labor also is a commodity whose use value to employers is less than its exchange value in wages. Thus, class is defined by relations to production and not to consumption. Marxism suggests that the extraction of surplus value by employers inevitably leads to underconsumption by the working class and the tendency toward crisis.

A third perspective on consumption focuses on the semiotic dimensions. Rather than a simple act of utility maximization, as represented by neoclassical economics, this body of work points to shopping and consumption as social and spatial practices that emanate from, and in turn reinforce, existing structures of power, culture, and ideology.

In highly individualized societies such as the United States, much personal status is achieved through the consumption of commodities. Indeed, self-identity and even self-esteem are frequently linked to owning the "right" brands of goods. Thus, what one may call the sensuous nature of consumption includes the complex social and psychological motivations that underpin the urge to buy, including consumers' egos, sense of self, status definition, and alleged individuality that comes from the purchase of mass-produced commodities.

Early-20th-century cultural theorist Walter Benjamin extended historical materialism to include the bourgeois infatuation with the commodity. He sought to uncover the ways in which the commodity penetrated into the consciousness of buyers, charting the growth of bourgeois consciousness in the emerging malls and stores of early-20th-century Europe. Working in Paris and Berlin during the 1920s, Benjamin's Arcades Project examined the linkages among the urban environment, experience, history, and memory, portraying cities as labyrinths in which individual subjectivity was swept aside by modernity and its impersonal relations, bureaucracies, and markets. Perception, Benjamin maintained, was itself historically specific. Commodities, in this reading, were far more than embodiments of labor power; they were also visual aesthetics with a significance above and beyond the narrow realm of the economic. In this light, Benjamin revealed that commodities are as much distillations of signs as they are embodiments of use and exchange values. Thus, Benjamin's Arcades Project captured the commodified nature of modernity, the deep linkages between seeing and knowing, on the one hand, and money and the commodity, on the other. This step effectively opened up the analysis of consumption as a social process, noting its mounting autonomy from production and the pervasive role of symbols in the construction and manipulation of consumer consciousness.

This line of thought reemerged in postmodern analyses of consumption, particularly the astute critique of contemporary capitalism offered by the political economy of signs. For Jean Baudrillard, the mass media have made the sign more important than its referents, creating a world of the *simulacra* in which we can no longer distinguish between simulations and reality or between true and false. In the context of post-Fordism, postmodern consumption centers as much on the symbolic value of commodities as on their use value. Thus, pseudo-Irish bars are more Irish than Ireland. Baudrillard's dissection of DisneyWorld and its Main Street reveals it to be just such a simulacrum—a giant shopping mall—and for Baudrillard the United States is essentially DisneyWorld writ large. Television carries this process of abstraction to new heights, reflecting and shaping the material world in complex and highly stylized ways.

GEOGRAPHIES OF CONSUMPTION

Drawing on the work of sociologists, historians, philosophers, and anthropologists, geographers have engaged in numerous lines of thought that suture commodities to their social and spatial origins. This body of work has tended to fall into three major categories.

First, drawing on the tradition of humanistic geography, some geographers have examined the relations among consumption, the body, and individual experience. A considerable literature, for example, has looked at food, its origins and cultural meanings in different geographic contexts, and its role in the unfolding of daily life. Similarly, geographers have examined the shopping mall not only as an economic phenomenon but also as a cultural site pregnant with meanings. Jon Goss, for example, studied the Mall of America in Bloomington, Minnesota, which has 520 stores, chapels, a roller coaster, an aquarium, and a rain forest. In this environment, fantasy, fun, and the commodity are merged into a seamless whole.

Second, many geographers have turned to consumption in the context of economic landscapes, including the pivotal role played by retail trade and consumer services. Traditionally, economic geography focused on production and the role of the so-called export base in economic development. When geographers turned to consumption, it was through the static and ahistorical lens of central place theory. More recent work has called attention to so-called nonbasic functions, including retail trade and personal services, and has shed light on their potential for job generation and economic change. Some geographers have studied enormous chains and franchises such as McDonald's and Wal-Mart. Studies of the geography of tourism are a burgeoning part of the discipline.

Third, geographers have focused on consumption in the context of the global economy, particularly the manner in which commodities are produced, distributed, and consumed via commodity chains. By embedding this sector within wider circles of finance, investment, trade, and consumption, this literature notes the ways in which globalization has unleashed a tidal wave of cheap imports that has propelled the high rates of consumer spending in societies such as the United States. This body of work traces the commodity through complex contingent lines of causality linking sellers and buyers across multiple spatial scales. Variations of this theme point to the highly gendered nature of consumption as well as to the moral and environmental dimensions that surround how commodities are consumed, including the sacrifices made by low-wage labor trapped in sweatshops in the developing world to provide American consumers with cheap goods. Such a perspective reveals consumption as being an economic, cultural, psychological, and environmental act that simultaneously reproduces both the world's most abstract space (the global economy) and the most intimate one (the individual subject and body).

—Barney Warf

See also Applied Geography; Body, Geography of; Class; Commodity; Cultural Landscape; Economic Geography; Food, Geography of; Geodemographics; Humanistic Geography; Identity, Geography and; Labor Theory of Value; New Urbanism; Phenomenology; Sense of Place; Spaces of Representation; Sport, Geography of; Subject and Subjectivity; Symbols and Symbolism; Tourism, Geography and/of; Uneven Development; Urban Geography; Urban Sprawl

Suggested Reading

Crewe, L., & Lowe, M. (1995). Gap on the map? Towards a geography of consumption and identity. *Environment and Planning A, 27,* 1877–1885.

De Graaf, J., Wann, D., & Naylor, T. (2001). *Affluenza: The all-consuming epidemic.* San Francisco: Berrett–Koehler.

Gereffi, G., & Korzeniewicz, M. (Eds.). (1994). *Commodity chains and global capitalism.* Westport, CT: Greenwood.

Goss, J. (1993). "The magic of the mall": An analysis of form, function, and meaning in the contemporary retail built environment. *Annals of the Association of American Geographers, 83,* 18–47.

Goss, J. (1999). Once upon a time in the commodity world: An unofficial guide to the Mall of America. *Annals of the Association of American Geographers, 98,* 45–75.

Gregson, N., Crewe, L., & Brooks, K. (2002). Shopping, space, and practice. *Environment and Planning D, 20,* 597–617.

Hartwick, E. (1998). Geographies of consumption: A commodity-chain approach. *Environment and Planning D, 16,* 423–437.

Hartwick, E. (2000). Towards a geographical politics of consumption. *Environment and Planning A, 32,* 1177–1192.

Lee, M. (Ed.). (2000). *The consumer society reader.* Oxford, UK: Blackwell.

Marsden, T., & Wrigley, N. (1999). Regulation, retailing, and consumption. *Environment and Planning A, 27,* 1899–1912.

Miller, D. (Ed.). (1995). *Acknowledging consumption.* New York: Routledge.

Stearns, P. (2001). *Consumerism in world history: The global transformation of desire.* London: Routledge.

Valentine, G. (1999). A corporeal geography of consumption. *Environment and Planning D, 17,* 329–341.

Wilk, R. (2002). Consumption, human needs, and global environmental change. *Global Environmental Change, 12,* 5–13.

Williams, C. (1997). *Consumer services and economic development.* London: Routledge.

CORE–PERIPHERY MODELS

A simplified view of economic space that assumes places can be categorized as belonging to an economic core (i.e., wealthy and possessing the means of production) or an economic periphery (i.e., poor and dependent on the core for the means to produce). The model is based on the observation of sharp economic development contrasts within and between nearly all territorial divisions. The core–periphery distinction can be found at any scale, from the local to the global. The specific characteristics of the core are vague but are generally thought to include the concentration of power, financial capital, human capital, research, innovation, diversified employment, and steady economic growth. Conversely, the periphery is characterized by low wages, low levels of diversification, volatile economic conditions, low levels of education, and little investment.

This method of classifying places is particularly useful to Marxist economists because it emphasizes the necessity of uneven development in market economies. The uneven development described by the core–periphery model springs from the Marxist assertion that the accumulation of wealth in the core is a product of the exploitation of resources obtained from the periphery. In addition, core systems construct patterns of trade that force the continued dependence of the periphery on the core. These patterns of uneven trade, wage minimization, multinational corporate structure, and migration encourage the departure of capital (both human and financial) from the periphery, thereby preventing less developed regions from altering their dependent status.

Although the core–periphery model is one of the most widely accepted conceptions in economic geography, it has faced criticism based on its simplistic reliance on trade as a causal mechanism and the vague treatment of power relations in the model. The pervasiveness of this core–periphery relationship is also a matter of considerable debate. Adherents of dependency theory consider the core–periphery relationship to be a necessary element of capitalism and thus a perpetual condition in market economies. Adherents to equilibrium economics assert that the reduced cost of operating in the periphery will encourage a diffusion of economic activity toward these areas, thereby ending uneven development. Much of the current research in economic geography is focused on either identifying the mechanisms that create and maintain the core–periphery dichotomy or on investigating the merits of the dependency and diffusionist arguments.

—*William Graves*

See also Dependency Theory; Economic Geography; Uneven Development; World Systems Theory

Suggested Reading

Knox, P., Agnew, J., & McCarthy, L. (2003). *The geography of the world economy* (4th ed.). New York: Oxford University Press.

Wolff, R., & Resnick, S. (1987). *Economics: Marxian vs. neoclassical.* Baltimore, MD: Johns Hopkins University Press.

CRIME, GEOGRAPHY OF

The geography of crime is the study of the spatial arrangement of criminals and crime. Geographic research on the location of criminals began in Western Europe during the early 1800s. Termed the cartographic school, researchers mapped the homes of criminals and related these maps to the socioeconomic environments of the country. At the turn of the 20th century, the geography of crime became more focused on urban areas when the urban ecologists at the University of Chicago related the home addresses of delinquents with characteristics of urban neighborhoods thought to spawn delinquency. The spatial pattern that they discovered is termed the *urban crime gradient*, which is the tendency for the number of criminals living in a neighborhood to decline with distance from the center of the city.

In contemporary times, the focus of geography of crime has shifted to the location of crime rather than the location of criminals. Because criminals must leave their homes to commit most crimes, analysts have focused on the three geographic concepts that describe this movement to a crime site: distance, direction, and reference point. These are three concepts that are used to locate anything in space.

Distance research has discovered that crime trips tend to follow a definite distance decay function. Criminals tend not to travel any farther than necessary to locate a crime site. However, if the crime is confrontive, they tend to avoid a buffer zone around the home. Therefore, the probability of a confrontive crime, such as rape or robbery, tends to increase with distance from the home until the edge of the buffer zone is reached and then decreases rapidly, following the principle of expending the least effort necessary to identify an opportunity for crime while avoiding recognition. Recently, these principles have been applied in geographic profiling where analysts attempt to predict the likely location of the home of a serial offender from the spatial arrangement of the offenses committed. Again, the assumption is that the offender lives in close proximity to the crimes committed.

Directional analysis focuses on the nature of places that have a lot of criminal activity surrounding them. These places are of two types: crime generators and criminal attractors. Crime generators are places such as high schools that cluster many people, some of whom have criminal tendencies, in one place on a daily basis. Criminal attractors are places to which criminals travel to identify a criminal opportunity. Examples include ATM machines for robbers, parking lots for auto thieves, and beaches for rapists. Because criminals will come from varying distances, the focus is on the directional bias toward one of these facilities. Directional bias is generally measured from the home of the criminal. Rather than using the four cardinal directions, bias is measured with respect to some anchor point such as the center of the city (toward or away from it in degrees on a protractor). Other anchor points that have been used in directional research of the spatial movement of criminals are workplaces, recreation areas, and illegal drug markets. Again, the spatial movement of the criminal is measured toward or away from the anchor point in degrees.

Finally, the reference point from which distance and direction are measured is an important determinant of crime patterns. Traditionally, the home of the criminal has been used as the reference point from which distance and direction of the crime trip is measured. This is because the home places constraints on how far the criminal can travel given that he or she must return in the evening.

Research on illegal drug offenders has determined that many do not have a regular home base to which they return every evening; many are homeless. In this case, other reference points are more important than a home in determining the spatial nature of crime trips. Drug marketplaces are more central to the lives of drug-dependent criminals than are home addresses.

Many criminals begin their crime trips from some place other than the home. This is the case for locations termed *crime generators.* Schools may be the beginning of crime trips for juveniles, and workplaces may be the origin points of crime trips for adults. Anchor points other than the home are important considerations when crime generators cluster crime in their locality.

The geography of crime has gained importance in police work with the advent of geographic information systems (GIS). Before real-time mapping was possible, there was a tendency for police to avoid persistent high-crime areas. This was termed *containment polity.* The reasoning was that it is nearly impossible to do effective police work without the cooperation of local residents. Therefore, if local residents oppose the police, why waste the time and resources to confront crime in their neighborhoods? Rather, police should try to stop the spatial spread of crime into the tipping-point neighborhoods surrounding the containment area.

Recent GIS analysis demonstrates that if persistent high-crime areas are not addressed vigorously, they will diffuse spatially into surrounding tipping-point communities. In this manner, whole sections of the city may be devastated. GIS analysis has allowed police administrators to turn containment policy on its head and to focus police resources on the most crime-ridden neighborhoods termed *hot spots.* COMSTAT (acronym for "computerized statistics") meetings of regional commanders use crime mapping to track the success of dampening hot spots and hold district commanders responsible for concentrations of crime that persist in their districts.

Geography has become increasingly important in crime analysis. Whether focusing on the homes of the criminals or the locations of the crimes, the GIS advancement in geography has allowed a much more sophisticated analysis than was possible when maps were drawn by hand and data tended to be restricted to those obtained from the police (Uniform Crime Reports of the Federal Bureau of Investigation) and the census (demographic, social, and economic data at the tract level). Features of GIS, such as buffering around places to determine whether they attract criminals, and overlays to aggregate rare events around features are just a couple of examples of how geography has advanced crime research.

—*George F. Rengert*

See also GIS; Law, Geography of; Social Geography; Urban Geography

Suggested Reading

Abeyie, D., & Harries, K. (Eds.). (1980). *Crime: A spatial perspective.* New York: Columbia University Press.

Brantingham, P., & Brantingham, P. (1981). *Patterns in crime.* New York: Macmillan.

Lowman, J. (1986). Conceptual issues in the geography of crime: Toward a geography of social control. *Annals of the Association of American Geographers, 76,* 81–94.

Smith, S. (1986). *Crime, space, and society.* Cambridge, UK: Cambridge University Press.

CRISIS

A crisis, in the lexicon of contemporary human geography, refers to a period of significant structural change and transformation. Typically, the term and the concept are used in various forms of Marxist analyses of capitalism. However, traditional economists occasionally refer to crises in the contexts of downturns in the business cycle. In both traditions, the notion of crisis speaks to the instability that lies at the core of capitalist development in time and space.

For Marx, capitalism's tendency toward crisis emanated directly from the extraction of surplus value in the production process. Because capitalists must extract surplus value to generate a profit, there is a long-run tendency for the system of production to overwhelm workers' capacity for consumption. The result is a chronic oversupply of goods, leading to declining prices and profits. As profits decline, firms are forced to react by cutting wages, restructuring production, or both. Marx predicted that capitalism's tendency in this regard eventually would so immiserate the proletariat that, in the final crisis, the working class would ultimately destroy that system and replace it with a socialist one better suited to working-class needs.

More recent views of crisis focus on changes that occur during periodic recessions and depressions. Economist Joseph Schumpeter argued famously that capitalist development is characterized by "creative destruction" as new technologies and markets destroy the old ones. In this reading, capitalism is in constant disequilibrium; indeed, much of the vitality and adaptability of the capitalist system arise directly from its continual processes of change. In the same vein, Simon Kuznets examined investment behavior as the motor that drives the business cycle. In his view, each firm must invest or disinvest in anticipation of future profits; thus, each company's individual rationality creates a collective irrationality; that is, the market is inherently unstable. Although crises are devastating for less competitive firms, often driving them into bankruptcy, they often make surviving firms even stronger. During downturns, when firms have relatively little to lose, they may experiment with new forms of production (i.e., technologies), new products, and/or new markets as well as seek out new geographic locations. Thus, crises are useful in reestablishing the conditions of profitability. For this reason, James O'Connor argued that crises are actually useful for capitalists as a whole even though they are fatal for some. Indeed, market-based systems would be deprived of their ingenuity without the periodic need to experiment and restructure; thus, capitalism is not only *crisis ridden* but also *crisis dependent*. Increasingly, therefore, as crises are seen as functionally necessary for the survival of capitalism, they have become viewed not as abnormal aberrations but rather as perfectly normal parts of the capitalist machinery.

A central contribution of Marxist geographers was to spatialize the notion of crisis. David Harvey played a profound role in this regard, particularly through his famous notion of the spatial fix. Harvey argued that the processes of competition and the extraction of surplus value led firms to accelerate the turnover rate of capital. Geographically, this process involved the search for more efficient transport systems. Capital tied up in transport is not directly realizing surplus value; therefore, reducing transport times accelerates the process of capital accumulation—what Harvey called time–space compression. However, reducing transport costs is difficult and expensive because the infrastructure needed to shuttle people and goods is expensive, is durable, and has a long depreciation time. Indeed, out-of-date transport systems (or the whole pattern of fixed capital investments in general) will inhibit future rounds of accumulation, eventually becoming a barrier to further accumulation. Thus, the spatial fix—the landscape that capitalism produces during temporary windows of stability—is periodically reworked during periods of crisis.

More broadly, crisis has become wrapped up with broader notions of restructuring in which capitalism undergoes periodic rounds of transformation. The late 19th century, for example, witnessed massive restructuring in the wake of the depression of 1893, including the ascent of large, well-capitalized, multiestablishment industrial firms; significant technological change; and the replacement of small local markets by a national market. Similarly, the late 20th century saw a crisis of profitability associated with the "petroshocks" of the 1970s, deindustrialization, the rise of the newly industrializing countries, the degeneration of the Rustbelt and the rise of the Sunbelt, the microelectronics revolution, and the ascendancy of neoliberalism worldwide.

Finally, these economic dimensions of crisis have been complemented by sociological and cultural ones. The most famous interpretation was offered by renowned sociologist and philosopher Jurgen Habermas in his concept of legitimation. Invoking Antonio Gramsci's notion of ideology, Habermas maintained that the state is continually torn between accommodating the needs of capital and production, on the one hand, and those of labor and social reproduction, on the other. To the extent that the system works smoothly, the state can serve the interests of capital under the guise that it operates on behalf of the general public good. However, during periods of traumatic realignment (e.g., the Great Depression of the 1930s), when its class bias is exposed, the state experiences a legitimation crisis, opening the door for alternative political movements.

—*Barney Warf*

See also Economic Geography; Labor Theory of Value; Marxism, Geography and; Restructuring; State; Time–Space Compression; Uneven Development

Suggested Reading

Berman, M. (1982). *All that is solid melts into air: The experience of modernity.* New York: Penguin Books.

Habermas, J. (1975). *Legitimation crisis.* Boston: Beacon.

Harvey, D. (1982). *The limits to capital.* Oxford, UK: Blackwell.

Harvey, D. (1985). The geopolitics of capitalism. In D. Gregory & J. Urry (Eds.), *Social relations and spatial structures* (pp. 128–163). New York: St. Martin's.

O'Connor, J. (1984). *Accumulation crisis.* Oxford, UK: Blackwell.

Skocpol, T. (1981). Political response to capitalist crisis: Neo-Marxist theories of the state and the case of the New Deal. *Politics & Society, 10,* 155–201.

Storper, M., & Walker, R. (1989). *The capitalist imperative.* Oxford, UK: Blackwell.

CRITICAL GEOPOLITICS

Critical geopolitics challenges conventional geopolitical accounts that posit an unproblematic use of geography as a causal or influential force in international politics. Based on poststructural theory, critical geopolitics has sought to subvert the taken-for-granted reasoning underlying geopolitics to insist, following Michel Foucault, that power and knowledge are always inseparable. There can be no apolitical or "natural" geographic influence on the practice of politics. Critical geopolitics has paid particular attention to the language of geopolitics (or geopolitical discourse). To critical geopolitics, language is not unproblematic, simply describing what is there. There is always a choice in the concepts that can be drawn on to make sense of a situation. Language is metaphorical, explaining through reference to other already known concepts. For instance, during the cold war, the domino metaphor simultaneously embodied a power political system where only two powers existed (the Soviet Union and the United States), where only force could oppose force, and where the unfolding of the process was inevitable; once started, the continuing fall of states was as unavoidable as stopping a line of dominoes from toppling once the first domino had been pushed. Disease metaphors were structurally very similar, relying on notions of contagion or the malign spread of infection, again depending on a simple notion of geographic proximity as the basis for social and political change.

Whereas traditional geopolitics regards geography as a set of facts and relationships "out there" in the world awaiting description, critical geopolitics believes that geographic orders are created by key individuals and institutions and are then imposed on the world as frameworks of understanding. Critical geopolitical approaches seek to examine how it is that

international politics are imagined spatially or geographically and in so doing to uncover the politics involved in writing the geography of global space. Gearoid Ó Tuathail called this process "geo-graphing" (or earth-writing). For geopoliticians, there is great power available to those whose maps and explanations of world politics are accepted as accurate due to the influence that these have on the way in which the world and its workings are understood and, therefore, the effects that this has on subsequent political practice. Critical geopolitics aims to challenge the objectivity of geopoliticians. For example, the privileging of sight (especially with the use of maps and diagrams) over other senses in geopolitical reasoning allows geopoliticians to write as if from afar—as if somehow unconnected to the world being surveyed. This reinforces the idea of an objective account rather than one written from a position grounded within the events being discussed. It hides the fact that geopoliticians have their own points of view and loyalties.

In arguing this, critical geopolitics suggests that geopolitics is not something simply linked to describing or predicting the shape of international politics; it is also central to the ways in which identity is formed and maintained in modern societies. National identity is not simply defined by what binds the members of the nation together; perhaps even more important, it is also defined by representing those who exist outside as different from members of the nation. Drawing borders around territory to produce "us" and "them" of the nation and those who are different does not simply reflect the divisions inherent in the world; it also helps to create these differences. Again, geopolitics does not simply reflect the facts of geography; dividing the world into domestic and international realms helps to form geographic orders and geographic relationships. Geopolitics reduces spaces and places to concepts or ideology. The complexity of global space is simplified to units that singularly display evidence of the characteristics that are used to define the spaces in the first place (e.g., Asia *is* exoticism, the Soviet Union *is* communism, Iran *is* fundamentalism, the United States *is* freedom and democracy).

The creation of a sense of difference, and particularly the sense of danger that this presents, has implications for the practice of domestic affairs in addition to how foreign policy is conducted. Thus, Simon Dalby suggested that geopolitics can justify limiting domestic political activity through the production of a greater enemy outside. At the same time, this presents a normative image of identity. So, for example, when the Soviet Union was imagined as being completely unlike the United States, any description of the Soviets as evil, aggressive, and unreasonable implies goodness, tolerance, and reason on the part of Americans.

The landscape of traditional geopolitics was populated by elite white men, a point explored by feminist and postcolonial critics. Cynthia Enloe suggested that women have been ignored in international politics, which traditionally has written a story of the spectacular confrontation of mighty states led by powerful statesmen, of the speeches and heroic acts of the elite, and of the specialist knowledge of "intellectuals of statecraft." Enloe refused to accept this story as covering the full extent of the workings of relations between states and instead focused on other actors and processes excluded and silenced by the conventional account—the role of international migration, the ideology of docile female labor for capitalist exploitation, the availability of sex workers for the global tourist industry, and so on. Enloe linked international geopolitics to everyday geographies of gender relations to highlight the constructed nature of scale and of state boundaries.

—Joanne Sharp

See also Geopolitics

Suggested Reading

Campbell, D. (1992). *Writing security: United States foreign policy and the politics of identity.* Minneapolis: University of Minnesota Press.

Dalby, S. (1990). American security discourse: The persistence of geopolitics. *Political Geography Quarterly, 9,* 171–188.

Enloe, C. (1989). *Bananas, beaches, and bases: Making feminist sense of international relations.* Berkeley: University of California Press.

Ó Tuathail, G. (1996). *Critical geopolitics.* Minneapolis: University of Minnesota Press.

Sharp, J. (2000). *Condensing the cold war:* Reader's Digest *and American identity, 1922–1994.* Minneapolis: University of Minnesota Press.

CRITICAL HUMAN GEOGRAPHY

A disciplinary trend, critical human geography is the result of the growing influence of—and interest in—critical theory in the social sciences. This paradigm change in scholarly thought must be understood in relation to, and as the result of, historical and social conditions. Although critical human geography is an emergent paradigm at a global scale, the discussion here focuses on its development in Anglo-American geography.

The emergence of critical human geography is tied closely to the social tensions of U.S. and British politics during the late 1960s. In the United States, it was especially the impact of the civil rights movement and the reaction to the Vietnam War that resulted in various forms of social critique and protest. In academia, this trend translated into the influence of a wide array of theoretical developments. Among them were Marxist critiques of capitalism, the critical theory of the Frankfurt School, French poststructuralism, postcolonial theory, feminist thought, and queer theory. A general theme uniting these different philosophical approaches is their use in reconceptualizing two aspects of human geography.

First, critical human geography seeks to provide a broad critique of the prevalent paradigms of scientific inquiry in the discipline. It is a reaction against positivism and its concern with objectivity and the scientific method. In addition, it undermines the assumptions of behaviorism and its emphasis on the goal-oriented decision-making models. Furthermore, it rallies against humanistic geography and its phenomenological approach to the lived world that often universalizes patterns of human behavior and meaning making. Last, it is a reaction against what it perceived as masculine models of science, and it contrasts these with distinctly feminist perspectives on science and knowledge acquisition. In summary, critical human geography intends to function as a potent critique of traditional scientific models in the discipline. It especially aims to deconstruct previously taken-for-granted scientific models by showing how scientific researchers, projects, data, and reports all are embedded in the power structures of a society and thus actively involved in socially constructing certain realities.

Second, critical human geography seeks to provide a powerful critique of the cultural, economic, social, and political geography of capitalist societies. Such endeavors have resulted in Marxist critiques of the capitalist logic behind urban design, expositions of the global patterns of exploitation in trade, studies on the increasing uniformity of cultural expression as a result of an emerging global culture industry, and much more. In addition, geographers have paid particular attention to the growing infringement on the public sphere, as evidenced by the number of studies addressing the surveillance and regulation of public space.

CRITICAL HUMAN GEOGRAPHY AND ALTERNATIVE MODELS OF SCIENCE

Scholarly work in critical human geography is an epistemological critique of the discipline. It calls into question the validity of the dominant positivist paradigm in geography and its use of the scientific method. It begins this critique by looking at the relationship between the researcher and his or her research objects or subjects.

Positivism idealized the concept of the objective researcher who, in the process of conducting a project, distanced himself or herself as much as possible from research subjects and the pressures or social forces within the discipline. It assumed and required that personal bias be left out of the research process and that disinterestedness guided the ethical conduct of a scholar who searches only for facts and not opinions. In contrast to such an objectivist model, Marxist geographers emphasize that scientific inquiry is always a product of the society in which it is produced and thus reflects and is influenced by power structures and dominant ideologies. It is almost impossible for a researcher to be unbiased and objective.

Moreover, feminist and queer geographers note that knowledge production in academia traditionally has been in the hands of heterosexual men. What constitutes knowledge and what counts as appropriate method was determined by these men. Feminists and queer theorists argue that men's research has been centered predominantly on goal-oriented, practical data gathering and has discounted intuitive and situational knowledge as invalid, subjective, and "soft." In addition to—or in place of—the hard science of gathering measurable results, feminism emphasizes that the collection of intuitive knowledge cannot be measured objectively. Other models of scientific inquiry that are more personal, subjective, qualitative, and even decisively political must be explored and permitted as valid research.

Critical human geography also questions the process of abstraction and reduction—of model making and its supposed objectivity. Traditionally, it was the behavioral geography of the 1960s that used the concept of "rational man" and his goal-oriented behavior as the starting point of inquiry and sought to model and standardize expressions of human spatial behavior. Both Marxist thought and feminist thought question the ability and reality of decision making on a purely rational basis and emphasize that behaviorist models are insufficient insofar as they do not account for spontaneous or intuitive behavior. Also, they question that actors are always aware or fully conscious of all their options. Instead, critical geographers argue, individuals often unconsciously choose from options that are predetermined or selected for them based on the society and ideology under which they live.

Much more than just a critique of scientific approaches, critical human geography offers a variety of methods to provide a critical analysis of society. Most important in the methodological approach is the argument that all knowledge and the spatial characteristics of reality are socially constructed. Marxist, but particularly poststructuralist, approaches in critical human geography seek to deconstruct taken-for-granted notions of space. The predominant tool of deconstruction is discourse analysis. Discourse analysis looks at the ways in which texts (e.g., speeches, articles, inscriptions) attach meaning to certain places and how these meanings are purposely created to represent certain positions of power. In other words, it links texts and the meaning they give to places with the people who created these texts and their positions of power. This is done to show how power is used to give meaning to places and to silence other texts and meanings.

An example of the interconnectedness of place, text, and power is the recently completed Memorial for the Murdered Jews of Europe in Berlin. From 1989 to 2000, various commissions and juries debated the design as well as the message of the memorial. Particular controversy arose as to who was supposed to be honored by the monument. Originally, it was intended as a site of remembrance for the millions of European Jews killed during the 12 years of the Nazi regime. Once the goal of the memorial was clarified, Sinti and Roma (European Gypsies) and other victim groups felt excluded from representation and argued for inclusion in the memorial. This argument was rejected by the organizing committees, which argued that a specifically Jewish memorial was needed to single out the Jewish victimhood in the Holocaust. German history historically has tried to blur the boundary between victims and perpetrators, and in the end the community of German Jews, supported by many allies in the German government, were successful in getting their specific memorial. The 11-year debate about the memorial was conducted by people representing various groups in different positions of power. Although the growing community of German Jews was given many opportunities to voice their position in the public media, the Sinti and Roma as a historically itinerant group have strong negative stereotypes attached to them even today. Subsequently, their position of power and their ability to influence public debate are significantly less pronounced.

In summary, discourse analysis carefully examines who plays what role in the ways in which place is invested with meaning. It analyzes who speaks in a public debate about place and who is silent, and it asks what it said and from what position of power, thereby dissecting public dialogue about the social construction of places.

CRITICAL HUMAN GEOGRAPHY AND ALTERNATIVES TO CAPITALISM

Rarely is the history of a scientific discipline and its shift in thinking so closely tied to one person as in the case of human geography and its critical turn. Especially when it comes to the development of critical human geography as a critique of capitalism, one scholar—David Harvey—clearly stands out as having defined this paradigm shift. His career began in England during the 1960s as an influential positivistic geographer interested in the development and advancement of the scientific method in geography. The publication of *Explanation in Geography* in 1969 marks the culmination of these efforts. Following his move to the United States, Harvey's career was marked by a radical shift of interest and the introduction of the first geographic work to explicitly draw on the writings of Karl Marx. Like no other work, *Social Justice and the City*, first published in 1973, set the tone for a critical Marxist analysis of the geographic structures and processes underlying the political–economic system of capitalism. Harvey's work focused on the analysis of the spatial logic of capitalism (i.e., the spatial requirements of profit generation and circulation). For example, he emphasized the necessity of capitalism to create cores and peripheries, or spatially manifested inequalities of participation, in a market economy. Capitalism depends on the accumulation of capital and its continuous circulation and reinvestment. This means that capitalists not only are looking for new opportunities to invest but also must

have new places to invest. Thus, capitalism depends on the constant growth of markets in both volume and spatial extent. Capitalism as a political and economic system not only happens in space but also actively produces it. In addition to examining the intricate details of this process, Harvey's work analyzed the role of the state and its foreign politics in the enforcement of the progress of global systems of capitalist production cycles.

Following Harvey's lead in the description of the intricate details of the geography of capitalism, several other scholars have contributed additional work on the aspects of space and place in capitalist societies. Most notably, feminist work has added a gender perspective to the spatial inequities created by capitalism and has shown how it is particularly women who are pushed to the margins of societies with fewer work opportunities, lower pay, and often longer distances to appropriate workplaces. Recently, there also has emerged a strong research tradition that pays attention to the diminishing role of public spaces in many capitalist societies. Don Mitchell, in particular, showed how large and often multinational corporations have been influential in the increased regulation and limitation of the right to assemble and to speak freely in public space. Mitchell and others interpret this as the result of the cooperation of state, county, and federal governments with capitalist interests. For example, many American cities now have in place by-laws that prohibit panhandling or even the prolonged presence of individuals in certain public spaces. Instead of addressing the roots of poverty in inner cities, governments now create rules that enforce the removal of people and activities that are considered harmful to local businesses. Thus, by increasingly regulating public space, the state and its institutions act in the interest of corporations and their desire to increase profits but neglect the needs of the citizenry.

—*Olaf Kuhlke*

See also Circuits of Capital; History of Geography; Marxism, Geography and; Social Geography

Suggested Reading

Harvey, D. (1973). *Social justice and the city.* Baltimore, MD: Johns Hopkins University Press.

Harvey, D. (2001). *Spaces of capital: Towards a critical geography.* London: Routledge.

Mitchell, D. (2003). *The right to the city: Social justice and the fight for public space.* New York: Guilford.

CULTURAL ECOLOGY

Cultural ecology is an approach within geography and anthropology to the study of people and the inter-relationships among cultures, resources, and environments. Cultural ecology and its practitioners hold that similar assemblages of environments and technologies demonstrate functional and causal relationships to concomitant forms of social organization. Therefore, it studies the patterns, practices, and processes whereby cultural groups adapt to their specific habitats or environmental conditions. In turn, cultural ecologists are primarily concerned with specifying and explaining subsistence activities or food production. These patterns and practices are seen as constituting the "culture core" of a given cultural formation. Cultural ecology can be seen as a subdomain within the larger and more diffuse field of human ecology and as a variant of anthropogeography in its nondeterministic expression.

Cultural ecology as a distinct and recognized subfield in geography and anthropology first emerged during the 1950s. Its older roots tap into the perennial concern to understand humans' cultural relationships with their ambient environments. More proximate antecedents can be seen in 19th-century geography's syntheses of human and physical geography, especially where concern for the cultural dimensions of humans' impact on specific environments was registered. Starting in the 1930s, anthropologist Julian Steward can be credited with naming the approach, theorizing its objectives, and providing examples of its practice with his studies of remnant Amerind hunter–gatherers in the American West. Much subsequent work in traditional cultural ecology has concentrated on hunter–gatherer populations and traditional agriculturalists.

Quite simply, the "simpler" or more direct the cultural group's interaction with the "natural" environment, the more tractable and accessible the analysis of adaptative processes (or at least this was the idea). Steward later applied his cultural ecological method to more complex societal formations, both prehistoric and contemporary. His method can be summarized as a tripartite exercise, namely (a) specifying the interrelations between environments and subsistence technologies, (b) studying the resulting behaviors, and (c) determining the effect of these behaviors on other aspects of culture. Collectively, these adaptational changes led to multilinear pathways of cultural evolution. Thus, Steward conceived his theory and method as counter to approaches that posited unilinear stage progression in cultural evolution. Rather than searching for universal principles of causation and uniform evolutionary trajectories as with dogmatic versions of historical materialism (e.g., Marxist) or cultural organicist cyclical models (e.g., Spenglerian), Steward looked for parallels in cultural causalities. The nondeterministic approach of cultural ecology has found widespread acceptance in the study of prehistoric societies.

Although Steward has been widely considered the intellectual author and first practitioner of cultural ecology, his ideas were actually incubated in the context of his graduate studies at the University of California, Berkeley. Steward sought to renew attention to culture–environment interactions and to rehabilitate cultural evolutionary perspectives within a more materialist anthropology. In the wake of the excesses of environmental determinism and unilinear evolutionism, Franz Boas and his followers moved anthropology away from these earlier tendencies. As a student, Steward had close contacts with Carl Sauer and his Berkeley School followers. In turn, much of what both Sauer and Steward proposed in terms of the study of subsistence systems and cultural adaptations was anticipated and articulated by British geographer C. D. Forde. Since Steward's initial 1950s launching, cultural ecology has undergone a number of adaptations to changing academic currents and critical concerns. These shifts and reorientations can be viewed in terms of decadal phases, each with its distinctive theoretical concerns, key practitioners, and prime texts. The 1950s phase, then, can be characterized by the concerns of Steward and others (both archaeologists and cultural anthropologists) to put questions of subsistence at the center of their investigations. For archaeologists, this included looking for comparative regularities in ancient civilizational processes rather than simply charting cultural–historical sequences or specifying civilizational patterns. For cultural anthropologists, it meant directing less attention to questions of kinship and ideational aspects of culture(s) and more to the material workings of cultural reproduction, especially food production and habitat appropriation.

The insights and methods of biological ecology also began to inform cultural ecologists' objectives. Anthropologist Frederick Barth demonstrated the utility of the ecological concept of *econiche* in his study of how farmers and pastoralists could symbiotically exploit the same environments. Steward's *Theory of Culture Change* remains a key text from this period.

The 1960s saw a rapid expansion in the numbers of practitioners, publications, and problems that came under the purview of cultural ecology. Differing approaches or traditions within cultural ecology also began to take form. Local studies focused on single groups, and limited areal extents were carried out in efforts to determine energy flows, subsistence techniques, and many other measurable dimensions of traditional people's ecological relationships with their environments. Geographer Harold Brookfield and his colleagues' work in highland New Guinea and other Pacific locales established methods and benchmarks that were replicated elsewhere in tropical contexts. Anthropologist Roy Rapport pioneered the emphasis on energetics that took off during the 1970s with his study of food rituals and energy flows in a New Guinean group. At the same time, a few cultural ecologists sought similar objectives at broader scales. Anthropologist Clifford Geertz's research on agricultural involution in Indonesia effectively showed how comparative study of differing ecosystems—in his case, antecedent rain forest, shifting cultivation, and wet rice farming—could enframe and elucidate the larger questions of Euro-expansion and its impacts on local peoples and their environments.

During the 1970s, earlier concerns were carried forward, some with new labels such as *adaptive dynamics,* while increased emphasis was put on recognizing the role that macropolitical economic factors played in shaping local cultural ecologies. The work of geographers Michael Watts in Africa, Lawrence Grossman in New Guinea, and Bernard Nietschmann in Central America laid the foundations for the emergence of political ecology as a distinct offshoot of cultural ecology. Other geographers with direct ties to the Sauerian tradition of landscape studies, especially William Denevan, David Harris, and James Parsons, and their students, including B. L. Turner, II, Gregory Knapp, and Kent Mathewson, enlisted cultural ecology for the study of ancient agricultural land forms (e.g., terraces, irrigation systems, raised fields). The past two and a half decades have seen an explosion of interest in political ecology and a maturation of some of cultural ecology's original objectives along with the abandonment of others. Much of the current work in cultural ecology shares many of the concerns, methods, and theoretical groundings of political ecology,

producing a productive blurring of boundaries. The sites and arenas of shared interest that engage both include questions of identity and social movements, pastoral–agricultural conflicts, ecopolitics and natural resource control, protected areas and indigenous presence, gender ecology, and environmental discourse and policy issues.

Perhaps the single largest change over the past half century within cultural ecology has been the theoretical shift from the older ecology based in cybernetic assumptions that natural systems tend toward equilibrium to the "new ecology's" recognition of the centrality of discontinuities, perturbations, and nonequilibrium dynamics in both "natural" and cultural systems.

—Kent Mathewson

See also Anthropogeography; Berkeley School; Environmental Determinism

Suggested Reading

Bassett, T., & Zimmerer, K. (2003). Cultural ecology. In G. Gaile & C. Willmott (Eds.), *Geography in America at the dawn of the 21st century* (pp. 97–112). Oxford, UK: Oxford University Press.

Butzer, K. (1989). Culture ecology. In G. Gaile & C. Willmott (Eds.), *Geography in America* (pp. 192–208). Columbus, OH: Merrill.

Steward, J. (1955). *The theory of culture change: The methodology of multilinear evolution.* Urbana: University of Illinois Press.

Turner, B., II. (1989). The specialist–synthesis approach to the revival of geography: The case of cultural ecology. *Annals of the Association of American Geographers, 79,* 88–100.

CULTURAL GEOGRAPHY

Anglo-American cultural geography has a long and rich history stretching back to the early 20th century in the United States and beyond that to late-19th-century German anthropogeography. Until the 1980s, there were few cultural geographers practicing in Britain. Two decades later, in a remarkable change that paralleled the "cultural turn" within the social sciences more generally, cultural geography has become one of the most popular areas of geography in Britain. We return to this development later, briefly tracing its genealogy, but first let us turn to the development of the subfield in the United States.

Cultural geography in the United States, from its founding during the 1920s through the 1970s, was dominated by Carl Sauer and his students at the University of California, Berkeley. Under the powerful influence of the charismatic Sauer, a coherent set of interests and approaches to research emerged under the unofficial name of the Berkeley School. Given the importance of Sauer to the foundation of the subfield, a few words are in order about his perspective. By the time he moved to Berkeley during the 1920s, he had rejected the still currently fashionable environmental determinism, which claimed that cultures were determined by nature (e.g., that hot climates produced less developed societies than did cold climates). Under the influence of cultural anthropologists A. Kroeber and R. Lowie, Sauer came to accept what was known as the "superorganic" notion of culture that treated culture as a kind of "black-boxed" entity that (rather mysteriously) shaped the behavior of different groups in different environments. He also developed a lifelong interest in Latin America and in prehistory. During the 1930s, Sauer fostered increasingly strong ties with biological scientists and pioneered research on the interaction between humans and the physical environment. He approached human–environment relations historically and focused on the human transformation of the earth.

Many of the most important ideas that Sauer introduced to the field (e.g., historical reconstruction of the impact of past cultures, the culture area or region, the diffusion of culture traits from region to region) were current at the time in German anthropogeography and American cultural anthropology. Sauer placed a greater emphasis on the human relationship with the physical environment than did the anthropologists, and perhaps this is where his most original contributions lie. His black-boxing of culture resulted from his view that geographers need not concern themselves with social, psychological, or political processes. (To be fair, much of his work was on historical cultures about which there were little data on the latter processes; nevertheless, many of his followers working on more contemporary issues had no such excuse for ignoring these.) Culture as a holistic entity was seen as a force that causes members of culture groups to act in culturally and historically specific ways. Such a broad formulation had the advantage of allowing the first generation of cultural geographers to describe the behavior of "cultural groups" without needing to invoke social–psychological or political

processes. It was assumed, perhaps heuristically, that people behaved as they did because their culture made them do so. Such a simplifying assumption allowed cultural geographers to focus on the abstracted processes in which they were most interested such as the historical diffusion of cultural traits across space and how particular culture traits work (e.g., how methods of cultivation work ecologically in particular types of places). As we will see, post-1980 cultural geographers called into question the assumptions of the Berkeley School, although more recently there has been a revaluing and greater appreciation of the environmental focus of the Berkeley School, which began to be lost with the first wave of what has been termed the *new cultural geography.*

Let us now turn to a consideration of the four principal themes in pre-1980s American cultural geography. The first is the diffusion of culture traits. Cultural geographers, like cultural anthropologists before the 1940s, sought to explain the development of cultures in terms of the diffusion of culture traits such as plants, domesticated animals, house types, and ideas rather than in terms of independent invention. In this manner, the movements of cultural groups could be traced by the cultural spoor they left behind. A second and related theme is the identification of culture regions through the plotting of material and nonmaterial culture traits. Attempts were made to correlate the incidence of such traits so as to identify relatively homogeneous cultural regions. A third theme was landscape interpretation, which attempted to trace the historical development of a particular landscape from its "natural" state into a cultural landscape. A fourth theme was historical cultural ecology. In this approach, attention was focused on how people's perception and use of their environment are culturally conditioned. Although these four themes continue to be active areas of research among North American cultural geographers, they no longer occupy the dominant position that they once did.

Traditional Berkeley School geography was supplanted by a new cultural geography originating in the 1980s. Simultaneously on both sides of the Atlantic, there arose a series of challenges to traditional cultural geography's modes of explanation. These challenges were the result of the rise of a Marxist-inspired geography in quest of contemporary social and political relevance, on the one hand, and a humanistic geography that sought to move to center stage the role played by individuals, on the other. Although these two approaches had very different models of explanation, both prioritized social and political theory. The result was a flurry of criticism of traditional cultural geography as antiquarian, overly simplistic, and deterministic in its explanation of social action and as incapable methodologically of handling the complexity of contemporary societies. Although such critiques occasionally were overdrawn, they pointed to some real problems with a deterministic superorganic conception of culture and with a romantic approach to peasant societies.

The rise of cultural geography in Britain was part of a broader cultural turn within the social sciences and had little in common with the genealogy of traditional American cultural geography. Because of the relative absence of a cultural geographic tradition in Britain and a lack of interest in the topics that inspired traditional American geography, British social and historical geographers sought to create a distinctively British cultural geography based on the study of culture as exemplified by the writings of Marxian literary critic Raymond Williams and the founders of the nascent field of cultural studies at the Birmingham Centre for Cultural Studies. In contrast to traditional American cultural geography's emphasis on cultural homogeneity and continuity over time, cultural studies celebrated diversity and change. And in contrast to the former's focus on rural developing societies, the latter sought to understand urban societies in the developed world. Whereas the former focused on the landscape and place as indicators of culture, the latter often sought to describe cultural practices and politics as they shape identities and lifestyles with more references to issues of class, race, gender, and sexuality than to landscape. Although during the 1980s and early 1990s the differences between traditional and new cultural geography were rather stark, they have softened over time, and in some areas of the subfield there has been blurring of the difference between the two. In particular, one can see this in the area of landscape interpretation and in a renewed interest in human–nature relations; however, as we will see, these similar foci have different genealogies. In the following section, a number of other areas of current interest to cultural geographers are discussed.

LANDSCAPE

Although landscape interpretation has tended to maintain important connections to traditional cultural

geography, some new directions have been charted that are more explicitly theoretical and show greater concern for the sociocultural and political processes that shape landscapes as well as the active role that landscapes play in these processes. Scholars have reconceptualized landscape in various ways. For those who draw on Marxian cultural criticism and the iconographic approach of art history, landscape is considered a "way of seeing" rather than a set of objects. Such a way of seeing, it is argued, is ideological, representing the ways in which a particular class has represented itself pictorially or the ways in which members of that class survey the landscape, especially as property. Others have applied poststructural notions of discourse, text, intertextuality, and power drawn from literary theory and Michel Foucault to understanding the ways in which landscapes are constructed and then read by those who inhabit or encounter them. This work often focuses on the political and social consequences of landscapes being taken-for-granted indicators of how societies are and should naturally be organized, especially within cultural regions. Another strand of poststructural landscape interpretation draws inspiration from Jean Baudrillard and explores landscapes as hyperreal simulacra (or simulations). Themed environments such as Las Vegas are seen as hyperreal, and philosophical questions surrounding the issue of authenticity are raised, for example, whether any landscape is any more or less authentic than any other. Disneyland and American malls are typical subjects of study. Yet another model of landscape interpretation is that of theater, drawing on dramaturgical approaches to the ways in which landscapes have an active role or a performative (constituting) role in social life. Important work has been conducted using combinations of these theoretical approaches to explore cultural memory, whether at the scale of individual monuments, urban landscapes, or nationalism.

Since the mid-1990s, there have been an increasing number of criticisms of the discursive turn within geography that is seen to overemphasize discourses and ideas about landscapes as opposed to the materiality of landscapes. Some have come from a psychoanalytic critique by feminist geographers unhappy with the masculinist "gaze" in landscape interpretation that emphasizes the visual pleasure of surveying the world from a particular privileged perspective (the "master of all he surveys" type of gaze). Other challenges have been inspired by the materialisms of

Marxism, actor–network, and nonrepresentational theory. These have tried to move landscape interpretation away from a focus on representation toward a greater stress on materiality, embodiment, and a rejection of the nature–culture dichotomy. Although these critiques have helped to correct an imbalance created by the discursive turn, important work continues to be done seeking a balance between representational and nonrepresentational practices as seen in an increase in studies of the cultural politics of landscapes.

NATURE

Nature–culture relations, which was one of the principal areas of research in the Berkeley School, has again become an important focus. Bruce Braun identified four strands of research present within this area. The first is cultural ecology, a perspective that was most popular several decades ago but still has adherents in geography. Cultural ecologists seek to connect cultural practices in "traditional" societies to local ecologies, seeing these as complex feedback loops between culture and nature. Although this position has much to offer in providing an integration of the cultural and the natural, it has been criticized as seeing culture as merely subsidiary and functional to natural processes.

A second strand, political ecology, can also be traced back to the Sauerian tradition with its interest in the relations of Third World peasants to the land. But unlike the cultural ecology approach, political ecology focuses squarely on the political and economic contexts and ecological consequences of peasant land use practices. Although this approach offers a welcome politicization of human relations to nature, often nature is treated as an inert entity manipulated by economic interests.

A third strand is the cultural approach to nature. Having been influenced by the cultural turn within the social sciences and poststructuralism, it argues that although nature is material, it is also socially constructed—that what counts as nature depends on linguistic and cultural meaning-giving practices. Cultural meanings conceptually divided up and thus constitute physical environments as objects of knowledge and power. Thus, researchers focus attention on the ways in which nature is represented by different cultural groups. An interesting body of feminist literature influenced in part by the work Donna Haraway has emerged under this approach. Some have made the criticism that this perspective substitutes a discursive

determinism for the environmental determinism of cultural ecology and the economic determinism of political ecology.

A fourth strand seeks to collapse the distinction between culture and nature that critics argue is central to these other approaches. Contemporary cultural geographers drawing on the work of nongeographers such as Haraway and Bruno Latour have begun to explore the ways in which culture and nature are mutually constitutive and inseparable. The work done under this nondualistic approach is broad and includes studies of animal geographies, including the history of domestication, demonstrating that genetic modification is not an entirely new phenomenon. Such arguments draw inspiration from Latour's notion of "hybrids" and "quasi-objects," everyday technologies that are neither cultural nor natural but rather both simultaneously. Cultural practices are not located in individuals. Individuals can be understood only as part of assemblages of the human and the nonhuman that cannot be explained by any simple causal model. It is a relational structure that some have termed an *extended organism*. The environment is intrinsic to any human or other organism; distinctions such as culture and nature debates or nature and nurture debates are seen to make no sense.

NONREPRESENTATIONAL GEOGRAPHIES

During the mid-1990s, there was a reaction against the overuse of notions of discourse, representation, and text that called itself nonrepresentational geography. One can identify two examples of nonrepresentational geography. The first, which draws on the writings of Latour and other sociologists of science, attempts to decenter the focus of analysis of social agency away from consciousness and cultural systems of meaning and toward nonhuman material agency. The second, which is associated largely with the work of Nigel Thrift, calls for a focusing of attention on forms of intuitive, noncontemplative embodied action such as song, dance, crying, and unarticulated human practices. Some question whether these latter practices are any more embodied than those that are normally thought of as "representational" such as seeing, speaking, and writing. However, these critiques are part of what has been termed a rematerialization of geography, and there is no question that the study of the body and the embodiment of culture are now important areas of research in cultural geography.

A second form of critique of representation comes from certain traditions within Marxism that have long been uncomfortable with placing consciousness at the center of analysis. Wishing to privilege deep structural economic forces and tending to understand consciousness as ideological, certain of the more economistic Marxists have raised the charge of "discursive determinism" and "idealism" against those who do not share their view that action is ultimately determined by the logic of economic relations. While acknowledging the critique that there is far more to the world than representation and that one must consider material (or bare) life and noncognitive or nondiscursive and embodied phenomena that may require new innovative methodologies, any model of cultural behavior that excludes discourse and representation as a central component is dangerously impoverished.

RACE AND POSTCOLONIALITY

Race has been prominent on the agenda of cultural geographers since the 1980s. During that first decade, most of the research and theorizing concerned what is called the racialized character of British and American urban societies, meaning that those societies are often understood in terms of their racial makeup—that although the concept of race may lack a scientific basis, it has important cultural meaning and thus social and political consequences (race is real because it has real social consequences). Since the 1990s, this interest has expanded into a concern with the centrality of race categories and erroneous theories about racial differences that have been central to the cultures of colonial and postcolonial societies. Important work has been undertaken by cultural geographers demonstrating how race is crosscut by gender and class and how whiteness must be considered as well rather than being treated as the unseen or unmarked norm. Linked to this has been an interest in the role played by the discipline of geography in the production of colonial knowledge and in cultural imperialism that results from a Eurocentric bias in geographic representations of the world, both academic and popular. Edward Said's well-known book *Orientalism* argued this point. Such work has added an important historiographic dimension to the work of cultural geographers.

CULTURE

There are many who find the contemporary definitions of culture to be problematic. Some still find culture to be too broad and deterministic a concept. One could argue that as the boundaries between notions of culture and other concepts such as nature, the economy, and politics become questioned or collapsed, the concept of culture needs to be rethought but not abandoned. As the idea of culture in the form of simplistic culturalist explanations and justifications is increasingly mobilized by political leaders, journalists, judges, managers in business, and policy advisers, it may be especially important to critically reexamine the concept. For example, some politicians and agencies concerned with economic development employ the concept of cultures of poverty by which they explain underdevelopment in terms of what they believe is "backwardness" or the unambitious nature of peasant cultures. Thus, just when the concept of culture is beginning to be widely used, often as a dangerous explanatory term in the world beyond the academy, it would be a very bad time for academics to abandon the concept rather than critically rethink it.

There are at least two persisting problems with the notion of culture. The first is that it tends to see populations of particular regions as having the same culture, thereby homogenizing and ignoring differences within societies. The second is that it posits a dualism between culture and nature. The first of these problems can be overcome by thinking of cultures as broad systems of understanding but not of agreement or shared values. This goes some way toward conceptualizing cultures as structured yet in no way homogeneous. The second problem—that culture is too focused on human agents—is resolved if culture no longer is seen as something apart from nature but rather is seen as embodied in humans that are a part of nature and whose bodies are essentially "open" to culture.

—*James S. Duncan*

See also Anthropogeography; Berkeley School; Body, Geography of; Chorology; Communications, Geography of; Consumption, Geography and; Critical Human Geography; Cultural Ecology; Cultural Landscape; Cultural Turn; Culture; Culture Hearth; Environmental Determinism; Environmental Perception; Epistemology; Ethics, Geography and; Ethnicity; Ethnocentrism; Eurocentrism; Feminist Geographies; Food, Geography of; Globalization; Human Agency; Humanistic Geography; Identity, Geography and; Ideology; Idiographic; Imaginative Geographies; Justice, Geography of; Languages, Geography of; Literature, Geography and; Modernity; Music and Sound, Geography and; Nation-State; Nationalism; Orientalism; Other/Otherness;

Phenomenology; Photography, Geography and; Place; Place Names; Political Ecology; Political Geography; Popular Culture, Geography and; Population, Geography of; Race and Racism; Radical Geography; Segregation; Sense of Place; Sequent Occupance; Sexuality, Geography and/of; Social Geography; Social Justice; Social Movement; Space, Human Geography and; Spaces of Representation; Sport, Geography of; Structuration Theory; Subaltern Studies; Symbols and Symbolism; Text and Textuality; Time Geography; Time–Space Compression; Travel Writing, Geography and; Urban Geography; Virtual Geographies; Vision; Whiteness; Writing

Suggested Reading

Duncan, J., Johnson, N., & Schein, R. (Eds.). (2004). *A companion to cultural geography.* Oxford, UK: Blackwell.

Foote, K., Hugill, P., Mathewson, K., & Smith, J. (Eds.). (1994). *Re-reading cultural geography.* Austin: University of Texas Press.

Leighley, J. (Ed.). (1963). *Land and life: A selection from the writings of Carl Ortwin Sauer.* Berkeley: University of California Press.

Mitchell, D. (2000). *Cultural geography: A critical introduction.* Oxford, UK: Blackwell.

Wagner, P., & Mikesell, M. (Eds.). (1962). *Readings in cultural geography.* Chicago: University of Chicago Press.

CULTURAL LANDSCAPE

A keyword in British and American human geography, cultural landscape is a multivalent concept that refers to the look or appearance of the earth's surface, to how that appearance is depicted in the visual arts, to the material objects that shape its appearance, and to an area of territory. For J. B. Jackson, one of cultural landscape's most significant interpreters, this complex term can be neatly summarized as a portion of the earth's surface that can be comprehended at a glance. Although deceptively simple, Jackson's traditional definition can provide a useful point of departure for a discussion that has expanded well beyond the boundaries of human geography as historians, architects, sociologists, anthropologists, literary critics, and social theorists all have found in cultural landscape a necessary concept to understand human-shaped environments.

EMERGENCE AND CHANGING DEFINITIONS OF A KEYWORD

Implicit in Jackson's definition is a tension that has long characterized discussions of landscape, a tension

that has a great deal to do with its etymology. The English word *landscape* contains within it two specific meanings that are at once complementary and, at times, contradictory: the human shaping of territorial space (the earth's surface) and mental or visual images of that space (that which can be comprehended at a glance). These two meanings—the material and the representational—entered the English language through different routes and eventually merged into the multi-faceted word that we know today.

During the Middle Ages in England, *landskipe* or *landscaef* referred to a specific portion of land occupied, managed, and controlled by an identifiable group of people—not natural scenery but rather land that had been modified by human interaction. This Old English sense of landscape as jurisdiction seems to have gradually disappeared from use when, by the 17th century, the related Dutch word *landschap* entered the English language. Here a landscape connoted the look or appearance of the land, especially in paintings of the rural scene. Landscape historian John Stilgoe described how, by 1630, *landscape* referred to both paintings and large-scale rural vistas that were pleasing to the eye—the hilltop views of villages, fields, woods, and church spires that so inspired England's emerging merchant class.

A third source came from continental Europe during the late 19th century, when universities in several countries—most notably France and Germany—developed influential scholarly traditions to examine the relationship between the natural environment and human intervention. In Germany, geographers began to define their new discipline as *landscape science,* whereby geography was most concerned with the form of landscapes in particular areas. These early German geographers strove to categorize with scientific precision the regions, settlements, village types, and agricultural systems throughout the country; thus, the word *landschaft* stood for a specific area defined by identifiable material features, both physical and cultural.

MORPHOLOGY OF LANDSCAPE

A young University of California, Berkeley, professor who had studied in Germany introduced the concept of landschaft into American geography. Carl Sauer began his long career as chair of the Department of Geography at Berkeley during the early 1920s and in 1925 published his landmark essay, "The Morphology of Landscape." More than perhaps any single work by a geographer,

Sauer's essay set the agenda for the discipline; it inspired a generation of scholars and set the stage for the innovative work conducted by him and his colleagues that came to be known as the Berkeley School of cultural geography.

For Sauer, landscape was not a pretty view to be seen; it was not a picture, a painting, or a vista. Rather, landscape meant an "area" or a "region" that was a product of natural attributes of climate, soil, and plant and animal life and of cultural attributes of population, housing, economics, and communication. It should be studied historically by examining how a natural landscape developed into a cultural landscape. This is how he famously put it: "The cultural landscape is fashioned from a natural landscape by a culture group. Culture is the agent, the natural area is the medium, the cultural landscape is the result" (Figure 1).

Such a concept of cultural landscape was meant, in large part, as a counter to the environmental determinism that had long dominated American human geography. Unlike that geographic theory, which aspired to enumerate the causal influences of the environment on humans, Sauer's landscape approach sought to show the interactions between people and the environment with an emphasis on human agency. More specifically, Sauer stressed the agency of culture as a shaper of the visible features of the earth's surface. This is not to suggest that the physical environment was of little importance; indeed, Sauer understood the physical environment to be the medium of cultural landscape modification. It suggests only that elements such as soil, topography, and climate should be incorporated into landscape study as the raw material for, and modified elements of, a deeply human place.

VERNACULAR AND ORDINARY LANDSCAPES

Whereas some scholars, such as geographer Richard Hartshorne, found Sauer's conception of landscape to be too close in meaning to an area or a region to be of much use, for many others it became the guiding principle for a wide array of studies of human–environment relations. Some of those studies were

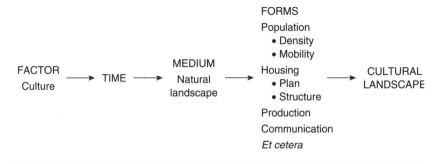

Figure 1   Carl Sauer's Morphology of Landscape. This diagrammatic representation of the morphology of landscape encapsulates Sauer's conception of cultural landscape as the product of the interaction between natural landscapes and human cultures. Figure by Joy Adams, adapted from Sauer's 1925 essay, "The Morphology of Landscape."

conducted by Berkeley School cultural geographers who charted the historical diffusion of ideas and practices from one region to another. For example, Wilbur Zelinsky correlated the occurrence of a particular form of town design with what he called the Pennsylvania Culture Area, and Fred Kniffen traced the spread of building types, settlement forms, and fences to distinguish cultural hearths and migration patterns. Other scholars not immediately affiliated with the Berkeley tradition also began studying the ordinary landscapes of everyday America, and none was more influential than Jackson.

A prolific writer, editor, and occasional teacher, Jackson founded *Landscape* magazine in 1951 as a forum for his ideas about landscape and, equally important, for a wide variety of young scholars interested in environmental concerns. Although Harvard University educated in the visual and literary arts, Jackson eschewed interest in aesthetics or "landscape beauty" in favor of what he called "vernacular landscapes"—the motels, fast-food franchises, mobile homes, garages, and strip malls of the workaday world. For Jackson, the true and lasting meaning of the word *landscape* is not something to look at but rather something to live in—with other people, not alone. The landscape is anchored in human society, in all its strange and wonderful variety.

Such an approach to vernacular landscapes, although still thriving today, received perhaps its most elegant treatment in D. W. Meinig's edited volume *The Interpretation of Ordinary Landscapes*. Bringing together essays by many well-known cultural

Figure 2   "Houses and Billboards in Atlanta" by Walker Evans. This 1936 photograph of an ordinary street scene in Atlanta can be interpreted in a variety of ways, each reflective of its viewer's subjective position. Reprinted Courtesy of the Library of Congress, Prints and Photographs Division, FSA-OWI Collection [LC-USF342-T01-008057-A DLC].

geographers, the book canonized Jackson's belief that all landscapes are expressions of cultural values and that cultural landscape study is a companion to the social history that seeks to understand the lives of ordinary people. In one of the book's most innovative essays, "The Beholding Eye," Meinig demonstrated how 10 different people, when looking at the same scene, could perceive it in 10 different ways. Take, for example, the 1936 photograph by Walker Evans of an ordinary street in Atlanta, Georgia (Figure 2). Viewers might see it as habitat, as an artifact of an earlier age, as a problem to be solved, as reflective of ideology, and so on. Meinig's central theme is that the interpretation of landscapes is far from an exact science and that our subjectivities inevitably shape those interpretations.

NEW DIRECTIONS IN CULTURAL LANDSCAPE STUDIES

One thing that is easy to neglect when looking at Evans's photograph of Atlanta (Figure 2) is that it is a

photograph—a representation of a three-dimensional reality. Drawing from European and British social and cultural theory, much recent work in human geography has examined precisely this relationship between the built environment and the media that depict it. Denis Cosgrove, for example, characterized landscape not as an object or a geographic area but rather as a "way of seeing"—a pictorial means of representing or structuring the world. With the explicit intention of theorizing the idea of landscape in a broadly Marxian understanding of culture and society, he described how this way of seeing was ideological because it represented how a privileged class depicted itself and its property. For scholars such as Cosgrove and Stephen Daniels, a landscape's power, as well as its duplicity, lies in its ability to project a sense of timelessness and coherency when in fact, as their work demonstrates, a landscape is anything but timeless and coherent. Surveying the picture-perfect landscape of California's fruit-growing regions, Don Mitchell made a similar argument, namely that such landscapes, beautiful as they might appear, "lie" to us when they obscure the often harsh social and labor conditions that went into their production.

This way of conceiving landscape has proved to be fruitful during recent years, generating a large number of diverse studies that emphasize the communicative and representational aspects of landscape; as methodological sources, art history and poststructuralist notions of text and textuality have become as important as cultural ecology. Furthermore, recent work has expanded well beyond the Berkeley School's near-exclusive focus on rural relic landscapes to encompass the urban environment and national mythologies. In her study of 19th-century landscapes in New York and Boston, Mona Domosh demonstrated how the upper-class leaders of those cities envisioned urban public culture in very different ways, resulting in varying representations and material, built forms unique to each place. Even more than cities, countries possess certain landscapes that are considered symbolic; nowhere is this more evident than in England. But as David Matless showed, underlying the entwined relationship between landscape and English identity are powerful social interests and historical actors that create or *construct* an organic sense of Englishness rooted in land and soil.

Whereas some feminist scholars, such as Gillian Rose, have objected to the visual emphasis of landscape as inherently masculinist, others, such as the writers in Vera Norwood and Janice Monk's edited volume *The Desert Is No Lady,* demonstrated that a consideration of landscape is essential to understanding matters of gender—as well as of race, ethnicity, class, and sexuality. Dolores Hayden's history of American suburbanization, unlike previous accounts that emphasized changing transportation technology, explored the interplay of natural and built environments and showed how resulting landscapes powerfully affect nearly every aspect of contemporary American life, including gender relations. James Duncan and Nancy Duncan also were wary of the visual appearance of landscape, but not on theoretical or methodological grounds; their study of environmental aesthetics in suburban New York illustrated that the physical presentation of a landscape carries with it a range of markers of inclusion and exclusion.

Interest in cultural landscape shows no sign of slowing down. Scholars such as William Cronon have helped to launch entire fields of study (e.g., environmental history) that rely on this central concept. No

less important, citizen groups concerned about the social and ecological costs of urban sprawl and environmental degradation increasingly describe their concerns with the language of landscape.

—Steve Hoelscher

See also Berkeley School; Cultural Ecology; Cultural Geography; Culture; Diffusion; Environmental Determinism; Feminist Geographies; Human Agency; Marxism, Geography and; Theory

Suggested Reading

Cosgrove, D. (1998). *Social formation and symbolic landscape* (2nd ed.). Madison: University of Wisconsin Press.

Cronon, W. (1983). *Changes in the land: Indians, colonists, and the ecology of New England.* New York: Hill & Wang.

Domosh, M. (1996). *Invented cities: The creation of landscape in nineteenth-century New York and Boston.* New Haven, CT: Yale University Press.

Duncan, J., & Duncan, N. (2003). *Landscapes of privilege: The politics of the aesthetic in an American suburb.* London: Routledge.

Hayden, D. (2003). *Building suburbia: Green fields and urban growth, 1820–2000.* New York: Vintage Books.

Jackson, J. (1997). *Landscape in sight: Looking at America* (H. L. Horowitz, Ed.). New Haven, CT: Yale University Press.

Matless, D. (1998). *Landscape and Englishness.* London: Reaktion.

Meinig, D. (Ed.). (1979). *The interpretation of ordinary landscapes: Geographical essays.* New York: Oxford University Press.

Mitchell, D. (1996). *The lie of the land: Migrant workers and the California landscape.* Minneapolis: University of Minnesota Press.

Sauer, C. (1963). The morphology of landscape. In J. Leighly (Ed.), *Land and life: A selection of writings of Carl Ortwin Sauer* (pp. 315–350). Berkeley: University of California Press. (Original work published 1925)

Stilgoe, J. (1982). *Common landscape of America, 1580–1845.* New Haven, CT: Yale University Press.

CULTURAL TURN

The cultural turn summarily represents a critique of traditional cultural geography by a group of scholars starting in the late 1970s. This group of geographers sought to redefine cultural geography with a critique of the so-called Berkeley School, the founding school of thought in cultural geography. The cultural geography of the 1980s and 1990s offered a two-step reassessment of core elements of traditional cultural geography. First, it sought to tackle and rewrite the problematic definition of culture as the central object of geographic analysis. Second, it supported the shift of cultural geography away from a static and empirical analysis of human–environment interactions and the examination material landscape to a more reflexive practice that involved a wider range of research techniques and more interaction between the researcher and his or her subjects of study.

Traditional cultural geography before the cultural turn is often referred to as based on the superorganic theory of culture. This means that cultures are composed of geographic units such as culture groups that are represented as collectives in which individuals have very little power or agency but rather are socially conditioned to act, behave, and express meaning in fixed ways. The superorganic concept of culture assumes that cultures (or aspects of culture such as religion and language) are largely independent of individuals and their behavior; individuals do not cause a culture to be formed, but culture is the agent that causes people to behave as they do. In such a concept of culture, values, beliefs, and meanings operate independently of individuals; they are not created and shaped by them, but individuals receive and are influenced by culture in the same fashion that their bodies and anatomies are the result of genetic codes. Whereas genetics controls people from within, culture controls people from without. Traditionally defined, then, cultural geography describes the spatial patterns of culture but not the cultural patterns of individuals within a culture. Culture as an agent—not individuals within a culture—is of primary interest to geographers. They examine cultural landscapes as if culture is the agent and nature is the medium that is geographically shaped, patterned, ordered, and transformed by it. Thus, the spatial organization of cultural patterns is the result of the collective internalization of cultural values by individuals who are characterized as passive recipients of information. Individuals and their cultural habits, beliefs, and traditions are simply representatives of cultural regions that express a certain character. Many prominent proponents of the superorganic view of culture often stereotype a region (and the individuals within it) as having a certain heart, soul, character, psyche, and/or personality. For example, stereotypical expressions such as "the soul

of Germany lies in its love of discipline" and "the core element of the American psyche is its strong sense of individualism" were characteristic of regional character types described by traditional cultural geographers.

The new cultural geography that instigated the cultural turn focused its critique on the limitations of superorganicism as it was promoted and popularized by the Berkeley School under Carl Sauer. Beginning in the late 1970s, two strands of critique of this limited concept of culture emerged. First, cultural geographers have increasingly moved away from the superorganic concept of culture and replaced it with one that takes into account the active role of humans as agents of cultural change. Culture no longer is conceptualized as something imposed on passive humans from without anymore; rather, it is delineated as a system of distinguishable practices, symbols, tools, and texts by which people attach meaning to experiences and events in their lives. Also, culture was redefined as socially constructed and malleable. Individuals were reconceptualized as agents of social change who could actively reshape cultural geographies; they were not helpless recipients of cultural traits. In short, culture emerged as a more or less coherent signifying system—as a set of ideas, texts, and symbols that give human lives meaning and that they express in public and private spaces. A second critique of traditional concepts of culture emerged in the work of Don Mitchell, who not only challenged the superorganic theory but also regarded the idea of culture as a signifying system as a problematic approach. Mitchell argued that the division of humanity into distinct, spatially recognizable culture regions is in itself a fallacy. Rather than a discrete reality, these geographic units of culture always affix a sense of uniformity to a certain group of people or a region as it does not really exist. Culture is multifarious and always changing; it cannot be defined simply along geographic lines. To capture the true diversity of cultural experience, and to decode the ways in which definitions of certain cultures have silenced such variety, the new cultural geography also includes a methodological critique. Whereas the Berkeley School was often accused of passive fieldwork, little interaction of researchers with their surroundings, and inadequate archive work, the cultural turn signified an embrace of a wider range of techniques. Influenced by feminist critique of geographic fieldwork, the 1990s saw a surge of scholarship that explored a wider variety of qualitative methods such as interviews, focus groups, participant observation, and discourse analysis.

—*Olaf Kuhlke*

See also Berkeley School; Cultural Geography

Suggested Reading

Duncan, J. (1980). The superorganic in American cultural geography. *Annals of the Association of American Geographers, 70,* 181–198.

Mitchell, D. (1995). There is no such thing as culture: Towards a reconceptualization of the idea of culture in geography. *Transactions of the Institute of British Geographers, 20*(1), 102–116.

Price, M., & Lewis, M. (1993). The reinvention of cultural geography. *Annals of the Association of American Geography, 83,* 1–17.

CULTURE

Arguably among the most contested and complex concepts ever discussed in the social sciences, culture is one of the significant ideas that scholarly classifications of society have created. In general, geographers and other social scientists have debated four separate aspects of the concept of culture over the past century and a half. First, the question arose as to whether culture is simply the sum total of all cultural expressions of a society or is some independent superorganic entity that, while profoundly influencing society, is still larger than and separate from it. Second, cultural geography in particular has seen a marked shift of focus from the study of material culture, such as tangible landscapes, tools, and other artifacts, to symbolic culture, such as religion, language, and other cultural texts. Third, social scientists have investigated the ways in which humans have constructed boundaries between culture and nature. Although these concepts traditionally were regarded as mutually exclusive, recent studies have shown how even natural landscapes have been consistently invested with cultural meanings and values. Fourth, and ultimately, some geographers have even questioned the use of the concept of culture as a whole and have called for an in-depth investigation of the social construction and use of the concept itself.

CULTURE AS A SUPERORGANIC OR SOCIALLY CONSTRUCTED ENTITY

The first question of importance for geographers is how culture affects society. Is culture imposed on a society as a superorganic entity from above and in a top-down fashion, or is it socially constructed from the bottom up and nothing more than the sum total of individual cultural expressions? Is culture static or constantly changing, and does it have a lasting, identity-building effect on individuals and communities, or is it rather fluent, with humans constantly reevaluating and redesigning their identities?

Superorganicism takes for granted that culture is an independent and stable entity. Culture is the agent that causes people to behave as they do. It superimposes behaviors and traits on them from beyond society. Values, beliefs, and meanings operate independently of individuals; they are not created and shaped by them, but individuals receive and are influenced by culture. Take the example of religion, where some groups argue that morals and dogma have been given to humans or were inspired by God rather than being constructed by humans as a consequence of their social interaction. Traditionally, cultural geography described the spatial form(s) (*morphos* in Greek) that culture imprinted on the landscape as an active agent, relegating individuals to passive recipients of information. Just as geomorphology described the natural formation of the landscape by the forces of nature, cultural geography illustrated the morphology (formation) of cultural landscapes by the force of culture. Carl Sauer's landmark essay on "The Morphology of Landscape" in 1925 represented this trend most clearly. In it, Sauer argued that geography ought to be concerned with the interactions of nature and culture as well as the influence that these entities have on each other. For Sauer, culture is an active agent that grows according to the natural landscape in which it is situated and remains attached to that landscape indefinitely. Nature provides a certain number of options for humans to transform and use it, and culture acts as the sum of all these potential options and not just the realized expressions. As humans realize certain cultural potentials that are given to them, they begin to transform nature and activate a set of cultural uses of the natural landscape. Thus, Sauer placed great emphasis on the use value of natural landscape and wanted geography to pay attention primarily to how productive human work makes natural landscapes permanently valuable for humans.

This rather static view of culture that more or less permanently tied cultural meaning to landscape has been exposed to considerable critique. From the late 1970s onward, cultural concepts have been reevaluated and the active and passive roles of culture and individuals have been reversed. Culture no longer is conceptualized as something impressed on passive humans but rather is defined as a system of distinguishable practices, symbols, tools, and texts by which people attach meaning to experiences and events in their lives. Also, culture has become more of a malleable concept, with individuals as agents of social change who can actively reshape cultural geographies. In summary, culture emerged as a coherent signifying system—as a set of ideas, texts, and symbols that give human lives meaning and that they express in public and private spaces. In geography, this trend was exemplified by the emergence of humanistic geography during the 1980s and later the linguistic turn that focused the object of cultural geographic inquiry not just on individuals but also on the variety of cultural texts they create (e.g., memorials, architecture, public art).

CULTURE AS A MATERIAL OR SYMBOLIC ENTITY

When studying cultures and their spatial components, geographers traditionally have examined the transformation of natural landscapes into cultural landscapes. The clear focus was on materials and their use value or the transformation of natural materials such as certain resources (e.g., wood, stone) into cultural artifacts for a useful purpose (e.g., shelter, housing, tool making). An example of this trend was the plethora of studies on North American house types that emerged from the 1940s to the 1980s.

At the same time, these landscape forms were said to have had an impact on the culture of humans in that area. Individual cultural habits were regarded as representatives of cultural regions that express a certain character. It was especially the prominent proponents of the superorganic view of culture that often created stereotypical representations of regions (and assigned the individuals within these as having a certain set of character or personality traits). Thus, material cultures were regarded as the results of a character-shaping and/or character-determining process that reflected humans' response to environmental conditions.

Geographers have also examined culture as a symbolic tradition. They have studied how people transform cultural and natural entities into symbolic objects not aimed at transforming the landscape but rather for the purposes of communication and meaning making. On the one hand, geographic scholarship sought to reveal the universal global patterns and practices of symbolic culture as they can be found in many belief systems. For example, the practice of pilgrimage or the establishment of sacred places according to stellar patterns and movements can be found in many world regions and cultures. On the other hand, geographers have focused on debunking myths of perceived cultural uniformity to uncover the culturally specific and diverse expressions of symbolic phenomena such as collective identities. In summary, the geographic study of symbolic culture has led to examinations of both culturally universal and diverse patterns of geographic expression of meaning; it has resulted in a broad scope of "maps of meaning" created by geographic scholarship.

THE CULTURE–NATURE BOUNDARY

Traditional definitions of culture clearly separated it as an entity distinct from nature. Originally, culture denoted the collective body of activities and knowledge that transform nature and natural landscapes. Hence, early anthropological and geographic studies focused on productive and transformative activities such as agriculture, horticulture, and viticulture. The culture–nature boundary was drawn to refer to human activities as culture that affects nonhuman nature and turns natural landscapes into cultural landscapes.

In contrast, the early examination of culture also included the determining role of the environment in the formation of human character traits. The evolution of human life was portrayed as having developed from a more primitive natural state to a more cultured restrained status often referred to as civilization. Geographers who embraced environmental determinism sought to portray an evolutionary development of cultural patterns as the result of environmental conditions. Charles Darwin's principles of evolution by natural selection were applied to the study of society and used to explain the development and superiority of certain cultures over others. Although largely discredited, even some modern scientific work repeatedly seeks to revive the principles of determinism to argue for a connection among environmental conditions, racial traits of humans, and humans' cultural achievements.

Recent geographic work goes even further in its analysis of the culture–nature boundary and argues that all untouched nature has been culturally constructed and invested with meaning by humans. Thus, while physically intact and in its original state, it has still been constructed and turned into something that it not necessarily is. For example, geographers have revealed how colonial powers imbued the natural landscapes of the African continent or even the Middle East and its inhabitants with moral values and cultural stereotypes. Such observations have led to the concept of social nature and the idea that untouched nature as we knew it does not exist anymore but that such pristine environments almost always are invested with social meanings.

QUESTIONING CULTURE AS A CONCEPT

Finally, geographers have questioned the use of the concept of culture as a whole. For example, Don Mitchell argued that the division of the ecumene (the inhabited surface of the earth) into culture regions with specific and more or less sharply defined boundaries is a misleading notion. These geographic units of culture always attach a sense of uniformity to a certain group of people or a region as it does not really exist. Culture is much too diverse and fluid to be mapped out along distinct lines, and when we do so we always somehow silence the true variety of cultural expressions. This critique reveals that the very definition of the concept of culture is a value-laden political act in which certain individuals define themselves and others along certain geographic lines; they begin to distinguish between "us" and "them." Thus, recent geographic work offers a potent ideological critique that emphasizes the role of power in the social construction of cultural landscapes. Under the influence of critical human geography, a distinct focus has now been placed on the actors that shape cultural representations and the ways in which cultural representations are influenced by societal elites. In addition, geographic work has now emphasized the role of class in the construction of cultural representations and shifted its attention to marginalized populations whose cultural expressions are often silenced by cultural elites and thus absent from public discourse. With regard to the concept of culture, it has been the ultimate goal of critical geography to deconstruct any notions of sameness in favor of uncovering the diversity of cultural experiences and the ways in which they are shaped by power.

—*Olaf Kuhlke*

See also Berkeley School; Critical Human Geography; Cultural Geography; Cultural Turn; History of Geography

Suggested Reading

Castree, N., & Braun, B. (2001). *Social nature: Theory, practice, and politics.* Oxford, UK: Blackwell.

Jackson, P. (1992). *Maps of meaning: An introduction to cultural geography.* London: Routledge.

Mitchell, D. (1995). There is no such thing as culture: Towards a reconceptualization of the idea of culture in geography. *Transactions of the Institute of British Geographers, 20*(1), 102–116.

Rushton, J. (2000). *Race, evolution, and behavior.* Port Huron, MI: Charles Darwin Research Institute.

Said, E. (1979). *Orientalism.* New York: Vintage Books.

Sauer, C. (1925). The morphology of landscape. *University of California Publications in Geography, 2*(2), 19–53.

CULTURE HEARTH

Though the overarching concept of culture hearth did not originate in geography per se, it has come to occupy a central place in traditional cultural geography's reconstructions of cultural origins and diffusions. Carl Sauer (1889–1975) seems to have introduced the term *culture hearth* in his 1952 Bowman Lecture, "Agricultural Origins and Dispersals." Hearth, with its ancient Indo-European cognates meaning charcoal and fire, well evokes Sauer's theory that agriculture's origins are to be found in contexts of leisured sedentary folk with sufficient diversity of sustenance and resources to explore natural processes imaginatively. Sauer also posited that control of fire was humanity's first great cultural acquisition and prepared the way for agriculture's inceptions many millennia later. Once kindled and tended, cultural traits such as plant domestication were then dispersed along avenues of adoption. The principles of cultural diffusion, and the notion of centers of innovation, can be traced back to earlier cultural and agricultural historians. Swiss botanist Alphonse de Candolle (1806–1893), in his *Origins of Domesticated Plants*, posed the question of global centers of plant domestication. During the 1920s and 1930s, Russian botanist Nikolai Vavilov (1887–1943) mounted dozens of plant-collecting expeditions to places that he believed were the original centers of plant domestication. He identified eight original centers in Asia, Africa, and the Americas. Botanists, archaeologists, and geographers all contributed to a vigorous research trajectory that continues the debate on agricultural origins and dispersals.

Friedrich Ratzel (1844–1904) can be credited with implanting the implicit idea of the culture hearth within the geographer's domain. Best remembered for laying the foundations for political geography and advancing environmental determinism in geography, Ratzel, in the second volume of *Anthropogeographie,* less conspicuously put locating culture centers (*hearths* in Sauer's poetic prose) and identifying culture traits and tracing their dispersals at the core of human geography. Ratzel also helped to make the delimitation of culture areas a major concern of anthropologists for the next half century. Ratzel inspired the development of the *Kulturkreise* (or culture circles) approach within anthropology. The object of Kulturkreise research was to reconstruct the diffusion of cultural traits from a few originating nodes or clusters and to map areas or regions of cultural cohesion. German anthropologists Leo Frobenius (1873–1938) and R. Fritz Graebner (1877–1934) were leading figures in this movement. American anthropologists found the culture area concept useful in their efforts to synthesize what was known about North American indigenous cultures. Anthropologist Clark Wissler (1870–1947) produced continental scale maps of native culture areas based on culture trait similarities and differences. Sauer's Berkeley School colleagues Alfred Kroeber (1876–1960) and Robert Lowie (1883–1957) were among the anthropologists who contributed to the debates and demonstrations of the concept. Sauer's interactions with Kroeber and Lowie, along with his own contributions to the culture area concept (especially his early work on plant domestication in Mexico), led to his formulation of the culture hearth idea. Sauer later proposed that plant domestication probably first occurred in tropical riverine contexts with root crops rather than seed crops. His favored hearth candidates were Southeast Asia and Northwest South America.

The culture hearth idea is not limited to questions of plant and animal domestication. Cultural–historical geographers have employed this construct to map a wide array of cultural traits and complexes. The work of Fred Kniffen (1900–1993) on the distribution and diffusion of material culture traits such as house types and Donald Meinig's (1924–) tripartite model (core, domain, and sphere) of dynamic culture regions offer good examples.

—Kent Mathewson

See also Berkeley School; Cultural Geography

Suggested Reading

Kniffen, F. (1965). Folk housing: Key to diffusion. *Annals of the Association of American Geographers, 55,* 549–577.

Meinig, D. (1986–2004). *The shaping of America: A geographical perspective on 500 years of history* (4 vols.). New Haven, CT: Yale University Press.

Sauer, C. (1952). *Agricultural origins and dispersals.* New York: American Geographical Society.

CYBERSPACE

Cyberspace is a context of human interaction constituted in and by digital signal flows. To interact with other people and machines in this digital environment, people must express their ideas in written words, codes, and graphic images without the use of gestures, contact, and physical presence. Thus, cyberspace is best understood as a virtual space or environment.

The term *cyberspace* is derived from a combination of *cybernetics* and *space,* with the former being the comparative study of computer operations and the human nervous system (a term coined in 1948 by Norbert Wiener). The term *cyberspace* can be traced to the science fiction writings of William Gibson, whose *Burning Chrome, Neuromancer,* and other "cyberpunk" novels popularized the idea of computer-mediated communication (CMC). Although Wiener believed that familiar patterns of human communication should serve as the model for CMC, Gibson and subsequent authors envisioned CMC as a radical form of new communication, thought, and experience. The motif of novelty has inspired far-flung speculation by novelists, artists, politicians, and social scientists, who have drawn on metaphors from architecture, space exploration, and the settlement of the American frontier to indicate innovation and unknown potentialities.

In a nonfictional sense, the origins of cyberspace date back to the 1960s when the U.S. Department of Defense funded the development of the first network of spatially dispersed computers. That network (ARPANET) started with 4 computers in 1969 and expanded to 18 computers concentrated on the Atlantic and Pacific coasts over the next 2 years. A robust network that would function even if random nodes and links were removed or out of service was made possible by the technique of packet switching, which broke up digital messages into packets and sent each of these separately, and by the transmission control protocol/Internet protocol (TCP/IP), which facilitated the interconnection of computers running at different speeds and sending different-sized packets of digital information. Through the efforts of the National Science Foundation (NSF) and state and local governments, a high-speed data transmission "backbone" for the United States was created by the early 1990s. Outside the United States, pioneering efforts began shortly after ARPANET and led to functioning networks by the 1970s and 1980s and a European Internet backbone by the 1990s. Local area networks (LANs) operating in workplaces and community access networks installed in certain towns and cities were connected to the Internet during the 1990s. The 1990s also brought the diffusion of the personal computer to the general population in the United States, and Internet service providers such as America Online developed consumer-oriented network content and services. The original uses for computer networks—military command and research—were quickly supplemented by gossip and debate and later by advertising, commerce, and entertainment. Because the diffusion of computer networking is now beginning to reach elites in the poor countries and the poor in wealthy countries, the network is incorporating an increasing range of devices, encoding systems, and types of social interaction.

Geographic discussions of cyberspace have challenged claims of "cyberenthusiasts" that (a) geographic space has been transcended through technology and (b) social relations in cyberspace will be radically different from those in physical space. Geographers take a critical position toward these ideas based on geography's interest in the material world and human–environment relations. People continue to occupy space, consume resources, engage in production and consumption, and enact the roles of embodied identities (displaying aspects of ethnicity, age, gender, and sexuality) no matter how much they "occupy" cyberspace. Space, place, distance, location, and embodied identity continue to be essential to social life.

Various geographic approaches demonstrate this basic idea. Studies of uneven development and economic dependency confirm Manuel Castells's idea that a "space of flows" is replacing the "space of places" and exacerbating the spatial centralization of economic and political power. Social constructivist approaches support assertions that communication technologies are defined differently in each place of use and by each group of users. Psychological theories led geographers to argue that people remain tied to embodied identities even as they take on disembodied roles, recapitulating hierarchical markers

such as gender and class. All of these approaches challenge the transcendentalism and aspatial character of mainstream views of cyberspace.

DIGITAL DIVIDE

The idea of a "digital divide" is central to geographic debates on networking; where Internet access is common, people generally have a high average income, and where Internet access is not common, people generally have a low average income. This situation reflects reciprocal causality; economic development spurs network access, and network access stimulates economic development. The percentage of the population with access to the Internet varies from less than 1% in the less developed countries of Africa, Asia, and Latin America to more than 50% in the United States, parts of Europe, Singapore, and South Korea. The digital divide is widening as Internet access climbs quickly in the developed countries, where individuals and businesses can easily afford computers and network access, and increases slowly in the rest of the world. Therefore, the economic benefits of information technologies accrue most quickly in places with a previous advantage, and technological and economic dominance are mutually reinforcing as factors of uneven development. Unevenness is manifested not only in access levels but also in the nature of computer use. People of lower- and middle-class backgrounds who use computers are more likely to perform routines such as data entry rather than guiding the production and diffusion of information throughout global networks by using information technologies for administrative, marketing, or management purposes.

SOCIAL MOBILIZATION

Complicating this picture is the fact that new information technologies greatly reduce the cost (in time, money, and labor) of progressive social mobilization. Organizations and movements at the margins of the global system have an international scope of action as soon as they invest a relatively small sum in computers and network access. This approach is often much cheaper than other forms of communication, such as physical travel and traditional postal communications, and the benefit is likely to increase over time due to technological development, competition in the information technology sector, and globalization. Environmental organizations, peasant movements, antiwar and antiarmament activists, human rights organizations, and campaigns for the recognition of stigmatized social groups all have benefited from cyberspace as an environment in which to mobilize. It is unclear, however, whether their ability to mobilize through cyberspace can keep pace with the increasing spatial concentration of the various forms of economic, political, and administrative power.

Representing the "layout" of cyberspace presents a special puzzle for geographers, and the methods for mapping cyberspace remain experimental and unconventional, reflecting only a few decades of progress. Representations of cyberspace include complex tree structures, flow charts, Venn diagrams, density gradients, maplike polygon arrays, and combinations of these.

It is clear that cyberspace is of growing importance as a context of experience and social interaction. As a person acts in and through information technologies, his or her identity evolves in response to the opportunities and constraints of this virtual space. One's community is less and less dependent on those who happen to be physically close, so one's sense of self comes to depend on an increasingly dispersed social network. This emerging "cyborg" identity may partially transcend geographic distance, but society as a whole shows no signs of losing its organizational pattern of centers and peripheries.

—*Paul C. Adams*

See also Communications, Geography of; Telecommunications, Geography and; Virtual Geographies

Suggested Reading

Crang, M., Crang, P., & May, J. (Eds.). (1999). *Virtual geographies: Bodies, space, and relations.* London: Routledge.

Dodge, M., & Kitchin, R. (2001). *Mapping cyberspace.* London: Routledge.

Graham, S. (Ed.). (2004). *The cybercities reader* (Urban Reader series). London: Routledge.

Janelle, D., & Hodge, D. (Eds.). (2000). *Information, place, and cyberspace: Issues in accessibility.* Berlin: Springer-Verlag.

Kitchin, R. (1998). *Cyberspace: The world in the wires.* Chichester, UK: Wiley.

DEBT AND DEBT CRISIS

The debt crisis is related to the emergence of an integrated global financial market and shifting capital flows to and from the "developing" world from the late 1960s to the 1980s. The shakeup of these capital markets during the early 1980s revealed the vulnerability of the global banking system. For the large commercial banks, the "crisis" diminished by the late 1980s as these banks wrote off their liabilities, sold their debts, or rescheduled debt payments. But for debtor nations, the debt crisis spurred deep cuts in public services throughout the 1980s and 1990s as part of broad economic and social restructuring programs.

Three underlying causes of the debt crisis were a ballooning of the global money supply during the 1970s, the changing structure of international debt, and a global economic recession that hit developing economies hard during the early 1980s. Beginning in the 1960s, the global money supply was influenced by the efforts of U.S. companies to finance overseas operations from U.S. and non-U.S. banks operating beyond the confines of U.S. banking regulation. U.S. deficits with the Vietnam War also increased the international supply of dollars. The looming crisis of competitiveness of the U.S. economy was temporarily resolved in 1971 when President Richard Nixon took the United States off the gold standard, suspending the rights of dollar holders to exchange dollars for gold and devaluing the dollar to encourage exports. This staved off an immediate crisis in the United States, but it destroyed the system of stable exchange rates. Speculators moved into international financial markets, leading to what some analysts have called the era

of "casino capitalism," marked by floating exchange rates and increasingly footloose capital flows.

Another part of the debt crisis was associated with the oil price hikes of the mid-1970s. As oil revenue rolled in, Oil Producing and Exporting Countries (OPEC) could not spend it all within their own economies. Because OPEC countries were not using the money to pay for goods and services, the threat was that this would withdraw money from the world economy and precipitate a global recession. Commercial banks began to recycle these "petrodollars" to developing countries, in some cases offering more money than the countries were seeking. For the nonindustrialized countries, especially Latin America, the heavy borrowing seemed to make possible the kind of development that was expected of them during this period.

Two factors combined to precipitate a financial crisis. First, the structure of the developing countries' debt started to change. In the case of Latin America, from the period following World War II through the early 1960s, nearly two thirds of the capital flowing into the region came in the form of official development assistance or public money from government aid and multilateral agencies. These funds were either direct transfers or favorable long-term, low-interest loans. By the end of the 1970s, however, more than 90% of the foreign capital coming into Latin America was private money in the form of direct foreign investment or private bank loans. Moreover, the character of private lending also changed. During the early 1970s, longer-term low-interest loans were the norm. Yet by the late 1970s, many banks began to shift to short-term lending at variable interest rates. Much of the developing world, especially Latin America, saw an explosion of short-term debt between 1978 and 1982. This changing debt structure would have dramatic consequences. When the United States, under President Ronald Reagan, began to tighten the money supply through higher interest rates during the early 1980s, many countries saw their debt multiply almost overnight.

A second factor precipitating the debt crisis was a general decline in the value of commodity exports from developing countries from 1979 to 1987. This meant that just at the time when debt service payments were rising steeply, many countries' ability to pay was collapsing. The result, for Latin America and other parts of the developing world, was a huge reversal from the net inflow of capital during the 1970s to a massive net outflow during the early 1980s.

The debt crisis surfaced in August 1982 when Mexico announced it could not meet its payment obligations. The "crisis" at that moment had to do with the overexposure of the international banking system given that many of the world's major banks had been lending way over their equity limits. If the major debtor nations had united in a payments moratorium, it could have forced a broad package of debt reduction. However, this did not happen. In Latin America, when President Alan Garcia announced in 1985 that Peru would limit its debt service to a proportion of export earnings, only Cuba and Nicaragua supported him. Other nations chose to negotiate bilaterally with international lenders to lower payments in exchange for more total debt. In 1989, the so-called Brady Plan, named for U.S. Treasury Secretary Nicholas Brady, allowed creditors to sell or trade their uncollected debts in secondary markets and distribute the financial risk more broadly in financial and investment markets. Raising interest rates on personal credit cards was one way in which these banks sought to mitigate their financial exposure.

Although the Brady Plan helped to end the crisis for the banks, developing countries faced forced fiscal reform, economic privatization policies, and sharp reductions in public spending on health and education. These shifts are known as "structural adjustment" policies and are mandated by international institutions, such as the World Bank, as conditions for future loans. Structural adjustment has been described as a "silent revolution" due to its long-term destructive effects on many countries in the global South.

—*Elizabeth Oglesby and Altha J. Cravey*

See also Dependency Theory; Developing World; Economic Geography; Globalization; Structural Adjustment; World Systems Theory

Suggested Reading

Boughton, J. (2001). *Silent revolution: The International Monetary Fund, 1979–1989.* Washington, DC: International Monetary Fund.

Corbridge, S. (1992). *Debt and development.* Cambridge, MA: Blackwell.

George, S. (1988). *A fate worse than debt: A radical analysis of the Third World debt crisis.* Harmondsworth, UK: Penguin Books.

Roddick, J. (1988). *The dance of the millions: Latin America and the debt crisis.* London: Latin America Bureau.

DECOLONIZATION

Decolonization technically refers to the breakup of empires, generally the European ones that took shape starting in the 16th century, and the formal independence of the former colonies. Just as colonialism began unevenly over the surface of the earth, so too did it end unevenly. World systems theorists argue that the opportunities for states on the global periphery to exert themselves against colonial powers are best when the core is in crisis. Thus, the Napoleonic Wars of the early 19th century afforded Latin America the opportunity to break away fairly early. Similarly, World Wars I and II proved to be the pivotal moments when Western control over much of Africa and Asia was finally broken.

The shift toward decolonization during the post–World War II era was complex. Often independence movements were composed of broad coalitions of nationalists, students, the intelligentsia, and peasants, frequently led by Western-educated intellectuals (e.g., Ho Chi Minh in Vietnam, Mohandas Gandhi in India). The cold war rivalry between the United States and the Soviet Union afforded such movements a political space that allowed them to play the superpowers off against each other because both the United States and the Soviet Union were eager to appear different from older European colonial conquerors and friendly to the masses of the emerging states. Often the struggle for independence was violent, involving protracted guerrilla conflicts and wars (e.g., Malaysia, Vietnam, Algeria). The relatively peaceful independence movement in India was the exception, although the division of South Asia into India and Pakistan involved extensive civil strife and the deaths of millions.

Independence movements gradually succeeded throughout the late 1950s, 1960s, and 1970s, leading to a proliferation of newly independent countries (from roughly 50 in 1945 to approximately 200 today in the United Nations). Major milestones in this process include India and Pakistan in 1947, Indonesia in 1949, and Angola and Mozambique in 1975. Virtually all parts of the globe have been decolonized, with a few small exceptions (e.g., French Guiana, Martinique, Gibraltar, Puerto Rico).

Decolonization involved political, economic, and ideological changes. Politically, this shift brought with it a new administrative and legal apparatus in the former colony, typically modeled after the colonial one. Indeed, often the very same people who served the foreign colonial power became leaders in the newly independent one. Ideologically, decolonization opened the door to challenges to long-standing racist notions of white inferiority (e.g., Ghana's Kwame Nkrumah), allowing a variety of experimental social projects (e.g., Tanzania's *ujamaa* [or African socialism]).

Although formal political independence inevitably brought with it the trappings of a new society—a new flag, currency, national airline, and so on—many observers question whether or not decolonization ended as simply as it appeared to end. Indeed, to dependency and world systems theorists, it is no accident that the former European colonies are inevitably part of the so-called Third World—the vast and diverse set of societies that encompass the bulk of the world's people but relatively little of its wealth. Despite ostensible political independence, such societies were often woefully unprepared for independence economically and remained heavily dependent on their former colonial powers for capital, trade assistance, and foreign aid, leading to widespread fears of neocolonialism, generally via multinational corporations. The dominant role of the United States as the world's leading neocolonial power, in both economic and military terms, made the contrast between nominal political independence and substantive economic independence all the more apparent. Most former colonies have inadequate infrastructure and human capital, with economies centered on raw materials (e.g., foodstuffs, mineral ores). Despite whatever measure is used—gross domestic product per capita, energy consumption, access to health or education services, and so on—former colonies almost always lag behind the industrialized world (although some, such as Singapore, rival it, and the newly industrializing countries have made rapid progress).

—*Barney Warf*

See also Colonialism; Dependency Theory; Neocolonialism; World Systems Theory

Suggested Reading

Betts, R. (1998). *Decolonization (The making of the contemporary world)*. London: Routledge.

Duara, P. (Ed.). (2004). *Decolonization (Rewriting histories)*. London: Routledge.

Le Sueur, J. (Ed.). (2003). *The decolonization reader*. London: Routledge.

DEINDUSTRIALIZATION

Deindustrialization refers to the large-scale loss of manufacturing jobs and subsequent labor market restructuring that has left many former manufacturing centers in ruins. Deindustrialization is most prevalent in the older industrial regions of Europe and the United States. Job loss is concentrated in heavy industrial sectors, particularly in steel and automobile manufacturing. The social and economic impacts of deindustrialization are dire. The unemployment caused by deindustrialization leads to increased poverty and a variety of other social ills such as crime, alcohol and drug abuse, suicide, and divorce.

The first and most influential work on deindustrialization was Bluestone and Harrison's *The Deindustrialization of America,* published in 1982. Bluestone and Harrison argued that deindustrialization is a deliberate corporate strategy to move capital out of manufacturing and reinvest it in more profitable (and speculative) activities such as financial services. The disinvestment in heavy manufacturing erodes national economic competitiveness in basic industry.

Deindustrialization is the flight of capital. Ignoring any sort of responsibility to the workers and locations that had been key elements during earlier rounds of accumulation, managers of industrial concerns let their factories become technologically obsolete. Corporations milked their older factories for as much profit as possible and then reinvested the funds elsewhere.

The mobility of capital and the willingness of corporate managers to use capital mobility as a way in which to gain concessions from workers and communities helped to disempower labor unions. Faced with massive job loss and eroding memberships, many unions gave in to corporate demands for wage and benefit rollbacks. Communities desperate to prevent plant closings readily negotiated tax breaks and other incentives to keep companies from leaving town. Corporations skillfully play workers and communities against each other, "whipsawing" them and gaining the greatest possible concessions. In many instances, companies that obtain worker concessions and community tax breaks remain only a few years and still leave town. Many corporate critics wonder whether any sense of corporate responsibility still exists.

Deindustrialization was part of a broader effort to reduce the power of organized labor. In Britain, Margaret Thatcher's government explicitly sought (successfully) to break the power of unions, particularly in the mining industry. In the United States, President Ronald Reagan actively signaled the federal government's support for union-busting efforts. The success at breaking the labor–management accord that had underlain the Fordist system signaled the ascent of neoliberal forms of economic governance and control.

The labor market effects of deindustrialization include the loss of well-paying manufacturing jobs. Many workers experience a permanent decline in income as they are forced to find employment in lower-paying service-sector jobs. Job losses as a result of deindustrialization are particularly pronounced among women and minority workers, but white male workers suffer the largest declines in income.

In manufacturing towns, the impact of deindustrialization is felt for many years. High levels of unemployment and accompanying social distress are common outcomes of large-scale deindustrialization. In many cases, people find it very difficult to adjust to the job loss associated with plant closures. Workers not only lose a way in which to make a living but also experience the loss of an entire set of social relationships that are based on the shared experience of work.

Manufacturing cities struggle to find a way out of the economic crisis provoked by deindustrialization. Often entire industrial landscapes disappear as companies demolish their closed factories. Cities pursue alternate economic opportunities, but it is difficult, if not impossible, to replace the jobs lost. Prisons, casinos, and entertainment districts all are economic development strategies used by cities anxious to find ways in which to move forward.

Youngstown, Ohio—once known as "Steel Town USA"—is the exemplar of a deindustrialized community. Economic crisis came to Youngstown on "Black Monday" (September 19, 1977) when the Campbell Steel Works announced its closing and the loss of 5,000 jobs. During the months that followed, another four major steel mills were closed. Residents of Youngstown and the surrounding Mahoning Valley actively contested the plant closings. In addition, a proposal to purchase the Campbell Steel Works and operate it as an employee-owned business was developed. Although local people donated funds for the purchase, the plan failed because the federal government refused to offer needed loan guarantees.

When U.S. Steel announced the closure of its Youngstown steel mill, angry workers occupied U.S. Steel's corporate headquarters in Pittsburgh, Pennsylvania. In addition, a coalition of local religious and union leaders filed a lawsuit that argued for a new form of eminent domain based on the idea that communities had a form of community property rights over industries located in their jurisdictions. Although this argument ultimately failed, a new concept of corporate community responsibility was introduced.

Youngstown's struggles to retain steel ultimately failed, and the city became an icon for the problems associated with deindustrialization. The struggles of displaced Youngstown steelworkers were publicized by Dale Maharidge and Michael Williamson in their book *Journey to Nowhere*. Inspired by that book, rock musician Bruce Springsteen wrote a song lamenting the loss of steel jobs and the destruction of the Jenny, one of the last surviving remnants of the steel industry.

A variety of economic development efforts have failed to provide a substitute for the high-paid employment offered by steel. Yet nearly 25 years after Black Monday, the residents of Youngstown retain the gritty determination to survive.

—*Jeff Crump*

See also Economic Geography; Industrial Revolution; Rustbelt; Uneven Development

Suggested Reading

Bluestone, B., & Harrison, B. (1982). *The deindustrialization of America: Plant closings, community abandonment, and the dismantling of basic industry.* New York: Basic Books.

Linkon, S., & Russo, J. (2002). *Steel Town USA: Work and memory in Youngstown.* Lawrence: University Press of Kansas.

Maharidge, D., & Williamson, M. (1996). *Journey to nowhere: The saga of the new underclass.* New York: Hyperion. (Original work published 1985)

Peck, J. (2002). Labor, zapped/growth, restored? The moments of neoliberal restructuring in the American labor market. *Journal of Economic Geography, 2,* 179–220.

DEMOCRACY

The word *democracy* originates from the Greek words meaning "the people" and "to rule." Thus, democracy means "rule by the people." The philosophy behind democracy is that all people have rights that cannot be taken away and that rulers and citizens have certain obligations to each other. The rulers have the obligation to protect citizens' rights, and citizens may take away the rulers' power if rulers do not fulfill their obligations.

Democracy is a form of government in which the voting citizenry, referred to as the people, have the power to alter the basic laws governing a state. There are several varieties of democracy, or means by which citizens may exercise this power, but the two most common forms are direct democracy and representative democracy. A direct democracy occurs when all citizens participate directly in governmental decision making, and representative democracy occurs when citizens elect officials to make decisions on their behalf. The term *democracy* may be used to assess how much influence people have over their government through elections and is demonstrated by the rule of law, that is, how much democracy exists.

Although in contemporary use, democracy is usually understood differently from the original use of the term by the ancient Greeks in their Athenian political system, present-day democracies may be characterized by these features:

- A constitution—written, unwritten, or both—that guides the formal operation of government, sets limits to government power, and outlines basic principles of legal rights that citizens may expect

- Election of officials

- Honest and equitable elections that are open to all citizens of voting age

- The right to vote and to stand for election (granted to all citizens of voting age)

- Freedom of expression, including the right to assemble peacefully and freedom of speech

- Freedom of association or the right to join with others either in personal relationships or in groups on the basis of shared views and beliefs

- Equal treatment under the law for all citizens, who have the right to appropriate legal procedures, safeguards, and established rules

- Access to alternative (nongovernment) sources of information

- An educated populace

Some of these rights may differ under certain conditions. For example, in the United States, a convicted felon may or may not have the right to vote depending on the state in which that person resides. In addition, in theory it is possible to have the problem of the tyranny of the majority. In that case, the right of all citizens to be treated equally under the law may be a concern if an elected majority opts to criminalize a particular minority, either directly or indirectly, on the basis of religion, sexual orientation, political beliefs, and so on. It might be argued that majority rule, despite such possible shortcomings, is better than minority rule, which has been shown through many cases historically to have overwhelmingly negative effects for large numbers of people. Some democratic systems use proportional representation as a way in which to ensure that minorities are represented fairly within government bodies, but other systems grant power predominantly to the two most popular political parties. However, the intention of a constitution, due process of law, and free elections is to curb the threat of a tyranny of the majority. Ideally, a democracy in practice entails majority rule with rights for minority groups.

Most countries that are considered to be democratic have government systems that are representative rather than direct democracies. The establishment of present-day democratic standards has been dominated by European countries and the United States, which share similar histories of industrialization and traits of economic development. Measuring other, less economically developed states by the standards of such Western democracies must be done with an awareness that those states may have different cultural values and historical contexts that influence the realities of how democracy may be practiced there. In addition, as new states are created (e.g., former Soviet republics, relatively new states in areas that were former colonies), democratic institutions such as those listed previously take time to evolve and to become established both in reality and in citizens' and rulers' expectations.

Elections are one feature of a democracy, but elections alone do not create a democracy. Other government systems have used elections to impart a sense of democracy, but dictatorships and totalitarian regimes, for example, may pressure citizens to vote in a particular way or restrict citizens' choices or their right to express their honest opinions. Elections are critical to democratic systems in that they allow citizens to remove rulers or administrations without altering the legal foundation of the government. This practice aims to maintain stability and reduce political uncertainty because the public is aware that it has the regular opportunity to change policies and to change who is in power.

—*Shannon O'Lear*

See also Political Geography

Suggested Reading

De Tocqueville, A. (2001). *Democracy in America.* New York: Signet. (Original work published 1835)

Sorensen, G. (1997). *Democracy and democratization: Processes and prospects in a changing world.* Boulder, CO: Westview.

DEMOGRAPHIC TRANSITION

Developed by several demographers during the 1920s, the demographic transition theory stands as an important alternative to Malthusian notions of population growth. Essentially, this is a model of a society's fertility (birth rate [BR]), mortality (death rate [DR]), and natural population growth rate (NGR) over time, using the simple relationship NGR = BR − DR. Because this approach is based explicitly on the historical experience of Western Europe and North America as they went through the Industrial Revolution, "time" in this conception is a proxy for industrialization. This approach can be demonstrated with a graph of birth, death, and natural growth rates over time that divides societies into four major stages (Figure 1).

STAGE I: PREINDUSTRIAL ECONOMY

In the first stage, a traditional, rural, preindustrial society and economy, fertility rates are high and families are large and extended. In agrarian economies, children are a vital source of farm labor, helping to plant and sow crops, tending to farm animals, performing chores, carrying water and messages, and helping with younger siblings. Children also take care of their elderly parents. In societies with high infant mortality rates, having many children is a form of insurance that some proportion will survive until adulthood. Thus, the distribution of birth rates around the world reveals

that the poorest societies have the highest rates in the world, particularly in Africa and most of the Middle East. In contrast, birth rates in North America, Europe, Russia, Japan, Australia, and New Zealand are relatively low.

However, in preindustrial societies, mortality rates also are typically quite high, meaning that average life expectancy is relatively low. The primary causes of death in poor rural contexts are the result of inadequate diets, unsanitary drinking water, and bacterial diseases. Thus, the world geography of death rates closely reflects the wealth or poverty of societies. Because both fertility and mortality rates are high, the *difference* between them—natural population growth—is relatively low, often fluctuating around zero. Although relatively few societies in the world live in these circumstances today, Stage I may describe certain tribes in parts of Central Africa, Brazil, and Papua New Guinea.

STAGE II: EARLY INDUSTRIAL ECONOMY

The second stage of the demographic transition pertains to societies in the earliest phases of industrialization, such as 19th-century Britain and the United States, or selected countries in the developing world today, such as Mexico. Early industrial societies retain some facets of the preindustrial world, particularly high fertility rates. Because most people still live in rural areas, children remain an important source of farm labor. The major difference is the decline in mortality rates, leading to longer life expectancies. Mortality rates decline as societies industrialize, not primarily because of better medical care but rather because of improved food supplies due to the industrialization of agriculture that played a major role in improving immune systems, including lowering infant mortality rates. Because the death rate has dropped but the birth rate has not, the natural growth rate grows explosively, a situation evident in a great number of countries in the developing world today (Figure 2).

STAGE III: DEVELOPED INDUSTRIAL ECONOMY

Societies in the throes of rapid industrialization, where a substantial share of people—if not the majority—live in cities, exhibit a markedly different pattern of birth, death, and growth rates from those earlier in the transition. Death rates remain relatively low, but in this stage fertility rates also exhibit a steady decline. Birth rates typically fall and families get smaller as societies become wealthier because urbanization and industrialization change the benefit-cost ratio of having children. In societies where large numbers of women enter the paid labor force—become commodified labor outside of the home rather than unpaid workers inside of it—mothers typically must drop out of the labor market, if only temporarily, to take care of their children. Economically, this process generates an opportunity cost to having children; the more children a couple has or the longer a mother refrains from working outside of the home, the greater the opportunity cost she and her family face. As women's incomes rise, either over time or comparatively within a society, the opportunity cost of having children rises accordingly, leading to lower fertility rates. As fertility rates decline, so too do natural growth rates. In short,

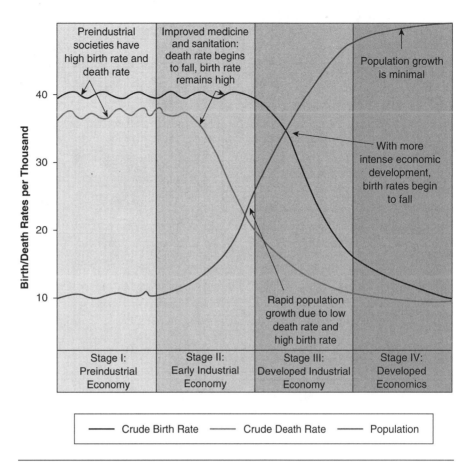

Figure 1   Demographic Transition

SOURCE: Stutz, F. and B. Warf. 2005. *The World Economy: Resources, Trade, and Development.* 4th edition. p. 82. Reprinted with permission from Prentice Hall.

relatively prosperous societies tend to have smaller families, and there is frequently a corresponding shift from extended to nuclear families in the process.

Historically, fertility levels fell first in Western Europe, followed quickly by North America, more recently by Japan, and then by Eastern Europe and Russia. In those areas, reproductive levels are near, or even below (in some countries), the level of generational replacement. Elsewhere, however, birth rates remain at much higher levels, although in China and Southeast Asia the birth rates are dropping quickly. There has been a modest decline in South Asia, the Middle East, much of Latin America, and parts of sub-Saharan Africa.

STAGE IV: DEVELOPED ECONOMICS

The fourth and final stage of the demographic transition depicts wealthy, highly urbanized worlds, a context indicative of Europe, Japan, and North America. Such

societies typically witness low death rates, the causes of which may change from infectious diseases to life-style-related ones, particularly those associated with smoking and obesity as well as, to a lesser extent, car accidents, suicides, and homicides. Birth rates also continue to fall in such contexts as many couples elect to go childless or to have only one child. When birth rates drop to the level of death rates, a society reaches zero population growth. When birth rates drop below death rates, as they have in virtually all of Europe and Japan, a society experiences negative population growth. Such countries are characterized by large numbers of the elderly, a high median age, and a relatively small number of children, all of which have dramatic implications for public services.

Globally, uneven economic development generates uneven patterns of natural population growth (Figure 2). The most rapid rates of increase are found throughout the poorer parts of the developing world, including

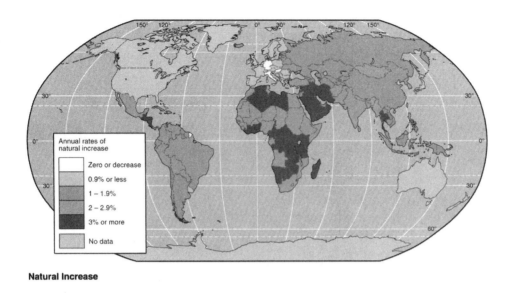

Natural Increase

Figure 2   Map of Natural Growth Rates Around the World

SOURCE: Stutz, F. and B. Warf. 2005. *The World Economy: Resources, Trade, and Development.* 4th edition. p. 91. Reprinted with permission from Prentice Hall.

countries the mortality rates have plunged in only one or two generations. Because mortality rates do not vary geographically as much as fertility rates, most of the spatial differences in natural growth around the world are due to differences in fertility.

—*Barney Warf*

See also Development Theory; Fertility Rates; Malthusianism; Mortality Rates; Natural Growth Rate; Population, Geography of

Suggested Reading

Jones, G., Caldwell, J., D'Souza, R., & Douglas, R. (Eds.). (1998). *The continuing demographic transition.* Oxford, UK: Clarendon.

Kirk, D. (1996). Demographic transition theory. *Population Studies, 50,* 361–388.

Peters, G., & Larkin, R. (2002). *Population geography: Problems, concepts, and prospects.* New York: Kendall/ Hunt.

Africa, the Arab and Muslim worlds, India, and Indonesia. In contrast, low rates of growth are found in the economically developed nations, including North America, Japan, Europe, Australia, and New Zealand.

CRITICISMS OF DEMOGRAPHIC TRANSITION THEORY

Although the demographic transition has wide appeal because it links fertility and mortality to changing socioeconomic circumstances, it has also been criticized on several grounds. Some critics point out that it is a model derived from the experience of the West and then applied to non-Western societies as if the latter are bound to repeat the exact sequence of fertility and mortality stages that occurred in Europe, Japan, and North America. There is no inevitability ensuring that the developing world must follow in the footsteps of the West. Some have pointed out that the developing world is in many ways qualitatively different from the West, in no small part because of the long history of colonialism. Furthermore, demographic changes in the developing world have been much more rapid than in the West. Whereas it took decades, or even centuries, for mortality rates in Europe to decline to their modern levels, in some developing

DEPENDENCY THEORY

The dependency approaches that emerged during the 1960s and 1970s represent an important and complex body of theory with Marxist and structuralist roots. Dependency theory first emerged as a critique of modernization theory toward the end of the 1960s as a growing disillusionment with the laissez-faire and diffusionist approach of modernization theory set in and as it became clear that there had been a failure to deliver the promised material benefits of becoming "modern." Thus, there evolved a more wide-ranging critique of development theory that was firmly rooted in the Third World and in certain traditions of "Third Worldism." The dependency school contended that dependency on a metropolitan "core" (e.g., Europe, North America) increases the "underdevelopment" of satellites in the "periphery" (e.g., Latin America, Africa). Third World poverty, it was argued, was not a result of local failures in the periphery but rather a direct consequence of the exploitative relations between First World and Third World—between Metropole and satellite.

According to the dependency scholars (or *dependistas*), economic dependency came about because these peripheral satellites were encouraged to produce what they did not consume (e.g., primary products) and to consume what they did not produce (e.g., manufactured/industrial goods). Where modernization theorists saw colonialism as part of an "awakening" of modernity, the dependency approach highlighted how colonialism underdeveloped the periphery and continued to do so, *neocolonially,* after the end of the empire. Unlike modernization approaches, dependency theorists sought to view development in historical context, arguing that colonialism had helped to put in place a set of dependent relations between core and periphery and highlighting the need to think about the forms of colonial and postcolonial incorporation into the world economy.

In Latin America, André Gunder Frank made the relations between "North" and "South" a key point of focus in his study of "The Development of Underdevelopment." Frank argued that the relations between the Metropole and the satellite countries were exploitative, pointing out that any surplus generated in the satellite countries was siphoned off to the North, breeding conditions of underdevelopment. Dependency theorists set out to oppose the modernization approach point by point to such an extent that, in a way, they ended up "checkmating" each other. Modernization theory had envisaged that Third World countries would gradually progress and evolve toward an urban-based, Western lifestyle of consumption, but the dependency scholars argued that unequal capitalist relations and a history of colonialism denied the Third World a chance of ever being fully industrialized. Unlike the modernization theorists, the dependency scholars focused more at the international and global scales and spaces of development, examining the structural relations of nation-states to the world economy.

Just as the modernization approach was adopted in a variety of ways by international institutes and bilateral donors, the dependency school was made up of all those opposed to U.S. policy and by groups of what were called "Third Worldists." Theorizing the manipulation of the periphery by the core was an important process given that by this time a variety of state socialisms had begun to appear (e.g., in Cuba, Angola, Mozambique, Tanzania, and Vietnam). The dependency approach had important roots in the United States, Brazil, Chile, and Colombia and later spread out into a variety of regions, including Africa, the Caribbean, and the Middle East. The economic program pursued by Iran during the 1980s, for example, reflected many of the ideas on "delinking" and self-sufficiency propagated by dependency theorists such as Frank and Samir Amin.

Theoretically, the dependency debate was an assault on the conventional wisdom concerning the relationship between international trade and the development process. The neo-Marxist aspects of its critique offered a revolution against capitalism as a way out, highlighting the weakness and vulnerability of Western capitalist economies and their dependence on the labor and resources of others as well as focusing on the political role of a local (or *comprador*) bourgeoisie in the process of underdevelopment. The development strategy of the dependency school was formed partly by an institution set up in 1948 known as the Economic Commission for Latin America (ECLA [or CEPAL in Spanish]). Raul Prebisch, who worked at ECLA, was an important figure in the dependency debates, arguing that the global economic system was divided structurally between rich and poor countries and urging greater regional cooperation in Latin America to counteract this scenario. For a few years, dependency approaches held the initiative, and eventually even the international development community was obliged to accommodate at least some of the critique; for example, the International Labour Office called for "redistribution with growth" in 1972.

Key criticisms directed at the dependency approach were that the theory represented a form of "economic determinism" and overlooked social and cultural variation within developed and underdeveloped regions. In addition, the term *dependence* had been used immoderately and led to oversimplification. It might have said much about the origins of underdevelopment, but a clear statement of what "development" itself might be was obscured by a rigid core–periphery model that some have read as a simple inversion of earlier binaries associated with modernization theory. Another point of contention was that the dependency theorists seemed to be calling for a delinking from the world capitalist economy at a time when the world economy was undergoing further globalization. Furthermore, the notion of underdevelopment in a way endorsed concepts of First World–Third World or core–periphery rather than seeking to fundamentally challenge this schema and beginning a search for alternative ways of differentiation.

The dependency framework also perhaps left the impression that there was an "evil genie" organizing the system, making sure that the same people win all of the time and somehow loading the dice. In addition, the economy (rather than the culture or politics of individual spaces and places) was still seen as being of primary importance by the dependency scholars in a way that lacked nuance and verged on the deterministic. Despite its appeals, dependency also did not inspire many policies of development except in Chile and Cuba for short periods. By the early 1980s, many commentators noted the diminishing returns of the dependency critique and pointed to an "impasse" because there seemed to be no way to go beyond the theoretical coordinates of these previous approaches. Nonetheless, in the context of growing inequalities between the North and the South, dependency approaches still continue to provide a rich source of ideas.

—*Marcus Power*

See also Development Theory; Modernization Theory; Neocolonialism; Underdevelopment; World Systems Theory

Suggested Reading

Frank, A. (1967). *Capitalism and underdevelopment in Latin America.* New York: Monthly Review.

Frank, A. (1996). The development of underdevelopment. In C. Wilber & K. Jameson (Eds.), *The political economy of development and underdevelopment* (pp. 105–115). New York: McGraw-Hill. (Original work published 1967)

Leys, C. (1996). *The rise and fall of development theory.* London: James Currey.

Peet, R. (1999). *Theories of development.* New York: Guilford.

Power, M. (2003). Development thinking and the mystical "kingdom of abundance." In M. Power, *Rethinking development geographies* (pp. 71–84). London: Routledge.

DERELICT ZONES

Abandoned buildings and vacant land are the hallmarks of derelict zones, that is, areas in which disinvestment, vacancy, and degradation are prevalent. The number and size of derelict zones in American cities have increased since the turn of the 20th century. The processes by which buildings and land become obsolete and undervalued in North American cities, towns, and rural areas have been hotly debated. Some view dereliction as an inevitable stage in an efficient land market, whereas others criticize the inequalities that give rise to derelict zones.

CAUSES OF DERELICTION

Newly developed locations have up-to-date buildings that become obsolescent as they age and deteriorate. Over time, buildings will be abandoned and land may be left vacant and ripe for redevelopment. Eventually, buildings will be renovated or replaced. According to this view, dereliction is a process through which obsolescent buildings and land uses of declining value are replaced by new structures and more valuable activities.

Other scholars also view derelict zones as the inevitable outcome of a capitalist market in which investment and its benefits are spatially uneven and socially unequal. Dereliction provides financial opportunities for investors who purchase deteriorating properties at depressed prices, hold them while deferring maintenance, and sell when property values increase. Investors disregard the high costs of disinvestment for nearby residents and business owners who live and work in a deteriorating environment that threatens their everyday lives and livelihoods. The presence of derelict buildings spreads a pall, stigmatizing nearby properties and their inhabitants. As a result, dereliction often spreads by contagion, with buildings and locations near derelict buildings being at much greater risk for reduced maintenance and deterioration.

UNEVEN AND UNEQUAL GEOGRAPHIES OF DERELICTION

Derelict zones are concentrated in places that lack the resources to resist them. Typically, central-city areas dominated by low-income populations, racial minorities, and public housing are the first to experience deterioration. Since the early 1970s, the loss of large manufacturing plants and their relocation to the suburbs and small towns hastened disinvestment in central-city neighborhoods. Those who could not follow their jobs remained behind, often isolated from well-paid employment. Discriminatory housing policies that facilitated the suburbanization of white Americans while hampering minority home ownership heightened racial segregation, putting minority areas at greater risk for disinvestment. Public housing programs concentrated the poor in high-rise developments, leading to the abandonment of aging but low-cost rental housing in central-city areas. Over time, many public housing developments also experienced disinvestment, becoming some of the most infamous derelict zones.

Recently, dereliction has increased in suburban areas, small towns, and rural areas where economic restructuring has rendered many locations less valuable for production. Faced with the economic burdens associated with job loss and a declining property tax base, local governments and communities have not always been able to reverse the subsequent deterioration. Where local residents are empowered and work closely with government agencies, evaluations of derelict zones as stigmatized places of little value can be resisted.

—*Valerie Preston*

See also Ghetto; Urban Geography

Suggested Reading

Jakle, J., & Wilson, D. (1992). *Derelict landscapes: The wasting of America's built environment.* Lanham, MD: Rowman & Littlefield.

DEVELOPING WORLD

Countries and regions that are described as part of the *developing world* typically are characterized by low levels of average income; high rates of poverty; wide social, economic, and spatial inequalities; and high levels of dependence on the markets and products of advanced industrial countries. Also referred to as *less developed countries, nonindustrialized countries,* the *Third World,* or the *South,* these countries also share similar histories of colonial rule or indirect domination. The term *developing world,* however, is more

than a simple way of classifying countries and regions. The term has operated historically as part of the discourse of development with its faith in Western notions of progress and modernity and its naturalization of the knowledges and social practices required to achieve it. The designation as *developing world,* therefore, has deep roots in a particular social imaginary of the world, one that ranks countries by the extent to which they differ economically, socially, and institutionally from Western countries considered at the apogee of modernity.

Defining regions as either developed or developing is a problematic exercise because, depending on the social or economic characteristics on which one is focusing, the numbers of countries and regions included can vary significantly. For example, although the United Nations (UN) does not have an established convention for designating countries as either developed or developing, it generally regards the 115 countries of Asia (excluding of Japan), Oceania (excluding Australia and New Zealand), the Americas (excluding Canada and the United States), Africa, and the Caribbean as the developing world. Even within this broad definition, however, exceptions exist. For example, Israel and the Southern African Customs Union are usually considered developed regions for international trade purposes, and the countries of the former Yugoslavia, although European, are treated as developing regions. Similarly, the countries of Eastern Europe and the former Soviet countries in Europe are not considered part of either the developing or developed world, even though many exhibit levels of poverty and inequality found in parts of Latin America and the Caribbean. Equally problematic to establishing a coherent definition is the fact that countries categorized as part of the developing world need not actually be in the process of increasing levels of wealth or social welfare and, in fact, may be in deep crisis. Often the clear contradictions between the category and reality are resolved through acknowledgment or by the creation of more detailed systems of classification. The World Bank, for example, warns that the use of the term *developing country* in their publications does not imply either that all of the economies belonging to the group are actually in the process of developing or that those not in the group have necessarily reached some preferred or final stage of development. However, the UN has developed additional categories, such as *least developed countries* (LDCs), *land-locked developing countries* (LLDCs), *small island economies* (SIDs), and *countries in transition from centrally planned to market economies,* to draw attention to specific constraints facing particular national territories. In 2005, 50 countries held the distinction of being LDCs given their extremely low levels of income per capita, social welfare, and high levels of economic instability.

Historically, the criteria used to define the developing world have focused largely on economic growth, with little concern for questions of equity, sustainability, productivity, or empowerment. Until the 1990s, for example, the most common way of differentiating the developing world from the developed world was through the use of gross domestic product (GDP) figures that measured the value of goods and services produced in a country in a given year. This focus on the economic performance of economies reflects the influence of modernization theories that proliferated throughout the developing world from the 1950s onward. The emphasis that modernization theorists placed on industrialization and economic growth was a product of the underlying belief that because the countries of Asia, Africa, and Latin America were backward and undeveloped, they needed to emulate the Western experience of industrialization to attain the assumed superior standards of living found in Europe and North America. Within modernization theory, therefore, these economic growth indicators functioned as a way of hierarchically organizing countries based on Western definitions of progress, humanity, and civilization.

Even though alternative measures of development currently exist, it is still common for international institutions to rely on measures of average income to define the developing world. For example, based on the gross national income (GNI) statistics in 2005, the World Bank categorized 154 countries, including 27 in Europe and Central Asia, as part of the developing world. In general discussions, for example, the World Bank defines countries of the developing world in terms of their levels of GNI per capita. The GNI is a measure of the worldwide income earned by a country divided by the population. Although recognized by the World Bank as insufficient on its own as a measure of welfare or success in development, the GNI has remained the most popular single indicator of economic capacity and progress over the past 50 years. On the basis of this indicator, countries are considered to be part of the developing world if they have GNI per capita levels below a benchmark, which in 2003 was set at less than

U.S. $9,385. This benchmark between middle-income and high-income countries was first established in 1989 when it became clear that many of the countries included in the middle-income group (a category based on earlier listings of what constituted *developing* and *industrial* countries) no longer met the criteria.

Beyond the issue of the appropriateness of the categories, the use of GNI per capita as a proxy for level of development remains deeply problematic. This is because national income statistics are measures of economic activity rather than welfare and, therefore, tell us little about the relative differences in the quality of life enjoyed across population groups. In fact, the GNI statistic does not even differentiate between economic activities that increase welfare costs and those that boost welfare benefits. Thus, natural disasters, war, and increasing levels of crime may actually increase the value of GNI, making a country's economic situation look better than it is, if they generate increases in market activity. As a way of defining the developing world, the GNI per capita statistic is also problematic because it provides no indication of the distribution of income. Thus, in a country with a highly unequal distribution of income, a $1 million increase in national income benefiting the richest 20% of the population would be valued in the same way as a $1 million increase benefiting the poorest 20%. Equally problematic is that national income statistics do not measure value generated by services and goods that are not traded. For example, the value derived from the unpaid work that many women perform is not included in the national income accounts, even though studies indicate that such labor is often crucial to maintaining levels of social welfare, particularly during periods of economic crisis.

The use of largely economic indicators to define the developing world was challenged during the early 1990s when the UN began to use the Human Development Index (HDI), a composite index measuring a country's achievements in three areas of human development—longevity, knowledge, and standard of living—to rank countries accordingly. Created by economist Mahbub ul Haq as a measure of social well-being, the HDI focuses on the choices that people have to live the lives they value rather than on the economy (which is viewed as only one means of enlarging human choices and capabilities). The HDI uses life expectancy at birth to measure longevity; a combination of the adult literacy rates and the combined primary, secondary, and tertiary gross enrollment ratios to measure knowledge; and GDP per capita to measure standards of living. The preferential use of human development indicators as a way of identifying places in the world where human choices, freedoms, and (consequently) levels of development are most constrained has been an important part of the UN's millennium development campaign to halve by the year 2015 the number of people living on less than $1 per day. Currently, the UN has set eight millennium development goals (MDGs) to be accomplished by 2015: (1) cutting poverty and hunger rates recorded in 1999 by half; (2) achieving universal primary education and (3) achieving gender equity at all educational levels; (4) cutting the mortality rate of children under 5 years of age by two thirds and (5) cutting the maternal mortality rate by three quarters; (6) halting and reversing the spread of major diseases, including HIV/AIDS; (7) cutting the number of people without sustainable access to safe drinking water by half, reversing the loss of environmental resources, and significantly improving the lives of at least 100 million slum dwellers; and (8) developing a global partnership for development, including more effective aid, more sustainable debt relief, and fairer trade rules. To monitor progress toward the MDGs, the UN has also constructed 48 social and economic development indicators for tracking development, including the proportion of a population living on less that $1 per day, HIV prevalence among pregnant women between 15 and 24 years of age, the proportion of land area covered with forest, and the proportion of seats held by women in national parliament.

While widespread dissatisfaction with the economistic and Euro-centric assumptions embedded in the term *developing world* during the 1990s led the UN to develop more human-centered definitions based on the constraints to human choices and capabilities, as early as the 1950s alternative terms had been offered emphasizing the historical–political and geographic factors that produced low levels of income, poverty, and inequality rather than existing levels of economic growth. Alfred Sauvy, a French demographer and historian, introduced the term *Third World* (or *tiers monde*) in 1952 to collectively define the countries of Latin America and the Caribbean, Africa, Asia, and the Pacific (excluding Australia and New Zealand). He argued that like the commoners and peasants who comprised the "third estate" at the time of the French Revolution (with the first and second estates being the church and the aristocracy, respectively), these

countries occupied a position in the world economy that was equally exploited and devalued in the global economy. Given their position, Sauvy argued that Third World countries could not be considered part of either the industrialized capitalist world (the First World) or the industrialized Communist Bloc (the Second World) because the Third World was always going to seek to become "something" and, as implied by the analogy, would probably need to do so by revolutionary means.

The term *Third World* was adopted by many countries during the cold war to highlight their desire to be liberated from external oppression and control through alignment with either the United States and Western European capitalist countries or the communist countries of Eastern Europe. During this period, the term became not only symbolic of the growing political awareness among countries of their shared histories of domination through direct and indirect colonial rule but also an emblem of an imagined community of poor and peripheralized places that was beginning to form in the world economy. At a meeting in 1955 in Bandung, Indonesia, 29 African and Asian states committed themselves to greater economic and cultural cooperation as well as opposition to colonialism. This group, which subsequently became the non-aligned movement, represented a moment in history when countries such as India, Ghana, Yugoslavia, Indonesia, and Egypt, recognizing the commonalities in their positioning in the world capitalist system, sought to cooperate and act as a political bloc for mutual benefit. As proclaimed in the Havana Declaration of 1979, the nonaligned movement aimed to ensure "the national independence, sovereignty, territorial integrity, and security of nonaligned countries" in their "struggle against imperialism, colonialism, neo-colonialism, apartheid, racism, including Zionism, and all forms of foreign aggression, occupation, domination, interference, or hegemony as well as against great power and bloc politics." The potential power of the Third World appeared to be evident during the early postwar years as the number of newly independent nations at the UN surpassed that of European and European-dominated nations. However, the ability of Third World countries to maintain solidarity and remain nonaligned to Western European and Soviet power blocs was weak. Given the poor levels of infrastructure and resources for investment in healthcare and education that many newly independent countries had to confront after independence, many nonaligned states succumbed to the offers of much-needed development aid made by major superpowers seeking to extend their spheres of influence. In the end, poor countries ultimately found themselves too reliant on First World aid to be able to effectively challenge their dominance. Thus, in 2005, although the nonaligned movement was still in existence, many questioned the continued usefulness of this association and the ability of its 116 members to represent the interests of the developing world given the collapse of the Soviet Union during the 1990s and the increasing presence of impoverished and equally historically dominated Eastern European countries, few of which were members of the group. In addition, internal conflicts among members made it difficult for the group to maintain a single unified voice. Conflicts between members such as India and Pakistan, Ghana and Togo, and Indonesia and Malaysia, as well as the internal conflicts of Yugoslavia that eventually led to its demise, all limited the ability of the movement to effectively challenge the richer nations over problems such as debt, poverty, imbalanced trade relations, and political representation that the nonaligned movement members face collectively.

During the 1980s, the *South* became a common way of referring to the developing world, popularized by the independently commissioned international development report titled *North–South: A Program for Survival,* also called the *Brandt Report.* Commissioned by Robert McNamara, then president of the World Bank, and chaired by Willy Brandt, a former West German chancellor, the report sought to revitalize negotiations between the poor and rich countries by formulating a basic proposal for balancing the inequalities in wealth, trade, finance, and money between the developing world and the developed world. The commission broadly defined the developing world as countries (with a few exceptions) that occupied the Southern Hemisphere. By geographically dividing the map into North and South, the report highlighted the stark differentials in the quality of life experienced between human populations located in the powerful industrialized states of the temperate zones of the Northern Hemisphere and those in the impoverished states of the tropical and semitropical zones of the Southern Hemisphere. The report argued that the underlying reason for these differentials lay in the North's domination of the international economic system, its rules and regulations, and its international institutions of trade money and

finance. Arguing that relative prosperity in the South could promote prosperity in the North and that economic trouble in the South could wreak havoc in the North as well, the Brandt commission sought to influence public opinion and ultimately government efforts to avert widespread economic crisis in the poorest countries. The Brandt Report called for changes to be made to the global economy to make it more democratic, fair, and equitable. It called for a restructuring of the World Bank and the International Monetary Fund, which were viewed as unrepresentative of many of the countries that they served, and for the rich industrial countries of the North to share their means and power with the countries of the South. Although *the South* became popular in the development circles as a way of describing the developing world, it neither acquired the radicalism associated with the term *Third World* nor acquired the sense of obligation associated with the idea of a global community that it sought to cultivate. It is perhaps for this reason that despite the widespread and public acceptance of the Brandt Report, very few (if any) of its recommendations for reducing the growing economic disparity between the North and South were ever adopted.

The challenge of finding a way in which to both define and articulate the collective and mutual interests of the developing world has grown significantly since the collapse of the Soviet Union during the 1990s. During the cold war, the ideological divisions between the First World and the Second World, or between the East and the West, made it easy to construct an imagined community of the South or Third World states bound together by common experiences of peripheralization and poverty within the global economy. After the cold war, however, the experiences of colonialism, peripherality, and poverty no longer could shape the strategic interests of either former colonies or postsocialist countries because the ensuing global spread of neoliberalism significantly reorganized national economies such that the meanings of old development categories no longer could remain stable. For example, by the 1990s countries such as Singapore, Chile, Botswana, and India were beginning to experience levels of economic growth and investment that made their strategic interests potentially different from those of crisis-ridden states. These countries, euphemistically described as *emerging markets,* were viewed to have opened up their markets and to have "emerged" onto the global scene. During the 1990s, the number of countries classified as emerging market economies expanded to include many of the countries of Eastern Europe as they began to change from a socialist system of production to a free market–based capitalist system. Also described as transition economies, post-Soviet economies are defined more in relation to their economic performance, and the opportunities they present for foreign investment, than on the basis of the quality of life enjoyed by their citizens. So despite the fact that countries such as Jamaica, the Dominican Republic, Romania, and Albania all rank as lower-middle-income countries with 2% or less of their populations living on less than $1 per day, they are not considered in the same way. Jamaica and the Dominican Republic remain part of the Third World, whereas Romania and Albania are considered to be transition economies. The constraints that these newly constructed categories impose on the ability of the world's populations to articulate common concerns over poverty and marginalization are significant because they maintain, rather than challenge, the idea that the path to progress, prosperity, and development is located largely in neoliberal market-based strategies to increase economic growth.

With the global spread of neoliberalism, it has become even more difficult to define the developing world in territorial terms. Many argue that the deregulation and liberalization of markets, combined with a diminished role for states, have increased disparities between the rich and the poor to such an extent that it is now possible to define a *Fourth World* composed of nations living within or across territorial state boundaries whose interests are not represented by them. Even though these groups may be located in the developed world, the levels of poverty, exploitation, and violence that they experience have a limiting effect on their human rights and capabilities that is much like that experienced historically by Third World states within the global economy.

It is clear that the concept of the developing world is riddled with inconsistencies and remains difficult to define, but the need to create an imagined community of people concerned with the marginalized within the global economy remains an important first step toward challenging domination itself. Since the 1990s, there has been an increasing movement among various groups in civil society toward articulating alternative ways of solving the problems of exclusion and social inequality without resorting to familiar categories such as the Third World, the developing world, and the South. Best seen in the networks of people's movements that meet each year at the World Social Forum, there is an increasing commitment to developing solidarities across borders in ways that are focused less on the histories that define the developing world territorially and more on the processes through which human rights and environmental responsibilities have been exploited and marginalized. By seeking to develop a common framework for understanding and mobilizing for social and economic justice, more accurate ways of articulating the common goals, aspirations, and obligations of the earth's populations will emerge.

—*Beverley Mullings*

See also Colonialism; Dependency Theory; Development Theory; Economic Geography; Globalization; World Systems Theory

Suggested Reading

Payne, R., & Nassar, J. (2005). *Politics and culture in the developing world: The impact of globalization.* New York: Longman.

Weatherby, J., Evans, E., Gooden, R., Long, D., Reed, I., & Carter, O. (2004). *The other world: Issues and politics of the developing world.* New York: Longman.

DEVELOPMENT THEORY

Most contemporary development theories seek to define the social, economic, or political conditions under which humans, both individually and collectively, are able to realize their potential, build self-confidence, and live with dignity and fulfillment. Such a simple definition, however, obscures the range and conflicting nature of practices that theorists have identified as significant in this pursuit.

The earliest theories of development emerged during the 1950s after the end of World War II. Before this time, there was limited concern for the levels of inequalities in existence among the world's human populations, most of whom were subjects of colonial rule. With decolonization and the emergence of a host of newly independent nation-states that appeared shortly after World War II, inequality, poverty, and standards of living materialized as issues of concern to political leaders and scholars in both the colonizing and postcolonial worlds. Arturo Escobar argued that the emergence of development theories at the end of World War II, however, must also be understood in the context of the cold war, particularly the competition between the United States and the Soviet Union for ideological and political supremacy. In this context, Escobar argued that development theory represents more than a humanist desire to spread peace and abundance throughout the world; it is also a social imaginary that historically has been used to justify a host of interventions that have at times been of greater benefit to the First World than to the Third World.

At the heart of all development debates continue to lie two fundamental disagreements. The first disagreement is over the extent to which securing the social conditions for dignity and self-determination should take precedence over creating the conditions for increasing productivity and economic growth. The second disagreement is over the value and meaning of collective and individual dignity and fulfillment itself.

Early development theories were primarily extensions of conventional economic theory that equated development with economic growth and industrialization. Embedded in both Western notions of progress and colonial constructions of race, early development theories viewed the countries of Africa, Asia, and Latin America as undeveloped and in need of significant economic change if they were to emulate the standards of living in existence in Europe and North America. Implicit in early theories was the assumption that development was a natural process that had reached its zenith in the industrialized countries of Europe and North America. As a naturalized process, the idea of development was closely linked to the associations made between science and progress in Anglo-American societies from the 18th century onward. Within this framework, poor countries could find dignity and fulfillment only if they emulated the economic experiences of the developed world and instituted policies to catch up with the West. Modernization theorist Walter Rostow, for example, argued that all countries passed through the same historical stages of economic development and that countries with less material wealth were merely at an earlier stage in this linear historical process. For Rostow, development required a five-step application of policies focused on investment, savings, and the encouragement of an entrepreneurial class. More sophisticated formulations, such as Nobel laureate Arthur Lewis's two-sector model, also constructed the problem of development as the consequence of the lack of accumulation of productive capital and a low savings rate in poor countries. In Lewis's model, the modern sector is typified as progressive, oriented toward industry, and more likely to be the vanguard of wealth creation. In contrast, the traditional sector was characterized by a large supply of unemployed and unproductive labor that was viewed as unlikely to contribute significantly to the development process given its inability to save. By establishing the conditions for capitalists within the modern sector to make profits, Lewis argued that there would be greater levels of reinvestment and available capital. Most modernization models of development made assumptions about the meaning of development that ultimately proved to be deeply problematic. Both Lewis and Rostow, for example, relied on binary differentiations between the modern world and the traditional world that not only were ethnocentric (the modern was seen as Western-like, integral, and progressive, whereas the traditional was seen as non-Western, backward, and residual) but also denied the histories that produced the patterns of uneven development between the First World and Third World and within the Third World itself.

The problematic assumptions of modernization theory, with its emphasis on dual sectors, economic growth, and a mechanistic set of stages, were challenged during the early 1970s by scholars who emphasized the fact that there were distinct structural constraints facing the decolonizing world that modernization theories did not take into account. Raul Prebisch, for example, argued that unlike the First World, Third World countries were constrained by the fact that they were tied, through trade, to an already industrialized First World. This made Third World pathways to industrialization different from those experienced earlier by Europe. Prebisch further observed that the trade relationship between the First World and Third World created a center–periphery relationship that made it impossible for Third World countries to function in the world economy in ways other than as dependent producers of raw materials for First World manufacturing industries. For Prebisch, without significant trade protection, the Third World would not break out of either its dependence on the First World or its peripheral resource role. The ideas of these scholars, who became known as structuralists, paved the way for a much more radical critique and theorization of development.

The failure to recognize the part played by the First World in the patterns of development found in the Third World became the basis of a new set of development theories from scholars who drew on elements of Marxist political economy to critically examine the underlying structures and relations that created inequalities in income, infrastructure, and quality of life between the First World and Third World. These so-called dependency theorists, and later world systems theorists, posed a significant challenge to the way in which development was imagined and practiced. Contrary to the modernization theorists, these scholars, many of whom came from Latin America and other parts of the global South, viewed underdevelopment as a situation that was actively produced by the capitalist system itself.

The early writings of scholars such as Celso Furtado, Samir Amin, and later Immanuel Wallerstein characterized the Third World as the product of the history of Euro-American expansion and the incorporation of colonized countries into the world economy. These scholars argued that development theories could not be viewed as a set of endogenous practices. Development theory needed to take into account the past and present forms of political and economic domination between the advanced capitalist countries and the poor countries to determine the solution to the problem of poverty. A variety of theories based on this argument were advanced by dependency theorists, but most supported the view that the inequalities among states and the unequal nature of interactions among them under capitalism created binaries of power variously described as dominant–dependent, center–periphery, or metropolitan–satellite that were self-reinforcing and detrimental to the less powerful states.

André Gunder Frank, in his early writings, argued that the root of the problem of underdevelopment lay in the way in which the wealthy advanced industrialized countries expropriated the surpluses created by poor peripheral ones. He argued that through trade, surplus was systematically extracted from peripheral areas and appropriated by more affluent centers. At the international level, this was manifest in the way in which economic surpluses generated in Latin America tended to benefit the affluent capitalist countries where foreign corporations were based rather than these poor countries themselves. Frank argued that, in fact, poor countries were likely to experience their greatest levels of growth when the level of their incorporation into the capitalist system was at its lowest. Implicit in Frank's initial formulation was the belief that it was only by delinking from the capitalist system that countries could hope to develop. Focusing on

the structures underlying the production process in Africa, Amin similarly theorized that so long as the Third World countries maintained their asymmetrical relationship to the First World, they would never become autocentric and self-determining. True development required as a prerequisite a weakening of the links between the First World and Third World through socialist transformation and the fostering of greater regional ties.

While dependency theorists challenged the ahistoricism of modernization theories and their formulation of catching up development, many of their ideas were also criticized. One major criticism, for example, lay in the way in which dependency theorists defined capitalism as a mode of exchange rather than as a mode of production. For dependency theorists, it was through the process of unequal exchange that centers were able to extract the surplus generated by peripheries. However, critics argued that without a definition of capitalism based on the relationship between labor and capital in the production of surplus, capitalism became a useless concept. As Ernesto Laclau argued, if the mode of exchange was the most important characteristic of capitalism, it would be possible to claim that capitalist systems existed as early as the time of the ancient Greeks. This critique raised the important issue of the scale at which the question of development was framed and the explanatory power given to the external factors (unequal exchange) relative to internal factors (class relations) in explaining the process of development. Other critics argued that the dependency argument was circular in logic and hence was fatally flawed. For example, Deepak Lal claimed that in arguing that countries were dependent because they were poor and were poor because they were dependent, theorists were locked into a circular argument that by definition could not be resolved. Most damning for the dependency theories, however, was the growing evidence that growth and development were possible among peripheral countries with ties to the industrialized countries in ways that challenged the economic superiority of the core countries. The emergence of newly industrializing countries (NICs) in East Asia significantly challenged the necessity of delinking from the capitalist system as advocated by many *dependistas*.

Derived from a much longer historical examination of the nature of the global economy, the work of the world systems theorists during the mid-1970s shared many of the ideas found in dependency theory such as the role of unequal trade in the exploitation of the periphery by the core and the framing of the question of development at the international scale. Unlike the classical dependency position, however, world systems theorists such as Wallerstein, and later Frank and Amin themselves, argued that it was possible for development to occur in the periphery and for countries to become semiperipheral and part of the core. They proposed that countries such as Brazil and South Korea functioned within the global economy as semiperipheral buffers, importing the high-tech products from the core and exporting semimanufactured goods to the periphery. Although these theories recognized the possibility of development within capitalism, they still viewed mechanisms underlying the process as deeply exploitative of the periphery.

By the mid-1980s, development theory was argued to be at an impasse. The emergence of the NICs, the collapse of Soviet state socialism, and the passage of a decade of economic crisis in the Third World in the wake of the 1974 and 1979 oil price hikes created a theoretical vacuum. The period corresponded with a general crisis within social theory itself, one described by many as a crisis of representation. For many theoreticians in the social sciences, the world of the late 20th century was changing in such a way that traditional ways of representing reality were becoming dissatisfactory—and, for some, impossible. Within development theory, this crisis took the form of a critique of the Marxist metatheory that shaped much of the neo-Marxist development theories of the 1970s. David Booth, for example, argued during the mid-1980s that neo-Marxist and Marxist development theories were at an impasse because they were too generalized, economistic, and excessive in their commitment to proving that the structures and processes found in poor countries were the necessary outcomes of their participation in the capitalist system. He argued that without greater attention to the diversity and complexity of the real world, these theories could contribute little to the practical issues facing the Third World. The retreat of many Third World states from socialist-inspired development strategies, combined with the theoretical void created within development studies, was filled quickly by the political and economic development ideology of neoliberalism that emerged during the 1980s as the economic strategy employed by many Western industrialized countries to get out of the monetary crises that had followed the oil-induced world recession. Initially formulated as an economic strategy to open up markets for the circulation of capital, neoliberalism quickly became associated with a set of political and development ideologies aimed at reducing the role of the state and opening up poor countries to global flows of capital. For neoliberals, free markets were the key to maximizing human welfare because they were the most efficient way of distributing capital and informational resources. If markets were allowed to freely determine the distribution of resources, individuals ultimately would be able to maximize their own economic and political and social needs and wants. Therefore, free market capitalism was a necessary first step toward political freedom and required a reduction in the role played by states. States no longer were encouraged to regulate or intervene in markets. Their role now was primarily to enable the free operation of markets by creating the legal systems needed to facilitate individual consumption and the movement of capital.

The global spread of the neoliberal economic model also had a significant effect on the discourse and idea of development. By the mid-1980s, the idea of development and the need to consider the Third World differently were eclipsed by the discourse of globalization, which was presented as an inevitable stage in the history of the world economy. Neoliberals argued that the need to create theories and strategies that specifically recognized the differences in the structures and historical experiences for Third World countries was fundamentally wrong and distorting. Lal, for example, argued that the distortions to markets produced when Third World countries sought to become self-reliant by protecting their markets from global competition and encouraging industrial strategies of import substitution were the cause of, rather than the cure for, much of the poverty in the global South. Views such as Lal's became highly influential within the Bretton Woods institutions of the International Monetary Fund and the World Bank during the 1980s and formed the basis for the structural adjustment policies (SAPs) that continue to define the conditions under which loans are made to Third World economies and, more recently, transition economies of the former Eastern Bloc. Under SAPs, recipients of loans are required to carry out a standard set of macroeconomic reforms that usually include the devaluation of national currencies, the raising of interest rates, the reduction of budget deficits, and the removal of price distortions such as subsidies and quotas. These policies are viewed by international lending agencies as crucial to creating the conditions for economic stability.

Longer-term strategies focus on the deregulation and liberalization of national economies and include policies such as the privatization of industry and resources that may once have been nationally owned. Other policies have revolved around creating the conditions to attract international investment such as the establishment of export processing zones.

The dominance of neoliberalism in development theory and practice has been heavily criticized over the past 20 years. Many have argued that the importance placed on creating suitable conditions for international capital not only failed to increase levels of productivity and growth in many poor countries but also heightened spatial, social, and economic patterns of unequal development with devastating consequences for historically marginalized groups. Some of the most vigorous critiques of neoliberal development theory have come from feminist and environmental scholars. They have shown how the failure of neoliberal theorists to adequately examine how women's labor or environment resources are negatively affected by free and unfettered markets has contributed to the increasing levels of poverty and violence experienced by many women and girls as well as the heightened levels of environmental destruction. For example, Diane Elson demonstrated how the neoliberal model places an additional burden on women by failing to take into account the ways in which the gender division of labor devalues the paid and unpaid work of women and girls. Drawing attention to the burden that the withdrawal of the state from collective welfare provision has placed on women charged with primary responsibility for social welfare in the home, Elson not only called for the gendered effects of seemingly genderless macro-level policies to be made clear but also, and importantly, called for development theorizing to be led by the consistent consideration of these effects. Both of these groups have also significantly challenged the nature of development theorizing itself by not seeking to create an all-encompassing macro-level theory of transformation but rather providing contextualized understandings of the relationships among the macro, the meso, and the micro in processes of social and economic change. In paying greater attention to multiple scales, there has emerged a greater recognition of the myriad ways in which needs, wants, and capabilities are expressed and valued across regions and among groups differentiated by gender, ethnicity, income, culture, and religion.

Since the 1990s, the search for a better understanding of the mechanisms, at multiple scales, that influence the behavior of participants in a society, and ultimately the nature of development, has generated a growing interest in the role of institutions (formal and informal rules, enforcement mechanisms, and organizations). Scholars from both Marxist and neoclassical traditions have begun to develop sociopolitical and economic frameworks to explain long-run institutional change. In these frameworks, described as the new institutional economics, scholars have drawn on neoclassical economic theories to identify how both the existence and absence of particular institutions influence both individual and collective behavior and ultimately the outcome of a particular development intervention. Scholars from the regulation school, alternatively, have used Marxist theory to examine how specific matrices of social, economic, and political institutions interact to produce long periods of economic stability despite the continued appropriation of the surplus value created by workers by capitalists. Although there appears to be much similarity in the ways in which these two approaches use multiple disciplinary perspectives to examine social, economic, and political institutions at multiple scales, significant differences exist in the questions they seek to answer. Whereas new institutionalists have focused primarily on identifying the institutions required for efficient markets to be established, regulationists have sought to avoid such policy prescriptives and instead concentrated on the reasons why particular institutions persist even when they generate contradictions. Although few of the early institutional approaches explicitly addressed issues of development, there has been a steady integration of the ideas of scholars associated with the new institutional economics (e.g., Douglass North) throughout the 1990s into the development policies of international agencies such as the World Bank. This has resulted in a shifting in the emphasis within neoliberal development theory from the minimal state toward a more active role for the state in creating the institutions necessary for the efficient operation of markets.

The 1990s also gave rise to a more sustained critique of both the neoliberal development model and the association between development and continuous consumption-led growth. Recognition of widening gaps in economic and political power between and within countries has led many to argue for a reorientation of the focus of development away from the enrichment of the economies within which people live and toward the enrichment of human life. For example, Amartya Sen argued for a people-centered vision development where individuals and groups define and direct for themselves the institutional mechanisms necessary for human capabilities to be maximized. From the debates regarding the ways in which diverse local communities might begin to redefine development, there has also emerged a growing focus on the need for global institutions dedicated to the redistribution of resources. Development theory today no longer can be considered as a single narrative about what it means to live an abundant human life; rather, it should be considered as a set of contested forms of knowledge. Yet as there emerge multiple understandings of the complexity of the basic elements required for each human to live a fulfilled life, there is a growing recognition of the need for global institutions to defend each individual's right to define and access them.

—Beverley Mullings

See also Colonialism; Debt and Debt Crisis; Demographic Transition; Dependency Theory; Developing World; Economic Geography; Export Processing Zones; Flexible Production; Fordism; Geopolitics; Globalization; Gross Domestic Product; Growth Pole; Hunger and Famine, Geography of; Import Substitution Industrialization; Industrial Revolution; Informal Economy; Infrastructure; Innovation, Geography of; Labor, Geography of; Labor Theory of Value; Marxism, Geography and; Modernity; Modernization Theory; New International Division of Labor; Neoliberalism; Newly Industrializing Countries; Peasants; Political Geography; Population, Geography of; Postcolonialism; Postindustrial Society; Poverty; Product Cycle; Rural Development; Rustbelt; Spatial Inequality; Squatter Settlement; Structural Adjustment; Sustainable Development; Terms of Trade; Transnational Corporations; Underdevelopment; Uneven Development; Urban and Regional Planning; World Economy; World Systems Theory

Suggested Reading

Amin, S. (1976). *Imperialism and unequal development.* Hassocks, UK: Harvester.

Corbridge, S. (1990). Post-Marxism and development studies: Beyond the impasse. *World Development, 18,* 623–640.

Elson, D. (1994). Micro, meso, macro: Gender and economic analysis in the context of policy reform. In I. Bakker (Ed.), *The strategic silence: Gender and economic policy* (pp. 33–45). London: Zed Books.

Elson, D. (2003). Gender justice, human rights, and neo-liberal economic policies. In M. Molyneux & S. Razavi (Eds.), *Gender justice, development, and rights* (pp. 78–114). Oxford, UK: Oxford University Press.

Escobar, A. (1995). *Encountering development: The making and unmaking of the Third World.* Princeton, NJ: Princeton University Press.

Frank, A. (1996). The development of underdevelopment. In C. Wilber & K. Jameson (Eds.), *The political economy of development and underdevelopment* (pp. 105–115). New York: McGraw-Hill. (Original work published 1967)

Laclau, E. (1971, May–June). Feudalism and capitalism in Latin America. *New Left Review,* pp. 19–55.

Lal, D. (1983). *The poverty of development economics.* London: Institute of Foreign Affairs.

Lewis, W. (1954). Economic development with unlimited supplies of labour. *Manchester School of Economics and Social Studies, 22*(2), 139–191.

Lipietz, A. (1987). *Mirages and miracles: The crisis of global Fordism.* London: Verso.

North, D. (1989). Institutions and economic growth: An historical introduction. *World Development, 17,* 1319–1332.

Prebisch, P. (1962). The economic development of Latin America and its principal problems. *Economic Review of Latin America, 7*(1), 1–22.

Rostow, W. (1960). *The stages of economic growth: A non-communist manifesto.* Cambridge, UK: Cambridge University Press.

Sen, A. (2000). *Development as freedom.* New York: Anchor Books.

Watts, M. (1995). A new deal in emotions: Theory and practice and the crisis of development. In J. Crush (Ed.), *Power of development* (pp. 44–62). New York: Routledge.

DIASPORA

Originating from ancient Greek and meaning dispersion, the term *diaspora* traditionally was associated with the Jews to describe their traumatic uprooting from ancient Israel, their forced exile throughout the world, their feelings of alienation in the host countries, their collective memory of their homeland, and their desire to return home. The Greeks and Armenians constituted two other examples of archetypal diasporas. These diasporic communities are generally characterized by a high level of ethnic organization in their host countries that usually includes cultural associations, political parties, schools, and other institutions with the goal of preserving a group identity.

However, with the emergence of globalization, the term *diaspora* has been used more widely. At the same time, there was a reconceptualization of the term to encompass phenomena of increased international population mobility unleashed by globalization such as augmented emigration to the developed countries, the telecommunication and transportation revolution, and the development of a cosmopolitan global culture. In this context, the 1980s and 1990s witnessed increased interest in the phenomenon of diaspora, and minorities whose experiences met the classical diaspora paradigm only in part began to be called diasporas, thereby blurring the lines among ethnic minorities, refugee flows, migrations, and diaspora. Moreover, the media's use of the term *diaspora* has played a significant role in ascribing new meaning to the notion of diaspora, including a so-called rock-and-roll diaspora and a soccer player diaspora.

Most contemporary scholars conceive diaspora broadly, arguing that even classical diasporas, such as that of the Jews, are socially constructed. These authors stress that diasporic identities are not innate; rather, they arise from the complex relationship among an ethnic minority, its host state, and its homeland. This understanding of diaspora opens the door to understanding the contemporary rise of complex multiple ethnic and national identities that various individuals or groups of people display.

Central to an understanding of diaspora is its tenuous position in between the host and home countries. Historically, and stereotypically, the nation-state viewed the diaspora as a threat. Diasporas were perceived as a menace to a host state's organic unity, and they were seen at best as just tolerated minorities who often were abused and forced to assimilate. Also, home states often have been ambivalent toward their diasporas. They perceived their diasporas as not authentic, as impure, and as having a hybrid identity.

However, today the discourse about diasporas has been redefined. Diasporas are now perceived in a much more favorable light, and their influence in shaping both home and host state politics has increased as nation-states realized that they could benefit from diasporas' services. This is because diasporas are, in some instances, better positioned than their host or home countries as transnational actors in a global world. Diasporas are transnational phenomena that escape the integrationist tendency of the nation-states and that continue to manifest dual or multiple national identities and allegiances.

—*Gabriel Popescu*

See also Globalization; Migration; Other/Otherness; Population, Geography of; Social Geography

Suggested Reading

Cohen, R. (1997). *Global diasporas: An introduction.* London: UCL Press.

Shain, Y. (1999). *Marketing the American creed abroad: Diasporas in the U.S. and their homelands.* New York: Cambridge University Press.

Tololyan, K. (1996). Rethinking diaspora(s): Stateless power in the transnational moment. *Diaspora, 5*(1), 3–36.

DIFFUSION

Geographic diffusion is the dispersal of information or objects throughout a geographic region. Classic studies on diffusion originated during the early 20th century and focused on topics such as the spread of new, or "modern," agricultural techniques. This emphasis suited the condition of the United States and other Western countries, which were transitioning from an agrarian society to an industrial society. Over time, research on diffusion began to explore other social attributes, particularly those features that were prevalent in urban environments. The ongoing process of globalization has added new complexities to this process.

In general, there are two types of geographic diffusion. The first type of diffusion is called *contagious diffusion.* As the name indicates, this conception of diffusion is borrowed from the science of epidemiology. In this type of diffusion, a characteristic is transmitted from one person to his or her nearest neighbor. Accordingly, contagious diffusion produces a wave-like pattern that gradually spreads outward from the site of origin. This process has been noted in the spread of architectural characteristics in the Midwest. The second type of diffusion is *hierarchical diffusion,* which involves the spread of an attribute from one city to another city. The assumption underlying hierarchical diffusion is that large urban centers function as sources of social and technological innovation. These cities retain a primary position within the hierarchy of human settlements. Accordingly, hierarchical diffusion first involves the transmission of information and objects of major cities (whose inhabitants often have similar attributes and interests) before spreading (or trickling down) to smaller and smaller human settlements. Historically, such a process was seen in the advent of industrialization and more recently in production and dissemination of music styles. As such, hierarchical diffusion produces a different geographic pattern than does contagious diffusion. Hierarchical diffusion "leapfrogs" from one urban location to another, thereby leaving substantial gaps. The intervening spaces remain unaffected until the attribute becomes pervasive throughout a given society. During recent years, a third type of diffusion has been articulated, one that is a variant, or inversion, of hierarchical diffusion. This latter type has been referred to as *reverse hierarchical diffusion.* As the name indicates, this type of diffusion originates in rural locations and spreads to larger urban centers. The most prominent example of this phenomenon in recent times is the growth and diffusion of Wal-Mart. In contrast to most other retailers (and the principles articulated in neoclassical economics), Wal-Mart began by establishing stores in rural locations that had been ignored (and underserviced) by other companies. Over time, Wal-Mart eventually began to set up operations in more densely populated locations.

From an analytical perspective, geographers have taken different approaches toward an explanation of diffusion. As in other areas of geographic investigation, issues of scale are prominent. Whereas some researchers emphasize the role of individual actors, other researchers emphasize the role of global economic systems or cultural orthodoxies. Thus, a critical theoretical distinction has emerged between those researchers who prioritize micro-scale phenomena and other researchers who accentuate macro-scale phenomena.

In micro-scale approaches, researchers often focus on the decision-making process of individuals. In such theoretical formats, individuals often are classified into one of three categories. Early adopters are those individuals who were willing to try new technologies. This amounts to a small segment of a population because the adoption of new innovations usually involves a certain degree of financial or personal risk. A second set of individuals also adopts innovative technologies, but only after these innovations have been adequately tested and their utility has been verified. By adopting such technologies at a later date, the inherent risk of innovation is reduced. At this point in time, the innovation becomes an attribute of mainstream society. The third category of individuals is classified as resisters. These are individuals who continue to engage in traditional practices and are skeptical of new innovations. In most cases, these individuals are considered to be a small percentage of a given population and one that might never assimilate into the dominant society. In geography, the most prominent examples of a micro-scale approach are the early writings of Torsten

Hagerstrand. Hagerstrand used a Monte Carlo approach, which assumed that individuals in closer proximity to an innovation were more likely to adopt that innovation. The complexities added by early adopters and resisters were accounted for by probabilities.

Although this approach clearly provides insight, many researchers have criticized its basic assumptions. Most notably, critics contend that the majority of this research has unduly focused on the economic utility and efficiency of innovations. As such, this theoretical approach conforms to neoclassical perspectives, which narrowly portray individuals as economic entities. Accordingly, this approach tends to homogenize the interests of individuals by suggesting that one standard (e.g., efficiency/profitability) determines whether an innovation will be adopted. It does not acknowledge that individuals have multiple concerns and interests that may influence the perceived value of an innovation. Perhaps more problematic is that in portraying nonadopters as resisters, this theoretical stance often is antagonistic to traditional or non-Western cultures. Indeed, in contrast to progressive adopters of innovation, resisters sometimes are portrayed as irrational, backward, or ignorant. This theoretical position is particularly problematic when dealing with non-Western societies that have suffered from colonialism and neocolonialism.

In contrast to micro-scale approaches, other researchers have emphasized processes that operate at larger scales. In particular, some researchers highlight the role of capital and transnational corporations. From this stance, the capacities of transnational corporations direct the process of geographic diffusion. In this vein, a classic of such phenomena is the so-called Green Revolution, which involved the diffusion of modern agricultural innovations (e.g., high-yielding seeds, fertilizers, pesticides, herbicides) from North America to Mexico, India, and Southeast Asia. The corporations involved in the creation of these products were central to their diffusion.

In addition, in the global context, cultural priorities vary considerably from one region to another. Secular priorities frequently conflict with religious worldviews. These influences have deep historical roots and are embedded in languages and practices that have a broad yet intricate reach within different societies. Recent writings on postmodernity have attempted to express the extent of this diversity. Conceptions of such diversity implicitly critique the homogenizing assumptions of neoclassical economics.

In reality, diffusion is most likely a combination of all these factors. To some extent, these theoretical positions are inextricable from one another. As a result, any effort to understand diffusion must account for the different networks (or sets of relations) that operate at different geographic scales.

—*Christa Stutz*

See also Location Theory

Suggested Reading

Hägerstrand, T. (1965). A Monte Carlo approach to diffusion. *European Journal of Sociology, 6*, 43–67.

Haggett, P. (1965). *Locational analysis in human geography.* London: Edward Arnold.

Rodger, E. (1995). *Diffusion of innovations.* New York: Free Press.

DIGITAL EARTH

The term *Digital Earth* was coined by then–U.S. Senator Al Gore in his book *Earth in the Balance,* published in 1992, to describe a future technology that would allow anyone to access digital information about the state of the earth through a single portal. The concept was fleshed out in a speech written for the opening of the California Science Center in early 1998, when Gore was vice president. By then, the Internet and Web had become spectacularly popular, and Gore sketched a vision of a future in which a child would be able to don a head-mounted device and enter a virtual environment that would offer a "magic carpet ride" over the earth's surface, zooming to sufficient resolution to see trees, buildings, and cars, and would be able to visualize past landscapes and predicted futures, all based on access to data distributed over the Internet. The Clinton administration assigned responsibility for coordinating the development of Digital Earth to the National Aeronautics and Space Administration (NASA), and several activities were initiated through collaboration among the government, universities, and the private sector (www.digitalearth.gov). International interest in the concept was strong, and a series of international symposia on Digital Earth have been held, beginning in Beijing, China, in 1999.

Political interest in Digital Earth waned with the outcome of the U.S. presidential election of 2000, but activities continue aimed at a similar vision, often under other names such as Virtual Earth and Digital Planet. The technical ability to generate global views, to zoom from resolutions of tens of kilometers to meters, and to simulate magic carpet rides is now available from several sources. Environmental Systems Research Institute (ESRI), the market leader in geographic information systems (GIS) software, now offers ArcGlobe as part of its ArcGIS package together with data sets at 30-m resolution. A Web-based visualization, developed by Keyhole, Inc. (purchased by Google in 2004), is available at www.earthviewer.com. NASA offers World Wind, its own public domain analog of Earthviewer (learn.arc.nasa.gov/worldwind/).

The vision of Digital Earth proposes that a complete digital replica of the planet—a mirror world—can be created. Such a replica would be of immense value in science because it would enable experiments to investigate the impacts of proposed human activities (e.g., the large-scale burning of hydrocarbons, the destruction of forests). This would require integration of data with models of process, something that is not yet part of any of the Digital Earth prototypes. Much research is needed on the characterization of processes before the full dream of Digital Earth can be realized. Meanwhile, the technology appears to be limited to virtual exploration of the planet's current and past physical appearance. Inevitably, there will be an emphasis on those aspects of the earth that are characterized by widely available data sets and that can be easily rendered in visual form. Thus, Digital Earth seems bound to privilege relatively static physical aspects of geography over dynamic social aspects.

—*Michael F. Goodchild*

See also GIS; Humanistic GIScience; Spaces of Representation

Suggested Reading

Gore, A. (1992). *Earth in the balance: Ecology and the human spirit.* Boston: Houghton Mifflin.

DISABILITY, GEOGRAPHY OF

The term *disability* is contested, used in many different ways in different contexts, and increasingly narrowly defined in legal terms with recent legislation. In general, disability is the study of people with mind and body differences, commonly referred to as physical and/or mental impairments, and the interactions between society and the capacity of disabled people to function as independent individuals.

Geography of disability explores disabled peoples' experiences of space and place, investigating the relationships among the geographic environment, the nature of individuals' impairments, and the role of society as a mechanism for including or marginalizing people with disabilities. Geography of disability refers to the landscape (in its widest sense) of disabled experience, from the urban to the rural, from the micro scale of household mobility to the accessibility of transportation networks across cities and countries. Research addresses not only the visible components of disability, such as wheelchair ramps (or lack thereof) in the built environment, but also a range of sociospatial processes that surround issues of disablement; a range of social, political, and cultural factors; and the complex interactions among power, space, and materiality.

During the 1990s, geographers began to examine their role and interaction with people with disabilities, paralleling changes in other social science disciplines, including sociology, cultural studies, anthropology, urban geography, planning architecture, and political science, leading to the formation of a distinctive subdiscipline—geography of disability.

CONCEPTUALIZING DISABILITY

There has been a historical continuum of meaning of disability—from the *moral* (disability is a sin or shameful) to the *medical* (disability is a defect or sickness to be cured by medical research), *rehabilitation* (disability is a deficiency to be fixed by rehabilitation science), and the *social* (disability is caused by society's barriers to including a disabled person as a fully integrated citizen). The most noticeable direction in contemporary geographic studies of disability has been the influence of the social model of disability, which stresses that disabled people are marginalized by social attitudes and normative ideas of the naturalness of being able-bodied that are written into the landscape to produce countless physical and social barriers to their full participation in society. The barriers were socially constructed rather than an inevitable result of people's impairments.

DEFINITIONS

The United Nations uses the following definitions. An *impairment* is any loss or abnormality of psychological or anatomical structure or function. A *disability* is any restriction or lack of ability (resulting from an impairment) to perform an activity in the manner or within the range considered normal for a human. A *handicap* is a disadvantage for a given individual resulting from an impairment or a disability that limits or prevents the fulfillment of a role that is normal—depending on age, sex, social, and cultural factors—for that individual. Therefore, a handicap is a function of the relationship between disabled persons and their environment. It occurs when they encounter cultural, physical, or social barriers that prevent their access to the various systems of society that are available to other citizens. Thus, a handicap is the loss or limitation of opportunities to take part in the life of the community on an equal level with others.

THE MEDICAL–SOCIAL CONTINUUM

The medical model has evolved most notably over the past two centuries, when the "expert" knowledge of medicine became embedded and realized through certain specific institutional practices such as hospitals, special schools, and asylums. The social model notes that a different understanding of "normality" exists when it is placed in a different context, that is, not in disabled people but rather in the society that fails to meet their needs.

For many people, the medical model of disability does not fully represent the role of society in disabling people with impairments or their personal experiences. The social model moves the focus of disability away from the individual to the environment of a person and structural factors, addressing the societal and geographic factors that led to the "disabling" of an impaired individual. It proposes a radical split between thinking about impairment and thinking about disability; people are considered disabled by society and the environments it produces rather than by their impairments. It situates disability in wider, more general sociocultural practices and structures (e.g., the media, planning, politics, education). From the position adopted by the social model, it is apparent that disabled people are excluded and marginalized from mainstream society through practices of exploitation, marginalization, powerlessness, cultural imperialism, and violence. However, the social model detracts from the experience of being disabled because much of this experience *is* derived from impairment. The social model offers a specific or particular explanation of the social oppression of disabled people; however, it is unable to fully explain it. The "reality" is perhaps a continuum between the two; that is, being "able" or "disabled" is not a fixed state but rather one set within the context of society, medicine, culture, politics, and economics. It is a complex interaction, not one determined solely by an embodied experience or by society. There is also a continuum of experience between the medical conception and the social construction of disability. In other words, a disabled person's experience is not derived purely from his or her impairment or from society alone. The individual and social models are logical duals that are interdependent; no one disabled person is oppressed without impairment or is impaired without oppression. For example, even if blind people were entirely accepted by and into society, they still would be unable to see or read nonverbal cues.

GEOGRAPHIC RESEARCH ON DISABLEMENT

Prior to around 1990, there was relatively little engagement with disability issues and geographers. However, research at that time focused on the following themes:

1. The ecological analysis and mapping of disability, mainly psychiatric geographies (attempts were made at identifying the ecological correlates of mental disorder to shed light on disease–environment relationships)

2. The location of mental health facilities and community reactions to such sitings and their socioeconomic effects

3. A historical geography of mental health asylums

4. The impact of healthcare reforms and the subsequent availability and quality of the services provided

5. The deinstitutionalization of disabled people with mental problems into the community

6. Investigations into the spatial learning of people with severe vision impairments in the physical environment (including route and environmental learning, spatial cognition, and research into raised-line [tactile] maps)

NEW GEOGRAPHIES OF DISABLEMENT

New approaches to disability, impairment, and chronic illness have emerged, expanding the scope, methodologies, and focus of geographic research to include more diversity and experiences of living in, as well as interacting in, urban and rural environments, including people's experiences of chronic illness such as HIV/AIDS, psychiatric illness, and multiple sclerosis. Disabled participants have created graphical representations of the inaccessibility of urban centers, highlighting the contested and political nature of cartographic representation and disability. Historical geographies of disability have expanded. The transport and mobility needs of disabled and elderly people have been explored generally and used to explore the spatial analysis of travel patterns. Planning and design issues have been investigated with respect to recent legislation in the United States and the United Kingdom. The role of disabled people as subjects, objects, or active participants engaged in critical research has been investigated. The number of disability studies within geography is continuing to grow. The debates around these contested meanings and the formulation of models to represent have very real consequences for disabled people. Political policymaking, urban planning, and educational provision affect all of us and the geographic space within which we interact.

—*Daniel Jacobson*

Suggested Reading

Butler, R., & Parr, H. (Eds.). (1999). *Mind and body spaces: Geographies of illness, impairment, and disability.* London: Routledge.

Gleeson, B. (1999). *Geographies of disability.* New York: Routledge.

Imrie, R. (1996). *Disability and the city.* London: Paul Chapman.

DISCOURSE

Increasingly being used as part of developments in cultural, social, and political geography from the late 1980s to the present, the term *discourse* relates to the ways in which meanings and identities exist and are created within modes of communication and language. Discourse analysis (the process of examining how we communicate meaning) suggests that modes of communication are not neutral but rather are embedded within the specific social and spatial relations they seek to describe. Within human geographic research, the words and practices used to describe places and people have come under closer scrutiny as theorists have shown not only that language helps to communicate our knowledge and research findings but also that this language—or discourse—itself shapes the kinds of results or experiences that can be described.

Discourse can refer to a range of communicative media—verbal interactions, written materials, media and artistic images, music, abstract symbolic icons, and so on—that come together to fit within a common understanding that makes these practices meaningful. For example, the image of a cross is meaningful as a religious symbol (e.g., signifying a church or a personal religious belief) or as a road safety guide (e.g., signifying a traffic intersection) only if it is recognized as being part of a religious belief system or a set of legal safety rules. If not part of a recognized set of meaningful discourses, this image could simply be viewed as two lines crossing. In this sense, the discursive field—or context—in which words, symbols, images, and gestures take place provides rules for meaningful communication.

The work of French theorist Michel Foucault has been particularly influential in geographic research examining the relationships among discourse, identity, and space. Foucault explored the links among knowledge, power, and discursive formations. Through his writing, Foucault attempted to illustrate that identities and truths that come to be viewed as common knowledge are culturally, historically, and geographically specific; that is, something that is considered "true" about certain people, social practices, or places in one context might not be considered so in another context—depending on what is seen as relatively typical, normal, or natural. The notion of normality is illustrated as being one in which subjective decisions are made about practices and identities that are considered culturally acceptable and thus part of mainstream everyday life. Such an approach toward understanding cultural practices challenges notions of an allegedly universal truth that transcends time and space. By examining the discussion and policing of subjects such as disease, gender, sexuality, and capital punishment, Foucault highlighted that the notion of practices (e.g., discipline, nationalism) or identities (e.g., criminal, hysteric, authority figure) could not exist outside of discourse; they had meaning, and actually came into existence, by being a part of the ways in which knowledge about them was produced, discursively constructed, and monitored through specific social practices. One of the most obviously useful ways of applying these ideas to human geography is illustrated through Foucault's examination of a reformulation of punishment through surveillance and the internalization of outside control

in the context of a prison and the ways in which this becomes a space in which symbolic discipline is as important as physical limitations.

Another particularly geographic example can be noted in relation to national identity and nationalist discourses. Benedict Anderson's popular concept of our ties to a particular national identity as being similar to belonging to an imagined community is based partly on the idea that as residents (or citizens) of a nation, specific symbols—parliament buildings, currency, the image of the crown, languages, and so on—can act as unifiers, linking a populace through their symbolic meaning. This meaning connotes not only a political identity but also one that is geographic and that relates to a specific place and territory. This means that although individuals who consider themselves to be citizens of a particular nation-state might never meet all of the other citizens or visit all of the other territories within the national jurisdiction, their sense of belonging *with* those people and places is fostered through social practices (e.g., the distribution of national newspapers, education in an official language, the use of a national currency that is recognized as having a specific value in a particular place) that help to foster the discursive creation of a national "community" that is linked to a specific locale.

The examples just discussed illustrate the spatial character of discourse and discursive constructions. Although some discursive studies have been critiqued as being less concerned with the material conditions that people and cultures negotiate, the works of Foucault and of other social theorists and geographers have largely been grounded in specific locales and contexts and have explored the embodiment of discursive identities as a central concern (e.g., how these identities and cultural practices are enacted and lived through bodies or buildings, how they in turn influence discourse).

Geographers have also noted that discursive frameworks can limit the diversity of viewpoints, place images, and/or identities that can actively engage in, and be engaged with, systems of communication. Discursive practices can exclude minority views and new alternative meanings because powerful groups often attempt to fix the meanings of particular words and symbols in ways that privilege dominant viewpoints. To understand discourse, therefore, we also must understand that it is intricately intertwined with power. In *Orientalism,* for example, Edward Said highlighted how Western colonial representations of Arabic and Islamic cultures as "exotic" and less civilized were as much about the construction of a white European identity as they were about the creation of an undermined "other" that could be easily stereotyped and dismissed. More recently, geographers have explored the impacts that racist and sexist discourses have had on the negotiation and experience of space and place at a variety of scales (e.g., in relation to representations of mobility, depictions of crime in cities, spaces of potential harassment, and restrictive immigration policies) and challenges that have been posed to exclusionary practices (e.g., through activist media, community groups, and reflexive research). Discourse, therefore, is something that is constantly changing and, when critically engaged, can help us to understand how we know, what we can articulate, and how we represent the places and cultures that we explore.

—*Susan P. Mains*

See also Epistemology; Identity, Geography and; Ideology; Other/Otherness; Spaces of Representation

Suggested Reading

Foucault, M. (1977). *Discipline and punish.* London: Tavistock.

Hall, S. (Ed.). (1997). *Representation: Cultural representations and signifying practices.* London: Sage.

Jones, J., III, & Natter, W. (1999). Space and representation. In A. Buttimer, S. Brunn, & U. Wardenga (Eds.), *Text and image: Social construction of regional knowledges* (pp. 239–247). Leipzig, Germany: Institut für Landerkunde Leipzig.

Rose, G. (2001). *Visual methodologies: An introduction to the interpretation of visual materials.* London: Sage.

Said, E. (1979). *Orientalism.* New York: Vintage Books.

DIVISION OF LABOR

The division of labor refers to the specialization in different stages of work that occurs within firms, among firms, and among regions and countries. Although it is not unique to capitalism, the division of labor is most pronounced under commodity production and profit maximization. Rudimentary divisions of labor based on gender appear in hunter–gatherer societies. Similarly, household divisions of labor typically are based on gender, but these vary widely among societies; feminists often point to gender-based divisions of labor in the context of patriarchy. The discovery/invention of agriculture led to a division of labor based on class, that is, slavery. Under feudalism, a rough differentiation between rural areas and towns began to emerge, as evidenced by the rise of the guild system. However, it is under capitalism that the division of labor reaches its most explicit level. It forms one of the core notions of contemporary economics and economic geography.

Eighteenth-century Scottish economist Adam Smith often is credited with originating the idea of (and the term) the division of labor in his book *The Wealth of Nations,* published in 1776. Smith noted that in many firms during the Industrial Revolution, different workers engaged in different steps in the production process, allowing each worker to learn his or her task in great detail and become experienced in it. The division of labor results when workers do not attempt to do all tasks but instead are limited to one task that they perform repeatedly. Smith illustrated this process through his famous example of pin manufacturing, where different workers engaged in 18 different steps such as drawing the wire, adding a head to the pin, and sharpening the point. Workers could produce far more collectively than they could as individuals working independently. In short, specialization leads to greater efficiency and allows firms to be as productive as possible. Smith further observed that the ability to specialize was contingent on how large a market firms served; larger markets allowed companies to become more specialized because they were more likely to find relatively rare clients, leading to Smith's maxim that the division of labor is governed by the size of the market. Thus, larger economies sustained more specialization and usually exhibited higher productivity than did smaller ones.

David Ricardo took Smith's line of thought further, adding a geographic dimension to this process in the form of the *spatial* division of labor as regions and countries specialized around their comparative advantage. Ricardo's contribution was to illustrate how the division of labor was inherently geographic and how countries and regions benefited through unfettered trade among places. Thus, the division of labor was made possible only when regions and countries become interdependent on one another, a process that runs throughout the historical geography of capitalism. The division of labor and the gains from trade are intimately interrelated at the scales of the individual, the firm, and the region or country. The spatial division of labor ranges in scale from the city (e.g., the distinction between central business districts and suburbs) to the global (i.e., the international division of labor). Historically, as groups of firms in similar industries located in proximity to one another to produce and benefit from the fruits of agglomeration, entire districts began to acquire distinct positions within the national and international divisions of labor.

A profound version of the division of labor emerged under Fordism during the late 19th or early 20th century, when the production process became highly specialized within large corporations. Fordism replaced the artisanal, labor-intensive divisions of labor common under mercantile capitalism, where workers were relatively skilled and performed many different steps in the production of goods. Fordism, in contrast, emphasized specialization of work tasks, a goal augmented by Taylorist time and motion studies as well as the reliance on economies of scale that these firms acquired. The result was that complex skilled jobs were decomposed into many simpler tasks, a process that made them not only efficient but also feasible for the waves of unskilled and semiskilled immigrants arriving at the time. More recent investigations of post-Fordist flexible production systems argue that a new division of labor within and among companies emerged during the late 20th century, a process that dramatically reconfigured labor markets in light of associated waves of technological change and globalization. Post-Fordist divisions of labor tend to be characterized by detailed differentiations of tasks among firms (as well as within them), a process that leads to intricate networks of input and output relations.

Marxism was also heavily affected by the notion of the division of labor. The unfolding of the division of labor, and its relations to the forces and relations of production, was one of the great motors of history, replete with numerous political and ideological contradictions. In the context of industrial capitalism, Marx argued that specialization reduced workers to being cogs in a machine, depriving them of control over the production process and alienating them deeply. Geographically, the spatial division of labor perpetually produced and reproduced by the flow of capitalism maintained a system of permanent uneven development, a differentiation sustained by interregional flows of surplus value extracted by wealthy regions or countries from less prosperous ones.

International development theory was highly affected by the idea of the division of labor. Modernization theory, for example, maintained that each country optimally occupied a niche within the global division of labor based on its comparative advantage, a view heavily criticized by dependency and world systems analysts as masking the exploitation inherent in capitalist production. Transnational corporations often engage in an intracorporate division of labor where the headquarters is located in the country of origin, typically in a large city, and less skilled assembly functions and branch plants are positioned in lower-wage countries in the developing world.

—*Barney Warf*

See also Agglomeration Economies; Class; Comparative Advantage; Dependency Theory; Development Theory; Economic Geography; Economies of Scale; Flexible Production; Fordism; Input–Output Models; Labor, Geography of; Marxism, Geography and; Modernization Theory; Transnational Corporations; World Systems Theory

Suggested Reading

Beneria, L. (1985). *Women and development: The sexual division of labor in rural societies.* New York: Praeger.

Coser, L., & Durkheim, É. (1997). *The division of labor in society.* New York: Free Press.

Massey, D. (1989). *Spatial divisions of labor* (2nd ed.). New York: Routledge.

Smith, A. (2003). *The wealth of nations.* New York: Bantam Classics. (Original work published 1776)

DIVISION OF LABOR, NEW INTERNATIONAL

SEE NEW INTERNATIONAL DIVISION OF LABOR

DOMESTIC SPHERE

With regard to labor, the term *domestic sphere* usually is used to refer to two quite different arenas of social life: the nation-state and the home. With regard to the nation-state, the term is used in contradistinction to those events or processes that take place in the foreign sphere or international realm. Hence, trade union policy may be categorized as that which applies domestically and that which applies overseas. In such use of the term, the nation-state is privileged and its boundaries serve as a kind of spatial marker between the domestic and nondomestic spheres. Two issues arise, however, with regard to this conceptualization of the domestic sphere. First, it assumes that the nation-state is in fact a relatively coherent spatial entity whose boundaries circumscribe particular absolute spaces (e.g., the spaces of France vs. those of Italy). Yet contemporary processes of economic globalization challenge this assumption, such that the phenomenon whereby a company such as General Motors produces parts in Mexico for its vehicles assembled in the United States sometimes make it quite difficult to determine where the domestic sphere ends and the nondomestic sphere begins. Second, a strict division of planetary space into the domestic sphere (i.e., the space "within" each nation-state) and the nondomestic sphere (i.e., the global/international space beyond any individual nation-state) relies on an areal view of geographic scale—that is, a view in which scales such as the "national" and the "global" are seen to contain different absolute spaces of varying size. If, however, the scales of social life are seen not as hierarchies of discrete areal units but instead as ropelike or capillary-like and connected in much the same way as a spider's web, it becomes much harder to determine what is "inside" the nation-state (i.e., what the domestic sphere is) and what is "outside" it (continuing the analogy, we might ask where a spider's web begins and ends and what it encompasses). Nevertheless, and despite such issues, the view of the domestic sphere as that which relates to things occurring within the boundaries of various nation-states still has wide commonsensical appeal, even if the nation-state's boundaries seem more porous today than at past historical moments.

The second way in which the term *domestic sphere* is used in the geographic literature is with regard to activities taking place within the home, whether these are paid (e.g., industrial homeworking) or unpaid (e.g., child rearing). In this regard, a number of writers have sought to make a distinction between "labor" and "work"—a distinction that has bearing on the activities taking place in the home and that also is common in European languages (e.g., *ponein* and *ergazesthai* [Koine Greek], *laborare* and *facere* [Latin], *lavorare* and *faticare* [Italian], *travailler* and *ouvrer* [French], *trud* and *rabota* [Russian], *arbeiten* and *wirken* [German]). Thus, 17th-century philosopher John Locke delineated between "the labor of our body and the work of our hands," with labor being the activity through which humans purposively create property out of the world that nature has provided—a delineation that can tend toward seeing labor as a public activity and work as a private one conducted within the home. Karl Marx, on the other hand,

defined labor as activity that helps generate surplus value within the capitalist system, with all other types of activity being seen as work. For her part, 20th-century philosopher Hannah Arendt differentiated between those biological processes necessary to sustain life (labor) and human activity (e.g., art) that has no utilitarian purpose but is an end in itself through which people freely pursue self-realization independent of biological necessity (work). In Arendt's view, then, work is associated with freedom and labor is associated with biological requirements.

These distinctions have relevance for debates in human geography. With the growth in influence of Marxist theory during the early 1970s, many geographers tended to see activities carried out in the home—activities such as cleaning, cooking, and child rearing—as part of the domestic "sphere of reproduction" because they were seen as necessary to ensuring that the capitalist system could reproduce itself on both a daily basis (workers could go to work each day clothed and fed) and a generational basis (new generations of workers were reared). Although important overall, this work of social reproduction was not viewed as "productive" because it did not relate directly to the generation of surplus value, which was Marx's definition of productive labor. Many feminists, however, criticized this conceptualization, arguing that it privileged activities directly associated with commodity production (often done by men outside the home) and denigrated activities such as child rearing and cooking (activities overwhelmingly conducted by women). This, they argued, played into theoretical frameworks that considered unpaid activities typically done by women (e.g., housework) to be "noneconomic" because of their failure to contribute directly to the totality of monetized relations in the economy or to processes of capital accumulation. (It is important to note that such conceptual attitudes toward female domestic labor were not new but rather stretch back to early-19th-century economists such as Nassau William Senior, who included female labor market activity within his definition of "the economic" but excluded housework and child rearing (done within the home, usually by women) because the latter did not result directly in objects that could be exchanged in the marketplace for money.) The exclusion from conceptions of "the economic" of such domestic activities has resulted, many theorists argue, in a significant misunderstanding of the world of work/labor because nonmonetized activities such as cooking and housecleaning are not reflected in statistical measures such as the gross domestic product (GDP), by which economic structure is gauged (e.g., hiring a housekeeper is seen to contribute to the GDP, but cleaning the house yourself is not). Indeed, the limitations of such a conceptualization are shown by the fact that in some industrialized countries the value of unpaid domestic labor is estimated to be 70% of the reported GDP.

—Andrew Herod

See also Economic Geography; Home; Labor, Geography of; Marxism, Geography and

Suggested Reading

Valentine, G. (2001). *Social geographies: Space and society.* Harlow, UK: Prentice Hall.

E

ECOLOGICAL FALLACY

Ecological fallacy can be defined as the erroneous inference about individuals from data or information representing the group or a geographic region. This is an important methodological problem among several disciplines in social sciences, including economics, geography, political science, and sociology, because these disciplines rely on many aggregate-level data sets such as census data. In geographic research, individual-level data may be aggregated to geographic units of different sizes or scales, such as census tract and block group, to infer individual behavior. Then the ecological fallacy is related to the scale effect, which refers to the inconsistency of analytical results when data aggregated to different geographic scales are used.

The problem of ecological fallacy is composed of two parts: how the data are aggregated and how the data are used. All data are gathered from individuals. But due to many reasons, such as privacy and security issues, individual-level data usually are not released; rather, they are aggregated through various ways in which to represent the overall situation of a group of individuals. The group of individuals can be defined by socioeconomic demographic criteria (e.g., below the poverty line, whites) such that individuals within the group should share similar characteristics. The group may also be defined geographically (e.g., within a county or a census tract) such that individuals are in the vicinity of a given location.

There are many ways in which to aggregate individual-level data. One of the most common methods is to report the summary statistics of central tendency such as the mean or median of the individual-level data. Another way is to report the number of observations possessing a specific characteristic such as the population within a given range of ages. Surely, these statistics can represent the general characteristics of the observations given that individuals have similar characteristics, but they definitely fail to describe precisely the situations of all individual observations. Not all individuals are identical, and the statistics might be good to describe only some.

Although aggregated data represent the overall situation of a group of individuals, there is nothing wrong with using the data for analysis so long as one recognizes the limitations of the data. Ecological fallacy emerges when one using the aggregated data does not recognize the limitation of using the data to infer individual situations and ignores the variability among individuals within the group.

This is a well-recognized but stubborn problem in social sciences. Many researchers attempt to solve this problem. Gary King claimed that his error-bound approach can handle the problem, but geographers are skeptical that his method can deal with the scale effect. Ideally, using individual-level data will not commit ecological fallacy. In general, less aggregated data, or data for smaller groups, are more desirable. Geographically, data representing smaller areas will be less likely to generate serious problems. Standard deviation or variance, which indicates the variability within the group, can potentially reflect the likelihood of committing ecological fallacy.

—*David W. Wong*

See also Quantitative Methods

Suggested Reading

Fotheringham, A. (2000). A bluffer's guide to a solution to the ecological inference problem. *Annals of the Association of American Geographers, 90*, 582–586.

King, G. (1997). *A solution to the ecological inference problem*. Princeton, NJ: Princeton University Press.

Wong, D. (2003). The modifiable areal unit problem (MAUP). In D. Janelle, B. Warf, & K. Hansen (Eds.), *WorldMinds: Geographical perspectives on 100 problems* (pp. 571–575). Dordrecht, Netherlands: Kluwer Academic.

ECONOMIC GEOGRAPHY

Economic geography is the study of how economic activities are stretched over the earth's surface at various spatial scales, ranging from the local to the global, and how they change over time and space. Defining the "economic," however, is no simple task; although this domain obviously includes production, transportation, and consumption, more recent analyses have blurred the distinctions between the economy and related social, cultural, and political spheres (e.g., through studies of the household and the informal economy). Whereas traditional approaches during the early 20th century were almost entirely empirical and descriptive in nature, by the 1950s the subdiscipline had become increasingly theoretical. Over time, economic geography has been characterized by different viewpoints and paradigms that exhibit different assumptions, foci, methods, and conclusions. Thus, there is not *one* economic geography but rather many economic geographies.

Economic geography today is concerned with explicating the spatial structure of capitalism. However, this focus should not exclude the important observation that capitalism is a relatively new phenomenon historically, originating in the 16th and 17th centuries and expanding to dominate the globe. Capitalism may be defined as a market-based system dominated by private ownership of the means of production and profit maximization. Long before capitalism, however, there were numerous other economic systems such as hunting and gathering, slave-based social formations, and feudalism. Finally, capitalism itself is a complex and multifaceted social and economic system that varies significantly over time and space.

NEOCLASSICAL ECONOMICS AND ECONOMIC GEOGRAPHY

Until the 1970s, the dominant approach to economic geography was neoclassical economics, emphasizing supply and demand; marginal analyses of costs, revenues, and utility; and a sharp distinction between economics and politics. This tradition dates back to the early 19th century, when Johann Von Thunen formulated the first model of land use in 1826, demonstrating that land values decline with distance from the market and that competing land uses generate a profit-maximizing surface. By 1909, Alfred Weber had developed a highly influential model of transportation costs, emphasizing that firms locate in places that minimize their total transportation costs. This school was concerned with developing theories of corporate organization and change, including the various "factors of location" that firms juggled in deciding where to invest. Neoclassical economics brought to geography a mathematical rigor that raised its level of analytical sophistication. Geographers came to appreciate the complex workings of investment behavior, uncertainty, utility maximization, and the power of the market in rewarding profit-maximizing decisions and punishing irrational behavior. This paradigm led to the widespread use of models, including the following:

1. Walter Christaller's enormously influential central place theory, a model of city systems that posits them as retail centers (central places) that distribute goods and services to their surrounding hinterlands, was enormously influential. A hierarchy of goods and services leads to a hierarchy of central places, with lower-order ones nested within the hinterlands of higher-order ones.

2. Gravity models became a way of modeling spatial interaction and predicting (but not explaining) patterns of interurban migration and traffic flows. They are still widely used in studies of commuting, shopping, and traffic planning. Combined with geographic information systems, this approach is widely used in engineering and site location studies.

3. Spatial diffusion, launched by Torsten Hagerstrand, was concerned with the ways in which innovations (e.g., technologies, information, diseases) moved through time and space. Use of such models introduced probability theory into models of the innovation adoption process and has been helpful to epidemiologists and in marketing.

4. Input–output models, invented by economist Wassily Leontief, used matrix algebra to simulate regional and national economies. In revealing the structure of linkages among firms and industries, it has been widely deployed in a variety of impact analyses and yields the multiplier effects of different activities.

Neoclassical economics also gave rise to product cycle conceptions of industrial change that linked changing markets for goods, as they moved from being innovations to mature goods, with associated changes in the production process. The product cycle explained the locational dynamics of firms as they grew from small, labor-intensive, vertically disintegrated entities into large, capital-intensive, vertically integrated ones, a shift that corresponded with the decentralization of firms away from core regions to the periphery. Although the product cycle originally was developed to explain the movement of corporations from the First World to the Third World, it was adapted to understand the decentralization of firms down the urban hierarchy.

Other geographers inspired by neoclassical economics plunged into analyses of international trade. Inspired by David Ricardo's extremely influential notion of comparative advantage, which explained differences in trade and production systems on the basis of the geography of production costs and factor productivity, they turned to more complex issues such as the impacts of government policies on trade, the role of multinational corporations, intracorporate trade, changing terms of trade (prices of exports and imports), technological change, and trade in services. The work of economist Paul Krugman led geographers to incorporate economies of scale and their spatial effects into trade theory.

Although neoclassical economics is internally self-consistent, is methodologically rigorous, and made great contributions, it has been criticized for its ahistorical nature and lack of context and for its silence about social relations, ignoring class and gender, power, politics, struggle, and contradiction. Location theory, it has been argued, is uncritical and limited in its relevance. By reducing the social to the individual, neoclassical economics engages in methodological individualism, missing the bonds that tie people together into wholes. Moreover, its model of the human subject—the asocial, utility-maximizing "Homo economicus"—proved to be unrealistic.

Because of these faults, many economic geographers began to look to other approaches.

ECONOMIC GEOGRAPHY AND HISTORICAL MATERIALISM

Drawing on the intellectual legacy of Karl Marx, geographers during the 1970s began to formulate views that centered on the role of social relations, historical context, and political power—not simply individual decision making—in the construction of geographies. Marxism represented space as a social product, constructed and reconstructed over time rather than as a passive platform on which firms made location decisions.

The political economy approach took as its point of departure the centrality of labor. Through the labor process, people enter into social relations, change nature, and materialize ideas. Marxism relied on the labor theory of value, which holds that all value ultimately is derived from the socially necessary labor time embodied in the production of goods. All societies have historically specific ways of organizing labor, that is, a division of labor that forms the "economic base." Ownership of the means of production (or lack thereof) establishes the basis of class as a fundamental category of social and spatial analysis. Class relations vary among different modes of production, but under capitalism they consist primarily of a small ruling class and a large number of workers, with other classes (e.g., professionals) situated in between.

Marxism distinguishes between use values (the qualitative subjective aspects that fill human needs) and exchange values (their quantitative price on the market). Under capitalism, labor is a commodity, that is, something bought and sold on a labor market. Because the use value of the worker to the capitalist exceeds the exchange value in terms of wages, capitalists appropriate surplus value from workers. Thus, the labor contract is held to be an inherently unfair and exploitative exchange, and profits are a theft from workers. Such a view gets beneath the view of commodities as mere things, revealing them to be embodiments of social relations. Because capitalists must hold down wages to maximize profits, there is a contradiction between production (maximizing the extraction of surplus value) and consumption (paying workers enough to purchase the goods they produce). This contradiction leads to an excess of surplus value, which lowers the prices of goods and thus the overall rate of profit, leading capitalism into chronic crisis. Thus, crises are held to be a normal—not an abnormal—part of capitalism. Over time, the overaccumulation of surplus value leads capitalists to seek new markets and new ways in which to export crises, making capitalist accumulation inherently expansionary, as manifested historically in the 16th-century "Voyages of Discovery" to the restless geographies of the modern multinational corporation.

David Harvey's great contribution to this process was to spatialize Marxist theory. Capitalists, under the constant pressure of competition, must increase the rate of surplus value extraction by accelerating the turnover rate of capital. Harvey argued that capitalists generate economic landscapes that reflect particular constellations of investments, as captured by the popular notion of the "spatial fix." However, transportation and communications infrastructures are expensive, are durable, and have long depreciation times. Because the creation of landscapes takes long periods of time, and because capitalist production tends to change quickly, geographies of investment may inhibit the formation of newer, more profitable landscapes. Thus, the spatial fix of each age is simultaneously its "crowning glory and prison." Capitalists are caught between fixity and motion; that is, they must continually negotiate a balance between old investments and the creation of new spaces. The creation of new transport and communications systems leads, in turn, to time–space compression, in which relative distances have been steadily conquered through transport and communications technologies. Time and space, which appear to be "natural" and outside of society, are in fact social constructions.

Similarly, Doreen Massey noted that the unique historical trajectories of regions as they occupied different roles within the changing division of labor created layer on layer of investments over time, generating palimpsests in which the residues from each layer shape the nature and impacts of subsequent ones. Massey's approach allowed for broad notions of uneven development to be reconciled with the specifics of individual areas.

Marxist urban political economy portrayed cities as systems of production and labor reproduction structured around lines of class and the division of labor. For example, Marxists pointed to the class politics of urban land use and the social (not just technical) nature of urban planning, opening the door to the understanding of urban regimes and growth coalitions. Suburbanization was seen as a manifestation of the changing urban division of labor, not simply the

product of new transport technologies and the desire for single-family homes. Inner-city poverty is the product of class exploitation and racism, which conspire to reduce ghettos to neocolonies providing low-wage unskilled labor. Similarly, gentrification reflected the corporate recapturing of the inner city during the late-20th-century wave of growth in producer services and globalization, not simply the desires of yuppies to live downtown.

In international political economy, Marxism shed light on colonialism and the international division of labor. During the 1960s, dependency theory argued that poverty in the developing world is the product of centuries of capitalist exploitation. Dependency theory portrayed the global economy as a zero-sum game in which the world's core (Europe, North America, and Japan) gained at the expense of the periphery (Latin America, Asia, and Africa). Bluntly, the West is held to have made the rest of the world poor. Unequal exchange was central to this process, and multinational corporations were the agents responsible. Dependency theory illustrated that poverty is not some natural state but rather an active process—a notion captured in the phrase "development of underdevelopment." Similarly, world systems theory, initiated by Immanuel Wallerstein, focused on the formation of a single world market but multiple political units (including the nation-state); thus, the political geography of capitalism is the interstate system, which allows capital to play multiple localities off one another. Typically, the world system is governed by a single hegemon, a leading power that "sets the rules of the game" such as Spain, the Netherlands, Britain, and the United States. World systems theory divided the world into a core, a periphery, and a semiperiphery, with the latter being countries that share aspects of both the core and periphery (e.g., East Asian newly industrialized countries [NICs]). In contrast to dependency, in world systems theory there remains the possibility that some countries could advance economically, permitting it to be more flexible and realistic in light of the advances made by many NICs during the late 20th century.

FORDISM AND POST-FORDISM

One outgrowth of Marxism was regime theory, which noted the different forms that capitalism took at different historical and geographic contexts. Regime theorists argued that capitalism exhibited temporary windows of stability marked by relatively stable patterns of production, consumption, and state intervention, followed by crises and restructuring. Regime theory became particularly relevant to the dramatic changes that followed the realignments of the 1970s, including the collapse of the older system of mass production, Fordism, and the rise of a newer system of "flexible production."

Fordism was associated with the mass production of homogeneous goods in which capital-intensive companies relied heavily on economies of scale to keep production costs low. Typically, firms working in this context were large, capital-intensive, and vertically integrated, controlling the chain of goods from raw material to final product. Fordism included highly refined divisions of labor within the factory, so that each worker engaged in highly repetitive tasks. It was closely associated with the work of Frederick Taylor, who applied time-and-motion studies to workers' jobs to organize them efficiently. Mass consumption and advertising came into being as the demand side of this process. As a social contract, Fordism tolerated labor unions and politically was associated with the Keynesian state. Geographically, Fordism generated manufacturing complexes such as the North American Manufacturing Belt, the British Midlands, the German Ruhr region, the Inland Sea area of Japan, and similar agglomerations of industrial firms. Historically, Fordism formed the backbone of the great economic boom during the three decades following World War II.

Ultimately, Fordism reached its social and technical limits. Following the petrocrises and competition from the Asian NICs, productivity growth during the 1970s slowed dramatically and waves of plant closures washed over the United States. Rates of profit in manufacturing dropped, and many firms closed down, moved overseas, or restructured themselves with a new set of production techniques. Fordism gave way to post-Fordist flexible production, which became widespread. Post-Fordism allows goods to be manufactured efficiently in small volumes as well as large ones and appeared at the historical moment when the microelectronics revolution was having a great impact on manufacturing. Post-Fordism reflected the imperative of firms to increase their productivity in the face of intense international competition. In contrast to the large, vertically integrated firms typical of Fordism, under flexible production firms tend to be relatively small, relying on computerized production techniques to generate small quantities of goods sold in relatively specialized markets. Microelectronics, in essence, circumvented the need for economies of scale. The classic technologies of post-Fordism include robots and "just-in-time" inventory systems that obviated the need for large expensive warehouses of parts (the "just-in-case" inventory system).

Many firms during the late 20th century engaged in "downsizing," that is, ridding themselves of whole divisions to focus on their "core competencies." Under the relatively stable system afforded by Fordism, most firms produced their own parts, justifying the cost with economies of scale. Given the uncertainty generated by the rapid technological and political changes of the late 20th century, many firms opted to "buy" rather than "make," that is, to purchase inputs from specialized companies. The use of subcontracts accelerated rapidly. As interfirm linkages grew rapidly, many firms found themselves compelled to enter into dense urban networks of interactions, including many face-to-face linkages, ties in which "noneconomic" factors such as tacit knowledge, learning, reflexivity, conventions, expectations, trust, uncertainty, and reputation were critical and cooperative agreements were common. Post-Fordist approaches came into economic geography during the 1980s, focusing on new manufacturing spaces such as California's Silicon Valley, Italy's Emilia–Romagna, and Germany's Baden–Württemberg.

Politically, post-Fordism was tied closely to the ascendancy of neoliberalism, which emphasized the ostensibly free market, deregulation, and privatization. Climbing out of the crisis of Fordism, global capital replaced the Keynesian national "spatial fix" with a highly fluid, globalized neoliberal counterpart. The post-Keynesian, post-Fordist state enhanced, and was in turn enhanced by, the greatly accelerated capacity of finance capital to move effortlessly across the globe, the latest chapter in the "annihilation of space by time" that has defined the historical geography of capitalism. At the local level, the globalization heightened competition among places for capital, a process manifested in popular calls for a good business climate, tax concessions, subsidies, and relaxed environmental controls.

SERVICES

During the 1980s, economic geography also turned to the study of services, which encompass an enormous diversity of occupations and industries. Services may be understood as the production and consumption of intangible inputs and outputs. It is impossible to measure these quantities accurately, yet they are real nonetheless.

The traditional perspective on services, postindustrial theory (now out of date), viewed services as information-processing activities and as a qualitatively new form of capitalism. Although many service jobs do involve the collection, processing, and transmission of large quantities of data, clearly others do not (e.g., trash collectors, security guards). More recent theorizations stress services as another form of commodity production, not as a qualitatively new phenomenon but rather as an extension of market relations into new domains of output and activity.

There is a broad consensus as to the major components of the service sector, including (a) finance, insurance, and real estate (FIRE) such as commercial and investment banking, insurance, and real estate; (b) business services that subsume legal services, advertising, engineering and architecture, public relations, accounting, research and development, and consulting; (c) transportation and communications, including the electronic media, trucking, shipping, railroads, airlines, and local transportation; (d) wholesale and retail trade firms, the intermediaries between producers and consumers; (e) consumer services such as eating and drinking establishments, personal services, repair and maintenance services, entertainment, hotels and motels, and tourism (the world's largest industry); (f) government at the national, state, and local levels, including public servants, the military, and all those involved in the provision of public services (e.g., public education, healthcare, police, fire departments); and (g) nonprofit agencies such as charities, churches, museums, and private nonprofit healthcare agencies.

Services comprise the vast bulk (75%) of output and employment in most economically developed countries of the world. Furthermore, services comprise the vast majority of all *new* jobs generated in these economies. In economically developed countries, services employment has increased steadily in the face of low rates of population growth, slowly rising rates of productivity and income growth, and manufacturing job losses. Even in much of the developing world, services comprise a large share of the labor force, including much of the "informal" (untaxed and unregulated) economy; this fact belies earlier assertions that all economies inevitably are transformed in a series of rigid stages (i.e., agricultural to industrial to postindustrial).

Services are traded interregionally and internationally by cities and countries. Many urban areas export services to clients located in other parts of the same nation (e.g., New York; Washington, DC). Services are also traded on a global basis, comprising roughly 20% of international trade. Internationally, the United States is a net exporter of services but runs major trade deficits in manufactured goods.

Services also include telecommunications. Because the circulation of information is critical to the operation of large complex economies, the history of capitalism has been accompanied by waves of innovation in communications. The telegraph and telephone allowed multiestablishment firms to centralize their headquarters functions while they spun off branch plants to smaller towns. Despite the proliferation of new technologies, the telephone remains the most commonly used form of telecommunications for businesses and households. The microelectronics revolution was particularly important in the telecommunications industry (arguably the most dynamic sector today), including the Internet and fiber optics. Large firms operating in multiple national markets require such systems to coordinate thousands of employees within highly specialized corporate divisions of labor.

Popular confusion about telecommunications includes simplistic notions that they entail "the end of geography." Often such views hinge on a utopian technological determinism that ignores the complex relations between telecommunications and local economic, social, and political circumstances. Predictions that telecommunications would allow everyone to work at home via telecommuting, spelling the obsolescence of cities, have fallen flat in the face of the persistent growth in densely inhabited global cities. Telecommunications usually is a poor substitute for face-to-face meetings, the medium through which most sensitive corporate interactions occur, particularly when the information involved is irregular, proprietary, and/or unstandardized in nature. For this reason, a century of telecommunications has left most high-wage, white-collar, administrative command-and-control functions clustered in large cities. Telecommunications is ideally suited for the transmission of standardized forms of data, facilitating the dispersal of functions involved with their processing to low-wage regions (e.g., back offices). In short, by allowing the decentralization of routinized processes, information technology enhances the comparative advantage of inner cities for nonroutinized, high-value-added functions that are performed face-to-face. Thus, telecommunications facilitates the simultaneous concentration and deconcentration of economic activities.

In the current round of globalization, heralded by the marked expansion in the scope, volume, and velocity of international linkages, worldwide telecommunications networks reflect what Manuel Castells labeled the rise of the "network society" dominated by a "space of flows." In a Fordist world system, national monetary control over exchange, interest, and inflation rates is essential; however, in the post-Fordist system, those same national regulations appear as a drag on competitiveness, a factor underpinning worldwide moves toward deregulation and privatization. For example, as large sums of funds flowed with mounting ease across national borders, national monetary policies have become increasingly ineffective.

THE CULTURAL TURN

During the 1990s, economic geographers became increasingly sensitized to the need to incorporate a more flexible understanding of culture. Post-Fordism directed attention to the critical roles played by noneconomic factors such as tacit knowledge, learning, reflexivity, conventions, expectations, trust, uncertainty, and reputation in the interactions of actors. Economic geographers emphasized culture as a complex contingent set of relations every bit as important as putatively economic factors in the structuring of economic landscapes, humanizing abstract economic processes by showing them to be the products of agents enmeshed in webs of power and meaning. In this light, culture enters into issues such as the accepted definitions of "work" (e.g., paid or unpaid, at home or outside of the home) as well as actors' understandings of what is normal or abnormal, proper or improper, and legal or illegal. Such factors shape corporate attitudes regarding loyalty, obedience, duty, and reciprocity and also shape informal linkages in which tacit knowledge circulates, the politics within firms and bureaucracies, and settlements of disputes.

This "cultural turn" took geographers from matters of production to the arena of consumption, a topic long dominated by neoclassical economics. This reading portrays consumption not only as an economic act but also simultaneously as a social act that is embedded in local and national relations of production, class, gender, and power; a psychological act that reproduces identity; and an ecological act that forms the end of value-added

chains stretching across the planet. The spatiality of consumption is thus a multiscalar process. Geographies of consumption allowed for the incorporation of consumer tastes, fashion, and lifestyle issues.

One manifestation of the cultural turn was actor–network theory, which views actors as drawing on networks of rules, resources, information, and power, with actors and networks being mutually presupposing and coevolving. To overcome the artificial boundaries between culture and nature, this view holds that actors need not be human but may include inanimate objects. Actor–network theory allowed an escape from the conventional focus on spatial scale in that networks operate across many scales simultaneously, folding space and time in a series of differential power geometries. Thus, spatial scale is not pregiven but rather is produced through, and is constitutive of, social relationships.

The cultural turn was central to the recognition of the contingency of economic systems; that is, actors can always shape them differently from those expected by overarching social "laws." Borrowing from institutional economics, this line of thought emphasized the capacity of human actors to construct their world, humanizing abstract processes. Thus, corporations and entire industrial districts were viewed as having trajectories of growth and decline over time and space that are path dependent; that is, choices and decisions made at one moment in time shaped their subsequent structure at later moments. This view brought to the fore the role of historical and spatial contexts in economic behavior.

Finally, the cultural turn revealed that far from constituting an unstoppable force, global processes are in fact embodied, interpreted, contingent, and contested. Thus, local regions do more than simply receive globally generated changes; they also produce them. Hence, globalization entailed different outcomes in different regions. By revealing how the global and the local are shot through with one another (or "glocalized"), this literature generated nuanced understandings of how globalization is manifested differently in different places, thereby helping to dispel simplistic assertions that globalization simply erases geographic specificity.

—Barney Warf

See also Agglomeration Economies; Agriculture, Industrialized; Circuits of Capital; Class; Colonialism; Commodity; Comparative Advantage; Competitive Advantage; Consumption,

Geography and; Deindustrialization; Dependency Theory; Development Theory; Diffusion; Division of Labor; Economies of Scale; Economies of Scope; Factors of Production; Flexible Production; Fordism; Globalization; Gross Domestic Product; Imperialism; Import Substitution Industrialization; Incubator Zone; Industrial Districts; Industrial Revolution; Informal Economy; Infrastructure; Innovation, Geography of; Input–Output Models; Labor, Geography of; Labor Theory of Value; Location Theory; Malthusianism; Marxism, Geography and; Mode of Production; Modernization Theory; Neocolonialism; Neoliberalism; New International Division of Labor; Newly Industrializing Countries; Political Ecology; Postindustrial Society; Poverty; Producer Services; Product Cycle; Profit; Rural Development; Rustbelt; State; Sunbelt; Terms of Trade; Time–Space Compression; Tourism, Geography and/ of; Trade; Transnational Corporations; Transportation Geography; Underdevelopment; Uneven Development; Urban and Regional Planning; Urban Entrepreneurialism; Urban Geography; Urban Sprawl; Urban Underclass; Urbanization; World Economy; World Systems Theory; Zoning

Suggested Reading

Amin, A. (Ed.). (1994). *Post-Fordism: A reader.* Oxford, UK: Blackwell.

Bryson, J., Daniels, P., & Warf, B. (2004). *Service worlds: People, organizations, and technologies.* London: Routledge.

Castells, M. (1996). *The rise of the network society.* Oxford, UK: Blackwell.

Dicken, P. (2003). *Global shift: The internationalization of economic activity* (4th ed.). New York: Guilford.

Gibb, R. (1994). Regionalism in the world economy. In R. Gibb & W. Michalak (Eds.), *Continental trading blocs: The growth of regionalism in the world economy.* New York: John Wiley.

Harvey, D. (1982). *The limits to capital.* Chicago: University of Chicago Press.

Knox, P., Agnew, J., & McCarthy, L. (2003). *The geography of the world economy* (4th ed.). London: Edward Arnold.

Krugman, P. (1991). *Geography and trade.* Cambridge: MIT Press.

Massey, D. (1984). *Spatial divisions of labor: Social structures and the geography of production.* New York: Methuen.

Sheppard, E., & Barnes, T. (Eds.). (2000). *A companion to economic geography.* Oxford, UK: Blackwell.

Stutz, F., & Warf, B. (2005). *The world economy: Resources, location, trade, and development* (4th ed.). Upper Saddle River, NJ: Pearson/Prentice Hall.

Thrift, N., & Olds, K. (1996). Reconfiguring the economic in economic geography. *Progress in Human Geography, 20,* 311–337.

Vernon, R. (1966). International investment and international trade in the product cycle. *Quarterly Journal of Economics, 80,* 190–207.

Walker, R. (1985, Spring). Is there a service economy? The changing capitalist division of labor. *Science and Society,* pp. 42–83.

ECONOMIES OF SCALE

Economies of scale refer to the reductions in cost that firms achieve by producing in larger rather than smaller volumes of output. *Economies* in this context refers to the benefits incurred by reducing costs, largely by spreading fixed costs over a larger quantity of output.

Mass production occurs through the standardization of parts and a detailed division of labor. Specialized divisions of labor, however, require a relatively large scale of output because generally a large pool of workers is necessary. Scale economies operate when increases in factor inputs generate disproportionately larger increases in output; in more technical terms, the production function in not linear. For example, if a firm increases its inputs of labor and capital by 20% but sees its output rise by 30%, it enjoys economies of scale. Thus, they represent the opposite of diminishing returns in the production process.

Economists portray scale economies as a curve of long-run average costs (Figure 1) that graphs the unit costs as a function of scale. As unit costs decrease, they reach an optimal point and ultimately began to increase, reflecting diseconomies of scale (diminishing marginal returns to scale) that occur when a firm becomes too large to manage and operate efficiently.

Economies of scale tend to favor the formation of larger firms and hence relatively oligopolistic market structures (those dominated by a few giant companies). Large firms generally pay much less for material inputs than do small firms because the former buy in bulk and often enjoy economies of scale in transportation as well as in the production process. The presence of economies of scale varies widely among firms and industries. It is indisputable in sectors such as industrial agriculture and capital-intensive forms of manufacturing such as steel and automobiles. The degree to which services with intangible outputs enjoy economies of scale is less clear.

Economic scale is closely intertwined with geographic location. Indeed, the choice of location cannot be considered in isolation from scale and production technique. Different scales of operation may require

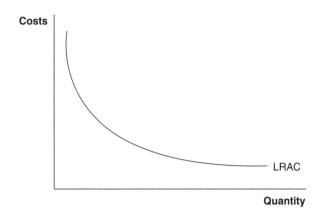

Figure 1   Economies of Scale

different locations to give access to markets of different sizes. Conversely, location itself can influence the combination of inputs and hence the technique adopted. Economies of scale tend to favor a select group of geographic locations over dispersed production patterns.

—Barney Warf

See also Agglomeration Economies; Economic Geography

Suggested Reading

Krugman, P. (1992). *Geography and trade.* Cambridge: MIT Press.

Stutz, F., & Warf, B. (2005). *The world economy: Resources, location, trade, and development* (4th ed.). Upper Saddle River, NJ: Pearson/Prentice Hall.

ECONOMIES OF SCOPE

The term *economies of scope* refers to the economies, or benefits, that firms derive by producing particular combinations of output. These exist if one firm can produce two separate products more efficiently than two firms can independently produce them separately. Economies of scope resemble economies of scale but operate in a different manner. Whereas economies of scale refer to the lower costs involved in producing larger quantities of a single type of good, economies of scope refer to the benefits generated from producing a mix of goods and thus are common in multiproduct firms. These arise essentially because a firm can use a given stock of factor inputs to generate a variety of related or complementary outputs. Average costs decline due to the *mix* of output between two or more products. Thus, economies of scope refer to the potential cost savings that result from the joint production of products not directly related to one another. If a firm can produce multiple goods and services more efficiently than several firms could produce them independently, it enjoys economies of scope. Otherwise, it suffers from diseconomies of scope (increases in price that accompany the production of different goods).

For example, a fast-food franchise may produce two types of foods more quickly than it could produce both in isolation, largely by using the same storage and preparation facilities. As another example, a firm's administration and management may carry out services necessary to the production of a variety of different goods, including research, marketing, and financing. Economies of scope are common in the learning process when the knowledge and experience gained in the production or sale of one good are useful in the production and sale of other goods. One warehouse may be used to store a range of products. The salaries and transportation costs of a sales force may be used effectively to sell a variety of different goods or services. A publishing firm may realize cost savings by using staff members to produce more than one magazine. These benefits are frequently found in marketing and distribution, where they underscore strategies such as product bundling.

Economies of scope also occur when there are cost savings generated by the production of by-products, that is, when the production of one good automatically triggers the production of another. For example, a beef producer may also generate leather, or a lumber company may also create sawdust.

Potential economies of scope underlie the mergers, acquisitions, and takeovers of firms seeking to diversify into new product lines and markets. In this case, the attraction is the ability to reduce costs by operating two or more businesses under the same corporate umbrella.

Economies of scope and economies of scale are often inversely related, particularly as firms face a "make" or "buy" decision regarding their inputs. For example, firms may subcontract (buy) inputs rather than make them "in-house" when they face rising uncertainty or rapid change in products or technology, when the labor process resists easy automation, or when the optimal scales of operation of production processes are markedly different. Thus, from the transaction costs perspective, externalization allows external economies of scale to replace internal economies of scope. By externalizing, firms substitute variable costs for fixed ones and spread the risks of production over their subcontractors, a particularly vital role during peak periods of demand.

—*Barney Warf*

See also Agglomeration Economies; Economic Geography; Economies of Scale

Suggested Reading

Stutz, F., & Warf, B. (2005). *The world economy: Resources, location, trade, and development* (4th ed.). Upper Saddle River, NJ: Pearson/Prentice Hall.

EDGE CITIES

Joel Garreau's 1991 book *Edge City* gave the first critical in-depth account of a phenomenon that had been gradually appearing on the metropolitan landscape since the 1960s. That phenomenon was the clustering of commercial development outside of the central city at the intersection of major highway interchanges, predominantly consisting of office buildings, shopping, and entertainment complexes. Garreau wrote about edge cities in a matter-of-fact way, postulating that the agglomeration of high-rises on the fringes of major metropolitan areas was merely the inevitable future of metropolitan form. Indeed, the edge city phenomenon can be interpreted as logically following on the heels of suburban residential expansion that occurred after World War II. Widespread population dispersal out of the central city was necessarily followed by places that could accommodate individual consumption (shopping malls) and then by jobs for suburban employees in the form of office high-rises. Edge cities now are easily visible as rising clusters of steel and glass buildings outside of virtually every major American city (as well as many cities outside the United States), particularly in northern and central New Jersey, southern California, the San Francisco Bay area, Boston, Detroit, Atlanta, and Phoenix. Well-known edge cities include Tyson's Corner, Virginia (outside of Washington, DC), and Plano, Texas

(outside of Dallas). There are roughly 200 edge cities in the United States.

There are negative consequences to edge city growth, and many see edge cities as an unfortunate outcome of postmodern economic processes. Not only do they detract from the vitality of the original downtown core, pulling jobs and services away from needy populations, but they also represent a form of urban development that is visually confusing, land consumptive, and largely unplanned in any meaningful way. Furthermore, they exist without representational government, except perhaps some regional or county-level control, and thus have been described as "stealth" cities. In addition, edge cities are generally automobile dependent and thus have been attacked as representative of the antithesis of the walkable, diverse, compact urban form characteristic of sustainable metropolitan form. Whereas the notion of an edge city is now a generally recognized phenomenon, recent scholarship has documented alternative suburban forms, notably the *edgeless city,* which does not go as far as the edge city in terms of agglomeration. Such patterns represent an unorganized and diffuse composition of office space that will be even more difficult to redress than the edge city. In that respect, the edge city may offer some potential for redevelopment and revitalization. Some interpret the edge city as comprising a framework for potential new urban cores. The edge city can be valued as representing a maturation of suburban sprawl, where at least the peripheral spread of post–World War II growth patterns is channeled into dense clusters of regionally distributed office space and accompanying shopping opportunities. All that is needed to make the edge city an actual city, albeit a satellite one, would be to incorporate housing and public services in a more integrative way. This transformation will not be straightforward given that the infrastructure of the edge city, particularly its position relative to highway interchanges, will make such a transformation very challenging.

—*Emily Talen*

See also Exurbs; Suburbs and Suburbanization; Urban Fringe; Urban Geography

Suggested Reading

Garreau, J. (1991). *Edge city: Life on the new frontier.* New York: Anchor.

Lang, R. (2003). *Edgeless cities: Exploring the elusive metropolis.* Washington, DC: Brookings Institution.

ELECTORAL GEOGRAPHY

Electoral geography is the systematic investigation of geographically disaggregated information about elections. Electoral geographers investigate distributions of election outcomes and the geographic implications of electoral systems in democratic societies to draw inferences about underlying patterns, causes, and impacts of geographic differences in voting behavior.

Electoral geography is closely associated with democracy. Over the course of the 20th century, democracy diffused rapidly throughout the world. Only a small minority of the world's population lived in independent democracies in 1900. The large majority lived under colonial rule or in autocratic societies where most people had little voice in government. Today, all developed countries and many less developed countries are democracies. Moreover, the 20th century saw continued extension of the right to vote. In 1900, only a few countries gave women the right to vote, and many also denied voting rights to racial, ethnic, or religious minorities. Today, universal adult suffrage is the norm in most democracies.

The diffusion of democracy between and within countries has enhanced the intellectual value of electoral geography. In contemporary democracies, elections are held at regular intervals. Voters face free, albeit sometimes constrained, choices among alternative candidates or proposals. Data representing the outcomes of elections are collected and published for geographically disaggregated spatial subnational units such as states, provinces, counties, cities, and local governments. These data are at least reasonably accurate representations of actual voters' preferences. They are in the public domain and can be mapped, and electoral maps can be analyzed statistically and cartographically. During recent years, the statistical and cartographic analysis of geographically disaggregated electoral data has been enhanced greatly by the development of geographic information science. Through analysis of electoral data, electoral geographers draw inferences about underlying economic, cultural, ethnic, social, and environmental factors influencing differences in election outcomes between places and between individual elections over the course of time.

There are two types of democracy: direct and representative. Under direct democracy, voters express their preferences on specific issues. For example, several European countries recently held referenda on whether to endorse the European Union's constitution, and several U.S. states held referenda on state constitutional amendments constraining or banning gay marriage. Analysis of spatial patterns of *yes* and *no* votes on such issues provides information about underlying cultural, economic, and political processes. Representative democracy involves choices between candidates for public executive or legislative offices. Voters elect candidates, who in turn bear direct responsibility for making decisions concerning public policy. Frequently such analysis involves geographic comparison in levels of support for competing political parties. Statistical and cartographic analysis is undertaken to identify economic, cultural, demographic, and other factors associated with observed differences in levels of support between parties or candidates.

The analysis of sequences of elections in representative democracies is often an especially fruitful line of inquiry. In many countries, elections have been held for many years, with data available for the same areal units in numerous elections. For example, elections for president of the United States have taken place since 1789, with reasonably accurate public records of popular votes available by state and county going back to the 1830s. Since 1860, the Democratic and Republican parties have been the two major political parties in the United States. The geographic pattern of support for the two major parties has shifted both between and within states. For example, between the 1860s and the 1940s, the South was dependably Democratic, with the Republicans winning popular vote majorities in only a handful of states in a few elections. The Democrats' dominance of the South meant that both parties ignored this region and took it for granted. Since the 1950s, however, the Republicans have become dominant in the South. Moreover, the South has now become a dominant region in determining electoral outcomes; no candidate has won the presidency without winning significant electoral vote support in the South since 1924. Meanwhile, parts of New England and the upper Midwest that were reliably Republican for many years are now dominated by the Democrats. Researchers have identified and interpreted analogous shifts in party preference at the state level throughout the country as well.

Electoral geographers also analyze the nature of the electoral process and its geographic implications. The specific rules and procedures used to aggregate votes and determine the outcomes of elections vary from one democracy to another. For example, many countries, such as the United Kingdom and Canada, use a parliamentary system in which the party or coalition of parties with the most legislative seats forms a government. The head of state or prime minister is simultaneously a legislator. In the United States, on the other hand, legislative power and executive power are separated. Presidents and state governors are elected independently of federal and state legislators. In many countries, the chief executive is elected by direct popular vote. The United States, however, chooses its president by an Electoral College whose membership is determined by separate counts of popular votes in each state and the District of Columbia. Thus, a president can win an Electoral College majority without winning a plurality of popular votes, as happened in 2000.

In most representative democracies, legislators are elected to represent districts that are defined territorially. Since the 19th century, it has been recognized that the delineation of electoral districts, and the processes by which they are delineated, can and often does affect election outcomes and the eventual direction of public policy. The term *gerrymandering* has often been used to describe deliberate bias associated with the drawing of electoral district boundaries.

Electoral geographers have long been interested in questions associated with gerrymandering. Resolving issues associated with gerrymandering involves recognizing that districting inevitably involves compromises between incompatible objectives. Few would deny that districts should be drawn in a fair manner. But what constitutes fairness? The meaning of fairness is problematic in drawing electoral districts and more generally in aggregating individual voter preferences into collective outcomes. It is often difficult to reconcile the objective that majority wishes should prevail with the also important objective of protecting minority rights. For example, in a country where 60% of the electorate supports the majority party and the other 40% supports the minority party, the process of drawing each district to preserve this 60/40 ratio would mean that the minority party would end up with no legislative seats at all. The issue of preserving minority rights in representation has been especially controversial with respect to the representation of racial, ethnic, and religious minority groups in legislative bodies.

—*Fred Shelley*

See also Political Geography

Suggested Reading

Archer, J., Lavin, S., Martis, K., & Shelley, F. (2002). *Atlas of American politics, 1960–2000*. Washington, DC: Congressional Quarterly Press.

Archer, J., Shelley, F., Taylor, P., & White, E. (1988). The changing geography of America's presidential elections. *Scientific American, 268*, 44–51.

Flint, C. (2001). A timespace for electoral geography: Economic restructuring, political agency, and the rise of the Nazi party. *Political Geography, 20*, 301–329.

Johnston, R., Taylor, P., & Shelley, F. (Eds.). (1990). *Developments in electoral geography*. London: Routledge.

Shelley, F., Archer, J., Davidson, F., & Brunn, S. (1996). *Discovering America's political geography*. New York: Guilford.

EMOTIONS, GEOGRAPHY AND

In the history of the discipline, despite there having been no *explicit* body of work (until recently) related to *emotional geographies* (geographic knowledge written with and/or on emotions) or the *geography of emotions* (a mapping of different emotional states), much work in human geography speaks about how people emotionally embody space and place. Forming a wide-ranging and sophisticated intellectual backlash against the silence of positivistic and masculinist human geography on questions of embodied human, humanistic, feminist, psychoanalytic, and nonrepresentational approaches in the discipline all have conceptualized diverse emotional relations and spaces in everyday life.

In general, references to *emotions, feelings,* and *affect* in academic work indicate an interest in intense physical and social experiences (commonly represented as *love, happiness, sadness,* etc.) that have profound impacts on individuals, their relations with others, and their relations with things in the world. To be more precise in defining emotions, we can say that these are simultaneously *embodied, psychological, social, cultural, and physical states of being*. There are debates in academia as to whether these states are universal or culturally specific phenomena, but there is agreement that they are multifaceted and complicated and cannot be easily explained by either biological or social determinants alone. A further explanatory distinction refers to the difference between *emotions* and *feelings*. It is possible, for example, to associate feelings with intense and immediate bodily sensations (e.g., people refer to shivering with excitement) and to regard emotions as cultural productions whereby we consciously understand such bodily responses as particular sorts of emotions. Hence, conscious emotional experiences make sense to us through the cultural resources available as we learn what it means to feel certain sensations. Thinking through how such embodied intensities are encountered, recognized, reflected, and acted on through different sorts of spaces is the key work of the geographer interested in emotions.

To use a topical example that might demonstrate the importance of thinking through emotional geographies, we could highlight the range of human intensities felt as a result of the events on September 11, 2001, in New York. Due to a mixture of political, terrorist, and military maneuvers, people in both the United States and Iraq, for example, embody a range of everyday emotional states in relation to a range of spatial scales from the body to the home, to the city, to the nation, and beyond. Intense embodied feelings of fear, insecurity, and threat might accompany emotional anger, for example, and be manifest through these scales in different or similar ways for the people who live in these two places. Attention to these states of being and the geographies through which they are configured could take different directions and could include changing senses of place (perhaps using a humanistic approach), women's fears about public spaces (using a feminist approach), feelings about Western and non-Western difference (using a psychoanalytic approach), and remapping bodily consciousness during times of war (using a nonrepresentational approach). There is insufficient space here to unfold the nuances of these possibilities in terms of approaches to this example, but this list indicates something of the orientation that each might have toward the emotional geographies in question. Overall, geographers are contributing to the study of emotions by paying attention to how emotions are thoroughly implicated in our everyday spatial experiences, and such work tells us that although emotions are difficult to define and complex, they are nonetheless central to the ways in which we live in the world.

—Hester Parr

See also Body, Geography of; Human Agency; Identity, Geography and; Ideology; Subject and Subjectivity

Suggested Reading

Bondi, L., Davidson, J., & Smith, M. (2006). *Emotional geographies.* London: Ashgate.

Davidson, J., & Milligan, C. (2004). Embodying emotion, sensing space: Introducing emotional geographies. *Social and Cultural Geography, 5,* 523–532.

Parr, H. (2005). Emotional geographies. In P. Cloke, P. Crang, & M. Goodwin (Eds.), *Introducing human geographies.* London: Edward Arnold.

Pringle, R. (1999). Emotions. In L. McDowell & J. Sharp (Eds.), *A feminist glossary of human geography* (pp. 68–69). London: Arnold.

EMPIRICISM

Empiricism is the philosophical doctrine that knowledge and understanding originate in experience, especially sensory experience. Thus, the empiricist philosopher John Locke likened the mind of an infant to a "blank slate" that contained nothing but the potential to register the facts and concepts that the experiences of a lifetime would write on it. Of ancient provenance, empiricism was formalized during the 17th century and is a major force in modern ideology. Its most significant expression is experimental science, with an experiment being a more or less controlled experience. Empiricism is the epistemology of science; it specifies the grounds on which most scientists and geographers justify claims to factual knowledge.

Empiricism takes weak and strong forms. The weak form is the everyday habit of using sense data to answer questions about contingent facts. If one wishes to know whether or not there is milk in the refrigerator, for instance, one probably will look in the refrigerator. The question could be answered deductively (today is Wednesday, one's spouse shops on Tuesday, etc.), but such reasoning could lead at best to supposition. A weak empiricist believes that looking is the only way in which to verify or have positive knowledge that milk, or anything else, is present; hence, empiricism is also called verificationism (or positivism). Weak empiricism is very common in human geography.

Strong empiricism elevates this method to the epistemological doctrine that experience not only is a ground for knowledge but also is the best or only ground. There are not, and cannot be, meaningful assertions other than those verified, or verifiable, by positive empirical evidence. This is a revolutionary doctrine because its demand that all alleged truths be put to the test of experience undermines metaphysical beliefs grounded in intuition, pure reason, faith, tradition, and authority. Doctrinal empiricism sometimes is content to invalidate the epistemological claims of these alternative routes to knowledge and pronounce itself agnostic with respect to the objects these methods allegedly apprehend (e.g., God, value, beauty, freedom, the soul). Frequently, however, strong empiricism affirms the ontological doctrine that extraempirical, nonobservable things do not exist. Advocates call this ontological doctrine *naturalism,* whereas dissenters call it *scientism* or *positivism.*

Dissenters come in many varieties but share a belief that there is a knowable reality that exceeds and conditions the phenomena of sensory experience. This metaphysical or transcendental realm is not one that experimental science will one day discover with further investigations and improved techniques because it can be apprehended only by nonempirical methods. Antiempiricists sometimes are called rationalists because they believe that the mind (or at least the acute minds of rational souls) can directly apprehend transcendental truths intuitively and without the aid of the senses. (The name is confusing because their intuitions and inferences often affirm the very beings—God, the soul, value—that empiricists expunged in the name of rationalization.) In contemporary human geography, rationalist antiempiricism usually is called antipositivism, and it is evident in humanistic introspection and the theorizing of social theorists.

Curiously, one might say that the limitations of empiricism (and the naturalistic ontology that follows from it) have grown apparent as empiricism has prospered—as more and more nonempirical truths have been put to the test of experience, failed, and been discarded. The radical doctrine of empiricism does not verify the objective existence of values such as justice, beauty, and the good, and so it removes these rational ends from human calculation even as it, through empirical investigations, engrosses the means that humans may employ to achieve what now appear as arbitrary and emotive ends. Thus, widespread adoption of the doctrine of empiricism at least partly explains a postmodern predicament, that is, our endlessly increasing ability to do just about everything except agree just what it is we ought to do.

—*Jonathan Smith*

See also Epistemology; Logical Positivism; Ontology; Phenomenology; Realism

Suggested Reading

Joad, C. (1950). *A critique of logical positivism.* Chicago: University of Chicago Press.
MacIntyre, A. (1984). *After virtue* (2nd ed.). Notre Dame, IN: University of Notre Dame Press.

ENLIGHTENMENT, THE

The Enlightenment is that period of intellectual enquiry, broadly synonymous with the "long" 18th century (circa 1680–1820), when modern ideas of rationality, public criticism, and the emancipation of civil society through reasoned reform took shape. During the Enlightenment, ideas of "ancient authority" and "tradition" were challenged. Earlier, classical and Renaissance conceptions of the world and humanist scholarship had been rejected. Philosophical inquiry and science were widely believed to be the basis to socially useful goals. Religious restrictions would diminish in the face of secular tolerance. Humankind would be free from ignorance and error. Since then, and even during its development, the Enlightenment has been the subject of detailed scrutiny as to what it was, why it happened, and what its consequences have been. Conventional views of the Enlightenment as an essential, largely philosophical phenomenon evident in urban Europe, especially in the lives and writings of great men, have been challenged decisively during recent years. Questions of geography are central to these revised and revitalized notions of the Enlightenment.

In conventional interpretations, little attention was paid to the geography of the Enlightenment. Where it was, emphasis was given to its distinctive features and differences at the level of the nation-state, chiefly within Europe. Attention concentrated on the idea of the Enlightenment's originating "hearth" or "core" nations (e.g., France, England, Scotland, Holland, Germany) and to a "periphery" where the Enlightenment was evident later or in different form (e.g., Russia, the Scandinavian countries). Relatively limited attention was given to the Enlightenment in the Americas and to its presence and making in Portugal, Spain, or the countries of Eastern Europe. As Enlightenment studies in general have become more diverse—embracing, for instance, medical knowledge and questions of gender, exoticism, race, and sexuality—studies of the Enlightenment and geography have diversified beyond the scale of the nation and rejected simplistic distinctions between an Enlightenment core and periphery. Three distinct but interrelated themes may be noted.

The first theme is geographic knowledge and the Enlightenment. Geographic knowledge, gleaned through oceanic navigation, terrestrial exploration, mapping, and natural history survey, was crucial during the Enlightenment to new ideas about the shape and size of the earth, the richness of its natural diversity, and the nature of its human cultures. In this first sense, the Enlightenment as a philosophical movement depended on new geographic knowledge about the extent of what contemporaries then called the "fourth world" (the Americas) and, crucially, about the "fifth division" of the world (the Pacific world or, in modern terms, Australasia). One distinctively Enlightenment idea, that of society's development through a series of stages, was profoundly shaped by the "discovery" of new peoples on the islands of the Southern Ocean, for example, and by the extent of human cultural difference. Contemporaries referred to these geographies of human difference as "The Great Map of Mankind" and devoted considerable time to theories explaining the development of human society in relation to factors such as climate, the role of custom, and commercial capacity.

Second, we may think in terms of geography during the Enlightenment. Geography as one form of modern intellectual endeavor was itself shaped by the evolving encounter with new peoples and lands during the Enlightenment. This was apparent in terms of emphases on realism in description, systematic classification in collection, and comparative method in explanation. Geography during the Enlightenment was a discourse, a set of practices by which the world was revealed and ordered. It was also a discipline in which formal study was possible in schools and universities. It was likewise a popular subject, taught in academies and public lectures alongside history, astronomy, and mathematics, to educate citizens about the extent and content of the globe. In these ways, geography during the Enlightenment was part of what thinkers then called the "science of man"—that concern to understand the human world through the same observational and methodological principles as the natural world.

Finally, it is commonplace to refer to the different geographies of the Enlightenment. These different geographies are distinguished by their attention to the intrinsic diversity of the Enlightenment—to the social processes and contradictions underlying its intellectual and practical claims—and, above all, by a sensitivity to the importance of geographic scale in mapping and explaining the Enlightenment. Although the Enlightenment is still much studied in national context,

greater attention is paid to its global expression and consequences, to the local institutional sites and social settings in which the Enlightenment's defining ideas were produced and debated, to the uneven transmission of those ideas across geographic space, and to the variant nature of their reception. Thus, ideas of *the* Enlightenment as a uniform intellectual movement with particular national expression have been challenged by work that stresses Enlightenment—even enlightenments—as a social process or processes with diverse geographic expression.

Questions concerning the "where" of the Enlightenment are as important as those concerning the movement's "what" and its "why." Postmodernism has speculated on the end of the "Enlightenment Project." Although initially critical of Enlightenment writers' emphases on rationality, reform, and the power of critical argument, many postmodern theorists now would confirm the enduring significance of the Enlightenment as a set of political issues and as an object of historical and geographic study.

—*Charles W. J. Withers*

See also Discourse; Exploration, Geography and; History of Geography; Postmodernism; Spaces of Representation

Suggested Reading

Kors, A. (Ed.). (2003). *Encyclopedia of the Enlightenment* (4 vols.). New York: Oxford University Press.

Livingstone, D., & Withers, C. (Eds.). (1999). *Geography and Enlightenment.* Chicago: University of Chicago Press.

Porter, R., & Teich, M. (Eds.). (1981). *The Enlightenment in national context.* Cambridge, UK: Cambridge University Press.

Schmidt, J. (Ed.). (1996). *What is Enlightenment? Eighteenth-century answers and twentieth-century questions.* Berkeley: University of California Press.

ENVIRONMENTAL DETERMINISM

Environmental determinism is the doctrine that individual human actions, beliefs, and values are controlled or determined by the ambient environment. Accordingly, when applied to aggregates, as the doctrine normally is, societies, cultures, and civilizations are also held to be the product of their environments.

This doctrine or perspective on human–environment relations is among the oldest and most enduring ways of looking at, and conceptualizing, humans' place and condition in the world. Within geography conceived as the formal study of the earth's surface, environmental determinism has been an evident, and at times dominant, approach. Within geography's broader scope—that of humanity's individual and collective knowledge of the earth's surface (its places, patterns, and processes)—environmental determinism has been an evidential and fundamental feature of this thinking. No doubt, flickers of environmental determinist thought were part of our earliest cognitive awareness. Primitive cosmologies and religions are largely constituted on premises of natural forces' agency over human thought and action. Much of their ritual practice revolves around propitiating these forces of nature and environment. Over the past half century or more, there have been concerted campaigns within formal geography to counter and discredit environmental or geographic determinism. Nevertheless, it is a doctrine that continues to be retooled and deployed in cognate fields, and its persistence in popular thought is pervasive and seemingly permanent.

The ancient Greeks theorized and authored the first formal expressions of environmental determinism that are clearly part of geography's own scholarly past productions. The Greeks had well-developed ideas about the relation between climate, hydrology, vegetation, soil, relief features, and local to global locations and their controls or influences on individual behavior as well as on collective cultural attributes, attitudes, and actions. Climatic conditions in particular were linked to psychophysiological states of well-being or malady. In turn, from Hippocrates (fifth century BC) onward, the doctrine of the four humors (blood, phlegm, yellow and black biles) held that bodily balances and imbalances, and hence health itself, were determined largely by environmental factors. The Hippocratic treatise *Airs, Waters, Places* served as one of the main bases of Western medical theory for the next two millennia or more. Implicitly, it also served as a fountainhead of environmentalist thought for an equal duration. Whereas many Greek philosophers and scholars expressed elements of this psychophysiological body of thought, virtually all subscribed to the notion that environmental factors influenced the lives and ways of different peoples and their cultures. Herodotus's ethnographic and historical observations on differing peoples through the ancient ecumene or inhabited world are encyclopedic and offer many instances of environmental–cultural influences. Aristotle is perhaps the most cited proponent of climate plus location equaling potential for civilizational fruition. People of cold regions, particularly Europe, were spirited and freedom loving but lacked skill and intelligence. People of hot regions, particularly Asia, lacked spirit but had skills and intelligence despite their tendencies toward subjection and despotism. Greeks occupied the intermediate climatic and location regimes most suited to achieving the golden mean. Aristotle and other philosophers commented in detail on specific environmental effects such as the insalubrity of marshes, the qualities of different winds, and alluvial soils versus stony soils. In addition to specific local or regional environmental effects, the Greeks conceptualized global controls by latitudinal zones. Aristotle posited that the torrid zone, or the lands closest to the equator, were uninhabitable and that those most distant, the frigid zone, were equally uninhabitable. Thus, the lands between these extremes, the temperate zone, were the only ones suited for human habitation. Within the temperate zone, the Mediterranean littorals were ideally situated. Most Greek thinkers, and later the Romans, agreed with Aristotle on his appraisal of the temperate latitudes as being most suited for the perfection of human habitation and civilization.

From the fall of Rome (ca. 500 AD), through the European Middle Ages (500–1500 AD), and on into the Renaissance (ca. 1500–1650 AD), environmentalist doctrines at both the body–physiological and regional–locational scales were staples of geographic thought and theory. Afro-Asiatic traditions of geographic thought and practice also embodied environmental determinist concepts and outlooks. During the Middle Ages, Muslim scholars such as Ibn Khaldun (1332–1406) accepted much of the environmentalist theory of Greco–Roman antiquity but also inflected their own compendious geographies with environmentalist observations. The Chinese philosophy of Feng Shui, which guides the placing of buildings in relation to environmental features, reflects environmentalist currents within Chinese geographic practice during this time as well as before and subsequently. Medieval Christian scholars such as Thomas Aquinas and Albertus Magnus gave detailed attention to environmentalist explanations of human behavior according to geographic features and location. With the onset of the Age of Discovery and the Renaissance (ca.

1500–1650 AD), encounters with new lands and new and rediscovered ideas created new contexts for environmental determinist thought. French political philosopher Jean Bodin (1530–1596) was the key environmental theorist of this period. He drew heavily on Greco–Roman environmentalist theory but modified it with the flood of new geographic knowledge that inundated Europe during the 16th century. His robust conceptions of both history and politics as determined by geographic factors, especially climate, laid the foundations for important expressions of environmentalism during the Enlightenment (ca. 1650–1800 AD).

The Enlightenment, or the "Age of Reason," was also a high point of environmental determinist theorizing. Charles de Secondat Baron de Montesquieu (1689–1755) was the most influential and celebrated of these theorists. His *The Spirit of Laws* (1748) offered a universal treatise on politics and governance. Its elaborate arguments are based foremost on environmentalist reasoning and examples. Climatic conditions offer an overarching explanation for cultural and historical differences and customs. Environmental factors such as relief features, soils, and relative locations account for subvariations and outcomes. Although environmentalist arguments were voiced by many Enlightenment figures, including Jean-Jacques Rousseau and Denis Diderot, they also were challenged. Voltaire, David Hume, and Count Buffon all found fault with the doctrines of climatic causation. This debate carried over into the 19th century and formed one of the central axes on which modern geography was constructed. Nor was there a clear separation between the opposing camps until the early 20th century, when the refutation of environmental determinism became one of the central organizing principles of several different strands or schools of geography. For example, in the work of geography's modern founders, Alexander von Humboldt and Carl Ritter, one can find both appeals to environmentalist explanation and counterevidence. Ritter, however, relied on it heavily, whereas Humboldt resorted to it only occasionally. Toward the end of the 19th century, Friedrich Ratzel attempted to put environmentalist theory at the center of a new and scientific human geography. The first volume of his *Anthropogeographie* made the case for this approach. His training in zoology and his evolutionary perspective—more neo-Lamarkian than strictly Darwinian— underwrote the concepts he formulated. Chief among these were the ideas of *Lebensraum* (or "living space") and the state-as-organism. Ratzel's followers, particularly Ellen Churchill Semple, popularized Ratzel's concepts and introduced an explicitly geographic environmentalism to a broad Anglophone audience. Other American geographers, such as Nathaniel Shaler, William Morris Davis, Albert Brigham, and especially Ellsworth Huntington, helped to make environmentalism the main mode of explanation in American geography during the period circa 1890 to 1920. In its strong forms, turn-of-the-century environmental determinism helped ideologically to legitimate social Darwinism, racism, eugenics, colonialism, and other manifestations of the European and North American drive for global supremacy.

Despite Ratzel's initial influence, by the 1920s many European geographers found simplistic environmentalism wanting of both substance and relevance. French historians and geographers counterposed *possibilism* as a more nuanced understanding of human–environment relations. Carl Sauer was among the first American geographers to subject environmentalism to sharp critique and to reject it as either a theoretical or a methodological program for geography. By the 1930s, chorology or the regional approach had largely replaced environmentalism as the main focus of academic geography. By the 1950s, environmental determinism in geography had been largely discredited. Variants of what might be interpreted as environmental determinism continued to be advanced under the banner of Soviet Marxist geography into the 1960s. At the same time, scholars in some of geography's cognate disciplines have resuscitated environmental determinism to explain the uneven developmental trajectories of societies in widely differing historical periods, but particularly in modern times. Even today, in many of these appraisals, the world's tropical lands are inherently doomed to marginal roles and returns compared with those of temperate climes and latitudes.

—*Kent Mathewson*

See also Anthropogeography; Berkeley School; Cultural Geography; History of Geography

Suggested Reading

Glacken, C. (1967). *Traces on the Rhodian shore: Nature and culture in Western thought from ancient times to the end of the eighteenth century.* Berkeley: University of California Press.

Peet, R. (1985). The social origins of environmental determinism. *Annals of the Association of American Geographers, 75,* 309–333.

Sauer, C. (1925). The morphology of landscape. *University of California Publications in Geography, 2*(2), 19–53.

Tatham, G. (1951). Environmentalism and possibilism. In G. Taylor (Ed.), *Geography in the twentieth century* (pp. 128–164). London: Methuen.

ENVIRONMENTAL JUSTICE

During recent years, a large body of research has emerged suggesting that poor people and people of color suffer a disproportionate burden with exposure to environmental hazards and in particular accompanying the siting of waste management facilities. The term *environmental justice* itself is contested. Many proponents, be they community-based activists or public agencies (e.g., the U.S. Environmental Protection Agency), define it as a situation where no people, regardless of race, national origin, or income, are forced to shoulder an unequal burden and all are treated fairly with regard to the enforcement of environmental regulations.

In response to community demands for greater participation in the decision-making process, public agencies recently have begun to accept environmental justice as also entailing meaningful involvement by potentially affected communities in siting decisions affecting their health. As a result, environmental equity, consisting of both distributional equity and procedural equity, is commonly accepted as necessary to attain environmental justice. However, combining insight from the pollution prevention movement, and its focus on upfront toxic use reduction in place of pollution management, with organized labor's quest for greater democracy in the workplace, a number of scholars and community activists challenge this conventional understanding of environmental justice. For them, reliance on liberal notions of procedural and distributional equity, typically implemented through negotiation, mitigation, and fair share allocation among targeted communities, merely perpetuates the current production system that is, by its very structure, discriminatory and nonsustainable. These environmental justice advocates reject environmental equity as sufficient to attain environmental justice. In its place, they propose production justice, where the very structure of the production system itself is changed through democratic control over the decision to pollute and, by extension, over the decision to produce. Insofar as this requires class unity across political and national borders, the quest for progressive environmental justice has been inspired, but also severely hampered, by the globalization of capital.

Although the struggle against environmental contamination and dislocation by external forces began, at least in North America, with the arrival of Europeans and the subsequent war on native people and continued through slavery, resource extraction, and industrialization where both race and class were determining factors in risk exposure and community resistance, most observers date the modern environmental justice movement (EJM) back to the late 1970s and early 1980s. The class- and race-based components of the EJM can be traced, respectively, to local resistance at the predominantly white working-class community of Love Canal—America's most famous Superfund site—and, a few years later, to the arrests of more than 500 people for protesting the siting of a storage facility for PCB-contaminated soil in a poor African American community in Warren County, North Carolina. The latter protest was particularly important for development of the EJM because it led to the first nationwide survey of the demographic determinants of hazardous waste facility siting. Sponsored by the United Church of Christ (UCC), the study suggested that race, rather than income, was the single most significant determinant when accounting for disproportionate siting and, furthermore, that this was not mere coincidence but rather the result of what the report termed *environmental racism.*

The landmark UCC report, in turn, led to a flurry of research, much of it conducted by geographers, attempting to determine whether a disproportionate siting burden occurred at various levels of analysis, be this with county, census track, or zip code units. Researchers also considered whether it could be considered environmental injustice when the offending activity was in place before the poor and people of color moved in—a determination derided by critics as a meaningless "chicken or egg" debate when one acknowledges the institutional racism limiting free choice. At the grassroots level, the UCC report in turn provided justification for the growing EJM, eventually encompassing hundreds of communities (e.g., Kettleman City, California; Sierra Blanca, Texas; Chester, Pennsylvania; Geismer, Louisiana; and on many Native American reservations). Here the struggle against what was experienced as environmental racism often drew strength and guidance from the earlier civil rights movement and the ongoing American Indian movement.

At the regulatory level, following President Bill Clinton's 1994 Executive Order 12898, federal agencies are required to identify and address disproportionate and adverse human and environmental health impacts of their programs and activities. By extension, through federal funding and permitting requirements, many state agencies have followed suit, although the results have been uneven, with equity remaining the professed goal and yet rarely being achieved in practice.

Subsequent research has demonstrated that there is no universal explanation for the siting outcome that can be applied across time and space. Broad surveys of environmental racism suffer from lack of agreement on the proper scale of analysis, characterization of both the risk and affected population, and uncertainty over intentionality in siting decisions. Laura Pulido demonstrated that racism itself is a dynamic social and spatial process that cannot be reduced to simple overt action. Thus, uneven impact is likely due to the dominant social structures, practices, and ideologies that actually reproduce the privileged status of white people on a broad scale over time. Furthermore, insofar as geography matters when conceptualizing environmental justice issues—what one sees depends on where one looks—disproportionate impact in rural white areas, particularly with the location of a new generation of large regional solid waste repositories, has been on the rise. Hence, class discrimination, and lack of access to the actual decision to pollute in the first place, also enters the siting equation, as has been suggested through a number of recent studies in places such as rural Pennsylvania and Kentucky.

—Michael Heiman

See also Justice, Geography of; Race and Racism

Suggested Reading

Cutter, S., Holm, D., & Clark, L. (1996). The role of scale in monitoring environmental justice. *Risk Analysis, 16,* 517–526.

Farber, D. (Ed.). (1998). *The struggle for ecological democracy: Environmental justice movements in the United States.* New York: Guilford.

Heiman, M. (1966). Waste, race, and class: New perspectives on environmental justice. *Antipode, 28,* 111–121.

Pulido, L. (2000). Rethinking environmental racism: White privilege and urban development in Southern California. *Annals of the Association of American Geographers, 90,* 12–40.

United Church of Christ Commission for Racial Justice. (1987). *Toxic waste and race in the United States.* New York: United Church of Christ.

ENVIRONMENTAL PERCEPTION

Environmental perception refers to the subjective ways in which groups and individuals perceive and evaluate their environment. As a subfield of cultural and behavioral geography, environmental perception is not limited to the natural environment; rather, it includes factors such as built structures, customs, values, and other individuals or groups. Thus, studies of environmental perception highlight the discrepancies between individual and group choices based on their perceived environment and their actual environment. Geographers who study environmental perception assume that an understanding of space and place is fundamental to how individuals and groups perceive and experience their particular environment and the resulting behaviors in which they engage as a product of this understanding.

Initially conceived from the desire to situate empiricist methodology within a theoretical framework and from the view of geography as a spatial science, the concept of environmental perception draws from a multitude of disciplines, including (but not limited to) experimental psychology, neoclassical economics, anthropology, history, and computer science. Kevin Lynch's famous book *The Image of the City,* published in 1960, is often cited as one of the seminal works of environmental perception. Lynch discussed mental maps of urban landscapes in Boston, Jersey City, and Los Angeles and argued that an individual's perception of a city is linked closely to his or her relationship with the city—the individual's age, gender, ethnicity, educational level, ability to drive, length of residence in the area, and so on. This led to a tradition of cognitive mapping and spatial perception. Similarly, William Ittelson argued that environments surround individuals at multiple scales and that those individuals do not observe the environment so much as they explore it. More recently, geographers have evaluated the role of environmental perception in environmental policy and risk assessment.

—Micheala Denny

See also Behavioral Geography; Cognitive Models of Space; Cultural Geography; Humanistic Geography; Phenomenology

Suggested Reading

Aitken, S., Cutter, S., Foote, K., & Sell, J. (1989). Environmental perception and behavioral geography. In G. Gaile & C. Willmott (Eds.), *Geography in America* (pp. 218–238). Columbus, OH: Merrill.

Brookfield, H. (1969). On the environment as perceived. *Progress in Geography, 1,* 51–80.

Ittelson, W. (Ed.). (1973). *Environment and cognition.* New York: Seminar Press.

Liverman, D. (1999). Geography and the global environment. *Annals of the American Association of Geographers, 89,* 107–120.

Lynch, K. (1960). *The image of the city.* Cambridge: MIT Press.

Saarinen, T. (1999). The Euro-centric nature of mental maps of the world. *Research in Geographic Education, 1*(2), 136–178.

Tuan, Y.-F. (1974). *Topophilia.* Englewood Cliffs, NJ: Prentice Hall.

EPISTEMOLOGY

Epistemology is an area of inquiry in the discipline of philosophy. It is generally concerned with the study of the sources, forms, and conditions of knowledge. Whereas ontology seeks to contemplate the question of the nature and modes of being and asks questions about what exists and in what form it exists, epistemology addresses the problem of how we can get to know these different possibilities of existence. Thus, epistemology is concerned with the relationship between "what there is in the world that we can get to know" (objects, materials, etc.) and "how we can get to know it" (the methods of acquiring knowledge).

Numerous approaches to explaining the relationship between knowledge and the world have been established, most notably empiricism, rationalism, realism, pragmatism, and constructivism, all of which had significant impacts on the discipline of geography and the conceptualization of its core concepts such as space and place. In general, these approaches can be divided into foundational and nonfoundational epistemologies. These two approaches have different consequences for the ways in which geographic knowledge is acquired and disseminated.

Until recently, foundationalism was the dominant epistemological underpinning of all geographic inquiry.

It begins with the assumption that knowledge develops in the human mind and that a tangible reality exists in the world "out there." It argues that there are correct (legitimate) and incorrect (illegitimate) ways in which the human mind can gain knowledge of the world—of how it can get the world into the mind. In general, the most accepted approach to bringing the world into the mind has been the scientific method. It is the process by which scholars collectively and repeatedly attempt to construct a reliable, coherent, and nonarbitrary representation of the world. To do so, scientists seek to lay aside personal beliefs and traditions in their interpretation of nature and culture and to instead use standardized and widely accepted methods to examine reality and develop an abstract theoretical model of real entities. Ultimately, the scientific method aims at reducing bias or preconceptions by the researcher when developing a theory or testing a hypothesis. It assumes that undeniable facts and clear and distinct ideas and concepts—a certain detectable order—exist in the world and that humans can bring this world into the mind by continuously examining their surroundings with their senses. Furthermore, this knowledge is consistently reevaluated, questioned, and updated. Knowledge of the world is composed of what individuals gather, compare, exchange, and combine into a logical, testable, and transparent apprehension or model of this world. Space is conceptualized as a tangible, objective, quantifiable, qualifiable, and verifiable entity that can be described and measured. All foundational epistemologies argue that geographic theories, abstractions, descriptions, or models of reality can directly represent the reality of the world out there—outside of the human mind. Although present throughout most of geography's disciplinary history and still of notable importance, foundational epistemologies had the greatest influence on geography following the quantitative revolution of the 1950s and the subsequent influence of positivism as the dominant mode of scientific enquiry.

In contrast to such scientific approaches that assume reality to be a knowable objective entity out there and open to inquiry by the human mind, nonfoundational epistemologies negate such possibility of knowledge acquisition. Whereas foundational epistemologies seek to establish grand theories that can universally explain human and natural phenomena, nonfoundational epistemologies not only regard such attempts as impossible but also deny the objective and unbiased character of the scientific method. Feminism,

Marxism, poststructuralism, and postmodernism all offer different critiques of foundational epistemologies that especially target the role of the researcher in the scientific process and emphasize the limitations of language providing adequate representations of the world. For example, both feminist and Marxist philosophies stress that the process of generating scientific knowledge is far from being governed by objective, value-free research; rather, it is guided by political ideologies that represent the interests of dominating groups such as men, political majorities, and ethnic groups. Subsequently, the main goal of feminist and Marxist geographies is to decode the ways in which spaces and places have been inscribed with ideological representations. They show that space is not innocent but rather always a value-laden entity that is guided by the vested interests of certain groups within a society and that must be examined as such. Furthermore, these critical approaches denounce foundational epistemologies, especially positivism and the scientific method, as a decidedly male invention that overlooks the different ways in which knowledge can be gained. In the past, male-dominated and supposedly objective research based on the scientific method often has left unexamined the voices of marginalized and oppressed people such as women, certain racial groups, and sexual minorities. Poststructuralist and postmodern critiques of foundationalist epistemologies then add another point of contention. Whereas foundationalist epistemologies assume that the language, texts, and visual representations of reality can supply us with adequate models and theories about the world around us, nonfoundationalist approaches argue that such an a priori assumption is illusory. Language as a tool of communicating geographic knowledge is always providing us with incomplete representations of the world; whenever something is said, something else is left silent or silenced. All nonfoundational epistemologies argue that geographic theories, abstractions, and descriptions or models of reality can never directly represent the reality of the world out there—outside of the human mind. Instead of the construction of grand theories and universal models of geographic patterns and behaviors, they argue for the deconstruction of such models to reveal their often biased and ideologically colored representations of the world. In addition, nonfoundationalist epistemologies favor the construction and representation of a diversity of local knowledges; they prefer microexplanations and acknowledge the

perpetual incompleteness of scientific explanations. Ultimately, nonfoundational epistemologies seek to overcome the need for an epistemology as a whole and reject the possibility of certainty and universality in scholarly inquiries.

—*Olaf Kuhlke*

See also History of Geography; Ideology; Ontology; Postmodernism; Poststructuralism

Suggested Reading

Cloke, P., Philo, C., & Sadler, D. (1991). *Approaching human geography.* New York: Guilford.

Dear, M. (1994). Postmodern human geography. *Erdkunde, 48,* 2–12.

Gregory, D. (1994). *Geographical imaginations.* Oxford, UK: Blackwell.

ETHICS, GEOGRAPHY AND

Early definitions of ethics within geography focused on ethics as a means of distinguishing between good and bad and between right and wrong. Later definitions focused on ethics as the study of morality and of making moral judgments. This shift from prescriptive to relational definitions reflects the influence of the cultural turn within geography. Although the emphasis has changed, the definitions share a common core. This is an understanding of ethics as the evaluation of human conduct. The conceptualization of ethics within geography works in a range of ways—in broad theoretical debates about the relationship between geography and ethics and in debates about geography as a discipline and about individual behavior and choices.

Broad theoretical debates about the relationship between geography and ethics tend to focus on concepts such as space, place, nature, environment, development, and technology. With this focus, the emphasis is on the ways in which ethics and geography intersect in addressing these concepts and concerns. Some geographers are interested in the ontological basis of the intersection. In this context, a concern with ontology—theories of being—suggests that we consider ways of constructing and maintaining ethical relationships with ourselves, with others, with places, and with environments. Other geographers are interested in the epistemological basis of this intersection. In this context, a concern with epistemology—theories of knowing—suggests the need to develop ethical ways of knowing about ourselves, others, places, and environments. A range of different theoretical approaches are used to consider these questions, including realism, relativism, and (most recently) poststructuralism. The distinctions between ontology and epistemology, however, are not always clear. From the perspective of geographers concerned with questions of ethics, ontology and epistemology often are considered as interdependent. Thus, these broad theoretical debates have been operationalized through two main arenas. The first is in relation to the discipline of geography, and the second is in relation to the ethical behavior of individual geographers.

In considering the discipline of geography, two strands of inquiry are apparent. The first relates to the place of ethics in geography, and the second relates to the place of geography in ethics. Early attempts to consider the place of ethics in geography were instigated during the 1960s and 1970s. These included the work of Marxist and humanist geographers concerned with issues of social relevance, social justice, and values in geography. These concerns have continued to be of importance to geographers interested in issues of ethics, but the range of concerns has since expanded. During recent times, there has been particular interest in issues of development ethics and environmental ethics, and there is an emerging concern with the ethics of the relationship between human and nonhuman entities. The increasing significance of considerations of environmental ethics is illustrated by the establishment in 1998 of *Ethics, Place and Environment,* a journal devoted to the study of geographic and environmental ethics. In considering the place of geography in ethics, there has been a particular emphasis on what has been termed *descriptive ethics.* This refers to detailed descriptions of the ways in which the relationships between people and places construct, reinforce, or challenge ethical (or unethical) beliefs and practices. This set of relationships often is described as a moral geography and often is concerned with identifying the right and wrong places for particular kinds of actions. Both strands of inquiry tend to have normative aspects, particularly in relation to the ethics of spatial and social inequalities and injustices and in relation to their identification and amelioration.

When geographers write about ethics at the level of the individual, they usually are concerned with ethical behavior in relation to research and teaching. Feminist geographers have been at the forefront in highlighting the need for ethical behavior in relation to research projects and research subjects. In so doing, they draw attention to questions of power relationships—between researchers and the "researched," among groups of research subjects, and between the academy and the wider community. They encourage a reflexive approach to the research process and the development of collaborative and transformative research projects. Feminists also have been at the forefront in considering the ethics of teaching, both in terms of the relationships between teachers and students and in terms of pedagogy. However, the formalization of research and teaching guidelines in institutional settings such as universities means that ethical behavior at the level of the individual is increasingly proscribed in terms of risk mitigation. This leads to a narrow understanding of ethics and ethical behavior as adherence to a set of institutionally defined rules rather than as a set of moral judgments.

Thus, the consideration of ethics in geography takes a variety of forms, from debates about individual behavior to debates about theoretical stances. Discussions of ethics permeate much work in geography, although not always in an explicit form. Many geographers engage implicitly with ethical issues through their consideration of terms such as *justice, responsibility,* and *rights.* Other geographers engage implicitly with ethical issues through their consideration of or involvement in activism. Although geographers today are less likely to discuss ethics in terms of right and wrong, concern with a normative vision for geography still permeates geography's engagement with issues of ethics.

—*Mary Gilmartin*

See also Environmental Justice; Existentialism; Feminist Geographies; Feminist Methodologies; Humanistic Geography; Justice, Geography of; Marxism, Geography and; NIMBY; Poststructuralism; Radical Geography; Social Movement; Spatial Inequality; Urban Social Movements

Suggested Reading

Fuller, D., & Kitchin, R. (Eds.). (2004). *Radical theory/ Critical praxis: Making a difference beyond the academy?* [Online]. Available: www.praxis-epress.org/availablebooks/ radicaltheorycriticalpraxis.html

Nast, H. (1994). Women in the field: Critical feminist methodologies and theoretical perspectives. *The Professional Geographer, 46,* 54–102.

Popke, E. (2003). Poststructuralist ethics: Subjectivity, responsibility, and the space of community. *Progress in Human Geography, 27,* 298–316.

Proctor, J., & Smith, D. (Eds.). (1999). *Geography and ethics: Journeys in a moral terrain.* London: Routledge.

Smith, D. (2000). *Moral geographies: Ethics in a world of difference.* Edinburgh, UK: Edinburgh University Press.

ETHNICITY

Ethnicity is a difficult concept to define. It is a relatively recent term; the first recorded use of the term was during the 1940s, and it first appeared in a dictionary in 1972. It is, however, linked to *ethnic,* which has a significantly longer history. The term *ethnic* originally referred to people who were neither Christian nor Jewish, but by the 19th century it had come to refer to the (often racialized) characteristics of particular groups. Ethnicity refers, therefore, to the characteristics of groups that allow those groups to be understood or perceived as distinct. However, ethnicity also refers to how individuals understand their participation in, and identity in relation to, those particular groups. As such, ethnicity refers to individual *and* collective senses of identity.

There is disagreement over the characteristics that identify ethnicity. Within geography, earlier understandings of ethnicity focused on biological and cultural aspects of group identity. These included race, religion, language, similar cultural practices, and a sense of a common or shared history. Ethnicity, in these early definitions, was also associated with minority status, particularly within national boundaries. Over time, the focus shifted from ethnicity as defined in terms of shared attributes to ethnicity as the *perception* of common identity. Now the idea of ethnicity as a social construct is prevalent, with some geographers theorizing the various ways in which ethnicity and ethnic identity work by constructing and maintaining difference. In addition, some geographers have started to take issue with the idea of ethnicity as linked to minority status, arguing that everyone—including members of majority groups—has an ethnicity. These changing definitions of ethnicity within geography draw liberally on the work of other social scientists, such as sociologists and anthropologists, as well as on the cultural turn within geography.

There is much debate on the form of the relationship between ethnicity and race. In some instances, commentators see ethnicity as a subset of race, where each category among a small number of racial categories contains a greater number of ethnic categories. In other instances, commentators describe ethnicity and race as virtually interchangeable. There are difficulties with both of these approaches. With the first, it is problematic to assume that humans can be neatly categorized in such a way, particularly with the general acceptance that there is no biological basis to the category of race. With the second, describing ethnicity and race as interchangeable ignores the fact that most societies tend to treat ethnic groups quite differently from races and that their treatment is often more benign. So, although there are similarities in the concepts of ethnicity and race, particularly because they represent interactions between diverse populations, it is important to realize that ethnic groups are not necessarily racial groups, that racial groups are not necessarily ethnic groups, and that both ethnicity and race are social constructs that have different meanings in different contexts. Some commentators distinguish further, arguing that racial categories are imposed, whereas ethnicity is a process of group self-definition.

The relationship between ethnicity and nationalism has also received attention from geographers. Nationalism may be broadly defined both as a feeling of belonging to a nation and as a desire for that nation to have sovereignty over a specific territory. There are obvious similarities between ethnicity and nationalism. Nationalism, like ethnicity, is concerned with establishing and maintaining a sense of collective identity. Nationalism is also concerned with drawing and enforcing boundaries. Nationalism also may be based on, reinforce, or reinvent ethnic ties, particularly in the event of the conquest of a territory by an external power. However, it is possible to distinguish between ethnicity and nationalism by arguing that nationalism is intrinsically concerned with territory and sovereignty, whereas ethnicity may be but is not necessarily so. It is also necessary to recognize the often conflicted status of minority ethnic groups within nationalist movements.

Geographers interested in issues of ethnicity have paid particular attention to processes of spatial segregation, integration, and assimilation, especially in urban areas. The concern with ethnic spatial segregation can be traced, to a large extent, to the work of the Chicago School of urban sociology, most notably that of Robert Park, who argued, in his theory of human ecology, that social distance and spatial distance were interrelated. Drawing on the work of Park and others, and influenced by the techniques of the quantitative revolution and by a concern with the need for social relevance, some geographers sought to identify and delineate ethnic ghettoes to devise measures of segregation and integration such as indexes of dissimilarity and to use this mapping and measurement to inform policies directed toward ethnic integration and assimilation. American geographers initially used these techniques and approaches to consider the place of African Americans within American society, whereas geographers in the United Kingdom were initially interested in the extent of segregation between Catholic and Protestant communities in Northern Ireland, particularly Belfast. Within Britain, the focus extended to include black and Asian social and spatial segregation, particularly in cities such as London, Leicester, and Bradford. Since the 1970s, however, there has been growing dissatisfaction with this approach to the study of ethnicity. Its critics argue that (a) the concept of a ghetto is a pejorative "ethnic stereotype" that fails to adequately capture the complexities of ethnic social and spatial identities and that (b) measures of segregation are based on census data that often are incomplete and flawed.

Studies of ethnicity that aim to map and measure tend to have the unintended effect of fixing ethnic groups in place. In contrast, studies that are concerned with ethnicity as a social construct are interested in the conflict between fixity and mobility. Some studies aim to show the ways in which migration alters ethnicities, both for those who are migrating and for those who are living in the places to which migrants are moving. Other studies aim to show the ways in which ethnicities change in place. Still other studies aim to show how the construction of ethnic identities, particularly when those ethnicities are described in minority terms, is as much about the ethnicities of dominant or hegemonic groups as it is about minority groups. These three concerns come together in recent work in North America. Research on migrant groups in the United States, such as Irish and Italians, has shown how the construction of ethnic identity served as a form of protection for new immigrants but also served to reinforce ideas of racial superiority as migrant groups reinvented themselves as white. And within the discipline of geography, Kay Anderson's research highlighted the ways in which Chinatown in Vancouver, British Columbia, rather than being an organic expression of Chinese identity, worked instead as a (white) European construction. As such, for whites in Vancouver, Chinatown became the site onto which disease and depravity were displaced. Research on the social construction of ethnicity, particularly in connection with the relationship between ethnicity and place, is also being applied in other national settings, most notably Australia. In many instances, ethnicity and race are conflated as researchers seek to understand the various ways in which these concepts serve to distinguish between and discriminate against particular groups of people.

More recently, the issue of ethnic conflict has started to receive attention, particularly in relation to the practice of *ethnic cleansing*. Ethnic cleansing, a form of genocide, refers to the forced removal of an ethnic group from a territory claimed by another ethnic group or state. Bosnian Muslims, during the breakup of the former Yugoslavia, were victims of ethnic cleansing that included murder, starvation, and sexual assault such as rape. The term is also used to describe the treatment of European Jews during the Holocaust and the treatment of Tutsi by Hutu during the Rwandan genocide. However, interest in ethnic conflict is not confined to this issue. Other areas of interest include ethnic mobilization and ethnic politicization, often among minority or migrant groups and in a range of national and urban settings. Thus, geographers interested in the topic of ethnic conflict address the range of ways in which concepts of ethnicity are mobilized in the interests of warfare, power, politics, and territoriality.

Despite the move to social constructionist understandings of ethnicity, there remains a strong commitment to the study of ethnicity using the more traditional mapping and measurement techniques. This approach currently is undergoing a resurgence, in part as a response to new patterns of migration to Europe, North America, and Australia. As a consequence, some geographers are again involved in identifying patterns of residential and social segregation and integration from the perspective of migrant groups. Others are moving beyond census categories to other markers of ethnicity, most notably religion. Religion has, once again, become a significant area of study for geographers interested in ethnicity. The focus of these new research projects is generally on religions regarded as minorities within the Western societies where research is taking place. In particular, there has been a noticeable increase in research on Muslim communities, with a minor focus on Hindu, Sikh, and Jewish communities. Western Christianity receives limited attention. A small number of geographers are addressing issues of language. Again, however, the focus often is on minority languages, such as Welsh, rather than on the relationship between ethnicity and more widely spoken languages such as English. Thus, although some geographers argue that ethnicity should not be used only to describe minority status, in practice many geographers continue to conduct research as though ethnicity and minority status are implicitly connected. The study of ethnicity within geography continues to be marked by this tension and related tensions—whether to study ethnicity in terms of clearly defined categories that can be quantified or in terms of the complexity of the concept of ethnicity, and the interconnectedness of different ethnicities and of ethnicity with race, class, gender, and other markers of identity.

—*Mary Gilmartin*

See also Chicago School; Cultural Geography; Cultural Turn; Diaspora; Ethnocentrism; Eurocentrism; Ghetto; Identity, Geography and; Migration; Nationalism; Race and Racism; Religion, Geography and/of; Segregation; Social Geography; Urban Underclass; Whiteness

Suggested Reading

Anderson, K. (1991). *Vancouver's Chinatown: Racial discourse in Canada, 1875–1980.* Montreal: McGill–Queen's University Press.

Boal, F., & Douglas, N. (Eds.). (1982). *Integration and division: Geographical perspectives on the Northern Ireland problem.* London: Academic Press.

Dwyer, C. (2002). "Where are you from?" Young British Muslim women and the making of home. In A. Blunt & C. McEwan (Eds.), *Postcolonial geographies* (pp. 184–199). New York: Continuum.

Jackson, P., & Smith, S. (Eds.). (1981). *Social interaction and ethnic segregation.* London: Academic Press.

Peach, C. (2000). Discovering white ethnicity and parachuted plurality. *Progress in Human Geography, 24,* 620–626.

Roediger, D. (1991). *The wages of whiteness: Race and the making of the American working class.* London: Verso.

Zelinsky, W. (2001). *The enigma of ethnicity: Another American dilemma.* Iowa City: University of Iowa Press.

ETHNOCENTRISM

Ethnocentrism is the process by which people understand other cultures using their own culture as the norm. The term was coined by William Graham Sumner, a late-19th-century Yale University sociology professor. Ethnocentrism involves making assumptions about other cultures based on a limited experience of them. All groups can be ethnocentric, but it is often most obvious among Western academics who privilege a Western viewpoint without acknowledging how this limits their work. Often this process involves valuing familiar cultures more than unfamiliar ones and can lead to discrimination against cultures different from one's own. People are usually not aware that they are being ethnocentric because it is very difficult to identify the assumptions on which the behavior is based. Our realities are built on our experiences, and when we have new and different experiences, it is only normal to evaluate them based on our own realities. But using the standards of one culture to judge another culture does not work. Ethnocentrism is problematic because it usually leads to misunderstandings. To study and interact with different cultures, it is necessary to develop an awareness of ethnocentrism.

Ethnocentrism is an important concept in the history of both geography and anthropology. Throughout the 19th and early 20th centuries, both disciplines tended toward ethnocentrism, and studies of non-Western cultures saw these cultures as primitive or in need of development that would make them more like Western cultures. During the mid-20th century, social scientists from both Western and non-Western cultures began to question the ethnocentric manner in which much research was carried out. To counter this trend, the concept of cultural relativism was developed so that all cultures would be treated in similar ways and not prejudged based on familiarity or difference. This theory holds that with time, patience, and an open mind, we can learn to understand other cultures as they understand themselves. Ethnography, or the study of daily life, is considered a very good way in which to counter ethnocentrism. When we study other cultures in their own context, it is easier to understand their differences from our own cultures. The concept of situated knowledge, popular in feminist geography, helps us to identify our particular political, economic, and social positions to better understand how and from where we produce knowledge about others.

Examples of ethnocentric behavior include seemingly innocent comments such as "the British drive on the wrong side of the road" and "the Spanish are lazy and that's why they take a siesta every day." A good way in which to avoid ethnocentrism is to stay away from generalizations that are judgmental and do little to help understand differences in cultural practices.

—*Rebecca Dolhinow*

See also Cultural Geography; Ethics, Geography and; Ethnicity; Eurocentrism; Orientalism; Other/Otherness; Race and Racism; Situated Knowledge

Suggested Reading

Said, E. (1978). *Orientalism.* New York: Pantheon.

EUROCENTRISM

History and the social sciences have a long history of viewing the West, whatever that might be, as the so-called motor of history, that is, as the dynamic power that instigates change and progress while the rest of the world passively waits for the benefits of its wisdom and wealth. The doctrine that upholds the West as inherently superior to non-Western cultures is Eurocentrism, which has a long historical record. The very definition of Europe, for example, may be traced back to Vasilli Tatischev, Peter the Great's geographer during the 18th century, who defined the east end of Europe at the Ural Mountains as part of the Russian elite's desire to differentiate Europe from Asia as meaningful entities. Similarly, Georg Hegel, who viewed world history as the product of a spirit (or *geist*), argued that it reached its zenith in the European nation-state. Karl Marx viewed Western capitalism as alive and dynamic, in contrast to the static Asiatic mode of production. Max Weber ascribed rationality to European cultures, especially Protestant ones, dismissing Islam, Hinduism, and Confucianism, a view that undergirded modernization theory. In 20th-century environmental determinism, Karl Wittfogel advocated a hydraulic theory that dismissed the possibility of democracy in Asia as Oriental despotism. All of these perspectives put the West on center stage as the engine that drives the world economy, with everyone else allegedly hanging on.

The assumption of European and, by extension, Western superiority takes a variety of forms. In earlier versions, it hinged on a crude racism (e.g., the "white man's burden"). Martin Bernal, in *Black Athena,* noted that 19th-century European historians constructed a mythology in which Europe invented itself without reliance on earlier, wealthier, and darker-skinned cultures such as the Egyptians and Phoenicians.

James Blaut traced a model of history he called the Orient Express, in which the locus of progress moves from Southeast Europe (classical Greece) to the Northwest. Subsequently, via colonialism, everything good, progressive, innovative, and productive is held to diffuse out of Europe.

Edward Said's highly influential book *Orientalism* opened new ground in the discovery of Eurocentrism. In this view, colonialism was every bit as much a cultural and ideological project as it was economic and political in nature. Orientalism was the flip side of Eurocentrism, that is, the symbolic construction of unrealistic mythologized Orient that bore little relation to the complex societies of Asia or the Middle East but revealed much about Western views of themselves and their biases. Eurocentrism led to a conceptual reordering of the world through forms of knowledge that legitimized Western dominance. Typically, this move is organized around binary divisions; the West is white, progressive, powerful, rational, democratic, and superior, whereas the Orient is nonwhite, feminine, traditional, static, mysterious, irrational, despotic, and inferior. (Said himself was criticized for essentializing the West, i.e., stereotyping Europeans as racists as if they lacked diversity among them.) Coupled with modernization theory, Eurocentrism constructed a sense of historical time that represented the West as the present and the future, whereas the Orient was relegated to the past; thus, beyond Europe was before Europe. Eurocentrism and Orientalism revealed that every regionalization is a power relation—a way of representing the world in ways that serve some interests and not others.

Other geographers analyzed Eurocentrism and Orientalism in light of the European conquest and penetration of non-Western spaces. Geography as a way of knowing space—the active geographing of various parts of the globe—was part and parcel of the Western administrative control of such regions, which included the inventory of use values as well as the ideological legitimation of these relations. Western notions of space were vital parts of the colonial imaginary; the ways in which space was demarcated and brought into Western frames of understanding drew critical boundaries between identities, self and other, and underpinned particular regimes of power and knowledge. For example, Egypt occupied an important geographic and ideological position in the evolving colonial self-conception of the West—the ancient, stagnant, senile culture, simultaneously proximate and distant, that could be rendered sensible through the application of Western rationality, the empire of the gaze. The imaginative geographies of Western administrators and travel writers revolved around an often racist imagery that pervaded Western views of the Arabic "other." Thus, space, power, and identity were fused in an inseparable skein as Egypt was geographed by a foreign authority. Similarly, Africa was rendered a dark continent until Europeans shed light on it. It is important to note that Eurocentric and Orientalist ideas are not dead and forgotten but rather very much alive, as in Hollywood movies starring the great white hero in combat with hordes of brown-skinned natives (e.g., *Indiana Jones*).

—*Barney Warf*

See also Colonialism; Cultural Geography; Imaginative Geographies; Imperialism; Orientalism; Other/Otherness; Postcolonialism; Race and Racism

Suggested Reading

Blaut, J. (1993). *The colonizer's model of the world.* New York: Guilford.

Blaut, J. (2000). *Eight Eurocentric historians.* New York: Guilford.

Clayton, D. (2003). Critical imperial and colonial geographies. In K. Anderson, M. Domosh, S. Pile, & N. Thrift (Eds.), *Handbook of cultural geography* (pp. 354–368). London: Sage.

Driver, F. (1992). Geography's empire: Histories of geographical knowledge. *Environment and Planning D, 10,* 23–40.

Gregory, D. (2004). *The colonial present: Afghanistan, Palestine, and Iraq.* Oxford, UK: Blackwell.

Lewis, M., & Wigen, K. (1997). *The myth of continents.* Berkeley: University of California Press.

Said, E. (1979). *Orientalism.* New York: Vintage Books.

EXISTENTIALISM

Existentialism is a modern philosophical position that has its roots in the 19th- and 20th-century writings of Friedrich Nietzsche, Søren Kierkegaard, Martin Heidegger, and Jean-Paul Sartre. These writings share a general concern to reject systematic forms of reasoning and behavior in favor of individual expression and action.

Existentialism argues that although the human system of perception and cognition enables people to reason and reflect and marks them out as distinct from all other animals, the manifold experiences that each person goes through ensure that his or her personality remains unique. In this sense of the term, people create their own nature. Moreover, this nature will continue to change as time wears on and other contexts are experienced. The term *existence,* then, refers to this continual re-creation of the self through experience.

This existence is, however, fraught with anxiety and even dread. This is because people, it is argued, long for some kind of external or independent confirmation that the choices they make in life are indeed the right ones. In the absence of just such a confirmation, there is a feeling of what Sartre called "nausea," that is, the recognition of the fundamental lack of order to the universe. In a similar vein, Heidegger referred to the "anguish" that people feel when they realize that at each moment there is no overarching set of rules as to how they should proceed to live as humans; instead, there is a range of choices that can be made. Indeed, a common reference point in existential thought is this notion of *freedom* in that people will always have the opportunity, but also the responsibility, to choose their actions. Even the decision not to choose is itself a choice.

This philosophical position has major implications for academic analysis in which research questions are posed and methods of data collection and analysis are chosen. It follows that because each choice has been made by an individual with his or her own unique personality, no one else can truly comprehend the reasoning behind a particular decision, nor can they grasp the impact of that decision on other people. Hence, one cannot be an existentialist and claim to "explain" social events. One can, however, attempt to empathize with the experiences of another person. In this way, analysis becomes more of a dialogue with others than an objective series of hypotheses and observations.

Although existentialism can be traced back to the 19th century, it was not until the latter half of the 20th century that *humanistic geographers* began to engage with this body of thought. Not all of humanist geographic inquiry follows an existentialist path, yet the emphasis on the everyday, often mundane experiences of people has had a significant impact. The notion of *intersubjectivity,* for example, draws in large part from the empathic understanding of the experiences of others. This has been used by humanistic geographers to delve into the world of otherwise marginalized groups, including the elderly and children, using in-depth ethnographic methods. Moreover, the notion of *reflexivity,* which now cuts across the field of human geography more broadly, underscores the conviction that an explicitly *subjective* understanding of a situation by someone experiencing it is actually preferable to that of someone claiming to be a detached objective observer.

—*Deborah P. Dixon*

See also Body, Geography of; Emotions, Geography and; Ethics, Geography and; History of Geography; Humanistic Geography; Identity, Geography and; Ideology; Phenomenology; Sense of Place; Structuration Theory

Suggested Reading

Rowles, G. (1978). *Prisoners of space? Exploring the geographical experience of older people.* Boulder, CO: Westview.

Samuels, M. (1978). Existentialism and human geography. In M. Samuels & D. Ley (Eds.), *Humanistic geography: Prospects and problems* (pp. 22–40). London: Croon Helm.

EXPLORATION, GEOGRAPHY AND

Exploration, as an individual act or event and as a process that involves discovery, examination, recording, and reporting, is a key constituent of geography and geographic understanding. In terms of knowledge in the past about the earth's dimensions and content and about geography's development as a science, exploration is associated with oceanic voyaging and global circumnavigation—with the penetration of continental interiors, imperial expansion, tales of discovery and heroic endeavor, and major mapping projects. Yet exploration at a variety of scales and in different ways is also commonplace of modern life. Traveling in space, examining the world's oceans, using a tourist map, examining familiar places through educational fieldwork, and even poring over maps or consulting encyclopedia entries are all geographic

explorations of various sorts. In these senses, geography has an origin and a continuing existence as a science of action through exploration. Because this is the case, exploration embraces important methodological questions about the making of reliable geographic knowledge, the disciplining of the senses through fieldwork, and the authority of different knowledge claims. And its history may be told from different perspectives.

The term the *Age of Exploration* is conventionally applied to that period between the late 15th and late 17th centuries when the world was discovered and geographically "enlarged" by European navigators. The achievements of Bartholomeu Diaz in rounding what is now known as the Cape of Good Hope in 1488, of Christopher Columbus in discovering the Americas in 1492, and of Vasco da Gama in establishing trade connections with the Orient in 1497 changed forever previous conceptions of the earth. Ferdinand Magellan and Sebsatian del Cano were the first to circumnavigate the globe between 1519 and 1522, repaying the expense of the voyage with spices. Where the Portuguese led, the English, Spanish, Dutch, and French followed. Trade routes to the Far East and to the Americas and voyages of global circumnavigation were paralleled by exploration in search of the Northeast Passage and the Northwest Passage. These hoped-for trade routes between Europe and the Orient that would eliminate the need to voyage around the southernmost capes of South America or Africa did not materialize.

In the Age of Exploration, exploration was rooted in a belief in the supremacy of Christianity and in the unquestioned benefits to Europe of global trade, principally in spices and precious metals. Geographic knowledge was advanced because ancient views about the world were overturned. Columbus and those who followed in his wake demonstrated the existence of a continent hitherto unknown to Europeans (if well known to its inhabitants). The earth was shown to have more land than was previously believed, to be habitable at and beyond the equatorial regions, and to have great human and natural diversity. By the mid-1640s, Dutch navigator Abel Tasman reached the southwest coast of modern Australia and parts of what is today New Zealand but did not recognize the full extent of the lands in the Southern Ocean.

By the late 18th century, British navigator James Cook had added to the knowledge derived from men such as Tasman and William Dampier and decisively challenged the belief in *Terra Australis Incognita* (or unknown southern lands). Cook's three voyages between 1768 and 1780 added significantly to understanding of the world's continents and of the North and South Pacific. Enlightenment exploration by the British and others, French *voyageurs–naturalistes* Louis Antoine de Bougainville and Jean-François de Galaup, Comte de La Pérouse, and Alejandro Malaspina (the Genoan working for the Spanish), to name only a few, provided texts, specimens, and illustrations of people and landscapes. Thus, exploration provided the very "stuff" of geography, new material for natural philosophers, and accounts of exotic novelty for European audiences. By the later 18th century, oceanic navigators had charted the shape of the world's continents but not revealed their content. When the Association for Promoting the Discovery of the Interior Parts of Africa was established in London in 1788, the view held at that time was that nothing worthy of research by sea, except the poles themselves, remained to be examined. But by land, the objects of discovery still were so vast as to include at least a third of the habitable surface of the earth. Much of Asia, a still larger proportion of America, and nearly the whole of Africa were unknown.

Exploration of the earth's continents was a significant feature of 19th-century geographic enquiry. With their 1804–1806 Missouri River expedition, Meriwether Lewis and William Clark effectively began the exploration of North America, to be followed by J. C. Frémont during the 1840s and John Wesley Powell during the 1860s. In Latin America, Alexander von Humboldt made important contributions to geography through exploration. In Africa in particular, exploration went hand-in-hand with the imperial expansion of European nations and a determination to spread Christianity and extract natural resources. The Scot Mungo Park confirmed the course of the Niger River in 1796. In 1828, Réné Caillié was the first European to reach and return safely from the fabled desert city of Timbuctoo. As North and West Africa were revealed to the gaze of outsiders, the British, the Germans, the Portuguese, and the Dutch led the exploration of East, Central, and South Africa. Men such as David Livingstone, Richard Burton, J. H. Speke, Henry Morton Stanley, Heinrich Barth, G. Schweinfurth, James Augustus Grant, Keith Johnston, and Joseph Thomson mapped Africa's river systems, botany, and geology; debated the continent's economic potential; and collected natural history and ethnographic specimens.

Women explorers there and elsewhere often took a more humanitarian view. Modern exploration in the East Indies began with the work of Alfred Russel Wallace between 1854 and 1860. The coastal margins of Australia were first determined in 1803–1804 by Matthew Flinders, and the first transcontinental crossing was made by S. J. Eyre in 1841.

From the end of the 19th century, and particularly between 1895 and 1918, exploration centered on the polar regions. Polar exploration, often represented as the "Race for the Poles," was prompted by the first International Geographical Congress in London in 1895; longer-run traditions of polar voyaging and oceanographic science led by the Norwegians, Russians, British, Americans, and Swedish; and the commercial importance of whaling. Robert Peary and Matthew Henson, two Americans, were the first to reach the North Pole on April 6, 1909. Roald Amundsen, a Norwegian, first reached the South Pole on December 14, 1911. The death of British explorer Robert Falcon Scott, who reached the South Pole on January 17, 1912, but perished with his companions on the return journey, has become an icon of exploration-as-heroic-failure.

The history of exploration can be read as a chronology of accomplishment, exploration interpreted as both a means to and the result of the empirical quest for geographic knowledge through fieldwork, often undertaken in association with the imperatives of imperialism. Exploration and geography and empire were closely connected, not least because many explorers served the state as commercial agents or as military men. Yet if considered uncritically, exploration conveys simplistic notions of discovery and contact, the supremacy of the explorers' knowledge over others' knowledge, and the unproblematic achievement of geographic certainty. Exploration is not just something that geographers did and do. It is a practice shared by numerous sciences, part of what scientific fields do to claim legitimacy. Terms such as the "Age of Exploration" must be used carefully. Arab geographers knew much of Africa before the Portuguese did. Muslim Chinese navigator Zheng He undertook seven major expeditions to India and the West between 1405 and 1433. The Polynesian Voyaging Society, begun in 1973, sees its modern voyages as affirming Hawaiian identity through exploration. To portray exploration without reference to the native inhabitants of places "discovered" by Europeans and others is to offer a distinctively

Eurocentric view. Considered more fully as a process, exploration and geography and geography-as-exploration involve questions of cross-cultural liaison and translation in the field, the role of native agency, different knowledge systems and ethical responsibilities, and narratives less of heroism and dominance than of hesitancy and negotiation.

Exploration was often dangerous. Magellan was killed in the Philippines in 1521. Cook was killed by Pacific Islanders having unwittingly flouted their customary conventions. La Pérouse's expedition was lost at sea. Polar exploration is a record of failed expeditions, geography thwarted, lives lost in icy wastes. On land, disease, fatigue, hostile tribesmen, the facts of distance, and the failure of instruments acted to hinder safe passage or to reduce the value of the information secured. Many narratives of exploration minimize or omit altogether the key role of the native inhabitants. They are considered (if they are discussed at all) not as guides, translators, map makers, intermediaries, or sources of information (often passed off as the explorers' own) but rather as "objects" of inquiry or subject peoples. For these reasons, the study of exploration narratives and maps, of landscape depictions, and (since the mid-19th century) of photographs as forms of geographic visualization must be sensitive to context, to the intended audiences, and to the fact, whether intended or not, that exploration often silences the native voice.

Exploration is not an unproblematic route to geographic truth. Some explorers lied about their achievements. Exploration depends on trust in (and tolerance of) one's informants and companions, one's instruments, and one's self. It demands rules—even ethical codes—if it is to produce reliable knowledge from credible sources. From the later 17th century, learned academic societies such as the Royal Society in London published what were effectively "how to explore" guides—methodological manuals designed to help secure reliable information. By the 19th century, more specialist guides to exploration were available, including in Britain the Royal Geographical Society's *Hints to Travellers,* first published in 1854, and the *Admiralty Handbooks*. These exploration manuals stressed the importance of accurate observation and provided advice on instruments, mapping practices, appropriate clothing, and standards of behavior. The Royal Geographical Society (RGS), formally established in London on July 16, 1830, had its roots in the African Association and the Raleigh Dining Club, begun in 1827. The RGS was a leading institution for the promotion of exploration, particularly between 1830 and 1933 and notably in Africa, India, and the polar regions, and today it maintains an active role in advising on and supporting exploration.

The RGS was one of many geographic bodies and societies founded during the 19th century, several of which supported explorations of their national territories and colonial possessions. The Paris Geographical Society, begun in 1821, was influential in Caillié's Timbuctoo expedition and other French exploration in Africa. Other bodies were begun in Berlin (1828), in Mexico City (1833), and in Rio de Janeiro (1838), to list only a few. The American Geographical Society of New York was established in 1851. The Palestine Exploration Fund, established in 1865, reflected British interests in the geography of the Holy Land and had active German and French counterparts. The Society of Women Geographers was founded in 1925 by four American women explorers. The connections between exploration and geography are the focus of the Hakluyt Society, named after Elizabethan travel compiler Richard Hakluyt and founded in 1848 to publish travel and exploration narratives, and of the Society for the History of Discoveries, begun in 1960 to stimulate interest in the history of geographic exploration.

—*Charles W. J. Withers*

See also Cartography; Cultural Geography; Enlightenment, the; Epistemology; Fieldwork; Historical Geography; History of Geography; Imaginative Geographies; Orientalism; Subaltern Studies; Travel Writing, Geography and

Suggested Reading

Allen, B. (2002). *The Faber book of exploration: An anthology of worlds revealed by explorers through the ages.* London: Faber & Faber.

Baker, J. (1931). *A history of geographical discovery and exploration.* London: Harrap.

Bourguet, M.-N. (1999). The explorer. In M. Vovelle (Ed.), *Enlightenment portraits* (pp. 257–315). Chicago: University of Chicago Press.

Buisseret, D. (Ed.). (2005). *The Oxford companion to exploration.* New York: Oxford University Press.

Driver, F. (2001). *Geography militant: Cultures of exploration and empire.* Oxford, UK: Blackwell.

Grimbly, S. (Ed.). (2001). *Atlas of exploration.* Chicago: Fitzroy Dearborn.

Riffenburgh, B. (1994). *The myth of the explorer.* Oxford, UK: Oxford University Press.

EXPORT PROCESSING ZONES

Export processing zones (EPZs) are geographically defined areas where goods are produced for export to other countries. They are known by a variety of different names, including free zones, free trade zones, special economic zones, maquiladoras, and export platforms. Political leaders in host countries treat these zones as enclaves that have different rules and regulations from the rest of the country. In this way, the zones are outside the customs territory of the country. Private investors are encouraged to set up factories and other export activities via nonpayment of duties, favorable tax regimes, special laws, exemptions, and infrastructure subsidies that make the zones attractive. The number of EPZs worldwide has expanded rapidly since 1960, when the first one (the Shannon Free Zone) was built in Ireland. A decade later there were 10 host countries, and by 1995 there were 70.

EPZs have been especially important and highly criticized in developing countries, where they are created for a variety of reasons. The central motivation of government officials may be to create jobs; attract flows of foreign direct investment (FDI); increase foreign exchange earnings; or promote export-led industrial development, technology transfer and linkages to other economic activities, or some combination of these outcomes. By these metrics, EPZs have had success in some developing contexts.

Many different kinds of manufacturing and service activities exist side by side in EPZs, although labor-intensive operations such as electronics assembly and apparel are quintessential activities. Such labor-intensive sectors often hire a predominantly female workforce, so that many EPZs end up with a gender division of labor in which women occupy many of the rank-and-file jobs and form a majority of the workforce, sometimes as high as 80%. Unions are banned in some EPZs and discouraged in many others; in this way, wages are held artificially low and working conditions may be poor. Critics also argue that competition between EPZs creates a "race to the bottom" for workers and communities that depend of the wages of assembly plant workers.

Many developing countries first designed EPZs that looked like industrial parks with subsidized infrastructure such as factory space, communications systems, water/sewer networks, and electrical power.

More recent examples in Zimbabwe and Guatemala use innovative spatial arrangements that are geographically flexible and use stand-alone factories as EPZs. This approach allows companies on which EPZ status is conferred to locate wherever they desire. This is a highly flexible tool, enabling companies to operate from convenient locations while enjoying EPZ benefits. The result is a patchwork pattern of customs regulation. From a geographic perspective, the spatial flexibility will be an interesting phenomenon to assess and analyze. In all contexts, EPZs are inherently spatial. Therefore, bringing a geographic perspective to bear on understanding the social outcomes, labor practices, and governmental regulations associated with EPZs may be useful.

Although there is much controversy about the benefits of EPZs to host economies and the impacts on quality of life, EPZs remain an attractive option for many policymakers wishing to promote employment, inward FDI, and export-led industrial development. These policymakers use a geographic approach and carve out special districts in their countries for export expansion.

—*Altha Cravey*

See also Comparative Advantage; Developing World; Development Theory; Economic Geography; Industrial Revolution

Suggested Reading

World Bank. (1991). *Export processing zones.* Washington, DC: Author.

EXTERNALITIES

Broadly speaking, externalities (also known as *spillovers* and *neighborhood effects*) refer to "uncompensated welfare impacts," that is, actions and events that affect the welfare (positively or negatively) of one party or person by another without some type of remuneration. These arise when decision makers do not reap all of the rewards or bear all of the costs of their actions and can occur in both the production and consumption of goods and services.

Positive externalities improve the welfare of an individual or a group without a cost. For example, if one's neighbor has an attractive garden or plays music that one enjoys, the receiving party derives benefits without incurring costs. Most positive externalities are relatively trivial. However, network externalities, which reflect the rising utility of systems such as telephone networks and the Internet, are important; the more people use a system, the greater the value it has to each user.

However, negative externalities, which diminish the welfare of a person or group, are a different story. Examples of negative externalities include the reduction in real estate values created by the location nearby of an unwanted land use (e.g., a toxic waste plant). If a developer erects a high-rise that annihilates a homeowner's view, the affected party suffers a negative externality. More general cases involve the creation of air and water pollution, acid rain, noise pollution, and traffic congestion. Because the producers of negative externalities do not have an incentive to worry about the impacts of their actions on others, they generate social and market inefficiencies.

Negative externalities occur when the social costs of an action are not captured in the private costs in the form of the market price; thus, they are a prime example of market failure. For example, the true costs of operating an automobile include its impacts on highways, the environment, and public health, few or none of which are included in the price of gasoline or car insurance. As another example, a logging company may deprive a neighborhood of shade. In this case and similar ones, the social costs are greater than the sum of individual costs and lead to the overproduction of goods with high social costs. These often are seen as violations of individual rights and lead to serious ethical and political problems. Thus, negative externalities are commonly cited as necessary instances of government intervention such as zoning ordinances, health and safety regulations, and environmental conservation.

Geographers study the spatial location, frequency, and magnitude of negative externalities that are unevenly distributed. The presence of a sports stadium, for example, may generate a field of noise that negatively affects local residents. Negative externalities are particularly important in the analysis of transportation (e.g., congestion), land use (e.g., rural-to-urban land conversion), and natural resource conservation (e.g., hydroelectric dams and their impacts).

—*Barney Warf*

See also Agglomeration Economies; Economic Geography

Suggested Reading

Cornes, R., & Sandler, T. (1996). *The theory of externalities, public goods, and club goods.* Cambridge, UK: Cambridge University Press.

EXURBS

Exurbs are a particular kind of pseudo-urban settlement—disjointed fragments of urban form surrounded by countryside but not really belonging to it, noncontiguous residential, or other functional dependencies of a town or city.

An exurb is situated at some distance from the recognizable urban or metropolitan fringe, usually consisting of one or more housing tracts surrounded by open land still rural in character, as well as scattered employment and commercial sites, but few (if any) social services. Exurbs house urbanites who mostly commute to jobs in outer metropolitan suburbs or other exurban areas, sometimes over considerable distances. These residents depend on the service infrastructure provided by communities closer to or on the metropolitan fringe (to which the exurbanites likely contribute no direct taxes) and increasingly on Internet purchases brought to their home or office doors by parcel delivery services. Exurbs represent the leading edge of modern urban sprawl, driven by prolonged metropolitan decentralization, and may in time become outer suburbs of an expanding metropolitan built-up area.

ORIGIN AND ENLARGEMENT OF THE CONCEPT

The preconditions for exurbs have existed since automobiles became widespread, particularly in regions without complicated rural land tenure patterns, where urban market pressures could easily convert land to urban use. Introduced in 1955 by Auguste Spectorsky, the term *exurb* gained broader currency in the United States during the 1970s, when isolated but rapidly developed tract housing and custom home subdivisions began appearing in large numbers well beyond the existing metropolitan fringe, thanks to intercity superhighway construction and state and local road upgrading in the rural hinterlands. The concept was expanded to accommodate the increasing diversity of land use patterns and community types found beyond the fringe of many metropolitan areas during the 20th century. Occasionally, it is misapplied to locales better described by other terms such as *outer suburbs, satellite towns,* and *edge cities.*

Exurbs are found in some form in many regions of the world, but especially in highly urbanized countries with permissive land use regulations. They are most common in the United States, where abundant land, a highly mobile population, and fragmented governmental jurisdictions encourage their proliferation.

MEASUREMENT

Exurban zones typically stretch 30 miles beyond the suburban edges of towns of at least 50,000 residents and up to 70 miles beyond the fringes of cities of at least 500,000 residents. Exurban communities as a whole contain a mixture of established and more recent residents, with the latter commuting at least a half hour to work. In the United States, they are generally measured as aggregations reported at the county level because their geographic scatter has become so vast that it spreads over many counties surrounding even single metropolitan centers. Above all, the measures rely on population density but can also include employment data and occasionally physical attributes. The average population density of American exurban zones is 93 persons per square mile, in contrast to rural and urban densities of 4 and 1,149 persons per square mile, respectively. Whereas fully urban development has placed fully 55% of the U.S. population on less than 2% of the country's land area, exurban settlement has appropriated more than 14% of the total land area for only 37% of the national population.

TYPES OF EXURBS

Exurbs differ from their suburban counterparts in their land use mixture and population, containing much-lower-density residential development of varying ages and types such as dispersed roadside homes, housing subdivisions, country estates, mobile homes, hobby farms, and other recreational sites as well as converted farmhouses and functioning farms. Exurbs also contain a scatter of recent retail, commercial, office, and light industrial developments, often attached loosely to established villages and small towns spread throughout the rural area. With sustained metropolitan

decentralization and the rise of an invisible web of telecommunications and advanced technology, exurbs have grown diverse. Set within the galaxy of small residential clusters dotting the countryside, one can find quite specialized concentrations that range from "technoburbs" (office parks and research facilities) and their associated "nerdistans" (self-contained upscale living quarters for their high-tech workers) to elaborate recreational and retirement developments (with and without golf courses and other landscape amenities such as lakes and marinas).

TYPES OF EXURBANITES

Residents of exurbs comprise four main types. First, economy-minded residents, comprising nearly one third of the population, are not affluent, live in modest housing more remote from urban centers, and have relatively few children. Second, family-oriented residents, nearly as numerous among exurbanites, have several children, usually have two incomes, live closer to urban centers, and own average-cost homes on lots larger than those available in suburbs. Third, affluent residents, comprising roughly a quarter of exurbanites, are generally two-career managerial and professional couples without children, have large homes on large lots, and live closer to the urban centers where they work. Fourth, long-distance commuters, comprising roughly one in six exurbanites, are usually modest-income, one-worker families in blue-collar, technical, or sales jobs; have children; and paid the least for the largest lots but with the longest commutes to work.

Collectively, exurbanites place a high value on rural living, driven by a vision of the "pastoral ideal"—living in the "middle landscape." They are attracted by large lot sizes, low house prices, low crime rates, a good environment for raising children, low taxes, and (frequently) lack of urban controls such as land use regulation. Conversely, they are not put off by the low level of services or the distances to their jobs.

PLANNING ISSUES

Exurbs are widely seen as the most challenging form of urban sprawl. Because they comprise a rapidly rising proportion of urban residents in developed countries, proliferating exurbs pose daunting problems of land use management, service provision, and resource conservation. Paying for modern residential services, emergency response facilities, schooling, and other needs over such large expanses of territory at such low densities pressures the local taxing systems and strains natural and human resources. The ultimate costs of pursuing the spacious rural idyll on this scale are kept in abeyance only by the continued availability of historically cheap domestic and foreign oil. If that relationship were to change, today's networked exurbs could become tomorrow's geographic orphans, largely cut off from their sources of sustenance.

—*Michael P. Conzen*

See also Edge Cities; Rural Geography; Rural–Urban Continuum; Suburbs and Suburbanization; Urban Fringe; Urban Sprawl; Urbanization

Suggested Reading

Daniels, T. (1999). *When city and country collide: Managing growth in the metropolitan fringe.* Washington, DC: Island Press.

Lang, R. (2003). *Edgeless cities: Exploring the elusive metropolis.* Washington, DC: Brookings Institution.

Patel, D. (1980). *Exurbs: Residential development in the countryside.* Washington, DC: University Press of America.

Spectorsky, A. (1955). *The exurbanites.* Philadelphia: J. B. Lippincott.

FACTORS OF PRODUCTION

There are numerous variables that influence the location of firms and of industries, which are aggregations of firms. The locational decision of a firm is complex, and companies spend considerable time and effort in choosing their optimal locations. Investments in inappropriate locations can be disastrous. Thus, firm decision making is a rational process, if an imperfect one, and is subject to the laws of market competition. Although personal considerations such as climate and the owner's preferences occasionally may be important on the margins, firms cannot choose arbitrarily because if they do they will be forced out of business by their more rational competitors. The major factors of production that shape firms' locations include labor, land, capital, and managerial and technical skills. (Other factors, such as transport costs, are considered elsewhere in this volume.) All of these are necessary for production, and all exhibit spatial variations in both quantity and quality.

LABOR

For most industries, labor is the most important determinant of location, especially at the regional, national, and global scales. When firms make location decisions, they often begin by examining the geography of labor availability, productivity, and skills. The degree to which firms rely on labor, however, varies considerably among different sectors of the economy and even among different firms, which may adopt different production techniques.

The relative importance of labor varies considerably among industries. The demand for labor depends on the size of the firm involved and how labor intensive or capital intensive a given production process is as well as the cost. In very capital-intensive industries (e.g., petroleum), labor costs may be irrelevant. Thus, it is a mistake, but a common one, to assume that all industries seek out low-cost labor. Over time, most industries have become increasingly capital intensive; that is, they have substituted capital for labor, particularly when production in large quantities justifies the investments involved.

The supply of labor in a given region greatly affects the cost. In countries with high birth rates, the supply tends to be relatively high and labor costs are low. In economically advanced countries, the birth rate is low and labor is relatively expensive. Because some firms demand particular types of workers in terms of their age and/or sex, the demographic structure of a region also shapes the supply of certain types of employees (e.g., teenagers). Finally, because labor is mobile over space (but not perfectly so), migration (or immigration if international) also shapes the supply of labor. In regions that can attract labor easily, wage rates will tend to be low, all else held constant. When the supply is limited by, say, immigration restrictions, wage rates tend to go up. At the local level, housing costs also can constrain the supply of labor if they are so high that workers cannot find affordable places to live.

Under capitalism, the real cost of labor is determined by the relative productivity of labor rather than the cost of wages and fringe benefits. Thus, cost is hardly the only factor considering labor. Productivity is largely a function of the skills present in the local labor force (or human capital) that, in turn, are derived from formal and informal educational systems, on-the-job training, and years of experience. Firms will pay relatively high wages for skilled productive labor. Consider that if labor costs were central to the location of all firms, very low-wage countries, such as Mozambique, should attract vast quantities of capital—which they do not—and high-wage countries, such as Germany and the United States, should see a rapid exodus of jobs. The reality of the geography of labor is much more complex and involves labor markets in which jobs are constantly created and destroyed, skills are produced/reproduced and change, new technologies come into play, and other cultural, economic, and social forces can be important factors.

Moreover, the skill level of a given occupation greatly affects the size of its labor market. In general, skilled labor markets tend to be geographically larger than unskilled ones. Workers may migrate long distances for well-paying positions, and the market for many skilled jobs is global in reach. Unskilled positions, in contrast, typically draw from a relatively small labor shed; for example, few people would travel cross-country to take a job as a janitor or a retail trade cashier.

The labor process is saturated with politics. Labor is the only "factor input" that is able to resist the conditions of exploitation—to go on strike, to engage in slowdowns or sabotage, or to unionize. Unionization rates vary widely, adding to differentials in the cost of labor. Thus, in addition to the cost of labor, firms must consider the length of the work day, working conditions, health and safety standards, pensions and health benefits, vacations and holidays, demands for worker training, subsidized housing, and the role of labor unions, all of which shape wage rates and productivity levels.

LAND

At the local scale (i.e., within a particular metropolitan commuting area where labor costs are relatively constant), land availability and cost are the single most important locational factors affecting firms' location decisions. The cost of land reflects the supply and demand, and different types of firms require different quantities in the production process. In general, larger firms, particularly in manufacturing, require more land and thus are more sensitive to the costs, although in some sectors, such as producer services, firms pay very high costs (in rent or by purchasing a site). Firms often engage in intensive examination of several

selected possible sites before settling on an optimal location.

The cost of land is influenced heavily by its accessibility. Transport costs determine the location rent of parcels at different distances from the city. Thus, because land downtown is the most accessible, it is by far the most expensive; in most cities, land costs decline exponentially away from the city center. However, not all firms necessarily seek out low-cost land. The imperative to do so depends on the trade-off between land and transportation costs that firms make to maximize their profits. Firms that must have accessible land—generally labor-intensive firms that must maximize their accessibility to labor, to each other, and to urban services—will pay very high rents to locate near the city center. On the other hand, firms that do not require access to clients, suppliers, and services, such as large manufacturing firms in suburban industrial parks, make a different trade-off, choosing to locate on the urban periphery where land costs are low but transport costs are higher.

Since World War II, there has been a centrifugal drift of manufacturing to suburban properties. Large parcels of industrialized land are more likely to be available in the suburbs than in central-city locations, where accessibility makes land relatively expensive. Other reasons why industrial properties have expanded into the suburbs include locations that are easily accessible to motor freight by interstate highway as well as access to suburban services and infrastructure, including ample sewer, water, parking, and electricity. Industries may also be attracted to the suburbs because of nearness to amenities and residential neighborhoods. Suburban locations minimize labor's journey to work.

CAPITAL

Under capitalism, capital plays a major role in structuring the production process. Capital takes one of two major forms: fixed capital or liquid (variable) capital. Fixed capital includes machinery, equipment, and plant buildings. Besides the installation and construction costs, firms must budget for maintenance and repair and depreciation. The age of the capital stock of a region greatly affects its overall productivity levels. Liquid capital includes intangible revenues such as corporate profits, savings, loans, stocks, bonds, and other financial instruments. The rate of capital formation reflects variables such as corporate profitability (e.g., market prices, production costs), savings rates, interest rates, and taxation levels.

Liquid capital is theoretically the most mobile production factor. The cost of transporting liquid capital is almost zero, and liquid capital can be transmitted almost instantaneously in an electronically wired world. Fixed capital is much less mobile than liquid capital; for example, capital invested in buildings and equipment obviously is immobile and is a primary reason for industrial inertia. Any type of manufacturing that is profitable has an ensured supply of liquid capital from revenues or borrowing (depending on credit rating), and interest rates hardly vary within individual countries. Most types of manufacturing, however, initially require large amounts of fixed capital to establish the operation or periodically to expand, retool, or replace outdated equipment or to branch out into new products. The cost of this capital, which is interest, must be paid from future revenues. Investment capital has a variety of sources—personal funds, family and friends, lending institutions (e.g., banks, and savings and loan associations), the sale of stocks and bonds, and so on. Most capital in advanced industrial countries is raised from the sale of stocks and bonds, although American firms rely on this approach more than do firms in Europe, where banks play a larger role in industrial financing. The total supply of investment capital is a function of total national wealth and the proportion of total income that is saved. Savings become the investment capital for future expansion.

Whether a particular type of manufacturing, or a given entrepreneur, can secure an adequate amount of capital depends on several factors. One factor is the supply of and demand for capital, which varies from place to place and from time to time. Of course, capital always can be obtained if users are willing to pay high enough interest rates. Beyond supply-and-demand considerations, investor confidence is the prime determinant of whether capital can be obtained at an acceptable rate.

Capital also is important because firms can substitute capital for labor in a process of capital intensification. The history of capitalism is largely one of capital intensification in different industries, particularly in agriculture, where only a very small fragment of the labor force in industrialized countries now works. Capital intensification can increase productivity, but it may also displace workers. Only if the cost of goods drops sufficiently to increase real incomes

and worker expenditures can it generate job growth in the long run.

MANAGEMENT

Management involves the nuts and bolts of corporate decision making, including allocating the firm's resources, raising investment capital, keeping abreast of the competition and government rules and policies, making investment decisions, hiring and firing workers, and making marketing and public relations decisions. Corporate management reflects and shapes the organizational structure of firms, including the pattern of ownership and how decisions are made. Firm management forms may range from sole proprietorships to partnerships and may be either publicly or privately owned.

Within firms, management forms an important part of the corporate division of labor (i.e., headquarters vs. branch plants). Corporate headquarters determine a firm's overall competitive strategy, what markets and products to focus on, a firm's labor policies, whether to engage in mergers and acquisitions, and types of financing. Thus, these tend to be skilled, well-paying, white-collar jobs. Most are in large urbanized areas.

Technical skills are the skills necessary for the continued innovation of new products and processes. These skills are generally categorized as research and development (R&D). The R&D required for new products typically is a large and expensive process involving long lead times between invention and production, a process that is often beyond the scope of small firms.

—Barney Warf

See also Agglomeration Economies; Capital; Economic Geography; Economies of Scale; Labor, Geography of; Location Theory

Suggested Reading

Stutz, F., & Warf, B. (2005). *The world economy: Resources, location, trade, and development* (4th ed.). Upper Saddle River, NJ: Pearson/Prentice Hall.

FEMININITY

Femininity is the quality of being feminine or the trait of being female. The term *femininity* in particular evokes the normative assumption that women should embody and reflect feminine qualities such as being private, domestic, gentle, graceful, delicate, ladylike, passive, sensual, and emotional. These qualities depend on a defining opposition to masculine traits such as public, professional, strong, powerful, hard, aggressive, and rational. Broadly, gender differences and the definition of male and female depend on the distinction between femininity and masculinity and on an abiding heterosexuality between men and women that further emphasizes male and female traits; without femininity there would be no masculinity and vice versa. Feminine traits such as those just listed also are coded through Western race and class politics that emphasize whiteness, refinement, education, and wealth.

Importantly, feminists have noted that qualities of femininity derive from devaluation of women in patriarchal sexist societies. Although femininity relies on its binary relationship to masculinity, this binary is at once qualitatively unequal and essentially false. It is unequal because femininity reflects weakness and thus is depreciated in relation to the strength and superiority of masculinity. It is essentially false because such categories of difference do not hold up to scrutiny; binaries cannot encompass the diversity and difference of actual attributes and identities of women and men. The variability of women's identities illustrates the precarious condition of femininity given that femininity never exists purely or ideally in the material world.

Feminists insist that because femininity is idealized, it is not essential or biological to women. They argue that women reproduce femininity, albeit imperfectly, through their identification with feminine ideals and representations and through the daily practices of feminine identity. This reproduction of femininity is itself normative as it relates to idealized and socially sanctioned gender categories, symbols, practices, and representations. Feminine girls and women gain social, economic, and political recognition and reward that masculine women do not. This is evident through practices and attitudes regarding homophobia, violence against women who do not properly take on feminine identities or symbols, and the devaluation of effeminate men or androgynous individuals. Psychoanalytic approaches insist that the reproduction of feminine identity even entails the unconscious adoption of feminine gender norms by women through disciplining spaces of masculine power. Psychoanalysis also suggests that subjects must take on gendered symbolism to form psyches in the first place and thus to enter the social world.

During recent years, reactions to stereotypes of second-wave feminists as butch, antifeminine, rough, and aggressive have led to new embraces of femininity. So-called power feminists, in addition to some third-wave feminists and women more generally, insist that girls and women can be both strong and feminine. Mainstream movements such as "girl power" movements reflect these goals to redefine femininity as powerful and valuable.

—*Mary Thomas*

See also Feminist Geographies; Lesbians, Geography and; Masculinities; Sexuality, Geography and/of

Suggested Reading

Butler, J. (1993). *Bodies that matter: On the discursive limits of "sex."* New York: Routledge.
Laurie, N., Dwyer, C., Holloway, S., & Smith, F. (1999). *Geographies of new femininities.* Essex, UK: Longman.

FEMINISMS

Feminisms refer to political and academic movements that confront misogyny, sexism, and their compounded and unique effects on different women and girls. By deciphering and evaluating the oppression of women and girls, feminist scholars seek to alleviate sexism's various and harmful impacts. Here feminism is examined in the plural because common goals and a collective political identity have increasingly become difficult to delineate and justify given disparate theoretical explanations of gender, sexism, and patriarchy. Furthermore, feminist scholars use different methods to develop these models and often embrace dissimilar political affiliations. If there is one goal that unites feminisms, it is to improve the status and material conditions of all women. Political and academic feminisms, often but not always one and the same, have sought this goal through different trajectories over time. Often 19th- and 20th-century Western feminisms are referred to by their historical era and include first, second, and third waves.

FIRST-WAVE FEMINISMS

Western feminist movements first sought the inclusion of women in the everyday world of men. Thus, they demanded equal opportunity in politics, law, work and the economy, and public space. This stage of Western feminism often is referred to as liberal feminism because feminists drew on liberalism's ideal of individual liberty, freedom, and egalitarianism to make their claims. Indeed, feminists of the late 19th and early 20th centuries exposed the false promises of Western democratic states by showing that the value of human freedom did not extend to all subjects and specifically not to women. In general, first-wave feminists demanded the universal rights offered by liberal democracies such as the vote, education, and social welfare.

With suffrage, first-wave feminists in the West celebrated a great victory. However, the extent of women's inclusion in democracies subsequently remained extremely limited, as was evidenced by low levels of women's higher education participation and political office holding as well as high occupational segregation that kept women's labor force participation restricted to low-paying dead-end jobs. Coupled with these was the normative middle-class pressure for women to remain homemakers that, in turn, promoted the restriction of women's mobility and their confinement to primarily private spaces such as the household. Thus, after World War II, feminists began to question the ideal of universality and argued that it was patriarchal (male controlled) and masculinist (favoring men and their prerogatives). The ensuing critique of universalism and liberalism marks second-wave feminism.

SECOND-WAVE FEMINISMS

A significant contingent of second-wave feminisms continued to rely on liberalist claims to equality and civil rights, a similarity they shared with antiracist civil rights movements of the 1960s. However, second-wave feminists suggested that women needed to base political activism and claims for future rights on their own experiences rather than on those of men. In other words, feminists (primarily during the 1970s and 1980s) rejected the ideal of inclusion because, the claim went, they would only be vying for inclusion in a man's world built on men's values. Fundamental to this movement was an insistence that dominant political and economic institutions were essentially patriarchal and masculinist. Rather than merely struggling to gain access to such institutions, feminists also sought to critique the masculinism inherent in them. Feminists

advocated for the construction of new institutions based on what they labeled as feminine ideals and qualities such as care, communalism, and nurturing.

Feminist epistemology was divergent, but its overarching concern was to situate women's gender position as one that was unique to men's and that therefore could offer an alternative way of life to patriarchy. Thus, second-wave academic feminisms elaborated and connected theories of gender difference, gender oppression, patriarchy, and feminine identity construction and linked these to societal structures and institutions. A key attempt of this type of theory building in the second wave was standpoint theory. Standpoint theorists argued that the struggle against sexism and oppression should begin with women's ways of knowing and experiencing the world. These ways of knowing referred, first and foremost, to those that developed through the gendered division of labor such as reproduction, family and child care, and mothering.

Thus, standpoint epistemology emphasized inclusion and care rather than liberal universalism's paradoxical exclusion of women. Standpoint theorists explored psychology, science, political economy, methodology, race, and other areas to apply their critique and their feminist epistemology. Some standpoint theorists working in science studies also insisted that women's views of the world claim to be partial, as opposed to the masculinist assumption that scientific inquiry is objective, complete, and infallible. In this way, standpoint theorists made an important critique of universal values and traditional methodologies of physical and social science.

Second-wave feminism was marked by remarkable plurality, especially as it developed over time; several movements predominated and still have legacies in today's feminisms. Socialist feminists concentrated their critique on the gender division of capitalism that extracts women's reproductive labor without remuneration while it rewards male labor with wages; they insisted that capitalism and patriarchy are dually made and reinforced, so that a critique of one must entail a critique of the other. Radical feminism explained women's oppression as stemming singularly from male power, as seen through traditional family structures and values that reproduce men's control over women's bodies, labor, and sexuality. Some radical feminists supported women's social and spatial separatism, especially lesbian feminists who said that women's separation from men was the only means to achieve a break from patriarchy and to establish an essentially women's culture. Ecofeminists related nature and women through the suggestion that both are objectified and dominated by patriarchy. Although ecofeminists differed on their prescriptions for overcoming these harmful positions, most agreed that because women are socialized to be caregivers and nurturers, they would serve as more compassionate custodians of the land. Finally, psychoanalytic feminism looked to women's subjectivity to explore how women come to be subjects and take on gendered identities through patriarchical family structures and cultures.

However, feminist critics, especially feminists of color, challenged the assumption that there existed common feminine traits and questioned whether all women even share a common gender identity. This question points to the central issue of how to conceptualize gender, the category "woman," and the male/female gender division. Typical definitions of gender place it as a cultural construction in opposition to biological sex. Thus, female/male is a sexual division, whereas feminine/masculine is a cultural or social distinction. Such a distinction implies the learned qualities of gender and the idea that gender is expressed by men and women through their identities and behaviors. Feminine attributes are unique and opposite to masculine ones, for example, through feminine behavior such as emotionality and passivity. This social construction theory of gender maintains that these behaviors are taught to girls rather than maintaining that they are essential biological effects of being female. It also relies on a binary relationship between male and female.

Increasingly heated debates about the ability of the category "woman" to synthesize feminist claims placed the social construction theory of gender in jeopardy. Relatedly, the second wave's avowal of universalism often failed to recognize its own uneven application of particularity. Although second-wave feminists worked hard to establish effective politics to advance women's rights, they subsequently neglected the importance of social differences in the lives of all women. Second-wave feminists, through their prioritizing of gender and sexism, often failed to account for racism and ethnocentrism. Women of color and Third World feminists insisted that gender was not the only social relation of import to women. They also argued that women of color could not separate their gender identity from their other racial, ethnic, sexual, postcolonial, location-based identities. Thus, by the 1980s, second-wave feminisms had to confront

charges of racism, Eurocentrism, and the exclusion of different women in political movements and leadership. The struggle to include all woman regardless of race, ethnicity, sexuality, location, and so on continued to involve the rejection of an essential notion of "female" or "woman," so that feminist identity politics no longer could take for granted a collective and common gender identity.

THIRD-WAVE AND 21ST-CENTURY FEMINISMS

Over time, third-wave feminisms developed through this critique of identity politics. Third-wave feminists of the 1990s and now into the 21st century consider how to continue antisexist politics without its founding category. Third-wave feminisms instead situate politics around coalitions of different women in their attempts to maintain a fractured movement marked by contradictory definitions of gender and women's alliances to different social groups.

The theoretical movement of poststructuralism has played a central role in the development of the third wave, and both share an antifoundationalist epistemology. So-called poststructuralist feminism rejects the ideal of collective identity and advocates a politics of difference. Identity no longer is seen as liberatory or a model for future societies and institutions, as with standpoint theory. Deconstruction is a central method of poststructuralist feminists and is used to show that the category "woman" relies on a defining opposition to "man." The defining binary is suppressed during use; thus, deconstructionists insist that such categories must be disavowed.

Performativity theory has been an influential component of third-wave poststructural feminism. Performativity theory argues that gender is unnatural rather than an essential biological trait of a feminine body or sex. Instead, gender and gendered identities are given form through the repeated practices of gendered subjects; gender is a doing, and it gains visibility because people engage in gendered practices such as feminine behaviors and dress. Thus, gender is an effect of the repeated practices of people, but it is very important to remember that performativity theorists insist that such practices are powerfully enforced by normative pressures to behave properly. Furthermore, they claim that gender distinctions between men and women rely on heterosexuality because one can be a feminine subject properly only through a defining desiring relationship to a male. Of course, there are masculine women and feminine men, and there are homosexuals and bisexuals, but according to performativity, these just show us how it is possible to disrupt the normative relationship between gender and sexuality. A significant aspect of performativity theory is psychoanalytic theory because identification and subjectivity are central issues to its ontological questioning of gendered subjects and identities. Psychoanalysis helps performativity theory to explain how subjects are influenced by social pressures and norms and how they form identities through powerful influences.

In part due to performativity theory and in part due to the ongoing critique of a gender binary, third-wave feminist inquiries have spread beyond the second wave's primary focus on women and women's ways of knowing. Feminists now investigate a myriad of gender matters, including masculinity and male studies, sexuality and queer theory, and questions of materiality and the body.

Finally, postcolonialism has had a pivotal influence on third-wave feminism as well by critiquing identity hierarchies that, over time and space, have situated white European men, culture, and society as central and definitive norms to which others must aspire. Postcolonial feminists have been especially interested in how women were represented during colonialism and how masculinism, Eurocentrism, and imperialism continue to influence representations of women in Third World and diasporic communities—and, indeed, how a term like *Third World* even comes to stand in for diverse social locations in the first place. A central question of postcolonial feminism is who has the right to speak for whom. This is connected to the issues of representing women, defining feminist issues, and creating political movements that serve the interests of all women (and men) regardless of their divergent social and geographic locations.

—*Mary Thomas*

See also Femininity; Feminist Geographies; Feminist Methodologies; Gender and Geography; Masculinities

Suggested Reading

Butler, J. (1990). *Gender trouble: Feminism and the subversion of identity.* New York: Routledge.

Haraway, D. (1991). *Simians, cyborgs, and women: The reinvention of nature.* New York: Routledge.

Harding, S. (1986). *The science question in feminism.* Ithaca, NY: Cornell University Press.

McClintock, A. (1995). *Imperial leather: Race, gender, and sexuality in the colonial contest.* New York: Routledge.

Mohanty, C., Russo, A., & Torres, L. (Eds.). (1991). *Third World women and the politics of feminism.* Bloomington: Indiana University Press.

Moraga, C., & Anzaldúa, G. (Eds.). (1983). *This bridge called my back: Writings by radical women of color.* Brooklyn, NY: Kitchen Table Press.

FEMINIST GEOGRAPHIES

Feminist geographers argue that the discipline of geography has inadequately considered and theorized the gendered power relations that significantly influence everyday lives, institutions, environments, economies, and politics. Since its origin in the 1970s, feminist geography has grown into a significant force and has fundamentally affected all fields of human geography. In fact, it is impossible to review any subdiscipline or journal in human geography without seeing feminist geographers' contributions and the ways in which their inquiries have altered geography's theories, practices, and methods. This widespread effect makes an inclusive assessment of feminist geographies difficult, but here a few examples can serve to highlight feminist geographies' contributions to the practice and study of human geography.

WHERE ARE THE WOMEN IN GEOGRAPHY?

Feminist geographers have drawn attention to the paltry numbers of women in the profession of academic human geography, especially in comparison with some other social sciences such as sociology and anthropology. They suggest that the overwhelmingly male composition of human geography, especially in the past but also continuing to the present, has influenced what has been studied in geography. Many claim that this has meant that women's particular issues and the study of gender relations more broadly have been neglected in geographic scholarship. The research and writing in human geography may profess to be universally appropriate, but under scrutiny feminists find that seemingly universal claims about how the world works really refer to men's worlds. Furthermore, feminists claim that male dominance in geography and its scarce scholastic attention to women's lives and issues only further discourage women from entering the field.

Feminist geographers carry out their critique of the profession in part by exposing its masculinism. This refers to how the discipline has primarily served the interests of men and worked within masculine paradigms that emphasize objectivity over situated accounts of the world, that engage only certain methodologies, that observe an ideal of professional distance over political activism, and that call for a separation between personal life and public life. Feminists suggest that these have dissuaded women from becoming professional geographers and have restricted scholarship and geographic theory. Therefore, feminist geographers tie the doing of geography—that is, the practices of academics—with knowledges produced. They want to ensure that such bias cannot happen in the future.

Feminist geographers have been successful in their attempts. Nowadays it is more difficult to ignore feminists' arguments and disregard their contributions. Unfortunately, their successes do not equal a complete triumph over masculinism, and feminist geographers continue to struggle to remold geography so that women will be drawn to the discipline. Feminist geographers advocate good mentoring of female students and colleagues to help them succeed. Feminist educators also insist that teaching feminist topics will show all students, both male and female, how important it is to consider their approaches—and, hopefully, to inspire more inclusive disciplinary practices in the future.

THE WORK OF WOMEN

The research that feminist geographers conduct has been influenced by wider trends in feminist and social theory, so it is important to place feminist geographies in a context of feminisms' development over time. Feminist geographers have, however, made significant contributions through their unique geographic perspective on gender and women's lives. In particular, feminist geographies have investigated the different material worlds of women and men. Their motivation is to show that women live in, produce, and negotiate very different spaces than do men. They also argue that women's identities are made through gendered social meanings of space and environment.

For example, during the 1970s and 1980s, feminist geographers asked how women were adversely affected by economic change and restructuring. This was an important development because previous economic study neglected the differences between women's and men's lives while simultaneously accepting men's lives to be the norm. Feminist geographers, on the other hand, argued that a gender division of labor restricted women's ability to enter high-paying sectors of employment. Women traditionally have been segregated in "ghettos" of employment represented as and labeled "feminine" or "women's work" such as pink-collar jobs, low-end service work (e.g., housekeeping), informal sector work, and caring labor (e.g., teaching, child care). The types of jobs open to women also are determined by race and ethnicity, so that minority women are restricted to the most menial and poor-paying jobs. Certainly even the availability of women's work itself is regionally variable given that underemployment marks one characteristic of developing countries. Many of these early studies were influenced by socialist feminism, which tied women's oppression to both gender and class relations of power at different scales. An activist interest in exposing the inequalities of the labor market also motivated feminist scholarship.

Feminist geographies also sought to connect the gendered economy with social life and cultural spaces, that is, to argue that each buttressed the other. Thus, the example of the gendered division of labor extends to the ways in which regions, cities, and Western suburbs themselves are built to reflect and reestablish gender difference. Historical geographers have shown that the separation of work from home under capitalism reflects dominant gender norms that place women in private spaces at home (ideally in suburbs in the West) and men in public spaces such as the workplace (in urban spaces). All over the world, women are supposed to be supporters of men's labor and creators of home, yet rarely are they credited economically with heading households. Thus, when women do engage in labor for wages, they are represented as more capable of reproductive caring work or work that is viewed as inferior to men's work, in turn reinforcing poor wages and dead-end jobs. Therefore, feminists argue that women's labor force participation is shaped by social relations of gender, class, race, location, and space as well as the sexist and racist ideals constructing spaces and practices.

This example highlights how feminist geographies show the interconnections between different spheres and social relations. Neat simplifications and theories obscure the complicated processes that tie representations of gender to material life, such as through places and institutions, the built environment, and the way in which we think about nature, to the jobs that are available to women. Thus, feminist geographies often cross the borders that delineate subfields, and their contributions span economic, political, urban, social, environmental, and cultural geographies.

The example of women's work shows how places are made through broad and complex processes and problems. But it would be a mistake to assume that all of women's places are local like the home; for instance, the case of work also has offered feminist geographers a way in which to critique traditional approaches to global spaces and larger-scale processes of economies, resource use, and politics. Feminist geographers tie the disposability of women's low-wage work to global economic systems that rely on cheap labor to ensure profitability and that relegate women's labor to the lowest rungs. There is a regional component to this as well, as with the case of the movement of production to the global South or, more specifically, to places such as maquiladoras in northern Mexico, sweatshop labor in Indonesia or China, cash crop versus food crop farming in western Africa, and resulting cheap service labor in the United States or the United Kingdom. Feminist geographers also insist that the additional burden of women's informal labor subsidizes formal economic accumulation because, for instance, women's reproductive labor ensures the supply of workers, the nutrition of workers' bodies, the growing of food and tending of land, and the caring of the sick, elderly, and young. Development projects, feminist geographers argue, must consider the labors of women who are not necessarily participating in formal markets. The lack of women's political participation in many states also exacerbates the neglect of women's concerns in development and national and international economies.

On the other hand, feminist geographers have detailed the ways in which women and girls cope with economic hardship and development schemes by migrating for work, forming alliances with other women and families, rearranging household labor, and insisting on more "appropriate" development programs. Here appropriateness refers to how development and planning must consider women's varying needs according to location, religion, gender, sexuality, and other differences. Some feminist geographers even suggest that the idea of scales of regional development (i.e., the developing world) and the simplification of global process as primarily political–economic obscure women's multiple contributions to the world and their rich subjectivities. Such reductions lead to representations of Third World women as merely workers and laborers in a global economy aimed at

First World consumption. Women's lives all over the world are complex and rich despite the obstacles women face.

SPACE, PLACE, AND IDENTITY

Feminist geographers also study subjectivity or how women come to take on identities and think of themselves. Given that feminist geographers have shown that place, space, and gender are interrelated, surely place and space matter to how women form identities and how their identities in turn affect their production of place and space. Moreover, feminist geographers ask how women resist the dominant, and often sexist and racist, relations that largely influence the production of identity and place.

Feminist geographers have examined identities by looking at particular gendered sites, such as the home, the city, and the workplace, and at the spaces that contextualize women's lives such as nation, nature, urban, private, and public. These approaches often ask how women experience places and spaces in unique ways, such as women's fear of crime and perceptions of urban space, or how women create spaces for themselves such as homes and city neighborhoods. They are also engaged with the question of how boundaries are created, asking whether women are pressured to be in one space and not the other (e.g., the home and not the city street). Borders between spaces and places also interest geographers, and feminists have asked how mobility and movement across borders affect women's identities such as through international migration.

Feminist geographies do not just concentrate on women's differences from men or at investigations of specific sites. Feminist geographies also are interested in how identities come to have social meaning and relevance to individual women; the issue of subjectivity includes a consideration of how women learn about and practice their identities. Geographers specifically contribute to interdisciplinary feminist theories by considering how women's personal identities are made geographically through social space. In these attempts, feminist geographies show that women's identities are constantly changing in and through space and are mutable and spatially contingent, providing substance to the argument against an essential gender identity or idealized femininity.

As wider debates in feminism have highlighted, femininity is only one identity that women claim. Therefore, feminist geographies examine other identity practices of race, age, ethnicity, and sexuality as well as how different women must contend with spaces of racism, ageism, and homophobia. Feminist geographers also have begun to think about men's gender identities and masculinity and how embodiment matters to identity. For example, studies of health, disability, and illness explore women's and men's thoughts on their bodies, how people cope with the demands placed on their bodies, how ill and disabled bodies are or are not accepted in different places, and how women and men react to such discriminations.

It is important to remember that social spaces always are socially constructed, raising the question of how, when, and why women resist dominant norms and ideals with which they might not agree. The focus on resistance in feminist geographies is politically motivated given that feminists not only are interested in explaining gender oppression but also are intent on making women's lives better by fighting sexism. But resistance is conceptually tricky because most of us do not resist gender per se. Figuring out when women are resisting gender norms, as opposed to when they themselves are reproducing gender relations and meanings, presents feminisms with one of its most difficult and contentious tasks.

—*Mary Thomas*

See also Feminisms; Feminist Methodologies; Gender and Geography; Lesbians, Geography and; Masculinities; Situated Knowledge

Suggested Reading

McDowell, L. (1999). *Gender, identity, and place: Understanding feminist geographies.* Minneapolis: University of Minnesota Press.

Nagar, R., Lawson, V., McDowell, L., & Hanson, S. (2002). Locating globalization: Feminist (re)readings of the subjects and spaces of globalization. *Economic Geography, 78,* 257–284.

Rose, G. (1993). *Feminism and geography: The limits of geographical knowledge.* Minneapolis: University of Minnesota Press.

Women and Geography Study Group. (1997). *Feminist geographies: Explorations in diversity and difference.* Harlow, UK: Addison-Wesley Longman.

FEMINIST METHODOLOGIES

Feminist methodologies in geography are the research practices through which feminist geography is enacted. Methodology refers to an approach to research, including the practices of research design, data collection strategies, the conduct of research, and modes of analysis. Feminist methodologies are informed by a feminist epistemology or theory of knowledge. Drawing attention to how gender influences what counts as knowledge and how knowledge is produced, feminist methodologies put into practice an epistemological critique of the masculinist underpinnings of scientific positivism. Feminist methodologies have informed feminist geographers' choices of methods, although they have not constrained them. Feminist geographers continue to conduct both quantitative and qualitative research. Finally, feminist methodologies are distinguished by an attention to relations of power within the research process and by a commitment to feminist politics in all its diversity.

Although feminist critiques of positivist empiricist methodological practices were first articulated during the 1970s, it was not until the 1990s that feminist methodological discussions were first published in geography. Feminist methodology arose from a critique of the claims to objectivity and the assumptions of authority in social scientific research. Biases in research methodologies, feminists argued, not only have led to the exclusion of certain voices from the privileged ground of scientific knowledge but also have promoted exploitive research practices. Feminist geographers have shown how the methods historically associated with geographic research, such as exploring, charting, surveying, and mapping, have contributed to imperial practices, colonial ideology, and military conquest.

Feminist methodologies aim to break down the hierarchies of research, to shift the power relations between the researcher and the researched, and to cultivate relational, engaged, and emancipatory research practices. Adopting a stance of "reflexivity" in feminist research means considering how the shifting sociospatial constructs of identity and subjectivity affect relations between researchers and participants and how the *inequality* embedded in these relations affects the production of knowledge. Feminist geographers work both to destabilize this power relation in the research process and to acknowledge its impact on the practices and products of geographic research. For feminists, objectivity not only is impossible but also is an undesirable goal in that it disavows the intersubjectivity of the research process.

Feminist geographers have debated the question of whether working with a feminist methodology should give rise to a particular set of methods—that is, to particular techniques of data collection and analysis.

Critiques of dominant research practices have led some feminists to reject quantitative methods, arguing that these methods are tied to patriarchal structures, both in their claims to objectivity and in their reliance on predetermined inflexible categories. Indeed, the rise of feminist geography has been associated with the resurgence of qualitative methodologies in geography, including interviews, oral histories, focus groups, and ethnographies. Qualitative methods may offer particular strengths for addressing feminist concerns. For example, life histories may enable a researcher to bring forward the voices of those who have been silenced by dominant discourses, discussion (or "focus") groups may foreground the contextual construction of meaning, and participant observation may enable interactive research in which relationships, friendship, and connection are central to the research process. These methods may also be more conducive to practices of collaboration whereby researchers return the products of their studies to the communities in which they conducted their fieldwork, often soliciting feedback from research participants.

Despite this association between feminist geography and qualitative methods, feminist geographers have largely rejected the idea that only certain methods may be part of a feminist methodology. Although there is a diversity of positions among feminist geographers, many believe that feminists should avail themselves of the full toolbox of methods of geographic inquiry. Quantitative methods may be useful for the documentation of forms of discrimination or oppression, for the identification of trends or patterns, and for inserting feminist issues into public agendas. Furthermore, qualitative methods should not be assumed to be inherently less exploitive or power laden than quantitative methods. In fact, the entangling of friendship, political engagement, and ethnography presents an equally thorny set of issues for feminist researchers. In choosing their research methods, the critical questions that feminist geographers must ask themselves are, first, whether the methods are appropriate to their research questions and, second, whether the methods (not only the projected results of the research) further feminist goals. Keeping these questions in sight, feminist geographers not only have continued to use both quantitative and qualitative methods but also have contributed trenchant critiques of the quantitative/qualitative dualism itself.

Whatever methods are used, certain practices, approaches, and concerns characterize a feminist methodology. Feminist research arises from a feminist epistemology that guides the formation of research questions, the application of methods, and the interpretation of results. Feminist geographic research reflects a commitment to challenging the sociospatial imagination that supports patriarchal structures and relations of inequality in society. Just as feminist geographers have destabilized the binaries of public/private, work/home, and culture/nature in their research, feminist methodologies interrogate the relations of power embedded in research practices by breaking apart hierarchically coded relationships such as researcher/researched, objective/subjective, rational/emotional, and field/home. Finally, feminist research practice in geography aims to enact feminist politics through the research process. However, there is not one singular feminist politics. Because feminists are concerned with the politics of the production of knowledge, the question of who produces knowledge about whom continues to challenge any tendencies to the centralization of feminist identity or politics. The heterogeneity of feminist epistemologies means that there are multiple feminist methodologies and multiple feminist projects.

—*Anna Secor*

See also Epistemology; Feminist Geographies; Qualitative Research; Quantitative Methods

Suggested Reading

Jones, J., III, Nast, H., & Roberts, S. (Eds.). (1997). *Thresholds in feminist geography.* Lanham, MD: Rowman & Littlefield.

Mattingly, D., & Falconer-Al-Hindi, K. (1995). Should women count? A context for the debate. *The Professional Geographer, 47,* 427–436.

McDowell, L. (1992). Doing gender: Feminism, feminists, and research methods in human geography. *Transactions of the Institute of British Geographers, 17,* 399–416.

Moss, P. (Ed.). (2002). *Feminist geography in practice.* Oxford, UK: Blackwell.

Rose, D. (1993). On feminism, method, and methods in human geography: An idiosyncratic overview. *The Canadian Geographer, 37,* 57–61.

Staeheli, L., & Lawson, V. (1994). A discussion of "women in the field": The politics of feminist fieldwork. *The Professional Geographer, 46,* 96–102.

FERTILITY RATES

Technically, fertility rates refer to the number of children a woman gives birth to during her lifetime and typically ranges between 0 and 12, although in some cases women may have considerably more children. Demographers often use age-specific fertility rates, which depict births to women in given 5-year age categories, because women tend to have the highest fertility rates during their main childbearing years between 15 and 40 years of age. Thus, fertility rates are closely related to birth rates (i.e., the number of babies born per 1,000 people per year) and to fecundity rates (i.e., the rate at which women biologically are potentially able to have children). Fertility is the major means by which most populations increase in size (in addition to in-migration) and is the counterpart to mortality.

However, fertility is much more than just a dry statistical measure in that it encompasses a diverse and complex set of social, cultural, economic, political, and psychological phenomena that lead it to vary widely among and within societies as well as over time and space. Although there undoubtedly are some biological factors that shape fertility rates, including genetics and diets, fertility is for the most part a reflection of the circumstances that lead women to either have children or not have children. In the world today, fertility rates range from a low of 1.3 children per woman in Spain and Portugal to roughly 8.3 children per woman in Rwanda.

Traditionally, fertility rates in preindustrial societies have been very high for a variety of reasons. In many hunting and gathering societies, fertility (and women) was celebrated as a sign of nature's benevolence. In agrarian economies, children are a vital source of farm labor, helping to plant and sow crops, tending to farm animals, performing chores, carrying water and messages, and helping with younger siblings. Children are important resources for aging parents in the absence of government programs such as social security. Finally, in such societies with high infant mortality rates, high fertility rates ensure that some proportion of children will survive until adulthood. In short, high fertility rates are rational demographic responses to agrarian poverty.

As societies industrialize, fertility rates tend to decline, a process that usually occurs slowly, over several generations, and typically well after mortality rates have fallen. Essentially, industrialization and urbanization lower fertility rates because they alter the motivations to have many children and large families. The need for child labor typically declines, whereas the costs, especially as measured in terms of forgone income in a commodified labor market, rise steadily.

As women's incomes rise, either over time or comparatively within a society, the opportunity cost of having children rises accordingly, leading to lower fertility rates. Thus, national wealth and fertility rates almost always are inversely related. If fertility rates drop to the level of mortality rates, a society reaches zero population growth, a situation characteristic of most of Europe and Japan. Societies under such conditions are characterized by large numbers of the elderly, a high median age, and a relatively small number of children, all of which have dramatic implications for public services.

—*Barney Warf*

See also Demographic Transition; Mortality Rates

Suggested Reading

Yaukey, D., & Anderton, D. (2001). *Demography: The study of human population* (2nd ed.). Prospect Heights, IL: Waveland.

FIELDWORK

A corporate boardroom. A refugee encampment. Everyday life in Bordeaux. Sacred spaces in Australia's outback. Andean labor movements. Shared memories of a public space. All of these are fields in today's human geography, and *fieldwork* is simply the sum of experiences by which a researcher engages these social spaces to generate original information. For most human geographers, fieldwork means some form of physical displacement, usually referred to as getting "out there." This means listening to people, experiencing sights and sounds, and interacting with particular groups. Therefore, fieldwork represents a stimulating complement and contrast to forms of research that rely on data that are not explicitly constructed with the researcher's own questions in mind, that is, with "secondary" or "preconstructed" data (e.g., national census data). Fieldwork spans all types of human geographic inquiry and is a dynamic and exciting element of research in the discipline.

FIELDWORK: CONSTRUCTING DATA

Human geographers draw on a variety of field methods or techniques to construct data, usually through direct engagement with people. *Interviews* span a range of interactive forms, from impromptu chats and story-telling to highly structured one-on-one interviews. In a focus group interview, the researcher convenes several people to initiate discussion or activities around pre-selected themes, sometimes with the aid of visual imagery such as maps or photographs. In contrast to interviews, door-to-door-style *surveys* typically are designed to generate responses from large and representative samples of people, usually with the aid of a scripted questionnaire. *Participant observation* relies on the researcher's active engagement with particular places, people, and processes. This might entail attending a community meeting, hanging out in a mall, or accompanying individuals through a normal day. *Ethnography* is participant observation sustained through long-term multimethod interaction with a community. The ethnographer might live for months or years in a foreign settlement or take a job that offers immersion in a particular industry or retail space. For geographers interested in historical phenomena, *archival research* (e.g., examining old diaries or parish records) often is considered fieldwork because it offers one of the most direct ways of engaging the past.

In practice, most field-workers combine multiple methods. Participatory mapping projects, for example, usually incorporate interviews, group sketch-mapping exercises, and walks with residents to locate significant landscape features. In fact, one of the challenges of doing fieldwork is to figure out what combination of methods is the most effective and practical, especially in light of the unanticipated hurdles and new avenues of inquiry that inevitably arise during the fieldwork process. For many geographers, it is the promise of serendipity and discovery that makes fieldwork so personally stimulating, memorable, and attractive as a research mode.

Data generated during fieldwork can be stored in any number of ways, including as a sheaf of completed questionnaires, tape or video recordings, photographs, drawings, e-mails, handwritten notes, tally sheets, click meter readings, and sketch maps. Therefore, these field data may be in numerical, digital, visual, or textual form and may emphasize qualitative or quantitative aspects of the research topic.

Researchers sometimes refer to fieldwork as data gathering. But this is misleading because data are not lying around waiting to be collected. The term *data construction* is preferable because it conveys the extent to which data are the product of innumerable decisions and interpretations made by the researcher. After all, a photograph is framed by the researcher's choices about what to exclude from view, a tape recorder is turned off based on someone's assessment that a conversation is over, and the contents of a field journal reflect what the researcher considers to be worth jotting down. Thus, fieldwork—as data construction—simultaneously is composed of data acquisition *and* some degree of data interpretation.

FIELDWORK IN HUMAN GEOGRAPHY: 1900s–1970s

Fieldwork is not unique to geography; for example, geology and anthropology are deeply fieldwork oriented. But primary data collection has long been central to the geographic enterprise. During Europe's Age of Enlightenment, scientific observation was considered to be essential to learning about the world. This spirit was exemplified by proto-geographer Alexander Von Humboldt, whose famous work *Cosmos* grew from his exhaustive travel and observations, particularly in the Americas. Fieldwork and exploration were largely inseparable throughout the 18th and 19th centuries, when nascent geographic societies across Europe and the United States funded expeditions to visit and report back on far-flung places. Although ostensibly scientific, their detailed reports were unabashed in describing exploitable resources and potentials for European settlement. Ultimately, early geographic fieldwork proved to be essential in promoting and facilitating European imperial expansion. To this day, many people associate the image of a pith-helmeted scientific explorer with the colonization and exploitation of southern regions.

As academic geography developed during the 20th century in Europe and North America, fieldwork traditions that developed in Britain and Germany emphasized dispassionate observation, field measurement, and the physical experience of landscape. Cultural geographer Carl Sauer was arguably the greatest proponent of such methods in the United States. He encouraged students to identify suitable vantage points from which to assess visible manifestations of human activity in a landscape and to enhance their understanding of the visual scene by talking with informed inhabitants.

Generations of human geographers followed the model set by Sauer and other early-20th-century human geographers, and the years following World War II saw a deepening of fieldwork training in geography curricula. This was accompanied by growing interest in foreign fieldwork, much of it facilitated by cold war concerns to develop world–regional expertise. This period saw the development of human geographers' proficiencies in a range of field skills, including foreign language competence, survey and mapping techniques, aerial photo interpretation, and questionnaire administration. Beginning in the 1960s, however, this field-based tool kit began to be eclipsed by the positivist revolution's shift from data construction to data analysis. Quantitative geographers saw little merit in field-based observation beyond post hoc ground truthing to validate model results, and fieldwork training in human geography subsequently waned markedly.

Noticeably absent from virtually all published human geography research prior to the 1970s is any detailed reflection on personal or political issues that arose in the course of fieldwork. For readers, the effect was to remove the researcher from the objective "facts" that he or she had distilled from a place and to present fieldwork as an unproblematic, innate, and inherently objective way of accessing information about the world. Yet many human geographers of the period *did* feel the need to write about the triumphs and hurdles of their fieldwork experiences. Most resorted to confiding in letters to family and colleagues, but some published their reflections as separate books or memoirs. These products typically were considered to be unscholarly—albeit fascinating—asides that had little bearing on the "scientific" insights the fieldwork had produced.

CRITIQUES OF TRADITIONAL FIELDWORK PRACTICE

How human geographers conduct, think about, and write up fieldwork has since been transformed by insights from humanistic, feminist, and other critical geographies. During the 1970s, humanistic geographers were frustrated by the way in which quantitative approaches to geography ignored the richness and contingency of human agency. They saw in anthropology's increasingly sophisticated ethnographic methods a means to engage with the lived realities of everyday people that were hidden in generalizing spatial models. Furthermore, humanistic geographers who worked with underprivileged societies felt compelled to address the root causes of social inequality, and their fieldwork often melded with advocacy and activism. Although neither ethnography nor activism was new in human geography, humanistic geographers'

activist ethnography dramatically enlivened the practice of fieldwork as a tool of political commitment and social change.

By the 1980s, feminist geographers were using fieldwork to explore how men and women experienced, shaped, and were influenced by spatial processes in profoundly different ways. They showed how traditional fieldwork reflected a dominant masculinist gaze that was blind to women's lives and labors and overlooked the ways in which gendered power relationships influenced human and social geographies. Instead, they strove to solicit the narratives of women, particularly those whose stories had long been hidden from academic scrutiny because they inhabited jobs or spaces (e.g., as explorers, miners, factory workers, or labor organizers) that were conventionally considered masculine. They also drew on ethnography to record the performance of gender dynamics in multiple spaces. Feminist insights also contributed to understanding the power-laden context of interviews, as when women are interviewed with male kin. Researchers subsequently have extended these practices to accommodate the multiple other axes along which people align such as those of race, class, and sexuality.

Human geographers also have questioned the unequal power relations that underlie the ability of white educated researchers to conduct fieldwork, especially in cross-cultural settings at home and abroad. These postcolonial critiques argue that the funding and practice of fieldwork too often has been tied to agendas that served elite interests and supported unequal political–economic structures. This critique shined a particularly harsh light on fieldwork associated with modernist development projects in the global South such as when seemingly benign data construction among rural producers helped justify international interventions that exacerbated rural inequalities. In response, human geographers have since become much more careful to understand the power structures in which they themselves are embedded and to struggle to formulate fieldwork practices that contribute to positive social change. For example, participatory methods seek to transform research participants/informants into collaborators who contribute to project design, execution, and the interpretation and dissemination of results.

Finally, fieldwork practice has been increasingly influenced by poststructural and postmodern ideas that question the notion of a single "objective" truth or the authority of any researcher to speak about phenomena beyond his or her immediate world. In extreme cases, these critiques have led some human geographers to abandon fieldwork altogether. Others have worked hard to diversify and expand their field approaches so as to elicit and represent multiple perspectives on a given topic. Most significantly, however, human geographers have become much more critically aware of the way in which their own positionalities—their economic backgrounds, cultural biases, politics, and emotions—shape the data they construct. These insights inform fieldwork practice when, for example, a researcher offers multiple opportunities for an interviewee to ask his or her own questions about the researcher's motives, background, and interests. They also inform how human geographers write about their research. The use of the first-person writing viewpoint is now common, and much greater textual space is allotted to explicit evaluation of the choice and effectiveness of field methods, particularly with respect to researcher positionality.

THE FUTURE OF FIELDWORK IN HUMAN GEOGRAPHY

Human geographers have worried about fieldwork's role within the discipline. But at least two sources of evidence suggest that fieldwork in human geography is vibrant and thriving. First, the recent increase in publications on the history and modern practice of fieldwork in human geography suggests a new interest in understanding the past and future roles of data construction in the field. Second, there is a renewed emphasis on fieldwork in human geography curricula, with new texts devoted to helping students think about, conduct, and write about primary research methods and with particular attention to the political and ethical issues involved. Once common assumptions that students could figure it out as they went along have given way to consensus that becoming an effective and responsible field-worker requires training and practice. Essential skills include recognition that engagement with people and places is a privilege, not a right; that methods must be flexible enough to capture multiple perspectives on an issue; that the anonymity, rights, and well-being of collaborators must be ensured; and that those most likely to be affected by the fieldwork must be able to negotiate the use and mobilization of results.

Ultimately, fieldwork in human geography contributes tremendously to our ability to understand the world and our place in it. As an academic or applied skill, fieldwork is a profoundly important part of the human geographic enterprise; as a personal experience, it is arguably one of the most challenging and satisfying research modes.

—*Kendra McSweeney*

See also Empiricism; Exploration, Geography and; Feminist Methodologies; Interviewing; Participant Observation; Qualitative Research; Vision

Suggested Reading

Clifford, N., & Valentine, G. (Eds.). (2003). *Key methods in geography.* London: Sage.

Cloke, P., Cook, I., Crang, P., Goodwin, M., Painter, J., & Philo, C. (2004). *Practising human geography.* London: Sage.

Hoggart, K., Lees, L., & Davies, A. (2002). *Researching human geography.* London: Edward Arnold.

Scheyvens, R., & Storey, D. (Eds.). (2003). *Development fieldwork: A practical guide.* London: Sage.

Selections on "Women in the Field." (1994). *The Professional Geographer, 46*(1).

Starrs, P., & Delyser, D. (Eds.). (2001). Doing fieldwork [special issue]. *Geographical Review. 91*(1–2).

FILM, GEOGRAPHY AND

Although many geographers have used, and still do use, film as part of their teaching strategy, it is only within the past 10 years or so that sufficient quantities of research on this popular medium have been produced to allow for a disciplinary subfield called *film geography*. Indeed, the earliest disciplinary writings on film, produced for *The Geographical Magazine* during the 1950s, were aimed at elucidating their usefulness as teaching tools; geographers should be encouraged to use clips from those films that, it was argued, represented landscapes in as faithful a manner as possible, such that students could gain a sense of what it would be like to experience these places firsthand as if they were "in the field." In this regard, film was regarded as more successful in its mimeticism than were other media, such as photography, in that film managed to capture movement as well as form and so could be used to represent natural and social processes.

The publication of Aitken and Zonn's *Place, Power, Situation, and Spectacle: A Geography of Film*, however, heralded a new era that addressed precisely this assumption—that film should and could provide a transparent "window" to the real world. As geographers generally engaged with broader-scale academic debates over the "crisis of representation," so film geographers began to address the relationship between the "real" (that which the camera has filmed) and the "reel" (the image on screen).

Two major lines of geographic research have ensued from this examination of the real and the reel. The first stems from Marxist analysis—particularly the work of the Frankfurt School in 1930s Germany—and emphasizes the ways in which film serves to support capitalism ideologically. Emphasis is placed on the fact that film is the product of a highly successful industry; as such, one can trace the form and impact of successive rounds of capital investment and disinvestment, as well as relations of exploitation, across the globe. Moreover, it is argued, the content of such films more often than not serves to divert attention from the broader effects of the global capitalist system in the form of poverty, crime, and environmental degradation. Instead, film content panders to a voyeuristic interest in sex and violence or a more benign concern with the melodramatic plight of the individual. Last but not least, film works to commodify both people and place in the sense that as each becomes part of a film project, the individuality and uniqueness of each are reduced to a standard marketable package. The sum effect is to destroy critical thinking in film watching as well as filmmaking.

And yet not all Marxist analyses are so pessimistic. Frankfurt School member Walter Benjamin argued famously that film offers escapism for audiences; it offers new horizons outside of their day-to-day experiences. This feeling of liberation might not be realized through the actual overthrow of the capitalist system but is nevertheless indicative of the dreams, myths, and expectations that are an integral part of the complex and subtle process of film spectatorship.

The second line of research to investigate the relationship between the real and the reel ensues from a social constructivist (sometimes called poststructuralist) perspective. Social constructivism emphasizes the fact that objects do not exist in a vacuum; instead, they are given meaning through the actions and thoughts of people. Because of the complexity of the life world of each person, we can expect an infinite number of meanings to accrue around an object. In contrast to a scientific understanding of the world, social constructivism holds that one cannot point to one particular meaning as being somehow correct even if there are commonalities across a range of viewpoints as to the nature of a particular object or if one particular way of looking at an object seems to work better than another when attempting to manipulate that object for a particular purpose. Because we cannot escape our own subjectivity, we can merely hypothesize, but never can realize, what an object is like outside of those meanings ascribed to it. In this sense, we cannot point to the "real" nature of anything. Moreover, this means that we as individuals cannot actually "access" someone else's view of the world and so can make no claim toward representing his or her views in an authentic manner.

The significance of this for film geography is that one no longer can talk about film as representing, or mimicking, reality simply because there is no single coherent reality waiting "out there" to be filmed. To be sure, the camera records mass and motion, but the nature of those objects that appear off the screen is firmly located in the world of meaning. Similarly, the nature of those objects that appear on the screen are just as embedded in social meaning. Accordingly, film geographers have begun to look at (a) how particular meanings are indeed ascribed to people and place as they appear on the screen (this requires an appraisal of how cinematic techniques are used to present action, narrative, and emotion) and (b) how the meanings of those people and places on the screen interconnect with meanings established through other media, including actually being "in" a particular place. Underpinning both of these research areas is an interest in the power relations behind the production of meaning, as some notions of what people and places are like have become much more "taken for granted" than others. The pervasive characterization of nature as feminine and an encroaching society as masculine, for example, has been the subject of much research in film geography, as has the association of indigenous groups with the notion of noble savagery. And yet it is well to remember that these are held to be very much social constructions. As such, these meanings are much more complex than our descriptions of them and, moreover, are open to continual transformation. In a move that harkens back to the work of Benjamin, noted earlier, much is now being made of the engagement between film and its audience, particularly the question of how the meanings of those diverse places within which film watching occurs—the cinema, the home, the car, and even the bus/train via the mobile phone—are themselves transformed though the practice of film watching, a practice that is just as much about hearing, tasting, touching, and smelling as about spectatorship. From a simple exercise in pedagogy, then, film has become one of the key media through which geographers have explored a host of research questions regarding the ways in which we understand and reflect on the meanings ascribed to ourselves, others, and the world at large.

—*Deborah Dixon*

See also Spaces of Representation; Vision

Suggested Reading

Aitken, S., & Zonn, L. (Eds.). (1994). *Place, power, situation, and spectacle: A geography of film.* Lanham, MD: Rowman & Littlefield.

Cresswell, T., & Dixon, D. (Eds.). (2002). *Engaging film: Geographies of identity and mobility.* Lanham, MD: Rowman & Littlefield.

FIRST LAW OF GEOGRAPHY

SEE TOBLER'S FIRST LAW OF GEOGRAPHY

FLEXIBLE PRODUCTION

Flexible production, also called *post-Fordism,* refers to the forms of manufacturing that began to take shape and eventually became dominant throughout the late 20th century. In contrast to Fordism, flexible production allows goods to be manufactured cheaply but in small volumes as well as large volumes. A flexible automation system can turn out a small batch, or even a single item, of a product as efficiently as a mass-assembled commodity. It appeared, not accidentally, at the particular historical moment (1970s) when the microelectronics revolution began to revolutionize manufacturing; indeed, the changes associated with the computerization of production in some respects may be seen as capitalists' response to the crisis of profitability that accompanied the petrocrises. Flexible production also reflected the imperative of firms to increase their productivity in the face of rapidly accelerating, intense international competition.

The most important aspect of this new, or lean, system is flexibility of the production process itself, including the organization and management within the factory, and the flexibility of relationships among customers, supplier firms, and the assembly plant. In contrast to the large, vertically integrated firms typical of the Fordist economy, under flexible production firms tend to be relatively small, relying on highly computerized production techniques to generate small quantities of goods sold in relatively specialized markets. Microelectronics, in essence, circumvented the need for economies of scale.

The classic technologies and organizational forms of post-Fordism include robots and just-in-time inventory systems. The Japanese developed just-in-time manufacturing systems shortly after World War II to adapt U.S. practices to car manufacturing. The technique was pioneered by the Toyota Corporation (and hence sometimes is called "Toyotaism") and obviated the need for large expensive warehouses of parts (the "just-in-case" inventory system). *Just-in-time* refers to a method of organizing immediate manufacturing and supply relationships among companies to reduce inefficiency and increase time economy. Stages of the manufacturing process are completed exactly when needed according to the market—not before and not later—and parts required in the manufacturing process are supplied with little storage or warehousing time. This system reduces idle capital and allows minimal investment so that capital can be used elsewhere. The manufacturing run proceeds only as far as the market demands. Inventories are very small and are replenished only to replace parts removed downstream. Workers at the end of the line are given output instructions on the basis of short-term order forecasts. They instruct workers immediately upstream to produce the part they will need just-in-time, and those workers in turn instruct workers upstream to produce just-in-time, and so on. Post-Fordist approaches to production came to dominate much of the electronics industry, automobiles, and the minimills of the steel industry.

Flexible production is closely associated with vertical disintegration and increased subcontracting rather than "in-house" production. During the 1980s and 1990s, many firms engaged in significant "downsizing," often ridding themselves of whole divisions of their companies to focus on their "core competencies." A large number of companies reversed their old principles of hierarchical, bureaucratic, assembly-line (Fordist) processes as they switched to customized, flexible, consumer-focused processes that can deliver personal service through niche markets at lower costs and faster speeds.

In the process, the use of subcontracts accelerated rapidly. Firms always face "make or buy" decisions, that is, deciding whether to purchase inputs such as semifinished parts from another firm or to produce those goods themselves. Under the relatively stable system afforded by Fordism, most firms produced their own parts, justifying the cost with economies of scale, which lowered their long-run average cost curves. Under post-Fordism, however, this strategy no longer is optimal; given the uncertainty generated by the rapid technological and political changes of the late 20th century, many firms opted to "buy" rather than "make," that is, to purchase inputs from specialized companies. This strategy reduces risk for the buyer by pushing it onto the subcontractor, who must invest in the capital and hire the necessary labor.

A key to production flexibility lies in the use of information technologies in machines and operations, allowing more sophisticated control over the process. With the increasing sophistication of automated processes and, especially, the new flexibility of the new electronically controlled technology, far-reaching changes in the process of production need not be associated with an increased scale of production. Indeed, one of the major results of the new electronic computer-aided production technology is that it allows rapid switching from one process to another and allows the tailoring of production to the requirements of individual customers. Traditional automation is geared to high-volume standardized production; flexible manufacturing systems are quite different. In the business world, flexibility is based on customization of output to individual needs and wants, higher quality and higher value added, rapid response and delivery, and improved service and follow-up. Instead of greater capital investments in infrastructure and machines, software and marketing databases allow firms to estimate the needs of customers and identify niche markets. Such software allows firms to produce in "short runs" for market niches, quickly changing markets without setup costs and avoiding delays of assembly-line systems.

As interfirm linkages grew rapidly during the 1980s, many firms found themselves compelled to enter into cooperative agreements with one another (e.g., joint ventures). Quality control (i.e., minimizing defect rates) became very important. Many firms

succeed in this environment by entering into dense urban networks of interactions, including many face-to-face linkages and the roles played by "noneconomic" factors such as tacit knowledge, learning, reflexivity, conventions, expectations, trust, uncertainty, and reputation in the interactions of economic actors. Geographically, therefore, flexible production is closely associated with the dense concentrations of "high-technology" firms that emerged during the late 20th century, including California's Silicon Valley, Italy's Emilia–Romagna region, Germany's Baden–Württemberg, the Danish Jutland, and the British electronics region centered around Cambridge. In such contexts, firms can substitute agglomeration economies (or external economies of scale) for (internal) economies of scale achieved by producing in large quantities.

—*Barney Warf*

See also Economic Geography; Fordism

Suggested Reading

Amin, A. (1994). Post-Fordism: Models, fantasies, and phantoms of transition. In A. Amin (Ed.), *Post-Fordism: A reader* (pp. 1–39). Oxford, UK: Blackwell.

Gertler, M. (1992). Flexibility revisited: Districts, nation-states, and the forces of production. *Transactions of the Institute of British Geographers, 17,* 259–278.

Linge, G. (1991). Just-in-time: More or less flexible? *Economic Geography, 67,* 316–332.

Peck, J. (1992). Labor and agglomeration: Control and flexibility in local labor markets. *Economic Geography, 68,* 325–347.

FOOD, GEOGRAPHY OF

What humans eat varies dramatically across space. The food we consume, or even the style of cooking called a "cuisine," can tell us much about the site and situation of a place. The particularities of the place's site, such as its climate, soils, plant and animal life, and location, provide the options from which people choose to eat in that place. But beyond this, what people eat is also shaped by a place's situation or a place's ever-changing relationships with other places. As people migrate, they bring with them their food preferences, cooking techniques, and sometimes also the plants and animals required to make their cuisine.

This interaction changes both the migrants and the host society's food choices, producing a hybrid cuisine. Food is often used to describe globalization, a term that captures the most recent consequences of the changing situations of places. Geographers often illustrate globalization by describing the ubiquitous Chinese restaurant in American suburbs, the worldwide spread of American-style fast food, and globalization's backlash expressed in the French's protest against "le hamburger."

The examination of food is also a useful entry point into the discussion of the uniqueness of place. The French word *terroir* is used to express the qualities of an agricultural product, usually wine, that come from the characteristics of the place where it is produced—the soil, bedrock, local climate, and so on. Eating and drinking can be transformed into a geographic exercise as one considers where the product comes from, what makes that place unique, and how that uniqueness is expressed in the taste of the foodstuff.

There is not a single geography of food but rather many geographies of food. One of the most essential ones is the geography of plant and animal domestication. Domestication occurred as early humans switched from hunting and gathering their food to sedentary societies where they planted their food nearby. Through interaction—the selection and encouragement of certain characteristics through control of the reproduction of plants and animals over time—these plants and animals changed. For example, the modern tomato plant has been modified over time to have large fruit as growers saved and propagated the seeds from plants producing large fruit while discarding seeds from small fruit-bearing plants. Animals have been shaped similarly by human interaction. For example, horses have been bred for specific purposes such as pulling heavy loads (draft horses) and galloping fast but only for short distances (thoroughbreds). By breeding and selection, humans have reshaped nature.

The places of early domestication allowed communities to settle in one place, and in these places new social, political, cultural, and economic structures arose. These domesticated plants and animals and the new social forms diffused outward, changing our world dramatically. We can map where many of our agricultural products were first domesticated. For example, coffee and cotton in East Africa, sugarcane and rice in Southeast Asia, the potato and the tomato in the Andean Uplands, peanuts and the pineapple in present-day eastern Brazil, rice and pigs in West Africa, cattle and grapes in the Mediterranean, the horse and the dog in Southwest Asia, and the blueberry and cranberry in North America. These places of domestication are still important today because they are where most genetic variations in those species are located and where a wealth of indigenous knowledge on how to grow and use those plants and animals still resides in these local communities. For example, although Idaho and Maine grow large amounts of potatoes in the United States, there is little genetic diversity in what they grow. It is in the Andes Mountains where the greatest genetic diversity of the potato exists. In a germ plasm bank (i.e., a seed bank) in Lima, Peru, is stored 151 species of wild potatoes and approximately 4,000 varieties of cultivated potatoes.

Of course, the production of these agricultural foodstuffs has expanded much beyond their places of domestication. This geography changed dramatically with the so-called discovery of the New World and the Columbian Exchange. In fact, foods associated with Europe, such as the potato (Ireland) and the tomato (Italy), came late to these cuisines and only after Columbus's voyage. A major component of colonialism was to move commercially viable plants and animals away from their places of domestication to colonized places where they could be produced in plantation systems with forced or cheap labor while distancing them from their traditional pests.

Another important distinction in the geography of food is between places devoted to capitalist industrial agriculture and those people and places not integrated into the capitalist system and continuing to use indigenous provisioning practices. Capitalist agriculture is characterized by the use of industrial inputs of machines, fertilizers, pesticides, fungicides, and irrigation as well as, increasingly, by genetic modifications. Farmers specialize in a crop produced for cash and not for subsistence. This cash crop specialization also means the loss of local food self-sufficiency as food is imported to replace the food formerly grown and consumed locally. Noncapitalist agriculture is more diverse, with local production meeting local needs. Agricultural surplus is stored and shared locally. However, the places of noncapitalist food production are becoming fewer as capitalism continues to spread.

A geography of food is also sensitive to the special meanings produced by the production and consumption of food. For example, to "break bread" with someone is not only to share a meal but also to form a social bond. Consider the meanings associated with holiday meals such as Thanksgiving and Christmas. Meals on these special occasions remind us of the importance of ritual and community that has been mostly lost in our day-to-day eating practices. Food may well be valued for what price it can receive in the market (called a commodity), but it still retains some meanings defined outside of capitalism. As a commodity, however, food goes to those who can pay for it and not to those who are hungry. This has led to confounding situations where places export foods at the same time as segments of their populations starve. This occurred in Ireland during the potato famine and in India at various times in its history.

—*John Grimes*

Suggested Reading

Bell, D., & Valentine, G. (1997). *Consuming geographies: We are where we eat.* London: Routledge.

Cook, I., & Crange, P. (1996). The world on a plate: Culinary culture, displacement, and geographical knowledges. *Journal of Material Culture, 1,* 131–153.

Mintz, S. (1985). *Sweetness and power: The place of sugar in modern history.* New York: Penguin Books.

FORDISM

Fordism is named after American industrialist Henry Ford, who pioneered the mass production of automobiles during the early 20th century using standardized job tasks, interchangeable parts (which date back to gun maker Eli Whitney), and the moving assembly line. Ford's methods, which were very successful, were widely imitated by other industries and soon became almost universal throughout the North American, European, and Japanese economies.

The precise moment when Fordism became the dominant form of production in the United States is open to debate. Some argue that it began as early as the late 19th century during the wave of technological change of the 1880s and 1890s, when mass production first made its appearance, displacing the older, more labor-intensive, and less profitable forms of artisanal production. For example, during this period, glass blowing, barrel making, and the production of rubber goods such as bicycle tires became steadily

standardized, and the Bessemer process for fabricating steel was invented. Fordism, however, elevated this process to a whole new level, including highly refined divisions of labor within the factory, so that each worker engaged in highly repetitive tasks. Ford engaged the services of Frederick Taylor, the founder of industrial psychology, who applied time-and-motion studies to workers' jobs to organize them in the most efficient and cost-effective manner. By breaking down complex jobs into many small ones, Fordism made many tasks suitable for unskilled workers, including the waves of immigrants then arriving to the United States, and increased productivity greatly.

Others argue that Fordism was a particular kind of contract between capital and labor, one that tolerated labor unions (e.g., the Congress of Industrial Organizations [CIO]) that came into being during the 1930s, and so Fordism should be seen as beginning during the crisis years of the Great Depression. Yet others make the case that Fordism was the backbone of the great economic boom during the three decades following World War II, when the United States emerged as the undisputed superpower in the West, and that it should be dated back only to the 1950s. Whenever its origins, Fordism is reflective of a historically specific form of capitalism that dominated during most of the 20th century.

Fordism came to stand for the mass production of homogeneous goods in which capital-intensive companies relied heavily on economies of scale to keep production costs low and keep profits high. Thus, mass consumption and advertising would also come into being as the demand side of Fordism. Typically, firms working in this context were large and vertically integrated, controlling the chain of goods from raw materials to final products. Ford's plants, for example, saw coal and iron ore enter one part of the factory and saw cars come out the other end. Well suited to large, capital-intensive production methods, this system of production and labor control was largely responsible for the great manufacturing complexes of the North American Manufacturing Belt, the British Midlands, the German Ruhr region, the Inland Sea area of Japan, and similar agglomerations of industrial firms around the world.

While Fordism "worked" quite successfully for nearly a century, ultimately it began to reach its social and technical limits. Productivity growth during the 1970s began to slow dramatically, and the petrocrises and rise of the newly industrializing countries unleashed wave on wave of plant closures in the United States. Because wages and salaries are often tied to the overall growth of productivity, these changes led not only to widespread layoffs but also to declining earning power of American workers. Rates of profit in manufacturing began to drop during the 1970s and 1980s, and many firms faced the choice of either closing down, moving overseas, or reconstructing themselves with a new set of production techniques. It is in this context that Fordism began to implode, giving way to post-Fordist, flexible production techniques, which have become widespread throughout the economy today.

—*Barney Warf*

See also Economic Geography; Economies of Scale; Flexible Production

Suggested Reading

Amin, A. (Ed.). (1994). *Post-Fordism: A reader.* Oxford, UK: Blackwell.

FRACTAL

Fractals, a term coined by their originator Benoit Mandelbrot, are objects of any kind whose spatial form is nowhere smooth (i.e., they are irregular) and whose irregularity repeats itself geometrically across many scales. The irregularity of form is similar from scale to scale, and the objects are said to possess the property of self-similarity or scale invariance. A classic fractal structure is the Koch Island or snowflake (Figure 1). It can be described as follows:

(a) Draw an equilateral triangle (an initial shape or *initiator* [Figure 1A]).

(b) Divide each line that makes up the figure into three parts and "glue" a smaller equilateral triangle (a *generator*) onto the middle of each of the three parts (Figure 1B).

(c) Repeat procedure (b) on each of the 12 resulting parts (4 per side of the original triangle).

(d) Repeat procedure (b) on each of the 48 resulting parts (16 per side of the original triangle) and so on.

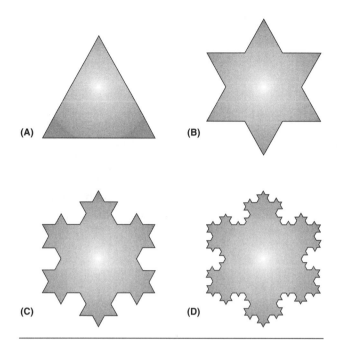

Figure 1   The Koch Island or Snowflake

This can ultimately result in an infinitely complex shape.

We use the term *fractal dimension* to measure fractals. The Euclidean dimensions are composed of 0 (points), 1 (straight lines), 2 (areas), and 3 (volumes). Fractal dimensions lie between these dimensions. Thus, a wiggly coastline (perhaps like each side of the Koch Island in Figure 1) fills more space than a straight line (Dimension 1) but is not so wiggly as to fill an area (Dimension 2). Its fractal dimension thus lies between 1 and 2. (The fractal dimension of each side of Figure 1 is actually approximately 1.262; the dimension of a more intricate, fjordlike coastline would be higher, closer to 2.) The tower blocks on the skyline of a city fill part of, but not all of, the vertical dimension, and so we can think of cities as having dimensions between 2 and 3.

The Koch Island shown in Figure 1 is a pure fractal shape because the shapes that are glued onto the island at each level of recursion are exact replicas of the initiator. The kinds of features and shapes that characterize our rather messier real world rarely exhibit perfect regularity, yet self-similarity over successive levels of recursion often can be established statistically. Just because recursion is not observed to be perfectly regular does not mean that the ideas of self-similarity are irrelevant. Christaller's central place theory provides one good example of a theory of idealized landscapes of nested hexagons that is applicable even though it is never observed exactly in real-world retail and settlement hierarchies.

Fractal ideas are important, and measures of fractal dimension can be as useful as spatial autocorrelation statistics or of medians and modes. Paul Longley and his colleagues described how the fractal dimension of an object may be ascertained by identifying the scaling relation between its length or extent and the yardstick that is used to measure it. Regression analysis provides one (of many) means of establishing this relationship. If we can demonstrate that an object is fractal, this can help us to identify the processes that give rise to different forms.

—*Paul Longley*

See also GIS; Humanistic GIScience

Suggested Reading

Longley, P., Goodchild, M., Maguire, D., & Rhind, D. (2005). *Geographic information systems and science* (2nd ed.). Chichester, UK: Wiley.

Mandelbrot, B. (1983). *The fractal geometry of nature.* San Francisco: W. H. Freeman.

GATED COMMUNITY

A gated community is a portion of public space that has, through the construction of physical walls or any other barrier, become separated out and privately controlled. Initially described by Richard Sennett in 1970 as purified communities, these enclosures subsequently have been classified by Edward Blakely and Mary Snyder in accordance with their function. Lifestyle communities refer to developments that create enclaves with all the amenities of the city being placed within predominantly residential land use enclosures outside of cities. The second category of gated communities refers to high-cost residential developments that are affordable only to a particular elite. The third category of gated communities is referred to as the security zone community, where enclosure is fitted to existing roads and infrastructure to create a controllable space. As the private–public separation of space has become increasingly blurred, these different categories of gated communities have become less distinguishable.

Gated communities frequently are sold as a panacea to the social ills of the city. Developers often play on fears in their marketing of gated communities. In particular, gated communities are a response to fear of crime and are characterized by the core features of surveillance and security. Other features may include environmental design interventions such as lighting and panoptic devices. The symbolic functioning of gated communities is summarized in Newman's discussion of defensible space. Newman's suggestions for safer environmental design include an enhanced sense of territoriality and the increased potential for surveillance of public space. Gated communities are highly contested. On one side of the issue is a set of urban residents who feel increasingly compelled to secure themselves against real and imagined burgeoning crime. On the other side is a series of human rights issues around the exclusion of marginalized urban citizens. In his seminal work "Fortress Los Angeles," Mike Davis led a series of critiques of gated communities. For these authors, gated communities are less about controlling violent crime (to which poorer communities are more vulnerable) and more about controlling the types of people who have access to or can be included in communities. Gated communities attract homogeneous groups of people and promote insularity. With their barriers and buffers and their security, surveillance, and control, gated communities represent an emergent urban spatial apartheid. Through these enclosures, the city becomes increasingly hostile to marginalized people, who become increasingly demonized because there is so little public interaction. The constant watch of private security services also skews the implementation of public security to the broader urban community and creates a false sense of security. Gated communities are not found only in First World cities; a number of authors have commented on the emergence of gated communities as a means of ensuring the safety of a privileged elite in sanitized spaces divorced from the pervasive poverty of Third World cities. Much of this work has been centered on the cities of São Paulo in Brazil and Johannesburg in South Africa, where gated communities have represented the strengthening of urban segregation.

—Teresa Dirsuweit

See also Gentrification; Housing and Housing Markets; Neighborhood; Segregation; Urban Geography; Urban Sprawl

Suggested Reading

Falzon, M.-A. (2004). Paragons of lifestyle: Gated communities and the politics of space in Bombay. *City and Society, 16*(2), 145–167.

Grant, J., & Mittelsteadt, L. (2004). Types of gated communities. *Environment and Planning B, 31,* 913–930.

Jürgens, U., & Gnad, M. (2002). Gated communities in South Africa: Experiences from Johannesburg. *Environment and Planning B, 29,* 337–353.

Le Goix, R. (2005). Gated communities: Sprawl and social segregation in Southern California. *Housing Studies, 20,* 323–343.

Manzi, T., & Smith-Bowers, B. (2005). Gated communities as club goods: Segregation or social cohesion? *Housing Studies, 20,* 345–359.

Wu, F., & Webber, K. (2004). The rise of "foreign gated communities" in Beijing: Between economic globalization and local institutions. *Cities, 21,* 203–213.

GAYS, GEOGRAPHY AND/OF

The geography of gay people encompasses three key ideas. First, the geography of gay people represents the spatial expression of an *individual enactment* of desire for the same sex. Second, beyond the individual, gay sexualities are *social phenomena* whose expressions must contend with social mores, laws, attitudes, and traditions. Third, such expression—both individual and social—raises questions about *rights* because for thousands of years the oppression of gay people has been regulated and bound by law and science, forcing gay people to take to the streets for political representation.

GEOGRAPHY OF SEXUAL IDENTITY

Being gay means having or expressing sexual and emotional desire for the same sex. Gays negotiate their identity by revealing their identity to others by "coming out" and by dealing with how society views gay identity. The "closet" represents both a personal space and a physical space where gays perform their sexual identity in secret to avoid society's judgment. Living in the closet limits interactions with others because it prevents honest and open relationships, yet such behavior often is normalized by the pressures of social life, family, and/or work.

Historically, oppressive ideas and practices have defined gay people in negative ways. During the 19th century, the mistaken clinical diagnosis of homosexuality as an illness put an allegedly objective spin on gay oppression, reinforcing restrictive laws on the lives of gay people. Such intolerant and hateful attitudes have caused gay people to struggle with heteronormative social structures (e.g., churches, marriage); unfortunately, these ideas remain today. *Heteronormativity* maintains that heterosexuality is normal and that any other practice is abnormal. *Homophobia* is the condition of irrational uncontrolled rage toward gay people that often leads to violence.

THE SOCIAL GEOGRAPHY OF SEXUALITY

Although sexuality's individual dimensions are very important to consider, sexuality extends beyond the individual simply because people are social beings. As an oppressed people, gay people must struggle with defining the contours of the spaces in which they can exercise desires and needs. In the places where they work, live, and socialize, they face discrimination in getting healthcare and suffer, as other groups do, from racial, gender, and class divisions. Although it is true that gay people often live and work in urban centers where they are close to resources such as friends, healthcare, protection, and social events, they also have moved increasingly to less populous, and traditionally less tolerant, places such as rural areas, small towns, and middle-class suburbs.

RIGHTS

The question of human rights sits crucially at the intersection of gay sexuality and geography. The legal definition and enforcement of space—who is allowed in it and what is allowed—constrains our ability to express sexual identity, both personally and publicly, making it a human rights issue. Gay sexualities always have been *political* expressions. As such, the geography of gay people has been closely linked to social movements. The Stonewall uprising in June 1969 in New York City represents *the* watershed moment in gay (men's) liberation in the United States, fueling other social movements for sexuality liberation. During the 1980s and 1990s, discriminatory healthcare practices in treating acquired immunodeficiency syndrome (AIDS) empowered gay people everywhere to take over the streets and to turn to the courts to gain recognition and rights.

Keeping in mind the important spaces for the expression of gay sexuality—in the home, among friends and family, at work or school, in the community, at social events, and in the mass media—one must remember that such expression is a human right. Because sexuality has been both scientifically and legally regulated, and gay expression has been prohibited and criminalized throughout history, the geography of gay people always has been a question of where and how they can live and work openly without fear of violence or repression.

—*Richard Van Deusen*

See also Justice, Geography of; Lesbians, Geography of/and; Queer Theory; Sexuality, Geography and/of

Suggested Reading

Bell, D., & Valentine, G. (Eds.). (1995). *Mapping desire: Geographies of sexualities.* London: Routledge.

Brown, M. (2000). *Closet geographies: Geographies of metaphor from the body to the globe.* New York: Routledge.

Chauncey, G. (1995). *Gay New York.* New York: Basic Books.

D'Emilio, J. (1983). *Sexual politics, sexual communities: The making of a homosexual minority in the United States, 1940–1970.* Chicago: University of Chicago Press.

Fone, B. (2000). *Homophobia: A history.* New York: Picador.

Greenberg, D. (1990). *The construction of homosexuality.* Chicago: University of Chicago Press.

Higgs, D. (Ed.). (1999). *Queer sites: Gay urban histories since 1600.* London: Routledge.

Knopp, L. (1986). Social theory, social movements, and public policy: Recent accomplishments of the gay and lesbian movements in Minneapolis. *International Journal of Urban and Regional Research, 11,* 243–261.

Lauria, M., & Knopp, L. (1985). Towards an analysis of the role of gay communities in the urban Renaissance. *Urban Geography, 6,* 152–169.

Nast, H. (1996). Unsexy geographies. *Gender, Place, and Culture, 5,* 191–206.

Weeks, J. (1985). *Sexuality and its discontents: Myths, meanings, and modern sexualities.* London: Routledge and Kegan Paul.

GENDER AND GEOGRAPHY

Geographic considerations of gender have offered insights into the gendered construction of spaces and the importance of place to gendered lives. During the late 1970s, gender and geography emerged as an area of inquiry that contested the assumed figure of man as representing all of humanity in geography. Since that time, gender roles, relations, embodiments, and their interactions with (and constitutions through) place, spatial processes, landscape, and environment have become exciting, innovative, and vast areas of geographic investigation.

Gendered geographies have not been uniform and vary in relation to how gender/sex is understood (Table 1). These conceptualizations are central to how geographers employ gender/sex and use them to explore a diverse range of gendered geographies. To simplify, theories of gender/sex range from essentialism (the biological separation of men and women), to social constructionism (the societal construction of gender), to poststructuralism (gender/sex comes into being through what we do). Although these categories are problematic and messy and there are slippages and overlaps among them, Table 1 offers an introduction to various conceptualizations of gender/sex within geography.

To emphasize the meaning of gender and geography, it is necessary to draw (artificial) distinctions between gender and geography and feminist geographies. The diversity of gendered geographies is then explored through a brief outline of some of the ways in which geography has worked with, and informed, different understandings of gender/sex. Following this, the gender of geography is considered. Clearly, due to the diversity and breadth of gendered geographies, this is only one (limited) story of many that could be told.

FEMINISM, GENDER, AND GEOGRAPHY

It should be made clear that gender and geography, alongside feminist geographies, does not focus solely on women. Neither is the inclusion of men and other forms of gender/sex (e.g., transgendered/transsexual individuals) the segregating feature. For the purposes of this entry, the defining feature of feminist geographies is a political motivation/commitment to analyzing and addressing issues of gendered power. Consequently, one can do gender geography without addressing issues of politics, power, patriarchy, heteropatriarchy, or feminist methodologies. The differences drawn on here pertain to relations of gendered power, which in theory can separate gender and geography from feminist geographies. Although this distinction is employed for the purposes of this entry, gender analysis and gendered critiques of power rarely are separated within geographies.

GENDER GEOGRAPHIES

Rather than offer a trajectory of development, this section illustrates some of the plethora of ways in which geography has worked with gender/sex and, in turn, the contribution that geographies can make to understandings of gender/sex. Until the late 1970s, geographic inquiry did not recognize gender, homogenizing human and representing only men's geography. In recognizing the absence of women's lives in mainstream (or *malestream*) geography, studies sought to add women to geographic inquiry as a discrete group with distinct geographies. These studies created the "geography of women." Centralizing women arguably led to essentialist definitions of women and men (Table 1). Yet in emphasizing the absence of women in geography, the geography of women provided a place for women within malestream analysis of our world. This work contested the assumption that men's geography was universally applicable and that human geography should explore solely manmade geographies. Furthermore, it demonstrated some of the problematic assumptions on which geographers' categories depended such as work as paid employment. Atlases that map aspects of women's lives are good examples of women's geography. They can include information such as women's employment; incidences of domestic violence; availability of shelters; fertility, abortion, marriage, and divorce rates; and poverty and credit for women. These maps not only represent "women's worlds" but also simultaneously redraw what is seen as mappable.

Despite the importance of including women in geographic inquiry, other geographers contended that focusing only on women does not account for how men and women interact, use space differently, and/or create distinctly gendered places. They argued that human geographic inquiry focused on male-dominated public spaces such as work, "manmade" environments, and public transport. Thus, there was an absence not only of women but also of women's geographies that were distinct from male geographies and often located more within, or related to, the private sphere of the home. Women's geographies, in this understanding, included patterns of travel beyond home to work to home and included trips to school, to child care, and to supermarkets/shopping malls.

These gendered geographies contend not only that gender is constructed onto biological sex but also that these constructions vary locally, regionally, nationally, and internationally and that globalization is constructing gender roles and relations. Thus, gender roles (activities that are associated with masculinities/femininities) and gender relations (the interrelations between the socially constructed categories of male and female) vary through, and are informed by, place, space, and time. Geographic studies that investigate the construction and maintenance of gender roles and relations are diverse and explore an extensive variety of topics (from food to the environment), scales (from the global international division of labor to the everyday spaces of toilets), and spaces (from British clothing changing rooms to Indonesian factory floors).

Since the late 1990s, men and masculinities have also come under scrutiny within geographic inquiry. The inclusion of men and masculinities as legitimate sites of inquiry rather than universally accepted norms has led to an examination of how diverse masculinities are formed and hierarchized. In examining differences among men, as well as between men and women, geographers explored how particular masculinities become hegemonic/subordinate through space and time and investigated masculinities across places and scales. Therefore, these investigations offer valuable insights into the diversity of men's lives. For example, how working-class men in Sheffield experience the spaces of work differs vastly from the professional classes in New York's financial district, although both may be affected by globalization processes.

Discussions of gender roles and gender relations, while challenging the essentialist assumptions of man/woman, at times do not account for differences between women (and men) such as ethnicity, sexuality, class, and (dis)ability. Arguing that gender is mediated through categories of social difference alters the very foundations of woman/man on which gender geographies are based. If social differences are not simply added to gender, men/women are (re)constituted through these social differences such that genders can be (re)made differently in relation to class, ethnicity, sexuality, and (dis)ability. Postcolonial critiques and black feminist critiques have argued that the category of woman is differentiated by race/ethnicity and that there are huge variations and distinctions even within categories such as black and Asian. Similarly, lesbians and people with disabilities have contended that gender

Table 1 Outline of Different Understandings of Gender/Sex and Their Use in Geographies

Theoretical Perspective	Conceptualizations of Sex	Conceptualizations of Gender	Examples of Geographic Studies
Essentialism	Sex is natural, biological, and unchanging. This is grounded in an understanding of nature as destiny.	Men and women are biologically different, and gender roles and relations (e.g., housewife/breadwinner) arise from biological sex.	Using "women" and "men" as categories can be important in mapping gender differences across the world (at times, this can be described more aptly as strategic essentialism).
Social constructionism	Sex is biological and unchanging but underpins rather than determines gender. Arguments in this vein contend that aspects of gender often are considered to be sex (e.g., management abilities, caring traits).	Men and women are differentiated by their socially constructed gender (masculine or feminine) roles through nurture and socialization. Relations between these genders are also socially produced and often (re)produce heterosexuality.	Gender roles and relations can be used to examine how and why the spaces of home and work are different for men and women. They have also offered possibilities for exploring regional variations in women's and men's employment.
Poststructuralism	Sex is not a biological given but rather is made through gendered enactments. There is no preexisting sex on which gender is built; instead, the division of gender produces sexed bodies. Rather than biology making the social, what we do and our relations with each other (re)make the biological within intelligible frameworks.	Gender is socially produced through what we do (performativity) in relation to heterosexual power relations that dichotomize gender into masculine and feminine. This "genre" reintroduces the body as a legitimate site of inquiry but does not reduce the "body" to solely biological determinants.	Examining the fluidity of gender/sex enables investigations into how what we do mutually constitutes our gendered bodies, gendered spaces, and gendered places. Contesting the dichotomization of man/woman leads to an exploration of how these are (re)formed in and (re)make space and place.

is not uniformly experienced and that geography itself can be accused of ableism and heterosexism. Turning this argument around, not only should gendered analysis take account of other social differences, but also analyses of other social differences need to account for gender. The fracturing and fragmenting of identities, such that woman and man no longer are coherent categories, lie within poststructural understandings of gender (Table 1).

Within poststructural gendered geographies, the boundaries of, and distinctions between, man/woman and male/female have been questioned. This understanding of gender/sex sees gender/sex as something we do repeatedly to become men/women rather than something we are (Table 1). By problematizing man/woman dichotomies, the separation of sex (as biological) and gender (as socially constructed) is challenged. Gender transgressions, such as drag kings/queens, transgendered/transsexual individuals, and intersexed individuals, illustrate that gender and sex are not necessarily linked in terms of man = masculine and woman = feminine. Geographies of gender have illustrated the importance of spatial processes and place both in rendering the categories of sex/gender fluid and in rendering how man/woman and male/female are reformed in context. For example, drag kings/queens who can perform one gender onstage and another gender in their everyday lives illustrate the differential manifestations of gender/sex in time and space. Although genders can be fluid across space and time, they often are policed into particular codes and norms of man/woman. The policing of behaviors within gender norms and the altering of these norms across space and time demonstrate that space not only is sexed but also is sexing. For instance, when men and women are segregated in the separation of men's/women's toilets, by rendering some bodies as in-place and others as out-of-place, sociospatial relations (re)make who fits into the category man/woman. These categories appear to be natural, and bodies that do not fit are considered to be "abnormal" in these places.

THE GENDER OF GEOGRAPHY

Geography has a particular gender. Since the 1970s, a striking inequality has been noted in the representation of men/women in geography departments. During the 1970s, women occupied less than 5% of faculty positions in graduate departments in North American universities; in Britain during the late 1980s, less than 10% of permanent posts were occupied by women. This was despite the equal numbers of undergraduate students undertaking geography degrees. Clearly, geography had a gender that influenced the objects and subjects of geographic inquiry and contributed to the very nature of geography itself. More recent figures suggest that there has been a change in the gender of geography and particularly that women's participation in academic departments has increased. For example, in Irish universities and in North America, women now represent roughly 15% to 20% of all academic members of staff. Although this is an increase from the 1970s, it is clear that women remain underrepresented within the discipline of geography.

Not only is there a numerical gender division, but also the positions that men and women hold in geography departments are different. Men are much more likely to be in secure (tenured) and permanent positions than are women. There continues to be an absence of women in positions of seniority; for example, in Britain only four geography departments had women who held professorship chairs in 2005. Despite this low figure, it should be recognized that this indicates an increase from the 1970s. Although it may be that women are "working their way up" in the discipline, requiring time to achieve both parity

in numbers and seniority, it cannot be assumed that gender equity will necessarily occur through time. Academic merit and what counts as good geography also need critical exploration alongside the working demands of being a geographer.

In spite of increasing equality, the figures continue to suggest a male-dominated discipline, and this illustrates the importance of gendered analysis that accounts for male/female divisions. However, as suggested in the previous section, women and men are not homogeneous groups and the achievement of some women and men does not necessarily equate to success for all. If masculinity is not just about men and femininity is not just about women, the gendered practices of geography also require nuanced exploration. It is important to note *which* masculinities and femininities are being valorized and that disparities associated with ethnicity, class, sexuality, and gender persist in the (re)constitution of geographic knowledges and geographers. Not only are analyses of gender important, but also the gendered analysis of geography as a discipline offers insights into how geographic knowledge is made.

—*Kath Browne*

See also Body, Geography of; Disability, Geography of; Femininity; Feminisms; Feminist Geographies; Masculinities; Queer Theory; Race and Racism; Sexuality, Geography and/of

Suggested Reading

Domosh, M., & Seager, J. (2001). *Putting women in place: Feminist geographers make sense of the world.* New York: Guilford.

McDowell, L. (1999). *Gender, identity, and place.* Cambridge, UK: Polity.

Rose, G. (1993). *Feminism and geography: The limits of geographical knowledge.* Cambridge, UK: Polity.

Women and Geography Study Group. (1997). *Feminist geographies.* Harlow, UK: Addison-Wesley Longman.

Women and Geography Study Group. (2004). *Gender and geography reconsidered.* Glasgow, UK: Author.

GENTRIFICATION

Gentrification is an imprecise and elastic term referring to a wide range of processes associated with changes in land use and the built environment (especially in urban contexts). Typically, it refers to changes that are characterized by some combination of the following: attempts to increase the profitability (rent) of land through reinvestment and redevelopment; planning and economic development projects (both public and private) aimed at increasing tax bases through revitalization; renovation and/or rehabilitation of buildings and infrastructure; demographic changes in the direction of increased middle- to upper-class and often white populations; transformations in social and cultural practices oriented toward these same populations (especially leisure and consumer cultures); displacements or outright erasures of working- and lower-class businesses and residences; decreases in affordable housing and commercial opportunities; destruction of poor, minority, immigrant, and/or other marginalized communities; consumer landscapes full of symbols and images connoting cosmopolitan aesthetics, desires, and aspirations; antivagrancy and related campaigns to protect areas' images as safe and desirable; and fetishization of certain kinds of cultural difference and diversity (especially sexual).

First coined during the 1960s, the term *gentrification* referred specifically to the reclaiming of poor and working-class inner-city residential neighborhoods by more middle- and upper-class in-migrants (a new urban gentry) and the concomitant displacement of indigenous residents by higher rents and property taxes. The implicit theoretical imagination at work was one primarily of a demand- and consumption-driven process, although the *reasons* for the shifts in demand and consumption habits that supposedly explained gentrification were multiple and debated. Often artists, gays, and other supposed cultural nonconformists willing to expend considerable amounts of sweat equity in rehabilitating homes and small businesses—products of the social revolutions of the 1960s as well as changes in urban occupational structures, transportation costs, family structures, and political movements alleged to have shaped their consumer tastes and preferences—were seen as the initial risk takers in a process that proceeded through stages. The culmination of the process was seen as the (re)colonization of gentrifying areas by somewhat more risk-averse affluent consumers and the construction of a locally based landscape of consumption catering to them. Ironically, original risk takers themselves frequently were displaced in the process.

Subsequent empirical work, however, called into question not only the alleged sequence of events but also the typical characteristics of both areas that were ripe for gentrification and of gentrifying and displaced populations. Some areas, it seemed, never gentrified despite being ripe; New York City's South Bronx was an oft-cited example. Other areas, such as parts of New York City's Harlem, gentrified while still remaining solidly middle class, thereby stretching the definition of gentrification in ways that sometimes privileged racial transformations over class ones. Some instances entailed wholesale displacements of indigenous populations (e.g., San Francisco's Western Addition), whereas others seemed hardly to displace anybody (e.g., New Orleans's Marigny neighborhood). In such cases, gentrification seemed to be more about cross-class changes in cultures of consumption than about distinctively class-based or demographic transformations. Many instances, such as that of San Francisco's Castro District, did indeed involve early in-migrations by culturally nontraditional risk takers (frequently gays but also single women with children), whereas others immediately involved affluent, more culturally conservative groups right from the start (many European cases fit this description). And many seemed driven at least as much by processes influencing property *developers* (e.g., the availability and preferences of investment capital) as by those influencing consumers of land and housing (e.g., changing occupational structures, job locations, and transportation costs). Even transformations in *rural* contexts, such as the rise of affluent resort communities and exurban hobby farms, came to be referred to as gentrification.

Spurred in part by such wide-ranging and conflicting empirical evidence, attempts to *theorize* gentrification more carefully revealed serious problems with the way in which this diverse set of practices was being conceptualized and explained. It became increasingly difficult to determine what counted as gentrification and what did not, and explanations for the various processes referred to seemed increasingly inadequate. Demand- and consumption-oriented explanations, for example, seemed naive in light of a growing theoretical appreciation of the role played by producers of and investors in land and housing during an era of increasingly mobile capital and global investment. Meanwhile, macroeconomic and structural explanations focusing on, for example, the logic of capital accumulation (e.g., Neil Smith's famous rent gap formulation in which gentrification is seen as a rational response by investors and developers to an entirely predictable mismatch between the rent generated by parcels of land under current highest and best uses versus alternative ones) failed to account for many of the cultural and local peculiarities and contingencies of gentrification. Eventually a consensus, of sorts, emerged around the idea that gentrification is best understood as a diverse and imprecise set of material and symbolic practices—of diverse origins—that nonetheless have powerful material and symbolic consequences.

Perhaps due to a malaise brought about by these difficulties, as well as to the recession of the early 1990s (which was predicted—incorrectly—to bring about degentrification), studies of gentrification went somewhat out of fashion for a time. But rapid urban transformations of the sort alluded to by the term only intensified. Property markets in major cities around the world boomed during the 1990s and early 2000s—especially in the United States, where persistent low interest rates supported by massive foreign investment in U.S. currency and Treasury bonds prevailed. Displacement, demographic transformations, and cultural changes of the sort typically associated with gentrification resulted in the ethnic cleansing–like erasure of nearly all poor and minority populations of some cities and boroughs (e.g., San Francisco; Islington, London; Manhattan, New York). In most of these places, tax bases were increased substantially as a result, causing local governmental authorities elsewhere to adopt pro-gentrification policies and practices, including subsidies to inner-city property developers, historic preservation tax credits, and antivagrancy laws. Neoliberal policies at all levels of government and internationally (those encouraging privatization, free trade, and a minimalist state) have only intensified this trend. The lived experience of gentrification, in other words, remains as real—and devastating—as ever. Consequently, gentrification is back on the agendas of many geographers. Now, however, it is conceptualized much more carefully and in concert with new (and nuanced) understandings of the relationships between global-scale circulations of capital and culture and local-scale lived experiences. The issues of class, race, culture, capital, and scale that are the heart of the matter are now seen as not only intersecting but also as mutually constitutive and as producing distinct, highly contingent, and even unique material forms that, despite their diversity, constitute a family

of empirical phenomena recognizable to many as gentrification.

—*Larry Knopp*

See also Rent Gap; Urban Geography; Zoning

Suggested Reading

Bondi, L. (1991). Gender divisions and gentrification: A critique. *Transactions of the Institute of British Geographers, 16,* 190–198.

Lees, L. (1994). Rethinking gentrification: Beyond the positions of economics or culture. *Progress in Human Geography, 18,* 137–150.

Mills, C. (1993). Myths and meanings of gentrification. In J. Duncan & D. Ley (Eds.), *Place/Culture/Representation* (pp. 149–170). London: Routledge.

Palen, J., & London, B. (Eds.). (1984). *Gentrification, displacement and neighborhood revitalization.* Albany: State University of New York Press.

Schaffer, R., & Smith, N. (1986). The gentrification of Harlem? *Annals of the Association of American Geographers, 76,* 347–365.

Smith, N. (1979). Toward a theory of gentrification: A back to the city movement by capital, not people. *Journal of the American Planners Association, 45,* 538–548.

Smith, N. (1996). *The new urban frontier: Gentrification and the revanchist city.* New York: Routledge.

GEODEMOGRAPHICS

Geodemographics refers to computer-based systems that combine spatially referenced data about consumers with statistical analysis and mapping programs that are used primarily to identify potential targets for business purposes. Geodemographics depends on the oft-quoted assumption that birds of a feather flock together or, more specifically, that you are where you live and, in business applications, that you are what you buy; that is, individuals' characteristics can be inferred from knowledge of aggregate demographic, socioeconomic, and behavioral data that describe their places of residence, and such characteristics predict the likelihood of purchasing particular constellations of commodities and services. No doubt, consumers can benefit from information tailored more precisely to their desires and needs, and there undoubtedly is a spatiality to everyday life, but critics have raised concerns about invasions of privacy that these systems threaten and about effects of the so-called ecological fallacy, through which individual characteristics are erroneously inferred from area or group characteristics.

Although collection of customer information began during the late 19th century, and segmentation of consumers was performed during the 1930s, it was technological and institutional innovations of the 1970s that brought about this revolution in marketing. Sociologist Jonathan Robbin combined factor and cluster analysis of data for 240,000 block groups of the U.S. Bureau of the Census to create 40 lifestyle categories, and these data were then cross-referenced with 36,000 postal delivery areas of the new Zone Improvement Plan (ZIP) of the U.S. Postal Service to produce a system called PRIZM (Potential Rating Index for ZIP Markets). This kind of information allowed businesses to target potential customers through discounted bulk-mail marketing campaigns and to undertake retail site evaluation and trade area analysis.

Geodemographics became the fastest-growing segment of the marketing industry, and a wave of mergers and acquisitions led to development of information conglomerates that are increasingly global in scope. For example, Claritas, the company that Robbin founded in 1971, acquired National Planning Data Corporation, Donnelley Marketing Information Services, National Decision Systems, and Market Statistics and was itself purchased by VNU, a company that includes A. C. Nielsen, Nielsen Media Research, Spectra Marketing Systems, and Scarborough Research. Recently, in partnership with Toronto-based Environics Analytics, Claritas has developed PRIZM CE, a segmentation system for Canada that integrates with its updated PRIZM NE system for the United States to allow continent-wide marketing programs. Skipton Information Group combines its proprietary Euro-Direct CAMEO clustering system and Micro-Vision's Market Maker geographic information system to bring life to data for postcodes in Britain, ZIP+4 in the United States, postal codes in Canada, Ilots in France, and Cho-Mokus in Japan as well as for administrative units in 24 other countries. Experian, a subsidiary of GUS, offers a segmentation system called Global MOSAIC that classifies more than 800 million of the world's consumers for cross-border target marketing, and it offers proprietary MOSAIC segmentation schemes for 20 countries.

Collaboration between public agencies and private businesses in Britain, the United States, and Canada in particular has facilitated development of massive electronic databases that combine information from censuses, public records, consumer and panel surveys, and commercial transactions, often with more than 1 trillion records. CACI Marketing System's ACORN segmentation scheme is based on more than 125 demographic statistics and 287 lifestyle variables in its Consumer Register database of 40 million U.K. consumers covering the 1.9 million postcodes in the United Kingdom; Claritas's PRIZM NE includes behavior records for more than 890,000 households and list-based data for more than 200 million households as well as census data down to block group and ZIP+4 areas of the United States; and MapInfo, a subsidiary of R. L. Polk Canada, combines 250 demographic and consumer behavioral variables from more than 100 million households to produce its PSYTE consumer segmentation system for 90,000 geographic units.

Many data are collected and distributed with only implicit consent of consumers, who are perhaps unaware of the further uses of data gathered in a particular transactional context. Although protest and legal actions have forced businesses to provide opt-out options, such surveillance often is represented and perceived as a necessary quid pro quo for participation in commercial transactions, and in any case, marketers will use their segmentation schemes to infer identity and behavior if such information is unavailable or withheld.

Contemporary segmentation schemes also go beyond available objective data to provide psychological profiles (psychographics) that offer, in the words of one vendor, "character—not just characteristics." Pictures, imaginary vignettes, and likely first names typically accompany the residential- and lifestyle-based classifications; for example, Looking Glass calls its segments names such as Jules and Roz ("affluent and physically active urbanites with children") and Denise ("single mothers on a tight budget"), and CACI's PeopleUK includes categories such as Tabloids and TV, Bingo and Betting, and Educated and Aware. CACI also offers a complete segmentation system called Monica based on the most common associations between MOSAIC segments and first names in the electoral role. The rankings, weightings, and premiums charged for contact with affluent consumers clearly reveal the differential values attached to the clusters and addresses by marketers.

Addresses are a vital part of geodemographics for several reasons. First, addresses are the most convenient way in which to match database records. Second, spatial data can be manipulated to create customized geographies such as retail trade areas, sales territories, and media footprints. Third, maps represent complex data in simple images. Fourth, addresses are means by which consumers are reached with marketing messages or products. Finally, marketers infer characteristics of individual consumers from aggregate data and substitute missing values in their databases based on residential locations. This last function is important because in some forms of data (e.g., census and vehicle licensing) individual records are not available for confidentiality reasons, and where records are missing the costs of collection are generally prohibitive. Despite the sophistication of consumer surveillance, geodemographics still relies on inference based on the neighborhood effect, and although this might reflect real differences in consumer identity from place to place, there is also a danger that it reinforces or even creates differences through the selectivity of subsequent marketing and retailing strategies.

Promotional literature for geodemographic systems is hype ridden, and the effects of their applications are generally limited to modest increases in customer response such as 5% to 10% in direct-mail campaigns. Nevertheless, it is still unnerving to learn the intimate details of consumer lifestyles and values and the geographic scale of the data; ZIP+4 and U.K. postcodes typically contain only 15 or so households. More sinister yet is the persistent use of metaphors of strategy, particularly of military operations, sexual conquest, and psychological manipulation. Surely many consumers would be disconcerted on knowing that they were being systematically observed, profiled, and targeted with such pinpoint accuracy for instrumental purposes.

—*Jon Goss*

See also Census; Census Tracts; Consumption, Geography and; Economic Geography

Suggested Reading

Burnham, D. (1983). *The rise of the computer state.* New York: Random House.
Claritas. (2005). *A corporate history* [Online]. Available: www.claritas.com/claritas/default.jsp?ci=6&si=2
Goss, J. (1995). "We know where you are and we know where you live": The instrumental rationality of geodemographic information systems. *Economic Geography, 71,* 171–198.

Harris, R., Sleight, P., & Webber, R. (2005). *Geodemographics, GIS, and neighborhood targeting.* New York: John Wiley.
Lyon, D. (2003). *Surveillance as social sorting: Privacy, risk, and automated discrimination.* New York: Routledge.

GEOGRAPHIC INFORMATION SYSTEMS

SEE GIS

GEOGRAPHY EDUCATION

Geography education is a teaching and research subfield focused on educational purpose, practice, and theory in geography, from prekindergarten through the postgraduate life span, in both formal and informal contexts. Much of the research in geography education has tended to investigate problems in kindergarten through Grade 12 (K–12) curriculum and instruction, but during recent years researchers have pursued a wider range of studies in spatial cognition, computers and multimedia, curriculum, pedagogy, and assessment at all levels of education. As of 2004 in the United States, five universities offered Ph.D. programs in geography with an emphasis on geography education and eight universities offered Ed.D. or Ph.D. programs in education with an emphasis on geography education. Many more departments offered programs at the master's level, especially professional master's programs for prospective geography teachers.

GEOGRAPHY IN ELEMENTARY AND SECONDARY EDUCATION

During the past two decades, geography has made considerable progress in the American school curriculum, having gained a discernible presence apart from social studies and a growing cadre of teachers skilled in modern analytical concepts and technologies. Improving the quality of geography teaching and learning in American education was the focus of a national reform movement that dates back to the early 1980s and that marshaled the talents of some of geography's leading scholars and professional organizations. In 1984, the Association of American Geographers (AAG) and the

National Council for Geographic Education established a Joint Committee on Geographic Education to publish *Guidelines for Geographic Education,* a document that informed teachers, school districts, local and state education authorities, and the general public of the importance of geography in K–12 education through the use of "five fundamental themes" and associated learning activities. Since its publication, more than 100,000 copies of the *Guidelines* have been disseminated nationwide. This period also witnessed the establishment of the state geographic alliances, a network of organizations whose primary purpose is to promote collaboration between university geographers and in-service teachers. The majority of these alliances were established during the late 1980s and early 1990s with grants from the National Geographic Society to set up an alliance in each state. Most of these alliances are still in operation, in many cases funded through endowments that were matched by the National Geographic Society and other organizations.

Perhaps the crowning achievement of the infrastructure building during the 1980s was the designation of geography as one of five core subjects under the National Education Goals, formulated by the National Governors Association in 1990 and codified by the Goals 2000: Educate America Act of 1994. The Goals 2000 Act prompted geographers to conduct an intensive disciplinary review that resulted in the development of national standards for K–12 teaching and learning, namely *Geography for Life: The National Geography Standards.* By 2004, geography was present in the curriculum standards in every state except Iowa.

Although the reform movement was successful in raising the status of geography in American education, it has yet to make significant headway in raising student achievement and teacher preparation in the subject. The most recent geography assessment by the National Assessment of Educational Progress found that more than two thirds of American students in Grades 4, 8, and 12 did not meet proficient competency standards. This situation is likely to continue until concerted efforts are made to reform the pedagogic content knowledge of geography teachers, especially at the preservice stage of the professional continuum.

GEOGRAPHY IN POSTSECONDARY EDUCATION

Geographers have also dedicated a considerable amount of effort to the improvement of teaching and learning in higher education. Some of the earliest efforts in this regard include the AAG's Commission on College Geography, established during the 1960s to focus on faculty development and instructional issues in undergraduate education. It published an extensive paper series on the teaching and learning of geography and during the 1970s initiated a number of professional development programs to promote innovative practice in curriculum and instruction, including the Commission on Geographic Education and the Teaching and Learning in Graduate Geography projects. During the 1980s, an informal effort known as the Phoenix Project, led by some of geography's most influential scholars, was successful in providing support and networking for another cohort of early-career geography professors. The Committee on the Status of Women in Geography has a substantial record of organizing conference sessions to address career issues for women graduate students and faculty. The latest project to examine academic professionalization in geography is the Geography Faculty Development Alliance, a 5-year project funded by the National Science Foundation (NSF) to provide early-career faculty with the theoretical and practical knowledge needed to succeed in their careers of research, teaching, and service. Some departments have also instituted Preparing Future Faculty programs in geography.

Geographers have also pursued a wide range of federally funded projects in undergraduate education, including the Core Curriculum in GIScience and Core Curriculum in GIS for Technical Programs, the Geographer's Craft Project, the Virtual Geography Department Project, Hands-On: Active Learning Modules on the Human Dimensions of Global Change project, and the Online Center for Global Geography Education project. Many of these projects have explored strategies to internationalize teaching and learning. In 1999, an International Network on Learning and Teaching Geography in Higher Education was formed with the aim of building an international community of geographers dedicated to collaborative teaching and research on postsecondary education issues.

Academic geography seems to have benefited from accomplishments in K–12 education. During the 1990s, the number of undergraduate degrees awarded in geography at U.S. institutions grew by 57% to the current average of 4,000 degrees conferred annually. At the graduate level, the 2002 enrollment in M.A./M.S. and Ph.D. programs totaled 4,432 students, an increase of 3% over the previous year. Much of the growth in enrollment comes at a time when geographers enjoy broader recognition in scientific and academic communities and increased demand in the workforce, particularly in the area of geographic information systems (GIS) technology. A recent NSF report, *Complex Environmental Systems: Synthesis for Earth, Life, and Society in the 21st Century,* presented a 10-year outlook for environmental research and education and cited geography as a key source of concepts and technologies to synthesize research questions and data acquisition across spatial, temporal, and societal scales. In 2004, the U.S. Department of Labor and U.S. Department of Education listed GIS technology as one of the three most important emerging areas for job growth alongside nanotechnology and biotechnology.

—*Michael Solem*

See also Applied Geography; History of Geography; Paradigm

Suggested Reading

Bednarz, R., & Bednarz, S. (2004). Geography education: The glass is half full and it's getting fuller. *The Professional Geographer, 56,* 22–27.

Geography Education Standards Project. (1994). *Geography for life: National Geography Standards 1994.* Washington, DC: National Geographic Research and Exploration.

Gewin, V. (2004). Mapping opportunities. *Nature, 427,* 376–377.

Weiss, A., Lutkus, A., Hildebrant, B., & Johnson, M. (2002). *The nation's report card: Geography 2001* (Office of Educational Research and Improvement, National Center for Education Statistics, NCES 2002-484). Washington, DC: U.S. Department of Education.

GEOPOLITICS

The term *geopolitics* refers to the linkage of space, power, and political practice. The links between particular aspects of physical or human geographic patterns and potential advantages for a political entity have been important parts of several forms of geopolitical thought. Early geopoliticians invoked a variety of approaches to ordering a chaotic world, including the incorporation of biological metaphors that became the basis for the infamous German *geopolitik,* which provided rationales for Nazi genocide. Geopolitical practice and research has ebbed and flowed within the United States, partly in reaction to the connotations of Geopolitik. The revival of the geopolitical research and thought occurred during the 1970s in the United States and has continued to develop along the four lines of geopolitical thinking considered in this entry: realist, political economy, critical, and feminist.

HISTORY OF GEOPOLITICS

Rudolf Kjellen, a Swedish political scientist, initially put forth the ideas that would undergird geopolitics in a book published as *Introduction to Swedish Geography* in 1900. His major contribution to geopolitical thinking was the 1916 publication, *The State as a Living Form.* Kjellen outlined the defining characteristics of the term *geopolitics* most fully, describing links among the physical environment, boundaries, governance, economics, and political goals. Geopolitical thought is also found in the writings of Englishman Sir Halford Mackinder, who in 1904 introduced heartland–rimland theory, which argued that control of the Eurasian landmass was key to global power. Mackinder's view of power was predicated on shifts in transportation technology and the physical geography of landmasses, with control of Eurasia being the pivot of power, leaving a naval power such as Great Britain at a disadvantage. Criticisms of Mackinder focused on the European-centered nature of his proscriptions for policy and their emphasis on serving the British Empire (although Mackinder's contemporaries were also instrumental in crafting the geopolitical visions of their own countries).

The German school of *geopolitik* produced a form of racialized naturalized geopolitics based on the melding of organic views of the state and the penchant for grand geographic theories as demonstrated by Mackinder's works. At the end of the 19th century, biological metaphors were imported by Frederick Ratzel, who argued for an organic view of the state, where states were likened to organisms needing resources and space for growth (or *lebensraum*). Such thinking also invoked the ideas of competition between state organisms, thereby naturalizing conflict and war. This view was reformulated by Kjellen, whose work greatly influenced German geographer Karl Haushofer, who argued that environmental conditions determined human activities, including political activity. Haushofer's ideas influenced Nazi Germany, a regime that practiced a form of geopolitics based on racial superiority and invoked an organic view of the state and nation, thereby needing lebensraum in which to grow and thrive.

The term *geopolitics* became associated with Germany's actions in justifying World War II and the Holocaust. Thus, the term took on a decidedly negative connotation in Anglo-American geography and thus geopolitical work declined within geography despite charges from Germany during World War II that the United States also was practicing geopolitics. This affected political geography as well.

Geopolitics experienced a renaissance of sorts during the early 1970s with the embrace of the term by then–U.S. Secretary of State Henry Kissinger. Kissinger's view of geopolitics centered on the maintenance of a balance of power between the two cold war rivals: the United States and the Soviet Union. Perhaps most famous from this era were the use of containment (a reversal of Mackinder's heartland–rimland thought where the goal was to encircle the Soviet Union and keep Marxist ideology contained to the geographic pivot) and the domino theory, which argued that the spread of communism would destabilize neighboring states and facilitate the spread of communism unless checked. Within South America, military leaders such as Augusto Pinochet often were trained in geopolitics, although researchers point out that it also connects to the organic view of the state. Today several independent schools of thought have emerged in analyzing geopolitical activity and are described in what follows.

REALIST GEOPOLITICS

This form of geopolitical analysis dominates the public's perception of geopolitics. Work in this vein focuses on the analysis of relative levels of power vis-à-vis various states, primarily in terms of military and economic capacity. Authors often propose various hierarchies and groupings that can be created to simplify the world. Dependent on the realist school of international relations, the 1970s saw an emphasis on understanding hierarchies of power and the fluidity of allies. With the collapse of the Soviet Union and the end of the cold war, realist geopoliticians shifted to addressing a world with only one superpower and where threats to powers that were not so clear anymore, including considerations of whether the United

States could maintain its geopolitical power in the new millennium. Geographers working in this tradition have dwindled in number but are still working on refining strategic visions of the globe that reflect the post–cold war world and the challenges that new technologies present as well as a greater awareness of the geopolitical visions crafted by allies and potential threats.

POLITICAL ECONOMY APPROACHES

Authors writing from a political economy perspective emphasize the relationship between the state of the global economy and political activities undertaken by states, stressing that economic imperatives guide political decisions in regard to resources, markets, and trade. One example of this would be world systems analysis, which argues that there is a core area of the global economy, a semiperiphery of countries whose economies have shifted from being mere supplies of labor and resources, and a periphery of states that are poorly integrated into the global economy. The relative levels of power are clear here given that many peripheral states are suppliers of resources and commodities to core states that have little power to set prices and serve as markets for products produced by core states.

More nuanced versions of a political economy approach were produced during the 1990s, primarily moving away from an emphasis on states themselves and refocusing away from the assumption that states are wholly integrated into the global economy in the same way through the state's territory. These models incorporate regional differentiation within states, emphasizing circuits of political and economic power concentrated in particular places (e.g., the global cities New York, London, and Tokyo; regional centers; the capitals of peripheral countries) along with the power institutions and movements. In this model, the core is not composed of states but rather is composed of those areas within states that are most integrated into the global economy. Peripheral areas in the developed and underdeveloped worlds are seen in the same light with no clear differentiation made.

CRITICAL GEOPOLITICS

This view of geopolitical activity leaves behind realist conceptions of a system centered on states and focuses on how geopolitical priorities are created through the many conduits of power available in societies. The point of critical geopolitics is not to serve the state and dominant interests in justifying what is normal or appropriate but rather to point out that geopolitical situations are constructed, and thus contestable, in civil society and that there are alternative geopolitical views, demonstrating that none is natural or preordained.

This school of thought analyzes the discourses used to frame conflicts, national priorities, and the positioning of nation-states as each attempts to address strategic goals. The scholarship in this area addresses the power to define the world in which nation-states function such as axis of evil, globalization, and free trade. The critical geopolitics literature views the creation of geopolitical visions as a combination of three pillars. Authors in this school differentiate among three forms of geopolitical practice: formal geopolitics (that of academics and think tanks), practical geopolitics (the practices of governments and nongovernmental groups), and popular geopolitics (the ideas about geopolitical situations as articulated through various popular media and always influenced by the other two forms of geopolitical activity).

FEMINIST GEOPOLITICS

Feminist geopolitical thought draws heavily on feminist critiques of international relations theory and critical geopolitics. Authors in this vein argue that international relations theory, in particular realist conceptions of international relations and geopolitics, addresses states and power in too much abstraction. This leads to an invisibility of citizens and subjects of power who are often hurt or killed in geopolitical activities and then generally ignored or discounted with terminology such as *collateral damage*. These critiques center on the masculinization of geopolitical practices—and, at times, the masculine gazes of authors, both in the realist and critical geopolitics variants. Central to feminist geopolitical analysis is the concept of security and, more important, the question of security for whom. Supporting these ideas are the connections among body, space, and power and ultimately what forms of power are being brought to bear on bodies, especially those not incorporated into the defining of strategic interests and those who bear the costs of geopolitical actions. As such, feminist geopolitics places value on the inclusion of real people into geopolitical calculus, not conceptualized targets that are not considered at an individual level. Feminist critiques of critical geopolitics charge that

critical geopolitics fails to offer a viable alternative to current practice and that deconstruction of geopolitical views is not enough despite the importance of denaturalizing claims to geopolitical truths.

—*Darren Purcell*

See also Core–Periphery Models; Dependency Theory; Global Cities; Marxism, Geography and; Nation-State; Political Geography; World Systems Theory

Suggested Reading

Agnew, J. (2002). *Making political geography*. London: Edward Arnold.

Cohen, S. (2003). *Geopolitics of the world system*. Lanham, MD: Rowman & Littlefield.

Hyndman, J. (2004). Mind the gap: Bridging feminist and political geography through geopolitics. *Political Geography, 23*, 307–322.

Kofman, E. (1996). Feminism, gender relations, and geopolitics: Problematic closures and opening strategies. In E. Kofman & G. Youngs (Eds.), *Globalization: Theory and practice* (pp. 209–224). New York: Pinter.

Ó Tuathail, G. (1996). *Critical geopolitics*. Minneapolis: University of Minnesota Press.

Sir Halford Mackinder and "The geographical pivot of history" [special issue]. (2004). *The Geographical Journal, 170*, 291–383.

Taylor, P., & Flint, C. (2000). *Political geography: World-economy, nation, state, and locality* (4th ed.). Upper Saddle River, NJ: Prentice Hall.

GEOSLAVERY

Geoslavery is a radically new form of human bondage characterized by location control via electronic tracking devices. Formally, it is defined as a practice in which one entity (the master) coercively or surreptitiously monitors and exerts control over the physical location of another individual (the slave). Inherent in this concept is the potential for a master to routinely control time, location, speed, and direction for each and every movement of the slave or, indeed, of many slaves simultaneously. Enhanced surveillance and control may be attained through complementary monitoring of functional indicators such as body temperature, heart rate, and perspiration.

Once viewed as a futuristic nightmare, human tracking is now affordable and available without restriction. For $200 plus a monthly service fee of $20, anyone can purchase an electronic device that puts George Orwell's *1984* surveillance technology to shame. They are marketed as kid-tracking devices, although some advertisements also mention pets and senior citizens. In vivid tones of doublespeak, one company offers service plans named "Liberty, Independence, and Freedom," but surveillance and control are their purpose.

Human tracking is part of a broad category of location-based services (LBS) that depend on geographic information systems (GIS) enhanced by coordinates derived from the Global Positioning System (GPS), radio transmission of real-time locations of tagged objects or individuals, and Internet-based monitoring systems.

Consumers welcome GPS receivers for personal navigation, especially for travel and outdoor recreation. There is much good and certainly no harm so long as the coordinates go directly to the user and no one else. Current GPS devices display maps produced by GIS containing detailed information about businesses, residences, and individuals. Human-tracking devices add radio communication that reports location data to a service center with its own powerful geographic information system. Subscribers pay for the privilege of peeking in at will to check on the individual being tracked.

Most LBS applications, including automobile navigation, cargo tracking, and emergency response, are overwhelmingly beneficial. Others, such as precision-guided weapons, are more controversial. Even many human-tracking applications per se will be neither coercive nor surreptitious and thus will not constitute geoslavery. Some will be quite beneficial.

Still, human-tracking devices pose the greatest threat to personal freedom ever faced in human history. At the very least, they will alter social relationships between some parents and children, husbands and wives, and employers and employees more dramatically than any other product emerging from the information revolution. Whatever legitimate uses there may be—to safeguard a child or an incapacitated adult, for example—abuses will occur. Even full-blown geoslavery is inevitable: The only question is how many people will suffer from it—hundreds, thousands, or millions.

After decades of fretting over Orwell's vision, hardly a whimper has been heard since the devices went on sale. Media attention has focused entirely on the advertised case—parents of good intention watching over their own children—and no one seems to have asked the following questions. Will the practice really protect children? Or will it introduce new risks? How will children react, emotionally and behaviorally, to constant surveillance and control? Will tracking be confined to children and incapacitated adults? Even so, which applications will require informed consent, legal proceedings, or medical hearings? Should human-tracking companies be licensed? Should their employees undergo background checks? What other safeguards are needed? Will human tracking become a ubiquitous tool of control throughout society? If so, which applications are acceptable and which are not? Which existing laws must be amended to place electronic means on a par with traditional means of branding, stalking, incarceration, and enslavement?

America's front line will be the workplace. Human tracking already has established a substantial foothold in many industries. How much will union leaders value their workers' freedom of movement? Will human tracking become a bargaining chip in future contract negotiations? Geographers have raised these and other crucial questions in scholarly journals and magazines, but questioning of any sort is strangely absent elsewhere. Far from critical review, news and talk show coverage amounts to little more than blind acceptance of manufacturers' claims.

It is time for an explicit national debate over human tracking and geoslavery that goes far beyond privacy per se. That will not occur, however, until citizens become alarmed, educate themselves, and demand answers. Currently, it is not clear whether they will resist. Recently, for instance, the U.S. Food and Drug Administration's easy approval of Verichip implants provoked no journalistic inquiry or public outcry. Neither did its approval, a few days later, of radio frequency identification tracking tags for Viagra.

—*Jerome Dobson*

See also Geodemographics; GIS; Humanistic GIScience; Location-Based Services

Suggested Reading

Dobson, J. (2000, May). What are the ethical limits of GIS? *GeoWorld*, pp. 24–25.

Dobson, J. (2003, May). Think twice about kid-tracking. *GeoWorld*, pp. 22–23.

Dobson, J., & Fisher, P. (2003). Geoslavery. *IEEE Technology and Society Magazine, 22*(1), 47–52.

Fisher, P., & Dobson, J. (2003). Who knows where you are, and who should, in the era of mobile geography? *Geography, 88,* 331–337.

Monmonier, M. (2002). *Spying with maps: Surveillance technologies and the future of privacy.* Chicago: University of Chicago Press.

GERRYMANDERING

Gerrymandering is the deliberate manipulation of spatial boundaries to provide a political advantage to a particular group. Gerrymandering links the political and the geographic in a very specific and material manner. American in origin, the term was first used to describe the 1812 creation of districts in Massachusetts that were designed to ensure a Republican majority over Federalists in the state legislature. Following Governor Elbridge Gerry's approval of the bill that created the districts, a contemporary cartoonist observed that the shape of the districts resembled a salamander. Responding to this comment, *Boston Gazette* editor Benjamin Russell, a Federalist, noted derisively that the district map should be called a "gerrymander." The term *gerrymandering* has since been used to describe the intentional distortion of electorate boundaries for political gain.

In territorially based representative democracies, gerrymandering is a powerful instrument expressed in multiple ways. A simple form of gerrymandering is the failing to redistrict as the population changes. Another type, opponent concentration or excess vote gerrymandering, occurs when boundaries are drawn so that one group is concentrated in the fewest number of districts so that this group may win there while its influence in other districts is restricted or negated. The complement to this, opponent dispersion or wasted vote gerrymandering, occurs when boundaries are drawn to split up or disperse a concentration of voters into several districts with the intention of preventing them from electing a candidate. An additional method, stacked gerrymandering, is intended to delineate boundaries in a meandering manner that encloses pockets of strength while avoiding areas of weakness.

Geographers and other researchers have investigated many issues involving gerrymandering. Geographers have noted that the nature of territorially based representation is such that the location of district boundaries can have a significant impact on election outcomes and, by extension, can also shape government policy and people's lives. Who these outcomes affect and how they affect them, both positively and negatively, is a point of departure for much research. Research has also focused on legal issues, including court-ordered redistricting that is informed and aided by geographic criteria, computer models, and the role of race and ethnicity in determining district boundaries. Finally, researchers have interrogated how gerrymandering infringes on the democratic process and why it is extremely difficult to eliminate from the political arena.

The simultaneous empowerment of one group and disempowerment of another group is inherent to gerrymandering and is a compelling aspect of this mechanism. The power and extent of gerrymandering cannot be understated. Indeed, nearly all voters, regardless of race, political affiliation, or location, have been affected by gerrymandering.

—Dean Beck

See also Electoral Geography; Political Geography

Suggested Reading

Archer, C., & Shelly, F. (1986). *American electoral mosaics.* Washington, DC: Association of American Geographers.

Glassner, M. (1993). *Political geography.* New York: John Wiley.

O'Loughlin, J. (1982). The identification and evaluation of racial gerrymandering. *Annals of the Association of American Geographers, 72,* 165–184.

Shotts, K. (2002). Gerrymandering, legislative composition, and national policy outcomes. *American Journal of Political Science, 46,* 398–414.

Webster, G. (1997). The potential impact of recent Supreme Court decisions on the use of race and ethnicity in the redistricting process. *Cities, 14*(1), 13–19.

GHETTO

Ghettos refer to sections of cities populated by minority ethnic or religious groups, that is, neighborhoods in which the minority is a majority. Some argue that their roots can be traced to Roman persecutions of Jews. During the Middle Ages in Europe, ghettos consisted of Jewish quarters in overwhelmingly Christian cities (e.g., the famous Warsaw Ghetto). Venice had a Jewish ghetto by the 14th century. Jews were forbidden from owning land outside the ghetto, and Jewish ghettos often had walls around them and were the subjects of vicious pogroms.

The nature of ghettos has changed over time. As Jewish ghettos were gradually disbanded during the 19th century, the term *ghetto* came to refer to other ethnic minorities such as Indian, Bangladeshi, and Jamaican immigrants to British cities. Although most ghettos tended to have below-average income levels, they were defined primarily in terms of ethnicity, not class.

The reasons for the formation of ghettos involve a combination of external constraints and internal motivations. External constraints include economic and political discrimination against the minority population, including formal or informal prohibitions against employment and the purchase of housing. Internal motivations that help underpin ghetto formation include the desire to be near one's ethnic group and language, the availability of marriage partners, access to culturally specific foods, and the informal webs of mutual assistance common among some ethnic groups.

In the United States, ghettos have taken a variety of ethnic forms, a theme well studied by urban geographers and sociologists (e.g., Chicago School social ecologists). During the waves of immigration from Southern and Eastern Europe during the late 19th century, many American cities had high-density ghettos composed of various ethnic groups (e.g., Italian, Irish, Polish). During the 1920s, immigrant Jews from Germany and Eastern Europe established a large ghetto in southern Manhattan centered around the garment industry. Often the cultural assimilation of one group over several generations and its dispersal into predominant Anglo-American communities, which varied from group to group, led to the group's replacement by another, less assimilated ethnicity. Thus, although the ethnic division of labor that underscores ghettos may be temporary so far as any individual group is concerned, it tends to be a permanent part of the urban landscape. The arrival of Chinese immigrants generated the first Chinatowns in New York and many West Coast cities. The migration of African Americans to northern cities circa World War I led to the formation of black ghettos, many of which were middle-class communities. Some ghettos (e.g., Harlem) became the center of rich artistic and political movements. The growth of the Latino or Hispanic population has generated the formation of Spanish-speaking barrios in cities such as Los Angeles, where distinct Mexican, Dominican, Salvadoran,

Colombian, Nicaraguan, and Cuban communities may be found. In many large cities that are the destination of migrant streams from around the world, it is not unusual to see ethnic communities of Armenians, Koreans, Thais, and Vietnamese, among others.

The transformation of American cities after World War II, particularly suburbanization and deindustrialization, changed the nature of American ghettos decisively. The intersections of class and race increasingly rendered minority-dominant neighborhoods poor with high levels of unemployment and crime. Increasingly, the term *ghetto* came to be associated with the black urban underclass and more generally with poverty, leading to the popular equation of ghettos with slums. However, some use the term to refer to gay or artistic enclaves in contemporary cities.

In short, ghettos reveal the intersecting dynamics of ethnicity, the urban division of labor, and residential segregation as they affect different groups under different historical circumstances.

—Barney Warf

See also Cultural Geography; Ethnicity; Urban Geography; Urban Underclass

Suggested Reading

Cutler, D., Glaeser, E., & Vigdor, J. (1999). The rise and decline of the American ghetto. *Journal of Political Economy, 107,* 455–506.

Wilson, W. (1987). *The truly disadvantaged: The inner city, the underclass, and public policy.* Chicago: University of Chicago Press.

GIS

Geography is fundamentally about building shared understandings of the world within and beyond its disciplinary boundaries and within and beyond the world of academia. Within academe, shared understanding is core to a dynamic and coherent discipline that is focused on robust, transparent, and (above all) usable representations of the real world. Beyond academe, geographic knowledge should not be the preserve of just the few, and a core mission of geography as a discipline is to reach out to other disciplines to provide a generalized understanding of space and spatiality—not least to provide a forum in which diverse views might be reconciled. In either setting, spatial representations should be accessible to the widest possible constituency in society. Much of human geography pays lip service to the need to *acknowledge* difference. At their best, geographic information systems (GIS) are not only accessible but also transparent and readily intelligible and so provide the only widely recognized formal spatial framework in the discipline for *reconciling* differences.

GIS are an applied problem-solving technology that allows us to create and share generalized representations of the world. Through real-world applications at geographic scales of measurement (i.e., from the architectural to the global), GIS can provide spatial representations that tell us the defining characteristics of large spaces and large numbers of individuals and are usable to a wide range of end users. They allow geography to address significant problems of society and the environment using explicitly spatial data, information, evidence, and knowledge. They not only tell us about how the world *looks* but also, through assembly of diverse sources of information, can lead us toward a generalized and explicitly geographic understanding of how the world *works*. As such, they lie at the heart of geography as a discipline, are pivotal to its ability to contribute to current real-world issues, and are core to its transferable skills base.

Beyond geography, the spatial dimension is viewed as inherently important by researchers and problem solvers working in a wide range of other academic and professional disciplines. In the world of business and commerce, for example, recent estimates suggest that global annual sales of GIS facilities and services may exceed $9 billion and are growing at a rate of 10% annually. The applications of GIS and their associated spatial data to which these figures relate range from local and national government departments; through banking, insurance, telecommunications, utility, and retail industries; to charities and voluntary organizations. In short, an enormous swathe of human activity is now touched, in some form or another, by this explicitly *geographic* technology and is increasingly reliant on it.

This line of thought illustrates the impact and significance that an inherently geographic endeavor is having on wider society and undoubtedly raises the external profile of geography as an academic discipline. Of course, a high level of economic activity does not necessarily equate with an increased likelihood of identifying scientific truth. Moreover, GIS-based representations of how the world works often

suggest how capital, human, and physical resources should be managed or how the will of the individual should be subjugated to the public good. This can raise important ethical, philosophical, and political questions in human geography such as questions of access to, and ownership of, information and the power relations that characterize different interest groups in civil society. Such general concerns about the use of technology should be used to inform issues of ethics and accountability, but they do not call into question their raison d'être or (in the case of GIS) their centrality to geography as a discipline.

DEFINING GIS

There are many definitions of GIS. Paul Longley and his colleagues defined GIS in relation to a number of component elements:

- A software product acquired to perform a set of well-defined functions (GIS software)
- Digital representations of aspects of the world (GIS data)
- A community of people who use these tools for various purposes (GIS community)
- The activity of using GIS to solve problems or advance science (geographic information science [GIScience])

GIS today are very much a background technology, and most citizens in developed countries interact with GIS, often unwittingly, throughout their daily lives. As members of the general public, we use GIS every time we open a map browser on the Internet, use real-time road and rail travel information systems for journey planning, or shop for regular or occasional purchases at outlets located by the decisions of store location planners. GIS have developed as a recognized area of activity because, although the range of geographic applications is diverse, they nevertheless share a common core of organizing principles and concepts. These include distance measurement, overlay analysis, buffering, optimal routing, and neighborhood analysis. These are straightforward spatial query operations to which may be added the wide range of transformations, manipulations, and techniques that form the bedrock to *spatial analysis*.

The first geographic information system was the Canada Geographic Information System, designed by

Roger Tomlinson during the mid-1960s as a computerized natural resource inventory system. At around the same time, the U.S. Bureau of the Census developed the DIME (Dual Independent Map Encoding) system to provide digital records of all U.S. streets and support automatic referencing and aggregation of census records. It was only a matter of time before early GIS developers recognized the core role of the same basic organizing concepts for these superficially different applications and GIS came to present a unifying focus for an ever wider range of application areas.

Any detailed review of GIS reveals that they did not develop as an entirely new area, and it is helpful to instead think of GIS as a rapidly developing focus for interdisciplinary applications that built on the different strengths of a number of disciplines in inventory and analysis. Mention should also be made of the activities of cartographers and national mapping agencies that led to the use of computers to support map editing during the late 1960s and the subsequent computerization of other mapping functions by the late 1970s. The science of earth observation and remote sensing also has contributed relevant instruments (sensors), platforms on which they are mounted (e.g., aircraft, satellite), and associated data processing techniques. Between the 1950s and the 1980s, these were used to derive information about the earth's physical, chemical, and biological properties (i.e., of its land, atmosphere, and oceans). The military is also a long-standing contributor to the development of GIS, not least through the development of the GPS, and many military applications subsequently have found use in the civilian sector. The modern history of GIS dates back to the early 1980s, when the price of sufficiently powerful computers fell below $250,000 and typical software costs fell below $100,000. In this sense, much of the history of GIS has been technology led.

A geographic information system today is a complex of software, hardware, databases, people, and procedures, all linked by computer networks (Figure 1). It brings together different data sets that may be scattered across space in very diverse data holdings, and in assembling them it is important that data quality issues are addressed during data integration. An effective network, such as the Internet or the intranet of a large organization, is essential for rapid communication or information sharing. The Internet has emerged as society's medium of information exchange and in a typical GIS application will be used to connect archives, clearinghouses, digital libraries, and data warehouses. New methods for trawling the Internet have been accompanied by the development of software that allows users to work with data in remote Internet locations. GIS hardware fosters user interaction via the WIMP (Windows, icons, menus, pointers) interface and takes the form of laptops, personal data assistants (PDAs), in-vehicle devices, and cellular telephones as well as conventional desktop computers.

In many contemporary applications, the user's device is the *client,* connected through the network to a *server.* Commercial GIS software is created by a number of vendors and is frequently packaged to suit a diverse set of needs, ranging from simple viewing and mapping applications, through software for supporting GIS-oriented Web sites, to fully fledged systems capable of advanced analysis functions. Some software is specifically designed for particular classes of applications such as utilities or defense applications. Geographic databases frequently constitute an important tradable commodity and strategic organizational resource and come in a range of sizes. Suitably qualified people are fundamental to the design, programming, and maintenance of GIS; they also supply the GIS with appropriate data and are responsible for interpreting outputs.

THE ROLE OF GIS

Even geographers can forget that their subject is important because everything that happens does so somewhere. In the broadest sense, geographic means pertaining to the earth's surface or near the surface, and in their most basic forms, GIS allow us to construct an inventory of where things (e.g., events, activities, policies, strategies, plans) happen on the earth's surface and when. They also provide tools to analyze events and occurrences across a range of spatial scales from the architectural to the global and over a range of time horizons from the operational to the strategic. GIS do this by providing an environment for the creation of *digital representations* that simplify the complexity of the real world using *data models.*

Fundamental to creation and interpretation of GIS representations is the first law of geography, often attributed to geographer Waldo Tobler. This can be stated succinctly as everything is related to everything else, but near things are more related than distant things. This statement of geographic regularity is key to understanding how events and occurrences are

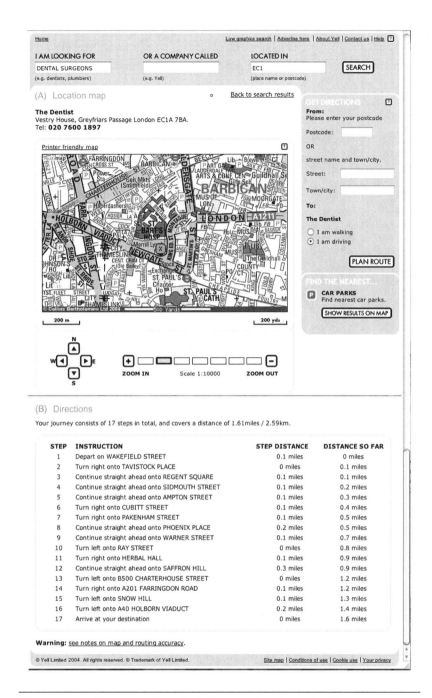

Figure 1   Example of Applied Geographic Information System

SOURCE: Longley, P., Goodchild, M., Maguire, D., & Rhind, D. *Geographic information systems and science* (2nd ed.). © 2005. Copyright John Wiley & Sons Limited. Reproduced with permission.

future events and occurrences. As human individuals, for example, our behavior in space often reflects our past spatial behavior.

Prediction implies regularity and the ability to devise a workable understanding of spatial processes. Yet regularities worthy of being described as *laws* are extremely rare in, if not entirely absent from, human geography. It is usually the case that the best we can hope for is to establish robust and defensible foundations on which to establish *generalizations* based on observed distributions of events and occurrences. The challenges of effective generalization are legion. Within human geography, for example, we may think of much of our own spatial behavior (e.g., the daily commute to work, shopping trips) as routine or nearly perfectly repetitive. Yet when we come to represent the spatial and temporal activity patterns of groups of individuals, the task becomes error prone and far from trivial. This is also true of spatial and temporal representations in general—be our interest in the representation of travel-to-work behavior, shopping, or disease diffusion. Good geography is, in part, about recording as many significant spatial and temporal events as possible without becoming mired in irrelevant detail. The geographer's *art* is fundamentally about understanding how and why significant events may be unevenly distributed across space and time; the geographer's *science* is fundamentally concerned with effective generalization about these events.

structured over space. It can be formally measured as the property of *spatial autocorrelation* and, along with the property of *temporal autocorrelation* (the past is the key to the present), makes possible a fundamental geographic statement, namely that the geographic context of past events and occurrences can be used to predict

It is in this way that GIS help us manage what we know about the world; hold it in forms that allow us to organize and store, access and retrieve, manipulate, and synthesize spatial data; and develop models that improve our understanding of underlying processes. Geographic data are raw facts that are neutral and nearly context free. It is helpful to think of GIS as a vehicle for adding value to such context-free bits and bytes by turning them into information through scientific procedures that are transparent and reproducible. In conceptual terms, this entails selection, organization, preparation for purpose, and integration. Human geography data sources are in practice often very diverse, but GIS provide an integrating environment in which they may be collated to support an *evidence base*. Through human interpretation, evidence is assembled into an individual's *knowledge base* of experience and expertise. In this way, geographic data can be related to specific problems in ways that are valid, consistent, and reproducible and, as such, can provide a cornerstone to *evidence-based policy*.

This is the cumulative manner in which GIS bring an *understanding of* general process to bear on the solution of specific problems that occur at unique points on the earth's surface. As such, GIS bring together the idiographic (the world as an assemblage of unique places, events, and occurrences) and the nomothetic (the quest to identify generalized processes) traditions in human geography—in the context of real-world practical problem solving. Many such problems involve multiple goals and objectives that often cannot be expressed in commensurate terms, yet a further strength of GIS is that they allow the formulation and application of explicit conventions for problem solving that are transparent and open to scrutiny. Analysis based around GIS is consistent with changes to scientific practice, specifically the challenges posed by mining today's enormous resources of information, the advent of interdisciplinary (as well as intradisciplinary) team collaboration, and the increasing rapidity of scientific discovery.

GEOGRAPHIC INFORMATION SCIENCE

Much of the *spatial analysis* that is core to GIS centers on pursuing specific hypotheses with respect to the data and information that are available—in the spirit of deductive reasoning. Induction also plays an (increasingly) important part in GIS-based analysis, whereby data "mining" is used to identify what to leave in (and, hence, what to take out of) a representation and what weight to assign that which is left in. Yet these complementary procedures of induction often can raise questions that are at the same time frustrating and profound. For example, how do GIS users know whether the results obtained are accurate?

How might the quality of the input data be ascertained with respect to other validatory sources that might be available to us? How can we be sure that the visual medium of a geographic information system does not obscure the underlying messages of a representation? What principles might help GIS users to design better maps? How can GIS be fine-tuned to assimilate the limits of human perception and cognition? Some of these are questions of GIS design, and others are questions about GIS data and methods. They all arise from practical use of GIS but relate to core underlying principles and techniques.

The term *geographic information science* was coined in an article published by Michael Goodchild in 1992. In it, the author argued that these questions and others like them were important and that their systematic study constituted a science in its own right. Information science studies the fundamental issues arising from the creation, handling, storage, and use of information. Similarly, GIScience should study the fundamental issues arising from geographic information as a well-defined class of information in general. Other terms have much the same meaning—*geomatics* and *geoinformatics, spatial information science, geoinformation engineering*. Each of these terms suggests a scientific approach to the fundamental issues raised by the use of GIS and related technologies, although they all have different disciplinary roots and emphasize different ways of thinking about problems.

GIScience has evolved significantly during recent years; one can get some idea of the range of current interests in the field by visiting the Web site of a biannual conference series on the subject (www.giscience .org). One disarmingly simple way of viewing the remit of GIScience is provided by the Varenius project (www.ncgia.org). Here GIScience is viewed as anchored by three concepts: the individual, the computer, and society. These form the vertices of a triangle, and GIScience lies at its core. The various terms that are used to describe GIScience activity can be used to populate this triangle. Thus, research about the individual is dominated by cognitive science, with its concern for understanding of spatial concepts, learning and reasoning about geographic data, and interaction with the computer. Research about the computer is dominated by issues of representation, the adaptation of new technologies, computation, and visualization. Finally, research about society addresses issues of impacts and societal context.

At its core, GIS are concerned with the development and transparent application of the explicitly spatial core organizing principles and techniques of GIScience in the context of appropriate management practices. These concerns provide an enduring intellectual nexus for the discipline of human geography. GIS are also a practical problem-solving tool for use by those geographers intent on practicing their vocation through solving real-world problems. The spatial dimension to problem solving is special because it poses a number of unique, complex, and difficult challenges that are investigated and researched through GIScience. Together, these provide a conduit for committed human geographers to pursue their interests through vocation in academic, industrial, and public service settings alike.

—Paul Longley

See also Applied Geography; Automated Geography; Cartography; Cellular Automata; Digital Earth; Geodemographics; Geoslavery; Humanistic GIScience; Information Ecology; Limits of Computation; Location-Based Services; Neural Computing; Ontology; Overlay; Tessellation; Tobler's First Law of Geography

Suggested Reading

Goodchild, M. (1992). Geographical information science. *International Journal of Geographical Information Systems, 6,* 31–45.

Goodchild, M., & Longley, P. (1999). The future of GIS and spatial analysis. In P. Longley, M. Goodchild, D. Maguire, & D. Rhind (Eds.), *Geographical information systems: Principles, techniques, management, and applications* (pp. 567–580). New York: John Wiley.

Longley, P., & Barnsley, M. (2004). The potential of remote sensing and geographical information systems. In J. Matthews & D. Herbert (Eds.), *Common heritage, shared future: Perspectives on the unity of geography* (pp. 62–80). London: Routledge.

Longley, P., Goodchild, M., Maguire, D., & Rhind, D. (2005). *Geographic information systems and science* (2nd ed.). Chichester, UK: Wiley.

GLOBAL CITIES

Global cities are the command and control centers of the world economy, the sites of vast complexes of skilled, high value-added activities with globe-spanning consequences. At the top of the international urban hierarchy, this handful of large specialized metropolises are simultaneously centers of creative innovation, news, fashion, and culture industries; metropoles for raising and managing investment capital; centers of specialized expertise in producer services such as advertising and marketing, legal services, accounting, and computer services; and the management, planning, and control centers for corporations and nongovernmental organizations that operate with increasing ease over the entire planet. The stereotypical global cities include the famous trio of New York, London, and Tokyo. To a lesser extent, they include cities such as Paris, Frankfurt, Toronto, Miami, San Francisco, Osaka, Hong Kong, Los Angeles, and Singapore. All of these lie at the core of a worldwide chain of value-added linkages that have steadily fostered a pronounced concentration of strategic headquarter functions in a few conglomerations and a persistent dispersal of unskilled functions to the world's periphery. This process reinforces the long-standing transition of employment in such regions from low-wage, low value-added, blue-collar occupations to high-wage, high value-added, white-collar employment. In short, global cities shape the world economy as much as they are shaped by it.

In one sense, global cities are as old as capitalism itself: Amsterdam, for example, played a key role in the 16th-century world economy. The current hegemonic position of these centers in the international economy may be interpreted as an outcome of the post-Fordist global division of labor that emerged during the 1970s, which was marked by the collapse of the Bretton–Woods agreement in 1971 and the shift to floating currency exchange rates; the oil crises of 1974 and 1979 and associated growth of Third World debt; the deindustrialization of much of Europe and North America and the concomitant rise of the East Asian newly industrializing countries; the steady growth of multinational corporations and their ability to shift vast resources across national boundaries; technological changes unleashed by the microelectronics revolution; the global wave of deregulation, privatization, and the lifting of government controls, all of which reflect the hegemony of neoliberalism worldwide; the integration of world financial markets through telecommunications systems; and the initiation of new trade agreements and trade blocs and agreements that accelerated the freedom of capital to transcend national borders. These changes produced a highly volatile, deregulated, globalized form of capitalism that greatly accentuated the position of global cities in the world space economy.

The strategic position of global cities is closely bound up with the ability to move vast quantities of money and information rapidly. Financial firms use an extensive worldwide web of electronic funds transfer networks that form the nervous system of the international economy, allowing them to move capital around at a moment's notice, arbitrage interest rate differentials, take advantage of favorable exchange rates, and avoid political unrest. Such networks create an ability to move money—by some estimates, more than $3 trillion daily—around the globe at the speed of light; subject to the process of digitization, information and capital became two sides of the same coin. A global web of fiber-optics lines firmly links New York securities traders to their counterparts in London and elsewhere, allowing money to be switched in enormous volumes. The volatility of trading, particularly in stocks, has also increased as hair-trigger computer trading programs allow fortunes to be made (and lost) by staying microseconds ahead of (or behind) other markets.

Despite their importance to worldwide financial markets, global cities also rely, paradoxically, on agglomeration economies, particularly face-to-face contacts saturated with trust and reciprocity. The core of such conglomerations allows for dense networks of interaction necessary to the performance of headquarters functions, including monitoring frequent changes in niche product markets, negotiating with labor unions, keeping abreast of new technologies and government regulations, keeping an eye on the competition, staying attuned to an increasingly complex financial environment, initiating or resisting leveraged buy-outs and hostile takeovers, and seeking new investment opportunities. Because their raison d'être cannot immediately be classified as economic but includes a vast variety of formal and informal cultural and political interactions such as tourism, the media, and fashion industries, global cities are more than simply poles for the production of corporate knowledge. The crux of global cities' role in the post-Fordist world economy is to serve as arenas of interaction, allowing face-to-face contact, political connections, and artistic and cultural activities as well as allowing elites to rub shoulders easily. At their core, global cities allow the generation of specialized expertise on which so much of the current producer services economy depends. The creation of expertise is no simple task, involving the transformation of information into useful knowledge. Despite

the enormous ability of telecommunications to transmit information instantaneously over vast distances, face-to-face contact remains the most efficient and effective means of obtaining and conveying irregular forms of information, particularly when it is highly sensitive (or even illegal). Thus, in the context of face-to-face meetings, actors monitor one another's intentions and behavior through observations of body language such as handshakes and eye contact, which are essential to establishing relations of trust and mutual understanding. Such interactions are simply not substitutable to the digital form required by telecommunications.

The analysis of global cities has been accompanied by a growing concern regarding mounting inequality within them. Saskia Sassen's famous volume *The Global City,* published in 1991, was hugely influential in maintaining that globalization leads directly to social polarization. She held that the growth of the financial sector, in particular, led to the formation of a cadre of well-paying positions typified by managers, executives, and stockbrokers, on the one hand, and large numbers of low-paying jobs typically filled by women and minorities in unskilled positions that cater to the elite, on the other. For the former, large annual bonuses are the norm; for the latter, who often struggle in minimum-wage jobs and with a steady supply of workers moving to the region from abroad, daily life becomes increasingly difficult. While a small elite earns millions buying and selling stocks, this argument holds, the spin-offs are to be found in low-paying unskilled jobs in retail trade, hotels, and personal services. For those at the bottom of the socioeconomic ladder, globalization can lead to diminished social mobility. Critics of Sassen's view focus on different causes of inequality, including the relative degree to which immigration, a polarized wage structure characteristic of many services, and public policy have contributed to the yawning gap between the poor and the wealthy in many such conurbations. The jobs–skills mismatch between employers who seek increasingly skilled labor and a workforce whose members possess insufficient human capital exacerbates central-city unemployment. More broadly, inequality reflects an entire system of social stratification—including occupational change, racial and ethnic segregation, poor educational systems, lack of affordable housing, and spatial isolation—that has evolved over time, fed by various waves of immigration. Sociologists often tie wage inequality to shifts in family structures, demographics, and educational levels. National-level policies, particularly the increasingly regressive income tax structure and the growth of unearned incomes, also contribute to this trend.

—*Barney Warf*

See also Agglomeration Economies; Flexible Production; Gentrification; Globalization; Producer Services; Urban Geography; World Economy

Suggested Reading

Fainstein, S., Gordon, I., & Harloe, M. (1992). *Divided cities: New York and London in the contemporary world.* Oxford, UK: Blackwell.

Hamnett, C. (1994). Social polarization in global cities: Theory and evidence. *Urban Studies, 31,* 401–424.

Knox, P. (1995). World cities and the organization of global space. In R. Johnston, P. Taylor, & M. Watts (Eds.), *Geographies of global change* (pp. 232–247). Oxford, UK: Blackwell.

Markusen, A., & Gwiasda, V. (1994). Multipolarity and the layering of functions in world cities: New York City's struggle to stay on top. *International Journal of Urban and Regional Research, 18,* 167–193.

Mollenkopf, J., & Castells, M. (Eds.). (1991). *Dual city: Restructuring New York.* New York: Russell Sage.

Sassen, S. (1991). *The global city: New York, London, Tokyo.* Princeton, NJ: Princeton University Press.

Taylor, P. (2000). World cities and territorial states under conditions of contemporary globalization. *Political Geography, 19,* 5–32.

GLOBAL POSITIONING SYSTEM

See GPS

GLOBALIZATION

Globalization typically is defined as the expansion in the scope, velocity, and impacts of international transactions such as trade, investment, migration, and communications. It is a complex subject that embraces many topics and can be approached from a wide variety of theoretical perspectives, but typically globalization entails the increased integration of different societies. There is no single process of globalization but rather a diversity of intertwined processes that reflect the persistent tendency of capitalism to stretch across national borders. Because it receives considerable media attention and lies at the core of many debates about economic trends and policies, globalization often has been surrounded by erroneous or simplistic misconceptions, both among those who advocate it and among those who fear it.

A common stereotype pertaining to globalization is that it is purely economic in nature. Much of the literature on this topic has focused on international trade and foreign investment, particularly the behavior of transnational corporations. Yet such a view is overly narrow and ignores the multiple ways in which globalization operates as a political, cultural, and ideological force as well. For example, immigration clearly is a topic pertinent to globalization, with many so-called noneconomic dimensions associated with it. Equally, one could point to the globalization of education, disease, or terrorism. Some of the aspects of globalization that are resisted most vehemently in parts of the world are its cultural dimensions, including the globalization of fast food, dress, and cinema, all of which are bound up with people's worldviews and daily lives.

A second simplistic view of this topic equates globalization with cultural homogenization, as if the world economy stamped a monoculture (typically American in nature) throughout the world. For much of the world, globalization is synonymous with Americanization. As the world's largest economic, military, and political power, the United States is simultaneously envied, imitated, and despised. Admiration for American culture typically is strongest among the young, so that globalization creates a generation gap in terms of outlook and preferences. However, although there can be no denying that cultural homogenization often takes place in the wake of globalization and frequently at the expense of old, deeply held traditions, it is equally true that globalization generally means different things in different places; that is, it is geographically specific. Global trends are mediated through national policies in different ways. The unique histories of individual places serve to impart local flavor to global trends, for example, when multinational corporations such as McDonald's must tailor their menus and advertising to local preferences. Thus, local regions not only undergo changes imparted to them by the global economy but also shape that global economy in turn. The global and the local are intimately intertwined, and geographers often use the term *glocalization* to capture this relationship.

A third frequent misconception about globalization is that it began, or reached its most prominent stage, only during the late 20th century. Clearly, there is little doubt that the world today is deeply globalized and becomes more so daily. However, the birth of capitalism on a global basis during the 16th century clearly marks an earlier epoch of globalization, as did colonialism during the following centuries. The Industrial Revolution unleashed waves of time–space compression that ushered in wave after wave of globalization. In terms of the relative magnitude of foreign investment, the late 19th century was at least as globalized as the present, if not more so. Moreover, globalization might even have earlier roots; work by Janet Abu-Lughod revealed the existence of a world system during the 14th century stretching throughout much of the Old World, and some world systems theorists have speculated on even earlier systems.

A fourth issue that is problematic in the study of globalization concerns its relations to the nation-state. Some argue that true globalization could not have occurred prior to the emergence of the modern nation-state during the 18th and 19th centuries; it is, after all, difficult to be international if there is nothing national. However, this view of globalization often is deemed to be too narrow and ignores the extensive evidence of premodern globalization. A related issue is the question as to whether globalization entails the end of the nation-state. Certainly, certain aspects of globalization have eroded the sovereignty of states in some matters; for example, the globalization of financial capital has made national monetary controls increasingly ineffective, and supranational organizations such as the European Union, the United Nations, the World Bank, and the International Monetary Fund have assumed some functions of the nation-state. The bluntest manifestation of this view is that globalization is boundary transcending and that localization is boundary heightening. However, it is simplistic to assume that globalization leads inevitably to the end of states as they currently are constituted, replacing them with some mythical, seamless integrated market that embraces the entire planet. Globalization always is refracted through national policies (e.g., concerning labor or the environment), and that is one reason why it has spatially uneven impacts across the world. Capitalism involves both markets and states, and the political geography of globalization is the interstate system, the existence of which is necessary for capital to play states and localities off one another.

A fifth stereotype about globalization is that it consists of some unstoppable teleological force that is independent of human intervention. In this reading, globalization is inevitable and countries can do little to stop it except accommodate its needs and requirements. Such a view denies the historical origins of globalization and the fact that people create it. In fact, globalization has experienced reversals, for example, during the trade wars of the 1930s. Moreover, globalization is resisted, sometimes successfully and sometimes not, often by those who believe that it presents a secular amoral threat to established local traditions and who view the market as a mechanism for reducing everyone to a consumer, annihilating all forms of identity except those that pertain to the commodity. For social segments with values that lie largely outside of the market, globalization can be deeply offensive morally. Thus, the more globalization has disrupted local value systems around the world, the greater has been the backlash against it.

Finally, a sixth frequent misconception about globalization holds that it always is beneficial. This claim, often advocated by those who hold to neoclassical notions of comparative advantage and the benefits of free trade, finds some empirical basis in the observation that the most globalized societies generally are among the world's wealthiest or most rapidly growing (e.g., the newly industrializing Asian countries). In this reading, globalization is associated with lower consumer prices, technology transfer, and improved efficiency. However, as dependency theorists and Marxists often point out, the history of capitalism is one characterized by uneven development, and globalization is no exception; that is, it represents capitalism at a global scale that reproduces poverty. Evidence for this argument may be found in local producers displaced by multinational firms, the exploitative labor conditions found in many sweatshops in the developing world, International Monetary Fund austerity programs, and international crises such as the East Asian financial crisis of the late 1990s. Among those who bear its costs but do not enjoy its benefits, globalization understandably breeds envy and resentment.

In sum, globalization is viewed most productively as a complex historical process that began before capitalism but was greatly accelerated by its worldwide expansion; that offers a mix of costs and benefits unevenly distributed around the world; that transforms but does not eliminate the role of the nation-state; that tends toward cultural homogenization, on the one hand, but simultaneously creates locally specific impacts, on the other; and that is not inevitable or unstoppable but rather can be, and has been, reversed and challenged.

—Barney Warf

See also Comparative Advantage; Dependency Theory; Developing World; Economic Geography; Marxism, Geography and; Modernization Theory; Nation-State; Newly Industrializing Countries; Trade; Uneven Development; World Economy; World Systems Theory

Suggested Reading

Abu-Lughod, J. (1989). *Before European hegemony: The world system A.D. 1250–1350.* New York: Oxford University Press.

Appadurai, A. (1996). *Modernity at large: Cultural dimensions of globalization.* Minneapolis: University of Minnesota Press.

Barber, B. (1995). *Jihad vs. McWorld: How globalism and tribalism are reshaping the world.* New York: Ballantine.

Blaut, J. (1993). *The colonizer's model of the world: Geographical diffusionism and Eurocentric history.* New York: Guilford.

Chase-Dunn, C. (1989). *Global formation: Structures in the world economy.* Oxford, UK: Blackwell.

Dicken, P. (2003). *Global shift: The internationalization of economic activity* (4th ed.). New York: Guilford.

Featherstone, M. (Ed.). (1990). *Global culture: Nationalism, globalization, and modernity.* London: Sage.

Frank, A. (1998). *ReOrient: Global economy in the Asian age.* Berkeley: University of California Press.

Friedman, T. (1999). *The lexus and the olive tree.* New York: Farrar, Straus and Giroux.

Friedman, T. (2005). *The world is flat: A brief history of the 21st century.* New York: Picador USA.

Herod, A., Ó Tuathail, G., & Roberts, S. (Eds.). (1998). *An unruly world? Globalization, governance, and geography.* London: Routledge.

Scott, A. (1997). *The limits of globalization.* London: Routledge.

Stiglitz, J. (2002). *Globalization and its discontents.* New York: Norton.

Wallerstein, I. (1974). *The modern world-system: Capitalist agriculture and the origins of the European world-economy in the sixteenth century.* London: Academic Press.

Wallerstein, I. (1979). *The capitalist world economy.* Cambridge, UK: Cambridge University Press.

Waters, M. (1995). *Globalization.* London: Routledge.

GPS

The Global Positioning System (GPS) is a worldwide navigation system developed by the U.S. Department of Defense. A constellation of 24 NAVSTAR satellites enables users to determine their position anywhere on the earth in three dimensions (latitude, longitude, and altitude) at any time of day and in any kind of weather. The GPS was declared fully operational on April 27, 1995.

Using the simple mathematical principle called *trilateration,* the location of a GPS receiver can be determined by computing its distance to at least four GPS satellites whose precise orbital positions are known. The distance to each satellite is determined accurately by measuring the time it takes for the signal transmitted by each GPS satellite to reach the receiver. Because these signals propagate at the speed of light, it is possible to convert the measured time to the distance between the receiver and each satellite. Because only one point on the earth can be at those precise distances from the satellites, the location of the GPS receiver can be determined.

The GPS consists of three major segments. The *space segment* consists of a constellation of 24 satellites circling the globe on four orbital paths and is designed so that, barring obstructions, a minimum of 5 satellites can be viewed to determine a position. The *control segment* tracks the satellites' orbits, monitors their status, and frequently relays updates to the satellites, including corrections to their on-board atomic clocks. The *user segment* consists of GPS receivers and auxiliary equipment such as antennas.

The GPS is designed for dual military and civilian use. The location accuracy available for civilian applications, called the Standard Positioning Service (SPS), originally was lower than the Precise Positioning Service (PPS) used for military applications. This was accomplished by intentionally degrading the signal using a procedure known as selective availability that ended on May 2, 2000. Since then, typical GPS accuracies are on the order of 15 meters horizontally and 25 meters vertically.

A number of factors, including satellite clock errors, orbital uncertainties, and atmospheric effects, contribute to GPS positional errors. These errors can be reduced using differential GPS (DGPS) techniques. Using DGPS, it is possible to determine a location to within 1 meter horizontally and within a few meters vertically. In DGPS, this is accomplished by continually comparing the position of a known location, called a *base station,* with its position determined by GPS that changes over time. The difference in these two positions measured at a base station can then be used to correct GPS positions measured at other locations. These differential corrections can be transmitted at radio frequencies to specially equipped

DGPS receivers and then can be applied in *real time* as the GPS coordinates are being collected or can be applied after GPS coordinates are collected using *postprocessing* techniques.

The GPS is used for a wide range of applications, including surveying, aircraft navigation, and even the game of geocaching. As GPS modernization continues and the European Galileo system is deployed, global navigation satellite systems will continue to improve.

—*Andrew Klein*

See also GIS; Humanistic GIScience

GRAVITY MODEL

The gravity model is a simple mathematical formulation used to model the interaction between two locations. It has been used to account for a wide variety of interactions such as telephone calls, automobile trips, and migration and merchandise flows. The model takes the form

$$I_{ij} = k\, P_i\, P_j\, /\, d_{ij}^{\,b},$$

where I_{ij} is the interaction between places i and j, k is an empirically determined constant, P_i and P_j are measures of the importance (or mass) of i and j (e.g., their populations), d_{ij} is the distance between i and j, and b is the friction of distance, an empirically derived parameter that represents the difficulty or cost of moving between i and j.

The gravity model is based on the law of physical gravitation, expressed as

$$F_{ij} = G\, M_i\, M_j\, /\, d_{ij}^{\,2},$$

where F_{ij} is the gravitational attraction between two objects with masses M_i and M_j separated by a distance of d_{ij} and G is the gravitational constant. The gravity model is an example of a model from the social physics school. The proponents of these models attempted to adapt the ideas and concepts of physical science, especially those of Newtonian physics and Darwinian ecology, to explain human patterns and processes. E. G. Ravenstein, writing about migration during the late 19th century, often is credited with being the first person to apply the gravity concept to social science when he observed that more migrants travel short distances than travel long distances and

that long-distance migrants tended to move to large centers.

The basic gravity model was incorporated into a variety of more complicated and sophisticated formulations used to describe the spatial extent of markets, the geographic distribution of demand, and transportation flows. W. J. Reilly used the gravity concept in the development of his "law of retail gravitation" that, among other things, delimited the market boundaries between cities. J. Q. Stewart and others developed potential models that characterized the interaction of a place with all other places. The results of these calculations typically were represented as potential surfaces that displayed the potential of all places in the study area simultaneously. That is, place i's population potential is calculated as

$$P_i = \sum (M_j\,/\,d_{ij}^{\,x})\ \text{for}\ j = 1\ \text{to}\ n,$$

where M_j is the population of j, d_{ij} is the distance between i and j, and x is the friction of distance. Other potential surfaces can be created to describe the spatial distribution of other phenomena. For example, by substituting a measure of disposable income or retail expenditures for population, market potential, an estimate of the spatial distribution of demand, can becalculated.

Today the gravity model is used most commonly in transportation geography and planning. Its flexibility and adaptability allow it to provide an accurate fit to data from a wide variety of situations and problems. Some have criticized the gravity model because they see little connection between the model's theoretical rationale and the problems to which it is applied. Critics also note that use of the model tends to support the status quo and the distribution of resources at the time the model is applied.

—*Robert S. Bednarz*

See also Transportation Geography

Suggested Reading

Ravenstein, E. (1885). The laws of migration. *Journal of the Statistical Society of London, 48*, 167–235.
Ravenstein, E. (1889). The laws of migration. *Journal of the Royal Statistical Society, 52*, 241–305.
Reilly, W. (1931). *The law of retail gravitation.* New York: Knickerbocker.
Stewart, J. (1947). Empirical mathematical rules concerning the distribution and equilibrium of population. *Geographical Review, 37*, 461–485.

GROSS DOMESTIC PRODUCT

Gross domestic product (GDP) is the total value of the economic output of a country. (In contrast, GNP is the total value of output of all members of a country regardless of where they might be.) To examine the economic productivity of a nation over time, the GDP usually is calculated on a yearly basis. To account for the effects of population size on the GDP, researchers frequently determine the per capita GDP of a country. The per capita GDP of a country gives an indication of the economic productivity of the average person. By assessing such statistics, geographers and other researchers are able to draw conclusions about the status and position of individual countries within the emerging global environment.

Around the world, GDP and per capita GDP vary quite substantially. In the United States, the per capita GDP is approximately $34,000. Per capita GDP levels in Western Europe, Japan, and Australia vary somewhat but are roughly similar to figures in the United States. In contrast, the per capita GDP levels of most countries in sub-Saharan Africa, South Asia, and Southeast Asia are less than $1,000. Countries in Eastern Europe (including Russia) and Latin America tend to have per capita GDP levels that fall within these extremes. In many respects, these geographic patterns reflect old social and geopolitical divisions. Based on these statistical parameters, a significant amount of disparity exists overall. Moreover, during recent decades, the size of this economic disparity has been increasing.

As a caveat, it has been noted that these statistics do not account for the relative value of different currencies in different economies. The value of a dollar, for example, is greater in some countries than it is in others. Therefore, some researchers have attempted to adjust statistics regarding GDP to account for these differences. By comparing the relative values of currencies, researchers have developed the notion of purchasing power parity. In general, by manipulating available data in this manner, the apparent economic gap between rich and poor countries is reduced (but not eliminated).

The total GDP of a nation is closely related to the type of economic activities that predominate within that country. In countries with low GDP levels, the majority of citizens usually are engaged in primary economic activities such as agriculture, forestry, and other forms resource extraction. In many of these countries, more than 75% of the population is engaged in agriculture. Secondary and tertiary economic activities play a comparatively minor role in these economies. Although the amount of manufacturing has increased somewhat in these countries during recent decades, it is still far less important than primary activities. Countries with high GDP levels typically are engaged in other types of economic activity. Tertiary activities (i.e., service-sector jobs) have become the main type of economic activity. Conversely, the percentage of people engaged in primary activities in these countries has dropped substantially over the past century and now represents a small minority. Due partly to the effects of globalization, secondary economic activities have also declined in importance in these countries during recent decades, but the drop in percentage has not been as precipitous.

These economic and geographic disparities are important because they have broad consequences. Not only do these conditions affect the economic affluence of citizens living within respective nations, but they also impact a host of other resources. For example, a lack of economic resources affects the medical infrastructure that is necessary to sustain communities. Consequently, life expectancy frequently is reduced and infant mortality often is increased in countries with low GDP levels. Similarly, educational systems suffer because of insufficient funds. More generally, other governmental resources often are unable to resolve emerging problems (e.g., public safety, habitat protection, transportation systems) because many countries simply do not have the financial means necessary to contend with them. These specific effects trickle down and affect the lifestyle and general welfare of the citizenry.

Geographers, economists, and other researchers are trying to better understand the cause of this widening economic gap among countries. Undoubtedly, innovations in technology have enabled countries to increase their productivity per person. Such innovations are likely to generate differences among countries, particularly during the early stages of development. Yet according to many theoretical models, these economic disparities eventually should diminish after these innovations have spread to other geographic regions. Walter Rostow's depiction of predictable stages of development is a prominent example of this theoretical perspective. However, due to the persistent nature of these economic inequalities, other researchers have formulated other theories. Perhaps the most prominent among these is dependency theory. Dependency theory suggests that economic disparities are not anomalies but rather an integral feature of global structural systems. That is, the current global structure, which emerged in the past out of colonial dependencies, is perpetuated by contemporary capitalist systems. From this perspective, countries with high rates of consumption require regions of relative poverty to persist in order to supply their material needs at a low cost. Thus, according to this perspective, a pattern of uneven development develops and endures.

Yet other researchers have suggested that the conceptual emphasis of GDP is misguided or inappropriate. In other words, some analysts contend that the economic focus of GDP distorts the condition of individual countries around the world. From this perspective, the GDP overemphasizes the centrality and importance of economic variables. These researchers argue that social life is not wholly determined by economics. As such, economic productivity should not be considered the sole index of a country's prosperity; rather, it should be considered one variable among many. For example, prevailing cultural systems have developed different means of adjusting to economic circumstances. Frequently, many of these systems, such as the organizational structure of families, do not show up in economic statistics. Distinctions between formal and informal economies may be important here as well. Accordingly, analysts have tried to develop other measures for assessing the status of countries that address the perceived deficiencies of the GDP. A prominent example is the Human Development Index (HDI), which incorporates variables such as life expectancy and literacy. Variables such as gender equality may also be integrated into such assessments. By including such variables, the analysis of individual countries becomes more rounded and perhaps more accurate.

GDP is the widely recognized means of representing and assessing the state of countries around the world. It is an analytical tool that is now deeply embedded in governmental institutions and economic systems. Yet as the forgoing suggests, GDP is a complicated concept. Although the notion of GDP certainly will continue to have theoretical value, its shortcomings must also be recognized in any analysis.

—*Christa Stutz*

See also Economic Geography; World Economy

Suggested Reading

McConnell, C., & Brue, S. (1993). *Macro-economics: Principals, problems, and policies.* New York: McGraw-Hill.

Meadows, D. (1974). *Dynamics of growth in a finite world.* Cambridge, MA: Wright-Allen Press.

Mishan, E. (1977). *The economic growth debate: An assessment.* London: Allen and Unwin.

Scott, A., & Storper, M. (Eds.). (1986). *Production, work, territory: The geographical anatomy of industrial capitalism.* Boston: Allen and Unwin.

Stutz, F., & Warf, B. (2005). *The world economy: Resources, location, trade, and development* (4th ed.). Upper Saddle River, NJ: Pearson/Prentice Hall.

GROWTH MACHINE

The term *growth machine* refers to the political and economic leadership of urban areas. Growth machine theorists claim that urban development is engineered to maximize profits of various local and nonlocal elite. Inherent in this thesis is that growth itself as an objective is never questioned and rarely debated among urban elites. Debates do not occur over whether growth is good or not; instead, they may occur over, for example, how to maximize growth and, in turn, profit. As such, the growth machine ascribes to value-free development and the ostensibly universal benefits of growth.

The thesis borrows from classical Marxism by making distinctions between use value and exchange value with regard to property. Whereas many urban residents attach some use value to their property, to the elites the value of a property is based solely on its exchange value via either rent or sale to the highest bidder.

The growth machine is composed of, and sustained by, both primary and auxiliary actors, all of whom benefit from urban growth. The primary actors are politicians, the media, and utility companies. The success of politicians often is linked to their ability to generate and sustain economic growth. Both the media (e.g., newspapers, television, radio) and utility companies have a vested interest in growth because they are tied to the local market and increase their revenues through a greater number of subscribers or customers.

The auxiliary actors are those that play a role in promoting and maintaining growth but generally are not as intimately connected to the growth process as the primary actors. Among these are cultural institutions (e.g., museums, theaters, universities, symphonies, professional sports teams) that often are dependent on the local growth machine to generate revenues. Organized labor is a proponent of growth because of the jobs that are generated despite labor's frequent confrontations with the capitalist class about the distribution of surplus value generated by this growth. Self-employed professionals and small retailers also have an interest in growth, although not so much for the increase in aggregate rents and potential displacement of their customer base as for the increase in customers and revenues.

Although the growth machine thesis is a useful starting point to understand the politics and economics of cities, it has limitations. It has been criticized as being weak methodologically, making it difficult to draw comparisons between cities. It also assumes that clear distinctions can be made between the rentier and nonrentier classes when in reality there are far fewer rentiers than is assumed in urban areas (rentiers are people who live off of returns from fixed assets). In addition, capital is not as footloose as was originally stipulated, and many other (noneconomic) factors go into locational decision making. As a result, the growth machine has largely been superseded by the more theoretically robust urban regime theory.

—*Ed Jackiewicz*

See also Urban and Regional Planning; Urban Entrepreneurialism; Urban Geography; Urban Social Movements

Suggested Reading

Fulton, W. (1997). *The reluctant metropolis: The politics of urban growth in Los Angeles.* Baltimore, MD: Johns Hopkins University Press.

Lauria, M. (Ed.). (1997). *Reconstructing urban regime theory: Regulating urban politics in a global economy.* Thousand Oaks, CA: Sage.

Logan, J., & Molotch, H. (1987). *Urban fortunes: The political economy of place.* Los Angeles: University of California Press.

Purcell, M. (1997). Ruling Los Angeles: Neighborhood movements, urban regimes, and the production of space in Southern California. *Urban Geography, 18,* 684–704.

GROWTH POLE

Growth poles are a spatial strategy for economic development centered around a dynamic industry that is geared to jump-start economic development in lagging areas that traditionally have not been major industrial or sector leaders. Unequal growth can be seen as useful in economic development planning because it provides an opportunity to bring what are regarded as unproductive areas into the production realm.

Simplicity is growth pole theory's great appeal, and the strategy became widely employed in urban and regional planning circles throughout the advanced industrial economies as well as in developing nations in Latin America and Asia. The idea originally was conceptualized by French scholar François Perroux in 1955 in an effort to decentralize the French automobile away from the Parisian basin. A decade later, J. R. Boudeville operationalized *la notion de pôle de croissance* so that slow-growth regions could develop quickly and extend multipliers throughout the rest of the nation's economy. In the United States during the 1960s, a presidential committee led by John D. Rockefeller studied a proposal to decentralize large metropolitan areas by building up medium-sized cities. Although the plan never was implemented because of population-based issues such as sex education and family planning, like the French efforts, it considered "trickle-down" effects to economic development by means of citing jobs and population within the hierarchy of a national urban system. When the Chilean automobile industry tried to decentralize production outside of Santiago into its northern cities during the 1960s and 1970s, many managers and engineers disliked the lack of cultural amenities around these new growth poles. In addition, growth pole development in the Chilean car industry failed to realize that a small domestic automobile market could not offset the vicissitudes of small international sales of European-brand automobiles.

Growth poles often require a judicious mix of public and private support for a propulsive industry. This usually entails supporting a basic industry that has significant forward and backward linkages. One difficult empirical question centers around what the balance between public and private investments should be to create such linkages. On the one hand, the state can encourage public or private investment through tax incentives. On the other hand, no guarantees exist about what those incentives might be, nor are there severe penalties imposed on the private sector if it deviates from industrial policies. Also complicated are the notions of developing appropriate forward and backward linkages in the production sphere so that significant economies of scale accrue to the producers.

Three stages characterize the implementation of a growth pole strategy. The first stage entails locating a propulsive industry within a targeted growth pole. The second (or polarization) stage anticipates backwash effects where the gaps between the center (growth pole) and hinterland (periphery) will actually widen. The third (or spread) stage should produce trickle-down benefits between the center and the periphery that ultimately will converge as economic development ensues. Like all stage models, these assumptions are inherently linear and normative and, therefore, will depart from reality in many settings.

Supporters of the growth pole strategy recognize initial drawbacks, including backwash or polarization effects in the hinterland region. Talented workers will leave small towns and rural settings in search of new opportunities in the new industry. Ironically, an undesirable effect may be to exacerbate urban–rural differences and augment the uneven development that growth pole strategies aim to remedy.

In conclusion, developing necessary infrastructure, attracting appropriate labor skills, and identifying the length of tax incentives are difficult empirical and policy questions for growth pole theorists and industrial planners. These difficulties underscore the challenges of taking abstract ideas and grafting them onto particular social, political, and economic units. Outcome measures become complicated when trying to determine the length of time to produce economic growth and to measure contributions to local economies. Therefore, growth poles respond to local geographic contexts that cannot be divorced from state governments, global forces, local culture, and political variables.

—*Joseph L. Scarpaci*

See also Development Theory; Economic Geography; Modernization Theory; Uneven Development

Suggested Reading

Boudeville, J. (1966). *Problems of regional economic planning.* Edinburgh, UK: Edinburgh University Press.

Gwynne, R. (1986). *Industrialization and urbanization in Latin America.* Baltimore, MD: Johns Hopkins University Press.

Jumper, S., Bell, T., & Ralston, B. (1980). *Economic growth and disparities.* Englewood Cliffs, NJ: Prentice Hall.

H

HEALTH AND HEALTHCARE, GEOGRAPHY OF

Geographers studying health and healthcare often employ concepts from contemporary social theory such as structuralism, humanism, poststructuralism, and postmodernism to examine relationships among health, health services, and places. The subdiscipline was developed during the mid-1990s as a reform of medical geography. Those calling themselves health geographers, while accepting the usefulness of medical geography, argued for an evolution away from a focus on the biomedical model and toward a socio-ecological model. At the same time, there was a call for a renewed focus on the reciprocal interactions between health and place. Places were seen not as locations within a spatial network with little individuality or meaning but rather as sites where health was negotiated among competing forces, where beliefs and feelings were expressed, and where differences in health and healthcare were manifest.

As opposed to the intensive quantitative methods favored by medical geographers, health geographers began to employ more extensive qualitative methods such as participant observation, semistructured interviews, and recording detailed narratives. They are sensitive to their role as the observer in relation to those being observed. Examples of some of the disciplines or subdisciplines that interested them include social geography, psychotherapy, environmental perception, medical anthropology, cultural studies, and comparative literature. Because the research of health geographers is informed by a mixture of social theories, it is difficult to categorize their work. However, in what follows, ongoing studies are related to several strands of thinking.

THE INFLUENCE OF STRUCTURALISM

A major concern in health geography is the inequalities in health and healthcare created by divisions within society such as class, gender, ethnicity, and income levels. Dominant groups (e.g., men, whites, corporations, the wealthy) are favored over their counterparts (e.g., women, minorities, factory workers, the poor). The less favored become the "other" and are marginalized. The medical profession *medicalizes* healthcare; that is, it imposes its explanatory models of disease and treatment on laypeople. The capitalist economic system and moves toward privatizing healthcare tend to increase disparities. The result is the creation of deprived unhealthy societies.

Health geographers seek to uncover the structural forces that lead to inequalities. Of particular concern in recent work is healthcare consumerism, that is, the move to appeal to the public to buy a range of healthcare products that include physician's services, expensive tests, and private rooms in hospitals outfitted like hotels. Those selling healthcare use sophisticated techniques to manipulate consumer preferences and value systems, as well as the structure of healthcare delivery systems, toward the consumption of their products.

THE INFLUENCE OF HUMANISM

Health geographers derive from humanistic geography notions about the importance of individuals' subjective experience of health and healthcare within specific places such as the home, the community, and a formal healthcare facility. They realize that place can have positive or negative meanings for people. People bring their beliefs about what causes illness and how it should be treated to healthcare situations. Healthcare geographers attempt to understand how people feel about their health and healthcare by spending a considerable amount of time with a few individuals to get inside their heads. Examples of situations that have been examined include questioning residents of a deprived inner-city neighborhood at length about their health beliefs and practices, asking users and staff of a mental hospital about specific design features that are conducive or not conducive to well-being, and finding out how guests at a respite center relate to the physical landscapes of the place.

Stemming from the humanistic tradition is an interest in the role of symbols in health and healthcare. Symbols create meaning as we relate them to our values and beliefs. Thus, our respect for physicians may be enhanced by their white coats, which are associated with purity, cleanliness, and honesty, or we may welcome or fear high-tech equipment in a hospital depending on our attitudes toward technology in general. Language is a symbol system that plays a very important role in, for example, encounters between patients and doctors. Health geographers need to pay close attention to the words people use in telling their stories about their health experiences.

THE INFLUENCE OF POSTSTRUCTURALISM AND POSTMODERNISM

An important area of thinking within health geography involves a focus on the identities that individuals attach to the human body. French philosopher Michel Foucault spoke of the body as an object of power relations; that is, people attempt to control bodies that have been marked or labeled as, for example, homosexual, disabled, or female. Biomedicine often constructs the deviant body, that is, one that falls outside its definition of what is healthy. Ideas such as these inform studies of topics such as staff attitudes toward patients in a mental hospital and the ways in which women with multiple sclerosis renegotiate their movements within their homes as their bodies undergo debilitating changes.

Poststructuralist concepts emphasize power relations that affect social relationships among individuals and between individuals and society. Within the area of health, power is manifested in the control of healthcare resources (e.g., drugs for human immunodeficiency virus/acquired immunodeficiency syndrome [HIV/AIDS] treatment), in the domination of healthcare spaces or territories (e.g., by excluding those without health insurance from a hospital), and in surveillance (e.g., locating a nursing station in a mental hospital ward so that patients are always in view). Health geographers are aware that medical knowledge is also power and can be used to dominate and control.

The importance of social and cultural differences is brought out in postmodern theory. Notions of difference have led health geographers to examine how people characterized by physical or mental disability, gender, sexuality, and ethnicity have different health beliefs, practices, and experiences. Informed also by feminist theory, women's health has become a focus of attention. Recent work has looked at the effects of women's roles and the physical environment on their health, gender discrimination in obtaining healthcare in developing countries, and attempts by women to have home births. In the area of sexual orientation, health geographers have carried out important studies that incorporate an understanding of the social construction of HIV/AIDS and the social and cultural networks in which gay men carry out their daily activities.

—*Wil Gesler*

See also AIDS; Body, Geography of; Critical Human Geography; Disability, Geography of; Ethnocentrism; Feminist Geographies; Humanistic Geography; Medical Geography; Other/Otherness; Postmodernism; Poststructuralism; Power; Qualitative Research; Social Geography; Structuralism

Suggested Reading

Curtis, S. (2004). *Health and inequality: Geographical perspectives.* Thousand Oaks, CA: Sage.

Gesler, W., & Kearns, R. (2002). *Culture/Place/Health.* London: Routledge.

HEGEMONY

Hegemony in common use means domination or authority over others. As the term was conceived by the Italian Marxist Antonio Gramsci, hegemony implies domination by consent, particularly the domination of subordinate classes by the ruling class. Hegemony stands in contrast to direct forms of domination such as force, persuasion, coercion, and intimidation. Instead, hegemony is achieved through cultural institutions whereby the interests of the dominant class are expressed as the interests of all classes. Hegemony conveys the power of social, political, and economic systems to produce the consent of subordinate classes to interests of the dominant class. The process by which dominant class interests become naturalized is known as hegemony, and it is through hegemony that the power of the dominant class is maintained.

Gramsci credited V. I. Lenin, the Russian leader and Marxist theorist, with the original conceptualization of hegemony, but it was Gramsci who explored the cultural aspects of the idea in his *Prison Notebooks.* Gramsci understood the control of civil society as deriving from the pairing of hegemony—meaning the political and cultural leadership of subordinate classes by the dominant class—with direct domination through the force of the state. Gramsci developed the notion of hegemony as an explanation for how dominant classes continue to further their own interests at the expense of, and with the participation of, subalterns (Gramsci's term for subordinate classes). He was interested in how a counterhegemony of subordinate classes might be sustained and eventually overthrow the hegemony of the dominant class.

Hegemony is achieved through individual participation in the activities of everyday life and culture. Subalterns come to accept ruling-class values and attitudes as natural and appropriate by engaging in ordinary political, social, and economic institutions (e.g., schools, media, markets, political parties). For Raymond Williams, a British social theorist, hegemony was culture in its deepest sense. As a form of control, hegemony encompasses the whole of lived practices. It is through everyday experiences that hegemonic processes are created and repeated. Hegemony is something in which individuals are fully immersed and which they re-create in their daily lives. The concept and process of hegemony may be understood as infiltrating all aspects of social life and relationships. Due to the extent to which hegemony penetrates everyday experiences, those who are immersed in it come to reproduce it.

Hegemony differs from active persuasion or coercion insofar as subjects come to accept it through their everyday activities and interactions. Ordinary social institutions and relationships are infused with the interests and values that validate the dominant position of the ruling class. By participating in social life, the interests of subordinate classes fall in line with those of the dominant class and subordinate classes becomes invested in maintaining and reproducing the interests of the dominant class as if they were their own. By accepting the worldviews and values of the dominant class, the subordinate classes acquiesce to their social and political leadership, allowing the ruling class to maintain its dominance.

However, care must be taken not to elevate hegemony to a totality (the whole of reality) or an ahistorical (timeless) form that denies its great flexibility and currency. Rather, hegemony is made and remade in everyday life through habitual practices. Daily activities are productive of and situated within hegemony; hegemony is realized as lived experience. All political, social, and economic activities comprise and re-create hegemonic processes that constitute individuals, their knowledge of the world, and their social relationships. In daily life, ways of knowing and acting in the world appear as common sense, but from a critical perspective these processes and relationships may be seen as both producing hegemony and being products of hegemony.

Hegemony is something actively renewed and recreated, just as it is continually resisted and limited. Significant oppositional forms to hegemony always exist in society. Gramsci allowed that hegemony includes elements working against it that he termed *counterhegemonies.* He argued that one of the strengths of hegemonic classes is their ability to co-opt, extend, and incorporate the demands of subordinate classes to continue to maintain their dominance. Hegemony is not passive but must constantly engage with emerging counterhegemonies that may be quite specific to time and place. Counterhegemonies indicate what hegemonic processes must control because hegemonic processes must respond to the oppositions that threaten them for hegemony to be maintained. Hegemony exists even as elements of counterhegemony circulate among subordinate classes.

Gramsci gave particular attention to the roles of intellectuals in the creation of hegemony and distinguished between (a) traditional intellectuals whose philosophy and work serve the ruling class and perpetuate unequal class relations and (b) organic intellectuals who emerge from the underclasses and work toward establishing a counterhegemony of marginalized groups. Gramsci was optimistic that through political education, subordinate classes would begin to act in their own self-interests. With the help of organic intellectuals, a subaltern counterhegemony would advance to undermine the power of the ruling class.

Hegemony as a concept shares commonalities with French Marxist Louis Althusser's conceptualization of ideology. However, hegemony contrasts with ideology (a system of ideas) in its wholeness as lived daily practices. The notion of hegemony as expressed in the *Prison Notebooks* was Gramsci's key contribution to Marxist thought. Scholars continue to struggle to understand thoroughly what Gramsci meant by his multiple uses of the term *hegemony* throughout his writing. Difficulties with translation from Gramsci's Italian, the reconditeness of the concept itself, and Gramsci's own need to avoid censorship of the *Prison Notebooks* all complicate the study of hegemony.

—*Kathleen O'Reilly*

See also Class; Class War; Colonialism; Discourse; Ideology; Marxism, Geography and; Radical Geography

Suggested Reading

Gramsci, A. (1971). *Selections from the prison notebooks of Antonio Gramsci* (Q. Hoare & G. Smith, Eds. & Trans.). New York: International Publishers.

Mouffe, C. (Ed.). (1979). *Gramsci and Marxist theory.* London: Routledge.

Williams, R. (1977). *Marxism and literature.* Oxford, UK: Oxford University Press.

HETEROSEXISM

Whereas homophobia has been understood as the overt hatred of, and discrimination against, gay men and lesbians, heterosexism can be understood as the assumption that heterosexuality is the only form of sexuality and/or is better than any other form of sexuality. Much of the literature on heterosexism has been developed in social psychology. Heterosexism is based on an understanding of the operations of power that contends that it is a result of social conditioning and upbringing rather than an individual pathology that is associated with homophobia. Heterosexism, heteronormativity (the normalization of gender and sexuality within heterosexual constructs of men and women), and homophobia tend to be conflated in

geographic inquiry. However, although these are interlocking forms of discrimination, they have developed in different disciplines and offer different insights into the workings of heterosexual power. This entry outlines the presumption of heterosexuality in geographic inquiry and practice and the spatial manifestations of heterosexism.

Geographic inquiry began, in a number of ways, from heterosexist presumptions, for example, the use of the stereotypical family (a woman, a man, and children) as a unit of analysis. In perpetuating the myth of universal heterosexuality through the use of such categories, these analyses have overlooked/ ignored alternative sexual lifestyles. By assuming that those who do geography are heterosexual, geographic practices can also render nonheterosexual lifestyles and identities invisible. For example, those going on field trips can be heterosexist in presupposing that all students/staff will be straight and, therefore, in segregating on the basis of gender/sex in room allocations.

Another key aspect of heterosexism is the hierarchization of sexualities, in particular the belief in the inherent superiority of heterosexuality and thus its inherent right to dominance. Spatial heterosexism can be seen where space is presumed to be straight and other forms of sexuality are (re)formed inferior and out of place so as to maintain the artifice of normalized heterosexual space. Heterosexism often works in subtle ways, and apparent tolerance of difference may hide the processes that (re)make the hegemony of heterosexuality. Heterosexism does not necessarily include the assumption that nonheterosexualites are deviant, yet it can be seen where the presence of alternative sexualities are acknowledged but rendered less important and less desirable than heterosexuality. The everyday nature of these taken-for-granted assumptions renders heterosexuality allegedly better than other forms of sexuality.

Heterosexism is manifest spatially where commonsense norms often regard space as implicitly heterosexual, validating displays of affection between men and women. In this way, the dominant sexuality in everyday spaces often is assumed to be heterosexuality. Nonheterosexual displays are policed through verbal comments, stares, and so on such that difference is noted and degraded (e.g., when nonheterosexual displays are considered flaunting it). Through these processes of surveillance, along with self-surveillance, space is *made* gay or straight through relations of power that hierarchize sexualized performances. The repetition of heterosexual performances creates the *illusion* of space as preexisting and as "naturally" heterosexual, thereby "invisiblizing" the sexualized power relations, such as heterosexism, that make it as such. Discourses and practices that (re)make heterosexuality superior to its other (nonheterosexuality) might not be named and might fall outside the remit of "heterosexism." However, heterosexuality continues to be naturalized and hierarchized in the (re-)creation of everyday spaces.

—*Kath Browne*

See also Feminisms; Gays, Geography and/of; Homophobia; Lesbians, Geography of/and; Queer Theory; Sexuality, Geography and/of

Suggested Reading

Bell, D., & Valentine, G. (Eds.). (1995). *Mapping desire: Geographies of sexualities.* London: Routledge.

HIGH TECHNOLOGY

High technology broadly refers to the advanced expertise and techniques that are used to directly address, examine, and solve specific tasks. It is manifested not only in products but also in processes. In general, the industries involved in high technology require higher levels of human capital than do other sectors given the high amounts of research and development required. In recent use, high technology generally has encompassed industrial categories such as electronics, telecommunications, information technology, and biotechnologies. However, the applications of high technology now extend to and influence a number of fields and industries, including those throughout manufacturing and services.

High-technology industries are viewed in many circles as drivers of regional economic development. Many centers of recent growth, such as Silicon Valley in the San Francisco Bay Area and the Route 128 area around Boston, have been based around concentrations of high-technology activities, including software research and development and (earlier) the manufacture of information technology hardware. Many researchers have suggested that positive externalities result when similarly focused firms, customers, workers, and other support activities are concentrated in proximity to one another. This colocation in turn spurs

competition and innovation, which drive progress in the field. In a sense, these regions of high technology often are viewed as the traditional industrial districts once were viewed.

Advances in high technology have made it possible for individuals, groups, and enterprises to communicate swiftly with each other across great distances. Technical progress and the cost reductions in information creation, management, and dissemination that often result from such progress have also been benefits of these advances. In most respects, high technology has led to many parts of the world essentially becoming closer together due to enhanced methods of communication. In many ways, the results of these new arrangements and relationships have led to a reevaluation of space and time as described by David Harvey. An issue to note is that those regions and groups with greater access to high technology can benefit, whereas those that do not have access to these technologies have, in effect, become more distant from the rest of the connected world. These spatial relationships and the impacts of technology are of interest to geographers.

—*Ronald Kalafsky*

See also Agglomeration Economies; Economic Geography; Flexible Production

Suggested Reading

Bresnahan, T., & Gambardella, A. (Eds.). (2004). *Building high-tech clusters: Silicon Valley and beyond.* New York: Cambridge University Press.

Saxenian, A. (1996). *Regional advantage: Culture and competition in Silicon Valley and Route 128.* Cambridge, MA: Harvard University Press.

HISTORIC PRESERVATION

Local and federal efforts to save historic properties, landscapes, and landmarks from demolition or dramatic alterations have become pervasive throughout contemporary American society. This historic preservation movement has coincided with a renewed interest in national and local heritage, perceived as threatened by fast-paced development since the 1960s. From initial interests in creating house museums during the 19th century, the realm of preservation now includes the conservation of entire neighborhoods and commercial districts, contrived outdoor museums, national historic landscapes, and festival marketplaces where derelict industrial structures are converted into venues for middle-class consumption. The consequential impacts of preservation practices on our human landscapes, economy, and society have attracted the attention of academics representing the disciplines of human geography, sociology, anthropology, history, architecture, urban design, and so on. Human geographers often must consider the preservation process in their understanding of tourism landscapes, downtown redevelopment, neighborhood gentrification, postmodern consumption, and economic globalization and localization.

Early federal involvement in preservation was manifested in the Antiquities Act of 1906 and in the creation of the National Park Service in 1916. The former allowed for the creation of national monuments to prevent the wholesale destruction of prehistoric remains in the American Southwest and elsewhere, whereas the latter added a newfound significance to preservation efforts in the West and Southwest and recognized the importance of environmental conservation. After intense pressure to do so, Congress in 1949 ultimately chartered the National Trust for Historic Preservation, the only national organization developed and sustained by the preservation movement. Subsequently, Congress passed the Historic Preservation Act of 1966, which provided for the creation of the National Register of Historic Places and encouraged the establishment of state historic preservation offices (SHPOs). In 2000, the National Register included some 70,000 listings, many of them historic districts, totaling more than 1 million individual properties.

SHPOs are responsible for conducting surveys of historic properties within their states. They also process nominations to the National Register and assist local communities and individuals with the nomination process. SHPOs further administer grants to projects and serve as a funding conduit from the federal level to the local level. Within municipalities, state-enabling legislation allows communities to establish local historic preservation ordinances, or zoning overlays, which can provide a variety of guidelines and restrictions on property alterations within the designated districts.

Local debate surrounding historic preservation initiatives is rooted in the American cultural value of maintaining private property rights, individualism, and the free market economy that assumes little government involvement. Preservation efforts typically are perceived as restrictive to individual ownership rights and economic growth. Conversely, preservation proponents highlight a consistent pattern of increased property values and capitalist investment within and around designated historic districts, indicating that preservation efforts can actually encourage economic redevelopment. The preservation movement is only increasing in strength as communities desire to remain in touch with their material heritage following decades of modern aversion to studying or appreciating the past.

—*Thomas Paradis*

See also Cultural Landscape; Postmodernism; Tourism, Geography and/of; Urban and Regional Planning; Zoning

Suggested Reading

Murtagh, W. (1993). *Keeping time: The history and theory of preservation in America.* New York: Sterling.

Tyler, N. (2000). *Historic preservation: An introduction to its history, principles, and practice.* New York: Norton.

HISTORICAL GEOGRAPHY

Historical geography, a branch of human geography that seeks to understand geographies of the past and often how the past impinges on the present, encompasses a broad range of scholarly activity. Practitioners of historical geography, deriving their theoretical perspectives, subject matter, and methodological tools from both history and geography, have long worked at the boundary of these two academic disciplines. Historical geography is, moreover, a hybrid approach and a series of concerns; it is an interdisciplinary field of inquiry that examines landscapes, environments, spaces, and places historically as well as how those geographies change over time.

The interdisciplinarity that always has characterized historical geography makes it especially relevant today. At a time when disciplinary borders are becoming ever more blurred and when dialogue across specializations is increasingly emphasized, historical geography is well positioned to reap the rewards of much fertile scholarship. Indeed, more and more scholars are identifying themselves as historical geographers, a trend seen in memberships of professional organizations, in presentations at learned meetings, and in scholarly publications. Equally important is the more general tendency to consider time as fundamental to geographers' craft. Time and space—conceptual siblings that some previously considered the sole preserves of history and geography, respectively—are thoroughly entwined social constructions that cannot exist independently. It is not surprising, then, that much of the most exciting work in the humanities and social sciences today occurs precisely at the borderlands of these two disciplines.

This work is marked by a liberal eclecticism that defies simplistic categorization. Over the past 20 years especially, the specific themes and approaches of historical geography have diversified along with human geography more generally. For some historical geographers who worry about a possible lack of intellectual coherence, this eclecticism is a source of concern, whereas for others, it is evidence of intellectual vigor and excitement.

HISTORICAL ROOTS OF CONTEMPORARY HISTORICAL GEOGRAPHY

The eclectic pluralism that characterizes historical geography today did not emerge out of a vacuum but instead arose from a century-long encounter between historical and geographic thinking. To understand the diverse nature of contemporary historical geography, it is first necessary to examine its historical roots, vestiges of which are still evident today. Those roots by no means follow a straightforward path, nor do they stem from one source. Rather, historical geography developed in a succession of overlapping periods of innovation from several intellectual strands. At times, historical work within human geography garnered attention as one of its central subfields; at other times, a historical perspective was dismissed for its perceived lack of explanatory power and antiquarianism. Such debates aside, each of these periods is marked by an impressive and steady increase in the work of historical geography.

Even before history and geography became fully professionalized in university settings during the latter half of the 19th century, scholars worked at their interface. None did so with greater insight or rhetorical force than the lawyer, manufacturer, congressman, and diplomat George Perkins Marsh. His 1864 book, *Man and Nature: Or, Physical Geography as Modified by Human Action,* profoundly reshaped Americans' attitudes toward the natural environment as he used world history as a tool for understanding environmental degradation in the United States. No less original than Marsh's arguments about the destructive effects of economic development on the American environment were those of historian Frederick Jackson Turner, who was also alarmed by the changes in the land. Environmental damage was less of a concern for Turner than were questions of national experience and character, which he insisted were founded on settlement geography. In his famous 1893 essay "The Significance of the Frontier in American History," and throughout his long career, Turner argued that American national character was forged in the settlement frontier—the ever retreating zone at the edge of the country's populated core.

Although neither Marsh nor Turner was a geographer, both worked at the borderlands of history and geography and both were concerned with the entangled nature of time and space. Arguably the first scholar trained as a professional geographer to approach the field in a distinctly historical manner was Ellen Churchill Semple, whose *American History and Its Geographic Conditions,* published in 1903, launched a new school of thought quite at odds with both Turner and Marsh. For Semple, the physical landscape demonstrably and decisively influenced American social and cultural life to such an extent that the environment became *the* determining factor in history. Although such a view may have been helpful in delineating the utility of the young discipline of geography, it was founded on pseudoscientific theories of racial hierarchy and more on speculation than on empirical research. Such work, at its extreme, offered blatant support for environmental determinism.

Although it is unclear how many historically inclined geographers became active proponents of environmental determinism, certainly enough did to make it the dominant paradigm during the 20th century's first three decades. By the 1930s, however, this environmentalist school of thought was losing ground as discussions of environmental control or influence gave way to descriptions of adaptation and response, terms that connoted a greater role of human agency in what was called man–land relations. A watershed moment in the refutation of environmental determinism, and in the maturation of historical geography, came with the 1932 publication of Charles Paullin's *Atlas of the Historical Geography of the United*

(a)   (b)

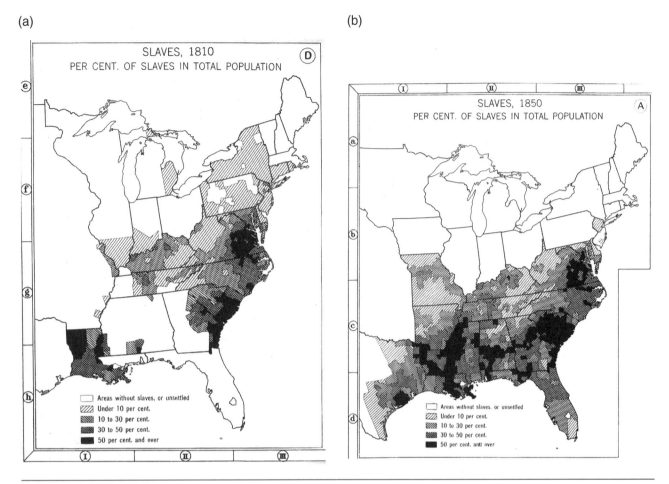

Figure 1   Percentages of Slaves in Total Population, 1810 (a) and 1850 (b)

SOURCE: Paullin, C. O. (1932). *Atlas of the historical geography of the United States* (J. Wright, Ed.). Washington, DC: Carnegie Institute; New York: American Geographical Society. Reprinted with permission of the American Geographical Society.

States, a tour de force nearly three decades in the making. With its emphasis on human initiative, social relations, and the dynamic interaction with the environment, Paullin's *Atlas* found more inspiration in the historian Turner than in the geographer Semple (Figure 1). Its maps and commentary emphasized social and economic life at the expense of traditional military history. Paving the way for future projects—most notably the three-volume *Historical Atlas of Canada* published between 1987 and 1993—the 1932 *Atlas* advanced the argument that historical cartography was not only a useful pedagogical tool but also a vital analytical method for historical geography.

The next several decades represented foundational, and controversial, years in the development of historical geography. On the one hand, important geographers such as Richard Hartshorne relegated historical geography to the outer fringes of the subject, insisting that

time, as the unique province of the historian, is unsuitable for geographic study. Others, such as English geographer Henry Clifford Darby, viewed historical geography as a pillar of the field. In the United States, Hartshorne's view probably held greater currency, at least until its persuasive refutation by Carl Sauer.

Already well known for his advocacy of anthropological methods and cultural landscape as central to geographic study, Sauer also emphasized the role of historical time in creating distinct cultural areas or regions. He used the occasion of his 1941 presidential address to the Association of American Geographers, "Foreword to Historical Geography," to deepen his commitment to an overtly historical approach. For Sauer, all geography was genetic; that is, it must be historical in the broadest sense. This argument positioned historical geography at the center of the field and underscored the importance of geographic diffusion as the best method of gauging the development of cultural traits. Although Sauer himself successfully applied his historical approach to geographic studies of Latin America, arguably his greatest legacy lay in its influence on his many students and on those geographers inspired by his interest in past geographies.

Fieldwork, a direct outgrowth of Sauer's empathy for anthropological methods, became the chosen way in which to gather historical–geographic data. In some cases, fieldwork consisted of archeological studies of human-induced vegetation change, particularly by indigenous peoples; at other times, fieldwork meant scouring the countryside for evidence of cultural implantation and transfer, especially in pioneer settlement. Among the students most active in promoting fieldwork for historical–geographic study was Fred Kniffen, who traced the origin and spread of material culture meticulously from rural house types and covered bridges to Native American groups and agricultural fairs (Figure 2).

Equally important for the long-term development of historical geography, but departing from an emphasis on fieldwork, was a growing interest in primary historical sources. Although some geographers, including Sauer, had used original documents previously, the first sustained commitment to their close study came from someone unconnected with the Berkeley School, namely Ralph Brown. In his 1943 book *Mirror for Americans* and his 1948 book *Historical Geography of the United States,* Brown eschewed any interest in relic material evidence in favor of primary historical sources. By adopting the methods and professional techniques of another discipline—history—to write historical geography, Brown cut a maverick path.

Despite the originality of Brown's work, his influence remained limited; much of it seems, in retrospect, antiquarian and narrow in scope. It took another historical geographer, Andrew Clark, to promote a more historically oriented historical geography, one that relied on the geographic concept of region but that also depended heavily on the methods and source materials of the discipline of history. But what really set Clark and his students at the University of Wisconsin apart was their emphasis on historiographic issues that lent themselves to geographic investigation. It was not enough, for instance, to merely describe past geographies and how they changed over time as Brown had done. Instead, Clark's method of studying the geography of change necessitated examining issues currently debated by historians, thereby ensuring that a great deal of historical geography took on a revisionist nature. The resulting scholarship—especially by Clark's many students and by scholars such as Carville Earle—often found its primary readership among professional historians as much as among geographers and frequently focused on pioneering settlement. James Lemon's 1972 book *The Best Poor Man's Country,* the winner of the American Historical Association's Beveridge Award for best book in American history, revised historical interpretations of community development and liberalism during the colonial period.

At a time when human geography moved strongly in the direction of model building, abstraction, and quantitative analysis, historical geography provided a haven for those who were uneasy with the positivist nomothetic paradigm of the day. This is not to say that all historical geographers eschewed quantitative methods; for example, many from the Wisconsin School found the techniques of spatial science helpful in mapping geographic change, and today some historical geographers are using sophisticated geographic information systems (GIS) mapping techniques to display historical data. But to understand human experience and intentionality, numerical data were of little use. Thus, many historical geographers found inspiration in humanistic geography's emphasis on subjectivity and worked to forge greater connections with other fields in the humanities, especially psychology, art history, and philosophy. A great deal of humanistic-inspired historical geography focused on the complexly woven tapestry of meaning inherent in ordinary landscapes.

THE INTERDISCIPLINARY ECLECTICISM OF CONTEMPORARY HISTORICAL GEOGRAPHY

If earlier moments of historical geography can be identified by significant schools of thought and roots of influence, today's eclectic pluralism resists such identification. Interdisciplinarity—long an implicit characteristic of the subfield—has become the self-conscious model for scholarship. Partly the result of trends more generally in the social sciences and humanities and partly the result of a thoroughgoing engagement with social theory, historical geography today is as diverse as the larger field of human geography; what has remained constant is the entwined coexistence of time and space in these studies. Among the intersecting themes that have emerged in this recent scholarship, four are especially prominent.

(a)

(b)

Figure 2   (a) "Built-in Porch Type House" Line Drawing (Figure 3) and (b) "Distribution of Built-in and Mid-western House Types" Map (Figure 13)

SOURCE: From Fred B. Kniffen, "Louisiana House Types," *Annals of the Association of American Geographers, 26* (1936): 179–193. Reprinted with permission of Blackwell Publishing.

Modernity and Power

Much recent work in historical geography is informed by a host of theoretical directions, including poststructuralism and postcolonialism, that signal concerns about questions of power. The most fundamental implication of this trend—much of which is indebted to the social theory of Michel Foucault—is that social power no longer can be conceived apart from its historical or geographic context; rather, it is integral to the formation of those historical geographies. Similarly, most scholars recognize that time and space dramatically alter those very power relations. Thus, discussions of European settlement in North America have given way to analyses of imperial conflict; Indian occupation and subjugation are given equal treatment with white colonialism. One of the most notable and ambitious scholars to rewrite American historical geography as a story of colonial encounter is Donald Meinig, whose four-volume *Shaping of America* interpreted 500 years of history not as benign westward expansion but rather as geographic change and tension predicated on the convulsive interaction between two unequal peoples. Writing at a much smaller geographic scale but with an equally long time frame, Cole Harris narrated the conquest of Canada's province of British Columbia as one of racism and imperialism and one implicated directly in larger global forces.

Extending beyond the regional frame to embrace both global and local scales is a hallmark of recent historical geography, especially as the modern world developed. Indeed, modernity has become a central organizing concept, as one finds in studies of the modern world economy and political power. Similarly, historical geographers have shown the importance of specific times and places, such as 18th-century London and 19th-century Paris, for the formation of modernity itself.

Identity

The interdisciplinary eclecticism so characteristic of contemporary historical geography is also found in the many historical–geographic investigations of identity and of the social or cultural construction of spaces implicit in those identity formations. Here the impact of theoretical positions derived from feminism, racial formation, and nationalism has been especially important. Indeed, the increasing importance of feminist theories and methods is quite striking, especially given the long and nearly complete absence of material on women in North American historical geography. Gender differences—especially as those differences are worked out through cultural, political, economic, and sexual differences—are increasingly seen as central to the creation of past geographies, whether in settlement and rural contexts, as Jeanne Kay showed,

or in the modernizing city, as Mona Domosh demonstrated. Thus, the historical geography of identity formation necessitates an examination of gender roles and differences. It also, and quite relatedly, must account for how race is understood and created. Wide-ranging studies, such as Kay Anderson's analysis of Chinatown in Vancouver, British Columbia, and Steven Hoelscher's analysis of African American segregation in Mississippi, demonstrated how historical racial categories and prejudices are created and sustained in geographic space.

If identity is located at the most personal level of one's gender, race, or sexuality, it can also be situated at the more distant or abstract level of the nation. Historical geographers have been especially active in showing how national identity has been created and sustained by museums and monuments, by the visual arts of painting and sculpture, by architectural design, and by the census. Especially potent for the successful realization of nationhood and economic development is the idea of heritage, that is, the contemporary use of the past for current purposes.

Representation

Although the material cultural landscape remains a vital subject for historical geographers, increasing numbers are forging entirely new paths by tracing the symbolic representation, or iconography, of past environments. Drawing from visual culture studies and on the theoretical perspectives of postmodernism, this work examines landscape as a cultural image, a pictorial way of representing, ordering, or symbolizing the world. For some, such as Denis Cosgrove and Stephen Daniels, landscape is a historically and culturally specific way of seeing and structuring the environment that, over the years, has helped to maintain class hierarchy. Such a perspective refuses to interpret landscape as a naively given object of external reality; rather, representations of landscape quite often serve distinct political and economic interests by hiding social relationships from view and by making what is represented seem natural.

Representations of landscapes come in a variety of forms, including travel narratives, photographs, maps, and paintings—all of which have been interrogated for the ideologies they communicate and the politics they support. Thus, J. B. Harley, in a series of pathbreaking essays, showed how maps can be deconstructed for their complicity in enabling projects of empire building and colonial rule. And Stephen Daniels examined famous landscape paintings of the American West, such as Francis Palmer's *Across the Continent* (Figure 3), for what they communicate about complex ideas of manifest destiny, American nationhood, and corporate expansionism.

Human–Environment Relations

The fourth area of critical interest among historical geographers is, in many ways, the one most directly connected with earlier traditions. Innovative work on indigenous and colonial land use, on the transformation of rural and wild environments during the Industrial Revolution, and on the roles of cultural attitudes, values, and ideas associated with ecological change owes much to a century of earlier research traditions. And yet here too new directions and new influences have enlivened the subfield, taking historical geography well beyond its previous focus on environmental influence and modification.

One of the most significant new influences comes from environmental history, an emerging subfield that shares many research problems, theoretical views, and methodologies with historical geography. Together, environmental history and an environmentally focused historical geography explore global expansion and the capitalist economy, natural resource exploitation and dependency, environmental justice, and urban ecological degradation. In foregrounding politically charged issues such as environmental quality, as in Craig Colten's study of drainage in New Orleans, and by examining the way in which differing political–economic systems create systemic environmental problems, historical geographers have mapped new ways in which to study past geographies.

—*Steve Hoelscher*

Figure 3   *Across the Continent: Westward the Course of Empire Takes Its Way*

SOURCE: Drawing by Francis F. Palmer and originally published by Currier and Ives in 1868.

See also Berkeley School; Cultural Geography; Cultural Landscape; Feminist Geographies; Fieldwork; Globalization; History of Geography; Humanistic Geography; Marxism, Geography and; Modernity; Nationalism; Postcolonialism; Power; Race and Racism; Theory

Suggested Reading

Baker, A. (2003). *Geography and history: Bridging the divide.* Cambridge, UK: Cambridge University Press.

Clark, A. (1954). Historical geography. In P. James & C. Jones (Eds.), *American geography: Inventory and prospect* (pp. 70–105). Syracuse, NY: Syracuse University Press.

Conzen, M. (1993). The historical impulse in geographical writing about the United States, 1850–1990. In M. Conzen, T. Rumney, & G. Wynn (Eds.), *A scholar's guide to geographical writing on the American and Canadian past* (pp. 3–90). Chicago: University of Chicago Press.

Cosgrove, D., & Daniels, S. (Eds.). (1988). *The iconography of landscape: Essays on the symbolic representation, design, and use of past environments.* Cambridge, UK: Cambridge University Press.

Domosh, M. (1998). Those "gorgeous incongruities": Polite politics and public space on the streets of nineteenth-century New York City. *Annals of the Association of American Geographers, 88,* 209–226.

Earle, C. (1992). *Geographic inquiry and American historical problems.* Stanford, CA: Stanford University Press.

Graham, B., & Nash, C. (Eds.). (1999). *Modern historical geographies.* Harlow, UK: Longman.

Harley, J. (2002). *The new nature of maps: Essays in the history of cartography.* Baltimore, MD: Johns Hopkins University Press.

Harris, C. (2002). *Making native space: Colonialism, resistance, and reserves in British Columbia.* Vancouver: University of British Columbia Press.

Harvey, D. (2003). *Paris, capital of modernity.* New York: Routledge.

Hoelscher, S. (2003). Making place, making race: Performances of whiteness in the Jim Crow South. *Annals of the Association of American Geographers, 93,* 657–686.

Kay, J. (1991). Landscapes of women and men: Rethinking the regional historical geography of the United States. *Journal of Historical Geography, 17,* 435–452.

Matthews, G. (1990). *Historical atlas of Canada* (Vol. 3). Toronto: University of Toronto Press.

Meinig, D. (1986–2004). *The shaping of America: A geographical perspective on 500 years of history* (Vols. 1–4). New Haven, CT: Yale University Press.

Ogborn, M. (1998). *Spaces of modernity: London's geographies, 1680–1780.* New York: Guilford.

Sauer, C. (1941). Foreword to historical geography. *Annals of the Association of American Geographers, 31,* 1–24.

HISTORY OF GEOGRAPHY

Geography has a long and complex history stretching back to prehistory. Although physical geography also has a history, this entry focuses on human geography. Every society can be said to create a geography, both in the sense of an ontology (i.e., as material landscapes and spatial distributions) and in the sense of epistemology (i.e., as a worldview). As long as there have been people, there have been geographies. Australian aborigines, for example, used so-called song lines to navigate the desert. Sumerians developed clay topographies of their cities, and Polynesians crossed the Pacific Ocean with maps of currents and winds made from sticks.

PREMODERN GEOGRAPHIES

Prior to the rise of modern Western capitalism, geography had roots that extended to classical Greece in the sixth century. In general, premodern geographies were empirical and inductive in nature, often consisting of encyclopedic compilations of place descriptions. Geography was a practical science often intertwined with geodesy, astronomy, surveying, exploration, trade, and military conquest. Classical Greece, from which Western culture ostensibly arose (but with numerous connections to older cultures), marked the first systematic attempts to describe the shape of the earth and map the known world (or the *ecumene*). For example, Thales (611–547 BC), who lived in Miletus, theorized that the earth floated on water and successfully predicted an eclipse on May 28, 585 BC. Anaximander (610–546 BC) constructed what might have been the first map of the world (since lost), invented the gnomon, and argued that the earth and all bodies were spherical. Herodotus (485–425 BC) was a historian who coupled history with geography in his famous studies of the Nile River. During the Athenian golden age, Aristotle (384–322 BC), a scientist and philosopher, theorized a geocentric astronomical system that held sway until the 17th century. He also advocated an early form of climatic determinism based on three belts of temperature and their ability to sustain civilization; that to the south of Greece was too hot and that to the north was too cold, leaving Greece alone ideal for civilization. At the famous library of Alexandria, Eratosthenes (276–194 BC) coined the term *geography* and estimated the circumference of the earth remarkably accurately. Later, Hipparchus (190–120 BC) theorized a grid about the world consisting of latitude and longitude lines.

Among the Romans, prominent geographers included Strabo (64 BC–24 AD), who was actually Greek and is best known for his 17-volume work *Geography.* Ptolemy (87–150 AD), also working in Alexandria, concluded that the task of geography is to describe the earth as a whole, and he did just that in his 8-volume *Guide to Geography.* He differentiated between geography (the study of universals) and topography (the study of localities) and defined chorography as integrating the two.

For a millennium under feudal Europe, geography suffered from the political and ideological dominance of theology. Exploration during this time was relatively rare, excluding perhaps the Vikings and the famous trips to Asia by Marco Polo. A well-known medieval geographic expression consisted of T-in-O maps, *oriented* to the east (i.e., Jerusalem), which depicted the continents of Europe, Africa, and Asia in crude, highly inaccurate terms.

In contrast, during the height of the Arab Empire from the 7th century to the 12th century, geography prospered in much the same way as did Arabic mathematics, astronomy, poetry, and medicine. The Arabs traded extensively with India and discovered the dynamics of the monsoons over the Indian Ocean. Arab cartography was so advanced that Christian kings (e.g., Roger II in Sicily) hired Arab mapmakers such as Idrisi (1099–1180). During the 9th century, Caliph Harun al-Rashid assembled scholars to translate Greek works into Arabic. Prominent Arab adventurers included Ibn Batuta (1304–1369), who traveled more than 70,000 miles throughout Eurasia over 30 years. Abu al-Raihan al Biruni (972–1050), a Persian born near the Aral Sea, wrote the *Kitab al-Hind,* which first examined the erosion of the Himalaya mountains. Ibn Khaldun (1332–1406), a scholar for the sultan of Egypt, wrote a multivolume world history, the *Muqaddimah,* which explicitly included human–environment interactions (possibly the first to do so).

THE RENAISSANCE AND ENLIGHTENMENT

The gradual expansion of capitalism during the 15th, 16th, and 17th centuries provided an enormous stimulus to geography that was highly useful in charting avenues of exploration and conquest, mapping land uses and distributions of resources, and helping Europeans understand the worlds they were conquering. Thus, Portuguese king Henry the Navigator sponsored a series of voyages down the west coast of Africa, including Vasco de Gama's trip around the Cape of Good Hope in 1497. Ferdinand Magellan set out to sail around the world but died en route (one of his ships completed the voyage). Christopher Columbus, sailing for the Spanish, inadvertently discovered the Americas in 1492, opening a vast new chapter in the rise of the West. Scientific voyagers included James Cook in three voyages throughout the Pacific during the 18th century. Explorers and scientists brought back to Europe vast quantities of data and samples to Europe, much of which were organized, stored, and displayed in universities, museums, zoos, and botanical gardens.

During this era, cartography exploded in scope and sophistication. Many of the first maps produced then were nautical sea charts, including portolan charts of coastal areas. During the 1400s, Martin Behaim in Nuremburg created the first globe. During the Dutch

golden age when Amsterdam was a major commercial center, the first atlases appeared; these were so important that they often were national secrets. During the 16th century, Martin Waldeensmuller made the first map to show the Americas. Gerardus Mercator (1512–1594), a Flemish engraver, in 1569 created a famous projection that allowed the use of straight rhumb lines or constant compass bearing, which was highly useful for long-distance navigation.

The Enlightenment also saw several prominent geographers make a break with theology. Bernhardus Varens (Varenius) was a 17th-century German who wrote the *Geographia Generalis* (published in 1650), which served as a major textbook for the next 150 years and was translated into English by Isaac Newton. Varens distinguished between specific geography, which was concerned with the unique character of places, and general geography, which was concerned with universal laws. Immanuel Kant, generally known as a philosopher, was also a professor of geography at the University of Konigsberg for 40 years (1756–1797). Kant made numerous contributions to geography. His philosophical orientation attempted to resolve the debate between British empiricists and Continental rationalists, arguing that the mind is predisposed to the rational organization of sense data. Kant maintained that people never experience things in themselves (*nuomena*), only their sense impressions of them (*phenomena*). Thus, the world has no inherent preexisting order but rather is constructed by the mind. Time and space are categories created by the mind to make sense of nature; they are not phenomena themselves but rather ways in which to organize phenomena. Kant's contributions to geography included papers in physical geography and his avocation of a view of space as an abstraction separate from nature. Importantly, he held that space was on a par with time in explanation, a notion that would be smothered by the rise of historicism during the 19th century.

THE 19TH CENTURY

The Industrial Revolution, growth of literacy, rise in universities, and pragmatic needs for geographic information gave rise to a steady growth of geographic knowledge, including the founding of geographic societies (e.g., in France [1821], in Germany [1828], in Britain [1830]). The two leading geographic figures during this period were Alexander Von Humboldt (1769–1859), one of the 19th century's most important scientists, and Carl Ritter

(1779–1859). Von Humboldt is best known for his extensive travels in Latin America (1799–1804) and Siberia (1837–1842). He accumulated vast amounts of empirical data, including plants, volcanoes, and climate; was the first European to see the Orinoco River; discovered the Humboldt Current; was the first to use isolines such as isobars and isotherms; theorized the notion of continentality, in which climates far from the ocean behave differently from those near the ocean; discovered that air temperatures decrease with altitude; and conducted extensive studies of Native American tribes. Many of these observations were contained in his 5-volume work, *Cosmos* (1830–1859), which offered an agnostic holistic view of nature, arguing that the earth must be viewed as a unified organic whole. In contrast, Ritter taught geography at Frankfurt and Berlin, traveling some in Europe but not beyond. He is best known for his massive 19-volume *Erdkunde,* which emphasized the comparison and synthesis of regions as an expression of a religious teleology. Humboldt and Ritter often are argued to be the founders of the regional approach and scientific geography; however, they differed on several fronts. First, Humboldt was secular, whereas Ritter was religious. Second, Humboldt focused on the physical environment, whereas Ritter focused on human geographies. Third, Humboldt worked at a fine level of areal differentiation, whereas Ritter worked at the scale of continents.

The 19th century is also known for two famous anarchist geographers. Elisee Reclus (1830–1905) participated in the Paris Commune in 1871 and subsequently was imprisoned and banished. In his 19-volume *Nouvelle Geographie Universelle* and 6-volume *L'Homme et la Terre,* he exhibited a concern with social inequality, preservation of the environment, and town planning. Peter Kropotkin (1842–1921) was a Russian aristocrat who became an anarchist after serving as a military officer in Siberia. In 1885, he wrote the essay "What Geography Ought to Be" in prison, arguing that spatial hierarchies mirror and reinforce social ones. His philosophy emphasized "mutual aid" among decentralized self-governing communities.

THE DEVELOPMENT OF AMERICAN GEOGRAPHY

From its inception, the United States developed its own tradition of geographic thought and practice. Among Thomas Jefferson's many accomplishments was his great interest in geography. Jefferson sponsored the Lewis and Clark expedition, which sought a Northwest Passage to the Pacific along the Missouri River, as well as the township and range system of land ordinance, helping to bring the territories of the expanding United States into a rationalized Enlightenment frame of consciousness. The late 18th century also witnessed vast numbers of atlases, textbooks, gazetteers, and travel guides. Jedediah Morse (1761–1826), a Connecticut Calvinist preacher often called the father of American geography, wrote several best-sellers to assist Americans in understanding their own territory, including Native Americans; his efforts formed a vital part of the construction of nationhood.

American geography exhibited several characteristics that differentiated it from its European counterpart. First, it emphasized the physical environment, with close ties to geology and meteorology. Second, there were strong links between American geography and the ongoing colonization of the West; geographers were active in the so-called opening up of the West through their surveys, charts, maps, and photographs throughout the 19th century, transforming what was for Europeans a blank space into a mappable, knowable, and hence controllable space. Third, geography in the United States placed a great emphasis on induction and empiricism (in contrast to European theorizing), fieldwork, and applied pragmatic applications.

A significant figure in the field at this time was George Perkins Marsh (1801–1882), who held many occupations. While serving as U.S. ambassador to Turkey and Italy, he observed the impacts of deforestation. On returning home, he became one of the first voices to protest the destruction of American woodlands and became an early advocate of resource conservation, particularly with his book *Man and Nature,* published in 1864. His work was widely influential during the late 19th century, assisting in the formation of the National Park Service and national forests. In the same vein, John Wesley Powell (1834–1902), originally a schoolteacher from Ohio who lost an arm in the Civil War, sailed down the Colorado River through the Grand Canyon in 1869, recording Indian customs, mapping the physiographic regions of the arid Southwest, and assessing its potential for development. Later, he directed the U.S. Geological Survey. Lake Powell in Utah is named after him.

During the late 19th century, geography became progressively institutionalized as a discipline in professional societies and universities. In 1851, the

American Geographical Society was formed in New York, sponsoring expeditions and the nation's oldest geography journal, the *Geographical Review.* In 1888, the National Geographic Society in Washington, DC, began to serve scientists working for the federal government. Under energetic editor Gilbert Grovesner, *National Geographic* grew rapidly in popularity and subscriptions, helping to bring the geographic imagination of American foreign conquests home to the expanding middle class during and after the Spanish-American War of 1898. In the process, it became the unofficial voice of American foreign policy. By the 1920s, with more than 1 million subscribers, it was the largest educational society in the world, and its photos and texts exercised considerable influence over popular conceptions of geography.

GEOPOLITICS AND ENVIRONMENTAL DETERMINISM

During the late 19th century, classical geopolitics and environmental determinism—as opposed to encyclopedic empiricism—arose as geography's first true paradigm, marking a shift in the discipline from a collection of facts to a body of theory. This move was closely associated with the aftermath of the Darwinian revolution and the secular challenge to Christian dogma, although it also drew on the earlier (and erroneous) ideas of French biologist Jean Baptiste Lamarck (1744–1829). The incorporation of the Lamarckian variant of Darwinism centered on the notion that culture is carried biologically, thereby removing the random chance central to Darwinian theory and positing evolution in much more rapid terms than did natural selection. Social Darwinism came to privilege the biological over the social in the explanation of history and geographies, often legitimating inequality as natural and necessary, as in the works of historian Herbert Spencer. Not coincidentally, this view of competition and hierarchy occurred in the midst of the waves of European colonialism.

Classical geopolitics, a term coined by Swedish political scientist Rudolf Kjellen, was formalized in the works of Friedrich Ratzel (1844–1904), who gave formal expression to German fears of encirclement at the hands of Russia, France, and Britain. Sometimes called the father of human geography, Ratzel's *Anthropogeographie* maintained that nation-states could be viewed as organic biological organisms; if they were not growing, they were dying. In their

search for living space (or *lebensraum*), states inevitably would come into conflict (or *kulturkampf*). Thus, Ratzel's views biologized international relations and were read eagerly by later German leaders, including Adolf Hitler. Nazi geographer Karl Haushofer (1869–1946) was instrumental in this regard.

A significant representative of classical geopolitics was Sir Halford Mackinder (1861–1947), who taught at Oxford University and served as a member of Parliament. In *The Geographical Pivot of History,* written in 1904, he was one of the first theorists to conceive of the world as an integrated political system. His argument centered on the notion that the age of sea power was over and that railroads, in opening up the interiors of continents, were generating new geographies of centrality and peripherality. The heartland, in this reading, consisted of Eastern Europe and western Russia, a pivot area relatively inaccessible to marine powers.

As the first ostensibly scientific paradigm in geography, environmental determinism framed the world in terms of a hierarchy of competing races in which climate, continents, and ethnic groups were flattened to dichotomies such as civilized versus uncivilized. Geography textbooks often described dark-skinned peoples as indolent and lazy and in need of Western intervention. Ellen Churchill Semple (1863–1932), who studied in Germany with Ratzel, was a prime proponent of this view, arguing in books such as *The Influences of Geographical Environment* (published in 1911) that climate and topography exerted essentially uncontrollable influences over people. Ellsworth Huntington (1876–1947) and Griffith Taylor (1880–1963) wrote in much the same vein, a line of thought that culminated in the racist program of eugenics.

By World War I, environmental determinism was increasingly falling into disfavor under the withering attacks of cultural geographers such as Carl Sauer. Principal errors of environmental determinism included its omission of countervailing facts and selective use of evidence, its inability to account for social relations, its ridiculous racism, and its simplistic understanding of nature that itself has been heavily modified by people. Advocates of the alternative doctrine of environmental possibilism, which gave much more weight to the capacity of people to shape their own worlds, began to gain legitimacy. The legacy of environmental determinism, however, included a widespread retreat from theory in general and a bifurcation between human geography and physical geography that put nature off-limits to social analysis until this schism began to heal during the 1990s.

AMERICAN GEOGRAPHY DURING THE EARLY 20TH CENTURY

A key figure in American geography at the end of the 19th century and during the early 20th century was William Morris Davis (1850–1934). Best known for his work in geomorphology, Davis taught geography at Harvard University, where he had a long list of influential disciples. He became the first president of the Association of American Geographers when it was founded in Philadelphia in 1904.

Geography was found throughout the Ivy League universities during the early 20th century. The first geography department, however, was at the University of California, Berkeley, beginning in 1897. The first Ph.D. program was located at the University of Chicago, starting in 1903, and this department produced a long stream of famous graduates, including J. Paul Goode, Harlan Barrows, Gilbert White, Homer Hoyt, Derwent Whittlesey, Carl Sauer, and Richard Hartshorne.

Carl Sauer (1889–1975) looms large as the father of American cultural geography. Teaching at Berkeley for three decades (1923–1954), he offered a view of culture that avoided the ugly ethnocentrism that was pervasive in environmental determinism, advocating instead a relativist reconception that uncoupled moral superiority from military or economic might. Sauer's accomplishments include the introduction of the concept of *landschaft* (or landscape) into American geography as both a cultural and natural product, the study of cultural hearths where innovations such as agriculture began, his advocacy of fieldwork (he was an expert on Mexico), and his insistence that cultural geography take historical context seriously. Sauer's department generated a large number of influential academic geographers such as John Leighly, Fred Kniffen, Joseph Spencer, Andrew Clark, James Parsons, Carville Earle, Wilbur Zelinsky, and Marvin Mikesell. At Harvard, Derwent Whittlesey (1890–1956) developed the notion of *sequent occupance,* a technique for analyzing landscapes as the product of successive groups over time, with each leaving an imprint and forming a palimpsest.

Isaiah Bowman (1878–1950) was a geographer known mostly for his prodigious administrative and political accomplishments. He studied under Davis at Harvard and participated in the American Geographical Society expedition to Peru that revealed Machu Picchu in 1911. Later, he served as director of the American Geographical Society (1915–1935), taking time in 1919 to work with the American delegation to the Paris Peace Conference in 1919, where he played a key role in drawing the boundaries of the new Europe. A founding member of the Council of Foreign Relations, he became well acquainted with Presidents Woodrow Wilson and Franklin Roosevelt. From 1935 to 1948, Bowman was president of Johns Hopkins University, during which time he also participated in discussions leading to the founding of the United Nations.

Because the 20th-century history of geography is described in detail elsewhere, including in several entries in this volume, the following offers only a brief overview of several major schools of thought that existed until the year 2000.

CHOROLOGY

In the wake of the collapse of environmental determinism, the discipline embraced a long-standing tradition of chorology, also known as areal differentiation. In Europe, a central figure in this vein was Paul Vidal de la Blache (1845–1918), considered the father of French geography and well known for his studies of small rural areas called *pays* and their associated styles of life (or *genres de vies*). He also played a central role in the introduction of possibilism to the discipline. His German counterpart was Alfred Hettner (1859–1941), who argued in the Kantian tradition that geography consisted of the art of regional synthesis, seeking relations among phenomena that other disciplines ignored.

In the United States, the most famous advocate of chorology was Richard Hartshorne (1899–1992), who studied under Hettner and then graduated from the University of Chicago in 1924. Chorologists maintained that the essence of the discipline was the description of regions in all of their glorious uniqueness and complexity, including cultural and physical phenomena. Hartshorne argued that the most productive analyses focused on small, relatively homogeneous regions, noting that any deployment of the regional concept was inherently a subjective tool to find meaning in the unwieldy complex of data found in the world. Thus, regions were mental tools to impose order on chaos. By eschewing theory, chorology found itself mired in empiricism and the discipline's theoretical and philosophical progress was halted.

Chorology drew to a close during the 1950s, beginning with a famous attack on Hartshorne's worldview by Frederick Schaefer in 1953. Essentially, Schaefer claimed that the view that geography is an integrative science concerned with the unique was naive and arrogant because such issues were common to many sciences. By refusing to search for explanatory laws, geography condemned itself to what Schaefer called an immature science. Rather than seeking idiographic regions, geographers should seek nomothetic regularities across regions. This critique helped open the door to the rise of positivism and the quantitative revolution.

POSITIVISM AND THE QUANTITATIVE REVOLUTION

Starting in the 1950s at the University of Iowa and particularly the University of Washington, geographers embraced a new theoretical paradigm, namely positivism. Epistemologically, this view embraced the scientific method. Drawing on older traditions such as applied geometry, physics, classical German location theory, and neoclassical economics, this school advocated a Cartesian view of space (i.e., geography as geometry) in which landscapes are reduced to isotropic planes. Central to this paradigm were the surfaces, nodes, and hierarchies as found in models, including central place approaches to urban networks, gravity models, diffusion models of disease and innovation, location–allocation models to optimize distributions, and input–output models. Appropriate models were held to distill the essence of the world, revealing its causal properties through the act of simplification.

Although highly successful within its own frame of reference, the positivist school also found itself facing heavy criticism. Humanists objected that such an approach might work well in the natural sciences but not in the social sciences. Positivism denied the existence of an observer, offering a false sense of objectivity. Rather, critics alleged, all data are value laden and filtered through theory. Positivism's celebration of quantitative techniques was criticized as sterile and restrictive; mathematics is but one way in which to understand reality and is not always optimal. Finally, Marxists argued that positivism's ahistorical approach lacked any account of social relations and thus focused on forms rather than on processes.

MARXISM

Drawing on the extensive tradition that began with Karl Marx (1818–1883), geographers during the 1960s and 1970s constructed a view of space and landscape centered on political economy. Although there are many variations of Marxism, they share in common a concern for the centrality of class, the production process, and the importance of labor; a historical understanding of how societies are constructed; and the power and politics that pervade all societies, including capitalism. Marxist geographers viewed space in explicitly social terms, that is, as a construction of historically contingent social structures and relations. Thus, the introduction of Marxism into geography was the spatialization of Marxism as an analytic tradition.

David Harvey (1935–) played a major role in this process through a series of books that introduced geographers to historical materialism and reworked it along geographic lines, including famous notions such as the cycle of commodities to money to commodities that lay at the heart of capitalism. In so doing, Harvey showed that time and space are not absolutes that lie outside of society but rather products of those societies. Every social formation under capitalism, he argued, constructs a *spatial fix,* that is, an optimal landscape that reflects the prerequisites of production yet simultaneously inhibits future rounds of production.

The contributions of Marxist geographers included shedding light on questions of uneven development and spatial divisions of labor; revealing the central role of the state in the formation and maintenance of capitalist relations (in contrast to the mythology of the free market); reworking urban theory around the issues of production, class, and social reproduction, including topics such as suburbanization, inner-city poverty, and gentrification; and adding great depth to international political economy, primarily through intersections with schools of thought such as dependency theory and world systems theory. Other Marxists initiated a tradition concerned with the social construction of nature and environmental politics, and yet others ventured into the domain of cultural geography, portraying culture and cultural landscapes not only as sets of ideas but also as power relations.

Critics of the Marxist tradition often pointed to its tendency to economic determinism, that is, the reduction of all social issues to labor and production. Others argued that Marxism suffered from an inadequate theorization of human agency and human consciousness.

HUMANISTIC GEOGRAPHY

During the 1970s, some geographers turned toward the philosophical traditions of phenomenology and existentialism. Originating in hermeneutics, or the study of textual meanings, this view drew on the works of philosophers such as Edmund Husserl (1859–1939) and Martin Heidegger (1889–1976) to explore the implications of human experience in a world without fixed meanings. Humanistic geographers were concerned with putting people back at the center of social analysis, intersecting with psychology, humanistic philosophy, literary analysis, and linguistics to offer a rich portrait of what it means to be human, that is, to construct webs of symbolic meaning. Jettisoning the myth of objectivity, humanistic geographers argued that understanding comes from shared interpretations and the clarification of values.

Central to the project of humanistic geographers was the work of Yi-Fu Tuan, who redefined the notion of place from a sterile tangible object to the intangible meanings that people give to it. Places, as embodied, erotic, and highly personal locales, were contrasted with space, a more abstract Cartesian notion. Humanistic geographers focused on how consciousness constructs places through understanding and interpretation, that is, how landscapes are authored. This line of thought led to the exploration of individual lifeworlds and the spaces of the body, the most intimate of geographies; bodies appear natural but are social constructions inscribed with social meanings. In so doing, humanism mounted a serious challenge to both positivism and Marxism, forcing social science to take seriously the implications of human consciousness, for example, in recognizing the role of contingency or the capacity of human actors to do otherwise.

Criticism of humanism maintained that this perspective lacked a systematic account of social relations, class, power, and production. Thus, humanistic thought centered on an asocial undersocialized view of the individual without reference to mechanisms of social reproduction.

GEOGRAPHY AND FEMINISM

Geography has long been a male-dominated discipline. As female academics grew in numbers and were inspired by the women's movement of the 1960s and 1970s, feminist geography grew in size, importance, and sophistication. Although there are many views within feminism, all of them assumed the common point of departure that focused on gender. Feminists maintain that social reality is pervasively gendered and that gender relations intersect (and often proceed) class, generally as manifested through patriarchy, that is, gender relations that favor men at the expense of women. Feminism as a political project seeks not to eliminate gender but rather to eliminate the inequality that arises from patriarchy.

Feminist geographers made numerous contributions to the field, focusing on how gender relations are intimately woven into existing allocations of resources and modes of thought in ways that generally perpetuate patriarchy. To ignore gender is to assume that men's lives are "the norm" and that there is no fundamental difference in the ways in which men and women experience and are constrained by social relations. Widely recognized as the first non-class-based form of social determination to acquire legitimacy, gender has been thoroughly denaturalized; although gender roles may appear to be natural (i.e., outside of society), they are socially constructed as webs of masculinity and femininity. Feminists opened up the family as an object of geographic scrutiny, questioning the schism between production and reproduction and between the economic and the social. In urban geography, feminists focused on gender differences in commuting. Methodologically, feminism helped to legitimate the use of qualitative methods such as participant observation, standpoint theory, and grounded theory, emphasizing that knowledge always is a view from somewhere and context bound.

POSTMODERNISM AND POSTSTRUCTURALISM

The 1980s and 1990s witnessed rising diversity in how geographers came to view the world, including the recognition that modernity, far from being a universal feature of human life, was one historical project among many. Diverse thinkers coagulated into a school of thought that came to be known as postmodernism and during the 1990s as the more politically assertive poststructuralism.

It is critical to note that there are many variations in this line of thought, but commonalities typically include placing great emphasis on the constitutive role of language and ideology in the formation of social life. Like humanists, postmodernists argue that every discourse interprets the world from a particular vantage point, that every view is a view from somewhere, and that what one sees depends on where one stands. Postmodernists maintain that for every topic, there are inevitably many competing discourses, none of which is inherently more correct than the others. Thus, there are no *a priori* grounds for deciding what is true or not. To postmodernists, everything is a discourse because there is no way in which to see the world outside of discourse. Such a view intersected with the literary tradition of deconstruction in which texts are pulled apart for their diverse and often contradictory meanings. It also borrowed heavily from the work of Michel Foucault (1926–1984), who emphasized that all knowledge is inescapably sutured to power; discourses were not simply reflective of the world but rather constitutive of its inhabitants. From this view, all grand sweeping views of reality (often labeled *metanarratives*) sweep a great deal under the carpet, obscuring as much as they reveal. Indeed, the essence of postmodernism is that reality is more complex than our languages allow us to admit and that every linguistic construction of order is a simplification and a distortion. Thus, every representation of reality—every discourse—is inescapably political; there can be no divorce between the ontology and epistemology of viewpoints. Like feminism, postmodernism welcomed views from subjugated minorities, the suppressed "other," and the subaltern in the developing world. Instead, postmodernists ask who it is that gets to decide what constitutes valid, legitimate, correct, or proper discourse. This approach draws from the work of Michel Foucault to argue that power lies at the heart of what is accepted as truth. As an emancipatory political project, this view forced a philosophical shift from the social core to its periphery, from the center to the margins; that is, it required intellectuals to recognize the importance of gender and race not only ontologically (i.e., as parts of social reality) but also epistemologically (i.e., as they pertain to the creation of knowledge).

Postmodern geographers are a diverse group. Marxists, such as Ed Soja and David Harvey, argue that postmodernism is the latest cultural expression of advanced capitalism and that so-called postmodern landscapes reflect the emergence of a globalized, hypermobile, post-Keynesian, post-Fordist regime of production. In this view, postmodernism essentially is a response to the enormous wave of time–space compression unleashed by contemporary capitalism. Others employ the postmodern celebration of difference to reassert the significance of localities and local uniqueness, arguing that when and where social events happen is fundamental to how they happen, a move that elevates geography to the level of epistemology.

Critics of postmodernism accuse it of degenerating into intellectual nihilism and endless relativism in which all positions come to have equal weight. Some Marxists argue that the postmodern focus on culture and discourse has submerged class as a meaningful social category. Moreover, an overemphasis on difference can be politically disempowering because without a common political project resting on common assumptions and values, it is difficult to forge alliances across a diverse social spectrum.

CONCLUSIONS

This brief review has done little more than sketch some of the broad contours of how geography as a discipline has existed for two millennia. Geography has served society in numerous ways, including exploration and trade, religious justification, naturalization of colonialism and racism, urban and regional planning efforts, and challenges to inequality. The central thesis here is that there is not one single way in which to view geography but rather a plethora of schools of thought about space that contend and intersect in creative ways. Moreover, geography's history is unfinished given that new ways of looking at space are ever likely to emerge.

—*Barney Warf*

See also Anthropogeography; Berkeley School; Chorology; Class; Critical Human Geography; Cultural Geography; Dependency Theory; Economic Geography; Empiricism; Ethics, Geography and; Exploration, Geography and; Feminisms; Flexible Production; Gender and Geography; Gentrification; Historical Geography; Humanistic Geography; Ideology; Idiographic; Imperialism Labor, Geography of; Labor Theory of Value; Location Theory; Logical Positivism; Marxism, Geography and; Nomothetic; Orientalism; Paradigm; Phenomenology; Political Ecology; Postmodernism; Poststructuralism; Qualitative Research; Quantitative Revolution; Radical Geography; Regional Geography; Rent Gap; Social Geography; Spatial Analysis; Structuralism; Structuration Theory; Theory; Time–Space Compression; Urban Geography; World Systems Theory

Suggested Reading

Cloke, P., Philo, C., & Sadler, D. (1991). *Approaching human geography.* New York: Guilford.

Dickinson, R. (1969). *The makers of modern geography.* New York: Praeger.

Harvey, D. (1984). On the history and present condition of geography: An historical materialist manifesto. *The Professional Geographer, 36,* 1–10.

Holt-Jensen, A. (2003). *Geography: History and concepts* (3rd ed.). London: Sage.

James, P. (1972). *All possible worlds: A history of geographical ideas.* Indianapolis, IN: Bobbs-Merrill.

Johnston, R., & Sidaway, J. (2004). *Geography and geographers: Anglo-American human geography since 1945* (6th ed.). London: Edward Arnold.

Livingstone, D. (1992). *The geographical tradition.* Oxford, UK: Blackwell.

Marcus, M. (1979). Coming full circle: Physical geography in the twentieth century. *Annals of the Association of American Geographers, 69,* 521–532.

Peet, R. (1985). The social origins of environmental determinism. *Annals of the Association of American Geographers, 75,* 309–333.

Schulten, S. (2001). *The geographical imagination in America, 1880–1950.* Chicago: University of Chicago Press.

HOME

The home is both a material place—a building, often with garden or yard attached, located in a particular neighborhood—and a space in which identities and meanings are constructed. Over the years, geographers have assumed the home to be a site of unchanging and stable social geographies, but more recently this assumption has been challenged on a number of fronts as conventional meanings of home have been scrutinized and deconstructed. As a result, the home has become a more fluid and contested space.

Traditionally, the home has been characterized in terms of key meanings that were summarized by Peter Somerville as (a) *shelter:* not only protection from the weather outside but also a place of physical security; (b) *hearth:* a place to relax and to be comfortable ("at home") and from which to offer welcoming hospitality; (c) *heart:* emotional security, with the home providing a site of love and affection; (d) *privacy:* a legal and sociocultural haven in which questions of "who enters" and "what are acceptable practices" can be regulated; (e) *roots:* a place in which to belong and which can be an expression of our identity; (f) *abode:* a place to stay and to sleep; and (g) *paradise:* an idealized expression of the emotional pleasures of belonging, being safe, and feeling secure. Interestingly, many of these meanings have also been related to the scale of nationhood, indicating the inclusions (and exclusions) of a home country.

These traditional assumptions about the nature and social relations of home have been challenged from at least three different perspectives. First, geographers have recognized home as a place of labor. Positive attributions of meaning about the home are predicated on the idea that home is separate from, and may be contrasted to, work. Feminist geographers in particular have critiqued any formula that separates out the "private" space of home from the "public" space of work, arguing that such a formulation unhelpfully conceives the home as a "woman's place" and that the domestic effort of making and maintaining the home cannot be distinguished from paid labor outside the home. With the increasing participation by women in the paid labor force; the growing mix of paid and unpaid labor in the performance of domestic duties such as child care; cleaning, and gardening in middle-class households, and the upsurge of home working and work brought home, previous distinctions between home and work have become increasingly untenable.

Second, home has been recognized as a space of oppression. Contrary to the idea that a home provides a safe, secure, warm, and loving haven, there is evidence that the home is a significant site of violence, fear, and male tyranny for women, children, and older people. Domestic abuse is the second-highest category of violent crime in the United Kingdom, and there now are specific refuges in every city for women forced to leave the home to escape such abuse. By its very nature, however, domestic violence often is carried out in private and concealed from public view, meaning that the scale of the problem is significantly underrepresented. Similarly, physical and psychological abuse of children contradicts the idea of home as haven. Geographers have provided evidence of how expressions of male domination, dysfunctional "family" circumstances (e.g., the introduction of new and unsympathetic stepparents), or issues relating to alternative sexuality can lead directly to oppression of young people in their home, and teenagers leaving home are accounting for increasing proportions of the homeless population. Perhaps even more hidden from public view is elder abuse—the physical, psychological, and/or financial abuse of older people, either in their own homes or in institutions of care that become their "homes" in later life.

Third, the home is a space of negotiation and contestation. As a domain in which different everyday lives are played out and different resources, emotions, and requirements are invested, it is unsurprising that paradisal ideas of home as a happy unified idyll are unrealistic. Geographers have emphasized that the home is a site where identity and meaning are constructed through consumption and where connections are made with global and local discourses about the home. In one sense, then, the home can be constructed as a place of distinction where taste, style, and sociocultural trophies of domesticity and family can be displayed in contrast to other homes. However, not only does the home represent a site of potential resistance to hegemonic taste and practice, but also all manner of negotiations take place *within* the home between different household members, rendering it a contested space. For example, geographers have emphasized the conflicts between children's desire for disorder and weak time–space boundaries within the home and parental preferences for order and stronger demarcations. These conflicts are lived out in the making and breaking of rules about what can be done and where as well as in the regulation and staging of performance and practice in the home.

—*Jon May and Paul Cloke*

See also Domestic Sphere; Feminist Geographies; Homelessness; Urban Geography

Suggested Reading

Hanson, S., & Pratt, G. (1988). Reconceptualizing the links between home and work in urban geography. *Economic Geography, 64,* 299–321.

Somerville, P. (1992). Homelessness and the meaning of home: Rooflessness or rootlessness. *International Journal of Urban and Regional Research, 16,* 529–539.

Valentine, G. (2001). *Social geographies: Space and society.* Harlow, UK: Prentice Hall.

HOMELESSNESS

Following an explosion in levels of homelessness of all kinds across the advanced capitalist countries over the past 20 years or so, there has been a significant increase in the number of academic studies of homelessness during recent years, including a growing body of work by geographers.

It is important first to define homelessness. This is by no means an easy task. Even if we confine our attention to the problems of homelessness in the advanced Western economies (rather than considering the much more extensive problems of homelessness and insecure housing in the global South), distinctions need to be drawn between different forms of homelessness, for example, between visible homelessness (or rooflessness) and hidden homelessness. Complicated legislative distinctions are also often drawn between different homeless groups, and these differ in different countries. These distinctions are important because they may determine whether or not a person is counted in official estimates of the homeless population or becomes eligible for (state) aid. Complicating this issue further, others have sought to think instead about the *experience* of homelessness, defining homelessness not in relation to the absence of accommodation but rather in relation to the absence of those feelings of security and belonging (usually) associated with a sense of home.

Hence, problems of definition make it very difficult to establish an accurate picture of the size of the homeless population because it must first be determined how homelessness is being defined. Such problems also make it extremely difficult to trace the distribution of (different) homeless populations or to compare levels of homelessness in different places.

In seeking to understand the geographies of homelessness, we can consider three key themes usefully. First, geographers have explored the causes of homelessness. Early work in the field often located the causes of homelessness with the individual, suggesting that people become homeless either because of some kind of pathology (e.g., alcoholism) or because they suffer from particular vulnerabilities (e.g., mental or physical ill health). As the traditional population of older, single homeless men has been supplemented by a new homeless population of younger people, women, and a growing proportion of people from ethnic minority groups, the current orthodoxy is to instead look to those broader structural changes that have had especially hard impacts on particular groups over the past couple of decades, thereby making them especially vulnerable to homelessness. Of these, particular attention has been focused on processes of economic restructuring (with deindustrialization and selective reindustrialization leading to a significant rise in the number of long-term unemployed and a more insecure labor market) and changes to state welfare regimes (notably reductions in state benefits, deinstitutionalization, and a sharp decline in the supply of affordable public housing). The best work in this field has sought to trace the impact of broader structural changes operating at a variety of scales (from processes of global economic restructuring to changes in national welfare regimes) to changes in levels of homelessness, and to the groups most affected by homelessness, in different localities.

Second, during the mid-1990s, American geographer Lois Takahashi called for more attention to be paid to the ways in which homeless people are represented and become *stigmatized,* and she developed a useful model setting out the various "axes of stigma" through which different homeless people are positioned—as vulnerable and deserving of help, as a threat and in need of control, and so on. As a result, geographers have begun to examine the ways in which homeless people, and the problems of homelessness more generally, are constructed in a range of media, including newspapers, novels and films, academic textbooks, census materials, and legislative systems. Perhaps the most interesting such work is that examining the ways in which particular spaces and places (e.g., city streets, alleyways, underpasses) come to be seen as "spaces of homelessness," whereas in other places and spaces (e.g., a "purified" rural "idyll"), even if problems of homelessness are evident, they tend to be ignored because homelessness there is literally "unimaginable." Such work is crucial. As authors in this field have argued, the ways in which the problems of homelessness, homeless people, and the geographies of homelessness are represented and defined have a significant impact on responses to the problems of homelessness.

Third, geographers have considered different spaces of homelessness at a variety of scales. Forging connections among homelessness, urban regeneration, and the politics of public space, at the city scale work has examined the growing tendency to try to sweep (visibly) homeless people from the streets of a revitalized urban core. Street clearance campaigns are increasingly common in cities across the United States, Western Europe, and Australasia, although beyond the United States this more punitive approach has tended to be tempered by programs aimed at increasing the supply of emergency accommodation and other services available to homeless people. Others have focused on the response of homeless people themselves to such campaigns and on the

(different) tactics that homeless men and women deploy to survive on the streets. At a finer scale, geographers have also examined the internal dynamics of the various institutional spaces (e.g., emergency shelters, drop-in centers, soup kitchens) that define the homeless city. For some, such spaces are best understood as an attempt to discipline homeless people (encouraging them to conform to normative constructions of home); for others, they are best understood as genuine spaces of care. With most such services being provided by nongovernmental organizations, attention has focused on the changing relationships between these organizations and the state, with some suggesting that the increasing supply of such services (the majority of which continue to rely on state funding) represents an expansion of the shadow state. Attention has also focused on the factors shaping the locations of such services. Here geographers have traced the implications both of the initial concentration of services in distinctive skid row districts and of the more recent dispersal of these services as skid row districts themselves have become subject to gentrification. Beyond the city, increasing attention now is being given to the specific dynamics of rural homelessness. Here geographers have examined the reasons behind the marked lack of provision for homeless people in rural areas and the difficulties this poses for those who become homeless in the countryside. Finally, providing a bridge between these different spaces, geographers have considered the complex mobilities that shape homeless people's lives at a variety of scales, examining the patterns of and reasons for these movements and the impact of such mobility on homeless people's ability to establish a sense of home when on the streets or moving between different places or institutional settings.

—Jon May and Paul Cloke

See also Home; Housing and Housing Markets; State; Urban Geography

Suggested Reading

Cloke, P., Milbourne, P., & Widdowfield, R. (2002). *Rural homelessness: Issues, experiences, and policy responses.* Bristol, UK: Policy Press.

Mitchell, D. (2003). *The right to the city: Social justice and the fight for public space.* New York: Guilford.

Ruddick, S. (1996). *Young and homeless in Hollywood.* New York: Routledge.

Takahashi, L. (1996). A decade of understanding homelessness in the USA: From characterization to representation. *Progress in Human Geography, 20,* 291–310.

HOMOPHOBIA

Homophobia is a negative attitude, indifference, or aversion toward homosexual persons or homosexuality in general. The term was first used in 1969 by American psychologist George Weinberg, who defined homophobia as the fear expressed by heterosexuals of being in the presence of homosexuals and the loathing that homosexual persons have for themselves. Homophobia stems from individual, social, and systemic prejudice and can result in hostility against, or an exclusion of, homosexual persons, both men and women, that has repercussions ranging from the daily to the lifelong.

The Fondation Émergence in Quebec created a useful typology describing nine ways in which homophobia manifests itself: (1) *mind-set homophobia:* a feeling of conviction that homosexual persons are abnormal or sick; (2) *heterosexist homophobia:* a belief that everybody is heterosexual and that heterosexuality is the only acceptable and legitimate form of intimate social organizing (this belief rests on the idea that majorities set norms); (3) *speech-based homophobia:* use of vocabulary and expressions that span from teasing to insulting; (4) *behavioral homophobia:* body language or attitude that shows discomfort, insecurity, or fear when in contact with homosexual persons; (5) *institutional homophobia:* institutional practices that put homosexuals at a disadvantage; (6) *opportunistic homophobia:* behavior of persons interested in homosexuality only for monetary or personal gain and who refuse all association with homosexual persons or organizations; (7) *internalized homophobia:* an unconscious form of homophobia that results from education and prevalent social values (homosexual persons are not sheltered from this form of homophobia because they receive the same education and are influenced by the same values as is everybody else); (8) *homophobia by omission:* a silent or passive attitude when faced with homophobia speech or behavior; and (9) *violent homophobia:* extreme manifestations of homophobia that lead to violence, from verbal aggression to hate crimes.

Geographers have demonstrated not only how homophobia shapes the human landscape but also how homophobia is a spatialized process. They have examined and documented the effects of homophobia at a variety of scales that range from the body to the global.

—*Glen Elder*

See also Gays, Geography and/of; Heterosexism; Lesbians, Geography of/and; Sexuality, Geography and/of

Suggested Reading

Brown, M. (2000). *Closet geographies: Geographies of metaphor from the body to the globe.* New York: Routledge.

Elder, G., Knopp, L., & Nast, L. (2003). Sexuality and space. In G. Gaile & C. Willmott (Eds.), *Geography in America at the dawn of the 21st century* (pp. 200–208). Oxford, UK: Oxford University Press.

Valentine, G. (1998). Sticks and stones may break my bones: A personal geography of harassment. *Antipode, 30,* 305–332.

HOUSING AND HOUSING MARKETS

Much more than shelter, housing is the address from which we launch our daily lives, the largest financial transaction of a lifetime for most of us, and the place most people call home. Governments, financial institutions, and the development industry contribute to the highly differentiated geographies of housing that result from its durability, fixed location, and heterogeneity.

HOUSING SERVICES

Each housing unit provides a bundle of diverse services beginning with shelter. The size of the dwelling, its style, its structural type (be it an apartment in a high-rise building, a townhouse, or a single-family dwelling), and the quality of the dwelling unit influence the quality of shelter. Aging apartment buildings with vacant units, broken windows, and poorly functioning elevators provide much poorer shelter than do new and well-maintained dwellings with air-conditioning, multiple bathrooms, and multicar garages.

Housing is also an important financial investment. The majority of the population aspires to achieve homeownership. Accumulating the capital to purchase a dwelling is an important challenge for many. The high rates of homeownership in Australia, Canada, and the United States, where the majority of households live in owner-occupied housing, attest to the popular belief that housing is a successful financial investment. The financial return on housing affects individual homeowners' lifetime wealth and income as well as the distribution of wealth and income among social groups. The financial benefits of homeownership vary over time and place, with various effects on different social groups in each region and metropolitan area. Maintenance is another significant financial aspect of housing. Currently, owners are investing heavily in maintenance and renovation, a sector of the economy that is expanding quickly in Canada, the United States, and much of Western Europe.

The residential area surrounding a housing unit offers access to the many locations pertinent to people's daily lives. The speed, reliability, and ease with which people reach schools, transportation routes, public transit, retail outlets, recreational facilities, and workplaces vary tremendously among residential locations. Neighborhood location may also be associated with proximity to undesirable land uses such as landfills and industrial activities.

The neighborhood is also an important social arena for many people. Despite recent technological changes in communication and transportation, many children, women caring for young children, elderly people, poor people, and disabled individuals still have social networks centered on the local area. These social relations may be important resources providing assistance with daily life, information about services and employment, and crucial social interaction. Social interaction also shapes residents' opinions, affecting voting behavior, other forms of civic involvement, and neighborhood reputations.

Where one lives also affects his or her social status. The finely textured geography of housing in each urban and rural area gives rise to geographies of status in which some locations are considered prestigious, whereas others are stigmatized. Prestige accrues to locations that are associated with elites and upwardly mobile social groups, whereas the residential areas where the poor and minorities concentrate often are stigmatized. Some neighborhoods retain their reputations and status across the decades, whereas the reputations of others decline as the neighborhoods age. Subsequent renovation and renewal may reverse the deteriorating fortunes and reputations of some neighborhoods, whereas the stigmatized reputations of others intensify their deterioration.

HOUSING PROVISION

The geography of housing is highly uneven, with marked spatial variations in housing types, values, and statuses. Constrained by their incomes and guided by their housing preferences, households strive to obtain satisfactory housing by renting and purchasing housing units from those available at any point in time in specific neighborhoods. Housing providers respond to households' preferences and incomes by adjusting the prices of vacant dwelling units, building new housing, and renovating existing housing to meet various tastes and incomes. In central-city neighborhoods, obsolete warehouses are converted to lofts that attract affluent professionals who value the convenience of living downtown near recreational and cultural amenities. High-rise apartment buildings catering to young singles replace aging walk-up apartments. In the suburbs, new housing caters to the preferences and budgets of first-time homeowners and people moving into new, large, and expensive housing.

Government policies influence the share of the housing stock that is provided outside the private housing market. In the public sector, regulations and norms, rather than a capitalist market, govern the construction, management, and allocation of housing units. Eligibility for publicly provided housing is based on specified criteria that often refer to social characteristics such as income, age, and household type and size. Since the 1970s, many governments have moved away from public provision of housing, preferring to provide rent supplements to low-income households and incentives to encourage the production of low-cost housing.

Governments also shape the geographies of private housing. In the United States, federal government policies stimulated new suburban development at the urban rural fringe and induced the deterioration and abandonment of older housing in central neighborhoods. Mortgage insurance policies that favored those purchasing new suburban housing heightened racial segregation in American cities, increased the cost of homeownership for many African Americans, and accelerated the deterioration of aging inner-city neighborhoods. At the state level, public housing policies led to the development of massive housing developments in which the poor, segregated racially, were concentrated and isolated from job opportunities and other communities. Municipal governments also shaped contemporary spatial patterns of housing with their responsibilities for the planning and regulation of land uses, maintenance and administration of public housing units, and provision of local services.

The mix of public and private housing at any location is also affected by the actions of property and housing developers and the financial institutions. The variety of new and renovated housing units within each neighborhood often is limited as developers cater to a narrow segment of the population. Financial institutions have reinforced the selectivity of development in the past by policies that favored mortgage lending for new suburban developments and restricted access to mortgage funds in older neighborhoods undergoing renewal, often reducing access to homeownership for minorities and other central-city residents.

The complex geography of housing provision and consumption affects each person's daily life and life chances. As the postwar experience in American cities also reveals, the uneven geography of housing contributes to current social inequalities.

—*Valerie Preston*

See also Home; HUD; Neighborhood; Suburbs and Suburbanization; Urban Geography

Suggested Reading

Bourne, L. (1981). *The geography of housing.* London: Edward Arnold.

HUD

The U.S. Department of Housing and Urban Development (HUD) was established as a Cabinet-level department in 1965 but identifies its origins in the creation of the Federal Housing Administration (FHA) in 1937. This view is appropriate given that HUD has significantly reduced its urban development activities to focus on housing and homeownership initiatives. This shift follows a broader federal policy movement throughout the 1990s that identified homeownership as a potential solution to a wide range of urban problems.

HUD is responsible for developing and administrating federal urban and housing policy. The department includes five program offices: Housing, the Government National Mortgage Association (Ginnie Mae), Public and Indian Housing, Community Planning and Development, and Fair Housing and Equal Opportunity.

The FHA is the largest program in the Housing office and insures mortgages issued by financial institutions based on specific criteria. In so doing, the FHA substantially reduces the lending risk for financial institutions and allows them to make more loans at lower interest rates to moderate-income homebuyers.

Ginnie Mae operates in the secondary mortgage market by purchasing mortgages from the originating financial institutions and then selling securities backed by those mortgages to private investors. Investors are able to make a profit on the securities, and Ginnie Mae returns a profit as the mortgages are paid back in a timely manner. The secondary market provides financial institutions with a new source of capital that allows them to make more home loans. In addition, those loans can be made to higher-risk borrowers (typically moderate-income, first-time borrowers) because the risk of the mortgage is spread out across a large number of investors.

The Public and Indian Housing office oversees hundreds of local public housing authorities (PHAs) and provides housing for the lowest-income households in the nation. This office has drawn substantial criticism over the years for its bureaucratic excess (a charge leveled at HUD in general). The HOPE VI program was a response to some of these criticisms. This program attempted to rectify two decades of neglect by providing $5.3 trillion in funding from 1993 to 2002 to address the problems of "severely distressed" public housing. HOPE VI focused on a new form of public housing with lower densities and, when possible, mixed-income residents. Residents were also encouraged to use the Housing Choice Voucher program (formerly termed Section 8 housing), which provides low-income households with vouchers that can be used to subsidize rent (or house payments) in private market housing. The vouchers are popular, but waiting lists for them often are long and the program is poorly funded. HOPE VI has won awards for its innovative approach and has clearly changed the physical landscape of public housing. However, it has also been the target of many criticisms for a variety of reasons. Most important, the original one-to-one replacement policy was dropped, and significantly more low-income housing units were demolished than were rebuilt.

The Community Planning and Development office implements HUD's community and economic development programs, including Community Development Block Grants (CDBG) and the Economic

Development Loan Guarantee Fund. Although some of these programs are sizable, most pale in comparison with the funds involved in the housing programs.

One of the most significant HUD activities involves the regulation of two organizations that are not part of HUD itself. Fannie Mae and Freddie Mac, massive government-sponsored enterprises, dominate the secondary mortgage market. In 2002, the two combined to issue 68% of the $3.7 trillion in mortgage-backed securities. Both entities, although privately held, are regulated directly by the secretary of HUD. In combination with Ginnie Mae, they give HUD influence on nearly 80% of all mortgage-backed securities.

—Daniel J. Hammel

See also Housing and Housing Markets

Suggested Reading

Gore, A. (1994). *Department of Housing and Urban Development: Accompanying report of the National Performance Review.* Washington, DC: Office of the Vice President.

Swope, C. (2002, December). HUD the Unlovable. *Governing*, pp. 26–30.

HUMAN AGENCY

The concept of human agency refers to the ability of humans to make conscious choices and communicate these to other people. It stands in contrast to the simplistic concept of *free will,* which argues that our choices are not the process of causal chains of social interaction and ultimately are undetermined by such relations. Views that center on human agency make no such claims. Instead, it openly acknowledges that humans make decisions and impress them on the world only through some form of interaction with a larger collective. This has several ethical implications. First, human agency relies on personal efficacy. Only if humans see that their actions show results and render the desired effects will they continue to consciously be part of a larger social collective. Second, the degree of efficacy and interaction then depends on the ways in which human subjectivity is modified by power. Human agency is formed only within social interaction that is the result of power discourses. For

example, an activist for an oppressed political minority party in a country dominated by a dictatorial regime probably has fewer possibilities to voice his or her opinion publicly (and to influence policy) than does the dictator. The activist is part of a hierarchical collective where one voice dominates all others. The activist's power in directly facing the dictator is very limited, and such action might even be life threatening. But in cooperation with other dissidents, and in covert gatherings as well as public ones, the activist might be able to gradually develop a social movement that, through its collective voice and manpower, challenges and upstages the dictator. This exemplary case shows how human agency is placed in a social context and how power mediates the ways in which people communicate and challenge one another. When addressing human agency, social scientists always examine the relationship between individuals and societies and focus on how institutions such as governments mediate the power relations between these entities.

In addition to reflecting on the relationship between individuals and societies as it is controlled by power, the concept of human agency has been debated in the philosophy of science. In particular, it has raised the questions of what the object of scientific inquiry should be and how the scientist should relate to this entity. These relationships can best be addressed by looking at the discipline of geography itself.

The emergence of a debate about the significance of human agency for the discipline of geography can be traced to the popularity of humanistic geography during the 1980s. At that time, a first serious examination of the role of the geographer in the research process occurred. Humanistic geographers spelled out the role of human agency in geography and provided a historical perspective and critique of how the discipline has dealt with this concept. They accused geography of too much focus on objects (inanimate materials) and too little emphasis on subjects (humans and their emotions, motives, and beliefs). The dominating philosophies of science that shaped geographic paradigms of the 20th century—most notably positivism and Marxism—undervalued or left aside the individual and collective power of human agency. Whereas the early human geography of the French school of Paul Vidal de la Blache focused on humans and their active role in transforming their environment within the constraints of nature, this interpretive understanding of people and their interactions with the landscape gave way to scientific inquiry based on positivist methodology.

Geographers increasingly adopted a new approach to science that focused on collecting social facts disconnected from individual and collective human consciousness. In many ways, this emulated the data collection process of the scientific method as it was (and still is) dominant in the natural sciences. Following the quantitative revolution of the 1950s, spatial analysis modeled the economic and social patterns created by humans in the natural landscape and paid little attention, if any, to individual human behavior and motivations. Beginning in the 1960s, behavioral geography remedied this problem only partially by focusing on repetitive quantifiable phenomena of human behavior that could be modeled and portrayed in universalizing models. For example, there emerged a larger number of studies in human migration behavior that examined social, economic, and cultural motives of individuals for changing their residences; families, households, and single individuals were asked where and why they moved. Although many of these studies actually involved interviews and large-scale survey research that actively included a population of research subjects, geographers nevertheless neglected to relate the motivations and patterns they observed to the social and political contexts in which they were formed. Statistical analyses summarized people's motivations, but to a large degree the results disregarded how these were formed and in what social context they developed.

The influence of Marxist structuralism is another philosophical paradigm that has been critiqued by geographers concerned about human agency. Marxian political–economic analysis reduces actors to puppets unconsciously acting out the conceptual logic of capitalism and attributes human behavior to some hidden force beyond people's awareness and control. In summary, geographers historically have situated the discipline at diametrically opposed starting points of analysis; whereas some approaches reduce human agency to measurable collective facts, others deny humans any form of agency and control. Recent theoretical approaches, such as feminism, structuration theory, poststructuralism, and psychoanalytical theory, have revived this debate and attempted to either merge structural and agency-based approaches in a unified theory or abandon the idea of human agency altogether.

—*Olaf Kuhlke*

See also Humanistic Geography; Identity, Geography and; Structuration Theory

Suggested Reading

Gregory, D. (1981). Human agency and human geography. *Transactions of the Institute of British Geographers, 6,* 1–18.

Pile, S. (1993). Human agency and human geography revisited: A critique of "new models" of the self. *Transactions of the Institute of British Geographers, 18,* 122–139.

HUMANISTIC GEOGRAPHY

Humanism is a term that encompasses a variety of philosophical positions that go back to the Renaissance, when scholars such as Erasmus and Petrarch offered views of the social world that put people in the center, in contrast to the prevailing religious interpretations. Closely associated with humanism is hermeneutics (from Hermes, the Greek messenger of the gods), which is essentially the study of meanings. Originating from medieval attempts to find the one "true" meaning of the Bible, hermeneutics became extended to include the multiplicity of meanings inherent within all literary texts and social actions.

Two closely related approaches to humanistic thought have characterized it over time: phenomenology and existentialism. Both are concerned with the shape of human experience—the nature of subjectivity—and there is considerable overlap. Whereas phenomenology tends to emphasize the nature of human experience and meaning, existentialism is more often concerned with the ethical conduct of life.

Several giants in the history of philosophy invoked these lines of thought. Danish Christian existentialist Søren Kierkegaard (1813–1855) offered a Romantic critique of the Enlightenment, claiming that objectivity is a myth and that all people faced an agonizing choice between faith and reason, between the sacred and the profane, between ethics and pleasure. Edmund Husserl (1859–1939) formed a transcendental phenomenology, noting that the view of science as an objective map of the outer world reduced the human observer to a passive receptor. He argued that objects do not have meanings in and of themselves; rather, meanings are constructed by the human mind. Husserl called for a science of phenomenology that would strip away the biases that the mind creates in its perceptions of the world in order to see essences—the reality of things in themselves. Martin Heidegger (1889–1976) asked the apparently simple question, "What does it mean to be?" and offered a very complex answer. His view rested on the notion of the hermeneutics of being (*Dasein*), the understanding of which meant an escape from abstract theorizing. Jean-Paul Sartre (1905–1980) attempted a merger of existentialism and Marxism, noting that in contemporary capitalism the human condition is depersonalized and alienated.

Essentially, all of these views maintain that objectivity is a hurdle to effective understanding and that there is no privileged conceptual vantage point; every view is a view from somewhere and is inescapably laden with biases. We cannot know the world except for the meanings that people give to it. Thus, human subjectivity is not a barrier to understanding the world but rather the only route to knowing it. Meanings are essentially arbitrary phenomena, and logic cannot inform our moral choices. Despite this predicament, as Sartre noted, humans are "condemned to freedom"; that is, they must make choices even if there are no firm grounds for doing so.

Thus, the project of a humanistic social science was to put people back in the center of social analysis, that is, to reveal the things that make people human (i.e., consciousness). Social science has long had a poor conception of the human subject—a flaw that humanism attempts to overcome. It is consciousness that makes us subjects rather than objects, that is, allows us to be actors in the world with will and volition. Mapping human consciousness allows us to move past the sterile models of human behavior such as *Homo economicus* to recover the sensuous nature of experience—the ways in which the self, the environment, and others are framed symbolically. This task involves some understanding of intentionality—our deeply human desires and motivations, anticipations and expectations.

Uncovering the multiple dimensions of human consciousness, however, is no simple task. It is essential to avoid simplistic and biologically reductionist notions of "human nature." In the broadest sense, consciousness is what makes us human. In some respects, human consciousness differs qualitatively from animal consciousness (e.g., in humans' sense of self, time, humor, and death), although with many primates this difference is a matter of degree. Constructing a humanistic understanding of consciousness has also invoked various psychological understandings of sensation, perception, and cognition, leading to intersections with behavioral approaches. Consciousness includes our emotions and

memories, pleasures and fears, sexuality, hopes for the future, and more—both rational and irrational. This view sees humans as active creative actors and stresses their constructive role in making the world. Social reality does not simply happen to individuals "behind their backs" or "above their heads"; individuals make the world that makes them. Thus, humanistic social science is unapologetically anthropocentric, antinaturalist (it objects to using the same means to understand the natural world and the social world), and antideterminist, noting that people's actions render social structures ever changing and contingent.

Humanistic thought emphasizes the central role of language as a set of signs that we use to negotiate the world and share meanings. Language is how we bring the world into consciousness, and thought is always linguistically structured. As linguists and philosophers such as Wittgenstein have demonstrated, language is an opaque medium of understanding with a structure of its own. There is no language-free theory, and language limits and constrains the ways in which meanings are constructed, at times letting them escape their authors. The intersections of humanistic thought and literature in the form of textual deconstruction allowed every system of signs (e.g., a text, a landscape) to be pulled apart.

Like positivism and empiricism, the humanistic approach begins with the individual and experience in the construction of knowledge. The task of social science is to enter into another's taken-for-granted world, to see reality through the other's eyes, and to acknowledge the other's view as a valid source. Truth is found in the subjective meanings that people assign to their worlds, and explanation is the recovery of their intentions. Thus, humanism advocated a self-consciously empathetic social science that did not strive for the holy grail of objectivity but rather confronted its own inevitable assumptions and biases. This approach forces researchers to acknowledge both the subjectivity of the observer and the subjectivity of the observed—to question their own assumptions and biases—rendering the old subject/object dichotomy false and inserting the researcher into the research process. In so doing, humanism confronted social science with the need to clarify the ethics and morals of the observer, making clear his or her positionality in the research process. It also legitimated the use of qualitative research methods, such as participant observation and case studies, that sought to uncover the views of subjects.

Humanistic thought has a long history in the discipline of geography. During the early 20th century, French cultural geography owed much to Paul Vidal de la Blache, who studied the unity of culture and landscape in terms of the lifestyles (*genres de vie*) in small rural areas called *pays,* uniting land and life through an understanding of how consciousness and the earth are deeply intertwined. During the 1960s and 1970s, several authors made major contributions to the American literature on humanistic geography. Edward Relph's *Place and Placelessness* was concerned with the cultural impacts of mass production and consumption, the homogenization of capitalist landscapes, and resulting alienation. Ann Buttimer introduced the notion of *lifeworlds,* a phenomenological recovery of *genres de vie* that took as its point of departure the multiple ways in which consciousness was preconsciously sutured to locales in the intimate rhythms of everyday life. David Lowenthal wrote on landscape tastes and perceptions and on the relationship between history and cultural heritage. David Ley offered richly detailed urban ethnographies of the inner city and social geographies of Canadian cities.

Yi-Fu Tuan, who coined the term *humanistic geography,* held pride of place in this pantheon. Tuan's contributions included the widely popular notion of "sense of place", that is, the highly subjective set of feelings and impressions that individuals attach to specific locales. In this reading, places are intangible webs of meaning, not simply physical points. Sense of place, for example, makes a house into a home, makes a church into a building with deeply religious meanings, or defines a gang's turf. Tuan applied this set of notions, broadly grouped under the label *topophilia,* to studies of nature versus wilderness, spaces of pain and torture, sacred places, patriotism, pets, and more.

This line of thought also differentiated between space and place. In part, the difference is a matter of scale; space generally concerns broader domains than the individual experiences on a daily basis. However, space in the Western tradition often is used in a highly abstract sense such as a Cartesian plane or isotropic plains used in mathematical models. In contrast, place tends to be smaller, localized, more intimately experienced, intangible depositories of experience. The shift from space to place—one of the major contributions of humanistic geography—saw a transition from the abstract disembodied space to the embodied, erotic, personal, pungent places of individual worlds. Such a move exhibited a concern with particularity and specificity rather than with generality and made little effort to search for "general laws." Humanistic geographers were interested in what makes places unique, how they enter human consciousness, and how that consciousness in turn constructs places through interpretation. In so doing, they opened to geography linkages to hitherto closed domains such as landscape architecture, cultural anthropology, the sociology of the self, and the arts and humanities.

Another topic legitimized through humanistic geography was the geography of identity and the body. Whereas classical theories of the human subject portrayed identities as unitary and stable, phenomenologists argued that identity is a multiplicity of unstable, context-dependent traits—sometimes contradictory—that change over time and space. Identities are both space forming and space formed, that is, inextricably intertwined with geographies in complex and contingent ways. Likewise, human geographers explored the multiple ways in which identity, subjectivity, the body, and place are sutured together. The interface between body and mind is an ancient topic of philosophical consideration; the fact that we both *have* bodies and *are* bodies confronts us with the nebulous intersections of mind and matter. The body is where the mind resides, the locus of consciousness, tangible and corporeal evidence of its existence, giving existential and phenomenological depth to lived experience. Although bodies typically appear as "natural," they are in fact social constructions deeply inscribed with multiple meanings—"embodiments" of class, gender, ethnic, and other relations. The body is the primary vehicle through which prevailing economic and political institutions inscribe the self, producing a bundle of signs that encodes, reproduces, and contests hegemonic notions of identity, order and discipline, morality and ethics, sensuality and sexuality.

In insisting on the primacy of the intentional subject, humanistic scholars were adamant that geographies and landscapes always are *authored,* that is, created by people who give meaning to them. This position was very much at odds with rival perspectives, including the impersonal geometries of positivism. Humanists challenged behavioral geographers to explore not only the actions of people but also their intentions, avoiding simplistic black-box models such as *Homo economicus.* Finally, humanist thought mounted a serious challenge to Marxism, pointing out its flawed conception of human subjectivity and questioning its economic determinism and teleological view of history and geography, where people are represented as finders of a world already made. Instead, humanists argued that the social world was open-ended and contingent, forever in the process of becoming.

Humanistic thought, however, also had its critics. Marxists and others pointed out that it offered no account of social relations—of class, power, and production. Moreover, humanism's notion of the subject, however rich, was a largely asocial undersocialized account of individuals in purely personal—not interpersonal—terms. For example, a dwelling is not just a site of caring and memories but also a locus of social reproduction, family relations, patriarchy, and power. Moreover, by being silent about social relations, humanistic thought lapsed into an uncritical view of the world as simply structured by choice, a mythologized vision of "free will" devoid of social constraints. Methodologically, humanism's critics argued that the approach was deeply flawed; for example, it offered no means of validating, confirming, or disproving its claims. Some even held that humanistic thought was opposed to science. These problems led Entrikin to conclude that humanism sufficed as a critique of other positions, a way of unmasking presuppositions, but not as an alternative.

Humanistic thought made great contributions to the discipline, helping to revive cultural geography and forcing researchers to take seriously the complex question of human consciousness. It jettisoned the myth of objective research and made explicit discussion of values and biases an integral part of the process. In the end, humanistic geography, faced with serious critiques of its own, was largely integrated into other paradigms such as structuration theory and various poststructuralist perspectives that arose during the 1980s and 1990s.

—*Barney Warf*

See also Body, Geography of; Ethics, Geography and; History of Geography; Human Agency; Identity, Geography and; Participant Observation; Phenomenology; Qualitative Research; Situated Knowledge; Social Geography; Space, Human Geography and; Symbols and Symbolism

Suggested Reading

Buttimer, A. (1976). Grasping the dynamism of lifeworld. *Annals of the Association of American Geographers, 66,* 277–292.

Duncan, J., & Ley, D. (1982). Structural Marxism and human geography. *Annals of the Association of American Geographers, 72,* 30–59.

Entrikin, J. (1976). Contemporary humanism in geography. *Annals of the Association of American Geographers, 66,* 615–632.

Ley, D., & Samuels, M. (1978). *Humanistic geography: Prospects and problems.* Chicago: Maaroufa Press.

Relph, E. (1976). *Place and placelessness.* London: Pion.

Tuan, Y.-F. (1974). *Topophilia.* Englewood Cliffs, NJ: Prentice Hall.

HUMANISTIC GIScience

Humanistic geographic information science (GIScience) aims to integrate all human knowledge derived from the humanistic tradition with key issues related to spatial data representation, analysis, and visualization in the context of GIScience. Instead of attempting to maximize accuracy by minimizing or even eliminating uncertainty, humanistic GIScience incorporates the human subjective and even imaginative dimensions of experience in the process of spatial data handling.

Recent developments in ubiquitous computing have made nearly everything computable and at the same time have sharpened our focus on the fundamental limits of computation. The goal of a humanistic GIScience is to build dialogues with a variety of different scholarly traditions to develop a better understanding of the complexities of reality (Table 1). It will be necessary to continue to pixelize the social and at the same time socialize the pixels. Recent works on the integration of naive geography, indigenous knowledge, feminist perspective, and public participation theory into the conventional geographic information systems (GIS) modeling processes are essentially examples of humanistic GIScience in practice. At the interface of computing technologies with humanistic scholarship, we can expect exciting groundbreaking development in the near future.

Because the quest for new means of analysis and modeling via computers has been increasingly entwined with a persistent search for the deeper meaning of such activities, we should expect to witness a revitalization of the aesthetic and humanistic traditions within GIS and cartography. Artists' renditions of reality in novels, poems, paintings, movies, music, and songs can be very rich sources of inspiration for geographers to explore alternative conceptualizations of space, place, time, environment, region, and scale.

Table 1   Defining Characteristics of Humanistic GIScience

Scientific Knowledge	Humanistic Knowledge	Humanistic GIScience: Synthetic Knowledge
Specialized, partial	General, holistic	General, holistic
Experimentation	Observation	Observation and experimentation
Immutable mobiles	Mutable immobiles	Mutable mobiles
Cultural disjunction	Culturally compatible	Culturally compatible

With the infusion of aesthetic traditions into GIS, we can anticipate that humanistic GIScience will flourish in the near future. In addition to representations of space framed by Euclidean geometry, humanistic GIScience attempts to find novel ways in which to handle the textures of place as articulated in the humanistic tradition as well as the structures of space.

Although the development of GIScience was concomitant with the development of GIS, GIScience no longer is reliant on the tools to exist and have meaning. With the development of a humanistic component, GIScience became more than a computational science in search of new algorithms. In fact, GIScience is also emerging as a humanistic science in search of meanings for its computations and an area for speculating about what lies beyond the limits of computation— GIScience's new *terrae incognitae*.

—*Daniel Sui*

See also Art, Geography and; Geoslavery; GIS; Humanistic Geography; Limits of Computation

Suggested Reading

Dreyfus, H. (1992). *What computers still can't do.* Cambridge: MIT Press.

Egenhofer, M., & Mark, D. (1995). Naive geography. In A. U. Frank & W. Kuhn (Eds.), *Spatial information theory: A theoretical basis for GIS* (pp. 1–15). Berlin: Springer-Verlag.

Flake, G. (1998). *The computational beauty of nature: Computer explorations of fractals, chaos, complex systems, and adaptation.* Cambridge: MIT Press.

Gelernter, D. (1997). *Machine beauty: Elegance and the heart of technology.* New York: Basic Books.

Kwan, M.-P. (2002). Feminist visualization: Re-envisioning GIS as a method in feminist geographic research. *Annals of the Association of American Geographers, 92,* 645–661.

Sui, D. (2004). GIS, cartography, and the third culture: Geographical imaginations in the computer age. *The Professional Geographer, 56,* 62–72.

HUNGER AND FAMINE, GEOGRAPHY OF

The persistence of hunger and famine represents a fundamental paradox of the contemporary world system. At the global scale, since the early 1970s the growth of food production and trade has easily outpaced population growth. Nevertheless, millions of people around the world, from New York City to New Delhi, experience hunger, while millions of others, from North Korea to Southern Africa, bear the impacts of famine.

UNDERNUTRITION AND MALNUTRITION

Hunger is a condition of involuntary food deprivation that takes a variety of forms, each having its own geography. The two most commonly recognized forms are undernutrition and malnutrition. Very basically, undernutrition occurs when a person is unable to consume enough calories to remain healthy and active on a continuing basis, depending on her or his age, sex, height, and daily physical exertion. Malnutrition specifies a person's inability to consume a nutritionally balanced diet. This can involve either deficient or excessive consumption of particular nutrients such as protein and carbohydrates. Both of these kinds of hunger result in a range of physical and mental effects on the body, most notably stunting (low height for age), wasting (low weight for height), underweight (low weight for age), impaired cognitive development, greater susceptibility to disease and injury, and (in the case of malnutrition) obesity.

World maps of undernutrition and malnutrition typically are drawn using estimates of deficient calorie, vitamin, and protein consumption; stunting; underweight children; and infant mortality. Deficient calorie consumption is the most common indicator used to examine the global distribution of hunger. The United Nations estimates that more than 800 million people, roughly 13% of the world's total population, do not consume enough calories on a daily basis. More than half of those live in just four countries: India, China, Bangladesh, and the Democratic Republic of the Congo (formerly Zaire). In relation to national populations, inadequate calorie consumption is highest in countries in Central, East, and Southern Africa, where more than one half of all people are undernourished. Especially high rates are also found in South Asia, the Caribbean, and Central America. Among all developing countries, the lowest rates of caloric deficiency are in Mexico and the countries of North Africa. Although a significant number of people in the United States, Canada, Western and Northern Europe, and Japan also experience this condition, the rate of undernutrition is extremely low in these countries.

FOOD INSECURITY

Hunger takes a third form, known as food insecurity, which occurs when people are unable to reliably obtain a healthy and socially acceptable diet in socially acceptable ways. This understanding of hunger rests on the acknowledgment that food is not just a source of energy and nutrients and that the food consumption process involves more than just the ingestion of food. It recognizes that food—what kind and how it is produced or acquired, prepared, served, and eaten—is central to cultural identity and social standing. Therefore, hunger disrupts not only physiological processes and mental development but also "normal" cultural and social functioning. As food insecurity, hunger represents a lack of control by people over what they eat and on what terms, and it is revealed in feelings of anxiety, frustration, guilt, and humiliation as well as "abnormal" behaviors or practices surrounding food consumption such as "dumpster diving" and trash picking, panhandling or begging, and using soup kitchens and food pantries.

Inasmuch as food insecurity is a subjective and distinctly contextual condition—that is, one for which objective and universal indicators do not exist—it is difficult to map at the global scale. The way it is experienced and the reasons why differ significantly from place to place. For example, food insecurity exists in the anguish of a mother in India who, without other recourse, feeds her children imported food that was processed outside their religious conventions; in the fears of a Mexican farmer that, by needing to grow genetically modified maize, he is subjecting his family to adverse health effects and undermining the sustainability of his farm; and in the social alienation of a Canadian man who lacks income and must obtain food through an emergency kitchen.

FAMINE

Famine is a complex process of social breakdown involving acute poverty, disruption of the food consumption process, and the occurrence of hunger in its various forms. Although it is typically associated with sudden mass mortality due to starvation, this is not an essential feature; those who have experienced famine tend to characterize it as a phenomenon of increasing vulnerability and the collapse of coping strategies rather than death. The potential for famine, which often is "triggered" by weather events, economic and political "shocks," or war and conflict, resides in a range of social relations and structures, including commodity and labor markets, land tenure systems, oppressive and unequal race and gender relations, political rights, systems of government, foreign debt, and class capacity. The major social effects of famine involve demographic (e.g., decrease in fertility, increase in deaths, delayed marriages, migration), economic (loss of income and assets), and public health (susceptibility to disease such as human immunodeficiency virus/acquired immunodeficiency syndrome [HIV/AIDS]) impacts.

It is not readily apparent when famines begin and end, but over the past two decades a clear geographic pattern is discernible, marked by occurrences or near occurrences in East Africa (e.g., Somalia, Ethiopia [both on repeated occasions]), Southern Africa (e.g., Zimbabwe), North Korea, Afghanistan, Central America (e.g., Nicaragua, Honduras), and Haiti.

GLOBALIZATION, DEPOLITICIZATION, HUNGER, AND FAMINE

The geography of hunger and famine exists not only in their spatial patterns of incidence but also in the causal processes underlying them. Two broad geographic processes are particularly relevant during the early 21st century. The first involves the tendency toward globalization in the agro-food system as a result of corporate concentration in food production, processing, distribution, and retailing and the opening of national food sectors to imports and investment capital from multinational firms. This trend is linked to a general decline in subsistence and locally oriented farming (a majority of the world's population still earns a livelihood by producing food) and to the increasing inability of consumers, particularly those with low incomes, to determine what they eat. The second process consists in the dislocation of responsibility for hunger and famine prevention and alleviation from national-level governments to international relief agencies and humanitarian organizations and to local communities, charities, and social networks. This simultaneous global–local shift represents a "depoliticization" of hunger and famine because it places responsibility for them in the hands of private individuals and organizations that ultimately are less accountable to those suffering the conditions and that are less able to provide fundamental solutions.

—*Andy Walter*

See also Food, Geography of

Suggested Reading

De Waal, A. (1998). *Famine crimes: Politics and the disaster relief industry in Africa.* Bloomington: Indiana University Press.

Food and Agriculture Organization of the United Nations. (2003). *The state of food insecurity in the world 2003.* Rome: Author. Available: www.fao.org/sof/sofi/index_en.htm

Riches, G. (1997). *First World hunger: Food security and welfare politics.* New York: St. Martin's.

Von Braun, J., Webb, P., & Tesfaye, T. (1999). *Famine in Africa: Causes, responses, and prevention.* Washington, DC: International Food Policy Research Institute.

I

IDENTITY, GEOGRAPHY AND

Identity is one of the most significant dimensions of social and spatial analysis. Loosely defined, identity concerns the psychological sense of self, its nature and importance, its relations to others, and the shape and boundaries of human experience. Hence, identity is simultaneously a deeply personal phenomenon and social phenomenon that reflects, and in turn shapes, individual and collective behavior. Individual and collective identities, such as nation-states, are mutually presupposing.

THEORIES OF IDENTITY

Classical theories of the human subject typically portrayed identities as stable in time and space. The classical Cartesian notion of the subject that emerged during the Renaissance and Enlightenment portrayed identities as a consistent bundle of traits that transcends contexts. Such a view emphasized the inherent rationality of humans—their predictability and universally shared qualities—and became the prevailing Western notion. This view of identity is replicated to one degree or another in schools of thought such as logical positivism, location theory, neoclassical economics, Weberian sociology, and (to a lesser degree) Marxism. However, traditional social science suffered from an impoverished sense of the subject.

Beginning with the entry of various humanistic perspectives (e.g., phenomenology) during the late 20th century, social science began to acquire or construct a much richer, more realistic, more human and humane understanding of identity. This transition drew deeply from the well of phenomenology and its concerns for the shape of human experience. Although this literature is varied and diverse, some of its central contentions include the following.

1. The notion exists that identities always are social—not simply individual—products that internalize roles such as class, gender, age, ethnicity, and sexuality. Identities both constitute and are constituted by the social world and always are historically specific. Identity is rooted in the routines of everyday life—in our performances as members of a given class, gender, ethnicity, and sexuality. Adopting an identity reproduces these relations in everyday life. The unity of individual and collective notions of identity was advanced by the introduction of structuration theory by sociologist Anthony Giddens during the 1980s, a maneuver that effectively overcame the long-standing division between "micro" approaches, which focused on individual humans (e.g., phenomenology, other humanistic views), and "macro" approaches, which began and ended with social structures but ignored the dynamics of individual behavior (e.g., structural Marxism).

2. Following the influential works of philosopher Michel Foucault, a consensus arose that identities always are tied to power relations. Power, in the forms of discourse and ideology, is manifested in relations of normality and marginality that often appear as natural or normal. Thus, those who hold power define normality and abnormality in a manner that advantages them, but these categories always are contested and subverted. This move gave the discipline a far more explicit and wide-ranging concern for the nature of politics that infuses the world of the everyday and the interior spaces of the individual. The emphasis on

239

power and politics drew greatly from traditions such as Marxism and feminism, and opened the door to a multiplicity of other forms of social determination, such as ethnicity and sexuality. More broadly, it infused geography and social science with a concern to uncouple truth from power, a step necessary in any theorization of difference without hierarchy. Much of the postmodern concern with difference and multiplicity reflects this philosophical change.

3. Identities are always *embodied*. Although the body typically appears as "natural," it is in fact a social construction deeply inscribed with multiple meanings—"embodiments" of class, gender, ethnic, and other relations. Likewise, the human body has become an inspirational topic for human geographers, particularly the multiple ways in which identity, subjectivity, the body, and place are sutured together. The body is the primary vehicle through which prevailing economic and political institutions inscribe the self, producing a bundle of signs that encodes, reproduces, and contests hegemonic notions of identity, order and discipline, morality and ethics, sensuality and sexuality. The body is also the most personalized form of politics; all power ultimately is power over the body.

4. Finally, postmodernism and poststructuralism injected a concern for the multiplicity of forms of identity and their relations to boundaries, noting that identities are constructed through difference, by defining what they are not; there is always an "other," and othering is a power relation. In addition, the view that a human being has only one consistent identity was abandoned in favor of a view holding that identity is a multiplicity of different, unstable, context-dependent traits—sometimes contradictory—that change over time and space as individuals move among what were once held to be fixed categories of meaning.

GEOGRAPHY AND IDENTITY

Human geographers have sought to understand the nature and meaning of identity and its place (both literally and figuratively) in the world. Identities are both space forming and space formed, that is, inextricably intertwined with geographies in complex and contingent ways. Space affects not only what we see in the world but also how we see it. This theme emerges, for example, in divisions between the public and the private, between front stage and back stage, and the phenomenological meanings we attach to places in everyday life.

Explorations of identity led geographers to analyze its significance to a multitude of topics such as diasporas, disabilities, citizenship, childhood, emotions, and insanity. Others turned to how identities formed, and were formed by, media such as art and literature. This body of work bore obvious relations to postcolonial analyses of ethnocentrism and Eurocentrism, for example, in constructions of categories such as the West. Yet others deployed identity in the understanding of cultural landscapes, including memorials and the selective use of the past.

The recognition of identity has been important to the cultural turn, in which economic geographers emphasized culture as a complex contingent set of relations every bit as important as putatively economic factors in the structuring of economic landscapes, humanizing abstract economic processes by showing them to be the products of agents enmeshed in webs of power and meaning. This move was also central to understanding topics such as consumption in a manner far more sophisticated than that offered by neoclassical economics and its obsession with the sterile self-interest of *Homo economicus*.

IDENTITY AND CYBERSPACE

Given the explosive growth of the Internet, geographers have joined others in examining how identity is changed via cyberspace. Electronic communications have contributed to a massive worldwide round of time–space compression that has reconfigured the structure of social relations and the rhythms of everyday life, including the growth of cybercommunities and their associated virtual selves, the digital divide that separates information "haves" and "have-nots" both globally and locally, and the political uses of the Internet. The Internet is a social product that is interwoven with relations of class, race, and gender and inescapably subject to the uses and misuses of power. In an age when ever broader domains of everyday life are increasingly mediated electronically, this literature has moved beyond simplistic dichotomies, such as online and offline, to suggest the ways in which the real and the virtual are increasingly shot through with one another.

IDENTITY AND GLOBALIZATION

A burgeoning literature has sought to document the complex ways in which the rapidly changing world system has brought in its wake rapid and thoroughgoing alterations of local subjectivities, including various forms of nationalism, ethnicity, and religion. This genre has ranged far and wide in exploring pressing issues such as Orientalism, local responses to globalization, and the resurgence of ethnic-based fundamentalism in the face of a seamlessly connected, international, information-based economy. Identities in the late 20th century, under postmodern, post-Fordist hypermobile capitalism, are rapidly being transformed by the time–space compression of hypermobile capitalism in which hyperreality becomes the norm. This process has accelerated the emergence of identity politics that emphasizes marginalized sources of subjectivity, including gender, sexuality, ethnicity, and postcolonial perspectives.

CONCLUSIONS

The analysis of identity has had far-reaching effects in human geography and comprises part of a broader transformation that refocused the discipline away from spatial patterns toward social processes, including the manner in which people represent themselves and are represented by one another. In so doing, geographers developed a richer understanding of what it means to be human as well as a markedly more sophisticated view of power, ideology, and how social categories are naturalized and denaturalized. Finally, the increased focus on identity has changed not only what geographers study but also how they study it, legitimizing the use of qualitative methods, breaking down the power relations between researchers and their subjects, emphasizing that knowledge always is a view from somewhere that it is context bound, and advocating new avenues such as participant observation, standpoint theory, and grounded theory.

—Barney Warf

See also Art, Geography and; Behavioral Geography; Body, Geography of; Children, Geography of; Consumption, Geography and; Cultural Geography; Cultural Landscape; Cultural Turn; Cyberspace; Disability, Geography of; Economic Geography; Emotions, Geography and; Ethics, Geography and; Ethnicity; Ethnocentrism; Eurocentrism; Existentialism; Femininity; Feminisms; Feminist Geographies; Gays, Geography and/of; Gender and Geography; Geodemographics; Homophobia; Human Agency; Humanistic Geography; Ideology; Imaginative Geographies; Marxism, Geography and; Masculinities; Mental Maps; Orientalism; Other/Otherness; Participant Observation; Phenomenology; Popular Culture, Geography and; Postmodernism; Poststructuralism; Qualitative Research; Race and Racism; Sense of Place; Sexuality, Geography and/of; Situated Knowledge; Social Geography; Spaces of Representation; Structuration Theory; Subaltern Studies; Subject and Subjectivity; Symbols and Symbolism; Time Geography; Time–Space Compression; Travel Writing, Geography and; Urban Geography; Virtual Geographies; Vision; Writing

Suggested Reading

Adams, P. (1995). A reconsideration of personal boundaries in space–time. *Annals of the Association of American Geographers, 85,* 267–285.

Crow, D. (Ed.). (1996). *Geography and identity.* Washington, DC: Maisonneuve Press.

Curry, M. (1997). The digital individual and the private realm. *Annals of the Association of American Geographers, 87,* 681–699.

Giddens, A. (1991). *Modernity and self-identity: Self and society in the late modern age.* Stanford, CA: Stanford University Press.

Keith, M., & Pile, S. (Eds.). (1993). *Place and the politics of identity.* London: Routledge.

Kirby, K. (1996). *Indifferent boundaries: Spatial concepts of human subjectivity.* New York: Guilford.

Pile, S., & Thrift, N. (Eds.). (1995). *Mapping the subject: Geographies of cultural transformation.* London: Routledge.

Turkle, S. (1997). *Life on the screen: Identity in the age of the Internet.* New York: Touchstone.

IDEOLOGY

In 1796, an erstwhile cavalry officer-turned-philosopher named Destutt de Tracy (1784–1836) coined the word *idéologie,* meaning the science of ideas. A concise and accurate definition of ideology today is "a powerful system of ideas." Ideologies and their impacts, both salient and subtle, manifest everywhere in geographic landscapes at all scales. Powerful ideas are invariably *political* ideologies. This is because all ideologies, whatever their provenances or manifestations, are involved to varying degrees in the political organization of social and spatial relationships involving authority.

IDEOLOGY, AUTHORITY, AND CRITICAL THINKING

An idea is whatever comes to mind. Any idea is potentially a component of ideology. Both animals and humans experience the world as sensations, but only humans can nurture their sensations as ideas through reflection and articulation and, in combination with other ideas, can empower them as ideology. Ideology is a human social tool capable of changing what is into what can be. Ideologies proliferate in many guises and often (but not always) are identifiable as words that have the suffix *–ism,* for example, patriotism.

Large-scale spatial expressions of ideology are likely to occur when an ideology becomes invested with authority. Authority is a legal or rightful power to command and act. Authority invests in ideology as a tool to justify its inalienable right to exercise power. Justification resides in doctrines and theories that claim confidence in their certainty of knowledge. For example, Thomas Jefferson, author of the Declaration of Independence and founder of an ideology called Jeffersonian liberalism, held certain truths to be self-evident. The Declaration of Independence was a doctrine that provided a detailed justification for his radical social revolution. Jefferson's face today appears in a massive profile on Mount Rushmore, a planned pilgrimage site for patriotic Americans. This political artifact carved in stone is a spatial expression of Jeffersonian ideology intended for civic educational purposes as a didactic (teaching) and mnemonic (memory) device.

Ideology, on close examination, is just rhetoric making truth claims. Perhaps for this reason, the closest synonyms for ideological in popular use today are *dogmatic* and *fanatic.* Thus, ideology and critical thinking—as critique—have a close, but adversarial, relationship. Critical thinking from the time of Socrates has been a critique of domination by an authority. Critical thinkers are able to advance arguments that successfully undermine ideological knowledge claims that authority makes to justify its right to rule. Despite their efforts, so long as there are ideas and authority, there will be ideology.

ORIGIN OF IDEOLOGY AND TRANSFORMATIONS OF ITS MEANING

Ideas are as ancient as humankind, but Tracy's invention of ideology did not occur until John Locke (1632–1704) had reformulated the concept of an idea in the context of a Cartesian universe—as the mind's immediate object of perception, thought, or understanding. Locke's intellectual precursors already had launched investigations into provocative topics such as human nature, freedom, religion, society, law, and art. Sir Francis Bacon (1561–1626), in his *Advancement of Learning* (published in 1605), argued specifically that the mind must be educated and disciplined in defense against bad habits of thought or else people would be led to believe what is false or misleading and thereby become complacent and too easily accepting of authority.

Following Bacon, Rene Descartes (1596–1650) wrote *cogito ergo sum* ("I am thinking therefore I exist"), which inspired Locke to propose that sensation and reflection were the source of ideas. Locke's proposition diffused to France with Voltaire (1694–1778), and Etienne Bonnot de Condillac (1750–1780) argued that sensation was the *only* source of ideas. It was at this point that Tracy, who had spent time in prison during the Reign of Terror reading Locke and Condillac, invented the term *ideology.* He subsequently elaborated on his science of ideas in *Elements of Ideology* (multiple volumes published 1801–1815). Jefferson translated and published those parts that dealt with political economy. Tracy's contributions inspired proponents of social revolution throughout Europe and the Americas and influenced the founders of socialism and communism.

Tracy considered his ideology to be a natural science, but he was also a social activist interested in educational reform and promoted the doctrines of liberalism against centralization of authority. During the 19th century, the Napoleonic era transformed ideology into an explicitly political concept rooted in the oppositional doctrines of dissidents and other legitimate political actors. It was with this negative connotation that ideology reemerged in the writings of Frederick Engels (1820–1895) as part of his theoretical collaboration with fellow socialist Karl Marx (1820–1895). It is primarily through this intellectual doorway of Marxism that ideology as term and concept emerged within human geography as part of a critique-of-dominance discourse among radical geographers.

IDEOLOGY AND HUMAN GEOGRAPHY

The history of geography is the history of geographic ideas; thus, critical thinking as a critique of dominance in human geography is inherently ideological. There have been two major currents of this sort of critique in North American human geography during the past 40 years. One is the scientific Marxist critique of capitalism, within which ideology has the narrow technical definition as false consciousness. By distracting workers from recognizing the conditions of their own existence, ideology enervates their revolutionary potential. The other current is a more general critique of empiricist–positivist ideological dominance of geography made by humanistic (hermeneutical) geographers and structuralist (radical) geographers.

The basis of the humanistic critique of empiricism–positivism is epistemological and irresolvable, involving different knowledges. The structuralist critique (apart from the false consciousness issue) measures Marxism's commitment to social relevance against empiricism–positivism's commitment to value neutrality.

The heyday of the structural Marxist movement in human geography began during the early 1960s, yet for several reasons its momentum peaked by the end of the decade. Sociologist Karl Mannheim (1893–1947) argued in *Ideology and Utopia* (published in 1936) that knowledge was disguised ideology and that Marxist knowledge was a systems-maintaining idea; therefore, Marxism itself was an ideology with its authority subject to critique. Mannheim's sociology of knowledge influenced Thomas Kuhn (1922–1996), who wrote of the structure of scientific revolutions and paradigm shifts. Philosopher-scientist Paul Feyerabend (1924–1994), in defense of the liberty of thought and critical thinking, castigated positivist science as the tyrannical ideology of scientism and thus as just another ideology full of rhetoric but no truth—no different from Marxism.

WHITHER IDEOLOGY?

By the 21st century, the revolutionary fervor of radical human geography had dissipated into a plurality of less radical Marxian intellectual initiatives—feminism, critical realism, bioregionalism, deconstructionism, and the broader poststructuralism, each with its own specialized theories and doctrines and transdisciplinary connections. Industrial economies were being rapidly replaced by knowledge economies that spawned appropriate new systems of powerful ideas that Marx and Engels never had anticipated. Critical social theory, and ideology along with it, took a cultural turn. Former radical social(ist) geographers rapidly repositioned themselves within a new cultural geography, amid trendy yet oddly juxtaposed intellectual movements that included critical literary theory, pragmatism, postmodernism, and technophilia.

Meanwhile, ideology slipped its moorings from within the rhetoric of critical social theory and drifted into the hitherto unexplored epistemological relativism of popular culture. Politicians, rap stars, talk show hosts, cartoon characters, vegetarian chefs, human geographers, and a multitude of others today use the term *ideology* both critically and uncritically and by doing so invest it with their own meanings. Human geographers as students, teachers, and researchers should find in these postmodern trends increased opportunities and enriching experiences. Indeed, modifying Tracy's original definition for ideology to read "the art and science of the spatial expression of ideas" offers human geographers new direction in their efforts.

—David Nemeth

See also Critical Geopolitics; Critical Human Geography; Cultural Turn; Discourse; Enlightenment, The; Hegemony; History of Geography; Humanistic Geography; Marxism, Geography and; Paradigm; Postmodernism; Poststructuralism; Power; Radical Geography

Suggested Reading

Gregory, D. (2000). Ideology. In R. Johnston, D. Gregory, G. Pratt, & M. Watts (Eds.), *The dictionary of human geography* (4th ed., pp. 369–370). Oxford, UK: Blackwell.

Taylor, R. (1974). *The word in stone: The role of architecture in national socialist ideology.* Berkeley: University of California Press.

Winner, L. (1980). Do artifacts have politics? *Daedalus, 109*(1), 121–136.

IDIOGRAPHIC

The term *idiographic* refers to the unique aspects of individual areas, that is, those that cannot be understood easily on the basis of general rules of inference or deduction. Much of geography traditionally has been concerned with the idiographic in the context of regions and places, long mapping the colorful and extraordinary. However, the uniqueness of places has also been at the center of significant philosophical debates about how to study geography.

The tradition of chorology or areal differentiation, which predominated during the early 20th century and was epitomized by Richard Hartshorne, maintained that geography is an integrative science that is concerned exclusively with the unique. In this perspective, regions form the highest form of understanding. Idiographic understanding holds that each region is a unique combination of physical and human elements in the landscape. Smaller regions are more likely to be more internally homogeneous, and broader ones can be understood through the accretion of small units. Upholding the idiographic in this manner essentially disregards the need for general themes or causal properties that transcend regions, the key point of nomothetic (law-seeking) approaches to geography. Thus, the idiographic has long been associated with empiricist and inductive forms of thought in geography, that is, generalization without explanation.

Beginning with Fred Schaefer's famous critique of regional geography in 1953, the idiographic began to wane in popularity. The move into a nomothetic science sought to subsume all the unique details of place under general laws of understanding that could be applied in all contexts. The attempt to make geography "scientific" entailed a shift from regions without theory to theory without regions. This shift corresponded with the decline in popularity of regional geography more broadly.

However, during the 1980s, beginning with Doreen Massey's famous work on regions in the changing spatial division of labor, geographers acquired a new respect for the idiographic. The so-called localities school attempted to resurrect the idiographic by approaching it in terms of the historical development of regions over time. Beginning with the observation that no social process unfolds in precisely the same way in different places, this view held that regions acquired unique combinations of imprints of different divisions of labor (e.g., investments, labor market practices, cultural forms). In such a view, general laws of explanation are observable only in unique idiographic contexts, and the local becomes more than some inexplicable phenomenon; it becomes an object of scientific understanding. In an age of globalization, the local always is shot through with the global, requiring a multiscalar approach. Unlike the earlier tradition of chorology, therefore, this approach is theoretically sophisticated and far from the empiricism that plagued earlier attempts.

—Barney Warf

See also Chorology; Locality; Nomothetic; Regional Geography

Suggested Reading

Hart, J. (1982). The highest form of the geographer's art. *Annals of the Association of American Geographers, 72,* 1–29.

Hartshorne, R. (1939). *The nature of geography.* Washington, DC: Association of American Geographers.

Massey, D. (1995). *Spatial divisions of labor* (2nd ed.). New York: Routledge.

Schaefer, F. (1953). Exceptionalism in geography: A methodological examination. *Annals of the Association of American Geographers, 43,* 226–249.

IMAGINATIVE GEOGRAPHIES

Imaginative geographies, the images of the world and its diverse people that help a group to define its identity, are cultural representations that carry both emotional and ideological weight. Within human geography, the study of imaginative geographies takes such representations seriously; images, as shapers of people's identities and understandings of the world, also shape the world itself. Thus, imaginative geographies blur distinctions between the "real" world and the "fictional" world. That is, they are real not because imaginative geographies accurately depict the world but rather because they have reflected and reinforced people's imagination of the world in tangible and concrete ways.

Human geographers often have looked to literature as a way in which to understand how a novelist such as William Faulkner uses geographic facts of a region and converts those facts into fiction. Such an approach, although interesting, is rather different from how cultural and historical geographers during recent years have used the concept of imaginative geographies, which was first proposed by Palestinian American literary critic Edward Said. In his influential 1978 book *Orientalism,* Said put forth a powerful argument about the ways in which Western colonial powers came to understand non-Western cultures in general—and the "Orient" in particular—as unchanging and primitive. Crucially and related, by casting non-European peoples in such a derogatory light, colonial powers were then able to define their own identities in relation to what the "others" supposedly were not—advanced, dynamic, and sophisticated. This essential opposition between the civilized American or European and the savage native became an essential element in the struggles for domination during the colonial era, vestiges of which remain evident today. Such imaginative geographies, Said posited, clearly are linked to political–economic power and to the asymmetrical social relations that are at the heart of racism.

Imaginative geographies are based, to a very large degree, on the circulation of textual and visual materials that give substance and meaning to such images. Intelligence reports, maps, and popular travel writing became extremely influential means by which people learned about distant places and people. During the 19th century, readers pored over the travel accounts of famous writers, such as Florence Nightingale and Gustave Flaubert, to discover the ruins of Egypt. Imperial Britain employed sophisticated mapping techniques not only to create the spatial image of its Indian empire but also to legitimate its colonialist activities as triumphs of rational science bringing "civilization" to allegedly irrational and despotic Indians. And from the mid-19th century through the early 20th century—during the "age of empire"—photographs played a decisive role in constructing imaginative geographies. With its heightened sense of realism, its ability to produce a feeling of "being there," and its widespread dissemination, the photograph became an active instrument in producing geographic knowledge of distant places, including those that came under colonial rule (Figure 1). Travel writing, maps, and photography remain important vehicles for producing and consuming imaginative geographies, but they have been supplemented by newer media such as television, movies, popular magazines, advertising, and mass tourism.

Two central points stand out from these observations. First, vision and visuality play a fundamental role in the making of imaginative geographies. Although neither Nightingale nor Flaubert spent much time sketching during their travels in Egypt, there is an intensely visual quality to their written images. In other words, viewing, looking, watching, and observing figure prominently in how distant geographies are imagined. Second, imaginative geographies might be something created or fabricated, but this does not mean

Figure 1   "The Geography Lesson" Showing the Merging of Maps, Literature, Art, and Photography in the Construction of Imaginative Geographies

SOURCE: Stereographic daguerreotype by Antoine Claudet, 1851.

that they are devoid of concreteness or substance. On the contrary, both their circulation in material form (e.g., the pictures, globe, and book depicted in Figure 1) and their effects (e.g., the governmental and economic policies undertaken by colonial powers) are very real. Thus, whether one regards imaginative geographies as essentially true or false, there can be no doubt that they have important consequences for how people live.

Human geographers have actively embraced Said's significant concept, especially in their examination of spaces produced by colonial and postcolonial encounters. Some have extended the concept to explore the ways in which children in different countries learn about each other while online, whereas others have considered how imaginary geographies in the Jim Crow South helped to maintain a culture of segregation and a political–economic system of racial oppression. In these cases and in many more, images of cultures and places might be something made or constructed, but those constructions never are made from entirely new cloth; rather, they draw on wide-ranging collective imaginations about geographic difference. The durability and longevity of those very images force geographers to question their own imaginative geographies in the writing they produce.

—Steve Hoelscher

See also Art, Geography and; Colonialism; Cultural Geography; Historical Geography; Identity, Geography and; Literature, Geography and; Orientalism; Other/ Otherness; Photography, Geography and; Postcolonialism; Power; Race and Racism; Text and Textuality; Vision

Suggested Reading

Aiken, C. (1977). Faulkner's Yoknapatawpha County: Geographical fact into fiction. *Geographical Journal, 67,* 1–21.

Driver, F. (1992). Geography's empire: Histories of geographical knowledge. *Environment and Planning D, 10,* 23–40.

Edney, M. (1997). *Mapping an empire: The geographical construction of British India, 1765–1843.* Chicago: University of Chicago Press.

Gregory, D. (1995). Between the book and the lamp: Imaginative geographies of Egypt, 1849–50. *Transactions, Institute of British Geographers, 20,* 29–57.

Hoelscher, S. (2003). Making place, making race: Performances of whiteness in the Jim Crow South. *Annals of the Association of American Geographers, 93,* 657–686.

Holloway, S., & Valentine, G. (2000). Corked hats and Coronation Street: British and New Zealand children's imaginative geographies of the other. *Childhood, 7,* 335–357.

Said, E. (1994). *Orientalism* (2nd ed.). New York: Vintage Books.

Schwartz, J. (1996). The geography lesson: Photographs and the construction of imaginative geographies. *Journal of Historical Geography, 22,* 16–45.

IMPERIALISM

Imperialism is a political relationship of dependency without involving territorial annexation and occupation. This relationship is distinguished by three necessary elements: inequality, domination, and a multiplicity of causes. An imperial relationship is one of effective domination or control over foreign entities through various direct and/or indirect means. Imperial domination includes control over the foreign entities' political, economic, and cultural practices. States, for example, may implement unfair trade practices and treaties. Actual military force is not a necessary condition; the implied use of force may be sufficient to compel other territories to acquiesce to the demands of the more dominant state. The spatial manifestation of imperial relations may be termed *empire.*

This use is contrasted with colonialism, usually defined as the direct physical control of one territory by another. It is possible, from this perspective, to have imperialism without colonialism but not vice versa; colonialism always includes imperialism. Many countries were affected by imperialism but not by colonialism. China and Siam (present-day Thailand), Iran, and Afghanistan, for example, were not colonies but did experience imperialism. In China, as a case in point, Britain established a string of "treaty ports" along the southeastern coast during the mid-19th century. Along with these ports, the British imposed a series of unequal arrangements, including extraterritoriality, which undermined Chinese sovereignty.

The term *imperialism,* derived from the Latin word *imperium,* was first introduced during the mid-18th century and originally used in reference to *Pax Britannica.* During this period, Britain occupied (in the form of colonies) many territories throughout the world; however, Britain also influenced many more kingdoms and states through imperial practices. Retrospectively, the term has been used to describe the early policies and practices of Rome and China. It has also been used to describe the Soviet Union and contemporary U.S. foreign policy (i.e., *Pax Americana*). Similar to colonialism, the term *imperialism* has been transformed to reflect changing global political–economic realities and our understanding of territorial relationships.

During the late 19th and early 20th centuries, the term *imperialism* was significantly transformed and modified through Marxist writings. Imperialism, in classical Marxian use, refers to a particular stage of capitalism. Karl Marx, in his critique of capitalism, premised that once the oppressed and exploited proletariat realized their actual conditions, they would rise up in a spontaneous socialist revolution. This revolt did not materialize among the laboring classes of Europe, however, leading some Marxists to reconsider the inner workings of capitalism. Vladimir Ilyich Lenin in particular set out to explain why the proletariat did not revolt; indeed, at the time of his writing, Lenin witnessed the growth of a privileged proletariat in industrialized states and the concurrent hyperexploitation of peasants in nonindustrialized territories. Lenin's definition of imperialism, therefore, was historically specific. He developed his ideas at a particular moment and in response to specific practices emanating from Europe, namely Europe's partitioning of Africa and Asia. This was a period of intensive colonial domination, one that also witnessed the emergence of corporate monopolies and finance capital. In his classic 1916 book *Imperialism, the Highest Stage of Capitalism,* Lenin agreed with Marx that capitalism was confronted with a series of inherent and inevitable crises and contradictions. Capital tended to "overaccumulate," with idle surpluses of labor coinciding with surpluses of capital. It was imperative, therefore, for capital to continuously expand and seek new sites of investment as well as new sources of materials, markets, and (cheaper) labor. According to Lenin, imperial states would identify other territories to exploit. This could be in the form of colonies or, more indirectly, through the imposition of unequal treaties. Geographer David Harvey later called this a *spatial fix.* Consequently, through the subordination and oppression of indigenous peoples, as well as the monopoly of production, exchange, and consumption, these colonies were hyperexploited. Capitalists could then afford to transfer some of the superprofits to the proletariat within their home countries. In effect, these superprofits were used to buy off and placate the laboring classes with marginally higher wages and better working conditions. In this way, the proletariat benefited from the hyperexploitation of foreign people and thus was less inclined to stage a revolution. Therefore, imperialism, from a Marxian perspective, refers to a situation where the domestic proletariat of highly developed capitalist countries are bought off with the profit of the exploitation of lesser developed territories. Nationalist and racist ideologies may be used to further justify the hyperexploitation of foreign territories.

In contemporary use, imperialism has been narrowly defined as the ideological justification of colonialism. This use is more culturally defined and is especially prevalent in the writings on postcolonialism. Literary critic Edward Said, for example, defined imperialism as the practice, theory, and attitudes of a dominant center ruling a distant territory. Thus, imperialist discourses are found in the writings, paintings, and photographs of foreign peoples and places by colonial settlers, missionaries, merchants, and so on. Most well known is Said's conception of *Orientalism.* Although Orientalism exhibits many different formations, it is simply, at one level, a primitive geopolitical division of the world into two halves: the Orient and the Occident. Therefore, Orientalist discourses encompass a style of thought based on an ontological and epistemological distinction made between the East and the West. This binary vision is oppositional and hierarchical, with the (primitive, savage, and uncivilized) East subordinate to the (modern, enlightened,

and civilized) West. Said maintained that this division was the starting point for many theories, epics, novels, scholarly texts, and political accounts of the Orient. Therefore, imperial discourses, from a postcolonial perspective, were and are used to legitimate and justify the exploitation, oppression, and subjugation of foreign entities.

Recently, scholars have also called into question the intentionality of empire building. For example, some researchers wonder whether there must be a conscious attempt on behalf of policymakers to extend a state's dominance over other territories. This is variously described as "empire light" or an "inadvertent" empire. A concern over intentionality has also appeared in debates surrounding the meaning of power. Political scientist Joseph Nye, for example, has distinguished between "soft" power (the capacity to influence and modify opinion and interests through noncoercive means) and "hard" power (the massive use of military and economic force to compel compliance).

Imperialism and *colonialism* have been used somewhat interchangeably during recent years. Indeed, the different uses of the terms are well illustrated by current writings on U.S. foreign policy. Both David Harvey's book *The New Imperialism* and Derek Gregory's book *The Colonial Present* ostensibly covered the same terrain—the establishment of a Pax Americana, the meanings of the U.S.-led war on terror, and the invasions and occupations of Afghanistan and Iraq. However, one (Harvey) concentrated on imperialism, whereas the other (Gregory) focused on colonialism. Indeed, Gregory was forthright in his use of the term: "I speak about the colonial—rather than the imperial—present because I want to retain the active sense of the verb 'to colonize.'" Both Harvey and Gregory also called into question the temporality of colonialism and imperialism. Harvey witnessed something "new," whereas Gregory identified the continuities between past and present practices. Gregory's employment of the phrase "colonial present," moreover, was in response to postcolonialism.

—*James Tyner*

See also Colonialism; Economic Geography; Globalization; Postcolonialism

Suggested Reading

Gregory, D. (2004). *The colonial present: Afghanistan, Palestine, and Iraq.* Malden, MA: Blackwell.

Harvey, D. (2003). *The new imperialism.* Oxford, UK: Oxford University Press.

Said, E. (1978). *Orientalism.* New York: Pantheon Books.

Said, E. (1993). *Culture and imperialism.* New York: Knopf.

IMPORT SUBSTITUTION INDUSTRIALIZATION

Import substitution industrialization (ISI) is a government strategy designed to encourage domestic industry to supply markets previously served by foreign imports. The goal is economic growth along with a measure of political and economic self-reliance. ISI policies aim to create and nurture infant industries to substitute for imports, as the phrase itself suggests. These new domestic industries may be locally owned (public or private) or foreign owned. The latter, foreign direct investment, usually requires a series of incentives to attract foreign firms to set up production facilities within national boundaries. ISI has been especially significant at one time or another in large economies such as China, India, Japan, Brazil, Mexico, and Argentina. As an inward-oriented strategy, ISI sometimes is contrasted with export-oriented industrialization, although the two are not mutually exclusive and many countries' policies combine elements of both.

ISI originated in Latin America, partly as a result of the twin shocks of the Great Depression and World War II and partly as an innovative and sophisticated set of policy prescriptions. ISI policies involve a complex mix of strategies, including the use of relatively high import tariffs, quota restrictions on imports, and controlled access to foreign exchange. Such protectionist measures to encourage domestic production usually are combined with disincentives to exporters. The idea is to increase industrial capacity and complexity over time by starting with simple consumer goods such as clothing, processed foodstuffs, and pharmaceuticals. ISI theorists prescribed four stages in this process. As firms gained stability and the nation became self-reliant in simple consumer goods, the country would move to more complex products such as stoves, radios and television sets, and automobiles—known as consumer durables. Countries at this second stage that did not have the capacity to manufacture every part of a given product, such as a stove or an automobile, were allowed by Latin American ISI theorists and planners to import the needed components. Thus, the importation of industrial parts was permitted, but that of assembled manufactured goods was not. With the industrial foundation gained during the second stage, nations would concentrate on the third stage, the production of intermediate goods, which specialized in the manufacture of the parts and components for consumer goods. The fourth and final stage emphasized the production of capital goods (e.g., machines to make parts and components), including steel and chemicals needed in the manufacture of heavy machinery. ISI theorists suggested that industrial self-sufficiency would be accomplished during these four stages.

To succeed at industrialization, ISI theorists advised national governments to protect local industry during its formative period from foreign competitors that, because of their head start, were capable of underselling local manufacturers. Latin American countries used several techniques to protect their new industries, including the establishment of import quotas that limited the number of competing foreign goods allowed into the country, the imposition of very high tariffs or duties on those products that were permitted into the country, and the maintenance of overvalued currencies that had the effect of discouraging sales of industrial raw materials abroad. Another common government response—and one that created some political fallout—was the outright nationalization of public utilities, airlines, railroads, mines, steel mills, and other basic industries.

The ISI strategy produced remarkable growth rates—and rapid industrialization—in certain countries in Latin America during the decades of the 1940s to the 1970s. Mexico's sustained high growth rates, for instance, were popularly understood and celebrated as the Mexican miracle during these years. Throughout Latin America and the Caribbean, the ISI strategy became increasingly codified during the 1940s and 1950s. Regional leaders embraced the strategy as a way of leaving behind the devastation of the Great Depression and avoiding these localized problems in the future. The idea was that any country, no matter how small or impoverished, could become more self-reliant and less vulnerable to price fluctuations and shocks in the world economy by industrializing. To pursue this end, government leaders would channel the bulk of the nation's spending into the development of a domestic industrial sector. The products of local industry would then replace imported manufactured goods, saving the country vast sums of money that could be used to raise the standards of living of its citizens.

By the late 1960s, the ISI strategy was being criticized for being inefficient, for failing to improve trade deficits, and for failing to provide sufficient employment. Subsequently, export-oriented industrialization was promoted by the World Bank and neoclassical economists and became a more popular model. On close examination, however, many countries combine elements of both strategies. Economists Raul Prebisch and Hans Singer developed ideas that spurred the implementation of ISI.

—*Altha J. Cravey*

See also Dependency Theory; Developing World; Development Theory; Economic Geography; Modernization Theory; Newly Industrializing Countries; World Systems Theory

Suggested Reading

Gereffi, G. (1990). *Manufacturing miracles: Paths of industrialization in Latin America and East Asia.* Princeton, NJ: Princeton University Press.

Kitching, G. (1989). *Development and underdevelopment in historical perspective.* London: Methuen.

INCUBATOR ZONE

The emergence of incubator zones can be traced to the concept of a business incubator, that is, a shared office space facility that seeks to provide its tenant companies with a strategic, value-adding intervention system of monitoring and business assistance. Business incubators promote early-stage firms' chances of survival with systematic internal support but also facilitate access to services from the external environment such as governments, venture capital firms, universities, and community advisory organizations. The whole networking apparatus constitutes an incubator zone in both a spatial sense and a function sense.

The Batavia Industrial Center, created in 1959 at Batavia, New York, was the first incubator. After more than 40 years, including a significant diffusion of the concept since the 1980s, the original single-dimensioned incubator that mostly provided internal assistance to tenants has evolved into a diversified concept called the *incubator zone* that strives to sustain economic development with a portfolio of methods such as propelling small business success and creating jobs, diversifying the local economic base and promoting regional entrepreneurship, supporting innovative research, and reducing the time from technology adaptation to commercialization. Simultaneously, the original incubator zones concept has diffused from industrialized countries, such as the United States, France, and Sweden, to industrializing countries, such as China, Brazil, India, and Israel.

Incubators may be classified in a variety of ways (e.g., in terms of their primary financial sponsorship or industry focus), yet all successful incubator zones must meet several criteria. First, there must be a satisfactory degree of fit between each incubator's service and needs of the local/regional economy. Otherwise, a lack of appropriate start-ups and networking partners would constrict the successful formation of the zone. Thus, an incubator zone may be viewed as a catalyst embedded in a local/regional economic development plan. Second, an incubator zone serves as a "one-stop shopping" framework in which there are numerous desired products, services, and resources, all conveniently and efficiently colocated and accessible through one organization for the benefit of the entrepreneur and the local/regional economy. Furthermore, the successful zone virtually underpins a consistent and sustainable knowledge spillover that is symbolized by frequent formal and informal communication among entrepreneurs, incubator leaders, incubator employees and tenants, investors, governors, and community leaders that ultimately transforms the incubator zone into an entrepreneurial community.

Incubator zones may be organized around different business models. Some are heavily, if not totally, subsidized by government or community groups. In general, such subsidies greatly enhance the success of incubator zones. Unless deep financial reserves exist to span the relatively long time period that is needed for most early-stage tenant companies to reach revenue or value levels necessary to pay for the full cost of the services provided, incubation efforts often fail. Also, incubator zones often are organized around prominent or propulsive industries in the local/regional economy (e.g., information technology, biotechnology and biomedicine, agriculture, retail specializations). Such initiatives also serve to enhance success because of the strong fit with local regional specializations and resulting synergistic spillover effects.

—*Roger Stough and Junbo Yu*

See also Economic Geography; Innovation, Geography of; Urban and Regional Planning; Urban Entrepreneurialism; Zoning

Suggested Reading

Autio, E., & Kloftsen, M. (1998). A comparative study of two European business incubators. *Journal of Small Business Management, 36,* 30–43.

Hackett, S., & Dilts, D. (2004). A systematic review of business incubation research. *Journal of Technology Transfer, 29,* 55–82.

Lewis, D. (2002). *Does technology incubation work? A critical review of the evidence.* Athens, OH: National Business Incubation Association.

Scaramuzzi, E. (2002). *Incubators in developing countries: Status and development perspectives.* Washington, DC: World Bank.

INDUSTRIAL DISTRICTS

Broadly defined as geographic concentrations of industry, the concept of the industrial district extends to a series of relations among firms, customers, suppliers, and local governance institutions. The concept was once limited to concentrations of manufacturing firms and related activities; however, the scope can be extended to companies in the service sector (e.g., finance, entertainment) and in research-related activities such as information technologies. Although this term has been applied to various industry-specific regions worldwide, it is rooted in descriptions by Marshall of the concentrations of small firms in Britain. Recently, the concept of the industrial district has been applied to other industry-specific regions such as Hollywood and Silicon Valley. The general idea is that the conditions and relationships within industrial districts foster competition, innovation, and subsequent economic growth.

There is no universal definition of what constitutes an industrial district, and this is a source of debate regarding the application of the concept to economic regions. However, there are a number of generally accepted characteristics that define such an economic and social entity. The industrial district consists of a spatially proximate group of firms in a particular industry or in a tightly related group of industries. In most cases, such a region is noted for a lack of vertically integrated large firms. Instead, an industrial district contains a large number of small and medium-sized enterprises involved in flexible specialization. Individual firms perform different functions within the supply, production, and distribution chains of an industry. Most industrial districts are marked by an atmosphere that is conducive to production, with a set of complex relationships among different actors within these regions. There are embedded relationships among firms in these regions, with varying degrees of social and professional networks among firms, in particular among the smaller companies. The benefits that can derive from locating in such a region include proximity to customers, access to skilled workers, nearby supporting industries, and (in most cases) proximity to competitors. Within the district, workers often move from firm to firm because their skills are interchangeable in this industry-focused region. There is marked competition among companies filling similar functions, yet there is a large degree of cooperation among companies. This arrangement in turn spurs innovation and competition among firms within this region. Firms can benefit from the agglomeration of industry, benefiting from external economies. In turn, this production drives economic development across the region.

Geographic research addresses the characteristics of industrial districts such as the local conditions that enable the growth and maintenance, and perhaps cause the potential decline, of these areas. Moreover, it examines spatial relationships among the actors (e.g., labor, firms, government) within the region and with entities outside of the industrial district.

—*Ronald Kalafsky*

See also Agglomeration Economies; Central Business District; Flexible Production

Suggested Reading

Marshall, A. (1890). *Principles of economics.* London: Macmillan.

Porter, M. (1990). *The competitive advantage of nations.* New York: Free Press.

Scott, A. (1988). *New industrial spaces.* London: Pion.

INDUSTRIAL REVOLUTION

Although capitalism has been a dynamic society since its inception, the pace of technological and social change accelerated greatly during the 18th and 19th centuries during the Industrial Revolution. It is important not to equate capitalism with industrialization. Historically, the Industrial Revolution occurred long after capitalism began; indeed, for most of capitalism's history, it involved preindustrial forms of manufacturing, including artisanal, mercantile, and household types of production. However, starting in the mid-1800s, an explosive increase in the speed and productivity of capitalist production occurred that transformed the worlds of work, everyday life, and the global economy. Industrialization is a complex process that involves multiple transformations in inputs, outputs, and technologies. Three dimensions are particularly important here: inanimate energy, technological innovation, and productivity growth.

INANIMATE ENERGY

If preindustrial societies relied on animate sources of energy (i.e., human and animal muscle power), industrialization can be defined loosely as the harnessing of inanimate sources of energy. The first of this type was running water in waterwheels, a source used since the late Middle Ages to grind corn and flour and to saw wood. Running water was a major source of energy during the earliest stages of the Industrial Revolution, but it required firms to locate near streams and rivers; moreover, most streams are ephemeral; that is, they do not flow all year long.

A more efficient source of inanimate energy involves the steam engine, the first of which was built by Scottish engineer James Watt in 1769 as part of an effort to expunge water from coal mines that reached under the ocean. The steam engine marked a turning point in the process of industrialization. Wood provided the first major source of fuel for this invention, which required heating water into steam to drive the engine's pistons. As producers began to cut down forests in Britain in large numbers, deforesting much of the country, wood supplies began to dwindle and the rising costs eroded profits. As wood became scarce, producers switched to coal. Thus, as Britain industrialized, several areas became major coal-producing centers, including Wales and Newcastle. As the Industrial Revolution spread across the face of Europe during the 19th century, the large coal deposits of the Northern European lowlands became increasingly important, fueling the growth of manufacturing complexes in Belgium, northern France, Germany,

Poland, and the Ukraine. In the United States as well, Appalachian coal played a key role in the nation's industrialization. During the late 19th century, coal was joined by other fossil fuels, particularly petroleum and natural gas. The abundance of cheap energy was the lifeblood of industrialization, and production became increasingly energy intensive as a result.

TECHNOLOGICAL INNOVATION

The Industrial Revolution witnessed an explosive jump in the number, diversity, and applications of new technologies. A technology is a means of converting inputs to outputs. These can range from extremely simple to highly sophisticated. As industrialization produced an increasingly sophisticated division of labor, opportunities for new inventions rose rapidly. These were employed in agriculture, manufacturing, transportation and communications, and services.

During the Industrial Revolution, a major reorganization in the nature of work occurred with the development of the factory system and a far more detailed division of labor. Prior to this era, industrial work was organized on a small-scale basis, including home-based work. By the late 18th century, firms in different industries were grouping large numbers of workers together under one roof, a process that effectively created the industrial working class. Inside factories, workers used vast amounts of capital, that is, many types of machines. The introduction of interchangeable parts, invented by American gun maker Eli Whitney, made machines more reliable. By the early 20th century, Henry Ford introduced the moving conveyor belt, which further accelerated the tempo of work and the ability of workers to produce.

PRODUCTIVITY GROWTH

As a consequence of the technological changes of the Industrial Revolution, productivity levels surged. *Productivity* refers to the level of output generated by a given volume of inputs; *productivity increases* refer to higher levels of efficiency, that is, greater levels of output per unit of input (e.g., labor hour or unit of land) or, conversely, fewer inputs per unit of output.

Productivity levels rose exponentially during the 19th century. As the cost of producing goods declined, standards of living rose. Most workers labored long hours and endured standards of living still quite low compared with those we enjoy today. Nonetheless, over several decades, industrialization saw many kinds of goods become increasingly affordable. Because wage rates historically have been linked to the marginal productivity of labor, the working class became better off. Most important in this regard concerns the industrialization of agriculture. As food became progressively cheaper, diets improved as more people ate more and better food than ever before. With the notorious exception of the Irish Potato Famine of the 1840s, hunger and malnutrition gradually declined throughout Europe.

GEOGRAPHY OF THE INDUSTRIAL REVOLUTION

The Industrial Revolution unfolded very unevenly over time and space. Whereas capitalism had its origins in Italy, industrialization was very much a product of northwestern Europe. Some put the first textile factories in Belgium in cities such as Liege and Flanders. However, Britain became the world's first industrialized nation. By the end of the 18th century, Britain stood virtually alone as the world's only industrial economy, a fact that gave it an enormous advantage over its rivals. Cities in the Midlands of Britain, such as Leeds and Manchester, were known as the "workhouses of the world" for their high concentrations of workers, capital, and output and were centers of the early textile industry. Others, such as London, Glasgow, and Liverpool, became shipbuilding centers. In many cities, networks of producers in guns, watches, and light industry formed dense industrial districts.

A half century after it began in Britain, the Industrial Revolution diffused to the European continent, North America, and Japan. In Europe, this saw the formation of industrial complexes in the lower Seine River Valley and Paris. In Italy, the Po River Valley became a major producer of textiles and shoes. In Scandinavia, cities such as Stockholm became major ship producers. In Germany, which was late to industrialize, the Ruhr region became a global center of steel, automobile, and petrochemical firms starting in the 1870s.

By the early 19th century, the Industrial Revolution leapfrogged across the Atlantic, igniting the industrialization of southern New England with the growth of the textile industry there. The Manufacturing Belt became a second great region during the late 19th century, largely on the basis of the steel, rubber, and automobile industries. However, industrialization also dramatically affected agriculture and meatpacking as well. Russia did not become industrialized until the 1920s, when the Soviet Union leaped to become the world's second-largest economy in the span of a decade. Starting in the 1870s, Japan became the first non-Western country to join the industrialized nations. During the 20th century, the process of industrialization diffused to many developing countries, particularly in East Asian newly industrializing countries. In a sense, the industrialization of the developing world, which is still very partial and incomplete, is a continuation of a long-standing historical process.

CYCLES OF INDUSTRIALIZATION

The nature and form of industrialization varied in successive historical periods. Capitalism is prone to long-term cyclical shifts in its composition, often in waves of roughly 50 to 75 years' duration. This saw the rise of different industries at different times. Thus, industrialization was not just one process but rather a series of processes that varied over time and space.

The first wave of the Industrial Revolution, from the 1770s to the 1820s, centered on the textiles industry. In Britain, as in the rest of Europe, North America, Japan, and the developing world today, textiles *always* have led industrialization. This sector, which is easy to enter with few requirements of capital or labor skills, initiated the industrial landscapes of most of the world. Because this wave was centered in Britain, it saw that nation become the leading economic power in the world, initiating the period of the *Pax Britannica.*

The second wave, from the 1820s to the 1880s, was a period of heavy industry. During the 19th century, sectors such as shipbuilding and iron plants were critical. These types of firms, which were large scale and capital intensive, differed markedly from the light industry of textiles. They required massive capital investments, were difficult to enter, and moved toward an oligopoly rather than a competitive market. This was the period during which the U.S. Manufacturing Belt began to form, although most of its growth occurred after the Civil War of the 1860s.

The third wave of industrialization, from the 1880s to the 1930s, saw the appearance of numerous heavy industries, including steel, rubber, glass, and automobiles. This was a period of massive technological change, including capital intensification and automation of work, as well as economic changes. As local markets gave way to national markets, most sectors experienced steady oligopolization, that is, concentration of output and ownership in the hands of a few large firms led by robber barons.

During the fourth wave of industrialization, which started during or immediately after the Great Depression of the 1930s and lasted until the oil shocks of the 1970s, the primary growth sectors were petrochemicals and automobiles. With a relatively stable global economy, this era saw the domination of the world system by the United States, which produced a huge share of the world's industrial output. The fifth wave of industrialization, often held to begin after the oil shocks of the 1970s, has been led by the electronics industry, which was powered by the microelectronics revolution and by the explosive growth of producer services.

During each era, the major propulsive industry was commonly featured as the "high-tech" sector of its day. Thus, just as electronics often is celebrated at this historical moment for its innovativeness and ability to sustain national competitiveness, so too were the textile industry of the 18th century and the steel industry of the 19th century associated with high levels of productivity and wages.

CONSEQUENCES OF THE INDUSTRIAL REVOLUTION

The Industrial Revolution permanently changed the social and spatial fabric of the planet, particularly in the societies that now form the economically developed world. Within a century of its inception, industrialization transformed a series of rural poverty-stricken societies into relatively prosperous urbanized ones.

The Industrial Revolution essentially created the modern working class. For the first time in human history, large numbers of workers labored together using machines. These conditions were quite different from those facing agricultural workers, who were dispersed over large spaces and relied on animate sources of energy. Industrialization gave rise to organized labor markets in which workers were paid by the hour, day, or week. This process was not easy given how brutally exploitative working conditions were during this time. Workers typically labored for 10, 12, or even 14 hours per day, 6 days per week, for relatively low wages. Often the work was unsanitary and dangerous—even lethal—as workers were subjected to accidents, poor

Figure 1   Child Labor in British Textile Factories During the 19th Century

SOURCE: Stutz, F., & Warf, B. (2005). *The world economy: Resources, location, trade, and development* (4th ed.). Upper Saddle River, NJ: Pearson/Prentice Hall. Reprinted with permission from Prentice Hall.

lighting, and poor air quality. Child labor was also common, subjecting those as young as 4 or 5 years of age to horrendous conditions (Figure 1).

As a result of this process, time—like space and so much else—became a commodity, something bought and sold. The transition from agricultural time to industrial time was important. Prior to the Industrial Revolution, people experienced time seasonally and rarely felt the need to be conscious of it. With industrialization, however, time was measured and divided into discrete units as signaled by the factory whistle, bell, and/or stopwatch. This change marked the commodification of time through the labor market.

Industrialization also produced labor unions. The first resistance to employers occurred during the 18th century and involved the British Luddites, who blamed their miserable working conditions on the machines they used and often destroyed them in attempts to halt their exploitation. By the late 19th century, organized labor had created a number of unions in the United States, including the Knights of Labor and the American Federation of Labor; during the 20th century, the Industrial Workers of the World and the Congress of Industrial Organizations were also created. Thus, industrialization often was a period of considerable class conflict.

Geographically, the Industrial Revolution was closely associated with the growth of cities. Almost everywhere, industrialization and urbanization have been virtually simultaneous processes. The reasons why firms concentrated in cities are important. Cities clearly were centers of capital as much as they were centers of labor. There are powerful reasons for firms to concentrate, or agglomerate, in cities. Most firms benefit by having close proximity to other firms, including suppliers of parts and ancillary services, and access to an infrastructure, information, and a labor force. Industrialization changed societies from predominantly rural to predominantly urban in character. In Europe, North America, and Japan, for the first time in history, the majority of people lived in cities. In the United States, for example, the first national census of 1790 showed that 95% of Americans lived in rural areas. The proportion of those living in urban areas increased throughout the 19th century, and by 1920 50% of the nation's population lived in cities. Today, it is roughly 85%.

Industrialization also shaped the population growth rates and demographic composition. On the eve of the Industrial Revolution, famous theorist Thomas Malthus predicted that rapid population growth would create widespread famine. Yet Malthus was soon shown to be wrong, at least in the short run. The industrialization of agriculture generated productivity increases greater than the rate of population growth, and the creation of a stable food supply improved most people's diets. As a result, life expectancy rose. Industrialization also lowered death rates, particularly as malnutrition declined and infant mortality rates dropped. Eventually, public health measures and cleaner water helped to control the spread of most infectious diseases. As death rates dropped, the populations of industrializing countries increased dramatically. This change was also accompanied by a shift from the extended family to the nuclear family. Eventually, industrialization also led to a decline in the birth rate; families had fewer children, and growth rates declined.

Yet another impact of the Industrial Revolution concerned the global economy. Capitalism had formed a loose network of international trade well before the 18th century. The harnessing of inanimate energy for transportation dramatically accelerated the speed of both land and water transportation, forming a significant round of time–space compression. New industrialized forms of transportation not only were faster but also cheaper, resulting in cost–space convergence as well.

These changes dramatically lowered the barriers to trade, and the volume of imports and exports internationally began to soar. Europe, starting with Britain, could import unprocessed raw materials, including cotton, sugar, and metal ores, and export high value-added finished goods, a process that generated large numbers of jobs in Europe and contributed to a steady rise in the standard of living.

Finally, the industrial world economy saw an explosion of international finance. For example, British banks, largely concentrated in London, began to extend their activities on an international basis, lending to clients and investing in markets overseas. Much of the capital that financed the American railroad network was from Britain. Thus, the globalization of production was accompanied by the steady globalization of money and credit.

—*Barney Warf*

See also Fordism; Time–Space Compression; Urbanization

Suggested Reading

Landes, D. (1969). *The unbound Prometheus: Technological change and industrial development in Western Europe from 1750 to the present.* Cambridge, UK: Cambridge University Press.

Stutz, F., & Warf, B. (2005). *The world economy: Resources, location, trade, and development* (4th ed.). Upper Saddle River, NJ: Pearson/Prentice Hall.

INFORMAL ECONOMY

The informal economy entered international development discourse during the early 1970s when an International Labour Organization (ILO) mission to Kenya noted the diversity of small-scale economic activities that were not recognized by public authorities. In addition to what were considered waning traditional occupations, it cited profitable and efficient enterprises. These activities were unrecorded, unregulated, and largely ignored by policymakers and government. Although the ILO study noted the dynamic nature of the informal economy, this economy originally was perceived as consisting of activities that were of little relevance to the formal or "real" economy of a country. Informal occupations were considered to be transitory in nature and were predicted to disappear as development, in the form of economic growth and industrialization, progressed. Today it is widely acknowledged that the informal economy is a fixed characteristic of most countries of the world. Although understandings of informal economic activity and employment arrangements have advanced considerably, full appreciation of this complex phenomenon is far from complete.

Contrary to predictions, informal employment is on the rise in all parts of the world. In 2002, the World Bank estimated that 40% of nonagriculture gross national product in low-income countries and 17% in high-income countries were generated in the informal economy. In developing regions, informal economic activity takes the shape of both self-employment and wage employment in small-scale informal enterprises, including street vending, home-based workshops, and personal services. In advanced capitalist economies, the advance of informal employment is seen in the shift to nonstandard employment relations in the form of hourly and piece-rate employment as well as subcontracting to small informal units and industrial out-workers. And the transition economies of the former Soviet Union and Central and Eastern Europe have seen informal economic arrangements of many types proliferate as economic well-being has faltered. Informal economic activities and employment arrangements in all parts of the world generally do not subscribe to official regulations or provide worker benefits and social protection.

Traditionally perceived as consisting of survivalist activities, the informal economy acquired negative connotations. However, because the majority of informal economic activities provide goods and services whose production and distribution are legal, they are considered "extralegal." The informal economy is widely characterized by low entry requirements (e.g., capital, qualifications), a small scale of operations, skills often acquired outside of formal education, labor-intensive methods of production, and adapted technology. The informal economy does not include the reproductive or care economies when they consist of unpaid domestic activities, nor does it include the criminal economy.

Due to the heterogeneous character of the informal economy, numerous definitions have been elaborated. The Swedish Agency for International Development Cooperation identifies four trends in defining the informal economy: (1) definition by economic unit and whether it complies with official regulations that apply to its trade (e.g., registration, tax payment, zoning); (2) definition by employment category whereby all work that takes place in an income-producing enterprise that is not recognized or regulated by existing legal frameworks is considered informal; (3) definition based on location or place of work, including home-based workers, street traders, itinerant or temporary job workers, and those who labor between home and the streets (e.g., trash sorters); and (4) definition by income or employment-enhancing potential, ranging from modern enterprises (e.g., data processing) to survivalist tasks (e.g., shoeshining, trash sorting, domestic servant work). Although definition by economic unit is the conventional method to define the informal economy, most observers favor definition by employment category whereby all nonstandard wage workers who work without minimum wage and without ensured work or benefits are considered to be informally employed. The array of definitions should be seen not as an obstacle but rather as an aid to understanding the varying nature and manifestations of the informal economy.

Different causal factors at work in different contexts explain the persistence and/or expansion of the informal economy. According to the ILO, factors relating to the pattern of economic growth experienced in many developing countries, widespread economic crisis or economic restructuring throughout the developing world, and the globalization of the world economy explain the persistence and expansion of the informal economy. Recent evidence suggests that the informal economy is less likely to shrink if economic growth does not entail improvements in employment levels and equitable distribution of income and assets. Evidence further suggests that people will voluntarily engage in informal economic activities because of excessive regulation from government or to supplement declining incomes in the formal economy.

Countering early perceptions of the informal economy as insignificant to the formal economies of most countries is evidence of many interdependencies between informal and formal economic activities. The economies are linked through the trade of goods, raw materials, and tools and equipment; the acquisition of skills; and subcontracting relationships. Many individuals participate in both the formal and informal economies. Within the current debate, some highlight the role of the informal economy in stimulating growth in the market economy by promoting flexible labor and/or absorbing labor displaced from the formal economy. However, most observers recognize that linkages and power relations among the informal economy, the formal economy, and the public sector are significant and differ depending on which segment of the informal sector is under consideration.

Although the informal economy is a significant and growing component of urban and rural life in most developing countries, regional differences are significant. In a recent fact-finding study, the Swedish International Development Agency found that in Africa, employment in the informal economy accounts for approximately 80% of nonagricultural jobs, 60% of urban employment, and 90% of all new jobs. In Latin America and the Caribbean, the informal economy accounts for more than 60% of urban employment. In Asia, it accounts for 45% to 85% of nonagricultural employment and 40% to 60% of urban employment. In contrast to the formal economy, self-employment in petty commerce and home-based production drastically exceeds wage employment in the informal economy. Major donors and multilateral agencies recognize that the informal economy holds significant potential for the creation of jobs and the generation of income.

Although not all workers in the informal economy are poor, the majority of the poor are found in the informal economy. The ILO found that the relationship between working in the informal economy and being poor appears when workers are classified by employment status, industry, or trade. Informal incomes decline worldwide as one moves across types of employment. Employers within informal enterprises fare better than owner-operators, who fare better than informal and casual wage workers; industrial out-workers fare the worst. Those who toil in the informal economy are among the most exposed and poor groups within the labor market. Most face inadequate labor legislation, labor protection, and social security. They have limited access to wage workers' organizations and limited bargaining power. Their incomes are low and irregular. Working conditions are poor; they have insecure contracts and few benefits.

The link between poverty and informal employment is strongest for women; according to the ILO, women account for 60% to 80% of the informally employed worldwide. Again, location matters. In India, 96% of all women workers are employed in the informal economy, and most are "invisible." In Mexico, the proportion is 58%, and in South Africa it is 45%. In advanced capitalist countries, women represent the majority of part-time workers, ranging from 98% in Sweden to 68% in Japan and the United States. Women are underrepresented in higher-income employment statuses (e.g., employer, self-employed) and are overrepresented in lower-income statuses (e.g., casual wage worker, industrial out-worker). Children who work in the informal economy predominate in the lowest-paying and most hazardous occupations such as trash sorting, domestic work, and apprenticeships where they face particular dangers and hardships.

Workers and producers in the informal economy are linked in various ways to the global economy. A large share of the workforce in global industries works in export-processing zones, sweatshops, or their own homes under informal employment arrangements. Global commodity chains link individual workers and enterprises, often operating under both formal and informal arrangements and spread across several countries, to one another. Economic globalization can lead to new opportunities in the form of new jobs for wage workers and new markets for the self-employed within the informal economy. In the aggregate, however, globalization tends to lead to shifts from secure forms of employment to insecure ones and to more precarious forms of self-employment, reinforcing the links among poverty, informality, and gender. Consensus exists that neoliberal economic policies promoting liberalization, privatization, industrial reorganization, and migration underlie the continued growth of the informal economy in industrialized, poor, and transition economies.

The informal economy is of increasing concern for policymakers because it is the part of the economy where the majority of populations, especially the poorest ones, support themselves. It accounts for a significant share of employment and output in all regions of the world, and it helps to meet the needs of poor consumers by making accessible low-priced goods and services. Major donors and multilateral agencies stress the importance of addressing the informal economy. In 1999, the ILO formulated a vision of "decent work" for all workers. It recognized that decent work deficits are more common in the informal economy and drew attention to reducing the employment gap, improving rights at work, providing social protection, and increasing the voice of workers.

Several responses to decent work deficits have arisen from grassroots sectors. The Self-Employed Women's Association, the oldest trade union of informal women workers in the world, was founded in 1972 in India. Women in Informal Employment Globalizing and Organizing, a global action research coalition, was formed in 1997. HomeNet, an international alliance of home-based workers, and StreetNet, a similar alliance of street vendors, were formed during the 1990s. The nature of the informal economy makes the right to decent work for all workers especially elusive.

—*Maureen Hays-Mitchell*

See also Developing World; Globalization; Urbanization; World Economy

Suggested Reading

Becker, K. (2004). *The informal economy: Fact finding study.* Stockholm: Swedish Agency for International Development Cooperation.

International Labour Organization. (2002). *Globalization and the informal economy: How global trade and investment impact on the working poor.* Geneva, Switzerland: Author.

International Labour Organization. (2002). *Women and men in the informal economy: A statistical picture.* Geneva, Switzerland: Author.

World Bank. (2002). *World Development Report 2002: Building institutions for markets.* Washington, DC: Author.

INFORMATION ECOLOGY

Information ecology focuses on the environmental impacts of the digital economy in the information age. The key questions to be addressed in information ecology are as follows. To what extent will information technology help improve the environment? Will information eventually substitute for materials and energy as the dominant commodity in the digital economy? What are the deeper connections between information flows and material and energy flows? From a broader perspective, information ecology should be regarded as an expansion of geography's human–environment interaction tradition in the age of ubiquitous computing and instantaneous communication.

Methodologically, information ecology is an offspring of industrial ecology, which focuses on the design and development of industrial manufacturing systems based on the principles and laws found in the ecosystems. Life cycle analysis for specific sectors and input–output analysis for given regions are two commonly used methodologies in information ecology.

The discussion of the interchangeability between information and energy is not a new topic; it can be traced back at least to Maxwell's Demon problem in physics during the late 19th century. Maxwell's Demon challenges the second law of thermodynamics by claiming that information is able to generate free energy starting from a state of maximum entropy. The problem has been approached primarily from three disciplines: physics (e.g., idealized models of our physical environment), engineering (e.g., descriptions of machines dedicated to particular tasks), and economics (e.g., national accounts and models of production and consumption).

In 1993, Daniel Spreng argued that the interchangeability among energy, time, and information can be observed in several instances in physics, engineering, and economics. He also argued that new information technology can be used to substitute time and energy to improve the quality of life without adding stress to the environment. In 1994, Xavier Chen contended that information could replace traditional production factors, such as capital, raw materials, and energy, via its embeddedness into the production factors and its combinations with them.

However, neoclassical economic theory is not capable of modeling information within its traditional framework. According to this theory, a production factor has four intrinsic properties: divisibility, substitutability by other factors, complementarity with other factors, and independence vis-à-vis the other factors. Nonmaterial information, on the other hand, is independent of any of these four properties. It is neither additive nor divisible, and it is neither easily quantifiable nor exhaustible. These characteristics of information make it difficult for scholars to analyze the interchangeability between information and other production factors in the context of traditional neoclassical economic theory.

Information ecology should be an integral component of the geography of the information society. Recent empirical studies have shown that the environmental implications of new information and communication technologies (ICTs) are complex. On the one hand, it has been shown that innovations in ICTs have several effects—dematerialization, decarbonization, and demobilization—on society, thereby saving energy, reducing emissions, and promoting sustainable development. On the other hand, new ICTs were also shown to play a major role in stimulating new rounds of global consumption, thereby increasing demands for both materials and energy—widely known as the *rebound effects*. The net effects of ICTs on the environment are uncertain. Information ecology is still in its infancy, and much more interdisciplinary research is needed.

—*Daniel Sui*

See also Humanistic GIScience; Innovation, Geography of; Political Ecology; Social Informatics; Urban Ecology

Suggested Reading

Chen, X. (1994). Substitution of information for energy: Conceptual background, realities, and limits. *Energy Policy, 22,* 13–23.

Ruth, M. (1995). Information, order, and knowledge in economic and ecological systems: Implications for material and energy use. *Ecological Economics, 13,* 99–114.

Spreng, D. (1993). Possibilities for substitution between energy, time, and information. *Energy Policy, 21,* 15–27.

Sui, D. (2004). GIS, cartography, and the third culture: Geographical imaginations in the computer age. *The Professional Geographer, 56,* 62–72.

Sui, D., & Rejeski, D. (2001). Environmental impacts of the emerging digital economy: The E-for-environment E-commerce? *Environmental Management, 29,* 155–163.

Weinberg, A. (1982). On the relation between information and energy systems: A family of Maxwell's demons. *Interdisciplinary Science Review, 7,* 47–52.

INFRASTRUCTURE

The infrastructure of a country or region refers to the vast network that makes possible the movement of goods, people, and information over time and space. The term generally includes roads, streets, and highways; seaports and airports; hydroelectric dams and power-generating plants; telephone and communications lines; natural gas pipelines; and aqueducts, sewers, and other water control facilities. The development of modern cities would not have been possible without well-developed infrastructures that sutured distant places together into a coherent whole and rolled out the landscapes on which production and circulation could occur. Although certain infrastructural forms existed even before capitalism (e.g., Roman roads), the modern infrastructure arose during the Industrial Revolution when roads, canals, sewers, and

other such networks made possible the circulation of people, capital, goods, and information on a rapid, sustained, and regular basis. Because they exert a huge influence over economic and social activity and land uses, infrastructures involve considerable efforts on the part of urban and regional planners.

Infrastructures are enormously expensive and almost always built by the state rather than by private firms in the market. The reasons for this are many and varied. Few private entities can mobilize the resources necessary to construct such phenomena, which involve large sunk costs and economies of scale. The U.S. national highway system, for example, was the largest single project ever undertaken by the federal government; started in the 1950s, it was not completed until the 1970s, with profound impacts on land uses and economic activity among and within cities.

Moreover, infrastructural goods face the rules of public economics rather than private economics, with very different conceptions of property rights, ownership, and the right of exclusion. In public goods, where there is no excludability and often there are universal service obligations, there always is a "free rider" problem whereby users do not directly pay for the costs of building or maintaining the infrastructure they use. More broadly, infrastructures socialize the costs of transportation for firms while keeping the benefits private. Under the global force of neoliberalism, however, many infrastructure components around the world (e.g., electrical generating systems) are being privatized and sold to private investors.

Infrastructure is a key part of David Harvey's *spatial fix*, that is, the landscapes constructed by capitalists to serve their interests at different historical junctures. The surplus value required to create these projects (the secondary circuit of capital) is huge, their costs must be amortized over long periods, and they change slowly.

Although infrastructures often are studied in technocratic apolitical terms, they are in fact highly political in nature. Who pays (and who does not), who benefits (and who does not), and where roads go (and where they do not) all create changing surfaces of inequality as the positive and negative externalities of the infrastructure are unevenly distributed over space and among different social groups.

—*Barney Warf*

See also Agglomeration Economies; Economic Geography

Suggested Reading

Graham, S., & Marvin, S. (2001). *Splintering urbanism: Networked infrastructures, technological mobilities, and the urban condition.* London: Routledge.

Harvey, D. (1982). *The limits to capital.* Chicago: University of Chicago Press.

INNOVATION, GEOGRAPHY OF

The geography of innovation is the study of human creativity, expressed by the creation and movement of new devices, products, and processes over space. It entails articulating the geographic location of innovation and/or places of innovation production. Such study can be undertaken at the individual, firm, organizational, global, or regional level. Prevailing models suggest that there is a complex association between the rise in technological infrastructure and the origin and production of innovation and that this affiliation is both self-reinforcing and mutually reinforcing. This process can help to reorganize a space economy and/or reinforce regional disparity. Reorganization is clearly evident in the United States, where many southern and western states are showing greater innovation potential than they did previously. The location of innovation between urban and rural areas, on the other hand, still is noticeably asymmetric.

Innovation location is shaped by varied spatial and organizational emphasis on research and development and human development/creativity. The geographic compartmentalization of innovation is strongly shaped by the location of (a) explicit (or codified) and tacit (or noncodified) knowledge; (b) globalization; (c) people, firm, and regional connectivity; (d) places that exhibit unique social and economic characteristics; and (e) those areas that noticeably support a technology-based infrastructure. Regions that possess and foster entrepreneurial and bohemian populations are more apt to be places of innovation. This cultural acquiescence can provide innovative regions with a unique social and cultural basis from which creativity can further develop. Place does matter, and issues of urban agglomeration and localization are important aspects of innovation location. Urban agglomeration can increase support for innovators that are in and around a group of cities that are within proximity to one another. A distance decay effect with respect to the geography of innovation benefits innovators that are

closer together. Localization enables actors in the innovation process to benefit from their proximity within and around a particular city, industrial park, or metropolitan area. This more localized effect on the innovation process contrasts with a trend toward global patterns. Early models on the geography of innovation concentrated on diffusion and location patterns. Mapping the process by which innovation is created, externalized, brought to market, and expanded has become important to understanding the geography of innovation. Previous insensitivities to space or location in the innovation process are being replaced with a greater appreciation of them.

—*Brian Ceh*

See also Agglomeration Economies; Economic Geography; Factors of Production; Flexible Production; Industrial Districts

Suggested Reading

Acs, Z. (2000). *Regional innovation, knowledge, and global change.* London: Pinter.

Acs, Z., & Varga, A. (2002). Geography, endogenous growth, and innovation. *International Regional Science Review, 25,* 132–148.

Audretsch, D., & Feldman, M. (1996). R&D spillovers and the geography of innovation and production. *American Economic Review, 86,* 630–640.

Feldman, M. (1994). *The geography of innovation.* Dordrecht, Netherlands: Kluwer.

Feldman, M., & Florida, R. (1994). The geographic sources of innovation: Technological infrastructure and product innovation in the United States. *Annals of the Association of American Geographers, 84,* 210–229.

INPUT–OUTPUT MODELS

Input–output models are general equilibrium models defined for national or regional economies. They are based on a tableaux of transactions by groups of industries that are referred to as *sectors*. Pioneered by Wassily Leontief, they build on systems of national accounts that track gross product and gross income and also include measures of intermediate transactions. These models decompose all industrial production in whatever region is being modeled into groupings of industries (sectors) that have distinctive production structures. Wassily Leontief's brilliant insight in developing these models was to recognize that economies classified according to these production structures could then be generalized so that the final production measures in national or regional accounts could be mathematically related to measures of industrial output in particular sectors.

Input–output models typically are built around a matrix of transactions. The sales (or output) of any given sector are distributed as sales to other sectors (intermediate sales) and sales to categories of final demand (consumption, investment, government, and exports to other regions). In any sector, the distribution of sales is paralleled by accounts that describe where a sector purchases the inputs needed to produce its output. These are divided among purchases from other sectors, purchases from sources of value added, and imports from other regions. The grouping of industries into sectors is based on an analysis of the structure of input (purchase) requirements such that each sector has a distinctive input structure.

The transactions matrix in an input–output model is generalized into a system of input coefficients, defined by dividing the purchases in a sector into proportions of total purchases. Through appropriate matrix arithmetic manipulations, these input coefficients can be modified to represent sector-to-sector specific "multipliers." The multiplier concept is at the heart of input–output models. These multipliers relate levels of sales (output) to levels of final demand in specific pairs of sectors. Input–output models frequently are specified so that the multiplier measures are expressed in multiple ways such as measures of industrial output, employment, and labor income.

Input–output models can be estimated by surveys of national or regional economies. However, they frequently are estimated for regions through nonsurvey approaches and have various degrees of "closure." In regional applications, it is common to include in the multiplier structure of these models components of the income and final demand accounts that are strongly tied to levels of local production such as personal consumption expenditures and labor income.

Input–output models are conceptually similar to several other widely used models of multipliers, including simple economic base models and regional econometric models. Simple economic base models do not distinguish the magnitude of multipliers of individual sectors; instead, they simply describe total output or employment as a function of export activity.

Regional econometric models develop more complex multiplier relations than those found in regional input–output models. They extend the mathematics of these models to deal with temporal patterns in multipliers and provide more sophisticated measures of impact than those provided by input–output models.

—*William Beyers*

Suggested Reading

Leontief, W. (1936). Quantitative input–output relations in the economic system of the United States. *American Economic Review, 64,* 823–834.

Miller, R., & Blair, P. (1985). *Input–output analysis: Foundations and extensions.* Englewood Cliffs, NJ: Prentice Hall.

INSTITUTIONS

The term *institutions* refers to assemblages of people and resources under socially recognized affiliations in private and public sectors. Institutions are widely recognized as important analytic objects in geography and the social sciences because of their production and control of resources (e.g., jobs, government subsidies, mortgage loans, access to public services and facilities). Institutions practice control of resources through rules of conduct, regulation of provider populations, and procedures of allocation. This science of allocation, far from being neutral and value free, flows out of an instrumentalist–political reality whose specifics are much debated. The two dominant schools of thought, Weberian and Marxist, posit different political realities. Whereas Weberians emphasize the drive of institutions to bureaucratically grow and self-perpetuate, Marxists speak of the drive of institutions to either directly accumulate capital or support the capital accumulation process of political–economic elites. In this context, each recognizes that institutions profoundly shape social and spatial patterns in their everyday functioning.

In their most powerful phase, institutions can individually or collectively evolve into bureaucracies whose hierarchies of managers, regulatory procedures, and modes of conduct organize an influential bureaucratic vision of social life and society. This vision reduces the complexities of subtle social issues to the arena of quick and efficient management and control. The result is the widespread acceptance of such narrowed issues (and how they should be dealt with) to the easy realms of right versus wrong, good versus bad, or desirable versus undesirable.

In human geography today, new institutional forms that transcend the simple public–private sector dichotomy are posited as emerging. Currently, the human landscape is seen to be populated by the likes of public–private partners, market-based local governments, and "quangos" (i.e., quasi-autonomous government organizations). Each represents the sense of a blurred public–private distinction as institutions carry important elements of both in their daily actions. These new institutional forms, unlike previous ones, embed the power of the public sector (e.g., sanctioned rules and regulations, subsidies, legitimacy affiliated with government intentions) and the private sector (e.g., capital, agglomerations of workers). In this context, many geographers today speak about the growing institutionalization of power in societies such as the United States and the United Kingdom.

—*David Wilson*

See also Marxism, Geography and; Political Geography; Power; Urban Social Movements

Suggested Reading

Giddens, A. (1981). Introduction. In M. Weber (Ed.), *The Protestant ethic and the spirit of capitalism* (pp. 1–12). London: Routledge.

INTERVIEWING

Interviewing is a research method in which information, opinions, and/or stories are gathered verbally from subjects in person or by remote means such as telephone or e-mail. Although interview data can be quantified, interviews typically are a technique used in qualitative research. Interviews are open-ended, meaning that the researcher has a list of questions with no predetermined range of answers (as one would have in a closed-ended survey) and so listens to and records the subjects' responses. Interviews can be highly structured, where a set list of questions is posed to each respondent, or semistructured, where the researcher follows the questions or themes more loosely, adding spontaneous follow-up questions depending on the responses and the flow of conversation. Indeed, many researchers stress the importance of viewing interviews as conversations with their subjects, rather than merely orally conducted surveys, so that the richness of individuals' experiences and thoughts can be explored. Interviews may involve a one-on-one conversation between the researcher and the subject, or they may be used in a focus group setting (although focus groups have somewhat different dynamics both in practical terms and in terms of research outcomes).

Interviewing is an important skill for qualitative researchers. First, interviewers should be knowledgeable about the topic or themes covered so that they do not waste respondents' time on trivia, yet they should maintain their scholarly curiosity to the extent that appropriate follow-up questions may be constructed spontaneously. Second, interviewers should be well organized, structuring questions with care (e.g., starting with a general question and moving toward greater detail), performing test runs with a small sample of respondents to evaluate the questions and practice asking them smoothly, and testing any recording equipment before the interview. Finally, interviewers should be aware that whereas good social skills will make interviewing more successful, practice does make interviews flow more smoothly and feel more comfortable. Some important things that interviewers should bear in mind are to be open-minded with respondents (even if they disagree with things people are saying), to really *listen* to what people are saying both to understand them and to construct spontaneous follow-up questions, and to contact respondents later with clarifications or further questions.

Interviews have been used increasingly by geographers during recent decades as qualitative research has experienced a resurgence, and critical perspectives have shaped the practice of geography. Here feminists have been especially important both in incorporating interviewing as a research method and in holding up the technique for critical evaluation. Interviews have been the focus of a great deal of critical discussion by feminists and others, in part because they often involve a power difference (by race, gender, class, age, etc.) between the researcher and the subject; sometimes the researcher has the more powerful social status, and sometimes the subject does. Particularly in research on marginalized social groups, researchers have been increasingly sensitive to ethical questions of exploitation of respondents for research and career gains, to issues of how knowledge produced from interviews can be represented fairly, and to the end result of research in terms of social change, advocacy, and/or empowerment for the subject groups.

—Meghan Cope

See also Feminist Methodologies; Participant Observation; Qualitative Research

Suggested Reading

Dunn, K. (2005). Interviewing. In I. Hay (Ed.), *Qualitative research methods in human geography* (2nd ed., pp. 79–105). Oxford, UK: Oxford University Press.

Fontana, A., & Frey, J. (2000). The interview: From structured questions to negotiated text. In N. Denzin & Y. Lincoln (Eds.), *Handbook of qualitative research* (2nd ed., pp. 645–672). Thousand Oaks, CA: Sage.

Kvale, S. (1996). *Interviews: An introduction to qualitative research interviewing.* Thousand Oaks, CA: Sage.

Limb, M., & Dwyer, C. (2001). *Qualitative methodologies for geographers: Issues and debates.* London: Edward Arnold.

INVASION–SUCCESSION

The terms *invasion* and *succession* in human geography reflect the influence of Charles Darwin and evolution as well as the influence of Ernst Haeckel and Frederick Clements and ecology. Important 19th-century thinkers such as Herbert Spencer adapted their principles to explain the changing social order. Social arrangements became the outcome of competition for advantageous locations. New entrants into an area disturbed old forms, setting off competition for dominance and generating a new set of land uses and new social organization.

Such explanations permeated understandings of American geography. Historian Frederick Jackson Turner incorporated invasion and succession into his influential frontier thesis developed in 1893. According to Turner, successive waves of settlement (or *invasion*), each pushing the frontier farther west, created a distinctive American personality—buoyant, self-reliant, and enterprising. Each group recapitulated the experience of those groups that came before, with each stage ultimately leading toward mature settlement (or *succession*).

Turner drew directly on the work of the U.S. Bureau of the Census's Francis Walker and Henry Gannett, who tracked America's population growth and viewed it as inexorably progressive. According to Walker's *Statistical Atlas of the United States,* published in 1874, the country was composed of three major settlement categories: wilderness, frontier, and fully settled. Wilderness would turn to frontier after invasion by early entrants, scouts, and trappers. Ranchers followed, and then farmers followed after that. Ultimately, cities would sprout—the final phase (or, in ecology terms, the climax phase) of human settlement. Each density level signaled a specific social configuration and associated economic form.

In urban geography, invasion–succession is most closely associated with the 1920s work of Chicago School urbanists Robert Park, Ernest Burgess, and Roderick McKenzie. The period was one of dynamic urban growth. The Chicago School used empirical methods to analyze urban spatial patterns. The researchers drew explicit analogies between behavior and processes shaping plant and human communities. Cities were composed of distinctive interdependent areas comparable to ecological communities, each dominated by its own characteristic land use and population. The city's zones ranged in function (e.g., central business district, residential zone) and desirability (e.g., better residences, slum areas). Invasion–succession was the process that explained change. Burgess's concentric zone model of the city placed the central business district at the center, the site for the most successful economic competitors. New immigrant groups had to vie for space with more advantaged groups; therefore, opportunistically, they could invade only the less desirable areas near their jobs—the vacant lots of ecology. Each new immigrant group would be followed by others like it, creating a mosaic of ethnic neighborhoods over time.

Invasion–succession explanations for development have been critiqued and changed. Turner's America never became fully settled in terms of density, yet high-tech cities are located in the midst of low-density settlement. Urban patterns are driven by capital flows in discussions of gentrification and abandonment, and the processes are viewed as anything but natural.

—Deborah Popper

See also Chicago School; Housing and Housing Markets; Neighborhood; Urban Geography

Suggested Reading

Park, R., Burgess, E., & McKenzie, R. (Eds.). (1925). *The city.* Chicago: University of Chicago Press.

Popper, D., Lang, R., & Popper, F. (2000). From maps to myth: The census, Turner, and the frontier. *Journal of American and Comparative Culture, 23,* 91–102.

Vasishth, A., & Sloane, D. (2002). Returning to ecology: An ecosystem approach to understanding the city. In M. Dear (Ed.), *From Chicago to L.A.: Making sense of urban theory* (pp. 343–366). Thousand Oaks, CA: Sage.

J

JUSTICE, GEOGRAPHY OF

Although there are multiple definitions of social justice, two leading notions deal with issues of *procedural* justice and *distributional* justice. Procedural justice issues in the United States traditionally have focused on libertarian ideals of personal rights, where individual freedoms and equality of opportunity are hallmarks of democracy and the free market system. These issues of justice are codified into documents such as the Declaration of Independence and the Bill of Rights, in which individuals are granted the opportunity to pursue goals in life with minimal interference from the government or the community. But even as these words were enshrined in the nation's founding documents, slaves were considered private property and women were not given the right to vote in presidential elections until the 1920s. Even today, widespread social injustice still occurs on the basis of race, gender, class, religion, and other social categories.

Although we expect to be treated equally and have the same opportunities as everyone else, we should not expect the *outcomes* in a democratic society to be equal or even fair. Notions of distributional justice focus on these inequalities that are inherent in society, where oppression of the weak by the powerful and discrimination against those who are deemed to be "different" is all too real. Distributional justice focuses on how resources and rights can be redistributed within society to combat these unjust outcomes such as poverty and prejudice. Concern for others, a focus on the community over the individual, and a redistribution of wealth all are important aspects of distributional justice. The argument here is that for

a democracy to function, individuals must also share the responsibility of sustaining society by eliminating social and economic injustices. In other words, there is a paradoxical element in social justice; we cannot enjoy our own individual rights without respecting the rights of others in our community.

Human geographers are concerned with social justice because social inequality is formed in geographic ways; people are discriminated against because they live in segregated places or are pushed into spaces at the periphery of society. For example, racial minorities sometimes are relegated to living in marginal neighborhoods, and homeless people are discriminated against in public spaces such as city parks and streets. Therefore, promoting social justice involves making geographic decisions and examining the geographic spaces in which these injustices occur. Only then can one formulate strategies on how to eliminate the barriers that cause these injustices.

For the most part, social justice issues in geography took root during the 1960s, when major cultural upheaval, such as the civil rights movement and massive protests against the Vietnam War, helped to make social justice an issue to be taken seriously. At the same time, much of social science was being criticized for being too preoccupied with using dehumanized quantitative measurements to explain complicated human behavior. Hence, attention was diverted to helping solve social problems and addressing injustice. As geography moved forward during the 1970s, much attention was paid to the processes responsible for spatial disparity in people's lives and how that created various "spatial injustices." In effect, social values became a more important element in the practice of human geography. During the 1980s and to the present day, notions of social justice are tied in with the shift to postmodern intellectual inquiry, which questions claims of truth by the dominant powers in society and emphasizes human diversity and human difference. There is also a reemphasis on studying localities by investigating how people in particular places hold particular views they deem to be good or bad, right or wrong, just or unjust. These issues are important for understanding and empowering social group identities and the impact they have on the local culture.

Geographic concepts of scale are important here because social injustice occurs at the scales of the body, the household, the community, the region, the nation, and even the globe. For example, discrimination and prejudice (and sometimes violence) take place through "othering" of bodies by dominant members of society such as ignoring the rights of the disabled because their bodies are deemed to be "unhealthy" or "unnatural." At the community level, a waste incinerator may be placed within a poor and racially segregated neighborhood, causing health problems for residents who, because of their marginal position in society, are powerless to stop it. At the international scale, large groups of immigrant workers may cross national borders, seeking to support their families, and become subject to oppression, discrimination, and fear. Whatever the scale of injustice, these issues cannot be separated from the geographies in which they occur.

These geographies also manifest themselves as social justice movements across many different landscapes and public spaces. These movements range from battles for civil rights in the American South during the 1960s to radical acquired immunodeficiency syndrome (AIDS) activism during the 1980s and 1990s. They are designed to upset the social order through disruption of established social space and to call attention to injustice, domination, and oppression. All of these contested cultural politics have one thing in common, namely that they create or contest "geographies of power" by simply asking the question, "Who has the right to this space?" These groups of people are bound together by common identities and common goals in which they seek to reconstruct the world in such a way that they may live on their own terms and not on the terms of those who seek to control them. Subjugated groups, such as women, gays, and people of color, want to represent themselves in the spaces of the countryside, the city, and the nation as well as on the global stage. Their struggles over how and what spaces they wish to represent are indeed played out through creation of geographies of social justice.

—*Thomas Chapman*

See also Environmental Justice; Ethics, Geography and; Political Geography

Suggested Reading

Adams, M., Blumenfeld, W., Castaneda, R., Hackman, H., Peters, M., & Zuniga, X. (Eds.). (2000). *Readings for diversity and social justice: An anthology on racism, anti-Semitism, sexism, heterosexism, ableism, and classism.* New York: Routledge.

Harvey, D. (2000). *Spaces of hope.* Berkeley: University of California Press.

Merrett, C. (2004). Social justice: What is it? Why teach it? *Journal of Geography, 103,* 93–101.

Mitchell, D. (2000). *Cultural geography: A critical introduction.* Oxford, UK: Blackwell.

Mitchell, D. (2003). *The right to the city: Social justice and the fight for public space.* New York: Guilford.

Smith, D. (1994). *Geography and social justice.* Oxford, UK: Blackwell.

Young, I. (1990). *Justice and the politics of difference.* Princeton, NJ: Princeton University Press.

L

LABOR, GEOGRAPHY OF

Historically, issues of workers and labor have been the domain of economic geographers and addressed largely by industrial location theory. Arguably, the most significant work in this subfield was based on the models of Alfred Weber, a German economist and sociologist and the brother of famous sociologist Max Weber. Principally, Weber tried to show how spatial variations in labor costs shape where industries locate. However, during the early 1970s, Weberian locational analysis was criticized by a new generation of geographers influenced by political economy, particularly Marxism. These geographers argued that Weber's approach did not account for the complexity of labor as a category; for instance, it ignored issues such as the gender makeup of the labor force and whether workers were unionized or not. Furthermore, it implicitly examined the making of economic landscapes from the perspective of industrialists/capitalists, who were seeking to locate new industrial facilities, rather than from the perspective of workers themselves.

Consequently, economic geographers began to examine the category of labor in different ways and in greater depth. Some, such as David Harvey, showed how under capitalism the need to extract surplus labor from workers shaped how economic landscapes are made. Thus, Harvey argued, if capitalists were to be profitable, they needed to arrange their operations in particular geographic ways, perhaps by relocating labor-intensive operations to developing countries overseas while keeping capital-intensive operations in economically core countries such as the United States and Britain. Others contended that although labor power

was a commodity that needed to be purchased by capitalists—that is, that workers needed to be paid for their time spent working—it was a commodity unlike other elements of production such as iron ore and electricity. Instead, they maintained, labor is a pseudocommodity given that worker behavior can drastically affect the labor process, unlike in the case of other commodities. Furthermore, a number of researchers showed how labor can be an important locational factor even in highly capital-intensive industries; for example, many companies choose to locate their research and development operations close to major universities, hoping to make use of the intellectual resources available in the labor pools located around such institutions.

By the 1990s, however, a new set of issues was being addressed with regard to labor. A number of authors argued that whereas these earlier critiques of Weberian locational theory were important elements in further deconstructing the category of labor, they nevertheless still viewed labor from the perspective of how industrialists make investment decisions—an approach termed the *geography of labor* by Andrew Herod. In contrast, Herod argued for what he called a *labor geography,* meaning an approach to theorizing labor that centered on how workers are embedded within particular geographic structures, how geographic considerations affect workers' political and economic decision-making processes, and how workers directly and indirectly shape the making of capitalism's geography. Rather than viewing labor from the perspective of how capitalists make locational decisions, this approach focused on workers as spatial actors. In this vein, a broad-ranging body of work developed showing how labor markets operate as spatial structures and how these serve to embed workers in particular locations geographically, how workers and employers struggle over the geographic scale at which things such as union contracts will operate, how traditions of labor militancy or quiescence may be diffused across space, and how practices of trade unionism are influenced by spatial considerations.

During recent years, there has been growing cross-disciplinary interaction between geographers and industrial relations specialists with regard to how workers' political and economic practices both shape space and are shaped by space. Equally, issues of labor now are being addressed beyond the subdiscipline of economic geography as researchers focus on cultural and political aspects of the geography of labor. For example, researchers influenced by poststructuralism have begun to question what it even means to talk of work and workers. Thus, is work only those activities for which we receive a wage, or do other types of non-paid activities (e.g., housework) also count as work? If so, what does this mean for understanding the category of labor, and how is it constituted geographically?

—Andrew Herod

See also Domestic Sphere; Economic Geography; Labor Theory of Value; Location Theory; Marxism, Geography and

Suggested Reading

Harvey, D. (1982). The limits to capital. Oxford, UK: Blackwell.

Herod, A. (1997). From a geography of labor to a labor geography: Labor's spatial fix and the geography of capitalism. Antipode, 29, 1–31.

Herod, A. (Ed.). (1998). Organizing the landscape: Geographical perspectives on labor unionism. Minneapolis: University of Minnesota Press.

Herod, A. (2001). Labor geographies: Workers and the landscapes of capitalism. New York: Guilford.

Martin, R., Sunley, P., & Wills, J. (1996). Union retreat and the regions: The shrinking landscape of organised labour. London: Jessica Kingsley.

Peck. J. (1996). Work–place: The social regulation of labor markets. New York: Guilford.

Wills, J. (1996). Geographies of trade unionism: Translating traditions across space and time. Antipode, 28, 352–378.

LABOR THEORY OF VALUE

The labor theory of value has long been one of the central pillars of political economy, particularly Marxism. Contrary to popular opinion, Karl Marx did not invent this notion—it was also used by Adam Smith and David Ricardo—but he gave it its most famous and thorough exposition.

This view privileges labor analytically above all other forms of human activity. To meet their needs, people must work, that is, engage in collective labor. Thus, through labor, people enter into social relations, reproduce society, materialize abstract ideas, and change nature. No good or service has value without labor time embodied in it. Thus, the labor theory of value is transhistorical; it applies to all societies. The organization of labor, however, varies from society to society (or among modes of production); in capitalism, it occurs primarily through the labor market.

The labor theory of value begins with the distinction between use values and exchange values. Use values refer to the qualitative, non-market-based qualities of a good (e.g., a house shelters someone, a drink quenches someone's thirst), whereas exchange values refer to the quantitative market price of the good. Marx argued that value was produced in the workplace and not in the market, where values were only realized. Thus, value was a reflection of the physical social process of labor and not the ethereal mental world of utility maximization. Marx maintained that the value of a good to a society reflected the socially necessary labor time that went into its production; for example, if one individual spends 20 years making a pair of shoes, they are not worth much because the labor time embodied in them would be far above the socially necessary amount. Over time, prices reflect the labor time embodied in the production of goods. Machinery and equipment, or organic capital in Marx's terms, represent embodied labor time of previous rounds of labor.

The centerpiece of the labor theory of value is the manner in which surplus value is extracted. Labor, like everything else in capitalism, is a commodity, and the cost of producing laborers reflects the amount of socially necessary labor time that goes into socializing them and keeping them able to produce. In the labor market, the employer (or capitalist) will seek to pay a worker only the amount necessary to reproduce the worker, which is less than the amount of output that the worker generates in a workday. Thus, a certain share of the worker's output is appropriated by the capitalist as surplus value; the worker's use value to the capitalist is greater than worker's exchange value (and so he or she is alienated from the production process). Far from being an exchange of equals (between labor time and wages) as in neoclassical economics, this view holds that all labor markets, by definition, are inherently exploitative. The contentious politics of this relation is manifested in struggles over wage rates, the length of the workday, the pace of work, and so on. This view also underpins Marx's claim that all profits are thefts from the workers. Viewing labor in this way is necessary to get behind the fetishization of commodities, revealing them as social products and not as isolated things.

The production process and the extraction of surplus value are collective processes and not individual ones. Structurally, firms and workers generate output in the workplace, and the output is sold on the market and converted to money form. Once the output is sold and firms recoup the costs of their capital investments, the resulting surplus is divided between workers and firms in the forms of wages and net profits, respectively. This division is decided via class conflict and struggle—not via supply and demand—and forms the basis for class formation and class war.

Firms, under the constant pressure of competition, threat of ruin, and lure of profits, must continually seek to restructure the production process. This may involve technological change that increases the productivity of labor and the creation of a surplus. Improvements in transportation times also accelerate the turnover of capital—its conversion from commodity to money form and back again—unleashing a wave of time–space compression. Or, restructuring may involve lowering the costs of reproduction of labor by offloading them to the state in the forms of collectively provided goods and services (e.g., infrastructure). Marx also held that capitalists maintain a reserve army of the unemployed—workers desperate for jobs even at low wages—and they serve to hold down wages and thus increase the rate of surplus profit extraction.

The existence of surplus value indicates that workers never can buy all that they produce; there is an inherent contradiction between production and consumption. Firms must seek to maximize production or else their competitors will drive them out of business; in lowering the costs of labor, they reduce the possibilities of consuming the goods they produce. Thus, capitalism has a chronic, inevitable, built-in tendency to overproduce, generating a surplus that cannot be consumed, leading to lower prices for outputs and thus lower profits and wages, in turn leading to bankruptcies and heightened class conflict.

The labor theory of value holds that markets are inherently anarchical and crisis ridden. Some of the surplus value can be siphoned off in the forms of investment in the built environment or accumulated in money form as financial capital. Because the production process is continually subject to overproduction of surplus value, it forms the basis for what Marx called capitalism's "wolf hunger," that is, the need to expand continually in search of new markets. Thus, the labor theory of value offers a useful means to understand the historical expansion of capitalism since the 16th century as it spilled out of Europe and conquered a large variety of precapitalist social formations around the world.

Geographers have used the labor theory of value in a variety of ways. David Harvey relied on it as the basis for his notion of a spatial fix, that is, the particular geographic constellation of production that serves the interest of capital for a given window of time until it becomes rendered obsolete by the ongoing need to restructure and expand. In this reading, capitalism's landscapes are simultaneously the crowning glory of past capitalist development and a prison that inhibits the further progress of accumulation, as Harvey put it, as the process creates barriers to its own further development. This view allows us to see landscapes as struggled over—the political products of class conflict—rather than as serene outcomes of perfectly functioning markets. The labor theory of value also forms a common means of approaching the spatial division of labor at multiple spatial scales, that is, the tendency of capitalism to segregate different types of labor in different regions. Wealth and poverty in this view are two sides of the same coin. Core regions (e.g., the traditional Manufacturing Belt, the world's First World countries) benefit because of their ability to extract surplus value from peripheral regions (e.g., the South, the developing world). Thus, the labor theory of value is central to views of global political economy such as dependency theory and world systems theory.

—*Barney Warf*

See also Economic Geography; Labor, Geography of; Marxism, Geography and; Uneven Development

Suggested Reading

Harvey, D. (1982). *The limits to capital.* Chicago: University of Chicago Press.

Harvey, D. (1985). The geopolitics of capitalism. In D. Gregory & J. Urry (Eds.), *Social relations and spatial structures* (pp. 128–163). New York: St. Martin's.

LANGUAGES, GEOGRAPHY OF

Language (in contrast to speech) may be understood in several ways: (a) as a means of organizing thought, (b) as a way of communicating for producing and sharing meaning, and (c) as a vehicle for bringing the world into consciousness. Thus, language is simultaneously a psychological, social, and cultural phenomenon. In many countries (e.g., Belgium, Canada), languages have deep political significance. Because languages are unevenly distributed across space, they are also inherently geographic as well.

Because languages are semantically and historically related to one another, it is common to group them into families of varying sizes. Linguists and cultural geographers typically maintain that there are roughly eight major language families as well as several others termed *isolates.*

By far the largest and most widespread of the major language families is the Indo-European, a group first identified by linguist William Jones during the 18th century. Starting with the migrations of the so-called Aryans circa 1500–2000 BC, perhaps as a result of their domestication of the horse, Indo-Europeans moved in two directions from their homeland near the Caucuses Mountains. One group moved east into northern India, becoming the basis of the Sanskrit-based Indic languages such as Hindi, Bengali, Urdu, Gujarati, Bihari, Marathi, and Nepali. Others remained in the Middle East, where they eventually became the Iranic family, including Farsi (formerly Persian), Kurdish, Armenian, and (in Afghanistan) Pashto. The other major branch of Indo-Europeans moved into Europe, where they diverged into several groups. These include the Latin-based Romance languages (Italian, Spanish, Portuguese, Catalan, French, Romansch, and Romanian) that arose during the disintegration of the Roman Empire. Greek and Albanian form separate categories in their own right. Farther north, the Germanic languages include German, Dutch, the Scandinavian tongues, and English. Celtic, an early branch once widespread throughout Western Europe, today consists of Scottish and Irish Gaelic, Welsh, Breton in western France, and extinct tongues such as Cornish; this branch is in danger of disappearing. In Eastern Europe and Russia, the Slavic branch includes Polish, Russian, Ukrainian, Czech, Slovak, Slovenian, Serbo-Croatian, and Bulgarian. The Baltic group of Lithuanian and Latvian is another.

With the expansion of Spanish, Portuguese, French, and British colonialism, Indo-European languages were carried throughout much of the world, becoming dominant throughout the New World, Australia, and New Zealand (Figure 1). Today roughly half of the world speaks an Indo-European tongue of one sort or another. English in particular, riding the heels of the British and American empires, has become the *lingua franca* spoken by more people than any other tongue (when second-language speakers are included). English is unquestionably the world's dominant language in commerce, trade, scholarly publications, airlines, international finance, and tourism.

A second major language family is Afro-Asiatic, which extends across the Middle East and North Africa (Figure 2). This group includes most of the extinct or nearly extinct languages of the ancient Middle East such as Canaanite, Phoenician, Assyrian, and Aramaic (of which pockets survive). The dominant branch of Afro-Asiatic is Semitic, which includes Arabic (with numerous dialects) and Hebrew, which was a nearly extinct language before it was revived by Zionists at the end of the 19th century. Other branches include Berber, widespread in the Mahgreb, and Kushitic, the dominant family of Ethiopia (i.e., Amharic) and Somalia.

Ural-Altaic, also sometimes called Finno-Ugric, comprises a third family. The origins of this group, probably near the Altai Mountains of Mongolia, are lost in prehistory. It is likely that speakers of this family are descendants of several waves of migration that generated populations that continue to speak loosely related tongues stretched across Eurasia (Figure 3). Finnish and Estonian are one example, and Hungarian, the language of the Magyar who settled in Eastern Europe during the 8th century, is another. A third branch is the Turkic languages, all of which emanated from the Turkish migrations into Central Asia and Anatolia during the 9th and 10th centuries; remaining Turkic languages include Turkish, Azerbaijani, Kazakh, Uzbek, Turkmen, Kyrghiz, and

Figure 1   Countries in Which Indo-European Languages Are Dominant

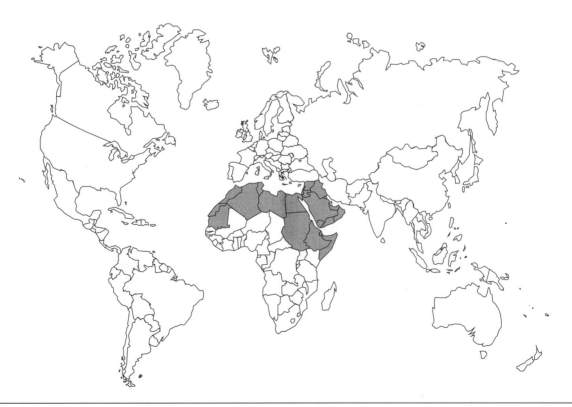

Figure 2   Countries in Which Afro-Asiatic Languages Are Dominant

Figure 3   Countries in Which Ural-Altaic Languages Are Dominant

Figure 4   Countries in Which Bantu or Niger-Kordofanian Languages Are Dominant

(in western China) Uighur. Yet another branch is Mongol and Manchu, which were formerly spoken in Manchuria but now are extinct, as are indigenous tongues of Siberia such as Samoyed and Tungic. Finally, many linguists assign Japanese and Korean to this family as well, although this is controversial.

Africa south of the Sahara desert is a complex mosaic of tongues from several language families. In addition to Afro-Asiatic languages in the north (Arabic and Berber) and in the Horn of Africa (Amharic and Somali), it has smaller families such as Nilo-Saharan and (in southwest Africa) the famous click languages of the Khoisan family (e.g., !Kung). However, the bulk of the many languages spoken throughout this vast region fall under the Bantu or Niger-Kordofanian language family, which includes thousands of tongues (Figure 4). Arising from the migrations of agriculturalists from Central Africa around the time of Christ, this family includes languages as diverse as Mande in West Africa; Kikuyu in Kenya; and Tswana, Nbele, and Zulu in Southern Africa. Along the eastern part of the continent, Swahili, a trade language that combined words of different languages (including some from Arabic), has long formed a lingua franca.

In East Asia, the Sino-Tibetan language family is the most commonly spoken group (Figure 5). Common to this group is the use of tones (although these are found in some African languages as well), in which pitch is part of the meaning of the word. This family includes Chinese, which embraces a variety of languages that are not mutually intelligible but that use a common writing system (a feat made possible only with a pictographic writing system, not an alphabet). Chinese includes Mandarin, the dominant language of northern China (and the most commonly spoken language at home in the world) and close to a national tongue (Szechwanese is a dialect), as well as Cantonese and lesser-known ones such as Shanghainese, Wu, Hakka, and Fukienese or Taiwanese. This group also includes Tibetan and Burmese (the latter because of Tibetan migrations down the Irrawaddy River).

A sixth major family is Malayo-Polynesian, a diverse group that extends across much of Southeast Asia into the islands of Polynesia and Micronesia (Figure 6), thereby including Hawaiian and Maori (in New Zealand). Originating among tribes in Taiwan, this group includes the dominant Bahasa languages of

Malaysia and Indonesia (each with countless dialects) and the numerous tongues of the Philippines, of which Tagalog is the best known. Circa 500 AD, Indonesian sailors crossed the Indian Ocean and settled in Madagascar, making the language Malagasy part of this family.

Several other families are worth noting. Southern India is home to a sizable population that does not speak Indo-European languages; instead, it speaks the Dravidian tongues such as Tamil, Telugu, Kannada, and Malayalam. The Indo-Chinese peninsula is home to two distinct language groups: Austro-Asiatic (Vietnamese and Cambodian) and Thai-Kadai (Thai and Lao). The aboriginal peoples of Australia and Papua New Guinea, who comprise 1% of the world's population, speak 20% of the world's languages in an enormously diverse group often called Indo-Pacific. The Americas were home to a huge range of indigenous languages prior to the mass extermination unleashed by the Europeans; more than a dozen families existed in North America (e.g., Iroquoian, Siouan, Salishan, Athabascan, Mayan) and in South America (e.g., Andean, Chibchan, Macro-Carib). Finally, isolate languages such as Basque, with no surviving relatives, and Kartvelian tongues such as Georgian continue to survive.

Today, there are roughly 5,000 to 6,500 languages remaining in the world. Most, however, have very few speakers, often numbering only in the dozens, and are not written. The total number was much larger in the past and has been declining steadily for centuries. The rise of nation-states often led to deliberate homogenizations of cultures and dialects, and today globalization and national school systems have contributed to the decline. In the Americas, disease, genocide, government-run boarding schools (in the United States), and cultural assimilation annihilated large numbers of languages. Today, 96% of the world's population speaks one of the top 20 languages, and many observers predict that 50% of all languages will disappear within the next century (i.e., one every 2 weeks). This decline represents a crisis in cultural diversity that deprives humanity of the rich ways of viewing the world inherent in having different languages.

Finally, it is worth noting that in addition to the geography of languages, geographers have been increasingly interested in the role of language in the representation of space. If language structures and

Figure 5   Countries in Which Sino-Tibetan Languages Are Dominant

Figure 6   Countries in Which Malayo-Polynesian Languages Are Dominant

mediates thought, it plays an enormous role in how discourses about the world are organized. Thus, the geography of language and the language of geography may be seen as deeply intertwined.

—Barney Warf

See also Cultural Geography; Religion, Geography and/of; Spaces of Representation

Suggested Reading

Comrie, B., Matthews, S., & Polinsky, M. (Eds.). (1996). *The atlas of languages.* New York: Quarto.

Dalby, A. (2003). *Language in danger: The loss of linguistic diversity and the threat to our future.* New York: Columbia University Press.

LAW, GEOGRAPHY OF

The geography of law concerns the analysis of the ways in which legal practice produces and shapes space, and the spatial representations and metaphors found within law, with an examination of the political and ethical consequences of such legal geographies.

Space is everywhere; we live in and through a geographic world. Consider any mundane activity such as a visit to a shopping mall. This includes spatial practices (the act of shopping), spatial representation (the advertisements and designs within the mall that often are designed to conjure up images of other exotic places), place (or perhaps the absence thereof given the ways in which corporate domination and mall design conjoin to produce so-called placeless spaces), and scale (marketing and distribution networks through which "global" products are channeled to our local stores). A lawyer, on the other hand, would point out that law is everywhere; there is no aspect of our lives that is beyond law, defined either formally (in the sense of statutes, judicial interpretation, etc.) or informally (the messier, but no less important, legal sensibilities that we carry in our heads and express in our daily practices). Thus, in an immediate sense, law makes things happen. At an extreme, people may be confined or even killed in its name. Law also shapes social identities and relations; to say that one is a student, a citizen, or an owner is, in part, to speak legally. From this perspective, our visit to the mall could also be conceived in terms of the ways in which

we bump into law, usually unconsciously but sometimes with surprising force. Our shopping trip is made possible by a network of legal arrangements such as property (e.g., ownership of the mall, leasehold arrangements, the transfer of property associated with our purchases) and contracts (e.g., the labor contracts into which workers enter).

Both the geographer and the lawyer would argue for the importance of space and law, respectively. The geographer, for example, might point out that the mall works, as an economic space, to the extent that its design and spatial layout induce people to spend money. The lawyer would argue that without thinking about law, we miss certain crucial qualities of the mall. For example, its legal designation as private property governs the sorts of political practices, such as pickets and protests, that can happen within it. Given the mall's contemporary importance to everyday life, the lawyer might argue, this has the effect of limiting political speech.

Separately, both of these accounts are persuasive. It is odd, then, that only recently have they been brought together. Social life, legal geographers argue, is both spatial *and* legal. The mall is both a space *and* a bundle of legal practices and discourses. So, for example, the mall's significance to speech is not an abstract one; it is the fact that people enter and use a physical space that gives it practical relevance. Thus, if law and space are important to social life separately, their combined effects may be worth examining. It is this realization that has prompted a number of scholars, in fields such as geography, anthropology, and law, to look more carefully at the geographies of law.

Although it is hard to characterize this diverse scholarship, which draws from a range of theoretical sources for inspiration, one useful distinction is that between law-in-space and space-in-law. Law-in-space concerns the ways in which law—statutes, judicial interpretations, the practice of legal officials, and so on—helps to produce space. Thus, legal actors, such as judges, can usefully be thought of as geographers to the extent that their interpretations and edicts make space. Property is an obvious example; the grid of land ownership (both public and private), which we all live in and negotiate on an everyday basis, has been made through generations of sedimented legal practice on the part of judges, surveyors, lawyers, police officers, municipal officials, and (of course) owners and nonowners. Space-in-law comes at the geographies of law from the other direction. Legal discourse is full of spatial tropes and metaphors. Legal actors frequently engage in spatial representation, being concerned with the delineation of zones, boundaries, and domains. When a judge designates a shopping center as private, from this perspective the judge draws on (and helps to reproduce) a particularly consequential "map" of the social world. When property rights are discussed in terms of autonomy, boundaries, and separation, we also rely on a geographic set of metaphors. Law-in-space and space-in-law clearly flow into each other; spatial metaphors within law, because of law's institutional power, can be imposed on the world. Partly for these reasons, as well as a desire to think past the law–space binary, some recent scholarship has tried to come up with a new vocabulary that recognizes the inseparability of law and space.

Legal geography offers important and valuable insights. Thinking the legal (e.g., property, crime) in terms of space, or thinking the spatial (e.g., boundary, place) in terms of law, promises new and enriched insight. Understandings of both space and law are changed, and new questions are posed. This is important in an analytical sense; for example, if we are interested in how spatial arrangements are put together or sustained over time, recognizing the role of law is important. However, it is also important politically. For better or worse, legal geographies have political and moral effects. Designating the mall as "private property" can close down certain possibilities. Yet workers, seeking to picket their employers within the mall, may also use a legal language of rights, freedom, and citizenship to challenge such geographies.

—*Nicholas Blomley*

See also Civil Society; Electoral Geography; Environmental Justice; Gerrymandering; Justice, Geography of; Local State; Nation-State; Political Ecology; Political Geography; Redistricting; Social Justice; Spatial Inequality; State; Urban and Regional Planning; Urban Managerialism

Suggested Reading

Blomley, N. (1994). *Law, space, and the geographies of power.* New York: Guilford.

Blomley, N., Delaney, D., & Ford, R. (Eds.). (2001). *The legal geographies reader: Law, power, and space.* Oxford, UK: Blackwell.

Holder, J., & Harrison, C. (Eds.). (2003). *Law and geography.* Oxford, UK: Oxford University Press.

Sarat, A., Douglas, L., & Umphrey, M. (Eds.). (2003). *The place of law.* Ann Arbor: University of Michigan Press.

LESBIANS, GEOGRAPHY OF/AND

Lesbian geographies draw together the gendered and (nonhetero)sexual dimension of geographies. They take a number of forms, from examining how lesbians appropriate urban space to looking at how lesbians negotiate and use space, including heterosexualized space (i.e., space that is made to be heterosexual). Writing about lesbians in geography challenges a heteronormative discipline that often marginalizes those who do not fit into the heterosexual masculine hegemony. In contesting the heteronormativity of the discipline, lesbian geographies also contest the trend to subsume lesbian and queer women within all-encompassing umbrella terms/research foci of queer or LGB (lesbian, gay, or bisexual). Lesbian geographies take gender as a salient axis of consideration alongside sexuality, yet they recognize that the category lesbian is constituted in time and space and is not uniform across other social differences. In taking gender and sexuality seriously, lesbians cannot simply be added to geography because these geographies contest the methodologies and ontologies of geography itself.

Research that maps gay territories has argued that women and lesbians tend not to concentrate in given areas and that, because of this, lesbians are less likely to achieve local political power. This is attributed to the gendered division of labor, male control of urban public spaces, and a gay male drive to appropriate place. However, lesbians do appropriate urban space and are involved in urban politics, but these things might not be done in the same ways as with (gay) men. For example, lesbian concentrations in particular neighborhoods might not be as visible as gay male appropriations of particular districts, yet they still point to an acquisition of territory. These concentrations, rather than being due to commercial enterprise or public visible presence on the urban landscape, can arise from social networking. Furthermore, lesbians are not confined only to urban areas; the appropriation of ruralities has a history within lesbian separatist communes. The rural is viewed as an escape from cities and as an opportunity to develop communities away from patriarchy and exist closer to nature.

Lesbians can appropriate and use public sites in visible ways yet not have a territorial presence. Julie Podmore, recognizing the tendency of lesbian geographies to focus on the absence of women in heterosexual space or their territorial clusterings, contended that how lesbians *use* public sites for interactions and communality might not fit within the territorial paradigms of urban geographies. Therefore, lesbian geographies can be excluded because of *what* is being examined and *how* these studies are being undertaken. The daily lives of women who exist outside heterosexuality and often use heterosexual spaces might not fit traditional geographic investigations. Yet lesbian geographies include an exciting plethora of ways in which space is appropriated, used, negotiated, resisted, and re-created. The (re)appropriation of space through music, clothes, and language, as well as more political activities (e.g., so-called dyke marches) and desiring enactments (i.e., expressions of same-sex desire), offers insights into the heterosexualization of space and its reworkings outside hegemonic norms.

Lesbian geographies might presuppose a coherence and unity within the category lesbian. However, relations of power within queer communities are also related to class, race, age, (dis)ability, and other social differences. Thus, lesbians are heterogeneous. Furthermore, the label *lesbian* can be seen as contingent and differentially appropriated across space and time. The identity lesbian might not be significant in particular cultures or historical periods; therefore, the category itself is problematic. Recognizing that lesbian is not a coherent or homogeneous category and that labeling is fluid across space and time, lesbian geographies have the potential to further contest the very nature of geography and how we do geography.

—*Kath Browne*

See also Feminisms; Feminist Geographies; Gays, Geography and/of; Gender and Geography; Queer Theory; Sexuality, Geography and/of

Suggested Reading

Bell, D., & Valentine, G. (Eds.). (1995). *Mapping desire: Geographies of sexualities.* London: Routledge.

Podmore, J. (2001). Lesbians in the crowd: Gender, sexuality, and visibility along Montreal's Boul. St-Laurent. *Gender, Place, and Culture, 8,* 333–355.

LIMITS OF COMPUTATION

Computer technology has advanced at a pace unprecedented by other technologies. Although such achievements have transformed the human experience, they have also fueled misconceptions about the limits of computation itself. From a popular perspective, the scope of problems that can be solved by computers may seem unlimited. In reality, computers are limited by several factors—some contingent and others more fundamental.

Contingent limitations apply to the current state of computer technology. Actual computers are artifacts of engineering; as such, they are subject to certain physical limitations and may lack sufficient resources (e.g., speed, memory) to solve certain problems. For example, to list all possible partitions of 100 countries into two even halves (the city partition problem) would require at least 2^{100} steps, which would take more than 30 trillion years on a computer at the speed of 1 billion steps per second. Note that this is a problem that is solvable by computers in principle but that we just cannot afford the waiting time given the speed of the current computers. Moreover, further improvement in computer power does not seem to help much. For example, even with the computer speed improved by 1,000 times, the city partition problem still would require more than 30 billion years of computer time.

Other contingent limitations apply to our current knowledge of computer algorithms. Computer algorithms have been a very active research area in computer science. Many powerful algorithmic techniques have been developed during the past four decades. For example, now it is a trivial task using computers to schedule a minimum cost trip from a given city to another given city (the shortest path problem). The problem on a map of 100 cities takes a computer only miniseconds. This algorithm has been used very popularly by travel agents. On the other hand, if we add the condition that the tour must pass through all cities on the map (the traveling salesman problem), a seemingly moderate condition, the problem becomes much more difficult. The problem is solvable in principle; we can simply enumerate all possible such tours and pick the one with the minimum cost. However, such an enumerating algorithm would take more than 2^{100} steps on a computer, and this, as already calculated for the city partition problem, would require an unaffordable waiting time.

On the other hand, in spite of much effort by computer scientists and mathematicians over the past four decades, no one has been able to develop an algorithm for the traveling salesman problem that is essentially better than the trivial enumerating algorithm. Therefore, currently no known computer program is able to solve the problem in a practical manner. Note that the traveling salesman problem is different from the city partition problem; the 2^{100} computer steps are *necessary* for solving the city partition problem, whereas much faster computer algorithms *might* exist for the traveling salesman problem but simply have not been discovered yet. During the past 40 years of research in a variety of areas in computer science, a class of more than 1,000 problems, all similar to the traveling salesman problem, has been identified. It is known that if any of these problems can be solved efficiently, all problems in the class can be solved efficiently using a technique of problem reductions. On the other hand, many of these problems have been studied for decades, and no efficient algorithms have been found for any of them. Thus, it is natural to conjecture that no efficient algorithms exist for these problems and that all of the problems in the class are computationally intractable. These problems have been named *NP-complete* problems. Many important computational problems in a large variety of areas are known to be NP-complete. Studying NP-completeness has been central in the research in theoretical computer science.

But this is only half of the story. Beyond the physical limitations on what can be computed *in practice,* there are deeper and more fundamental limitations on what can even be computed *in principle.* These limitations apply to the nature of computation itself. In fact, classic results in computational unsolvability were studied by mathematicians and philosophers prior to the construction of the first computers. After all, computers are built based on Boolean logic, and the execution of a computer program can be regarded as a process of mathematical logic reasoning. According to Gödel's incompleteness theorem, no formal mathematical system can verify the validity of all mathematical statements. Translated into the language of computer science, this says that no computer program can test the correctness of all other programs. In fact, it has been proven using formal mathematics that there are a large number of very simple computer programs whose correctness cannot be verified by any computer program. For example, suppose that a computer programming teacher has taught his first class and assigned his students to write their first computer program that simply prints "Hello, World." It would be quite natural for the teacher to expect to write a "testing program" that can test the correctness of the programs submitted by the students. However, fundamental research in computational unsolvability has shown that no such testing programs exist even for such a simple task. Note that the impossibility results here are outright and have nothing to do with the computational resource. These fundamental impossibility results have also found wide applications in science and engineering.

In summary, limitations of computation have been studied thoroughly from both practical and theoretical points of view. These studies may play important roles in the research in geographic information science (GIScience). In particular, research directions in GIScience, such as computable city, electropolis, Digital Earth, and virtual field trip, all are closely related to computer and information technology. A clear understanding of what is possible and what is impossible using computer technologies seems to be necessary for the research in these areas.

—*Jianer Chen and Scott M. Pike*

See also GIS; Humanistic GIScience

Suggested Reading

Franzen, T. (2005). *Gödel's theorem: An incomplete guide to its use and abuse.* Wellesley, MA: A. K. Peters.

Garey, M., & Johnson, D. (1979). *Computers and intractability: A guide to the theory of NP-completeness.* New York: Freeman.

Hopcroft, J., Motwani, R., & Ullman, J. (2001). *Introduction to automata theory, languages, and computation* (2nd ed.). Reading, MA: Addison-Wesley.

LITERATURE, GEOGRAPHY AND

For centuries, readers have flocked to the places described in works of literature—to Dickens's London, to Joyce's Dublin, to "Tara" or the home of Anne of Green Gables—and geographers have realized that works of fiction offer powerful interpretations of landscape, character, and memory that in turn influence the very real places the novels fictionalize.

Although a specialization called *literary geography* was not so named until 1977, it seems likely that as long as there has been literature, readers of such works have pondered literary geographies. "Where was that story set?" is a question that has drawn millions of people to travel and dozens of scholars to study literary geography. Indeed, the earliest works of literary geography in the United States and the United Kingdom generally set as their goals the answering of this question for various literary works; as early as 1907, geographers were attempting to map the areas described in literature.

Many literary geographers have believed that geographers' analyses could help to ground (and identify) the "real" roots of literary landscapes, even when those landscapes were expressed in highly symbolic fashion. Geographers have also seen literature as a source of geographic data and perceptions, preserving such information even from ancient times. Indeed, a special subfield of literary geography is engaged in the study of the Bible; scholars generally share the goal of using the Bible as a source of geographic information for the study of ancient Israel, and this highly specific focus has granted them a status separate from other geographers studying literature. But most literary geographers have concentrated on the "loftier" side of the canon; only more recently have scholars begun to examine works such as mysteries and science fiction.

In addition to identifying literary locations, another early emphasis was on how well particular works described the places where they (allegedly) were set and how well such writers interpreted their regions. Geographers looked to literature for evocative engagements with both physical–geographic and human–geographic phenomena—from hurricanes to house types—and then also sought to use such literature in the classroom. Novels became seen as a way in which to provide insights into landscape perception, into varying regions and the people who lived in them, and into physical processes on the earth's surface. Literature, in short, could be used as supplemental texts in geography classes—and it still is today.

Literary geographers have examined individual works, or the combined works of just one author, to explore how and why geography (or setting) is important in the plot or mood of the works and how the setting itself advances the plot. Some have been able to go beyond the identification of "actual" places where literary works were set to understand how some settings

are in fact complex composites of multiple times and multiple places (both real and imagined), forged in the writers' minds and representing only part of what other viewers of the same settings might see. They have understood, for example, how landscapes in literature can transcend their simple roles as settings to be used powerfully as metaphors and symbols.

Indeed, as early as the 1920s, some geographers began to find literary works revealing not just of geographic features or phenomena but also of the *meanings* of the natural and cultural worlds. With the rise during the 1970s of humanistic geography and its emphasis on subjective experience and meaning, literary geographers began to use literature to study how landscapes and places were perceived and interpreted by individuals and groups. Literary geography became an ancillary to traditional cultural/historical geography, and scholars looked to literature for rich data about landscape and place, particularly in regard to feelings, values, attitudes, and meanings.

More recent efforts have engaged literary works as cultural texts revealing not just idiosyncratic details of local landscapes but also normative geographies (descriptions of how a place should be) that could offer, for example, powerful (but often masked) lessons for readers, that could act as a draw for tourists, or both. Indeed, literary works, whether written for children or for adults, convey more than just ideas about landscapes and spatial perception. Fiction also encodes ideas about moral social geographies, conveying how things should (perhaps) be in themes with which not all would agree as both landscapes and social groups are stereotyped to convey particular messages.

Recent scholars have also focused on studies describing the impact that works of literature have had on particular places and groups, particularly in the case of literary tourism. Indeed, literary places, whether associated with famous authors or with their individual works, act as powerful draws for tourists as they flock to such places to "relive" the scenes from famous novels in the "real" landscapes in which they were set. It has been simple for scholars to criticize such tourists for foolishly conflating the fake and the real, for example, pointing out that despite elements of fact in fiction, novels always are fundamentally the creations of their authors' imaginations. But literary geographers have transcended such criticisms, recognizing that places need not be "real" to be meaningful and that, because of the potentially powerful emotional connections readers make with literary works, fictional places can be imbued with powerful meaning even when visitors know they are fictional. Such meanings may relate directly to the literary works the tourists engage, but they may also relate to circumstances in the visitors' lives that happened when they were reading the text, for example, evoking fond memories from their childhoods.

Thus, still today, issues around the authenticity of particular literary places remain everpresent. Visitors commonly question whether a particular site's identification with a work of fiction is in fact genuine. But contemporary literary geographers recognize that the question of a site's ultimate authenticity may remain unanswered because, first, literary places often are grounded more solidly in the writers' imaginations than in actual landscapes and, second, authenticity itself is a subjective experience that is grounded, at least in part, in the minds of the visitors themselves. Thus, although such worlds may be fictional—imagined—the meanings of literary places may be very real to their beholders, engendering potentially powerful and pleasurable experiences for those who visit and forever transforming the landscapes that the literary works described.

—*Dydia DeLyser*

See also Cultural Geography; Spaces of Representation; Text and Textuality

Suggested Reading

DeLyser, D. (2003). Ramona memories: Fiction, tourist practices, and placing the past in Southern California. *Annals of the Association of American Geographers, 93*, 886–908.

Herbert, D. (2001). Literary places and the tourism experience. *Annals of Tourism Research, 28*, 312–333.

Noble, A., & Dhussa, R. (1990). Image and substance: A review of literary geography. *Journal of Cultural Geography, 10*, 49–65.

Sandberg, L., & Marsh, J. (1988). Literary landscapes: Geography and literature. *The Canadian Geographer, 32*, 266–276.

LOCAL STATE

The local state has been used in geography and related disciplines since the late 1970s. The concept was first employed in a Marxist analysis of political practices in a borough of London. Although sometimes used synonymously with *local government*, it has broader meanings encompassing local judiciaries and other local authorities, QUANGOS (quasi-nongovernmental organizations that are financed by government but operate independently), local offices of central governments, and even (in certain cases) public–private partnerships and nonprofit organizations. Use of the term *local state* has become less common during recent years, and it has been partially superseded by the language of regulation theory. Some object to the term because it seems to imply local sovereignty, whereas the local state is ultimately limited in its capacity for independent action.

The local state serves in part as an agent of the central state. That there are very few central states without governmental units at smaller scales suggests the utility of local state structures. Placing authority for the provision of some public goods and services at the local level increases local participation in decision making and, at least theoretically, allows for local solutions to local problems. Local solutions can be more efficient than a "one size fits all" central solution. In a heterogeneous country with wide spatial variations, the local state can be seen to provide a measure of self-determination. Local electoral politics, regulations, and taxation provide a measure of local autonomy. But it is also argued that if problems and crises are segmented spatially into separate localities, blame for failure also rests locally, protecting the central state from criticism. Local government reorganizations (as occurred in the United Kingdom in 1968 and 1974) and shifting responsibilities between different levels of government are ways in which the central state maintains remote control of the local state.

Some see a significant role of the local state as legitimizing the actions of the central state and acting as the central state's agent. The local state is seen as an attenuation of the central state's political apparatus, enabling the central state to maintain social control. In fact, the relative autonomy of the local state to act independently of the central state varies from country to country and over time. In some cases, the local state merely enforces the rules of the central state and derives its funding from the same source. In the United States, the federal government's welfare policies enable it to retain a high degree of control through regulation of eligibility and accounting, even though the local state administers many of the programs. In other cases, local authorities have taxing, spending, and rule-making powers and can act relatively independently of the central state. Land use zoning is an example of mainly local control with few central government regulations. In still other cases, local nongovernmental entities, citizens groups, and other social movements may be influential enough to introduce local regulations—whether gay marriage in Massachusetts or vehicle entry fees to central London—that are counter to central government policies. Such actions by local states that conflict with central state policies are not uncommon when opposing political parties control local and central governments. For example, in the United Kingdom, when Conservatives in central government want to reduce public expenditure, Labor-run urban governments resist cuts in services. And in the United States, Republican cuts in federally funded welfare programs are opposed by local Democratic administrations.

The idea of the local state is associated mainly with leftist analysts within a Marxist or political economy tradition. They see the role of the central state as maintaining conditions for capital accumulation/capitalist production, whereas the local state supports capitalist *re*production. Thus, the local state typically provides public education, sanitation, leisure services and facilities, and local law enforcement and fire protection. This division of labor is not absolute, however, given that the local state also helps to maintain conditions appropriate for capital by, for example, providing infrastructure, job training, and tax breaks. Meanwhile, the central state also has reproduction roles in areas such as educational standards, healthcare funding, and environmental protection. Two significant related issues that receive increasing attention are the spatial mismatch between local revenue streams and service needs and the increasing tendency to privatize or outsource public services.

In the United States, as an example, the local state may experience diminishing fiscal resources at the same time as demand for local services is increasing. As U.S. central cities lost many taxpayers to suburbia and tax revenues dwindled, the demand for human services increased at the same time as the poverty population grew as a percentage of total population, and frequently overall, as immigrants and other low-income people moved to the city. The costs of services such as law enforcement and fire protection for both residents and suburban commuters continued. Bankrupt or struggling cities called on central government for assistance with varying levels of success. When federal aid was received, it usually came with strings attached that reduced local autonomy. With local states competing

for both central state funding and private-sector investments, it is perhaps surprising that "civil wars" between local jurisdictions do not occur more often. Such conflicts often are played out politically through local representative objections to pork-barrel funds for another local state. In the United States, the formula for distributing Homeland Security funding resulted in higher per capita allocations to unlikely terrorist targets such as North Dakota than to New York City, a target of the terrorist attacks on Sepember 11, 2001, causing even the Republican mayor to strongly object. But both the authority of the central state and the apparatus of electoral politics serve to legitimize a system that perpetuates sociospatial inequalities and inequities.

The privatization of public services, with local governments contracting with private-sector firms for everything from education (e.g., charter schools), to crime prevention (e.g., private security firms and prisons), to sanitation (e.g., for-profit garbage hauling), has numerous critics and implications. Although arguments can be (and are) made for the efficiency of such arrangements and for greater consumer choice, such as in the case of charter schools, accountability to the public may diminish. Privatized services often are provided by nonunionized workforces, thereby reducing wages and benefits for workers. As housing programs for low-income residents have shifted from federally funded, local authority–owned and –managed units to public subsidies for private builder and housing vouchers for recipients to purchase their own housing in the open market, the number of people served typically has been reduced even if the quality of housing has improved. Charter schools in the United States typically are publicly funded but privately run. They increase choices for parents who seek alternatives to standard public education. There are, for example, charter schools that focus on the arts, focus on the local community, or have single-gender enrollments. But they typically can select who they admit and so can "skim" the better students from the public schools, thereby reducing the standardized test scores in those public schools and jeopardizing their funding.

Since the late 1970s, when the term *local state* was introduced, changes in the economy away from Fordist models of production to a service economy in Europe and North America have been accompanied by changes in local government functions and alliances. Today, local governments are increasingly engaged in economic development and focus on attracting capital investment to their localities. Local governments are becoming increasingly entrepreneurial and are allying themselves with private-sector organizations to foster their local economies. Cities, states, and other local states market themselves as places for profit making while the local welfare state is reduced. In the interest of image enhancement, the homeless, beggars, and car window washers are regulated out of sight. This shift is accompanied by an expansion of the local political system to include a growing variety of nongovernmental entities. Public–private partnerships, improvement districts, consulting firms, universities, nonprofits, business roundtables, chambers of commerce, trade unions, and even churches and arts organizations play increasing roles in directing the functions of the local state. These groups are not elected, at least not by the citizenry as a whole, and although they allow particular groups to have a voice, they are not democratic in the traditional sense. However, the trend toward including such actors in local state decision making and service delivery is supported by both people on the right (because it reduces governmental spending and regulation) and people on the left (because it empowers community groups). In the United States, the federal government has increasingly focused on providing support for local nongovernmental entities, whether through programs such as the empowerment zones of the Clinton administration or through the faith-based social services of the Bush government. Although such programs may be administered through the local state, the regulations and funding are federal.

As the local state becomes more entrepreneurial, there is also more public-sector investment in trophy buildings such as sports stadiums and convention centers. Although the public investment usually is justified on the basis of eventually generating revenues for the city coffers, the underlying motivation often is image enhancement and ammunition for quality-of-life promotion. It can be useful to a city mayor to be home to a winning sports team even if he or she does not personally enjoy the games. The emphasis of the entrepreneurial state is on efficiency rather than equity and on wealth creation rather than redistribution. Thus, even the wealthy can obtain grants, loans, and tax abatements if their schemes are thought to induce economic growth. The local state must generate the aura of a good business climate and an amenity-rich environment to compete globally for investments and desirable (affluent) residents.

It can be argued that the local state, especially in the United States, has become a tool for protecting more affluent segments of society from the various disadvantages of proximity to the poor. The United States has myriad local states that often are minute and overlapping. This political fragmentation, together with local state control of zoning, public education, and other services, enables affluent communities to zone out the poor by large minimum-size lots, to maintain higher levels of per-pupil funding in schools, and to avoid the costs of subsidizing services even of adjacent cities to which suburbanites commute. This results in local states with few resources accepting locally unwanted land uses (LULUs) and "not in my backyard" projects (NIMBYs) such as landfills, refineries, prisons, and other facilities that can have negative effects on local residents. Thus, the mechanism of the local state facilitates socioeconomic spatial segregation reflected in the fact that, for example, in the New York metropolitan area some municipalities have double the average household income as do others not far away. Because of the strong correlations between income and race, the local state also reinforces racial segregation.

As noted earlier, the scale of the local state varies considerably, ranging from places the size of California (which would be the world's sixth-largest economy if it were an independent state) and New York City (a global city and world financial capital) to tiny parishes and sparsely populated Scottish counties. Unsurprisingly, the larger local states exert more influence on external affairs and are more affected by global flows. In maintaining conditions supportive of major corporations and banks, New York City acts as a command center for global finance capital at the same time as the local state copes (or not) with the human consequences of high levels of immigration. The local states within New York City (e.g., the five boroughs and their municipalities) have less autonomy and less influence over whether they will benefit from global flows or be devastated by them.

In summary, the concept of the local state has evolved considerably during the past three decades and now incorporates nongovernmental institutional entities and influences discussed by regulation theorists. Meanwhile, the relationships among the local state, its counterparts in other places, the central state, and global processes are in flux. The local state is changing, even if it has boundaries unchanged in 500 years. The local state is no island.

—Briavel Holcomb

See also Flexible Production; Political Geography; State; Urban and Regional Planning; Urban Entrepreneurialism; Urban Geography; Urban Managerialism; Urban Social Movements; Urbanization

Suggested Reading

Bounds, M. (2003). *Urban social theory: City, self, and society.* Oxford, UK: Oxford University Press.

Clark, S., & Gaile, G. (1998). *The work of cities.* Minneapolis: University of Minnesota Press.

Duncan, S., & Goodwin, M. (1987). *The local state and uneven development.* New York: St. Martin's.

Hall, T., & Hubbard, P. (1998). *The entrepreneurial city: Geographies of politics, regime, and representation.* Chichester, UK: Wiley.

Jones, M. (1998). Restructuring the local state: Economic governance or social regulation? *Political Geography, 17,* 959–988.

Taylor, P., & Flint, C. (2000). *Political geography: World-economy, nation-state, and locality.* Harlow, UK: Pearson Education.

LOCALITY

Locality is another word for place. If one thinks of different possible scales, locality is nearby. Within geography, it has stood for the smallest political designation, the place where one lives, the area contained within narrow political borders. Often viewed as having an obvious meaning, it reemerged during the 1980s as a significant term for the local level when those working with a variety of new theoretical approaches undertook to redefine the relationship between space and place and their relation to political, economic, and cultural forces. The word *locality* offered an alternative to *community* as it emphasized linkage and change as opposed to community studies' greater association with traditional culture or embattled resistance.

British geographers such as Doreen Massey and the Centre for Urban and Regional Studies (CURS) projects were especially important in adding analytic juice to the word *locality.* Locality studies arose in Britain during the early 1980s in response to the massive changes transforming the British economy and society. Localities research developed as a critique and/or enrichment of structural analyses of economic redevelopment that tended to view processes at the global or regional level. Work done under the rubric of locality studies required a more finely grained analysis of the interaction between large-scale structural forces that drove economic restructuring worldwide and more locally based, historically layered social arrangements. Structural analyses, for example, explained how capitalism's globalization of manufacturing production led to the growth of textile plants in low-wage countries. They captured how the overall process led to plant closures and unemployment for industrial countries and their industrial workers. But the structural analyses often provided little understanding of why some places got the new plants and others did not or of which places could ward off closures and which would elude the threat entirely. Thus, the point of the localities approach was not only to understand the specifics of place but also to understand how the processes articulated across space in creating the world's highly differentiated set of arrangements and outcomes.

Each locality has its own confluence of forces, mix of businesses and business organization, capital available for investment, attractiveness to outside capital, labor force, culture, and openness to change and risk. Locality studies of the textile industry in Europe, for example, found clearly different local patterns that depended on materials (e.g., cotton vs. wool), ownership patterns (e.g., large-scale, international, centrally controlled structures vs. smaller, local-scale, guild-like structures), technology (highly mechanized vs. labor intensive), and labor force (e.g., wage differentials and gender).

Some industries are less mobile than others. Those that depend on place-specific natural resources, a workforce with very distinctive training, or especially costly plant and infrastructure conditions may find themselves highly committed to their locations. Other industries may be relatively footloose, with requirements easily met in a number of places. These variations influence the locality, but localities are not merely passive responders to the industries' decisions. They mobilize their own forces in representing their attributes and in changing them.

Governments are instrumental in creating each set of local conditions. For example, particular choices in governmental regulation and governmental investment generate local transportation and communication advantages for some places—easy interstate highway access, guaranteed airline service, inexpensive broadband Internet availability, and so on—while marginalizing others. Differences in educational standards can mean the difference between a workforce that is technically trained and one that is not. Each locality exists within its own set governments operating at various scales. These interact—sometimes intentionally and sometimes inadvertently. Funding for particular projects, for example, may be authorized at one level and distributed at another level. For those living within a locality, the multiple governmental forms are an integral and sometimes unquestioned part of life. Yet formal arrangements are highly variable. For example, in strong home rule states, each locality has its own school district; in others, the locality has few formal powers and the county is responsible for education.

The workings of government, however, are much more complex than its formal structure and very much conditioned by its interaction with various constituencies. For each locality, local growth promotion mobilizes its cast of characters into coalitions. Governmental figures ally with local elites. But just as local industries may be more or less dependent on place, local elites may also vary in their commitment to place as well as in their mix of owners and workers, large and small enterprises, and residents and businesses. The specific coalitions vary depending on the issue at hand and the threat from outside forces. A locality studies approach disentangles the interplay between the external forces and local ones that generate each place's spot in, say, the infrastructural educational network or the plant or plant relocation competition.

Over time, the locality studies approach has put more emphasis on culture. Although gender and race were factored into localities work of the 1980s, they were considered mainly in the context of the workplace. In most communities, men dominated the manufacturing labor force, but women were more integrated into some than into others. In some places, the decline in manufacturing jobs shifted the employment structure to increase the dependence on part-time service-sector jobs, which are more often held by women, changing both the workforce and household dynamics.

The locality studies approach has been both academic and activist, assuming that this work should benefit real people in real places. Although it emerged as a critique, the approach has also been the object of numerous critiques. Some questioned whether the localities approach entirely avoided the theoretical weaknesses of the idiographic approach, relied too much on descriptions of individual locations, and paid too little attention to general principles. Was it too focused on economic questions? Was its integration of social, cultural, and political forces weighing each influence properly? For Marxists, its failings tend to be in its appreciation of social forces; for humanists, its failings tend to be in its understanding of cultural forces and human agency. Has it been able to integrate all of the geographic scales? Did the term itself lose meaning, becoming synonymous with *local* or *locale*? Flattening in meaning over time, the term and the debates around it have reinvigorated geography's understanding of the workings of place.

—*Deborah E. Popper*

See also Idiographic; Regional Geography

Suggested Reading

Cooke, P. (1996). The contested terrain of locality studies. In J. Agnew, D. Livingston, & A. Rodgers (Eds.), *Human geography: An essential anthology* (pp. 476–491). London: Blackwell.

Massey, D. (1994). The political place of locality studies. In D. Massey, *Space, place, and gender* (pp. 125–145). Cambridge, UK: Polity.

LOCALLY UNWANTED LAND USES

During the early 1980s, urban planner Frank Popper coined the term *locally unwanted land use* (LULU) to describe often-reviled land uses such as waste repositories, prisons, power lines, airports, highways, and mega–shopping complexes. LULUs—often associated with their sister acronym NIMBY (not in my backyard)—frequently are concentrated in poor and minority areas, and the perceived threat generally diminishes with distance from the LULU.

From the country's earliest days, Americans of means have sought to keep objectionable land uses out of their immediate locales. In 1916, this impulse was formalized when New York City enacted the nation's first zoning ordinance. More recently, citizen objections to local environmental intrusions have become increasingly public, frequent, and powerful.

Why did LULUs become such a prominent issue during the 1980s and beyond? Among the reasons are new types of LULUs (e.g., "big box" shopping complexes, nuclear waste disposal sites), increasing distrust of central government, growing awareness of pervasive environmental injustices, and increased capacity at the local level to organize and resist. Opposition to LULUs was inspired particularly by the high-profile Love Canal case of the late 1970s, when citizens of Niagara Falls, New York, succeeded in bringing national attention to toxic chemical contamination in their neighborhood.

LULU activism tends to be about pollution and social disruptions rather than about the nature protection issues embraced by many major national environmental groups. In contrast to traditional environmentalists, LULU activists are more likely to be poor, politically marginalized, and members of minority groups. Wealthier and more influential citizens are more adept at heading off LULUs from the outset; indeed, they are the beneficiaries of a political system that separates them from such problems quite effectively. Still, LULUs often do threaten the elites; highways and cell phone towers in wealthy exurbs, public recreation corridors (greenways) in affluent rural areas, and proposed Cape Cod wind farms are but a few examples.

Some see local resistance to LULUs as selfish obstructionism; in the most extreme cases, indeed, there are those citizens who would object to even the most benign changes to their neighborhood. But at a deeper level, LULU activism holds promise for bringing about more environmentally and socially sustainable solutions in public policy realms such as waste management, transportation, and urban planning. Geographer Michael Heiman argued that we are gradually moving from NIMBYism to "not in anybody's backyard," in other words, toward pollution prevention and broader citizen involvement in decision making. Recycling of formerly landfilled materials, elimination of toxic chemicals from production processes, and alternatives to major new expressways are examples of such outcomes. On the other hand, central governments, under pressure to solve pressing problems, will in many instances respond to LULU sentiments with legislation that overrides local objections to facility siting. And strategic lawsuits against public participation (SLAPPs) have met with some success in stymieing LULU opponents. Nonetheless, the local environmentalism collectively spawned by numerous LULU confrontations has become an enduring and important part of America's political–environmental landscape.

—*Robert J. Mason*

See also NIMBY; Urban and Regional Planning; Zoning

Suggested Reading

Heiman, M. (1990). From "not in my backyard!" to "not in anybody's backyard!" *Journal of the American Planning Association, 56,* 359–362.

Popper, F. (1985). The environmentalist and the LULU. *Environment, 27*(2), 7–11, 37–40.

LOCATION THEORY

Location theory refers to a conceptual perspective widely used in economic geography during the 1950s and 1960s, primarily during the reign of the philosophy of logical positivism. Location theory exemplified geography as a spatial science, primarily concerned about abstract laws applicable under all circumstances.

Location theory in various forms shared a common focus on the centrality of models as bridges between the empirical world and the theoretical world of prediction and explanation. Models distill the essence of the world, revealing causal properties via simplification. A good model is simple enough to be understood by its users, representative enough to be used in a wide variety of circumstances, and complex enough to capture the essence of the phenomenon under investigation. Typically, models were developed, tested, and applied using quantitative methods. This approach to geography relied on a Cartesian view of space, generally in the form of an isotropic plain in which space was reduced to distance and spatial variations occur only through transport costs. Essentially, location theory reduced geography to a form of geometry, a view in which spatiality is manifested as surfaces, nodes, networks, hierarchies, and diffusion processes. In short, this presented a view of space devoid of social relations.

Location theory developed close ties to neoclassical economics, spatializing economic relations through the use of cost minimization, profit maximization, and utility maximization models. Often it began with the assumption, and reached the conclusion, that unfettered markets were Pareto optimal in nature. In some cases, location theory turned to cognitive psychology and behavioral geography to incorporate probabilistic models of behavior under uncertainty in studies of spatial cognition, decision theory, and suboptimality. This line of thought was central to the discipline of regional science, a hybrid of economic geography and spatial economics.

Although it became popular during the 1950s and 1960s, location theory drew on an older tradition of German economic geography that extended back to the early 19th century. For example, Johann Henreich Von Thünen (1783–1850), a wealthy Prussian landlord, bought a 1,146-acre estate at Mecklenburg in 1810 and compiled data for *The Isolated State* in 1821, the first model of land markets, which subsequently had important effects on studies of early land use. In 1909, Alfred Weber (1868–1958), younger brother of famed sociologist Max Weber, inspired the first industrial location theory. Weberian analysis centered on transportation costs, arguing that firms located where the sum of transporting inputs and outputs was minimized, leading them to be resource or market oriented. In 1933, Walter Christaller (1893–1969) wrote his enormously influential dissertation *Central Places in Southern Germany,* which founded central place theory, the conception of city systems in terms of a hierarchy of market areas distributing goods and services. This view of city systems posits them as retail centers (central places) that distribute goods and services to their surrounding hinterlands. Each good has a threshold (minimum market size) and a range (maximum distance consumers will travel to purchase it). A hierarchy of goods and services leads to a hierarchy of central places, with lower-order ones nested within the hinterlands of higher-order ones. Assuming an isotropic plain, a hexagonal network of market areas should emerge. Central place theory became the basis of most urban geography during the late 1950s and 1960s. Finally, August Losch (1906–1945) contributed to location theory with the publication of *The Economics of Location* in 1939.

Location theorists developed and applied a variety of models to understand economic and demographic phenomena such as urban spatial structure, location of firms, influences of transportation costs, technological change, migration, and optimal location of facilities. One group, gravity models, originated in Newtonian physics and became a highly successful way of modeling spatial interaction and predicting (but not necessarily explaining) patterns of interurban migration. Gravity models are widely used in studies of commuting and shopping as well as for traffic planning and models of transportation and communication. Similarly, network and graph theory became a useful way of mathematically describing the essential properties of networks of any kind, noting the accessibility of different nodes. It became a useful way of approaching transportation problems (e.g., the famous "traveling salesman" problem). Increasingly sophisticated mathematics invoked linear programming algorithms to develop location allocation models, a convenient way in which to find optimal locations (e.g., for factories or stores) by minimizing transport costs subject to some constraint (e.g., cost, travel time). Combined with geographic information systems (GIS), this approach is widely used in engineering and site location studies such as for public services.

A somewhat different category of location theory concerned the process of spatial diffusion. Launched by famed Swedish geographer Torsten Hägerstrand (1916–2004), this approach was concerned with the ways in which innovations (e.g., new technologies, information, diseases) moved through time and space. Diffusion was held to be either contagious or hierarchical. Use of so-called Monte Carlo models introduced probability theory into models of the innovation adoption process. Originally designed to maximize the adoption of new techniques for development purposes, the topic is useful for epidemiologists and in marketing, where it is applied to analyses of consumer behavior.

There is no doubt that location theory made great contributions to human geography. It popularized the use of rigorous quantitative techniques such as multiple regression, log–linear modeling, factor analysis, discriminant functions, and entropy maximization, and it minimized armchair speculation by asserting the importance of testable hypotheses. Its models have found a wide array of applications, including traffic planning and retail trade analysis, and shaped development strategies during the heyday of modernization theory. Location theory uncovered a great deal of structure, pattern, and regularity in human spatial behavior, shedding considerable light on how markets function and allowing alternative scenarios to be envisioned and assessed. Coupled with GIS, this approach raised the analytical sophistication of economic and urban geography significantly.

Ultimately, however, location theory declined in popularity as logical positivism fell from favor. Its central problems included the fetishization of quantitative approaches, its use of an increasingly unrealistic assumption of an objective value-free observer, and its silence concerning social relations and social structures. The ontology of location theory is essentially atomistic (i.e., it reduces the social to the individual through a process often called *methodological individualism*), and in so doing it tears variables from their social context. The focus on models led location theorists to ignore wider issues of class, gender, power, and struggle, instead positing an antiseptic, ahistorical, and sterile world of social order but not social change. Because location theory did not take subjectivity seriously, it deprived itself of any way of including human consciousness in all its complexity and richness. Thus, location theory focused on appearances rather than causes, and forms rather than processes, thereby naturalizing the status quo and leaving itself incapable of being critical.

—*Barney Warf*

See also Agglomeration Economies; Applied Geography; Behavioral Geography; Comparative Advantage; Diffusion; Economic Geography; Economies of Scale; Factors of Production; Gravity Model; Growth Pole; History of Geography; Innovation, Geography of; Input–Output Models; Logical Positivism; Model; Nomothetic; Population, Geography of; Quantitative Methods; Quantitative Revolution; Spatial Analysis; Spatial Autocorrelation; Spatial Dependence; Spatiality; Time, Representation of; Tobler's First Law of Geography; Transportation Geography; Urban Geography

Suggested Reading

Harrington, J., & Warf, B. (1994). *Industrial location: Principles, practice, and policy.* London: Routledge.

Stutz, F., & Warf, B. (2005). *The world economy: Resources, location, trade, and development* (4th ed.). Upper Saddle River, NJ: Pearson/Prentice Hall.

LOCATION-BASED SERVICES

Location-based services (LBS) are information services dependent on the geographic location of users to deliver spatially adaptive content such as maps, routing instructions, and friend finding. Common applications include emergency response, personal navigation assistance, fleet management, and recreation. Typically, users are mobile and are located through the presence of location-aware devices, often wireless cell phones. These devices obtain positional estimates for individuals and distribute this information to a broader network through established communication protocols. If positional estimates are sampled with frequent regularity, movement traces are created, identifying all locations visited by individuals in space and time.

The development of LBS to this point has coincided with advances in geographic information technologies coupled with the proliferation of wireless communications. These efforts have been stimulated largely by the Federal Communications Commission's enhanced 911 (E911) mandate. E911 requires wireless cellular carriers to develop capabilities to position users according to predefined accuracy standards as a means of improving the delivery of emergency services. LBS development represents largely an attempt to leverage the required investment in positioning technologies to achieve commercial gain. Positioning efforts have focused primarily on satellite and terrestrial solutions. Satellite positioning occurs by embedding a Global Positioning System (GPS) client within a mobile device. Terrestrial solutions use radio location algorithms within the wireless network. Additional efforts, especially those initiated by the Open GIS Consortium, have focused on easing LBS development by creating standards for distributing and sharing locational resources across multiple entities.

Considerable attention has been given to using LBS as a mechanism for collecting disaggregate activity–travel data from users, with the possibility for data collection efforts unprecedented in both volume and detail. In cases with frequent sampling, an individual's movement trace can be measured as a sequential set of time-stamped locations (x, y, t). The absence of reliable data sources has been a continual impediment to the advancement of methods and theories for understanding human geographic processes. Furthermore, rich frameworks such as Torsten Hägerstrand's time geography traditionally have also been limited by data scarcity. Analyses of observed movement traces may provide insights into spatiotemporal patterns of human interaction and their relation to broader spatial structures.

Despite potential societal benefits, LBS often are criticized for potential losses in personal locational privacy. Locational privacy suggests that individuals have the right to control the observation, storage, and sharing of their movement traces to limit personal identification or inference into sensitive activities and behaviors. The implementation of LBS suggests near-continuous tracking situations in which individuals have less control over what is known of their whereabouts in the past, present, and future. Nonetheless, it is difficult to dismiss the benefits of LBS even in the face of these concerns. An ongoing challenge involves identifying privacy protection strategies that maintain the utility of these services.

—Scott Bridwell

See also GIS; Humanistic GIScience; Time Geography

Suggested Reading

Monmonier, M. (2002). *Spying with maps: Surveillance technologies and the future of privacy.* Chicago: University of Chicago Press.

Schiller, J., & Voisard, A. (Eds.). (2004). *Location-based services.* New York: Morgan Kaufmann.

Smyth, C. (2001). Mining mobile trajectories. In H. Miller & J. Han (Eds.), *Geographic data mining and knowledge discovery* (pp. 337–367). London: Taylor & Francis.

LOGICAL POSITIVISM

Logical positivism is a school of philosophy, originating in Vienna during the 1920s and 1930s, that claimed that "real" knowledge is based on logical consistency and empirical verifiability. If either condition is contravened, the claim to knowledge is spurious or "nonsense." But if both conditions are met, truth is guaranteed.

The search for clear criteria to separate true meaningful statements from false meaningless ones has preoccupied philosophers ever since there have been philosophers. Those who gathered around Moritz Schlick in Vienna during the interwar period (the "Vienna Circle") and who included some of the 20th century's most preeminent philosophers—Oscar Neurath, Rudolf Carnap, Herbert Feigl, and Friedrich Waismann as well as (on the fringes) Karl Popper and Ludwig Wittgenstein—believed they had the answer. Building on French philosopher and sociologist Auguste Comte's earlier 19th-century philosophy of science, positivism, logical positivists argued that there are just two types of true meaningful statements, each of which can be defined and delineated precisely.

- *Analytical statements* are true and meaningful by virtue of their definition and have a logical connection with one another. The paragon example is mathematics. Based on self-defined primitive axioms and the purest form of logical derivation, mathematical statements are tautological. That is, they are true by definition and therefore unassailable. Analytical statements are the basis of the formal sciences, logic, and mathematics.

- *Empirical* or *synthetic statements* are empirically verifiable; that is, they can be proven true or false unambiguously by comparing them with real-world observations. The most important of such statements are enduring empirical regularities—laws—which in turn are a necessary component of scientific explanation. Sciences in which the principle of verification is applicable are the factual sciences.

However, a lot is left out from these two categories. For example, aesthetic judgment, moral argument, political opinion, and various forms of metaphysical speculation all represent spurious knowledge from logical positivism's perspective. Such discourses might spark interesting debates, but they do not meet the criteria of scientific knowledge, the only standard for an epistemology of objective explanation.

Not surprisingly, the period when geography most endeavored to follow the strictures of logical positivism was when it strove to achieve objective explanation—to be a science. That episode, which was known as the quantitative revolution and began during the mid-1950s, was associated with importation into the discipline of both scientific theory and sophisticated statistical techniques of description and empirical verification. The first shot in that revolution was an article titled "Exceptionalism in Geography" by Fred Schaefer, a German political émigré at the University of Iowa and a friend and Iowa colleague of Gustav Bergmann, former member of the Vienna Circle. This article was a plea for pure logical positivism. Arguing against an older regional geography concerned with mere description of unique places, Schaefer urged objective explanation based on discovering empirically verified morphological laws. A morphological law is an empirically verified repeated association between one geographic event and another. For Schaefer, morphological laws are the geographic equivalent of the laws of natural science; following logical positivism, they are necessary for objective explanation.

After Schaefer's manifesto, much of the quantitative revolution was driven less by grand philosophical statements than by a desire for technical competence in specific theories and techniques. Ironically, the next important disciplinary statement about logical positivism, David Harvey's tome *Explanation in Geography* also functioned as a memorial for it.

Harvey's book was the most comprehensive and rigorous philosophical justification of logical positivism ever put forward in the discipline. It paid as much attention to the kinds of analytical statements that logical positivist geographers could make in mathematics, logic, and (especially) geometry as it did to their synthetic statements, particularly their use of the verification principle embodied in particular statistical techniques. But almost as soon as he wrote the book, and by some accounts even before he finished it, Harvey had doubts about the logical positivist project that lay at the volume's center. In many ways, it was the wrong book at the wrong place and wrong time. Even by the 1940s, people associated with the Vienna Circle, such as Wittgenstein and Popper, had become critics, arguing for contrary positions. And by the 1960s, the idea of law-driven, objective scientific explanation was unraveling quickly. Thomas Kuhn, in *The Structure of Scientific Revolutions,* undermined the notion of independent facts (synthetic statements) by arguing that they never are pure; they always are tainted by social and personal values. Popper claimed that scientific theories never are verified anyway; they are only falsified. And Imre Lakatos, in *Criticism and the Growth of Knowledge,* went one step further, arguing that the internal architecture of scientific theories prevents both verification and falsification. Scientific programs rise and fall, but not on the basis of the quality of their synthetic or analytic statements. Subsequently, science studies, a body of literature critically examining the practice and epistemology of science, has only deepened these objections to logical positivism's binary agenda.

It is unclear which of these criticisms was most crucial in Harvey's about-face. But within 2 years of the publication of *Explanation,* he publicly rejected the quantitative revolution, and the concomitant philosophy of logical positivism, instead propounding Marxism as an alternative. Certainly, a key problem with logical positivism for Harvey, but one of the rationales for its originators, was the exclusion of the political. Logical positivism strove for unbesmirched certainty anchored in either pure facts or pure logic; anything impure, such as politics, was excluded. For Harvey, however, impurity was fundamentally interesting. It was where life and its problems were lived. To omit impurity was to omit what made geography a relevant and compelling discipline.

—*Trevor Barnes*

See also Empiricism; Epistemology; History of Geography; Location Theory; Model; Quantitative Methods

Suggested Reading

Gregory, D. (1978). *Ideology, science, and human geography.* London: Hutchinson.

Guelke, L. (1978). Geography and logical positivism. In D. Herbert & R. Johnston (Eds.), *Geography and the urban environment: Progress in research and applications* (Vol. 1, pp. 35–61). Chichester, UK: Wiley.

Harvey, D. (1969). *Explanation in geography.* London: Edward Arnold.

Schaefer, F. (1953). Exceptionalism in geography: A methodological introduction. *Annals of the Association of American Geographers, 43,* 226–249.

M

MALTHUSIANISM

One of the first social scientists to tackle the matter of population growth and its consequences was the Englishman Reverend Thomas Robert Malthus (1766–1834), whose famous book *Essay on the Principles of Population Growth* (published in 1798) electrified the world. Malthus's ideas, contrived in the early days of the Industrial Revolution during the late 18th century, had an enormous impact on political economy and demography. Malthus was concerned with the growing poverty evident in British cities at the time, and his explanation was largely centered on the high rates of population growth that he observed and that are common to early industrializing societies. Thus, it is with Malthus that the theory of overpopulation originated. His pessimistic worldview earned economics the label of the "dismal science."

The essence of Malthus's line of thought is that human populations, like those of most animal species, grow exponentially (or, in the parlance of his times, geometrically). A geometric series of numbers increases at an increasing rate of time. For example, in the sequence 1, 2, 4, 8, 16, 32, and so on, the number doubles at each time period, and so the increase rises from 1 to 2, to 4, to 8, and so on. Exponential population growth, in the absence of significant constraints, is widely observed in bacteria and rodents, to take but a few examples from zoology. Note that there is an important assumption regarding fertility embedded in Malthus's analysis; he portrayed fertility as a biological inevitability, not as a social construction. This argument was in keeping with the large size of British families at the time. In short, in Malthus's view,

humans, like animals, always reproduced at the biological maximum; they were portrayed as prisoners of their genetic urges to reproduce. It is worth noting that Malthus's argument carried with it a strong moral dimension; it was not just anyone who reproduced rapidly, he observed, but most particularly the poor.

Malthus maintained that food supplies, or resources more generally, grew at a much slower rate than did population. Specifically, he held that the food supply grew linearly (or, in his terminology, arithmetically). An arithmetic sequence of numbers, in contrast to an exponential one, grows at a constant rate over time. For example, in the sequence 1, 2, 3, 4, 5, and so on, the difference from one number to the next is always the same. Malthus's view that agricultural outputs increased linearly over time reflected the preindustrial farming systems that characterized his world. In such circumstances, without economies of scale, an increase in outputs is accomplished only with a proportional increase in inputs such as labor, reflective of what economists call a *linear production function*. However, this view of agricultural output actually is rather optimistic by Malthus's reckoning. He argued that in the face of limited inputs of land and capital, agricultural output was likely to suffer from diminishing marginal returns. For example, as farmers moved into areas that were only marginally hospitable for crops, perhaps because they were too dry, too wet, too cold, or too steep, farmers would need increases in inputs that were proportionately much larger than the increases in output. Diminishing returns, he held, would actually lead to increases in agricultural output that were smaller than those a production system lacking economies of scale would generate.

When one plots the exponential growth of population against the linear growth of food supplies, it is

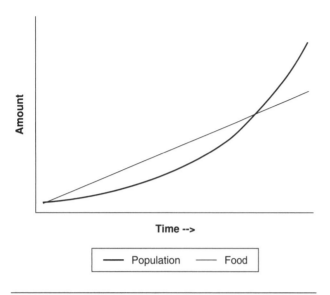

Figure 1   Geometric and Arithmetic Growth

clear that sooner or later the former must exceed the latter (Figure 1). Thus, in the Malthusian reading, populations always and inevitably outstrip their resource bases, and people are condemned to suffering and misery as a result. Malthus blamed much of the world's problems on rapid population growth, and subsequent generations of theorists influenced by his thought have invoked overpopulation to explain everything from famine to crime rates to deviant social behavior.

Malthus refined his argument to include "checks" to population growth. Given that natural population growth is the difference between fertility and mortality, "positive checks" are factors that reduce the fertility rate. Contraceptives are an obvious case in point, although Malthus objected to their use on religious grounds, instead advocating "moral restraint" or abstinence. Other positive checks include delayed marriage and prolonged lactation, which inhibits pregnancy. If positive checks failed, as he predicted they would, population growth ultimately would be curbed by "negative checks" that increased the mortality rate, particularly the familiar horsemen of the Apocalypse—death, disease, famine, and war. Thus, Malthusianism attributes to rapid population growth a variety of social ills, including poverty, hunger, and disease.

Malthus's ideas became widely popular during the late 19th century, particularly as they were incorporated into the prevailing social Darwinism of the time,

which represented social change in biological terms, often naturalizing competition as a result. However, to many observers, it became increasingly apparent that his predictions of widespread famine were wrong. The 19th century saw the food supply improve, prices decline, and famine and malnutrition virtually disappear from Europe (except for the Irish potato famine of the 1840s). By the early 20th century, Malthusianism was in ill repute. Critics noted that Malthus made three major errors. First, he did not foresee, and probably could not have foreseen, the impacts of the Industrial Revolution on agriculture; the mechanization of food production simply rendered the assumption of a linear increase untenable. Indeed, the world's supply of food has consistently outpaced population growth, meaning that productivity growth in agriculture has been higher than the rate of increase in the number of people. This observation implies that there is plenty of food to feed everyone in the world and that hunger is not caused only by overpopulation but also by a variety of other factors such as politics.

Second, Malthus did not foresee the impacts of the opening up of midlatitude grasslands in much of the world, particularly in North America, Argentina, and Australia, which increased the world's wheat supplies during the formation of a global market in agricultural goods. Third, and perhaps most important, Malthus's analysis of fertility was deeply flawed. During the Industrial Revolution, fertility rates declined and family sizes decreased. Thus, contrary to his expectation, humans are not mere prisoners of their genes and the birth rate is a socially constructed phenomenon, not a biological destiny.

During the 1960s, Malthusianism underwent a revival in the form of neo-Malthusianism. Neo-Malthusians acknowledged the errors that Malthus made but maintained that although Malthus may have been wrong in the short run, much of his argument was correct in the long run. Neo-Malthusians also added an ecological twist to Malthus's original argument. The most famous expression of neo-Malthusian thought was the Club of Rome, an international organization of policymakers, business executives, scholars, and others concerned with the fate of the planet. The Club of Rome funded a famous study of the planet's future, published in 1972 as *The Limits to Growth,* which modeled the earth's population growth, economic expansion, and resource consumption as well as energy and environmental impacts. It concluded that the rapid population and economic growth rates of the post–World

War II boom could not be sustained indefinitely and that ultimately there would be profound worldwide economic, environmental, and demographic crises. Much of this argument was framed in terms of the exhaustion of nonrenewable resources and ecological catastrophe. Unlike Malthus, neo-Malthusians advocated sharply curtailing population growth through the use of birth control and had an important impact on international programs promoting contraceptives and family planning.

Although neo-Malthusianism retains some credibility that the original Malthusian doctrine does not, it too suffered from a simplistic understanding of how resources are produced (e.g., when the price of oil rises, corporations find more oil). In addition, family planning programs in the developing world often have failed to live up to expectations, frequently for the simple reason that advocating contraception to curb population growth ignores the reasons why people in impoverished countries have large families and many children.

—*Barney Warf*

See also Demographic Transition; Hunger and Famine, Geography of; Political Ecology; Population, Geography of

Suggested Reading

Leisinger, K., Schmitt, K., & Pandya-Lorch, R. (2002). *Six billion and counting: Population growth and food security in the 21st century.* Washington, DC: International Food Policy Research Institute.

MARXISM, GEOGRAPHY AND

As a political philosophy and as a framework for understanding how society works, Marxism has its origins in the theories of Karl Marx (1818–1883), a German exile who lived much of his adult life in Britain. Marx can fairly be described as a philosopher, an economist, a political theorist, a historian, a journalist, and a revolutionary. Indeed, it was his revolutionary politics—and the threat of arrest by the authorities—that led him to flee his native land in May 1849. Politically, Marx was involved in several organizations, such as the International Workingmen's Association (also known as the First International), and wrote political pamphlets about historical and contemporary events, such as the Paris Commune of 1871, in which workers established an independent government to oppose Emperor Napoleon III. Marx penned many works, but some of his most famous include *Grundrisse (Outlines)* on wage labor, the state, and the world market; *Capital,* a three-volume work examining the economic structure of capitalism; *The Eighteenth Brumaire of Louis Napoleon,* a critique of the establishment of a dictatorship in France in 1851 by followers of President Louis Bonaparte, Napoleon's nephew (the Eighteenth Brumaire refers to November 9, 1799, in the French Revolutionary Calendar, the date on which Napoleon Bonaparte seized power in a coup d'état); *Critique of the Gotha Program,* an evaluation of the political manifesto of Germany's United Workers Party; and *The German Ideology,* an outline of his basic thesis concerning human nature, namely that the nature of individuals (their proclivities for cooperation or competition) are not inherent but rather depend on the material conditions determining their existence. With his friend and political associate Friedrich Engels, Marx also wrote *The Communist Manifesto* (published in 1848), which was designed to serve as the platform of the Communist League, a workers' organization.

Marx is perhaps best known for his theory of history and for his analysis of how capitalism operates as an economic system. During his student days, Marx had been a follower of German philosopher Georg Hegel, who had argued that what he called the "spirit of different ages" is embodied in various nation-states and that history is a record of the progression of human societies from less freedom to greater freedom and is shaped by the application of reason. Arguing that each society has its own personality that explains its level of development, Hegel believed that world events represent the necessary unfolding of the historical spirit through time. In this regard, he often has been accused of holding a teleological view of history, one in which history is thought to unfold according to its own internal logic. Hegel's model of history was based on a temporal progression in which concepts of reality unfold according to the outcome of dialectical reasoning. Such a system of dialectics argues that an existing element (or thesis) contains within itself inherent contradictions that unwittingly create its opposite (its antithesis). The result is a conflict between the two that ultimately results in the emergence of a new element (the synthesis), but this new element also contains its own internal contradictions, causing the process to begin anew. Hegel argued that each iteration of this dialectical conflict resulted in a level of development higher than the previous one and would continue until an end point where the highest possible state of freedom had been reached (a so-called end of history).

Although deeply influenced by Hegel (as well as by Adam Smith and David Ricardo), Marx reworked his ideas in significant ways, adopting a materialist vision of history—usually referred to as historical materialism—in which material economic and political interests are seen to drive history, in contrast to Hegel's idealist vision in which it is logic, reason, and ideas that do so. For Marx, dialectical contradictions have their origins in the relationships between social and economic classes. The beginning point of analysis, therefore, must be the economic circumstances within which humans find themselves. Furthermore, Marx argued that ideas are shaped by class relations and flow from the material conditions of the world, not the other way around; that is, the ideas of every age are essentially those of its ruling class. Hence, during the Middle Ages, kings conveniently argued that their authority was given to them by God, an idea that facilitated their hold on power, whereas in the contemporary period the ideology that ownership of private property is an innate, almost divinely or biologically given right serves to naturalize the unequal division of wealth in society between those who own property and those who do not. Thus, with his historical materialism, Marx metaphorically turned Hegel on his head, applying his ideas concerning dialectics to explore how conflicts over material conditions (rather than over ideas) between classes with different sets of economic interests—landholders and peasants, capitalists and workers—drive history forward. In such a theoretical framework, the purpose of knowledge for Marx is not only to understand the world but also to improve it. Indeed, Marx's most famous quote in this regard—"The philosophers have only interpreted the world, in various ways; the point, however, is to change it"—is one of two epitaphs carved on his tomb in Highgate Cemetery in London (the other is "Workers of all lands, unite").

In terms of Anglo-American geography, Marx's ideas were first widely explored during the early 1970s as a number of human geographers—radicalized by the Vietnam War and civil rights and antipoverty struggles—sought both to understand the economic and social geographies they saw around

them and to make their academic work address the big social justice issues of the day (e.g., racism, imperialism, environmental degradation, poverty). The intersection of Marx's ideas with geography (here geography is used to refer both to an academic discipline and to the production of landscapes) can be characterized as an effort to "Marxify" geography and to spatialize Marx. Thus, geographers adopted Marxist theoretical terms to explain the world around them, in the process challenging what they saw as a largely conservative discipline that they believed tacitly supported the status quo. Likewise, they sought to understand better how the geography of capitalism is implicated in the operation of capitalist social relations of accumulation, although there was considerable variation in the approaches they took to do so.

First, there was debate over the relative significance, for shaping the geography of capitalism, of the deep economic structures in capitalist society relative to the agency of humans embroiled within such structures—that is, between approaches that were more structuralist in nature and those that were more voluntarist in nature. Thus, some geographers tended toward a version of Marxism that was heavily influenced by French Marxists Louis Althusser and Etienne Balibar, a version that many critics argued portrayed humans as little more than automatons whose actions were determined by the deep economic structures of capitalist society and who could exercise little influence on these structures; in other words, the social structure (and hence geography) of capitalism was seen to change largely as a result of the playing out of its internal contradictions. Many critics argued that such an approach presented a base–superstructure model of capitalist society; that is, once the economic processes of accumulation were understood (the base), all other aspects of society such as politics and culture (the superstructure) were seen to be structured by it. Other geographers drew inspiration from Marxists such as British historian Edward Thompson, whose theoretical approach presented a view in which humans were able to transform the deep structures of society through political struggle, albeit within limits (i.e., they were not seen to have complete freedom of action). Many structuralists, however, argued that such voluntarist approaches allowed for too much agency and did not take into sufficient account how the economic structures of capitalist society constrained such agency.

This effort to explore the relative importance of structures and human agency was known during the

1980s as the structure/agency debate. Part of the reason for the debate lies in Marx's own writings given that he emphasized structure and agency in different contexts and at different times; his political pamphlets tended to stress agency more. Marx's economic texts analyzing the nature of capitalism were apt to focus more on the deep economic structures of capitalism, whereas Althusser argued that there was a significant epistemological break in Marx's life between his earlier, voluntaristic Hegelian writings and his later, more structuralist ones. Despite their differences, ultimately both sides in this debate were attempting to operationalize in their research Marx's famous maxim from *The Eighteenth Brumaire of Louis Napoleon* that "Men [and women] make their own history, but they do not make it as they please; they do not make it under self-selected circumstances, but under circumstances existing already, given and transmitted from the past."

Second, whereas many geographers drew on Marxist ideas to try to understand the economic geography of capitalism (and other types of society such as medieval Europe and centrally planned countries, e.g., the Soviet Union, which many criticized as simply an example of state capitalism rather than a truly Marxist society), others sought to understand the relationship of ideology and cultural practices to the structures of capitalist society. Many of these latter writers were influenced by the cultural Marxist writings of the Frankfurt School, which had been constituted by German Marxists who fled Adolf Hitler and came to the United States. Drawing on Marx's argument that the dominant ideas of a society are the ideas favored by its ruling class (and that they get inculcated into the population through things such as the educational and legal systems), such geographers explored how culture (broadly defined) can serve as a vehicle of ideological domination. They also developed critiques of science and technology, arguing that scientific and technological advances usually do not result from the pursuit of knowledge for its own sake but instead progress as adjuncts to processes of capitalist accumulation and social control; for example, government and the corporate world tend to fund scientific and technological endeavors that either promise to generate greater profits for private interests or serve to maintain the status quo (offering technical solutions to social problems such as environmental pollution, thereby hoping to reduce the necessity of a more fundamental economic and political restructuring of society).

Third, geographers attempted to work out the relationship between processes of capitalist accumulation and class dynamics, on the one hand, and how economic landscapes are made, on the other; that is, they sought to examine the relationship between social relations and a society's spatial structure. This effort has three important iterations to it. Initially, many geographers saw the geography of capitalism as simply a reflection of how capitalism is organized as a social and economic system. There are several reasons why this approach was adopted, including the desire to not repeat the mistakes of environmental determinism (in which geography was given a determining role in patterns of social development) and as a reaction to Hegel. Hegel had argued, after all, that the spirit of the age was encapsulated in the nation-state (a spatially defined entity); therefore, turning Hegel on his head paradoxically resulted in a downplaying of space, even as these geographers were intent on understanding how the spatiality of capitalism is made. Subsequently, in response to the idea that capitalism's geography is merely a reflection of its social relations, growing numbers of geographers began to explore how space was not only reflective of capitalist social relations but also constitutive of them.

Drawing on the ideas of, among others, French Marxist Henri Lefebvre, who had argued that capital must create particular spatial configurations of investment for capitalist social relations to be reproduced, this second iteration adopted a more dialectical approach to the relationship between space and social relations; social relations shape the geographies of capital accumulation, yet such geographies also shape how the social relations of capitalism operate. For instance, David Harvey argued that capitalists had to ensure that a particular configuration of the economic landscape was put in place—what he called a *spatial fix*—if accumulation were to occur successfully, suggesting that capital builds a landscape appropriate to its own condition at a particular moment in time, only to destroy it through disinvestment at a subsequent point in time. Others such as Neil Smith argued that the uneven development of economic landscapes under capitalism was not the result of the impossibility of even (as opposed to uneven) development but rather the outcome of contradictions within capital. On the one hand, extraction of surplus labor requires capital to be fixed in space and time (the process must occur somewhere); on the other hand, capital always must seek to remain sufficiently mobile so as to be able to relocate to more profitable locations, with the outcome of this tension being an economic landscape pockmarked by the scars of uneven development. At the same time, Doreen Massey outlined an argument for considering the geography of capitalism as being constituted by a geologic-type continuous depositing across the landscape of layers of investment— sometimes deep, sometimes shallow—that interacted with preexisting spatial arrangements of production and consumption to produce new geographies of uneven development. In this view, the geographic path dependence of places was shaped by what they had been in the past, so that new rounds of investment in the form of light manufacturing in Britain during the 1970s frequently went to areas that had been major steelmaking and coal-producing regions during the earlier 20th century but were, by the latter part of the century, suffering from deindustrialization (and so had lots of unemployment and thus cheap labor). The key here, then, was to understand how older economic landscapes continued to influence patterns of uneven development even after the social relations of capitalism that initially had produced them had been transformed by new political, economic, and technological forces. Put another way, the landscapes of the past continued to influence the making of the landscapes of the present and future, even as these latter landscapes gradually erased the older landscapes.

Despite the insights that such approaches provided for understanding how the geography of capitalism is made and for the integral part played by space in the functioning of capitalist social relations, this approach tended to be rather capital-centric in nature, explaining the geography of capitalism through exploring the internal contradictions of capital and the needs of capitalists to structure the economic landscape so as to facilitate accumulation. Consequently, this led to the development of a third iteration in which the geography of capitalism was seen to result from the sociospatial struggles between capital and labor, struggles that themselves were shaped by the geographic organization of capitalism. Known as labor geography, this approach argued that workers and capitalists (together with various factions within these two groups) often have different interests with regard to shaping the economic landscape of capitalism; capitalists need to create landscapes of accumulation, whereas workers need to ensure the creation of landscapes of employment (which are not necessarily the same thing)—and that it is important to theorize how workers attempt to implement, in the economic landscape, spatial fixes that they see as being advantageous to their own interests. The ultimate goal in all of this theorizing, however, has been to develop a historical–geographic materialist approach to understanding society, one where social beings are seen to be active in both space *and* time.

—Andrew Herod

See also Capital; Circuits of Capital; Class; Class War; Critical Human Geography; Dependency Theory; Hegemony; History of Geography; Ideology; Justice, Geography of; Labor, Geography of; Labor Theory of Value; Mode of Production; Radical Geography; Resistance; Social Justice; Spaces of Representation; Spatial Inequality; State; Structuration Theory; Underdevelopment; Uneven Development; World Systems Theory

Suggested Reading

Harvey, D. (1973). *Social justice and the city.* London: Edward Arnold.

Harvey, D. (1976). Labor, capital, and class struggle around the built environment in advanced capitalist societies. *Politics & Society, 6,* 265–295.

Harvey, D. (1978). The urban process under capitalism: A framework for analysis. *International Journal of Urban and Regional Research, 2,* 101–131.

Harvey, D. (1982). *The limits to capital.* Oxford, UK: Blackwell.

Harvey, D. (1985). The geopolitics of capitalism. In D. Gregory & J. Urry (Eds.), *Social relations and spatial structures* (pp. 128–163). New York: St. Martin's.

Massey, D. (1984). *Spatial divisions of labor: Social structures and the geography of production.* New York: Methuen.

Smith, N. (1984). *Uneven development: Nature, capital, and the production of space.* Oxford, UK: Blackwell.

Soja, E. (1980). The socio-spatial dialectic. *Annals of the Association of American Geographers, 70,* 207–225.

Soja, E. (1989). *Postmodern geographies: The reassertion of space in critical social theory.* London: Verso.

MASCULINITIES

The concept of masculinity is complex, having various definitions, historical roots, and logical approaches. Robert Connell, one of the key theorists of Western masculinities, argued that commonsense definitions of the term tend to fall into one of four approaches: essentialist, positivist, normative, or semiotic.

Essentialist definitions single out a core characteristic, usually a biological trait such as sex, and develop their account of masculinity based on this essential characteristic. *Positivist* social science attempts to provide an objective account of masculinity based on what men actually "are." *Normative* definitions recognize some of the internal contradictions found in various forms of masculinity and instead posit a standard for men to attain—what men ought to be. *Semiotic* approaches focus on symbolic understandings of masculinity and place it in relation to femininity, in effect defining it as not femininity.

All four types of definition suffer from flaws. Essentialist approaches are problematic because the choice of what characteristic is essential to masculinity is arbitrary (and is not essential at all). Positivist approaches, while claiming to be neutral and objective, nevertheless assume that people already have been sorted into the categories "men" and "women" and then proceed to measure the differences between these putatively distinct groups. In so doing, they never ask the really important questions about how gender itself is constituted in and through gendered social relations. Normative approaches tend to think of masculinity as a role to which men aspire. However, men rarely meet the complex restrictions of such roles. Semiotic approaches are the most sophisticated and often form the basis for important works in feminist geography and cultural studies, but they still tend to focus too heavily on the textual and discursive aspects of social life.

In response to the pitfalls in these definitions, social theorists who study gender have developed a more critical definition of masculinity. Connell, for example, suggested that masculinity should be understood more broadly as a set of practices by which men and women locate themselves in gender relations, articulate with that place in gender, and produce gendered effects on others and themselves. This kind of definition is important because it allows us to understand the temporal contingency of masculinity. However, it fails to account for the *geographic specificity* of different masculinities.

Here is where geographers have made important contributions to understanding the concept. Indeed, given the importance of context in the construction of masculinity, it should be very clear that masculinity is both *temporally* and *geographically* contingent. In other words, time and space make for different masculinities.

Thus, given the many possible gendered practices, relationships, and contexts that come together in the making of identities in different times and spaces, it is much more helpful to think not of a singular masculinity but rather of multiple *masculinities.* In addition, any one masculinity, as a product of practice, can be simultaneously positioned in differently structured relationships. Accordingly, masculinities always are complex and contradictory; they are highly contingent, unstable contested spaces within gender relations. This is why we speak of hegemonic and subordinate masculinities. Hegemonic masculinities are constituted through and meet the restrictions of dominant social relations. Hegemony works to control individual actions, even when doing so is not in people's own best interest; for example, consider how the hierarchies of business are not in the interests of most men, yet men not only go along with them but also often defend the very practices that disempower men. Subordinate masculinities do not meet the strict codes of dominant ideals, and some even subvert and contest dominant ideologies. Media such as movies and television provide useful examples of hegemonic gender identities because to be popular with a large audience, they often reproduce the dominant ideals of society. Accordingly, current Hollywood "hunks" such as Brad Pitt and Colin Farrell can be seen to exemplify hegemonic masculinity. Socially constructed characteristics such as their strength, decisiveness, handsome features, muscular bodies, and obvious (hetero)sexuality both constitute and are constituted by hegemonic masculinity. Nevertheless, such masculinities always are under threat; their strength and beauty can be turned on their head such that these "manly" men also become objects of homosexual desire.

At the same time, subordinate masculinities, examples of which can be found among the protagonists in currently popular American television shows such as *Queer Eye for the Straight Guy* and *Will & Grace,* can also be seen to both subvert and reinforce hegemonic masculinity in America. The popularity of the campy gay characters in *Queer Eye for the Straight Guy,* for example, surely undermines the "tough guy" imaginaries that support hegemonic masculinities. At the same time, these gay characters provide a foil by which we judge the not masculine, thereby further reinforcing hegemonic masculinity. It is this paradoxical stability and instability of masculinities, and especially the importance of context for understanding how masculinities are constructed, that makes it important to understand their *geographic character*.

—*Lawrence D. Berg*

See also Feminisms; Gender and Geography; Sexuality, Geography and/of

Suggested Reading

Berg, L., & Longhurst, R. (2003). Placing masculinities and geography. *Gender, Place, and Culture, 10,* 351–360.

Connell, R. (1995). *Masculinities.* Berkeley: University of California Press.

Jackson, P. (1991). The cultural politics of masculinity: Towards a social geography. *Transactions of the Institute of British Geographers, 16,* 199–213.

Phillips, R. (1997). *Mapping men and empire: A geography of adventure.* London: Routledge.

MEDICAL GEOGRAPHY

Medical geography applies the concepts and techniques of geography to the study of disease and health. It is one of the most practical subdisciplines of human geography in that it seeks to address issues that are clearly of vital concern to people in their daily lives. It is concerned with issues such as the spatial distribution of breast cancer mortality rates, the pattern of diffusion of influenza, the effect of people's movements on the spread of human immunodeficiency virus/acquired immunodeficiency syndrome (HIV/AIDS), the geography of toxic hazards, why hospitals are located where they are, and comparisons of the distances people must travel to primary healthcare in urban and rural areas. Although medical geography traces its origins to the Greek physician Hippocrates (ca. 460–377 BC), it only became a recognized field of study during the 18th century. Its recent flourishing began with the publication of Jacques May's series of global disease and nutrition maps during the early 1950s.

Medical geographers examine health and disease in the context of both a variety of cultures around the world and a variety of physical environments. They must be familiar with other fields of study within the natural sciences, the social sciences, medicine, and public health. Depending on their specific interests, those aspiring to become medical geographers should have a thorough grounding in cultural ecology, climatology, virology, epidemiology, health policy, biostatistics, anthropology, and geographic information systems (GIS), among other areas. Traditionally, medical geography has been divided into two main subcategories, disease ecology and healthcare delivery, although these two concerns are interrelated. A third subcategory, spatial analysis, always has been vital to medical geography and has become more so with the use of GIS.

DISEASE ECOLOGY

Disease ecologists examine the effects of disease on the triangle of human ecology, which consists of (a) *habitat* (the physical environment in which people live), (b) *population* (human characteristics such as gender and age), and (c) *behavior* (the actions of people). Within this framework, a wide variety of human ailments have been examined. The study of transmissible diseases, most important in the developing world, entails looking at the complex relationships among organisms that create the disease (agents), humans who may be infected (hosts), organisms that transmit the disease (vectors), and animal hosts (reservoirs). For example, African *trypanosomiasis* (or sleeping sickness), endemic in large areas of tropical Africa, is caused by the agents *Trypanosoma gambiense* and *Trypanosoma rhodesiense,* vectored by the tsetse fly, and has wild game and cattle as a reservoir.

The most important diseases in most of the developed world are degenerative and chronic diseases, including various cancers, heart disease, and stroke. These diseases usually are caused by a wide range of factors that disease ecologists need to examine such as genetic predisposition, radiation levels, water pollution levels, diet and exercise regimes, income levels, migration patterns, and social integration levels. Disease ecologists have also been at the forefront in investigating the spread of diseases that are transmitted through close human contact or through the air or water. A good example is a geographic study of the diffusion of cholera in the United States in 1832.

HEALTHCARE DELIVERY

Medical geographers who specialize in healthcare delivery are interested in the geography of how disease is dealt with in formal and informal systems of healthcare. They are interested in demonstrating where geographic inequities exist in healthcare systems and suggest ways in which health planning can be allocated or used more equitably and effectively. Various types of healthcare systems—folk medicine, traditional medicine (e.g., Chinese, Ayurvedic or Indian, Unani or Islamic), biomedicine, and complementary and alternative medicine (CAM)—have come under their scrutiny. Studies by geographers have included determining where managed care systems are most viable, examining informal caregiving to the elderly, examining spatial patterns of opposition to the siting of mental health facilities, and tracing physician migration across regions.

The geography of healthcare delivery traditionally has had several foci. One concern is the spatial distribution or geographic availability of healthcare resources such as doctors, nurses, lab technicians, high-tech equipment, clinics, and hospitals. Availability may be examined by the use of dot maps to depict the distribution of acute-care hospitals or of choropleth maps of physician-to-population ratios. Another concern is the geographic accessibility of healthcare to various groups of people. Here the role of distance to care—measured as straight lines along maps or roads in terms of travel time or perceived travel time—plays a key role. Examples of accessibility studies include comparisons between blacks and whites of travel times to primary care physicians and the effect of people's activity spaces on their use of hospitals.

Geographers have had input into investigations dealing with defining medical service areas (e.g., the geographic areas from which most of a hospital's patients come), planning for the effective regionalization of healthcare resources (e.g., establishing regions for the delivery of prenatal services), and developing models that efficiently locate healthcare facilities and allocate populations to them.

SPATIAL ANALYSIS

At least since 1849, when John Snow mapped cholera cases in central London and showed that deaths from the disease were concentrated in homes that took their water from the Broad Street pump, maps have been indispensable to medical geographers. Maps are used by disease ecologists to examine the diffusion of HIV/AIDS and to compare the spatial distribution of disease rates (e.g., heart disease) with patterns of human behavior (e.g., dietary preferences). Those interested in healthcare delivery use maps to depict flows of patients to hospitals and to delineate regions served by emergency medical services. Medical geographers have several spatial analytic techniques in their arsenal. For example, point pattern analysis has been used to examine the degree of clustering of leukemia cases, and spatial autocorrelation has been used to examine the spatial extent of cancer deaths.

Spatial analysis has been greatly facilitated during recent years by the development of GIS. GIS enable medical geographers to geocode, store, and display the locations or spatial patterns of the kinds of data that are essential to their work such as the residences of disease victims, the locations of healthcare facilities, and the distribution of risk factors such as environmental pollution, mosquito habitats, and smoking rates. Furthermore, GIS software can integrate spatial data from different sources and perform the kinds of spatial analyses discussed earlier in this entry.

—*Wil Gesler*

See also AIDS; Behavioral Geography; Cultural Ecology; GIS; Health and Healthcare, Geography of; Humanistic GIScience; Location-Based Services; Migration; Mortality Rates

Suggested Reading

Cromley, E., & McLafferty, S. (2002). *GIS and public health*. New York: Guilford.

Meade, M., & Earickson, R. (2000). *Medical geography* (2nd ed.). New York: Guilford.

Ricketts, T., Savitz, L., Gesler, W., & Osborne, D. (Eds.). (1994). *Geographic methods for health services research: A focus on the rural–urban continuum*. Lanham, MD: University Press of America.

MENTAL MAPS

A mental map is psychological or internal representation of a place or places. The term arose during the psychological turn in human geography of the late 1960s as a key component of behavioral geography that concerned itself with addressing the role of environmental perception as a mediating factor between humans' actions and their environment. Mental maps were viewed as a tool, that is, a key to unlocking the connection between people's understanding of their environment and their spatial choices and behavior.

This was an explicit attempt to explain human spatial activities by understanding them in terms of behavioral processes.

Mental map is one of many terms now known by the umbrella term *cognitive mapping*. The rationale for the study of mental maps is straightforward; our quality of life is greatly dependent on our ability to make informed spatial decisions through the processing and synthesis of spatial information, within a variety of situations, at differing scales. While attempting to navigate or explore an environment, we all have a spatial awareness of our surroundings, to varying extents. Through the varied stimulation of our senses, we come to know about places in the world. These sensory inputs emerge directly from the environment, where our senses engage in direct perception with objects that are close enough to touch, taste, smell, hear, and see. We can also learn geographic information from a multitude of less direct sources such as books, television, radio, newspapers, maps, models, and conversations.

Integrating these raw materials of experience and learning is a complex process. Mediating sensory inputs of environmental and spatial information are cognitive processes, attitudes, and beliefs. Through contemplation, sensory experience is transformed into knowing and understanding. The process that a mental map attempts to capture has become more widely known as cognitive mapping, commonly defined as a process composed of a series of psychological transformations by which an individual acquires, stores, recalls, and decodes information about the relative locations and attributes of the phenomena in his or her everyday spatial environment.

ALTERNATIVE TERMINOLOGY

Because mental maps deal with an abstract construct—this mediation between the environment and representations of the environment—they have been the focus of wide-ranging interdisciplinary study. Studies have addressed exploring the implied knowledge about the configuration, structure, and relationships in space as well as a concern with the meaning, thoughts, and beliefs associated with that space. The term *cognitive maps* often is associated with wayfinding behavior. Other terms, reflective of the interdisciplinary nature of their focus, have included the following: image schemata, spatial schemata, abstract maps, cognitive configurations, cognitive schemata, cognitive space, cognitive systems, conceptual representations, configurational representations, environmental images, mental images, mental representations, orienting schemata, place schemata, spatial representations, topological representations, world graphs, and cognitive collages.

UTILITY OF MENTAL MAPS

Early work with mental maps suggested their utility for (a) acting as a common referent when exchanging and communicating spatial information (e.g., facilitating the description of a route to another location or to another person), (b) acting as a rehearsal mechanism for spatial behavior (e.g., to mentally prepare and preplan a journey), (c) acting as a mnemonic device used to facilitate the memorizing of information by spatially referencing the information (e.g., by locating items to be remembered within a certain spatial context), (d) acting as a shorthand device for structuring and storing spatial knowledge, and (e) acting as a device for constructing imaginary or other worlds in the mind's eye (e.g., the ability of an individual to mentally build an awareness of a landscape that he or she has never visited from a traveler's description).

PROBLEMS WITH THE TERM *MAP*

It is clear that mental or cognitive maps facilitate the thinking of, feelings about, and activities in space that underlie our behavior. They are a blend of space in thought and the structure of that space along with the awareness of, beliefs about, and images, emotions, meanings, and symbolic properties of the space. It should be stressed that the term *mental map* acts as a shorthand term for this complex internal storage and should not generally be thought of as equivalent to a cartographic map. In 1994, Rob Kitchin proposed four useful interpretations of the term *map* in this context. First, is a mental/cognitive map a cartographic map? In this explicit statement, a mental/cognitive map is stored in the hippocampus, a part of the brain associated with long-term memory. The cognitive map is a three-dimensional Euclidean model of the world with rigid geometry. Second, is a mental/cognitive map *like* a cartographic map? As an analogy, a mental map has cartographic maplike properties; however, it is sketchy, incomplete, and distorted but still is assumed to work as the "map in the head" inspected by the mind's eye as an equivalent to a graphical map being read by the optics of a physical eye. Third, does a mental map work as if it were a map? This metaphor is interpreted as if it were like a person with a map in his or her head. The powerful spatial metaphor map is so useful because it makes spatial arrangements explicit and guides the use of imagery to remember conceptual relationships that are hard to imagine alone. Fourth, is it that the term *map* is redundant? That is, is it a hypothetical construct, a convenient fiction, where the word *map* has no literal meaning but rather is a convenient shorthand term?

THE FUTURE OF MENTAL MAPS

From the heyday of the term during the 1970s, many geographers have remained interested in mental maps. A more recent focus within human geography has moved through qualitative assessments of the meaning of place through various philosophical and literary styles. The representation and meaning of places remain central to much postmodern, poststructuralist, postcolonial, and feminist-based research. However, here the focus has shifted from the individual's mental schemata to the involvement of social, cultural, and historical discourses. Yet mental or cognitive maps are persuasive factors underlying our actions and behaviors. Recent advanced neuroscience imaging studies have led to a resurgence in interest in mental maps as cartographic objects. Studies with London taxicab drivers have documented specific place cells in discrete brain locations firing when shown photographs of famous London landmarks. In addition, the rise of geographic information systems (GIS) as a tool for manipulating, researching, and representing geographic space has led to renewed interest in mental maps. The research agendas of many geographic information science (GIScience) communities include the cognition of geographic space, of which mental/cognitive maps play a pivotal role. If a geographic information system worked more closely with our own internal representations of the world, would GIS be more intuitive and easy to use? Could a geographic information system be adapted to the understanding of underrepresented and marginalized groups based on their mental maps?

—*Daniel Jacobson*

See also Behavioral Geography; Body, Geography of; Cartography; Cognitive Models of Space; Environmental Perception; Existentialism; Humanistic Geography; Humanistic GIScience; Identity, Geography and; Music and Sound, Geography and; Ontology; Phenomenology; Qualitative Research; Social Geography; Spaces of Representation; Subject and Subjectivity; Symbols and Symbolism; Time, Representation of; Vision

Suggested Reading

Gould, P., & White, R. (1974). *Mental maps.* London: Penguin.

Kitchin, R. (1994). Cognitive maps: What are they and why study them? *Journal of Environmental Psychology, 14,* 1–19.

Kitchin, R., & Blades, M. (2002). *The cognition of geographic space.* New York: I. B. Tauris.

MIGRATION

Migration is the movement of people from one geographic location to another. Migration may result from many different causes. In some cases, economic opportunities may motivate individuals to move. Algerian guest workers in France are an example of this situation. War and political conflict may also instigate large-scale movements of people. For instance, violence emerging from internal political conflicts in Rwanda and Sudan has created mass migrations during recent years. Political and/or religious oppression has also created the context for migrations. Early migrants to the United States are a historical example of this situation, and refugees from Cuba provide an example of ongoing developments. Environmental disasters, such as drought, flooding, hurricanes, and nuclear disasters, may also generate waves of migration. Finally, in some cases, economic development may force migration. For example, the construction of the Three Gorges Dam in China has forced the removal of more than 1 million people to make space for a large reservoir behind the dam. These diverse factors can be divided into two general categories: push factors and pull factors. Push factors are variables that cause people to migrate because the situation in their current location is unsatisfactory, inadequate, or dangerous. Pull factors are variables that attract people from other locations because they offer positive opportunities or amenities.

Migration can also be categorized in other ways to better describe the characteristics of specific geographic movements. Most notably, a migratory movement may be classified as international voluntary migration, international forced migration, internal voluntary migration, or internal forced migration. International voluntary migration involves the movement of individuals across international borders. Examples of this type of migration include the movement of Europeans to the United States in the late 19th century and the movement of migrant laborers during more recent times. Examples of international forced migration include the African slave trade of the past, the relocation of English prisoners onto the continent of Australia, and the movement of Jews out of Nazi Germany. Internal voluntary migration involves movements that remain within national boundaries. In the United States, examples of this include events such as the Gold Rush, the movement to the suburbs after World War II, and responses to deindustrialization in the Midwest. Examples of internal forced migration include the placement of Native Americans on Indian reservations (the so-called Trail of Tears) and the relocation of Japanese Americans in internment camps during World War II.

Importantly, each type of migration operates at a different geographic scale and presents a distinct set of opportunities and challenges. With regard to scale, migration may affect the urban ecology of localities (e.g., suburbanization, white flight), the characteristics of different regions within a country (e.g., the movement away from the Rustbelt and toward the Sunbelt in response to economic opportunity and climate), or the overall composition of a country (e.g., the declining importance of U.S. immigration emanating from Europe in comparison with Latin America and Asia). Depending on the topic under consideration, geographers must be aware of the complexities produced by different geographic scales.

Similarly, migration produces opportunities and challenges. Within the economic realm, migration may boost economic productivity by using labor in a more efficient manner. On the other hand, this migration may engender costs that must be paid for by the preexisting local population. In contrast to common representations, migration produces a series of costs and benefits. These costs and benefits are unequally distributed throughout a society. Whereas some economic sectors may benefit greatly from these movements (e.g., agriculture), others may experience adverse effects. Some of these effects are direct, whereas others are more indirect. For example, migration into a region often produces a need for greater infrastructure. The incoming population will require schools, fire departments,

housing, transportation facilities, and retail establishments. Some of these requirements benefit localities (e.g., in the marketplace), whereas others require additional expenditures (e.g., transportation). In addition to these complexities, migration may have a substantial impact on cultural identity. This is reflected in the changing nature of language and the nature of acceptable practices in public places. Due to the potential changes it might produce, international migration often is viewed with a great degree of ambivalence and antipathy. These dynamics are further complicated by the fact that migrants often assimilate in their new surroundings.

Furthermore, it must be noted that migration as a whole involves a heterogeneous group of people with very different characteristics. These specificities must be accounted for in any analysis. Most of the public attention is focused on low-wage (perhaps illegal) migration. Political refugees from other countries, such as Haiti, are another noted example. This marginalized population possesses a distinct set of skills and requirements that are distinct to these types of movements. On the other hand, highly skilled professionals may also move to other countries. These individuals bring a distinct knowledge base and set of financial resources that often distinguish them from other migrants. These and other differences must be clarified to better understand the impacts of migration.

Finally, it must be noted that migration affects not only the recipient region or country but also the region or country of origin. Most generally, migration may reduce the population in the country of origin, thereby transforming the existing relation between supply and demand in the labor force. This potentially can reduce the amount of poverty within a country and thus diminish the likelihood of political unrest. Migrants may also send money back to their country of origin. In some cases, the aggregate size of these remittances is quite large. Such remittances are important because they can help secure the livelihoods of many families and improve the general welfare of a country. One negative effect of migration on source countries is the possible loss of educated, highly skilled individuals. This process is referred to as "brain drain" and may include valuable personnel such as doctors, lawyers, and governmental administrators. This can bring substantial hardship on countries that have many economic and social problems to solve. In some cases, however, these professionals may return to their country of origin with new skills that may be used to improve the position of the country.

The nature of migration changes over time, and migration policies develop in response. For example, in the United States, immigration laws were first enacted in response to concerns regarding race and cultural identity. Concerns regarding economics and drug trafficking instigated the perceived need for stronger regulations later in the century. More recently, concerns about terrorism and national security have fostered the formation of stricter enforcement of national boundaries. Although these examples are specific to the United States, they indicate the changing nature of international relations. Each country must consider the particular issues facing it at a particular point in time.

—*Christa Stutz*

See also Labor, Geography of; Mobility

Suggested Reading

Guinness, P. (2002). *Migration.* London: Hodder Murray.

Ravenstein, E. (1885). The laws of migration. *Journal of the Royal Statistical Society, 48,* 167–227.

Rubenstein, J. (2005). *The cultural landscape: An introduction to human geography* (8th ed.). New York: Macmillan.

MOBILITY

Mobility refers to the ability of people to move from one location to another. Throughout most of human history, the mobility of people remained essentially unchanged. Yet within modern societies, the mobility of people has changed quite dramatically over the years, creating a host of opportunities and problems. Researchers in diverse disciplines must be aware of the effects of mobility because it affects the structure and distribution of contemporary civilization.

In large part, the changing nature of mobility is linked closely with the changing nature of transportation systems. Throughout the early stages of human history, the means of human mobility was limited to walking. This meant that mobility was constrained by the limitations of human endurance. With the advent of domestication and certain technological developments (e.g., harnesses), the use of animal power became a means of increasing the mobility of individuals. Thus, the scope of activity was enlarged because horses and other animals have greater physical endurance than do humans. By the beginning of the 20th century, the mobility of individuals was transformed once again by the invention of the automobile (the "horseless carriage"). In addition to these developments, railroads, shipping lines, and airplanes have expanded the geographic scope of personal mobility, potentially allowing individuals to become global citizens.

The effects of such mobility are substantial and can be seen at many spatial scales. From a cultural perspective, the issue of mobility is important because it can greatly affect the perspectives and worldviews of people living in particular locations. When mobility was constrained by the limits of human or animal power, many societies remained largely insular and "local." With the increase in mobility, however, interactions among distant regions became more common, thereby serving as the context for the exchange of information, ideas, and material items. In this way, cosmopolitan influences began to influence previously isolated localities and regions. Although it was an anomaly of its time, the Silk Road was an early example of this phenomenon. Years later, the age of shipbuilding and exploration continued this process in new venues. More recently, the age of jet travel, tourism, and consumerism has created an entirely new set of international dynamics.

A quick comparison of cities indicates the influence of changing mobility on the internal structure of human settlements. For instance, many of the European cities that emerged during the Middle Ages have a dense pattern of development that was suited to foot traffic and animal transportation. Similarly, older cities on the East Coast of the United States frequently were designed with a linear, gridlike street pattern to facilitate pedestrian traffic. Conversely, cities that emerged after the invention and diffusion of the automobile (e.g., cities in the southwestern United States such as Phoenix and Los Angeles) often have a different pattern of development. Typically, these cities extend over larger areas and are less densely populated. Perhaps the best examples of this phenomenon are modern suburbs that usually contain cul-de-sacs and other curvilinear street patterns that diverge considerably from older grid systems.

The issue of mobility is also closely associated with accessibility. Mobility is the means by which places become accessible. Accessibility has become increasingly important for many reasons. In the past, many individuals lived comparatively isolated lives. Farmers, for instance, could provide for most of their own needs. Similarly, artisans and craftsmen often produced their artifacts in the space of their own homes. In each of these cases, the workplace and the living space were one and the same. With the emergence of modern capitalism and commodity exchange, the workplace and home became separate realms, thereby increasing the importance of personal mobility. In the current social context, modern lifestyles are highly interdependent on many places and many resources. Thus, to maintain a certain lifestyle, mobility and accessibility are critical issues to be addressed. This potentially includes access to the workplace, the marketplace, educational facilities, healthcare facilities, and governmental agencies. All of these resources are needed to maintain individuals and a viable population.

The issue of mobility presents a series of pragmatic difficulties. Given the size of contemporary cities, planners must devise new transportation systems that are capable of transmitting large numbers of people. In addition, planners must develop such systems in a way that is not overly harmful to the surrounding environment. As such, mobility often requires a balancing of interests. Along these lines, the relative importance of public and private transportation must be considered. Private transportation often provides more flexibility (and mobility for individuals), but it also generates a number of other problems (e.g., traffic congestion, air pollution). Public transportation potentially can resolve some of these problems, but it is also a less flexible form of transportation that limits the mobility of individuals. In making these decisions about transportation systems (and thus mobility), governments must decide how to prioritize these concerns before they allocate development funds.

Concerns regarding mobility have also become wrapped up with concerns about social equity and justice. The status of disabled individuals within society is a case in point. During recent years, many societies have tried to make places more accessible to disabled individuals by constructing buildings, transportation systems, and landscapes that facilitate the mobility of this diverse population. Similarly, governments have attempted to develop transportation systems that meet the needs of low-income populations. Such developments attempt to ensure opportunities for disenfranchised segments of society.

Altogether, these considerations illustrate the importance of mobility and the challenges it presents. The changing nature of technology and the shifting composition of local populations mean that the issue of mobility will continue to be a dynamic area of research and investigation.

—*Christa Stutz*

See also Migration; Transportation Geography

Suggested Reading

Fellman, J., Getis, A., & Getis, J. (1992). *Human geography: Landscapes of human activities.* Dubuque, IA: W. C. Brown.

Haggett, P. (1965). *Locational analysis in human geography.* London: Edward Arnold.

Hensher, D., Button, K., Haynes, K., & Stopher, R. (2004). *Handbook of transport geography and spatial systems.* London: Elsevier.

Isard, W. (1956). *Location and space economy.* Cambridge: MIT Press.

MODE OF PRODUCTION

In the terminology of Marxism, the mode of production means the way of producing. This is a combination of the forces of production (or the raw materials, human labor power, tools, technology, and improved land that constitute a society's productive capacity) and the social relations of production (or the property, power and class relations, legal frameworks, and forms of association that govern how production takes place). The mode of production substantively shapes the mode of distribution and the mode of consumption, and all of these together constitute the totality of the economic sphere. For Karl Marx, the way in which people relate to the physical world (forces of production) and the way in which they relate to each other socially (relations of production) are bound together closely but not harmoniously. Change occurs when the forces of production develop to such a degree that they come into conflict with the existing social relations of production. For Marx, therefore, conflict between the means of production and the relations of production is the basis for social revolution.

MODES OF PRODUCTION IN HISTORY

Marx saw human history as divided into epochs shaped by the existing mode of production:

Primitive communism: Human society is organized into small tribal groups with shared production and consumption.

Slave mode of production: This mode, the first class-based society, was centered on the use of coerced labor—the direct possession of humans—as in ancient Greece and Rome.

Feudal mode of production: Feudalism refers to a set of reciprocal legal and military obligations among the warrior elite of Europe during the Middle Ages. It also refers to the manorial bonds tying the peasantry (serfs) to the land.

Capitalist mode of production: Surplus value (profit) accrues through the control over wage labor. The ruling class is the bourgeoisie, which exploits the proletariat. The key forces of production include the factory system accompanied by the development of a modern bureaucratic state.

Socialist mode of production: In Marx's framework, the socialist mode of production was a theoretical conception of a society based on workers' control over production. Forms of collective social organization, such as cooperatives, strike committees, and labor unions, could approximate a socialist mode of production.

Communist mode of production: This is a hypothetical stage at which social classes will cease to exist. Marx argued that the internal crises of capitalism would give rise to this stage.

ARTICULATION OF MODES OF PRODUCTION

By the late 1960s, scholars influenced by Marxist theory were increasingly concerned with understanding the uneven spread of capitalism throughout the world, especially in areas referred to as the Third World. The Latin American dependency school sought to show how the integration of colonized regions into the world economy did not produce patterns of development similar to Western Europe; rather, it produced underdevelopment or dependent development, a system of permanent economic extraction and unequal relations. Thus, the idea of successive modes of production was critiqued for its evolutionary cast, and many theorists turned to the idea of the articulation of distinct modes of production in explaining uneven development.

The articulation of modes of production approach to studying development was influential during the 1970s and early 1980s in the sociology of development literature and in what became known as the new economic anthropology. The contribution of this approach was to show how purportedly precapitalist social relations may persist within the broader development of a capitalist economy. Indeed, these relations may exist not as separate from capitalism or as pockets of decline but rather as intrinsic to the extension of capitalism. As Aidan Foster-Carter noted in 1978, capitalism does not necessarily dissolve what came before it; it only coexists with other types of societies.

By the late 1980s, the articulation of modes of production approach itself was facing criticism. Critiques noted the risk of promoting a static ideal-type model in the mode of production language. Another problem was analytical slipperiness. For example, just how many modes of production and articulations might exist? Was there a specifically colonial mode of production? Was there a peasant mode of production? Or were all of these simply part of the uneven development of capitalism? Narrow economism was another critique. Feminist critics especially argued that the exclusive attention to the productive sphere in the modes of production framework ignored the important ways in which production and social reproduction are mutually constitutive. Attention needed to be paid to the concrete social practices that bound together these interconnected spheres.

—*Elizabeth Oglesby*

See also Colonialism; Communism; Dependency Theory; Developing World; Labor Theory of Value; Marxism, Geography and; Uneven Development

Suggested Reading

Anderson, P. (1974). *Lineages of the absolutist state.* London: Verso.

Foster-Carter, A. (1978, January–February). The modes of production controversy. *New Left Review,* pp. 47–77.

Kay, C. (1989). *Latin American theories of development and underdevelopment.* London: Routledge.

Peet, R. (1991). *Global capitalism: Theories of societal development.* London: Routledge.

Wolf, E. (1982). *Europe and the people without history.* Berkeley: University of California Press.

MODEL

A model is an abstraction of the real world. Models are developed and used in geography to simplify the enormous complexity of the earth's surface and to look for analogues for geographic processes in the work of other disciplines. The most basic geographic model is a map—a simplification of the world that allows people to navigate around their neighborhoods and cities without getting lost in a sea of detail. Maps enable geographers to portray on paper or on a computer screen the physical and social processes they wish to study. Any individual map involves hundreds of choices about what to include and not to include, how to symbolize a particular feature, the scale most appropriate for the problem at hand, and the map projection to be used. Any individual map is the mapmaker's best judgment about how to simplify or model the real world. Similarly, any individual model represents what the model builder believes to be the essential features of the processes in which he or she is interested.

A classic 1967 volume edited by British geographers Richard Chorley and Peter Haggett, *Models in Geography,* linked model building and pattern seeking in geography to the growing scientific orientation of the discipline, a prevalent trend at that time. They argued that a good model is simple enough to be understood by its users, representative enough to be used in a wide variety of circumstances, and complex enough to capture the essence of the phenomenon under investigation. Models can be static depictions of the spatial structure of a system at one point in time or dynamic representations of systemic change over time. From a research perspective, they are bridges between the empirical world of what we observe and the theoretical world of prediction and explanation.

Models often are represented in statistical and mathematical terms. For example, the gravity model from physics was used extensively to represent spatial flows such as migration, trade, and commuting. According to the rules of the gravity model, spatial interaction increases with the population of the origin and destination and decreases with growing distance between them. Of course, we know that many other characteristics of places determine flows among them, but model builders identify population and intervening distances as essential features of movement systems. Optimization models evaluate the trade-offs people make among conflicting goals, for example, among efficiency, equity, and environmental protection. Facility siting often uses this type of modeling to evaluate the trade-offs involved in any particular spatial configuration of hospitals, child care centers, welfare offices, and (potentially) hydrogen refueling stations. Human geographers also have relied on graphic models to represent the spatial organization of agricultural and urban land uses, the geography of development, the diffusion of innovations, and spatial organization of the settlement system.

Statistical and computational models are experiencing a resurgence in human geography due to the enormous power and popularity of geographic information systems (GIS). Statistical models use statistical concepts to represent real-world elements and their interactions. Computational models use the computer as an integral part of the modeling process, and the outputs often are visual and highly interactive. Computational models often involve "what if" simulations, and their outputs often are visual and interactive in nature. The two are combined to produce dynamic landscape model simulations that formalize thinking about what is and is not important about the landscape, articulate how the landscape behaves in response to various stimuli, develop plans for how the landscape is used, and produce time–space simulations regarding potential futures of the landscape.

Advances in computing technology, the growing availability of frequently updated spatial data at increasingly finer levels of resolution, and the development of more sophisticated GIS have led to the development of spatial decision support systems (SDSS), which are in essence models of the landscape that allow decision makers to visualize the consequences of alternative decision making. They often involve integrating elements of human and physical geography to obtain as complete a picture of the landscape as possible. Unlike research models that are strongly process oriented and focused on hypothesis testing with the eventual aim of prediction, these policy-oriented SDSS models emphasize the integration of what already is known about key human and physical processes. Policy models are problem oriented in the sense that they must produce usable solutions within a reasonable time frame. The need for solutions determines the time horizons for the calculations performed as well as the temporal and spatial resolution at which processes are represented. Policy models often are quite complex in the sense that they reflect the interactions of government and other social spheres, but they are not a complete and accurate representation of the real world. Environmental decision support systems have been used to predict the effects of climate change and climate variability on land use and water resources and to assess flood risks. SDSS model-building efforts have aligned human geography more closely with their physical geography colleagues and have moved the discipline to a more central position in local, regional, national, and international policymaking.

—*Pat Gober*

See also GIS; Location Theory; Logical Positivism

Suggested Reading

Boots, B., Okabe, A., & Thomas, R. (Eds.). (2002). *Modeling geographic systems: Statistical and computational applications.* Boston: Kluwer Academic.

Chorley, R., & Haggett, P. (Eds.). (1967). *Models in geography.* London: Methuen.

MODERNITY

The concept of modernity has varied and complex uses but is generally used to designate one of three things: a historical era associated with a series of societal transformations that began in Europe and subsequently diffused to most of the globe, a distinctive form of consciousness or experience characteristic of that era, or an aesthetic or artistic movement (often referred to as "modernism") concerned with exploring or representing this modern experience.

MODERNITY AS HISTORICAL PERIOD

There is little agreement about the precise temporal boundaries of modernity, but its roots lay in the Renaissance of the 15th and 16th centuries and its development is linked to the rise of industrial capitalism in Europe during the 18th and 19th centuries. During this time, European societies witnessed a series of important political and economic transformations that are generally associated with the modern era, including the expansion of capitalist markets and the progressive development of a system of factory production coupled later with mass consumption; the growing importance of salaried employment and an increasingly sophisticated division of labor; a significant increase in urbanization and the growth of a cosmopolitan urban citizenry, spurred by the evolution of widespread literacy; the development of specialized forms of knowledge and expertise, including the familiar practices of governance, civil service, and planning and design; and, following the French and American Revolutions, the establishment of the modern state system, founded on principles of territorial sovereignty and some form of electoral democracy.

These social and institutional transformations culminated in the form of society that characterized Western Europe and North America from the late 19th century through the 20th century. Although this particular view of modernity emerged in a specific time and place, it has nevertheless been held up by many as a more or less universal model to be emulated by societies around the world. Thus, those parts of the world labeled as *traditional societies* have been encouraged to foster development through the adoption of Western-style political and economic institutions, a process generally referred to as *modernization.*

MODERNITY AS ETHOS

Many invocations of modernity refer not only to a series of events or a historical era but also, more broadly, to the form of consciousness or type of experience typical of that era. That is, modernity can be understood in part as a distinctive ethos or sensibility, the roots of which can be traced to the social and cultural transformations of the European Enlightenment of the 17th and 18th centuries. Chief among these changes was a decline in the importance of religious dogma and superstition in favor of a worldview that placed human subjects at the center of their own destiny. Thus, the culture of modernity is characterized by a growing belief in the power of reason, rationality, and truth as well as a faith in the ability of science and technology to harness the powers of humans and nature for the betterment of society.

Another defining feature of modern consciousness is a particular awareness of time and history. Indeed, as derived from the Latin *modernus,* to be modern means to be living "in one's own time" as opposed to the past. It has since evolved to describe a sense of being cognizant of one's place within the movement of history and of continually progressing to overcome the limitations of the past. In a broader sense, then, the ethos of modernity entails a sense of looking toward the future and of embracing—even celebrating—the flux of restless change that is the hallmark of modern society.

MODERNITY AS AESTHETIC MOVEMENT

Toward the latter half of the 19th century and into the 20th century, various aesthetic movements took this modern experience of progress and change as the object of their art. Modernist cultural forms are extremely diverse and defy easy characterization. In general, however, the modern movement in art, literature, and music placed emphasis on self-consciously exploring alternative forms of representation and on expressing the modern experience through the use of juxtaposition, ambiguity, and paradox. Examples include cubism in painting, surrealism in literature, and the experimental musical works of Arnold Schoenberg and John Cage.

MODERNITY'S CRITICS

As its many critics have pointed out, European modernity always has had its darker side. Despite the lofty ideals of freedom and progress inscribed at its core, modern society has produced various forms of totalitarianism, fascism, and genocide, and its tenets generally have been perfectly compatible with the institutions of slavery and patriarchy. The growth of industrial capitalism, which fueled modernity's spirit of innovation and change, also produced shameful inequalities and brought about a hollow and alienated form of commodified experience. The technological and scientific breakthroughs that promised an end to famine and suffering have also produced the horrific instruments of modern warfare, including the everpresent threat of nuclear annihilation, and have bequeathed to us a severely despoiled environment. And the modern institutions of governance and planning have threatened to become an "iron cage" of bureaucratic rationality and abstraction, with their emphasis on order and control through the modern state powers of discipline and surveillance.

Equally troubling, critics point out that the proponents of modernity sought to universalize what was really a very particular geographic and historical experience of societal change. By equating modernization with Westernization, the discourses of modernity served to dismiss non-Western societies as stagnant and unchanging and to denigrate alternative forms of knowledge and experience. For this reason, some observers suggest that it may be better to speak of *multiple modernities* or *alternative modernities* as a way of accounting for the diverse range of encounters with the institutions and attitudes of modern life (a stance commonly associated with *postcolonial* theory).

During recent decades, a significant debate has emerged between defenders of the legacy of modernity and those who challenge its claims to authority and universality. Some contemporary critics argue that we have entered a new era of *postmodernity* in which the singular vision of progress and optimistic faith in reason have given way to a celebration of multiple and different perspectives and a healthy skepticism toward any claims to absolute truth (an epistemological position that also characterizes *poststructuralist* thought). Others suggest that modernity's failures are no reason to turn our back on its many promises; they believe that modernity's principles can and should be redeemed as guides toward a better future.

—*Jeff Popke*

See also Enlightenment, The; Industrial Revolution; Modernization Theory; Postcolonialism; Postmodernism; Poststructuralism

Suggested Reading

Bauman, Z. (1991). *Modernity and ambivalence.* Ithaca, NY: Cornell University Press.
Berman, M. (1982). *All that is solid melts into air.* New York: Penguin.
Giddens, A. (1990). *The consequences of modernity.* Stanford, CA: Stanford University Press.
Harvey, D. (1989). *The condition of postmodernity.* Cambridge, MA: Blackwell.

MODERNIZATION THEORY

Modernization theory refers to series of interrelated theoretical claims that were put forward by social scientists from the 1950s through the 1970s. Many of these social scientists were from the United States, and modernization theory had much to do with the role of the U.S. government in the cold war, particularly in countries of the Third World.

MAJOR THEMES OF MODERNIZATION THEORY

The major themes of modernization theory are well exemplified by the writings of economic historian Walt Rostow, sociologist Talcott Parsons, and political scientist Samuel Huntington. Rostow, a top adviser to the U.S. administration of President John F. Kennedy and a leading architect of the Vietnam War, argued in his widely read 1960 book, *The Stages of Economic*

Growth, that all societies will tend to pass naturally through a series of six stages of development, going from undeveloped, predominantly rural, agricultural, and premodern to fully developed, predominantly urban, industrial, and modern. Rostow proposed this on the basis of an account of industrial development in Europe and North America, but he argued that the patterns he discerned in these regions of the world would also emerge in the developing countries of the post–World War II era provided that they did not succumb to the projects of Communist parties, which Rostow referred to as a "disease of the transition" to modern society. It was for this reason that Rostow saw military opposition to communism in Vietnam as a logical counterpart to the promotion of modernization in the developing world.

Parsons, in books such as his 1966 work *Societies,* elaborated sociocultural dimensions of modernization theory by developing an account of how societies move out of and back into equilibrium. Parsons's account implied that as societies change from less developed to more developed, they become more socially and economically complex, based less on personal relationships and inherited status and more open to individual efforts at attainment of wealth and status.

Huntington developed a detailed argument for a process of political development that would correspond roughly to the stages of economic growth proposed by Rostow. Taking up the notion that communism is a disease of the transition, Huntington argued in his 1968 book *Political Order in Changing Societies,* as well as in other works, for the utility of authoritarian government in countries that were in the midst of the transition to modernity. During this transition, Huntington argued, coalitions of displaced peasants and urban groups such as students might form around the promises of communism and thus attempt to derail the forms of modernization pursued by Western-oriented developmental elites. In this context, U.S. government support for authoritarian regimes was justifiable. As modernization proceeded under military governments, Huntington suggested, the growth of a large urban middle class eventually would undercut the threat of communism and allow democratization on the Western model.

MODERNIZATION THEORY IN GEOGRAPHY

Some ideas from modernization theory were taken up by geographers during the 1960s and 1970s. A specific focus of geographic research was the diffusion of social processes identified with modernity through the development of economic infrastructure in developing countries. Thus, geographers such as Edward Taaffe, Richard Morrill, and Peter Gould examined the development of transportation networks in developing countries, whereas Edward Soja put forward a fairly comprehensive view of the geography of modernization in sub-Saharan Africa. Other geographers, such as Brian Berry and Jonathan Friedmann, presented modernization theory–inspired arguments about patterns of urbanization in the developing world. Within the discipline of geography, however, many early advocates of modernization theory approaches turned in other directions by the 1970s and 1980s, as exemplified by Soja's turn toward postmodernism and Friedmann's development of world systems theory–inspired approaches to urbanization.

CRITICISMS OF MODERNIZATION THEORY

Modernization theory developed in a context where most Western economists accepted Keynesian arguments for specific state interventions to boost economic growth. As such, many modernization theorists accepted "pump priming"—state-led investments in large-scale infrastructure to promote economic "take-off"—as recommended by Rostow and development economists of the 1950s and 1960s. When the neoliberal transformation of the U.S. government began during the 1970s and 1980s, this aspect of modernization theory became less popular among state planners and economists, with a new emphasis on private sector–led investment replacing the Keynesian emphasis on state projects. However, central aspects of modernization-style thinking have remained within neoliberalism, including the preference for capitalist development, an emphasis on the standard character of the process of industrial transformation across different countries, the relative invariance of the appropriate development policies, and the necessity to "insulate" key development policy decisions from influence by popular forces that might insist on policies that the development experts consider inappropriate.

Because of its central connection to U.S. foreign policy initiatives in the developing world and to the U.S. war in Vietnam, modernization theory was embroiled in controversy from its inception. Marxist critics rejected its assessment of capitalism as the only appropriate path to development, whereas neo-Marxist critics such as dependency and world systems theorists went so far as to argue that capitalist development simply could not occur within developing countries in the same fashion as it had occurred in the developed countries precisely because the sorts of power relations between developed countries and developing countries exemplified by U.S. policies in the Third World would prevent developing countries from being able to generate autonomous growth and retain the wealth generated by industrial transformation to the same extent as the earlier industrializing countries of the global core had been able to do. Just as modernization theoretic ideas have remained important within neoliberalism, so too have these kinds of criticisms of modernization theory remained prominent within contemporary Marxist, postdevelopment, and feminist theories.

—*Jim Glassman*

See also Dependency Theory; Development Theory; Feminisms; Marxism, Geography and; Neoliberalism; World Systems Theory

Suggested Reading

Huntington, S. (1968). *Political order in changing societies.* New Haven, CT: Yale University Press.

Porter, P., & Sheppard, E. (1998). *A world of difference: Society, nature, development.* New York: Guilford.

Rostow, W. (1960). *The stages of economic growth: A non-communist manifesto.* Cambridge, UK: Cambridge University Press.

MONEY, GEOGRAPHY OF

Geography and money are no strangers to each other. A sizable literature has documented the complex, often contradictory ways in which finance and space are shot through with each other. This topic finds its origins in an earlier sociology of money; writers as diverse as Karl Marx, Max Weber, Émile Durkheim, and Georg Simmel all were concerned with the relations between modernity and commodification. For example, under industrial capitalism and the waves of urbanization it generated, cities arose as sites of new forms of social relations centered on money, leading to a widespread objectification of social relations in which everyone becomes a buyer or a seller. The Chicago School, particularly Louis Wirth, was appalled by the predatory relations and culture of calculation that pervaded capitalist societies as ever more people were drawn into a money economy.

In its broadest sense, therefore, money was instrumental in the time–space compression of capitalism, the formation of the nation-state, and the rise of a global economy. Capitalism without complex systems of finance to lubricate investment and trade is unthinkable. Because money is highly mobile, most attention has focused on the international geography of money and the ways in which money supplies are regulated at the global scale.

Under the Bretton Woods agreement erected by the United States, there was very little exchange rate fluctuation from 1947 to 1971; most currencies were pegged to the U.S. dollar, fluctuating only within 2% within a given year without International Monetary Fund intervention. The dollar, in turn, was pegged to gold (at $35/ounce). The fixed exchange rate system required the free international movement of gold as well as minimal government interventions to offset its effects such as changes in the money supply designed to change real interest rates. The regulations for exchange rates imposed by Bretton Woods were designed largely to avoid the rounds of depreciations that deepened the Great Depression of the 1930s. Under this system of international regulation, currency appreciations or depreciations reflected government fiscal and monetary policies within a system of *relatively* nationally contained financial markets in which central bank intervention was effective. Trade balances and foreign exchange markets tended to be strongly connected; rising imports caused a currency to decline in value as domestic buyers needed more foreign currency to finance purchases.

The system of stable currencies ended abruptly with the collapse of the Bretton Woods agreement in 1971 and the shift to floating exchange rates in 1973, reflecting U.S. trade imbalances with its European partners and the overvaluation of the dollar, whose strength was maintained only through a steady outflow of gold. The accumulation of U.S. dollars overseas, which significantly enhanced the growing Euromarket during the 1960s, contributed to an increasingly unviable trade imbalance. Finally, President Richard Nixon announced that the United States no longer would abide by the Bretton Woods rules governing the dollar's convertibility to gold, forcing a global switch to flexible exchange rates. Thereafter, supply and demand would dictate the value of a nation's currency, and currency trading became big business.

The global sea change in capitalism that began with the traumatic petrocrises of the 1970s and massive restructuring of industrialized economies included a fundamental renegotiation of the relations between financial capital and space. Freed from many of the technological and political barriers to movement, capital has become not only mobile but also *hypermobile*. A key part of this new order was the emergence of what might be called stateless money, which originated in its contemporary form through the Euromarket. Originally, the Euromarket comprised only trade in assets denominated in U.S. dollars but not located in the United States; today, it has spread far beyond Europe and includes all trade in financial assets outside of the country of issue (e.g., Eurobonds, Eurocurrencies). One of the Euromarket's prime advantages was its lack of national regulations; unfettered by national restrictions, it has been upheld by neoclassical economists as the model of market efficiency.

Capital markets worldwide were profoundly affected by the microelectronics revolution, which eliminated transactions and transmissions costs for the movement of capital in much the same way as deregulation and the abolition of capital controls decreased regulatory barriers. Banks, insurance companies, and securities firms, which are generally very information intensive in nature, have been at the forefront of the construction of an extensive network of leased and private telecommunications networks, particularly fiber-optics lines. Electronic funds transfer systems form the "nervous system" of the international financial economy, allowing banks to move capital around on a moment's notice, arbitraging interest rate differentials, taking advantage of favorable exchange rates, and avoiding political unrest. Such networks give banks an ability to move money—by some estimates, more than $3 trillion daily—around the globe at stupendous rates. Electronic networks provide the ability to move money around the globe at unprecedented rates (the average currency trade takes less than 25 seconds); subject to the process of digitization, information and capital become two sides of the same coin.

In the securities markets, global telecommunications systems facilitated the steady integration of national capital markets. Electronic trading frees stock analysts from the need for face-to-face interaction to gain information. The National Association of Securities Dealers Automated Quotation System (NASDAQ), the first fully automated electronic marketplace, is now the world's largest stock market and lacks a trading floor. Online trading allows small investors to trawl the Internet for information, including real-time prices (eroding the advantage once held by specialists such as Reuters), and to execute trades by pushing a few buttons.

The ascendancy of electronic money changed the function of finance from investing to speculating, institutionalizing volatility in the process. Foreign investments, for example, have increasingly shifted from foreign direct investment to intangible portfolio investments such as stocks and bonds, a process that reflects the securitization of global finance. Unlike foreign direct investment, which generates predictable levels of employment, facilitates technology transfer, and alters the material landscape over the long run, financial investments tend to create few jobs and are invisible to all but a few agents, acting in the short run with unpredictable consequences. Furthermore, such funds are provided by nontraditional suppliers; a large and rapidly rising share of private capital flows worldwide no longer is intermediated by banks. Thus, not only has the volume of capital flows increased, but also the composition and institutions involved have changed. Globalization and electronic money had particularly important impacts on currency markets. Since the shift to floating exchange rates, trading in currencies has become a big business driven by the need for foreign currency associated with rising levels of international trade, the abolition of exchange controls, and the growth of pension and mutual funds, insurance companies, and institutional investors.

The world of electronic money has changed not only the configuration and behavior of markets but also the relations of markets to the nation-state. Classic interpretations of the nation-state rested heavily on a clear distinction between the domestic sphere and the international sphere, that is, a world carved into mutually exclusive geographic jurisdictions. State control in this context implies control over territory. In contrast, the rise of electronic money has generated a fundamental asymmetry between the world's economic systems and political systems. Because finance has become so inextricably intertwined with electronic transfers of funds worldwide, it presents the global system of nation-states with unprecedented difficulties in attempting to reap the benefits of international finance while simultaneously attempting to avoid its risks.

—Barney Warf

See also Economic Geography; Globalization; Telecommunications, Geography and

Suggested Reading

Cohen, B. (1998). *The geography of money*. Ithaca, NY: Cornell University Press.

Corbridge, S., Martin, R., & Thrift, N. (1994). *Money, power, and space*. Oxford, UK: Blackwell.

Leyshon, A., & Thrift, N. (1992). Liberalisation and consolidation: The single European market and the remaking of European financial capital. *Environment and Planning A, 24,* 49–81.

Leyshon, A., & Thrift, N. (1997). *Money/Space: Geographies of monetary transformation*. London: Routledge.

Martin, R. (Ed.). (1999). *Money and the space economy*. New York: John Wiley.

Solomon, R. (1999). *Money on the move: The revolution in international finance since 1980*. Princeton, NJ: Princeton University Press.

Strange, S. (1998). *Mad money: When markets outgrow governments*. Ann Arbor: University of Michigan Press.

Thrift, N., & Leyshon, A. (1994). A phantom state? The de-traditionalization of money, the international financial system, and international financial centres. *Political Geography, 13,* 299–327.

MORTALITY RATES

Mortality refers to the incidence of death per 1,000 people among a given population and is essentially the same as the crude death rate; therefore, it is closely linked to life expectancy. Mortality should not be confused with morbidity, which is the incidence or prevalence of a given disease. Demographers typically rely on age- and sex-specific mortality rates, which measure the number of deaths of a given 5-year age group of males or females. Mortality rates vary considerably across the life cycle depending on the particular social circumstances in which people live. Typically, mortality rates tend to be relatively high for infants (especially in economically underdeveloped societies), tend to be low for children and young adults, and then rise steadily as people enter middle age, rising dramatically in old age. However, mortality is a complex phenomenon with multiple demographic, economic, sociological, psychological, cultural, and geographic dimensions.

Like fertility, mortality is a reflection of both biological circumstances (e.g., genetics, diet) and socioenvironmental context. The causes of mortality vary greatly among societies (as well as within them). Infant mortality rates (number of deaths of babies [less than 1 year old] per 1,000 infants) are an important measure of a society's health because infants are the most vulnerable members of any society. Typically, infant mortality rates are high in preindustrial societies both historically and currently; for example, in much of sub-Saharan Africa, infant mortality rates exceed 120 per 1,000 babies.

In preindustrial social contexts, the leading causes of mortality generally are infectious bacterial diseases (and, to a lesser extent, viral diseases), including respiratory infections (e.g., pneumonia), diarrheal diseases, cholera, malaria, tuberculosis, and measles; today, that list also includes acquired immunodeficiency syndrome (AIDS). Many of these are waterborne diseases. Thus, mortality rates in much of the developing world (Latin America, Africa, the Middle East, and Asia [excluding Japan]) tend to exceed 20 deaths per 1,000 people annually.

As societies industrialize, the mechanization of agriculture and the corresponding lower price of food tend to improve diets and thus lower mortality rates and raise life expectancies. Improved public health measures (particularly clean drinking water) and access to healthcare are also important. Thus, mortality rates in Europe, Japan, and North America generally are less than 7 deaths per 1,000 people annually. These changes are an integral part of the demographic transition.

Moreover, the decline in mortality rates is accompanied by a shift in the causes of mortality, a phenomenon often called the *epidemiological transition*. Essentially, mortality in economically advanced societies tends to result from environmental and behavioral causes, including smoking- and alcohol-related deaths, which produce proximate causes of death such as heart disease, strokes, and various forms of cancer. Excluding the middle-aged and elderly, other important causes include automobile accidents, homicide, suicide, and household accidents.

—*Barney Warf*

See also Fertility Rates; Medical Geography; Population, Geography of

Suggested Reading

Yaukey, D., & Anderton, D. (2001). *Demography: The study of human population* (2nd ed.). Prospect Heights, IL: Waveland.

MULTICRITERIA ANALYSIS

Multicriteria analysis is a set of analytical models of individual or group decision making when multiple considerations, possibly conflicting with one another, are at play. Systematic procedures for analyzing complex decision situations have been researched and developed since the 1960s, primarily in the fields of management science, military science, and operations research. Geography, particularly analytical economic geography, has become a major testing ground for the validation of multicriteria decision techniques owing to the fact that most data used by individuals, as well as by public and corporate managers, have a geographic component.

In some instances, models of multicriteria decisions serve to comprehend the reality of the process and factors leading to making decisions. Utility-based models of residential choice, in which trade-offs exist among multiple characteristics of housing units as well as the relative location of each choice alternative with respect to employment, retail, and service districts, are classical cases of such a descriptive framework. On the other hand, multicriteria analysis can also set a standard to which decision making ought to adhere. A state agency that is looking for a route to ship a truckload of hazardous materials to a processing center would rely on a normative approach to minimize the risk of population exposure while maintaining travel time and travel cost within acceptable bounds. Both classes of decision problems abound in spatial planning, business operations, and public policy and management.

A multicriteria decision problem typically involves multiple components, including a goal or set of goals, a decision maker or group of decision makers, choice or decision alternatives (e.g., actions, locations, policies, objects), and criteria against which alternatives are evaluated. A criterion is a rule on which to test the desirability of alternatives. When criteria are the attributes of geographic entities or decisions (e.g., distance, population density), the phrase *multiattribute decision making* (MADM) has been used. On the other hand, when criteria are identified to the decision makers' objectives with respect to the desired state of the outcome of the decision process, an analysis of *multiobjective decision making* (MODM), whereby objectives are functionally related to attributes of the alternatives, is conducted.

Even when multicriteria analysis is intended to replicate the outcome of human decision making, its methods are merely a simplification of the real processes of decision making used by humans. The decision process typically is decomposed in a small number of manageable tasks. A common framework for MADM would involve the following steps: definition of the decision problem (What is to be accomplished?), compilation of decision alternatives (What are the options?), identification of evaluation criteria and of their relative importance to the decision maker (What is the decision to be based on?), evaluation of each alternative on each criterion (How does each option fare from different perspectives?), integration of partial evaluations into an overall performance score of each alternative, and (finally) option selection. Multicriteria analysis now is routinely interfaced with geographic information systems (GIS) to fully incorporate the inherent spatial dependencies among alternatives, their attributes, and decision objectives.

—*Jean-Claude Thill*

See also Quantitative Methods

Suggested Reading

Malczewski, J. (1999). *GIS and multicriteria decision analysis.* New York: John Wiley.
Thill, J.-C. (Ed.). (1999). *Multicriteria decision-making and analysis: A geographic information sciences approach.* Brookfield, VT: Ashgate.

MUSIC AND SOUND, GEOGRAPHY AND

The geography of music and sound explores geographies beyond the discipline's traditionally visual investigations. This subfield of human geography examines the geography of music both as a cultural product such as literature and as a performance such as theater. It is the geography of where and how music is made and listened to, of music's function as a cultural product and an industry, and of the roles of sounds and silences in shaping space and place.

Although there were some attempts to pair aspects of sound with landscape settings as early as the 1920s, music geography really began during the late 1960s in part as a tool for explaining cultural geography concepts to introductory geography courses. Some geographers dismissed music geography as unscholarly, especially research dealing with popular music (little of the work done in music geography concentrates on music not considered to be "popular"). However, a small number of geographers in North America embraced the subfield, and the geography of music enjoyed limited but continuing success during the 1970s and 1980s. By the early 1990s, as music was finally being approached critically within the social sciences, more geographers working around the globe gained interest in the geography of music and began to expand the subfield's subjects and approaches. Today, research in the geography of music addresses several different key issues through a variety of methods, closely linking music geography to many other areas within cultural geography. For the foreseeable future, music research appears to be a subfield of growing interest for all of the social sciences and specifically within human geography.

Traditionally, key issues in North American music geography have addressed several themes based on the conception of music as a cultural trait similar to aspects of material culture such as architecture. These themes include distributions of music types at both the world and regional scales, diffusion of musical styles, cores of musical production and innovation, the spread of music through technological innovations (especially in telecommunications), and the music industry's effects on selected landscapes. All of these studies yield cartographies of production, diffusion, and/or consumption of music.

During the past decade and a half, other music geographers have also addressed connections between music and the global economy as well as several other topics. These include the very definition of music; the importance of music in culture at the local, national, and global scales; the role of soundscapes; the nature and workings of the music industry; and the role of music in experiencing space and place. All of these reflect a growing attempt to engage and situate the cartographies of music in social, political, and economic realities at scales from the global to the personal.

For a few decades, traditional music geographers have been deciphering the cartographies of music and sound. Whereas some have attempted to map musical aspects at a global scale, many regional and national styles of music have been addressed. Cartographies of instruments, genres, and subgenres are used to mark culture areas and regions of a musical trait's core and periphery.

As popular culture spread within academics as a viable field of critical study, so too did music within geography. What once was ridiculed as unscholarly is finding an increasingly open audience within cultural geography. With this new attitude today, exciting new directions within music geography have expanded the depth and bounds of the subfield, so much so that some are even seeing music geography and the geography of sound as an important critique of the discipline's traditional visually based studies.

These exciting developments include the recognition of music as an important aspect of everyday life. Music and sound have huge impacts on how individuals interact with the world, and now that geographers are acknowledging this fact, it is leading to interesting work that examines music's place in creating national or local pride, music's close and complex association with questions of authenticity of place and culture, and the impacts of music's role as a multi-billion-dollar culture industry. Music performance and listening practices create unique spaces and geographies that have only been recently explored in our research efforts.

—*Christopher J. Dalbom*

See also Cultural Geography

Suggested Reading

Connell, J., & Gibson, C. (2003). *Sound tracks: Popular music, identity, and place.* London: Routledge.

Leyshon, A., Matless, D., & Revill, G. (Eds.). (1998). *The place of music.* New York: Guilford.

Nash, P., & Carney, G. (1996). The seven themes of music geography. *The Canadian Geographer, 40,* 79–74.

NATION-STATE

The nation-state is an ideal concept rather than one that is an actual or real geographic phenomenon. The nation-state is the ideological belief that the population of one state consists entirely of the members of one national group. Nearly all states in the world contain multinational populations and so violate the nation-state ideal. For example, Great Britain is home to the English, Welsh, Scottish, Ulster, and Irish nations. Those pressing for Scottish independence from Great Britain are acting in the nationalist belief that the Scottish people have a right to their own state, a Scottish nation-state. Although nation-states are practically nonexistent in the world, nationalist politics, or the desire to create nation-states, has been the most effective and powerful ideology of modern times and is the cause of the ever changing boundaries of the world political map.

Despite the fact that the nation-state is an ideal concept, everyday language usually denies the problematic difference between political reality and rhetoric. Politicians usually convey the impression that their country is a nation-state, in other words, that their population shares a common national identity. However, contemporary geographic analysis is more focused on the national diversity within states and on how the geography of collective identity transcends state spaces in the form of networks.

The nation-state was the fundamental geographic unit of the modern period. The Treaty of Westphalia of 1648 established the principle of state sovereignty, that is, a singular authority to rule over a territorial area defined by state boundaries. In Europe, the system of nation-states replaced feudal empires in which a hierarchy of sovereignty through baronial and majestic rule was played out over a network of fuzzy political boundaries. In addition, the ideal of the nation-state emphasized the ideal that sovereignty lay within the people rather than within a royal divine right to rule. The head of state was granted authority through the will of the people rather than being ordained by God. The nation-state became the political home, as well as the political vehicle, for a national citizenry.

Identity with the nation-state needed to be nurtured. The nation-state political project replaced established local and regional identities and loyalties with a national identity as well as an investment in the state. The invention of national traditions, customs, and holidays has been an integral part of popular practice that legitimizes the ideal of the nation-state. The epitome of the success of the nation-state ideal was seen on the battlefields of World War I. Prior to the outbreak of war, Europe had been rife with talk of international socialist revolution. But between 1914 and 1918, millions of Europeans blew each other to pieces in the name of national security. The relationship between interstate politics and individual duty, bridged by the ideal of the nation-state, was captured by wartime poet Wilfred Owen's caustic reference to the British propaganda slogan, "Dulce et decorum est Pro patria mori" (It is sweet and fitting to die for one's country). Despite continued political challenge to this ideal, the notion of the nation-state demanding personal sacrifice in the name of interstate competition remains today.

European colonialism created states, with the ideal of the nation-state in mind, across the globe.

Achieving statehood was equated with becoming modern. State boundaries were imposed by colonial powers with little regard for the geography of ethnic and tribal identity. After the collapse of European empires during the 20th century, leaders of the newly independent states tried to forge a sense of national identity based on the state boundaries. In other words, they tried to create a sense of common nationality defined by citizenship in the newly independent states. This political project has had varying degrees of success. For example, British India quickly fragmented into India and Pakistan.

The nation-state is an ideal, but one to which all states aspire. Hence, each state propagates the belief that it meets the ideal. One of the aims of state education is to create a sense of common history and culture that embeds younger generations within a national identity. The goal is to create a belief that state citizenship is synonymous with national identity. All states practice this state construction of national identity, and they have differential success. Immigrant states (e.g., the United States, Canada, Australia, New Zealand) are multinational and use the story of the common experience of migration and settlement, even though the journey was started from many different parts of the globe, to construct a shared national identity. The practice of building states through immigration has resulted in the horrible decimation of indigenous populations and the marginalization of their cultures. Discrimination toward immigrant groups is also experienced, although the actual minorities being targeted change over time. Without diminishing the pain experienced by indigenous groups, the success and coherence of states created by immigration illustrate the ability of all states to create a national identity.

The state does not act alone in this endeavor. The media are also active in creating a sense of national unity. In everyday uses of terms such as "our weather" in creating enthusiasm for national sports teams and promoting national events (e.g., the Super Bowl), the media sublimely convey a notion that the "our" is unproblematic and natural. Moreover, the media's concentration on "us" helps construct feelings of "them," resulting in the normalization of the belief that the world political map is naturally made up of a mosaic of nation-states. The historical specificity of the nation-state as the geographic unit of society is unchallenged, and alternative forms of political organization are dismissed.

The ideal of the nation-state rests on the construction of an "imagined community." The propagation of the myth that all citizens share a communal national experience—a similar historical trajectory—is maintained despite the fact that there is no familial or physical connection among people. The imagined community of the U.S. nation-state, for example, connects the residents of Miami with those of Seattle despite the respective proximities of Havana, Cuba, and Vancouver, Canada.

Feminist scholars have critiqued the imagery used in the ideology of the nation-state. The notion of the motherland evokes a nurturing role for the nation that is translated into gender roles that limit women's task to the private realm of the household—the role of procreation and the socialization of children within the national community. In contrast, male gender roles are associated with defending the nation-state from external threat, a task that equates to a dominant role in public-sector tasks such as politics and the military.

Of course, not all states are able to create the sense that their citizens are members of a common nation. National separatism is the attempt to carve out a new state, one that better reflects the geography of national identity. The ideology of nationalist movements rests in the belief that multinational states wrong minority groups and that the rights of the minority nations will be better served if they create their own states. Nationalist movements challenge the legitimacy of particular states by arguing that they are not nation-states. However, nationalist movements simultaneously legitimize the ideal of the nation-state by claiming that they have a right to construct one, and it is the ultimate form of political organization.

The political geographic logic of the nation-state can easily lead to extremes of violence. The fact that the term *ethnic cleansing* is part of everyday vocabulary reflects the manner in which the ideal of the nation-state frames our understanding of the way in which the world political map is, and should be, organized. Violence in the former Yugoslavia, in Central Africa, and (more recently) in Sudan rests on the power of the ideal of the nation-state. Ethnic cleansing rests on the assumption that a state is somehow imperfect if more than one national or ethnic group resides within it. The ideal of the nation-state, one nation within the state, is held up to be essential for the future of the nation and the individuals within it. The belief in the nation-state relates individual fulfillment and national prosperity to residence within an allegedly pure nation-state. The political project of killing and expelling minority groups is motivated by faith in the power of the nation-state.

The construction of a nation-state may require a step beyond ethnic cleansing. Although all minority groups may be expelled from within a state's boundary, members of the nation may still reside in neighboring states. Hence, a further political geographic project motivated by the ideal of the nation-state is the expansion of boundaries so that all of the members of a particular nation are contained within one state. In the former Yugoslavia, for example, the project of a Greater Serbia was aimed at changing the political boundaries so that Serb residents of Croatia and Bosnia were encompassed within an enlarged Serbian state.

The politics of the nation-state resides in the power of the hyphen. The hyphen creates the ideal beliefs that the nation and the state are synonymous and that the former requires the latter. But during recent years, this geographic imagination has been challenged. Different geographies of identity are being created with varying degrees of success. Political projects such as the European Union are attempting to create supranational identities that transcend the scale of the nation-state. On the other hand, greater attachment to substate regions and more localized geographic attachments have diluted the dominance of national identity. Nonterritorial identities such as gender, race, and religion have also asserted a place in the political dialogue. The result has been the identification of nested identities—the realization that people define themselves in terms of attachment to geographic entities above, below, and including the scale of the nation-state.

Consideration of how the ideal of the nation-state has been challenged through identifying with other geographic scales has been complemented by consideration of the mobility of people and the fusion of identities. Migrants and their subsequent generations fuse old and new national cultures into hybrid identities that are a complex mixture of multicultural beliefs and practices. Identity is sustained by a network of connections to one set of national cultures (the mother country) that are experienced and given new meaning within the constraints and opportunities of the host country. In such a geographic condition, the hyphen in nation-state is severely challenged. State citizenship is negotiated through attachment to a hybrid of identities rather than to a singular national culture.

So what of the future of the nation-state? Nationalist conflicts, such as those in Israel/Palestine and Kashmir, provide harsh evidence that in some parts of the world belief in the nation-state as an ideal worth dying and killing for remains intense. Media coverage of any international sporting event is but one illustration that we are reminded of our attachment to the nation-state on a daily basis. However, the development of supranational political entities such as the European Union, as well as the diffusion of common cultural practices by the processes of globalization, offers substance to the argument that identity is being negotiated at a higher geographic scale, and in a less territorialized form, than the nation-state.

—*Colin Flint*

See also Critical Geopolitics; Diaspora; Ethnicity; Identity, Geography and; Migration; Nationalism; Other/Otherness; Political Geography; State; World Systems Theory

Suggested Reading

Anderson, B. (1991). *Imagined communities: Reflections on the origin and spread of nationalism.* London: Verso.

Appadurai, A. (1991). Global ethnoscapes: Notes and queries for a transnational anthropology. In R. Fox (Ed.), *Recapturing anthropology* (pp. 191–210). Santa Fe, NM: School of American Research Press.

Billig, M. (1995). *Banal nationalism.* Thousand Oaks, CA: Sage.

Herb, G., & Kaplan, D. (Eds.). (1999). *Nested identities: Nationalism, territory, and scale.* Lanham, MD: Rowman & Littlefield.

Hobsbawm, E., & Ranger, T. (Eds.). (1992). *The invention of tradition.* Cambridge, UK: Cambridge University Press.

Hooson, D. (Ed.). (1994). *Geography and national identity.* Oxford, UK: Blackwell.

Owen, W. (2004). Dulce et decorum est Pro patria mori. In W. Owen, *The poems of Wilfred Owen* [Online]. Available: www.pitt.edu/~pugachev/greatwar/owen.html

Smith, A. (1995). *Nations and nationalism in a global era.* Cambridge, UK: Polity.

NATIONALISM

Like culture, the terms nationalism, nationality, and nation are some of the most notoriously difficult to define. Although *nationalism* is derived from the past participle of the Latin verb *nasci* (meaning to be born) and the noun *nationem* (connoting breed or race), using etymological origins as the basis for understanding nationalism is not adequate. Most analysts agree that nationalism is a distinctly modern phenomenon. Whereas some focus on the *ideas* underlying a nationalist ideology, others stress the *material* conditions that acted as preconditions for the evolution of the nation-state. Geographers have drawn together the ideological, cultural, and territorial dimensions of nationalism.

Theorists interested in the ideas underlying nationalism focus on the emergence, during the late 18th century, of a notion that humans are autonomous, free-thinking individuals. With the diminishing role of the family, religious group, and/or community as the primary source of authority and identity, individuals moved toward a political ideal of their autonomy being best served through a larger cultural group—the nation. Those who have focused on the material changes that underpin the emergence of nationalism stress the link among modernity, the transformation of society from an agricultural one to an industrial one, and the nation. They argue that the "tidal wave of modernization" beginning in the 18th century necessitated the homogenization of cultures across space. Marxist thinkers have similarly focused on the material transformations of society and link the rise of nationalism with the development of a capitalist political economy. They suggest that nationalism is the response of peripheralized people to the uneven development of capitalism, where nationality is employed as a means of defense against core industrial economies.

By combining the cultural and ideological underpinnings of nationalism through the concept of an "imagined community," Benedict Anderson effectively linked together cultural, economic, ideological, and political processes. He claimed that the advent of the printing press had the effect of connecting populations over wide geographic areas. As the market for Latin printed books gradually became saturated, vernacular languages began to be printed. However, not each and every dialect was printed; rather, similar vernaculars were assembled together through a standardization of grammar and syntax. Accordingly, these new print languages were fundamental to the emergence of national consciousness, first, because they geographically connected speakers of, for example, huge varieties of English, Spanish, and German and made known to them the existence of those who shared the same language group. Second, print capitalism fixed language by having the capacity to produce and reproduce the same text not only over space but also through time. Third, print languages gave

priority to one dialect over another and thus awarded the chosen dialect political power (e.g., High German). Anderson's theory of nationalism has won widespread appeal both inside and outside geography. The cultural components that are important to establishing an imagined community of nationhood are outlined in the following sections.

LANGUAGE AND NATIONALISM

The cultural definition of identity frequently has rested on linguistic differentiation. There are four main reasons why language is useful as a basis of nation building. First, language arouses ideas of a common identity. Second, it forms a link with the past (e.g., earlier generations). Third, language becomes a link with authenticity. It provides a secular source of mass communication in modern society, yet it can lay claim to uniqueness. Fourth, a vernacular literature allows elites to become central to a nationalist movement. The politicization of language requires planning. The standardization of spelling and grammar and a mass education system achieves a degree of uniformity, at least so far as the written word is concerned. Language planning is crucial for the breaking down of old spatial barriers and the construction of new ones at the scale of the state. To take an example, the adoption of the Francien dialect of the Paris region as the national dialect evolved through various stages of language planning. The Edict of Villiers-Cottêrets by the Paris authorities in 1539 made Parisian French the official language of the royal domains, and the edict was facilitated by the first publication of a French dictionary and French grammar text in 1531. Where language planning is unsuccessful, tensions between linguistic communities forming a state can lead to separatist politics (e.g., Quebec in Canada).

HEROIC PASTS

Although nationalism has a territorial imperative in the acquisition of space or territory for a nation to inhabit, the imagined community of nationhood also has a temporal dimension in which periods of past glory are resurrected for current self-aggrandizement. This process is called the "invention of tradition," where the chronology of the historical imagination is eschewed in favor of a focus on particular historical or quasi-historical epochs when the cultural effervescence of the nation was particularly heroic. The geographic and the historical merge in the context of these imaginings as particular places and landscapes become centers for the rituals of collective cultural memory. In the context of Scottish nationalism, for instance, the tartanization of Scottish culture involved a selective reading of the "national" past that ignored the real political forces that produced such imaginings. The popularization of the tartan kilt and of tartan patterns associated with different clans corresponds with a period of changing economic fortunes in Scotland in relation to the British and the world economy.

The compelling appeal of invented traditions is not confined to minority groups within larger states; it is also found in well-established majority cultures. The contemporary nostalgia for the past tentatively has been linked to public dismay in the purported anonymity of high modernity, where identities are increasingly being fragmented in favor of more situated and flexible ones. The abandonment of some older myths and traditions associated with national cultures has seen their replacement with new or reconstructed ones. Tourism and the heritage industry frequently reinforce images of a heroic national past that is packaged for public popular consumption. New technologies, new forms of museum display, and the "theme-parking" of historical narratives may have replaced the older traditions of popular ritual, but the history purported to be represented through these installations often merely anchors conceptions of national identity on new terrains. Whereas in the 19th century national languages, school history textbooks, and religion formed the nexus of national imaginings, in the 21st century the heritage industry and the associated historicizing of interior design and architecture play a similar role.

The potency of a heroic history rejuvenated for current political reasons can be particularly skewed in the context of civil strife. For instance, the annual celebration of the 12th of July in Northern Ireland's commemorative calendar and the annual disputes surrounding the routing of Orange parades (e.g., at Drumcree) underscore the significance of representing history and the role of space in the parading of cultural memory. Although heroic histories form an important part in the establishment of an imagined community, these pasts are also intimately linked with heroic people in nationalist discourse, and these people are embodied in monuments and statuary that adorn the towns and cities of national states.

MONUMENTS AND NATIONALISM

If nationalism appropriates periods of the past to represent its continuity, it also appropriates historical persons, events, and allegorical figures to reinforce its cultural existence. During the 19th century, public statuary had firmly entered the public domain. Monuments dedicated to important figures in the nation's history began to emerge at a large scale. When the nation-state remembers its founding heroes, it also commemorates its wars—be they wars of independence, international conflict, or even civil strife. The war memorial varies in iconography depending on whether it is commemorating defeat or victory. War memorials that personify particular military leaders tend to be heroic in proportion and may use iconography from previous eras to convey the strength and national importance attached to individuals. The Nelson Column in London's Trafalgar Square adopts a design drawn from the iconography of ancient Roman imperial victories.

Memorials to World Wars I and II vary considerably in scale and iconography. They can be simple commemorative plaques placed in towns and villages listing local people who were killed, or they can be colossal national monuments located in prestigious positions within capital cities (e.g., the Cenotaph in Whitehall, London). Annual pilgrimages to these memorials and the laying of wreaths at their feet reinforces the state's recognition of the importance of war and the losses it incurs, but these occasions also enable the mass of the population to participate openly in a national event. The tone of the inscriptions on war memorials replicates those found on headstones in cemeteries but tends to conceal the class and gender divisions of the people they represent. In this sense, they function as "nationalized" monuments.

The nationalist discourse in which war memorials are conceived is confirmed by the fact that they rarely acknowledge the loss experienced by the "enemy." They are not memorials to all those who were lost in war; rather, they are interpreted in terms of national losses and national geopolitical considerations. War memorials commemorate *our* dead, not just *the* victims of conflict.

GENDER AND NATIONALISM

In many cases, analyses of nationalism are presented as gender-neutral discourses where the desire to create a national imagined community disguises other sources of identity such as gender, class, and "race." The imperatives for creating a unified voice frequently are seen to obfuscate other voices, especially those of women. The focus on ethnic identity around issues of language, tradition, and historical landscape has tended to underestimate the role of gender in the articulation of a nationalist politics. Nevertheless, the role of women and the gendering of nationalist discourse have been increasingly identified in human geography.

Women's participation in the development of ethnic and national consciousness has taken several specific forms. First, women have served as the biological reproducers of members of the nation. Encouragement to reproduce often uses religious and nationalistic arguments about women's duty to motherhood. Welfare policies, such as child benefits, may be introduced by the state to serve this end. Second, women can be seen to have reproduced the boundaries of a national group. Interethnic marriage may be discouraged so as not dilute the "purity" of the nation, and penalties for such behavior can be imposed by the state. Extreme forms of this racialized discourse could be seen in Germany under the Nazi regime. Third, women have played a role in the defining of the nation by acting as the ideological purveyors of the nation's culture through their role as transmitters of the culture. Women, as the main socializers of small children, may acculturate their offspring into the specified ethnic identity and may act as conveyors of the rich heritage of ethnic symbols associated with the nation. Fourth, the ideological construction of the nation around ideas of femininity has been embodied through women. The nation as a motherland to be protected from rape by a hostile group has figured large, particularly in national liberation struggles where men are asked to fight realistically in defense of actual women and metaphorically in defense of a gendered land. Finally, recent research has highlighted the very active role played by women in physical combat. Women play this role as makers of munitions, as nurses and carers for the injured, but also as actual combatants in guerrilla warfare. Although the role of women varies historically and geographically, and women's participation in nationalist discourse is in no way imposed on them totally, the purpose here is to highlight some of the ways in which gender can be incorporated into analyses of nation building.

NATIONALISM AND TERRITORY

Whereas issues of language, symbolic landscape, cultural icon, and gender all play a role in the articulation of a nationalist politics, they also serve to delimit, solidify, or negotiate the boundaries of national space. The marking of the territorial expanse of the nation has been crucial to both the definition and disputation of national borders. Connecting historical processes of nation building with a specific delineated territory has been central to the geography of nationalism. Confirming the link between a specific people and a place, however, has been a hotly disputed topic because the cultural geography of a place rarely represents an ethnically homogeneous piece of land. Consequently, the demarcation of national territory has been fraught with difficulties that at times have resulted in the most violent of territorial disputes. The redrawing of Europe's political map after World War I highlighted the problems of attempting to draw boundaries that would "contain" individual ethnic groups. The territorial dimension of nationalism cannot be underestimated in any discussions of the building of an imagined community of nationhood. The occupation of (and control over) space and the delineation of boundaries have been the source of many regional, national, and international conflicts, and geographers have attempted to explain and understand these disputes.

—*Nuala Johnson*

See also Ethnicity; Identity, Geography and; Nation-State; Political Geography

Suggested Reading

Agnew, J. (2004). Nationalism. In J. Duncan, N. Johnson, & R. Schein (Eds.), *A companion to cultural geography* (pp. 223–237). Oxford, UK: Blackwell.

Anderson, B. (1991). *Imagined communities: Reflections on the origin and spread of nationalism.* London: Verso.

Gruffudd, P. (1999). Nationalism. In P. Cloke, P. Crang, & M. Goodwin (Eds.), *Introducing human geographies* (pp. 199–207). London: Edward Arnold.

Hobsbawm, E., & Ranger, T. (Eds.). (1983). *The invention of tradition.* Cambridge, UK: Cambridge University Press.

Johnson, N. (1995). Cast in stone: Monuments, geography, and nationalism. *Environment and Planning D, 13,* 51–65.

Johnson, N. (2002). The renaissance of nationalism. In R. Johnston, P. Taylor, & M. Watts (Eds.), *Geographies of global change* (2nd ed., pp. 130–142). Oxford, UK: Blackwell.

Williams, C. (2003). Nationalism in a democratic context. In J. Agnew, K. Mitchell,
& G. Toal (Eds.), *A companion to
political geography* (pp. 356–377).
Oxford, UK: Blackwell.

NATURAL GROWTH RATE

Natural growth is the difference
between fertility and mortality rates,
expressed as a rate per 1,000 people
per year. Thus, in addition to net
migration, the natural growth rate
is one of the two major means by
which populations change in size
and composition. For most societies,
natural growth greatly exceeds
migration in importance.

The natural growth rate is a reflection of the complex forces that underlie the social and spatial dynamics of
both fertility and mortality. Some of
these are biological in nature, including genetics, which shapes birth and
death rates and life expectancy. Most,
however, are socially constructed and
include, among other factors, relative
access to food and healthcare, the
quality of drinking water, the structure
of employment, the costs and benefits
of having children, state policies, and
cultural preferences for large or small
families.

Natural growth rates vary widely
over time and space, reflecting the
particular social, political, cultural,
and economic relations of individual
societies. In general, in impoverished preindustrial societies prior to
the modern introduction of antibi-
otics, where both fertility and mortality rates are high,
natural growth—the difference between births and
deaths—tends to be low. Indeed, for most of human
prehistory, and world history until the Industrial
Revolution, natural growth hovered around zero.

Only with industrialization and, in particular, the
development of an adequate and reliable supply of food,

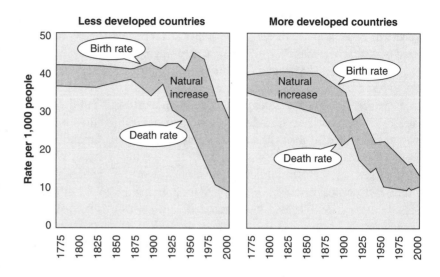

Figure 1   Natural Growth Rates of Developed and Less Developed
Countries

SOURCE: Stutz, Frederick P.; Warf, Barney, *The World Economy: Resources, Location,
Trade, and Development*, 4th Edition, ©2005. Reprinted by permission of Pearson
Education, Inc., Upper Saddle River, NJ.

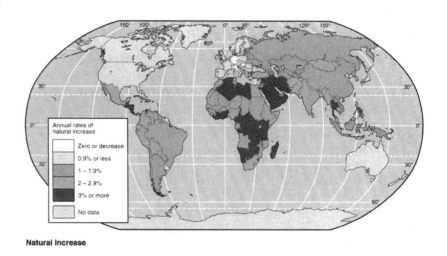

Figure 2   Natural Growth Rates, 2003

SOURCE: Stutz, Frederick P.; Warf, Barney, *The World Economy: Resources, Location,
Trade, and Development*, 4th Edition, ©2005. Reprinted by permission of Pearson
Education, Inc., Upper Saddle River, NJ.

as well as improved drinking water and public health
measures, did mortality rates begin to decline. Thus, natural growth rates are intimately associated with the
demographic transition (Figure 1). In societies where
fertility rates remain high but mortality rates have
dropped, natural growth rates tend to rise markedly.
Throughout the world today, where mortality rates vary

much less than do fertility rates, natural growth is highest in economically developing countries. Thus, in much of Africa, Asia, and Latin America, natural growth rates often exceed 2.5% annually; in some countries, they exceed 4.2% (Figure 2).

However, as fertility rates have declined worldwide over the past five decades, so too have natural growth rates. World average natural growth dropped from 2.6% annually during the 1950s to roughly 1.4% today. In the economically developed world, where birth rates often have dropped below death rates, natural growth rates often approach zero population growth or, as in the cases of Japan and most of Europe, are negative, leading to demographic decline. In such countries, net immigration becomes an important source of population growth and supply of labor.

—*Barney Warf*

See also Fertility Rates; Mortality Rates; Population, Geography of

Suggested Reading

Yaukey, D., & Anderton, D. (2001). *Demography: The study of human population* (2nd ed.). Prospect Heights: IL: Waveland.

NATURE AND CULTURE

Nature and culture have multiple meanings that are related to each other in complex ways. According to cultural critic Raymond Williams, they are two of the most complicated words in the English language. Major subfields and key concepts of human geography, such as cultural ecology, the landscape, environmental determinism, and human dimensions of global environmental change, focus on the intersections between these two concepts. Traditionally, geographers and others have struggled with the question of whether people are separate from nature, above and in control of nature, or subject to the rule of nature. Currently, some human geographers suggest that nature is socially constructed, so profoundly shaped by cultural perceptions, discourse, and physical manipulation that nature has no separate existence from culture. Others accept the idea that nature is partially constructed but explore ways in which to better understand the agency of nature and the materiality of

culture in shaping the social and environmental conditions of our countrysides, cities, regions, and planet.

When nature and culture are understood as separate entities, there are four general meanings of nature and two main approaches to understanding culture. First, nature refers to areas and processes that are separate from culture, external to people, and unaltered by human use, occupation, or transformation. Wilderness, in this sense, is a kind of pure nature. Second, nature is the essential quality or character of something, as in the statement, "By nature, kittens are playful." Third, it is also the whole material world and the processes that make nature function. Fourth, building on these meanings, people sometimes refer to nature as a force apart from human affairs that affects human affairs directly and that offers lessons for how people ought to behave. This is the idea in the concept of "Mother Nature."

The two main approaches to culture stress either its material aspects or its mental aspects, although in practice these ideas overlap. According to the first approach, culture refers to a set of human skills, technologies, and social organization that groups of people use to transform nature in ways that are useful and meaningful to them. For example, the words agri*culture,* api*culture,* and aqua*culture* imply the human ability to apply knowledge and technology to manipulate the natural processes of growth and reproduction of plants, bees, or fish to produce food and other useful items. In this sense, culture also means a set of nature-transforming skills or a particular way of life of a people, a period, a group, or humanity in general. Hunter-gatherers, the ancient Maya, and advanced industrial peoples have different cultures, according to this meaning. By extension, culture also refers to the products of intellectual and artistic activity such as art, music, literature, and architecture.

A second view of culture is less interested in its material aspects and instead stresses the way in which culture structures thought and behavior. According to a now mostly discredited superorganic view, culture exists above the individual, operates through its own internal logic, and strongly shapes individual behavior. A more common view accepts that culture results in patterns of thought and action among the individuals governed by it but sees culture as something malleable and expressive. Individuals create culture through their behavior, for example, hip-hop culture and corporate culture. A semiotic view of culture suggests that it is a collection of signs and symbols, a kind of grammar and vocabulary that makes it possible for people to interpret each other's behavior and to engage in meaningful behavior themselves. In the semiotic view, culture is an acted document, a web of meaning created by people interacting with each other, and a context through which people construct their individual identities and communicate those identities to each other.

The term *culture* sometimes is used synonymously with *civilization,* which is said to be a process of intellectual, spiritual, and aesthetic development in which people leave a state of savagery and progress through a hierarchy from low cultures to high culture. This idea in turn has been criticized for implying a unilinear process that falsely justifies the subjugation and domination of one people over another. A romantic version of the idea switches the low and high ends, such that cultures that are closest to nature are the most harmonious and good and civilization becomes a negative process.

RELATIONSHIPS BETWEEN NATURE AND CULTURE

Nature and culture exist in four main relationships with each other. According to views that maintain nature and culture as separate entities, there are three relationships. First, nature and culture can exist in a harmonious relationship in which nature provides for people's needs while people's activities do not damage nature. Second, nature can dominate culture. For example, environmental determinism is the view that environmental factors can exert an extreme influence on human culture. Availability of ports and wood supplies determines which nations will become seafaring powers. Similarly, the presence of frost-free winters, because of the implications for pest ecology, can be said to limit the productivity of tropical agriculture and contribute to international patterns of wealth and poverty.

In a third general relationship, culture dominates nature. The conquest of nature, in this sense, is a narrative in which the increasing material power of culture transforms nature in ways that make it ever more useful to people. Forest wastelands are transformed to fields and pastures. Footpaths become superhighways for automobiles. Pesticides, fertilizers, and improved seeds lead to higher yields in agriculture. However, the cultural domination of nature is not always seen as a positive thing. When human activities exceed natural limits or break a harmonious relationship, the conquest of nature narrative becomes a destruction of

nature narrative in which human activities cause a backlash. In this version of culture dominating nature, deforestation leads to flooding and soil erosion, automobile use leads to smog and global warming, and pesticides accumulate through ecosystems, resulting in the deaths of birds. Narratives of global environmental change, including species extinctions, ozone holes, global warming, and rising sea levels, often fall under this general understanding of the relationship.

The idea of evolution lends itself to an understanding of cultural influence on natural processes that defy deterministic explanations. Apples could be domesticated only because cultural selection interacted with the existing genes and reproductive characteristics of apples. The genetic variation and reproductive characteristics of oak trees, on the other hand, have precluded domestication to this point.

SOCIAL VIEWS OF NATURE AND THE MATERIAL BASES OF CULTURE

A fourth view of nature–culture relationships holds that nature is not external to people and culture; rather, nature is constructed discursively and materially. The social construction of nature view, for example, holds that all knowledge of nature is situated, contextual, and socially constituted, such that people can know nature only through their culture. The concept of "wilderness," for example, has clear historical and social roots in the United States at the time of the closing of the frontier.

Furthermore, nature and culture are entangled. The high culture of European classical music, for example, requires the continued existence of particular kinds of wood from tropical and temperate forests for the manufacture of musical instruments. On the other hand, even wilderness preserves require a set of social institutions, physical infrastructure, and human activities that "cultivate" endangered species by eliminating exotic species and incompatible human uses. Similarly, actor–network theory challenges the separation of nature and culture by stressing the importance of physical artifacts in networks of people and objects that need both physical and social components to accomplish anything.

Nature can also be said to be social because it is physically produced by people and their culture. This production of nature occurs, for example, when seed companies engineer improved seeds for sale to farmers. Genetic engineering, cloning, and global environmental change all are aspects of a seemingly definitive human subjugation of nature. Even the atmosphere is measurably altered by human activities and subject to international treaties. At an extreme, this production of nature view fits comfortably into the human domination of nature view, but the Marxist concept of the labor process can profoundly challenge the conventional separation of nature and culture. According to this concept, both people and nature participate in the labor process, and both are changed by it. Therefore, culture requires nature to reproduce itself and vice versa; the separation of nature and culture into opposites is an unwarranted abstraction.

These social views of nature emphasize the importance of the material (natural) context for culture. They establish nature and culture as interdependent concepts. Any given culture can exist only under specific natural conditions, including the way in which those conditions have been modified by current and past cultures. The "automobile culture" of the United States, for example, relies on a web of natural and cultural elements, including the existence and exploitation of petroleum reserves, a physical infrastructure of roads and filling stations, the ability of the atmosphere to absorb automobile emissions, and a set of institutions and physical artifacts governing people's use of automobiles ranging from traffic courts to metal stop signs. According to this view, people live in a state of nature even when surrounded by concrete, steel, glass, and plastic. The reverse is also true; even in wilderness areas, people can understand nature only through their own culturally constructed definitions of it.

In studies of robots, zoos, natural parks, river restoration, seed companies, genetic engineering, and the human body, human geography increasingly engages these difficult connections between nature and culture, and it challenges the way in which nature and culture usually are separated in thinking and writing. These challenges to the separation of the natural and the cultural confront conventional ontology—ideas about the character of the world and what things exist.

IDEOLOGIES OF NATURE AND CULTURE

Quite frequently, ideas of nature and culture are mobilized in ways that make specific social relations seem "natural," legitimate, and inevitable. Images of nature facilitate the exercise of power, and situated knowledge of nature expresses power relations. Human geographers frequently deconstruct such views, denaturalizing them by revealing them to be social products.

Although concepts of nature and culture are specific to a given culture, they seem to come from outside of that culture; therefore, these concepts provide powerful legitimating tools for existing social structures. For example, environmental determinism provides a seemingly scientific answer to the question, "Why is the Third World poor?" Environmental determinism explains this outcome as the result of the difficulties of tropical nature that preclude the kinds of agricultural systems and discourage the savings habits of temperate climates. History and international relationships drop out of the explanation. Similarly, concepts of static cultures arranged in hierarchies can be used to legitimate an invasion of another country so as to impose a "higher" culture on the people living in the other country. In the same way, recycling can be said to be natural and right because it mimics the functioning of ecosystems. The problem in these examples comes from the circular ways in which concepts of nature and culture that developed within specific societies are then mobilized to justify the way in which the society is arranged.

Human geographers show how these constructions of nature and culture are implicated in the exercise and justification of power, for example, pointing out how the image of a pristine rain forest erases a history of human occupation and use and, therefore, marginalizes the people currently living in it. Human geographers also point out the way in which understandings of nature and culture usually are gendered, with nature given a passive feminine character and culture given a masculine controlling character. They show how the domination of nature parallels the domination of women in society. In addition, animal geographies challenge the nature–culture dichotomy by suggesting that animals also have societies, have the ability to act and choose, and play roles in shaping history and geography.

Two main problems associated with social views of nature are, first, a possible overemphasis on the power of human society and the inability to appreciate the power of nature to influence social outcomes and, second, social views of nature that have little room for the aesthetic, spiritual, and moral values of nature. If nature is a social construct, why refrain from polluting it? Constructivist views of nature seem to deny that nature can be the grounds for value judgments about appropriate social organization and environmental behavior.

—Dan Klooster

See also Cultural Ecology; Cultural Geography; Environmental Determinism; Political Ecology; Wilderness

Suggested Reading

Castree, N. (2001). Socializing nature: Theory, practice, and politics. In N. Castree & B. Braun (Eds.), *Social nature: Theory, practice, and politics* (pp. 1–21). Malden, MA: Blackwell.

Geertz, C. (1973). *The interpretation of cultures.* New York: Basic Books.

Glacken, C. (1973). Environment and culture. In P. Wiener (Ed.), *Dictionary of the history of ideas* (Vol. 2, pp. 127–134). New York: Scribner.

Gold, M. (1984). A history of nature. In D. Massey & J. Allen (Eds.), *Geography matters! A reader* (pp. 12–33). New York: Cambridge University Press.

Williams, R. (1976). *Keywords: A vocabulary of culture and society.* New York: Oxford University Press.

NEIGHBORHOOD

Neighborhoods are a central institution in the organization of residential space in cities. They function simultaneously as institutional, sociological, economic, political, and geographic entities at multiple levels. At the scale of the individual and the family, neighborhoods exert a major influence over property values and the webs of social relations that tie people to those who live in close proximity to them. Neighborhoods, for example, are important to the spatial organization of friendships, the circulation of information and gossip, social support networks, and the socialization of children. Neighborhoods are important political actors in municipal governance, advocating for the rights of their residents through neighborhood associations of homeowners, coordinating decisions about public goods and services (e.g., tree planting, traffic lights, streetlights, sidewalks), the monitoring of crime, and the enforcement of zoning codes or covenants (e.g., regarding mandatory architectural details). In general, property values tend to be more similar within neighborhoods than among them.

Neighborhoods are both subjective and objective entities. Typically, neighborhoods consist of individuals who share some degree of social similarity in terms of their socioeconomic status, income, class, ethnicity, and (often) stage in the family life cycle. Thus, residents of neighborhoods frequently have similar worldviews and ideologies, political values,

and perceptions about the world. Proximity to like-minded neighbors is a significant part of the presentation of self and the construction of identity. Neighborhoods are psychological as well as socioeconomic entities, one means among several by which individuals are sutured into a community. The degree to which this unity of opinion occurs, of course, varies widely over time and space.

However, neighborhoods function in ways that are neither independent of nor reducible to the psychology of their inhabitants. Neighborhoods are a domain in which people may act as citizens—not simply consumers—and be involved in civic society (e.g., charity work, volunteering). Politically, neighborhoods are a major means through which local opposition to unwanted land uses is mobilized, for example, NIMBY ("not in my backyard") groups that may seek to prevent a land use deemed undesirable (e.g., a toxic waste dump, a home for mentally retarded people) from being located near their houses.

A classic perspective on neighborhoods articulated by Louis Wirth maintained that urban residential space was organized around three fundamental axes. First, neighborhoods reflect a common socioeconomic dimension. Their residents tend to belong to the same class and to have similar or related occupations, similar educational levels, and incomes that are not too different from one another. Through their ties to the urban division of labor, residents face similar restraints on housing affordability (whether high or low) and, having been socialized along similar lines, have similar preferences in terms of housing age, style, size, and location. Second, neighborhoods reflect similar stages in the family life cycle. For most people, the demand for residential space varies predictably with age. For example, young adults in the pre-child stage may prefer apartments, young parents typically require more room for young children and may desire single-family homes, and parents in the post-child stage may opt for smaller living quarters. Moreover, incomes often vary in tandem with family life cycle stage and thus affordability of housing.

Third, neighborhoods typically are marked by a similar ethnicity that operates in several ways to group together people of a similar racial background or cultural heritage. Historically, the Chicago School noted the formation of immigrant communities as a vital part of the American urban landscape, socializing newcomers to their new society and giving rise to a rich array of organizations concerned with jobs, entertainment, crime, and access to political resources. This pattern continues to one extent or another today. Ethnicity may reflect the desires of particular ethnic groups to share their culture in a common space, including access to culturally specific foods, language, friendships, and support networks. Ethnicity may also reflect external constraints on residential choices, including formal and informal discrimination in the housing and labor markets. People of particular ethnic groups often are channeled into particular income strata, a reflection of the historical experience of particular groups and their varying degrees of social mobility, education, access to public resources, and so forth. Jewish ghettos are an example of this tendency in European history, and African American communities exemplify this pattern in light of the long history of racial segregation characteristic of American cities.

Neighborhoods have changed markedly in nature and significance over time. Traditionally (i.e., until roughly the 1920s), they revolved heavily around face-to-face interaction and the sharing of tacit information in the forms of local friendships, acquaintances, and other networks. Because they were intimately associated with the rhythms of everyday life, neighborhoods often were well defined and very important to the social interactions of their residents. Most people were acquainted with their neighbors, watched one another's children, attended a common church, walked to work, and shared a common sense of identity as a community. For some, the world outside of the neighborhood often was a dimly visible horizon of experience.

By the 1920s, however, the meaning and role of neighborhoods began to change, particularly in the American context. The introduction of the automobile, the steady exodus of the middle class to single-family homes in the suburbs, and the use of telephones in lieu of face-to-face contact all gradually served to dissipate the traditional coherence of urban life. By the 1950s and 1960s, the steadily increasing trends toward commodification, individualism, and withdrawal from public life had begun to take their toll. The commodification of household functions tends to substitute formal networks for informal ones, decreasing the significance of neighborhood ties. Many central-city neighborhoods, particularly those suffering the strains of deindustrialization, rising unemployment, and poverty, began to dissolve and fragment. Some neighborhoods in suburban areas re-created the amenities and advantages originally found in the inner city, although generally in muted and subdued forms. There is no reasonable ground for suspecting that low-income neighborhoods cannot be as rich, functional, and supportive as higher-income ones, and indeed in some cases they may be more cohesive (e.g., when residents spend more time away from cramped living quarters and with one another). However, the social predicaments and stresses accentuated by poverty (e.g., crime) do tend to decrease the likelihood of residents mobilizing around common interests except under dire circumstances.

Neighborhoods never are fixed entities; they change constantly due to various circumstances. The aging of residents, changes in the price of housing, relative rates of investment in the housing stock, the influx of new residents, and shocks precipitated by the ever fluctuating urban division of labor all steadily rework the fabric of local communities. The Chicago School theorized ethnic change in neighborhoods as a process of invasion and succession similar to that found in plant ecosystems. More contemporary and realistic views hinge on the ways in which broad demographic forces, economic restructuring, and public policy conspire at multiple spatial scales to differentially alter the opportunities and constraints faced by people who occupy different parcels of the city.

—*Barney Warf*

See also Chicago School; Ghetto; Home; Housing and Housing Markets; Neighborhood Change; NIMBY; Segregation; Urban Geography

Suggested Reading

Keating, W., Krumholz, N., & Star, P. (Eds.). (1996). *Revitalizing urban neighborhoods.* Lawrence: University Press of Kansas.

Nelson, R. (2005). *Private neighborhoods and the transformation of local government.* Washington, DC: Urban Institute Press.

Wirth, L. (1938). Urbanism as a way of Life. *American Journal of Sociology, 44,* 1–24.

NEIGHBORHOOD CHANGE

Urban geographers are attuned to changes that occur at spatial scales smaller than that of the city. Observation and research have shown that neighborhoods often change quite rapidly and in a variety of ways. The type of change and the extent of change depend on the situation of the particular neighborhood in relationship to the metropolitan area as a whole and on any particular attributes that neighborhood may possess such as waterfront or higher terrain. Neighborhood changes demonstrate the impact of larger processes, whereas the sum of such changes modifies the pattern and structure of the entire metropolitan area.

Although threads of change often work together, it is useful to separate out changes in the physical structure, changes in socioeconomic conditions, and changes in population composition. This discussion concerns cities within the United States, although similar patterns can be observed in other contexts as well.

CHANGES IN PHYSICAL STRUCTURE

Urban and suburban neighborhoods are defined visually by their physical stock—infrastructure, commercial and industrial buildings, apartments, and houses. The urban land use model, associated primarily with economist William Alonso, saw the type, size, and density of physical structures as a function of the value of land. High-value land, traditionally near the center of the city, would attract offices and flagship stores in tall buildings. Lower-value land, traditionally near the periphery, would attract mostly low-density residential housing. As a city expanded and as access increased, the spreading zones of commercial, industrial, and residential structures would cause the physical structures within individual neighborhoods to change.

Historically, neighborhood change at the urban edge has been marked by high levels of new construction, particularly houses laid out at lower densities. More recently, outer suburbs have also attracted commercial, industrial, and apartment structures. This indicates that the nature of access has changed as superhighway intersections and airports define new locational advantages. Nonetheless, aspects of physical structure once associated with central business districts—high densities, tall commercial buildings—are found in these so-called edge cities.

Negative changes in land value have also influenced neighborhoods in the inner city. Here the process is one of increasing blight as existing physical structures are undermaintained and eventually may become abandoned. Sometimes the structures are burned down and the land is overgrown with weeds. Such neighborhood decay is particularly acute in those cities that are experiencing dramatic population

loss. A less dramatic effect is the process of filtering. Neighborhoods go through a life cycle whereby the housing progressively ages. Although some neighborhoods may retain or even increase their value with age, in many cases an aging housing stock attracts progressively poorer inhabitants as wealthier residents seek out better and newer housing. When the intrinsic value of the property exceeds the sum it is able to command in the marketplace, a rent gap emerges.

CHANGES IN SOCIOECONOMIC CONDITIONS

Closely associated with the changes in a neighborhood's physical structure are the changes in its social and economic situation. One famous model of urban form, developed by sociologist Ernest Burgess, related increases in neighborhood social status with distance from the city center. A second model, developed by Homer Hoyt, suggested that neighborhoods at separate economic levels were arrayed as wedges that began near the city center and then spread out to the city edge. Later, sociologists and geographers helped to spearhead the growth of social area analysis and factorial ecology that looked at how patterns related to income, life cycle, and ethnicity were layered on each other. Although these models are interesting generalizations, variations in socioeconomic status depend on many things. So too does the extent to which the socioeconomic condition of a neighborhood changes.

Geographers have been interested in tracing the patterns and reasons behind neighborhood socioeconomic change. One item of concern has been the increase in poverty and poverty-related indicators within many inner-city neighborhoods. Unlike during past decades, when poverty and increasing density went hand in hand, much of the current poverty is related to the exodus of population, especially of the middle class, from these same neighborhoods. The remaining population is characterized by high levels of social deprivation. Another factor that can affect the socioeconomic condition of a neighborhood pertains to the labor market or the accessibility of nearby employment opportunities. In many instances, there has also been an exodus of jobs from inner-city locations. Some other factors involved in shaping the socioeconomic conditions of neighborhoods include the placement of public housing and the lack of financial investment.

Situations of more positive socioeconomic change can occur when the existing population within a neighborhood becomes more prosperous or, much more likely, when a neighborhood begins to attract wealthier residents. This latter process, known as gentrification, has attracted a great deal of notice, although it still occurs in only a sprinkling of neighborhoods. By and large, the direction of migration, especially among wealthier residents, is from the city and inner suburbs to the outer edges of metropolitan areas.

CHANGES IN POPULATION COMPOSITION

Over the past 200 years, American cities have grown and also have become more ethnically diverse. The increased diversity has a great deal to do with national trends given that the United States has attracted immigrants from a wider array of places. It also has much to do with the decided proclivity among most such immigrant groups to settle within urban areas; throughout U.S. history, cities have been more diverse than rural areas and small towns. Within cities, particular neighborhoods have been most affected by both immigration and internal migration.

Urban America, during the early 1800s, was primarily a white, Anglo-Saxon Protestant domain. African American and Native American peoples were overwhelmingly rural. This began to change between 1820 and 1925, when between 33 million and 45 million people entered the United States from abroad. Although the influx was predominantly European, these newcomers brought with them different languages, religions, and lifestyles. The tendency of many immigrant groups was to settle in cities. Among the first wave of this European immigration, Irish Catholics and roughly half of the German arrivals found their way into cities. Late-19th- and early-20th-century European immigrants were overwhelmingly urban in their orientation. Roughly seven of eight Jews, Italians, and Poles resided in cities at a time when the United States was mainly rural. Moreover, the bulk of immigrants settled within inner-city neighborhoods that previously had been occupied by more established groups. Distinct residential quarters, marked by a particular ethnicity, began to take shape.

Following World War I, black Americans from the rural South moved to northern cities in what is termed the *Great Migration*. The black population of many northern cities soared, and African American newcomers often were channeled into a restricted set of inner-city neighborhoods, some of which had housed immigrant groups a generation prior. Since 1965, when many immigration restrictions were lifted, immigration has resumed. Now immigrants hail from countries in Latin America and Asia. Cities remain hugely popular, although now southern and western cities have joined northeastern cities as destinations of choice. Whereas many newcomers settle in inner-city neighborhoods, and in some cases this has led to a revival of urban growth, others have moved straight to the suburbs. What is remarkable today is the extent to which neighborhood ethnic change is occurring in places on the outskirts of the metropolitan area.

Concerns with ethnic neighborhood change have centered on the causes and consequences of segregation in both residential and business activities. Early theories posited a dominance and succession model whereby new ethnic groups came into a neighborhood and eventually displaced the existing population, whose members likely would move out into better areas. In such a scenario, segregation levels for particular groups likely would fall. The experience of the African American urban population altered this view somewhat. Segregation levels rose, and public housing—rather than dispersing the African American population— served mainly to reconcentrate it. With the newest groups, we are likely to see different outcomes. Locations are more varied, and even within inner-city neighborhoods, ethnic concentration has allowed for a total remaking of the neighborhood landscape.

—*David H. Kaplan*

See also Central Business District; Chicago School; Edge Cities; Exurbs; Housing and Housing Markets; Neighborhood; Suburbs and Suburbanization; Urban Ecology; Urban Fringe; Urban Geography; Urban Sprawl

Suggested Reading

Foner, N., & Fredrickson, G. (Eds.). (2004). *Not just black and white.* New York: Russell Sage.

Ford, L. (1994). *Cities and buildings: Skyscrapers, skid rows, and suburbs.* Baltimore, MD: Johns Hopkins University Press.

Kaplan, D., & Holloway, S. (1998). *Segregation in cities.* Washington, DC: Association of American Geographers.

Smith, N. (1996). *The new urban frontier: Gentrification and the revanchist city.* New York: Routledge.

NEOCOLONIALISM

Essentially, neocolonialism refers to the perpetuation of colonial economic relations after formal independence.

The long centuries of European colonialism drew to a rapid close after World War II, when the West's self-destruction provided a window of opportunity for nationalists in Asia and Africa. From the late 1940s to the mid-1970s, the number of ostensibly independent countries multiplied rapidly. (Latin America had achieved this state following the Napoleonic Wars of the early 19th century.)

However, despite the promises of modernization theorists that patience and the acceptance of Western capital ultimately would lead to social and economic development, many in the developing world, and some in the West, exhibited concern that long-standing colonial relations remained de facto, if not de jure, even though the former colonies had attained formal independence. Neocolonialism differs from colonialism, therefore, in the sense that the former colony has attained a nominal degree of political sovereignty. A new national flag or anthem in many respects did little to change the status for most of the population or the real relations of power between the former colony and the former colonizer. Poorly prepared for independence, many former colonies had little preparation to compete in the global economy and suffered from inadequate infrastructures, unskilled labor, a lack of managerial skills, and insufficient investment capital. Many had artificial borders that forced very different ethnic groups together, sewing the seeds for secessionism and tribal wars. In practice, most former colonies could offer the global economy little more than cheap raw materials (e.g., mineral ores, foodstuffs) and cheap unskilled labor. Thus, Western powers—typically the same ones that had once formally colonized them—retained effective economic control if not direct, formal political oversight.

The role of the United States is central to neocolonialism. Although the United States itself was a product of colonialism, by the late 19th century (following the Spanish-American War of 1898) it had increasingly become a colonial power in its own right. After World War II, when the United States remained unquestionably the leading economic superpower in the world, the age of the *Pax Americana* became virtually synonymous with neocolonialism. This relation reflected the dominant role of the United States politically and militarily, the enormous influence of American popular culture and media (which essentially rendered postwar globalization the same as Americanization), and the huge influence of American multinational corporations. Thus, despite its pretense to serve as a fount of world freedom, from

the perspective of neocolonialism, the American empire more closely resembles its European antecedents.

Multinational corporations or transnational corporations (TNCs) often are held to be the prime agents underlying neocolonialism. The vast bulk of them originate in the developed world, where large pools of investment capital exist, and many have output levels that exceed small or even intermediate-sized countries. The postwar rise of multinationals dramatically altered global investment and trade flows, as well as government policy, around the world. Today, some 30,000 TNCs control two thirds of world trade and employ 100 million people. Although some TNCs originate in Europe, Japan, and Canada, American ones are by far the largest group. Whereas most such firms invest in other developed countries, their presence in the developing world has mixed economic costs and benefits, including job generation and technology transfer but also displacement of local suppliers. Profits generated by the investments of such firms frequently are repatriated to their origin countries. Bank loans to developing countries, and the ensuing global debt crisis, are another form of foreign economic domination. Moreover, TNCs have a long and unsavory history of political interference, ranging from demands for tax relief to support for military coups d'état. States in the developing world that dared to confront multinationals often have faced the military wrath of the United States (e.g., Iran in 1953, Indonesia in 1965, Chile in 1973). Since the 1990s, the International Monetary Fund and the World Bank often have been accused of playing roles that aid and abet international capital at the expense of the citizenry in developing countries through neoliberal programs that tie debt restructuring to the privatization of public goods (e.g., airports, telecommunications firms, hydroelectric plants), deregulation, currency devaluations, and reductions in public assistance to low-income people. The local ruling class within developing countries often actively assists TNCs, global capital, and international institutions in this process.

The net effect of foreign economic domination, and often de facto political domination, is often to retard the economic growth of developing countries. Trapped in a world economy where the rules are stacked against them, many states in Asia, Africa, and Latin America continue to export raw materials and cheap labor and have experienced relatively little industrialization. The growth of some East Asian newly industrializing countries offers an exception to this trend. For most countries in the developing world, however, neocolonialism involves unequal exchange—the means by which surplus value is siphoned from poor countries to rich ones—in the form of poor terms of trade, that is, exporting low-priced outputs and importing expensive manufactured ones. Thus, what neoclassical economists typically label as *interdependence* may be seen as disguised exploitation and loss of economic sovereignty despite the nominal appearance of political independence. Efforts to promote free trade, for example, take on the appearance of promoting the interests of wealthy powerful countries at the expense of impoverished ones.

In sum, neocolonialism may be defined as colonialism in practice if not in name. The political mechanisms may be different, but the intent and outcome are identical to those of old-fashioned imperialism. In this light, colonialism never really ended; it only assumed a new form. The perpetuation of inequality between the world's core and periphery—between the developed and underdeveloped countries—attests to the general tendency under capitalism to reproduce uneven spatial development despite the nominal independence of the underdeveloped countries. Thus, the existence of neocolonial ties is a major impetus behind the growth of radical theories of global political economy such as dependency theory and world systems theory.

—*Barney Warf*

See also Anticolonialism; Capital; Colonialism; Core–Periphery Models; Debt and Debt Crisis; Decolonization; Dependency Theory; Marxism, Geography and; Neoliberalism; Transnational Corporations

Suggested Reading

Chomsky, N. (2003). *Hegemony or survival: America's quest for global dominance.* New York: Metropolitan Books.

Harvey, D. (2003). *The new imperialism.* Oxford, UK: Oxford University Press.

Johnson, C. (2004). *The sorrows of empire: Militarism, secrecy, and the end of the republic.* New York: Metropolitan Books.

NEOLIBERALISM

Neoliberalism is a political project that entails a set of particular economic and political ideologies. This theory is called neoliberalism because it is based, in part, on 18th-century liberal economic theory. Economics generally is concerned with how human needs and desires are satisfied when resources are not sufficient to meet everyone's needs. Based primarily on theoretical economic models, neoliberalism is a theory of how to structure economies through free markets. Traditionally, a market was the physical space, frequently a city square, where people gathered to trade goods. Markets work by bringing people who want to buy together with people who have something to sell. They can be literal, such as a corner store, or virtual, such as online shopping. The term can also be used to describe the daily processes of buying and selling in general without reference to any particular marketplace or commodity.

The basic tenet of both liberal and neoliberal economic theory is the strong belief in, and preference for, free markets that are allowed to self-regulate without outside interventions. Neoliberal ideology holds that the market should dictate the rules of society and not vice versa. From this premise, it follows that for markets to function properly, all state and collective interventions in the market must be removed. For example, state-instituted measures to lessen unemployment would be considered to be detrimental to the market because they would impede on the market's ability to self-regulate, with the idea being that unemployment may actually be in the best interest of the market. Neoliberalism currently is a very important and influential economic theory, one that shapes the daily lives of countless people across the globe.

KEY CHARACTERISTICS

The neoliberal economic philosophy is characterized by an absence of interventions in economic markets. Neoliberalism is also characterized by an emphasis on the individual and the importance of individual initiative and distaste for all collective efforts. The neoliberal social theory and policies that accompany neoliberal economics advocate for limited or no social interventions such as state-funded welfare and healthcare. A flexible labor force, one that is not permanent and cannot make claims for benefits but is readily available, is preferred. Part-time employment characterizes the neoliberal economy as many former full-time workers lose their jobs in favor of part-timers.

HISTORY

Adam Smith is considered the father of free market (or liberal) economics. In 1776, he published *The Wealth of Nations,* in which he wrote that the most efficient economies are created when all interventions are removed, thereby creating an unrestrained market. This free market ought then to be fostered by keeping it free of regulation. Smith's free market economics was very popular until the era of the Great Depression (1930s), when many people experienced intense poverty and called for a better way in which to control the economy and respond to the needs of an increasingly impoverished citizenry. Between the 1930s and the early 1980s, the liberal model was replaced by Keynesian economics, which was named for economist John Maynard Keynes and called for state regulation of the economy and constant interventions to control poverty and unemployment.

Keynesian economics was employed to rebuild the United States and Europe after World War II and represented the governing policies of the public works programs through the New Deal. As Keynesian policies worked to decrease the gap between the rich and the poor, they also put limits on trade of which corporations and the financial elite did not approve. It was in response to the large-scale market regulation of the Keynesian system that the current neoliberal order developed. Neoliberalism as an economic theory started at the University of Chicago with the work of philosopher and economist Friedrich von Hayek and his student Milton Friedman. From this beginning, neoliberalism has become a global presence supported by organizations such as the International Monetary Fund (IMF) and the World Bank. Two of the first proponents of neoliberalism were U.S. President Ronald Reagan and British Prime Minister Margaret Thatcher. Thatcher's TINA ("there is no alternative") campaign referred to neoliberal economic theory and its attending social policies as the only possible course for economic advancement. Internationally, neoliberal economics has had tremendous effects across the developing world. Embodied in structural adjustment programs, neoliberal policy is used as a requirement for procuring funds from both the World Bank and the IMF.

CRITIQUES

Many critics of neoliberalism view the economic and political ideologies of neoliberalism as based in theory and not in reality. They argue that on the ground, neoliberalism creates an ever widening gap between the rich and the poor and increasingly high levels of global poverty. Critics believe that if the market is left alone, it will only make the rich richer and the poor poorer; they believe that intervention is necessary to create equality. Neoliberalism is criticized for the manner in which it blames individuals for their economic problems and refuses to examine more systemic issues such as racism. Neoliberalism is important for geography because many of the current global flows of people and money are based on neoliberal theories of economics and politics and many powerful people believe neoliberalism to be the solution to many of the world's problems.

—*Rebecca Dolhinow*

See also Economic Geography; Globalization; Structural Adjustment

Suggested Reading

Ellwood, W. (2002). *No-nonsense guide to globalization.* London: Verso.

Klak, T. (Ed.). (1997). *Globalization and neoliberalism.* Lanham, MD: Rowman & Littlefield.

NEURAL COMPUTING

Neural computing refers to any of a number of computational methods that involve emulation of neurological function in human or animal brains. The most widely applied method of this kind is the artificial neural network (ANN), often simply referred to as neural net. Other methods include image processing, pattern recognition, and cellular automata. ANNs are modeled loosely on neurons in the brain and share an ability to generalize from specific information and to learn. They are applied principally to classification problems.

An ANN is a simple model of brain function consisting of an interconnected set of neurons. Each neuron has a number of inputs and outputs and converts a given combination of input signal levels to a defined signal output level. Typically, signal output levels are a weighted sum of input signal levels, often with a threshold applied so that the end result is a binary output regardless of the exact weighted sum of the inputs. This simple model of a neuron was first proposed by Warren McCulloch and Walter Pitts during the 1940s. The arrangement of interconnections between neurons usually is layered. An input layer receives signals derived from observational data. Each layer in a series of layers of neurons receives input from the previous layer, with its outputs feeding into neurons in the next layer. Many interconnection patterns are possible, but the details are relatively unimportant from an application perspective.

In classification tasks, a network may operate in either supervised mode or unsupervised mode. Supervised networks are trained on data where the desired classifications are known, often from ground truth empirical observations. In this mode, the weights associated with each input to each neuron are iteratively adjusted to produce the best possible match between the desired outputs and the network outputs. After training, the network can be applied to other data sets where the applicable classifications are not known. Unsupervised networks are similar to classical classification procedures such as clustering analysis.

The most common application of neural nets in geography is to the classification of remote sensed imagery. In this context, data from a number of channels form the input data, and land cover classifications form the desired output data. Training data are composed from regions of the study area for which reliable observational data are available. Although this is the most frequent application, it is possible to apply ANNs to classification problems in human geography or where the goal is estimating or predicting the likelihood of a particular outcome based on a number of factors.

In common with more traditional classification methods, an ANN maps each combination of input variables onto a classification. However, whereas traditional methods are restricted to simple mathematical combinations of input variables, ANN classifications assume nothing about their relative importance, enforce no distributional assumptions on data, and do not assume that linear combinations of variables are inherently more useful than complex nonanalytic functions. Nonreliance on analytic functions is both a strength and a weakness of ANNs. On the one hand, it admits the possibility of complex mappings from inputs to outputs that can reflect subtleties in observational data that are not readily represented by linear mathematical functions. On the other hand, the output from ANNs is not easily summarized. This has frequently led to ANNs being seen as "black box" solutions; that is, although they might work, they might not provide much insight into problems.

—*David O'Sullivan*

See also GIS; Humanistic GIScience; Limits of Computation

Suggested Reading

Hewitson, B., & Crane, R. (Eds.). (1994). *Neural networks: Applications in geography.* Dordrecht, Netherlands: Kluwer Academic.

Openshaw, S., & Openshaw, C. (1997). *Artificial intelligence in geography.* New York: John Wiley.

NEW INTERNATIONAL DIVISION OF LABOR

The new international division of labor (NIDL) refers to social and spatial changes in the demand for and organization of labor that began during the late 1960s, when patterns of investment and production shifted from being organized primarily at the national scale to being organized primarily at the global scale. The term was coined by Fröbel, Heinrichs, and Kreye during the 1980s to describe shifts in the location of German industrial investment away from specialized industrial zones within Germany toward regions within the global periphery with an abundance of low-cost labor. The NIDL marked a situation that was significantly different from the classical division of labor espoused by classical political economists Adam Smith and David Ricardo. Whereas Smith argued for the separation of the labor process into specialized tasks to maximize productivity in Britain's industrializing economy, Ricardo built on these ideas in his theory of comparative advantage to argue that countries should specialize in the production and trade of goods that were relatively (rather than absolutely) lower in cost. Together, these theories supported an international division of labor where workers in the Third World almost exclusively produced raw materials used in First World industries, whereas workers in the First World primarily produced the manufactured goods consumed in the Third World. Under the NIDL, the old international division of labor was transformed as an increasing number of transnational corporations (TNCs) began to establish production in the Third World that was geared primarily for markets in the First World. Whereas in some cases production represented a change in the firms' internal division of labor, in other cases it represented a shift in the location of the firms' production away from traditional First World locations.

Fröbel and colleagues identified three factors behind the shifting orientation of firm manufacturing production away from domestic markets toward global ones. First was the presence of a virtually inexhaustible supply of labor in the Third World that was extremely cheap, easily mobilized for production, and easily replaced. Second was the ability of firms to fragment the production process into tasks that could be carried out by labor with minimal skill. Third was the existence of transportation and communications technologies that could facilitate the partial production of manufactured goods in multiple geographic locations. Whereas technological innovations reduced the cost of transporting bulky goods from sites of final production and consumption, changes in communications also allowed firms to exert greater managerial control from a distance. These three factors were integral to the ability of firms to subdivide the commodity production process into fragments located in any part of the world with the most profitable combination of capital and labor. These factors, however, should also be understood as part of the general intensification of competition among corporations in response to the slowdown in productivity and declining profits during the 1960s. This is a situation that a number of writers have attributed to the general crisis of the Fordist system of production, consumption, and income distribution that characterized most advanced industrial economies from the 1950s through the 1970s.

The changing international division of labor has had a tremendous impact on labor in different geographic locations. Within the industrial heartlands of Europe and North America, the early shifts in investment triggered a process of relative economic decline in regions that once specialized in the production of manufactured goods. Throughout the 1980s, regions that were associated most closely with Fordist systems of production, such as the northeastern United States, northern England, and the Ruhr in Germany, tended to experience pronounced levels of deindustrialization as firms that once employed large numbers of skilled and semiskilled blue-collar workers terminated their operations. One of the immediate consequences of deindustrialization, therefore, was the reemergence of structural unemployment and stagnation that culminated in a decrease in real incomes in many industrial core countries. It is estimated, for example, that between 1969 and 1976, the mid-Atlantic region of the United States lost 1.5% (or 175,000) of its jobs, whereas between 1966 and 1976, net losses in the number of manufacturing jobs in the United Kingdom exceeded 1 million.

Within the global periphery, the shifting international division of labor produced mixed results. In much of East Asia and in a few Latin American countries, the NIDL generated opportunities for export-oriented capitalist development. Countries such as South Korea, Hong Kong, Singapore, and Taiwan all experienced increased flows of capital and investment across their borders as transnational manufacturing corporations sought new sources of low-cost labor. In many of these countries, this inflow of foreign investment was crucial to the "East Asian Miracle," the phrase used to describe the rapid industrialization of these economies. But in other countries, the growth of TNC factory production has been associated with the feminization of a poorly paid, non-unionized, exploited workforce.

The need for productive and easily disciplined labor has been explained as a reason why the changing spatial division of labor has simultaneously encompassed a change in the gender composition of the labor force. The feminization of labor in Third World factory production is a striking feature of the NIDL. From as early as the 1960s, when American corporations began to set up manufacturing and assembly plants in the Mexican border region under the state-run *maquila* program, a clear feature was the growing incorporation of young women into these factories. Young women often are considered ideal workers for assembly production and in many factories constitute the majority of the workforce. This is because they are viewed to be more likely to maintain high productivity and quality levels in the execution of repetitive tasks and are less likely to engage in organized labor action. Women are viewed to allegedly have such nimble fingers because the skills required to carry out the assembly tasks in garment and electronics manufacture are viewed to be similar to the everyday skills required in the home. It is argued that because the skills needed for factory production are perceived to be extensions of the skills used for unpaid work in the home, wages in these global factories are especially low. The International Labor Office, for example, noted that despite the fact that employment and output increased in Mexico between 1975 and 1993, the real incomes and purchasing power of workers fell by 20%. So although the globalization of factory production has brought new jobs for many women in Third World countries, these jobs often have done little to alleviate poor working and living conditions.

In explaining the reasons for the internationalization of production, Fröbel and colleagues emphasized the importance of low-cost labor, but this reason has been criticized as being overstated. To the extent that TNCs continue to locate production operations in places where labor is relatively costly, this suggests that additional factors may be at play. Thus, even in countries that appear to be sites of low-cost labor, other factors such as government incentives may play a role in explaining the location decisions of TNCs. In many Third World countries, for example, government provision of tax and customs duty holidays or subsidized plant and machinery within export processing zones may be even more attractive than the existence of a low-cost labor force. Similarly, in some places the presence of a specialized workforce might be an important factor in determining where investment is located. As services have grown in importance to economic growth in both First World and Third World economies, so too has the internationalization of service work. For example, the growth of India's software industry represents the latest phase in the NIDL. Today, both Third World and First World software TNCs are beginning to compete for highly skilled but relatively low-cost information technology workers in both the First World and the Third World, and this is creating opportunities for highly concentrated flows of investment into cities such as Bangalore, India, and Dublin, Ireland.

—*Beverley Mullings*

See also Dependency Theory; Development Theory; Division of Labor; Economic Geography; Export Processing Zones; Globalization; Import Substitution Industrialization; Modernization Theory; Neocolonialism; Transnational Corporations; Uneven Development; World Economy; World Systems Theory

Suggested Reading

Corbridge, S. (1986). *Capitalist world development*. Lanham, MD: Rowman & Littlefield.

Fröbel, F., Heinrichs, J., & Kreye, O. (1979). *The new international division of labour*. Cambridge, UK: Cambridge University Press.

Harvey, D. (2003). *The new imperialism*. Oxford, UK: Oxford University Press.

Pandit, K., & Casetti, E. (1989). The shifting pattern of sectoral labor allocation during development: Developed versus developing countries. *Annals of the Association of American Geographers, 97*, 329–344.

Peet, R. (1987). *International capitalism and industrial restructuring*. Boston: Allen & Unwin.

NEW URBANISM

New urbanism is a collective term for an array of planning and design practices originally developed to respond to dissatisfaction with contemporary urbanization processes, especially those linked to urban sprawl. This dissatisfaction included the separation of land uses (a precept of traditional zoning), the domination of daily life by the automobile, and the loss of a sense of community. New urbanism has been seen optimistically by some commentators as an answer to problems of giantism and a lack of attention to environmental and economic sustainability. In the early 21st century, new urbanism has become well known as one possible response to the impacts of automobile-driven suburbanization, edge cities, and alienation. It is, however, not without its critics, who have seen new urbanism as a veneer over the organization of space of capitalism that replicates race and class divisions within contemporary society, even as it embeds them in a utopian vision of communities of the future. These critics, many of whom are geographers, have cast doubt on the progressive visions of some advocates of new urbanism. It should be stated at the outset that new urbanism has evolved as a complex and nuanced phenomenon. Its status in contemporary urbanization processes continues that evolution.

The formal concept and practices of new urbanism developed in the United States during the past 25 years or so. The genesis occurred mainly in the professional practice of a small number of architects and designers struggling to overcome the perceived negative impacts of modernist land use planning and design. New urbanism has become an important facet of urban design and architecture and visions of contemporary community development. Indeed, advocates for, and practitioners of, new urbanism have influenced zoning and subdivision ordinances at the municipal scale. By the late 1990s, this influence had extended to the enabling legislation for planning and development in many states and had permeated federal urban policy in the context of both housing and urban redevelopment. New urbanism in a variety of guises has also become a feature of development and redevelopment processes in other core countries in the world economy.

Geographers and other social and behavioral scientists have been drawn into renewed and energetic discussions of the relationship between physical design and the processes by which people simultaneously create places and are influenced by them. In addition to these discussions, new urbanism has attracted attention in the academic literature of several other disciplines and in popular literature. Popular media have included several books and articles in widely read magazines. Peter Katz's book *The New Urbanism,* James Howard Kunstler's book *The Geography of Nowhere,* articles and discussions in *The Atlantic Monthly* and *Time,* and a cover story in a 1995 edition of *Newsweek* all are good examples of the latter trend. Whereas myriad developments in the United States and other countries are associated with new urbanism design principles, one community—Seaside Florida—attained iconic status when it formed the backdrop for the feature film *The Truman Show.*

ORIGINS AND INSTITUTIONS

Most observers see new urbanism emerging from critical discussions of planning and urban design by actual practitioners. The term *new urbanism* was predated by discussions of neotraditional neighborhood planning and transportation/pedestrian-oriented design. It was argued that both could address the much-lamented loss of community in the suburbs and, from some perspectives, in marginalized inner cities. Elements of these early practices included attempts to humanize the scale of new developments, increase their density and compactness, encourage the mixing of land uses and types of dwellings, expand public spaces and pedestrian environments, and thereby encourage social diversity and a sense of community. Many designs sought to replicate an earlier area of model towns and neighborhoods from the 1920s as well as an idealized image of community from some European countries. Many were strongly influenced by historic preservation, which also had become a force to be reckoned with in the evolution of architecture and planning during the 1970s.

Two widely acknowledged foundational influences on the evolution of new urbanism were the efforts of Peter Calthorpe in California and of Andreas Duany and Elizabeth Plater-Zyberk in Florida. Calthorpe is most associated with pedestrian- and transit-oriented developments, whereas Duany and Plater-Zyberk are best known for the development of private greenfield communities. Although clear ideological differences existed, few would disagree that there was a synthesis and agreement on some foundational thinking

regarding urban processes. This common ground led to significant and important momentum for the new urbanist critique of modernist planning and design. Key way stations were the development of the Ahwanee Principles in 1991 and later the establishment of the Congress for New Urbanism (CNU). The CNU has developed codes of practice and annual awards for best practices. It represents a clear and visible touchstone for those interested in new urbanism as a phenomenon. In broader terms, discussions and examples of new urbanism appear on myriad Web sites for architectural and design practices and in publications and on Web sites for advocacy organizations such as the National Association of Homebuilders. It should be made clear, however, that although new urbanism has gained an important niche in many dimensions of urban development and redevelopment, large-scale suburban and exurban housing and commercial developments that comply with traditional zoning and subdivision codes appear to continue to be the norm.

CONCLUSION

New urbanism will continue to be a focus of geographic analyses of the processes of urbanization. Empirical analyses of developments on the urban fringe, as well as in the transitional zones of inner cities, will be complemented by critical discussions of the interactions between design and planning professionals and the array of engaged parties, both state and nonstate, as part of the sociospatial dialectic of this facet of contemporary capitalism.

—*Richard Kujawa*

See also Urban Geography; Urbanization

Suggested Reading

Calthorpe, P., Corbett, M., Duany, A., Moule, E., Plater-Zyberk, E., & Polyzoides, S. (n.d.). *The Ahwahnee Principles* [Online]. Available: www.lgc.org/ahwahnee/principles.html

Congress for New Urbanism. (n.d.). [Online]. Available: www.cnu.org

Falconer Al-Hindi, K., & Till, K. (Eds.). (2001). The new urbanism and neotraditional town planning [special issue]. *Urban Geography, 22*(3).

Harvey, D. (1997, Winter–Spring). The new urbanism and the communitarian trap. *Harvard Design Magazine,* pp. 66–69.

NEWLY INDUSTRIALIZING COUNTRIES

The term *newly industrializing countries* (NICs), also known as *newly industrializing economies,* refers to a diverse set of Asian and Latin American nations as well as broader processes of global industrial restructuring that took hold during the latter half of the 20th century. NICs emerged during the 1950s and 1960s within a broader context of industrial promotion in developing nations. Such promotion shared a historical parallel with the Economic Commission on Latin America's call for import substitution industrialization (ISI). This strategy aimed to increase industrialization (mainly in Brazil, Argentina, Chile, and Mexico, where development strategies were called "inward-oriented" strategies) through a series of tariffs that would protect nascent industries from goods manufactured in Europe and the United States. ISI mapped out a production strategy that would evolve from the domestic production of consumer goods, then intermediate goods, and ultimately capital goods. Such a normative progression proved to be elusive for most developing nations, and the NICs' so-called miracle economies came to be the exception to remarkable industrial growth rather than the norm.

As a geographic category, the term *newly industrializing countries* refers to those developing nations whose industrial output increased quickly over a short period to contend with manufacturing and steel producers in North America and Europe. The late 1970s witnessed a significant increase in export-led growth from these nations that went beyond ISI by producing more advanced, technologically mature, and finished manufactured items in transport equipment and machinery categories. This production also signaled a shift away from processing raw materials, including basic manufactured goods and chemicals. By the new millennium, the NICs had developed a larger domestic and global market share that was focused increasingly less on Europe, Japan, and the United States and more on the Chinese and East and South Asian markets.

As a manifestation of industrial restructuring, the NICs reflect changes in the new international division of labor that unchained industrialization outside the North Atlantic region, Japan, and Australia. Many NICs specialize in communication and transportation technologies and rely on low-cost and mostly nonunionized labor. Initially, many relied on heavy state investment. Technology-enhanced production (e.g., computer-aided design), special public and private partnerships, and a focus on quality control have further secured the output of NICs in the global consumer durable markets. Increasingly, informatics products—especially computer chips and components and plasma (LCD) and high-density screens—constitute a key industrial sector.

Although no formal membership in the list of NICs exists, analysts during the 1960s began identifying the "Four Dragons" or "Asian Tigers" as South Korea, Hong Kong, Taiwan, and Singapore—nations where the governments often were intolerant of worker strikes or unionization and where ISI typified outward-oriented strategies. Accordingly, these rapidly growing economies accelerated after 1973 when the world economy was especially dynamic. However, their growth has not been driven by unfettered market forces. Instead, close government coordination of industrial production, as well has heavy-handed tactics in disciplining labor and other components of civil society, characterized these developmental states where the national governments continue to play a key role.

The locus of successful NIC development is concentrated in Asia. After Japanese rule and a grueling war with North Korea, South Korea changed from relying on imports during the 1950s to the local manufacture of basic consumer items such as food processing, clothing, footwear, and textiles. Its Economic Planning Board enlisted a series of economic and social plans that led to investment in large steel, shipbuilding, and automobile firms. By the 1970s, large private-sector consortia (*chaebols*) developed important vertical and horizontal linkages while emphasizing the exports of labor-intensive products. A period of secondary ISI that ensued during the following decade emphasized skill-intensive and higher value-added products.

Taiwan was also a Japanese colony that possessed a significant industrial base after World War II. As an "authoritarian corporatist" state, it tended to emphasize smaller firms than did South Korea. Taiwan passed through broad development phases that shifted from the light industries of ISI (e.g., food processing, textiles) to exports of electronics, textiles, and intermediate goods. Since the late 1980s, steel, telecommunications, and computers have dominated Taiwanese manufacturing.

Singapore transitioned in 1965 from a British colonial outpost and a city-state relatively deprived of natural resources to a nation of intense human capital investments in the financial, informatics, publishing, and manufacturing sectors. Labor-intensive production helped to reduce high unemployment and spur a parallel path of social and economic programs at an unprecedented level among East Asian nations. These industrial and social programs attracted foreign direct investment (FDI) and raised domestic savings (through a mandatory savings schema). It has become an international banking center that weathered the 1997 Asian crisis that plagued many of its neighbors. However, Singapore increasingly competes with China for new sources of capital.

China's centrally planned economy shifted course after the death of Mao Zedong in 1976, with the open policy of 1979, and then again with the revised 1997 constitution. Special economic zones receive targeted FDI from Europe and North America as well as from ethnic Chinese living outside of the mainland. Southern Guangdong Province, with its propinquity to Hong Kong, has especially benefited from these market liberalizations over the past decades. Whereas state-owned firms produced roughly three quarters of industrial output during the 1970s, they produce roughly one quarter of it today. Affiliation with the World Trade Organization has meant a drop in tariffs and will test China's production capacity on the world market.

Although the original classification of just four NICs remains a subject of debate, Indonesia, the People's Republic of China, and other nations occasionally are associated with the NICs. By the 1990s, there emerged references to a set of sub-NIC nations (e.g., Malaysia, the Philippines, India) whose low-cost industrial output started to rival the original Asian Tigers. The economic successes of the NICs in a rapidly globalizing economy call into question conventional free market assumptions about the role of private sector–led industrialization as well as labor mobility theses that suggest that workers' ability to organize and negotiate wages with management are prerequisites to industrialization.

NICs face the same challenges that industrial nations faced during the early 2000s as the global downturn in production and consumption slowed industrial growth. FDI strategies and a diverse global investment environment eagerly scout out new areas of manufacturing investment, especially in the less skilled sectors of textiles and food processing and in markets where labor is more docile. Sustaining growth equity continues to challenge NICs, although the Asian NICs have achieved a more equitable distribution of income than have their Latin American counterparts. More robust industrial production among the Asian NICs has also eliminated the level of debt rescheduling that has dogged Mexico, Brazil, and Argentina. The NICs demonstrate that their broad economic base offers an edge against many development problems that afflict countries outside of the traditional industrial spheres of North America and Australasia.

—*Joseph Scarpaci*

See also Dependency Theory; Development Theory; Globalization; Import Substitution Industrialization; Modernization Theory; New International Division of Labor; Transnational Corporations; Uneven Development; World Economy; World Systems Theory

Suggested Reading

Dicken, P. (2003). *Global shift: Transforming the world economy* (4th ed.). New York: Guilford.

Gereffi, G. (1990). International trade and the industrial upgrading in the apparel commodity chain. *Journal of International Economics, 48,* 37–70.

Smith, M., McLoughlin, J., Large, P., & Chapman, R. (1985). *Asia's new industrial world.* London: Methuen.

NIMBY

NIMBY is the acronym for "not in my backyard," a characteristic (if stereotyped) slogan of neighborhood and community groups opposed to locally unwanted land uses (LULUs). Typically, NIMBY movements arise in opposition to perceived environmental threats such as toxic waste dumps, trash incinerators, recycling centers, and landfills. At the state level, they may oppose nuclear power plants and potential transportation routes for trucks or trains carrying dangerous chemicals. During the 19th century, urban locational conflicts erupted over the siting of slaughterhouses, rendering plants, and saloons. NIMBY movements may also form as coalitions against social (as opposed to environmental) categories of land uses that they deem as undesirable such as shopping malls, prisons, bridges or tunnels, low-income housing projects, transit systems, homeless shelters, drug rehabilitation centers, and halfway homes for retarded people. NIMBY movements reflect the spatial distribution of undesirable effects on people's welfare, such as impingements on their health, noise, and fears of crime or visual and aesthetic blight (typical concerns of low-income NIMBY movements), or negative effects on their property values (a frequent motivator of middle-class NIMBY movements). Other NIMBY arguments are that LULUs will destroy small-town environments or strain local public resources. NIMBY tactics may include lawsuits, working the legislative and judicial machinery, protests and demonstrations, public relations campaigns in the media (e.g., letters to newspapers), and behind-the-scenes pressure on elected officials.

The strength of NIMBY movements varies largely in accordance with the socioeconomic status, educational level, and financial resources of its members. Low-income minorities generally have less access to political power in most municipalities, whereas white middle-class communities are more likely to have the ear of city government officials and be better positioned to oppose unwanted land uses through the courts or bureaucracies. Because decision makers siting noxious land uses (i.e., large real estate developers) are more likely to attempt to locate them in predominantly minority and low-income areas, such communities are more likely to give rise to NIMBY movements, often under the banner of environmental justice. Because local issues affect people's everyday lives, often in profound ways that speak to their deepest hopes for and fears about their future quality of life, NIMBY movements can attract members who are otherwise usually disengaged from formal politics such as housewives.

Critics of NIMBY movements maintain that they are elitist and parochial, hamper necessary development, exhibit a so-called drawbridge mentality, and/or covertly attempt to maintain neighborhood racial homogeneity under the banner of opposition to other unwanted aspects such as noise and pollution. Some argue that NIMBY movements are ethically inconsistent, simply attempting to displace LULUs to less politically powerful areas; that is, NIMBY movements displace the problems rather than solve them. In this reading, NIMBY movements are irrational, selfish, misguided, and obstructionist, and they prevent the attainment of societal goals by privileging local interests over social needs. After all, facilities such as waste incinerators must be built *somewhere*. Thus, critics of NIMBYs maintain that they elevate local benefits over broader social ones.

A structuralist interpretation of NIMBYs points to the inequalities inherent between the production and consumption of urban space, particularly the spatial distribution of negative externalities in the forms of locally concentrated costs and dispersed social benefits. Rather than viewing local resistance as irrational opposition to land developers, some view it as an inevitable outcome of the urban development process. NIMBYs represent consumers of land who are invested in a particular landscape under the threat of change, and their opposition to new facilities is a means of constraining the behavior of capitalists, who often enjoy the backing of the state. Thus, they can be no more annihilated than can private capital investments in the landscape, and they provide a necessary countervailing measure to land developers who would otherwise proceed unchecked.

—*Barney Warf*

See also Home; Locally Unwanted Land Uses; Resistance; Urban and Regional Planning; Urban Geography; Zoning

Suggested Reading

Heiman, M. (1990). From "not in my backyard!" to "not in anybody's backyard!" *Journal of the American Planning Association, 56,* 359–362.

Lake, R. (Ed.). (1987). *Resolving locational conflict.* New Brunswick, NJ: Center for Urban Policy Research.

Lake, R. (1993). Rethinking NIMBY. *Journal of the American Planning Association, 59,* 87–93.

Ley, D., & Mercer, J. (1980). Locational conflict and the politics of consumption. *Economic Geography, 56,* 89–109.

Meyer, W., & Brown, M. (1989). Locational conflict in a nineteenth century city. *Political Geography Quarterly, 8,* 107–122.

NOMADISM

Nomadism is a way of life in which peoples (nomads) move from place to place, often in a cyclic manner, without any fixed abode. They subsist as hunter-gatherers, pastoralists, or traders and move on when there is a reduction in the local resources. Hunter-gatherers, such as the Pigmies of the Congo and the Inuit of the Arctic, move as small extended-family groups within a well-defined area in which they are familiar with the location of resources such as food plants and potential food animals. Hunter-gatherers usually avoid places where they might come into conflict with dominant sedentary populations.

Pastoral nomads specialize in domesticated animal herding and move seasonally to locate places with good predictable pastures and a supply of water. The Rwala Bedouin, for example, move south in the winter to find pastures on the northern edge of the Nafud desert in Saudi Arabia, which is watered by rain-bearing winds tracking in from the Mediterranean. In the spring, they move back north into southern Iraq as the desert pastures become parched by the summer heat.

Pastoral nomads herd goats, sheep, camels, horses, and cattle, often in sensitive marginal environments not suitable for permanent settlement. Uncontrolled livestock numbers have led to widespread overgrazing and soil erosion around encampments, as in Saudi Arabia. In general, the more marginal the environment, the more often the herd must be moved on, although things are slowly changing. In the Middle East, the bedouin camp now has a water tanker that is filled from wells or local village supplies, and so good grazing becomes the single factor determining movement.

Although pastoral nomads depend on their animals for food, their animals are inextricably linked to social status and cultural practices. The number of livestock in the herd, for example, typically is a marker of social status and ultimately power. Pastoral nomads are not completely self-sufficient and come into contact with sedentary populations to trade animal products for goods such as cereals, oil, salt, and rice.

A third type of nomadic lifestyle is that of the peripatetic traders, that is, entertainers, craftsmen, fortune-tellers, acrobats, and casual workers who travel from village to village seeking customers. They meet the limited demand for services and goods that the settled community does not provide for itself. The gypsy travelers of Europe have mostly managed to maintain this type of nomadic existence in spite of the pressures to settle.

All nomads are opportunistic and resilient. Their mobility gives them advantages over settled communities in the face of environmental hazards, but in most parts of the world the nomadic lifestyle is in decline as more and more nomads seek jobs, education, and healthcare.

—*Peter Vincent*

See also Agriculture, Preindustrial; Cultural Geography; Developing World

Suggested Reading

Leonard, W., & Crawford, M. (2002). *Human biology of pastoral populations.* Cambridge, UK: Cambridge University Press.

Spooner, B. (1998). *The cultural ecology of pastoral nomads.* Reading, MA: Addison-Wesley.

NOMOTHETIC

Beginning with Fred Schaefer's famous attack on what he called "exceptionalism" in 1953, the central claim of the idiographic approach that history and geography are concerned only with the unique aspects of individual places, the discipline began a transition into a nomothetic body of knowledge, one that sought general laws of explanation independent of time and space. Whereas idiographic understanding sought to uncover all the aspects of one place, nomothetic understanding sought to reveal how one phenomenon varied among many places.

The idiographic–nomothetic debate, also known as the systematic versus regional geography debate, raged throughout the 1950s and early 1960s. This clash of views essentially involved the question as to whether geography should be involved in general laws of understanding (i.e., the relative emphasis on regional differences vs. regional similarities, the factors that differentiated places vs. the commonalities that ran through them). Thus, the shift from an idiographic geography to a nomothetic one was closely (but not exclusively) associated with the broader decline in the regional approach. This move involved the triumph of the abstract over the concrete, the general over the particular, and the universal over the specific. Nomothetic approaches valued abstraction and empirical regularities and held that the empirical world existed solely for the purpose of testing theory.

Rejecting the empiricism of the idiographic approach, nomothetic forms of geography have long been associated with the attempt to make the discipline more "scientific" in the same sense as the physical sciences. Thus, advocates of the nomothetic approach exhibited a disdain for induction and a sustained concern for the role of theory and explanation. This view centered on a sharp fact–value distinction, rigorous methods of data collection and sampling, deductive (also called nomological) logic, hypothesis testing, quantitative methods, reproducible results, and predictive ability. These tools are held to uncover the logical structures that underpin the play of empirical surface appearances. All of these were hallmarks of the philosophy of logical positivism. In this account, the process of explanation involves embedding the unique within the general, that is, showing an individual set of observations to be the outcome of wider principles at work in a variety of places. A law is held to be a hypothesis that is repeatedly confirmed under controlled conditions. Choosing among competing laws involves invoking Ockham's scalpel, that is, choosing the simplest approach. Methodologically, nomothetic approaches were closely associated with the rise of statistics and mathematical models that were central to the rise of spatial analysis and location theory. Epistemologically, nomothetic views relied on a Kantian view of absolute space rather than on later relativist notions of space as socially constructed.

Critics argued that this approach effectively reduced geography to geometry, ignoring historical context, social relations, and the role of human consciousness. Moreover, not all regional details could be swept under the carpet of general theory so easily, as the revived localities school maintained.

—*Barney Warf*

See also Idiographic; Logical Positivism; Model; Theory

Suggested Reading

Berry, B. (1964). Approaches to regional analysis: A synthesis. *Annals of the Association of American Geographers, 54,* 2–11.

Harvey, D. (1969). *Explanation in geography.* London: Edward Arnold.

Schaefer, F. (1953). Exceptionalism in geography: A methodological examination. *Annals of the Association of American Geographers, 43,* 226–249.

NOT IN MY BACKYARD

See NIMBY

ONTOLOGY

The term *ontology* is used in a large number of diverse fields. An ontology is, in its broadest sense, a systematic or formalized description of accepted properties and characteristics that relies on distinct institutional, social, and technical conventions. In geography, it represents a nexus of intellectual activities, most significantly philosophy and computer science. Its significance comes in no small measure through its role in the facilitation of information exchange and sharing in computer networks (interoperability). Examples of geographic ontologies include public transportation, noise emissions, and map features. Through systematization and formalization, these examples help facilitate the exchange of information about multiple activities that often are stored in different computer systems. Different disciplinary understandings of the term have led, in geography and geographic information science (GIScience), to people distinguishing "big Ontology" from "little ontology."

The general concept of ontology can be compared to a synopsis of a community's language and understanding of the world around the community. Each community's members pursue similar activities, but their environment influences how they conduct activities and what relationships they have with a variety of activities. An Arctic Inuit group understands the world its members experience in different ways from a Xhosa group in Southern Africa. Each group has an ontology that reflects the community's shared set of knowledge. The specifics of fishing activities are different (e.g., needing to cut a hole in the ice), but the systematization of the ontology leads to a clear definition of specific activities being part of fishing.

This "small *o*" concept of ontology has a practical orientation that accounts, to some degree, for different meanings of things but tends to describe functions. Functions offer much insight into what people do but make it hard to distinguish the means from the ends, especially means that are implicit such as the paper needed to write a letter and the sharpened stone needed to break through ice. For work with computers, this issue means that an ontology can oversimplify and possibly ignore key aspects of activities that give them significance. Rather than merely functions, small *o* ontologies should be taken as specification of a conceptualization. However, although this considers activities broadly, it remains limited to what is needed and desired for a particular conceptualization. Returning to the fishing example, an ontology of the Inuit conceptualization of fishing may be perfectly adequate for Inuit communities but not for Xhosa or other indigenous fishers dwelling in more temperate areas.

Philosophers have raised this question and focused on this "big *O*" issue, attempting to systematically, logically, and rationally describe the penultimate meaning of objects and activities. The philosophical concept of Ontology involves determining the essential characteristics and actions of fishing and describing the universal traits and activities of fishing. This concept originated with Plato's view that the human mind is chained in a cave and can only perceive reflections of the world outside of the cave. Big *O* ontology aims to describe that world.

In spite of these differences, each approach requires the systematic and formal representation of knowledge. In either approach, an ontology is independent of the agent's internal representation, for example, presented in a language and terms that a person outside of the agent's community can understand. Small *o* ontologies are called conceptualizations, which are abstract simplified views of the world to represent for a purpose and are explicit.

Because small *o* ontologies are far more prevalent in geography and GIScience, some relevant details are in order. First, most ontologies use a joint terminology that can be used by a number of disciplines and applications. An ontology usually uses a representation language based on first-order predicate logic, that is, logic suitable for algebraic expression in which predicates take only individuals as arguments and quantifiers bind only individual variables. Because a domain can be manipulated algebraically, terms and relationships in a domain can be described as axioms and put into relationship with each other.

Ontologies rely on commitments, especially agreements to use the shared vocabulary. These agreements are crucial to the success of ontologies. Different knowledge of the same activities or environmental processes and characteristics is possible, and different answers to queries are possible. An example of such a problematic query is the question, "What are all the towns in the Chicago area?" For some people, the Chicago area may be limited by political affiliations to any regional governing associations. Other people may consider the Chicago area to be economically defined. Without more knowledge about the context, these queries can be impossible to answer—a substantial problem for ontologies developed for particular conceptualizations.

A number of XML-based representation languages, including Ontolingua, Loom, and Frame Logic, are used for describing artificial intelligence ontologies that can address these problems. Other languages provide resources for developing and supporting Web services by connecting representation language to particular approaches to model, access, and construct relationships with other ontologies. Examples of these languages include Simple HTML Ontology Extensions (SHOE), Ontology Exchange Language (XOL), Ontology Markup Language (OML and CKML), and Resource Description Framework Schema Language (RDFS).

As many uses of geographic information systems (GIS) move more toward Web services' models (e.g., when a visitor to a national park using a car navigation system transparently loads map, road, and attraction information from separate Web servers), ontologies and these languages become critical to the successful combination of information. DAML Map is a sample application showing what is possible to do by merging different shareware or copyleft software packages and accessing different data sources.

In summary, ontologies define terms, query operations, and relationships. In networked environments with heterogeneous data sources, ontologies become crucial for the systematic and formalized description of geographic information. This makes it possible, using the Internet and semantic Web ideas and concepts, to access geographic information from multiple sources and to combine it as the need arises. Most ontologies are axiomatic frameworks for knowledge representation that require commitments and agreements to ensure that different groups use the ontologies' common vocabulary.

—*Francis Harvey*

See also Cartography; GIS; Humanistic GIScience; Spaces of Representation

Suggested Reading

Ashenhurst, R. (1996). Ontological aspects of information modeling. *Minds and Machines, 6,* 287–394.

Guarino, N., & Giaretta, P. (1995). Ontologies and knowledge bases: Towards a terminological clarification. In N. Mars (Ed.), *Towards very large knowledge bases* (pp. 25–32). Amsterdam: IOS Press.

Harvey, F. (2003). Knowledge and geography's technology: Politics, ontologies, representations in the changing ways we know. In K. Anderson, M. Domosh, S. Pile, & N. Thrift (Eds.), *Handbook of cultural geography* (pp. 532–543). London: Sage.

Kuhn, W. (2001). Ontologies in support of activities in geographical space. *International Journal of Geographic Information Science, 15,* 613–631.

OPEN SPACE

Open space, a concept employed to offset or counterbalance unchecked urban expansion, refers to the conservation of landscapes retaining characteristics of presettlement environments, pastoral agricultural lands, or restored areas meant to re-create or mirror such landscapes. Open spaces typically are formed or protected at the local level through the neglect of commons, the establishment of private land trusts, conservation planning, or active rehabilitation of abandoned or contaminated brownfields. The intent of advocates often is to keep these areas free of development in perpetuity.

Open spaces have been championed as having a wide array of benefits that not only accrue to the surrounding human population but also strengthen the integrity of local environmental and biotic systems. Keeping tracts of urbanized land undeveloped may allow for the preservation of environmental pockets reflective of the natural state of the environment or physical landscape. Ecologically, these areas have the potential to enhance biodiversity, aid in the conservation of endangered or threatened animal and plant species, and provide greenways or greenbelts that allow for the movement of larger, more mobile animal species. They are also valued for their ability to contribute to flood control, minimize erosion and mass movement, and provide protection from fire hazards. Open spaces often are used as hubs for recreation and sport by the nearby urban population as well.

Aesthetic considerations are also powerful motivating factors in the preservation of many open spaces. Open spaces are highly valued for their accessibility to the public, for their ability to provide a nearby "natural" experience in the middle of an otherwise concrete jungle, and as an important source of aesthetic beauty and scenic viewscapes that may provide affective and cognitive benefits to urban residents. Another valued asset of open spaces is their ability to buffer zones of dense development creating a patchwork or mosaic of urbanized landscapes interspersed with undeveloped areas. However, open spaces are not without their detractors. Critics claim that the removal of valuable urban land from development results in a weaker tax base and fragmentation of city services and also encourages higher-density development elsewhere in the urban landscape.

Although open space may be naively construed as natural environments, especially in contrast to the exaggerated built environment of cities, this categorization becomes problematic when the ideologies with which such landscapes are invested are examined closely. Preservation of pastoral or agricultural landscapes has been recognized by geographers as being as much a part of the cultural heritage of a landscape as a reflection of the natural environment. Restored, rehabilitated, or neglected landscapes are also invested with cultural meaning as their role, utility, and character are generated through interaction with the surrounding population. In this way, geographers may also view open spaces as unique urban places.

—*Michael Urban*

See also Conservation; Cultural Landscape; Nature and Culture; Political Ecology; Urban Sprawl; Wilderness

Suggested Reading

Garvin, A., & Berens, G. (1997). *Urban parks and open space.* Washington, DC: Urban Land Institute and Trust for Public Land.

ORIENTALISM

Orientalism has its origins in late literary theorist Edward Said's 1978 book by the same title. *Orientalism* is widely considered to be one of the most influential books of the 20th century, and its influences cross the humanities and social sciences. Said analyzed the writings and representations of Western European authorities on the region of the world they categorized as "The Orient," with his particular interest focused on what most geography books would label Southwest Asia, the Middle East, and the Near East. Said's central claim was that this broad body of work had scripted a notion of the Orient as an exotic "other," both repulsive and intriguing and unconnected to the long sweep of human cultural development that became—as a result of these Orientalists, Said argued—Europe's to claim.

What came to be referred to in this way as the Orientalizing of the Near East and its peoples was, Said contended, central to the imperialist projects of the West. By erasing the connectivity of civilizations and cultures of the region from the West's story and representing them as an exotic, bizarre, and inferior appendage, the Orientalists made colonial conquest a natural and logical extension of the rise of the West. Influenced by French philosopher Michel Foucault, Said sought to suggest that the Orientalists' discourse on the region had over time erased the real Orient or any alternative notion of regional identity.

Orientalism can be taken as one of the foundational texts of postcolonial studies, leading literary theorist Robert Young to consider Said to be one of the three main scholars (with Gayatri Chakrovarty Spivak and Homi Bhabha) to shape that field's development. It has created many imitators and spawned a growth industry of applications of its basic notions to other parts of the world where European imperialism has, these works claim, Orientalized the people it has conquered or erased. Scholars have taken the concept well beyond its literary origins and into the critical rereading of the representations of places found in postcards, art, architecture, and maps, among other devices. These latter categories captivate geographers, perhaps for obvious reasons, and it is no surprise to see many geographers influenced by this idea. The central ideas of Orientalism have, for instance, been deployed in important geography scholarship such as Derek Gregory's *The Colonial Present.* Gregory sought to show how the Western powers have reproduced and extended the Orientalist scripting of the Middle East in three contemporary conflict settings: Afghanistan, Palestine, and Iraq.

Orientalism has hardly been without its critics. Some critics seek to discredit the work and all that has come in its wake because they do not separate it from the political project of Palestinian human rights that was, by virtue of his long and bitter exile from his homeland, Said's life work. Other critics argue that his book is notably absent of the possibilities for voices of resistance within Middle Eastern countries to this othering process. Bhabha prominently extended Said's claims, even while criticizing them. In demonstrating that the connections between imperialist or colonialist rhetoric or discourse and realities on the ground often could be found wanting, Bhabha provoked a storm of interest in the muddled places in between the colonizer and the colonized. Bhabha pointed a host of scholars to the ambivalence, hybridity, and mimicry found in colonial representations of Orientalized places and, at the same time, to the same phenomena in the self-representations of colonized peoples. The latter thought is the jumping-off point for subaltern studies scholarship, commonly associated with Spivak. Subaltern studies scholars have problematized the question of the capacity for colonized peoples, such as those that the Orientalists critiqued by Said were writing about, to write back to the colonizers and subvert the discursive representations of them.

Although most of Said's work bears the profound influences not only of Foucault but also of cultural materialist theorists with strong Marxist credentials

such as Italian philosopher Antonio Gramsci, perhaps the harshest criticisms of *Orientalism* have come generally from Said's left, notably in the work of Aijaz Ahmad. Ahmad and other Marxist critics decry the work and the outpouring of scholarship it influenced for an alleged absence of grounded historical materialist class analysis. These critics bemoan the emphasis on deconstructing the discourse of texts rather than on the material consequences of European domination in the Middle East, alongside the sense that critics such as Ahmad have of Said's idealism about underlying causes in the region's subjugation.

Said himself replied to his critics through various expansions on the utility of the ideas he had crystallized in *Orientalism.* His responses are found in a number of publications, but *Culture and Imperialism,* published in 1993, is probably the most useful one for human geographers—in part because geography plays a crucial role in his arguments therein. Said acknowledged that the claims he made about the Middle East had now been extended to other parts of the world, even while he accepted that he had, in *Orientalism,* left out the whole story of the responses of colonized peoples to both the discursive and material tactics of imperialists. He defended the analysis of literature as a critical dimension of the materiality of imperialism, not only historically but also in contemporary times, because the discursive representations of Western outsiders, he argued, could be shown to have direct material links to policy outcomes such as the first Gulf War. Gregory's book has carried the claims of Said (now deceased) forward to the second Gulf War (the Iraq War).

—*Garth Myers*

See also Anticolonialism; Colonialism; Eurocentrism; Ideology; Imperialism; Nationalism; Other/Otherness; Postcolonialism; Subaltern Studies

Suggested Reading

Ahmad, A. (1992). *In theory: Classes, nations, literatures.* London: Verso.

Gregory, D. (2004). *The colonial present.* Oxford, UK: Blackwell.

Said, E. (1978). *Orientalism.* New York: Pantheon.

Said, E. (1993). *Culture and imperialism.* New York: Knopf.

Young, R. (1995). *Colonial desire: Hybridity in theory, culture, and race.* London: Routledge.

OTHER/OTHERNESS

The ideas of the "other" and "otherness" have associations with psychoanalysis, structuralist and post-structuralist theory, and postcolonial studies. Many scholars of psychoanalysis take note of the work of the human brain, alongside its own internal divisions (often seen as the conscious mind and unconscious mind), in making some form of a basic distinction between the self and the beings outside. In what is termed *object relations theory,* it is posited that a person learns as a very young child to see herself or himself as unconnected to the mother—as a distinct being. The fears and terrors that this realization brings with it cause the child to displace her or his feelings onto others. Psychoanalytic theories suggest that the outside segment of the binaries—self/other and same/different—often is feared, loathed, or held as inferior. Thus, people often seek to expel, reject, abject, or exclude what is taken as other, outsider, or different, for instance, people who are out of place from where the mind's prevailing order wants them. The term *othering* often is used for these exclusionary processes. These processes never quite succeed, according to many psychoanalysts, leading to a perpetual struggle for most selves between repulsion from otherness and desire for otherness.

Along with this inner world that in many ways remains geography's last terrain for exploration, it seems to be a fairly basic step of taking this work of unsuccessful policing of separations between binaries of self/other or us/them outside of the head. Surely, one of human geography's most fundamental reasons for existence lies in helping people to sort out areal difference—which places are the same or similar and which places are different or other. Societies often seek to separate same from other, whether the dividing lines be based on race, class, gender, or other categories. Like the processes in our heads, these social processes of separation have ambivalent outcomes. People in one place, of course, can be construed as different from people in another place; otherwise, human geographers would not have much work to do. The problems that keep human geographers employed hinge on how those differences are constructed, manipulated, and deployed at differing levels of a society's power structure and on just how incomplete or unsuccessful the constructions, manipulations, or deployments are.

Cultural anthropology, like human geography, relies on the basic idea that people differ from one place to another for a significant portion of its raison d'être. Johannes Fabian's study, *Time and the Other,* suggested that modern anthropology needed to contend with a considerable history of othering its objects of study by a pronounced focus on those who many Westerners conceive of as exotic or primitive. In Fabian's view, the field, by essentializing culture—boiling down differences to these supposedly exotic or primitive traits—reinforced and extended stereotypes that debilitated efforts toward cultural understanding.

Postcolonial studies, however, has troubled any neat separation between self and other, or between us and them, common to conventional theories in cultural anthropology or earlier human geography. On the one hand, Edward Said's *Orientalism* opened a whole field of analysis of how othering tactics served the interests of colonialism and imperialism; indeed, therein lies the impetus for Fabian's work discussed previously. Similarly, the group of Indian historians known as the subaltern studies collective sought, in effect, to rewrite South Asian history from the point of view of the other, albeit articulated in their works as the subaltern or subordinate classes. For another example, in the work of Timothy Mitchell, it can be seen that colonial cities defined themselves by what they were not or what they excluded from their midst. There is an other side to every city in the colonial imagination. Yet things are not nearly so simple, postcolonial studies scholars suggest, and as Mitchell's work shows, the exclusions seldom (if ever) produced the binaries that were intended. In particular, cultural studies scholar Homi Bhabha stressed the ambivalence of colonialism and mimicry that took place on both sides, leaving scars on either end of the encounter. The very otherness of the other often proves as desirable and alluring as it is alien and disgusting.

Poststructuralist or postmodern thought outside of the formerly colonized world also makes great use of the idea of the other or otherness. French philosopher and historian Michel Foucault helped to redirect a great deal of scholarly attention away from more conventional (i.e., elite) subjects of history toward marginalized groups in Western societies. Despite Foucault's very diverse subjects for investigation, each of his works generally shared his fresh approach to ideas of power that suggested that power did not flow in a neat top-down manner in societies but rather took diffuse capillary forms. Foucault and other post-structuralist French thinkers such as Jacques Derrida have been very influential in human geography, but these influences in this particular sphere of otherness have also been criticized. Notably, David Harvey worried that poststructuralism could lead to a kind of championing of difference that might take scholars down a thousand blind alleys on searches for power's diffusion—searches that would obscure what Harvey saw as the material sources of power located in the upper circuits of global capitalism. By focusing attention on inner worlds or the allure of difference, many critics (e.g., Harvey) charge, the power dynamics of political economy often can be shortchanged.

—*Garth Myers*

See also Postcolonialism; Subaltern Studies

Suggested Reading

Fabian, J. (1983). *Time and the other: How anthropology makes its object.* New York: Columbia University Press.

Mitchell, T. (1988). *Colonizing Egypt.* Cambridge, UK: Cambridge University Press.

Pile, S. (1996). *The body and the city: Psychoanalysis, space, and subjectivity.* London: Routledge.

Schecht, S., & Haggis, J. (2000). *Culture and development: A critical introduction.* Oxford, UK: Blackwell.

Sibley, D. (1995). *Geographies of exclusion: Society and difference in the West.* London: Routledge.

OVERLAY

Overlay is, for most people, one of the most definitive operations of geographic information systems (GIS). Overlay makes it possible to combine vector and/or raster data for any areas based on a common coordinate system. The analytical capabilities of overlay make a critical contribution to geographic analysis in many fields.

However, overlay is a late arrival to geographic techniques and methods. The history of overlay prior to the late 1960s remains veiled in ambiguity. The technical and conceptual basis finds a parallel in the development of offset printing introduced during the early 20th century. The first published reference to the technique is from a 1913 landscape architect's plan for Billerica, Massachusetts. The cost and complexity of conducting overlays by hand led to geographers using representative fractions to numerically indicate relationships. For the analytical integration of soil types and crops in Montfort, Wisconsin, the possibility of overlaying transparent maps of each property was discussed, but until GIS became widely available, the expense and cartographic complexity dissuaded the pursuit of overlay as a technique for geographic analysis.

Ian McHarg's presentation of overlay, published in *Design With Nature* in 1969, involves the use of overlay to superimpose transparent thematic maps drawn in darker colors as the value associated with the property changes over an area. Overlaying layers of natural features, social features, and engineering considerations for transportation planning produces a composite in which darker colors indicate higher values and more conflicts for planning. Lighter areas in the composite map show areas with lower values and fewer conflicts.

The concept of overlay remains complex. The colocation of two properties is invaluable for geographic analysis, but the significance of colocation is subject to many considerations. The ingenuity of McHarg's use of overlay masks problems with ensuring that measurements of properties can be meaningfully overlaid, which logical operations reflect geographic relationships, and reoccurring questions about how properties are valued. The accuracy of overlay is a constant concern.

Overlay techniques have been greatly advanced since McHarg's seminal work and are combined with logic selection operations to process geographic information. Raster overlay has become analytically richer, and vector overlay has been integrated into database systems. Overlay techniques remain an important interface metaphor operation for integration operations that now are processed entirely by database software with limited geometric intersection processing. For all of the advances in implementation, overlay still requires a substantial amount of interpretation. Geostatistical techniques have become important complements and alternatives to overlay.

—*Francis Harvey*

See also Cartography; GIS; Humanistic GIScience

Suggested Reading

Bailey, R. (1988). Problems with using overlay mapping for planning and their implications for geographic information systems. *Environmental Management, 12*(1), 11–17.

Finch, V. (1939). Geographical science and social philosophy. *Annals of the Association of American Geographers, 29*(1), 1–28.

Manning, W. (1913). The Billerica town plan. *Landscape Architecture, 3*(3), 108–118.

McHarg, I. (1969). *Design with nature.* New York: Natural History Press.

Steinitz, C., Parker, P., & Jordan, L. (1976). Hand-drawn overlays: Their history and prospective uses. *Landscape Architecture, 66,* 444–455.

P

PARADIGM

The concept of paradigms originally was developed by Thomas Kuhn in his landmark publication *The Structure of Scientific Revolutions,* first published in 1962. The term *paradigm* is generally used to describe (a) a normative framework that a science imposes on itself and (b) the instability and historical progression of theoretical assumptions and methodological procedures that a discipline experiences in more or less regular intervals (commonly referred to as a *paradigm shift*). The discipline of human geography experienced several paradigm shifts during the 20th century alone, and the concept of the paradigm has been widely accepted to illustrate and explain the evolution of human geographic thought. Yet some scholars argue that the concepts of the paradigm and paradigm shift are oversimplified and need to be critically examined.

PARADIGM AS A NORMATIVE FRAMEWORK

Scientific disciplines, according to Kuhn, establish for themselves a number of normative guidelines that regulate the activities of researchers within it as well as guard and distinguish it from others. Paradigms emerge as the dominant ways of thinking within a science and determine the accepted theoretical frameworks, the most commonly used methodologies, and the ways in which future scientists are trained. Thus, they describe the stable pattern of academic activity that provides rules about how research ought to be done, to be taught, and to be accepted by the largest number of specialists within a community of scholars.

PARADIGM SHIFT

What Kuhn called "normal science" typically proceeds for extended periods of time and accumulates theoretical knowledge and empirical data that fall within the dominant paradigm. Research continues to refine the basic constructs within a paradigm and applies the generally accepted models so long as problems can be solved according to a certain procedure and the discipline is making progress. At a certain point in the cumulative and historical development of a discipline, when it fails to make progress or falls short of providing solutions for contemporary problems, clusters of new ideas begin to emerge and actively challenge assumptions that were taken for granted previously. Fueled primarily by new scholars unwilling to conform to the theoretical and methodological standards of the existing paradigm, a period of "extraordinary science" emerges and ushers in a paradigm shift. A new set of questions and research problems that disprove the assumptions and predictions of existing theory surfaces. Previously accepted methodologies fail to provide solutions to new problems, and the moral and ethical underpinnings of certain research techniques are questioned. Small groups of researchers actively challenge the scientific establishment within a discipline and stimulate a long-lasting debate about the new concepts and ideas they introduced. Dominant beliefs are slowly dismantled by a growing group of new converts, and the scientists still embracing the "old" prevailing set of thoughts are slowly dwindling in numbers yet vigorously resist accepting the new set of ideas. The paradigm shift is completed when the supporters of the original paradigm retire, die, or are otherwise largely replaced by scholars adhering to the new alternative set of ideas that emerges as a new dominant paradigm.

PARADIGM SHIFT IN HUMAN GEOGRAPHY: AN EXAMPLE

Rather than reconstructing the history of human geographic thought and its multiple paradigm shifts during the past century, it is appropriate to illustrate the concepts of paradigm and paradigm shift in human geography by using one specific example. During the 1960s and early 1970s, inquiry in human geography was dominated by a paradigm based on positivist science. Geographers conceptualized their own role in the research process, and for that matter in the field, as that of detached, value-neutral observers out to collect empirical data that were to be fed into supposedly objective geographic models of society and space. These models of human spatial behavior were to be tested and continuously refined. For example, census data frequently were used to map out the complex sociodemographic patterns of cities. Multivariate statistical analysis, used in techniques such as factorial ecology, created maps representing the different social status or class of citizens that doubtlessly showed the spatial distribution of wealth and capital in an urban environment, but these representations could not answer the questions of how and why such patterns had developed and what exactly the causal mechanisms behind these patterns of distribution were. Such analysis could not be done by studying census data and mapping it out; rather, it required the researcher to get in touch with the community at hand and to get directly involved in understanding the political processes guiding the creation of urban neighborhoods. Subsequently, critical voices such as David Harvey emerged and created alternative models for understanding the complex lifeworlds of cities and the political power struggles within them. This initial push to understand cities used both humanistic and Marxist concepts, not positivist hypothesis testing and generalized model building, and led to a whole new paradigm of critical human geography that became a powerful toolbox for researchers.

CRITIQUES OF THE CONCEPT

As the preceding example suggests, the positivist paradigm of mapping and modeling urban patterns never ceased to exist completely and has even resurfaced

due to the arrival of computer technology and geographic information systems. Similarly, critical human geography now is a widely accepted approach to understanding spatial problems, yet by no means is it an overly dominating paradigm. This lack of dominance and the existence of multiple competing paradigms point to two central critiques of Kuhn's concept of the paradigm. First, consensus is rare within any one discipline at any one time. Second, competing paradigms often borrow their core ideas from other disciplines; thus, paradigm shifts are stimulated by sources outside of the discipline and are not triggered by internal forces as Kuhn suggested.

—*Olaf Kuhlke*

See also Epistemology; History of Geography; Logical Positivism; Marxism, Geography and; Radical Geography

Suggested Reading

Harvey, D. (1973). *Social justice and the city*. Baltimore, MD: Johns Hopkins University Press.
Kuhn, T. (1962). *The structure of scientific revolutions*. Chicago: University of Chicago Press.

PARTICIPANT OBSERVATION

Participant observation is an umbrella term for a range of methods to investigate the practices of individuals and groups in place. These can include participating and observing, engaging in informal conversations, and collecting documentary data. These methods facilitate an exploration of (meanings of) people's everyday actions and behavior, usually from the researched individuals' perspectives.

There is conceptual overlap between participant observation and ethnography. To make a distinction, ethnography may be seen as a research approach that incorporates the methodology, methods, interpretation, and dissemination of research. Participant observation is a (set of) method(s) central to ethnography. The use of participant observation does not necessitate an ethnographic approach, although to date most geographic studies that employ participant observation explicitly have been ethnographic.

Geography has a long, albeit marginalized, history of ethnographic studies based on participant observation. Ethnographic research within geography can be traced back to the origins of the discipline. Like early anthropologists, these ethnographies employed participant observation to understand "exotic" societies, regarded as spatially and temporally distinct, so as to make universal laws about humanity. A more recent and critical history can be traced to the Chicago School of human ecology and strands of humanistic geography.

The cultural turn within human geography arguably instigated a renaissance of qualitative methods, although participant observation remains relatively underused. Participant observation usually is an intensive method that is useful for conducting idiographic specific studies rather than for making generalizable knowledge claims or universal laws.

Because participant observation usually is relatively inductive and intensive, it can provide in-depth holistic understandings of the meaning of people's everyday practices in place. These benefits are also limitations. Intensive participant observation is time-consuming. Participant observation can, however, make a unique contribution to examining the materiality of everyday life along with discursive representations. Furthermore, although other methods give insights into people's interpretations of their practices, not all action is consciously rationalized. Observing people's practices, and questioning people about the significance of their actions, can lead to deep and holistic understandings.

Participant observation raises interconnected practical, methodological, and ethical issues. The practical issues can be divided into two key aspects: whether the research is overt or covert and what the level of participation and/or observation is. Overt research involves informing participants of the purpose and nature of the research, whereas covert research is undertaken without informing participants. Covert research raises a number of ethical dilemmas such as not giving participants the opportunity to (refuse to) consent to participating. However, this can be ethically tenable in particular circumstances. The reality of research often does not fall neatly into either side of the covert/overt dichotomy. For instance, it is not always possible to be overt with all research participants who may play a limited role in the research. Furthermore, it may be more ethical *not* to disclose the specifics of the research focus when this would reaffirm negative lines of difference. Levels of participation/observation are a sliding scale from the complete participant, to the participant-as-observer, to the observer-as-participant, to the complete observer.

Although different levels of participation are appropriate for various questions and settings, roles often shift during the research processes depending on context.

It is essential to provide an accurate record of the research processes. Often observations initially are recorded as brief notes and later are written into in-depth fieldnotes and/or diaries. The ease of recording varies according to the type of participant observation undertaken—overt/covert and level of participation. Increasingly, audiovisual technologies are used to record fieldwork. These technologies facilitate investigation of materialities that may escape initial observations. However, such technologies do not produce more "objective" research given that the gaze of the (video) camera is directed by the individual operating the machine, whether the researcher or the researched.

The knowledge produced by participant observation is situated, providing intersubjective interpretations of events rather than objective truths. This has two key impacts. First, discussions of the need to triangulate the data gained from participant observation are pervasive. Participant observation often involves an inherent triangulation as informal conversations enable the gathering of others' perceptions of events. Second, considerations of reflexivity and power are integral to participant observation. Researchers should be critical of the potential of research to reproduce/transform unequal societal power relations (a debate taken further within action research). We should question how who we are influences all aspects of the research—when we are constructing the field of study, when we are in the field, and when we are writing it up (while acknowledging that these are not distinct spaces). However, as researchers, we need to be aware that our identities are context specific and shifting and that not all of our actions are consciously reflexive.

Participant observation is an insightful method that enables the exploration of everyday practices in place and the construction of meaning. In common with all human geography research, it raises ethical ambiguities. Participant observation is the central method(s) of ethnography. However, it is not the sole preserve of ethnographic research. Arguably, all studies of society can be viewed as participant observation given that researchers are part of the social world they examine.

—*Louise Holt*

See also Chicago School; Epistemology; Humanistic Geography; Qualitative Research; Situated Knowledge

Suggested Reading

Atkinson, P., & Hammersley, M. (1998). Ethnography and participant observation. In N. Denzin & Y. Lincoln (Eds.), *Strategies of qualitative inquiry* (pp. 110–136). Thousand Oaks, CA: Sage.

Cloke, P., Cook, I., Crang, P., Goodwin, M., Painter, J., & Philo, C. (2004). *Practising human geography*. London: Sage.

Cook, I. (1997). Participant observation. In R. Flowerdew & D. Martin (Eds.), *Methods in human geography* (pp. 127–149). Harlow, UK: Longman.

Hoggart, K., Lees, L., & Davies, A. (2002). *Researching human geography*. London: Edward Arnold.

Laurier, E. (2003). Participant observation. In N. Clifford & G. Valentine (Eds.), *Key methods in geography* (pp. 133–159). Thousand Oaks, CA: Sage.

PEASANTS

The term *peasant* generally refers to anyone who cultivates the soil as a small landholder or an agricultural laborer. Originally, the *peasantry* referred to a class of people within the feudal systems of Europe who tilled the land and provided manual labor for a feudal estate. During the Middle Ages and as late as the 18th century, peasants constituted between 80% and 90% of the population in Europe and occupied the bottom rung of the social hierarchy. Peasants bore heavy rent, tax, and tithe burdens, and in some regions (particularly Eastern Europe and Russia) they were legally bound to a particular manor or landlord who ruled over virtually every aspect of their lives. Legally bound peasants were known as *serfs*. In some places such as Russia, serfs lived in virtual slavery because landowners could legally transfer them from one estate to another at will. Even where peasants were free—in Britain, northern Italy, Spain, most of France, and western Germany—their burdens and obligations were onerous and their living conditions were dismal. Peasants were required to work for their lords and pay a significant portion of their crops as rent in exchange for the right to cultivate their own pieces of land. Often they were also expected to provide free labor and to tithe (donate) 10% of their harvest to the Church. Producing enough food to feed their families under these conditions was a constant and uncertain struggle, and most peasants lived desperate impoverished lives.

More recently, social scientists have used the term *peasants* to refer to a broad range of groups engaged in subsistence-based or small-scale agriculture in Asia, Africa, and Central and South America, including small land-owning farmers, tenant farmers, sharecroppers, and landless agricultural laborers, who form the main agricultural labor force in a region. Some scholars argue that the term should refer only to those engaged in agriculture, whereas others argue that other groups such as rural artisans, traders, foresters, and fishermen should be considered members of the peasantry because many farmers also engage in these activities. There is also considerable debate about whether or not small-scale commercial farmers and landless agricultural laborers should be included in definitions of peasants. Most scholars agree, however, that nomadic and seminomadic pastoralists should not be included as such.

These debates surrounding the definition of peasant reflect the considerable diversity that exists in agricultural production systems both across world regions and historically and that has made establishing a precise definition of peasant so difficult. Furthermore, so-called modern-day peasants often do not refer to themselves as such, preferring instead identities such as *farmers* and *agricultural workers*. As a result, some scholars avoid using the term at all.

Despite these debates, consensus does exist on some basic features of peasant economies and societies. In general, peasant economies are characterized by a simple technology and a division of labor based on age and sex. The basic unit of production is the household. Peasant households have access to relatively small plots of land that they farm with their own labor, draught animals, and nonmechanized equipment. Production is small in scale; output per worker is relatively low, and peasant families generally consume what they produce. However, a portion of their output may be sold in the market or paid to a landlord. In fact, the extraction or transfer of surplus production, either freely or coercively, to dominant nonproducing rulers is a key tenet underlying classic definitions of peasant.

Because what peasants produce usually is appropriated in some form by nonproducing groups, peasant communities are not isolated, self-contained entities. Instead, they are connected to a larger society through economic linkages that can extend internationally. At the same time, considerable diversity exists between peasant communities in the degree and manner in which they are integrated into national and international markets. Within peasant communities, households usually are stratified by the amount of land they work, whether they own or rent land, and the extent to which they work their own land or work someone else's land. The degree to which peasant households are economically differentiated and face resource scarcity shapes the nature of social relations in peasant communities. Although peasant households do forge networks of social support, intracommunity relations can also be quite contentious. Thus, scholars have characterized peasant social relations variously as cooperative and competitive.

Peasants are relatively poor and so are relatively less powerful vis-à-vis wealthy farmers, landlords, and the state. Political relations tend to be vertically organized into patron–client networks that serve to provide certain forms of protection to peasant households at the same time that they stifle progressive innovation and reform and the formation of horizontal, class-based alliances.

As a result of their poverty and political disempowerment, peasants are especially vulnerable to external forces, including political imperatives and crises, ecological change, and the vagaries of the market. As societies undergo capitalist transformations, peasant groups are particularly adversely affected, and this often has led to peasant unrest and revolt. Despite frequent peasant uprisings around the world from the 14th through the 20th centuries, in only a few instances have peasant rebellions developed into sustained political movements achieving significant improvements in peasants' lives. Most often, peasant uprisings have been spontaneous, sporadic, and poorly organized—and thus easily suppressed.

—*Holly Hapke*

See also Agriculture, Preindustrial; Dependency Theory; World Systems Theory

Suggested Reading

Kincaid, D. (1993). Peasants into rebels: Community and class in rural El Salvador. In D. Levine (Ed.), *Constructing culture and power in Latin America* (pp. 119–154). Ann Arbor: University of Michigan Press.

Mintz, S. (1973). A note on the definition of peasantries. *Journal of Peasant Studies, 1,* 91–106.

Shanin, T. (Ed.). (1988). *Peasants and peasant societies: Selected readings* (2nd ed.). Oxford, UK: Blackwell.

Wolf, E. (1966). *Peasants.* Englewood Cliffs, NJ: Prentice Hall.

PHENOMENOLOGY

Phenomenology entered the field of geography during the late 1960s and early 1970s, when geographers began to draw from this philosophical tradition to critique spatial science. Phenomenologist geographers argued that a spatial scientific approach based on an objective epistemology failed to account for subjective human experiences of place. These geographers, often called humanistic geographers, wanted to shift geography's attention away from a spatial science of pattern and process and toward a deeper understanding of place meaning and, in some ways, back to analyses of human communities and the cultural landscape. Since its appropriation by geographers, some scholars in the field have debated the efficacy of this particular philosophical tradition in geography, its application to geographic inquiry, and geographers' interpretations of this tradition.

Phenomenology, in the simplest terms, can be defined as the study of both material and immaterial phenomena. Its emergence as a philosophy can be traced back to 19th-century critiques of positivism and empiricism. Phenomenologists argue that positivistic science and empiricism cannot account for normative questions, which are integral to the human condition, and thus positivists and empiricists fail to understand the conscious and unconscious self. Geographers drawn to this particular approach during the 1970s interpreted the work of philosophical phenomenologists, such as Edmund Hussrel, to mean that we should reject empirical science. Drawing also from the theories of Martin Heidegger and Maurice Merleau-Ponty, geographers set out to refocus geography's attention, perhaps naively, on self-defined *geographic experiences.* For example, Yi-Fu Tuan, Anne Buttimer, and Edward Relph argued that a geography based in phenomenology moved the field away from models of humans as rational actors that establish clear patterns and laws that could be measured mathematically across space; instead, geography should focus on the study of place as feeling, emotion, rootedness, and community, all of which are tied to an authentic (i.e., genuine, natural, and presocially constructed) experience of place. The challenge, of course, is that one's "sense of place" often is tied into unconscious thought. Situating phenomenological analyses of place in the unconscious raised the question of how to go about studying what often are taken-for-granted understandings of locale or community.

While some phenomenologists in the field have focused their attention on the unconscious essences that constitute place, others have turned their attention to the conscious phenomena, or experiences, that also construct and lend meaning to place. In so doing, it is possible to extend the dialogue of community beyond the flat notions of object–subject relations across space that are part of an objective scientific geography. Treating objects and subjects as oppositions in a binary fails to capture the intersubjective nature of subject–subject experience and interactions. This intersubjectivity is not only between human subjects but also between subjects and their place in the world. Thus, the focus of many phenomenologist (humanistic) geographers has been on subjective experiences while shying away from causal models, theory building, and scientific inquiry. As such, this approach to human geography has been critical of modernization and the rapidity with which many people experience space and spatial relations. They argue that place is breaking down as people are being made marginal and becoming detached from lived experience of their home, town, and community.

Not all geographers have been taken with the humanistic reading of phenomenology. John Pickles offered a critique of the ways in which humanistic geographers have appropriated this philosophical tradition. Pickles argued that phenomenology is both a descriptive and a transcendental philosophy that is intended not only to dismiss science and scientific practice but also to rethink how we do science. Phenomenology, in contradistinction to how humanistic geographers have interpreted it, is concerned not with individualized hypersubjective geographies but rather with the complexity of phenomena that constitute both space and place. Moreover, phenomenologists are interested in investigating how we construct our science in the first place and on what ontological basis that science is founded. Of interest, then, is how we can investigate what constitutes the lifeworld not simply as an individualized site of subjective experience but rather as a universal human condition and set of sociospatial structures (or the lifeworld's spatialities).

Despite the debates, phenomenology constitutes an important philosophical tradition in the history of geography. Humanistic geographers working through a particular reading of phenomenology have had a strong influence on some, and a connection to other, postpositivist geographers whose interests in studying place and landscape during the 1970s and 1980s were aided by the powerful critique of spatial science that phenomenological-based geography offered. Phenomenology generally is a complex philosophical

approach and remains an important tradition within a larger history of science and scientific thought.

—*Vincent Del Casino*

See also Humanistic Geography; Theory

Suggested Reading

Buttimer, A. (1976). Grasping the dynamism of lifeworld. *Annals of the Association of American Geographers, 66,* 277–292.

Pickles, J. (1985). *Phenomenology, science, and geography: Spatiality and the human sciences.* Cambridge, UK: Cambridge University Press.

Relph, E. (1976). *Place and placelessness.* London: Pion.

Tuan, Y.-F. (1971). Geography, phenomenology, and the study of human nature. *The Canadian Geographer, 15,* 181–191.

PHOTOGRAPHY, GEOGRAPHY AND

For more than 150 years, the relationship between geography and photography has been complex, dynamic, and mutually influential. This relationship, in which geographic concerns have shaped photographic practices and photographic technologies have nurtured and documented geographic pursuits, can be studied from historical, practical, and theoretical perspectives. In 1839, two quite different processes for making permanent images "from nature" were announced. These early photographic technologies offered a new way of encountering the physical and human world. Louis Jacques Mandé Daguerre's method of producing a unique image on a silver-coated copper plate and William Henry Fox Talbot's paper-based negative–positive process were quickly harnessed to geographic purposes in the form of field observations, travel accounts, prints, book illustrations, and teaching aids—uses that have survived and become increasingly sophisticated in an age of geographic information systems (GIS) and digital imaging.

From the first mention of photographic pictures in Alexander von Humboldt's *Cosmos* in 1849, to Vaughan Cornish's appreciation of landscape through photography of scenery in 1946, to Denis Cosgrove's analysis of the whole-earth images from the Apollo space mission in 1994, photography has played a variety of roles in the data-gathering practices, ordering mechanisms, and myth-making processes by which people have come to know the world and situate themselves in it. When first introduced, photography's ability to record, store, and disseminate information in visual form made it a natural complement to geography's long-established emphasis on observation, description, and visualization.

Difficult, messy, and time-consuming processes did not stop the first photographers from carrying hundreds of pounds of equipment on the Grand Tour, up the Nile, and into the jungles of the Yucatan. Embraced as an accessory to travel and employed to produce so-called "man on the spot" accounts, photography presented the scientific traveler and the gentleman adventurer a way in which to bring the world home, in visual form, for contemplation, study, enjoyment, or analysis. Quickly, the camera became an instrument for acquiring or disseminating geographic knowledge within military operations, boundary and geological surveys, topographical mapping, immigration programs, tourism promotion, and ethnographic investigations. Practiced by travelers, photography mediated the personal encounter with unfamiliar places and peoples; collected by armchair travelers, photographs served as surrogates for travel and firsthand experience of place.

As photography got easier, cameras got small, processing became commercially available, and images in full color became a reality, three forms of photography increasingly became an integral part of human geography: aerial photography, repeat photography, and photogrammetry. Employing established standardized methods for producing and interpreting such images, these applications embrace photographs as scientific data to be read, measured, and manipulated in the process of studying cultural landscape remains, establishing spatial coordinates, or observing landscape change. Early on, photographers carried their cameras (and their portable darkrooms) to the tops of hills and tall buildings to record landscape views. True aerial photography was first attempted during the mid-19th century by French photographer Felix Nadar, who in 1858 ascended to the height of several hundred meters in a balloon to obtain a photographic bird's-eye view of the earth from which he planned to produce an exact topographic map. Two years later, James Wallace Black produced a view of Boston—as the eagle and the wild goose see it, according to Oliver Wendell Holmes—from a balloon 1,200 feet in the air. Since then, vertical and oblique aerial photography has been used to

inventory and map natural and human-made features on the surface of the earth. Today, geographic applications of aerial photography continue and multiply in the form of contemporary satellite imagery.

Chronologically replicated photographic views of the same subject are used to demonstrate change over time by exploiting the camera's ability to create a photographic baseline against which development or decay can be measured. Geographers are familiar with this practice through scholarly rephotography projects as well as popular "then and now" comparisons.

Photo-topography or photogrammetry is another form of data photography intended to convey specific geographic information in visual form. First used during the late 19th century for practical surveying purposes, this application of the camera to geographic endeavor produces data to be mapped, compared, or quantified. The production and interpretation of air photographs, although in many ways very much a cartographic endeavor, employs photographs in an effort to read surface characteristics and spatial relations.

Used more generally as a means of gathering geographic information, photography became an integral part of exploring expeditions and imperial administrations, opening distant or exotic regions to 19th-century European eyes. Photographers in London, Glasgow, and Paris documented the changing geographic face of urban renewal, just as series of stereoscopic views produced on both sides of the Atlantic celebrated urban development in newly settled areas of the New World. Similarly, various projects to inventory native peoples of North America, India, and the British Empire themselves were photographic investigations of cultural geographies as well as visual expressions of the 19th-century penchant for ordering and labeling.

Beyond their obvious use in fieldwork, mapping, lecturing, and publishing, photographs are increasingly attracting scholarly attention as primary sources for their role in expressing, reflecting, and shaping the relationship of people to place. Even scientifically conceived photographs are now subjected to critical scrutiny for the layers of meaning beyond the visual facts presented in the images themselves. In the wake of recent theoretical writings on representation and the relationship between knowledge and power, historical and cultural geographers have turned their attention to the role of photographs in the geographic imagination. In their studies of the ways in which photographs construct, confirm, and contest notions of landscape and identity, they have contributed to recent work on the "visual turn" in the humanities and social sciences.

At the same time that academic geographers have turned to photographs as primary sources and tools of cultural and historical analysis, lavishly illustrated magazines, photography exhibitions, Web sites, and coffee table books—from *National Geographic* to Arthus-Bertrand's *Earth From the Air*—continue to foster a more popular geographic agenda to see and know the world and continue to employ photographs as windows on a world beyond our doorstep. Perhaps the most obvious use of photographs as windows comes in teaching and learning, where lantern slides and 35mm slides have been the mainstay of classroom instruction and conference presentations. Yet even as a major technological and pedagogical shift takes place as slides are replaced by digital images, slide projectors give way to data projectors, and textbooks give way to WebCT, photography remains an indispensable part of our understanding and study of human geography.

—Joan Schwartz

See also Spaces of Representation; Vision

Suggested Reading

Cosgrove, D. (1994). Contested global visions: One-world, whole-earth, and the Apollo space photographs. *Annals of the Association of American Geographers, 84,* 270–294.

Hoelscher, S. (1998). The photographic construction of tourist space in Victorian America. *The Geographical Review, 88,* 548–570.

Kinsman, P. (1995). Landscape, race, and national identity: The photography of Ingrid Pollard. *Area, 27,* 300–310.

Ryan, J. (1997). *Picturing empire: Photography and the visualization of the British Empire.* London: Reaktion Books.

Schwartz, J. (1996). The geography lesson: Photographs and the construction of imaginative geographies. *Journal of Historical Geography, 22,* 16–45.

Schwartz, J., & Ryan, J. (Eds.). (2003). *Picturing place: Photography and the geographical imagination.* London: I. B. Tauris.

PLACE

Place typically refers to a particular segment of the earth's surface that is characterized by the unique sense of belonging and attachment that makes it different from other places around it. Thus, place is a meaningful portion of space. In many ways, the concept of place lies at the heart of human geography. This is because it is a concept that is widely used in everyday life; it has become part of common sense. This place is here and that place is over there. This taken-for-grantedness, however, hides the variety of ways in which the notion of place can be approached and the number of roles in play in the human experience of the world.

LOCATION, LANDSCAPE, AND SENSE OF PLACE

At its most basic level, place refers to a *location.* A location is either an absolute indicator of where something is in terms of an abstract set of measures (e.g., a map reference) or a relative indicator of where something is in relation to other things (e.g., A is 50 miles north of B). This is one way in which place is used both in human geography and in our everyday life. Here place simply refers to the "where" of something.

Place also refers to a physical landscape that makes one place unique from another. We might talk about our favorite place (e.g., somewhere we went on vacation as a child) and mean the total sum of material (both "natural" and "cultural") at a particular location. In this sense, place is much more than a reference to an absolute or relative location; instead, it refers to buildings, beaches, hills, forests—all the things that make one place different from the next place.

An important component of place is its *sense of place.* This term indicates all of the subjective meanings that become attached to a location and the physical landscape that is characteristic of that location. Some of these meanings may be personal and arise from individual biographies. It may be, for instance, that a certain smell makes us think back to a particular place, or perhaps the memory of somewhere we once lived fills us with emotions that are hard to explain—nostalgia, fear, sadness, hope. Some senses of place, however, are more shared and are likely to be the product of mediation. Places, after all, often are inhabited by a variety of people who have things in common—who share daily routines. In addition, places are objects of representation and appear in novels, music, and film as well as on television. So, when we mention a place such as New Orleans or Hong Kong, there are likely to be certain meanings that arise again and again among a diverse group of people. We can think of these shared senses of place as *place images.*

PLACE IN HUMAN GEOGRAPHY

Humanistic Geography

Although place seems to lie at the heart of geography, the idea of place was not examined in any detail until the 1970s. For most of the 20th century, the idea of *region* formed the basis of human geography. Only with the advent of *humanistic geography,* and the writings of Yi-Fu Tuan, Edward Relph, and Anne Buttimer in particular, was place subjected to critical scrutiny. Humanistic geographers have developed a notion of place that sees it as a way of "being-in-the-world" developed from the philosophy of phenomenology. Being-in-the-world through place involves subjective attachments to the world. Modes of expression, such as literature, film, art, gardening, and architecture, all can be seen as ways of making places through subjective immersion in the world and the assignment of meaning to portions of the world. Humanistic geographers looked to such practices to counter the perceived meaninglessness of concepts such as *space* and *location* that had been dominant in spatial science during the 1960s and 1970s. The concept of place also differed from the earlier concept of region in that it was less descriptive. Whereas regional geography sought to define regions as clear portions of the world, humanistic geography was more interested in the ways in which the world was made meaningful. Rather than seeking to map places and noting the differences, humanistic geographers wanted to explore what it was like to be human in a world of places. It was concerned with how places were created through personal attachment as well as the shared forms of meaning production such as art, literature, and the transformation of the material world itself.

The Politics of Place

Although humanistic geographers were happy, for the most part, to write about the creation of place by human individuals, more recent critical human geographers have pointed out that the meanings ascribed to place vary according to who exactly it is we are talking about. Feminist geographers, for instance, have taken issue with the claim that "home" is an ideal kind of place where people feel attached, rooted, and safe. They have noted how home does not always provide such a cozy feeling, especially for women who often feel unsafe and claustrophobic at home. Indeed, the public meanings of place often are constructed by relatively powerful people at the expense of alternative meanings that might be developed by relatively powerless people. Thus, places often are objects of power created to further particular forms of domination based on class, gender, race, and sexuality. How a woman perceives a place may be very different from how a man perceives it. Similarly, the way in which a wealthy white man sees a place may be very different from the meanings given to the same place by a poor black man (e.g., in the American South). Places, in other words, are not so easily shared.

A Global Sense of Place

The differences between places cannot simply be understood as originating from internal processes specific to a particular place. Places are not closed and sealed things with clear borders and pure histories. Places are the product of connections to other places through various forms of flow, including the mobility of people, things, and ideas through transport networks and forms of communication. Thus, we can think of a "global sense of place"—an idea developed by Doreen Massey. Places are the unique intersection of all manner of movements rather than the product of a particular form of rooted history that is unconnected to the rest of the world. This changes the way in which "insiders" relate to "outsiders" in place. One way of looking at place is to see places as inhabited by insiders who feel at home and have a sense of belonging. Insiders, however, depend on the existence of outsiders to make sense. Thus, the place of nation often is constructed in relation to foreigners. Likewise, the long-term inhabitants of small rural villages may be wary of city dwellers who visit their area. In this sense, place often can be the center of regressive and reactionary politics through which all kinds of outsiders are demonized. If we accept that places are the product of connections, it becomes harder to think of insiders and outsiders given that places are, in fact, produced by those outside as much as by those inside. No longer are there clear boundaries separating one from the other.

Place in a World of Movement

To some, however, the rapid increase in the rate and scope of global mobility and communication has meant a decrease in the importance of place in the world. They have noted how the world—particularly the modern

West—has become more homogeneous, with places becoming more and more alike. Concepts such as *placelessness* and *non-place* have been developed to suggest that contemporary society has eradicated the unique meanings of specific locations. They could be anywhere. Examples include airports, shopping malls, chain stores, motorways, tourist sites, and fast-food restaurants. Thus, more and more of our lives are spent in such non-places. Others, however, have argued that the increasing mobility of the modern world has led to the increased importance of place. As it becomes easier to move and to communicate at great speed over great distances, it is argued, individual places need to work ever harder to distinguish themselves from everywhere else to attract business, residents, and tourists. Thus, place marketing and the heritage industry have become particularly important in the modern world. Image consultants are hired by local governments to make their place stand out from the others.

PLACE AS A WAY OF KNOWING

Place most often is used to refer to a thing in the world, usually something at a fairly small scale such as a neighborhood, village, town, or city. But the combination of location, material landscape, and sense of place occurs at a much larger scale (e.g., the nation, the globe) as well as at a much smaller scale (e.g., the corner of a room, a favorite chair). To think of a nation or a room as a place is to understand it in a particular way. The placeness of a nation or a room refers to certain characteristics attributed to such disparate things. Here place is a way of thinking about the world as much as a reference to things in it.

Place as a way of knowing organizes the world geographically. People, objects, and forms of behavior are ascribed meaning through their place in the world. When we use phrases such as "a place for everything and everything in its place," we are suggesting a particular "proper" geographic ordering of the world around us. People, objects, and forms of behavior are said to be "in place" or "out of place." Indeed, a cursory glance at the media often will reveal a wide range of people and activities being described as out of place. These range from homeless people in city neighborhoods, young people on street corners, to mothers breast-feeding in restaurants, to gay people kissing in public. Place clearly is used to structure normative expectations of proper behavior and thus plays a key role in the construction of moral geographies.

The fact that people, objects, and behavior can be thought of as in place or out of place suggests that place is more than a description of particular "things" that can be identified, mapped, and described. In addition to using place to refer to things in the world across scales (ontology), place can be thought of as a way of understanding the world around us (epistemology). Place as a way of knowing highlights how some forms of knowledge rely on a tight connection between people and place, on the one hand, and a clear sense of boundedness and rootedness, on the other. Thinking in these terms can be described as "sedentarist" because it focuses attention on the "proper place" of people. It is this kind of thinking that leads many to assume that mobile people, such as nomads, immigrants, and refugees, are out of place and thus a threat to the good order of a place-based world. Such thinking clearly has pernicious consequences because people perceived in this way often are treated badly by those who have invested in place and rootedness. In Nazi Germany, for instance, a complex mythology was created that described Aryan Germans as deeply rooted in the soil of the nation, whereas it described others—Jews, gypsies, and gays—as rootless people of the desert and the city where there was no opportunity for roots to establish. This mythology helped to justify the murder of more than 6 million people.

—*Tim Cresswell*

Suggested Reading

Auge, M. (1995). *Non-places: Introduction to an anthropology of supermodernity.* London: Verso.

Cresswell, T. (2004). *Place: A short introduction.* Oxford, UK: Blackwell.

Tuan, Y.-F. (1977). *Space and place: The perspective of experience.* Minneapolis: University of Minnesota Press.

PLACE NAMES

Place names (or *toponyms*) are used to identify and differentiate geographic features, both human-made (e.g., countries, cities, streets) and physically based (e.g., mountains, lakes, rivers). Toponyms are systems of spatial reference and symbolic expression. They are one of the most fundamental ways in which people claim, categorize, and attach significance to places.

Traditionally, geographers have classified and mapped names as cultural artifacts; however, newer research emphasizes the cultural politics of place naming.

PLACE NAME AS SPATIAL REFERENCE

Place names facilitate the identification and physical navigation of the landscape, using a single word or series of words to distinguish one place from another. As part of the larger structure of language, toponyms permeate our daily vocabulary, both verbal and visual. They are found on road signs, advertising billboards, addresses, and (of course) maps. Place names are critical to the use of nearly any kind of map, contributing to geographic knowledge and the development of a sense of place.

Although names are important points of reference, misunderstanding can result from conflicts in their use and spelling. The same name sometimes is applied to different places, and a place may have more than one name. Seeking greater uniformity in naming, particularly in light of growing electronic data demands, applied human geographers have worked to standardize the toponyms on maps and other publications. They have also established administrative procedures for dealing with new place names, name changes, and naming controversies.

Several countries, including the United States, have set up boards to review and rule on place name issues. The United Nations Group of Experts on Geographical Names (UNGEGN) assists in exchanging the results of these national efforts and promoting the benefits of international name standardization. Despite technical justifications for standardization, the process can lead to public disputes as authorities try to choose a single name for a feature.

PLACE NAME AS SYMBOL

Place names are also symbols to which people attach meaning and from which they draw identity. Names evoke powerful images and connotations, as became evident in Fayetteville, North Carolina, when residents along Anthrax Street demanded a name change. The street name had existed for several years but was reinterpreted in light of the tragedy of September 11, 2001, and bioterrorism fears. Its new name, Allegiance Avenue, offers a more patriotic association. The image-generating power of toponyms has long played a role in place promotion, from the intentional misnaming of Greenland to the more contemporary practice of crafting idyllic-sounding monikers for subdivisions and apartment complexes.

Humans apply place names to create a sense of order and familiarity, frequently choosing names that reflect and project their cultural point of view. Even when time has erased other evidence, toponyms can provide insight into people's religious beliefs, ethnic origins, history, environmental perceptions, and political values. Interpreting such names requires reading their many layers of meaning. In 1916, the Canadian town of Berlin, Ontario, changed its name to Kitchener, honoring the British secretary of war who died at the beginning of World War I. On one level, the naming reflected Canada's support of the Allied effort and the strengthening of its ties with England. On another level, it symbolized a nativist rewriting of the landscape and an anti-German hysteria sweeping across Canada.

Naming also represents a means of claiming or taking ownership of places, both materially and symbolically. In many world regions, a renaming of geographic features accompanied European colonial exploration. Explorers and mapmakers not only projected their Western values onto the landscape but also devalued the perspectives of indigenous inhabitants, who had been naming places for hundreds of years. The toponymic process sheds light on power relations—which social groups have the authority to name—and the selective way in which naming may privilege one worldview over another.

Place names often are characterized by a permanence that outlives their creators; however, they are not static symbols. Major global shifts frequently underlie the renaming of features—decolonization, the decline in communism, the rise of Islamic fundamentalism, the recognition of minority rights, and so forth. In the United States, racial and ethnic groups have actively sought the removal of names containing racial slurs and the renaming of places to recognize their historical achievements. Hundreds of streets and schools bear the name of civil rights leader Martin Luther King, Jr. These names sometimes serve as arenas for debates about culture and identity, thereby exposing the divisions and ideologies within society.

PLACE NAME STUDY

Geographers traditionally have collected, classified, and mapped toponyms as artifacts, using them to reconstruct the environmental and human history of places. Names provide clues about the direction and timing of human migrations, the location of past settlements, the original vegetation of an area, historical patterns in nationalism and commemoration, indigenous perceptions of place, and the boundaries of vernacular regions.

Newer approaches take a more contemporary focus and stress the cultural politics of place naming. Instead of passive artifacts, toponyms are vehicles for reinforcing and challenging cultural and political ideologies. Studies have examined how government elites in countries such as Israel, Germany, Romania, and the former Yugoslavia have used place names—particularly commemorative street names—to advance reinvented notions of national history and identity. Yet other scholars have examined naming as resistance, that is, how marginalized groups such as African Americans and the Maori of Aotearoa/New Zealand engage in toponymic struggles to establish their legitimacy within society. Mapping remains important to place name study; however, it is also necessary to study the naming process, the participants involved, and the very language used in referring to and debating toponyms.

—*Derek Alderman*

Suggested Reading

Alderman, D. (2002). School names as cultural arenas: The naming of U.S. public schools after Martin Luther King, Jr. *Urban Geography, 23,* 601–626.

Azaryahu, M. (1997). German reunification and the politics of street names: The case of East Berlin. *Political Geography, 16,* 479–493.

Randall, R. (2001). *Place names: How they define the world—and more.* Lanham, MD: Scarecrow Press.

Zelinsky, W. (1983). Nationalism in the American place–name cover. *Names, 31,* 1–28.

POLITICAL ECOLOGY

Variously referred to as a field, an approach, a theory, and a conceptual tool kit, political ecology analyzes environmental phenomena in the context of society–environment relationships and with the tools of a broadly defined political economy. Political ecology is wide-ranging and interdisciplinary, including not only geography but also anthropology, development studies, sociology, environmental studies, and others. Common themes include the effects of protected areas on local livelihoods, the social and environmental effects of integration into capitalist markets, struggles over landscapes and livelihoods, and competing perceptions of the environment. Although it began with a focus on rural land-based issues in developing countries, the scope of political ecology has widened significantly to encompass new themes, theoretical orientations, and geographic foci. Despite the wide variation, the field has points of agreement that bind it together. These include recognition of the importance of multiple scales of analysis and history, the social and political complexity of environmental problems, and the rejection of simplistic and purely technical explanations of environmental problems.

ORIGINS OF POLITICAL ECOLOGY

Political ecology emerged during the 1970s and 1980s as a response to two different fields that were thinking about environment and society. First, it critiqued a number of popular but overly simple explanations for environmental problems, particularly in developing countries. These included neo-Malthusian arguments that overpopulation was the primary cause of resource depletion and environmental degradation, assertions that markets and proper technology would solve all environmental problems, and policies that blamed environmental degradation on lazy, backward, or irrational land users. More fundamentally, political ecology insisted that environmental problems are not merely technical issues requiring technical solutions. Instead, the definitions, origins, and solutions of environmental problems are necessarily social and political in nature.

A second major catalyst was research in cultural ecology from the 1940s to the 1970s. Political ecology drew on early cultural ecology's concern with the relationship between culture and the environment. However, it also critiqued some aspects of this work, including the assumption that cultural traditions served ecological functions that always maintained equilibrium between people and their environment. In addition, societies often were treated as if they were isolated and without history or internal differentiation. Political ecology retained concern with the culture–environment relationship, but instead of treating cultures as isolated, it also took into account larger forces, particularly those related to markets and the state. In addition, it called for attention to class and other forms of differentiation within communities.

BASIC THEMES

A number of conceptual tools are fundamental to political ecology. One is the analysis of access to and control over natural resources, including the factors that shape differential access—whether by class, gender, ethnicity, or race—and its social and environmental outcomes. This theme was developed in the context of the effects of both capitalist market expansion and colonial legacies of nature preservation on the enclosure of the commons. Political ecologists study how loss of access to customary resources can violate resource users' moral economy—their collective perception of what is moral in the economic sphere—and thus engender various forms of resistance.

Another building block is the emphasis on examining both place-based and non-place-based factors and on integrating multiple spatial scales of analysis. It is insufficient to study only how a particular resource user behaves; one also must analyze what larger-scale forces (whether village politics, national policies, or global trade) affect that behavior. Whereas early political ecology writings conceived of the relationship between different scales as boxes nested within each other, or as chains linking smaller scales to larger scales, more recent work has treated those scales themselves as actively produced.

The role of perception in defining environmental problems is a third building block. This means not only that terms such as *degradation* are very difficult to define but also that one's definition of what is degraded, like one's vision of what nature should look like, is influenced by history, culture, and politics. Science is no exception, and some political ecologists have explored the differences among the visions of scientists, local people, and policymakers vis-à-vis what a desirable landscape or environment would be. Two other important elements of political ecology are an emphasis on history—both environmental history and the institutional history of resource management—and attention to the interactions among various forms of marginality (e.g., environmental, social, geographic).

NEW DIRECTIONS

The 1990s saw many assessments of political ecology. Observers noted that the field focused too heavily on rural land issues, that it failed to adequately address the role of gender and household power relations in environmental decision making, that it put too much emphasis on poverty as a cause of environmental problems, and that it lacked a coherent theory. Some believed that the field did not adequately emphasize politics and power, whereas others complained that it slighted ecological processes. In response, political ecology expanded in a number of new directions. These included studies of urban and industrial settings, the development of a feminist political ecology, and attention to the implications of new theories of nonequilibrium ecosystem dynamics for society–environment interactions. The critiques also led to a debate about whether the uniqueness of the developing world meant that there should be a specific Third World political ecology or whether the basic concerns and methods of analysis apply just as well to resource conflicts in developed countries.

Poststructuralism and the discursive turn also had an effect, spurring new writing on the social construction of scientific knowledge about the environment, especially in relation to indigenous knowledge. Work on new social movements also focused attention on environmental activism, including partnerships between environmental nongovernmental organizations at different scales and research on how environmental claims are intertwined with claims about indigenous identities. Others brought culture back into the mix. Rather than treating culture as static and traditional, cultural meanings now are understood as always open for contestation and always needing to be actively reproduced—and thus intertwined with politics. Finally, research on environmental risk, environmental justice, and environmental violence sometimes is also considered part of political ecology today.

CRITIQUES AND STREAMS

Political ecology is not without its critics. Among the common critiques today are that the term is now too broad—lacking empirical and theoretical coherence—and that it continues to neglect the ecological aspects of environmental phenomena. A few critics have called for abandoning the term altogether, charging that research on social–environmental phenomena should not assume a priori that politics is important.

The field can best be described as having a number of different streams with different emphases. One places politics at its center and is tied most closely with the political economy of resource access. Another one emphasizes the investigation of biophysical nature in human–environmental interactions. A third one places attention on cultural politics, identity, and the construction of environmental knowledge. These various streams, of course, are not mutually exclusive.

—Emily Yeh

See also Conservation; Cultural Ecology; Environmental Justice; Malthusianism; Nature and Culture; Resource; Scale

Suggested Reading

Blaikie, P. (1985). *Political economy of soil erosion in developing countries.* New York: John Wiley.

McCarthy, J. (2002). First World political ecology: Lessons from the wise use movement. *Environment and Planning A, 34,* 1281–1302.

Watts, M. (2000). Political ecology. In T. Barnes & E. Sheppard (Eds.), *A companion to economic geography* (pp. 257–274). Oxford, UK: Blackwell.

Zimmerer, K., & Bassett, T. (Eds.). (2003). *Political ecology: An integrative approach to geography and environment–development studies.* New York: Guilford.

POLITICAL GEOGRAPHY

Political geography combines aspects of both the political and geographic to study how political power is created, maintained, and exerted over geographic space. Political analysis sheds light on how geographic space is divided and structured to facilitate or thwart political activities and functions. Likewise, geographic analysis can reveal how geography is used to shape political identities and political structures as well as to mediate the political views and behaviors of people in places. Today, political geographers study topics ranging from the development of neighborhood political organizations, to the sectionalism of national elections, to the political geography of global terrorism. As a result, political geography is implicitly interdisciplinary in its subject matter, theoretical perspectives, and methods of analysis. It is informed not only by the discipline of geography but also by the disciplines of political science, political history, international relations, and political sociology, among others.

BACKGROUND

Political geography as a formally identified topic of academic inquiry dates back to the late 19th century and the work of German geographer Friedrich Ratzel. But it is also true that some of political geography's conceptual heritage dates back to the ancient Greeks, including authors such as Aristotle. From its inception (however defined) through the 1950s, political geography frequently was fixated on analysis of the state as its sole unit of inquiry. During this period, political geography frequently viewed its preeminent purpose as the active furthering of state goals or at least the identification of characteristics facilitating the maintenance of the state if not the international system of states.

Efforts to further advance state goals were central to the development of geopolitics during the late 19th and early 20th centuries, particularly in Germany, Great Britain, and the United States. Such an outlook reached its zenith in the development of German *Geopolitik* prior to World War II. Promoted as an objective science through propaganda, German Geopolitik provided a quasi-academic justification for Nazi violence and imperialism during the 1930s and 1940s. Due to its association with geopolitics, political geography was perceived unfavorably by many after World War II, and the area of study atrophied. During the past four decades, political geography has experienced a renewal, including an increase in its topical breadth, theoretical perspectives, and the geographic scales at which political geography research is conducted.

In spite of the many changes political geography has undergone during the past few decades, some of its core concepts have remained largely constant. For example, political geographers continue to consider the division of geographic space into units of political space, territoriality, boundaries, location, distance, and core–periphery relations, among others. Central among these is the concept of territoriality. Territoriality suggests human action in the acquisition, delimitation, and defense of geographic spaces. Thus, sizable demographic groups defining themselves as distinct nations have pursued the establishment of sovereign territorial states for their occupation. Recent successful examples of such efforts have emerged from the collapse of Yugoslavia and the Soviet Union such as Croatia, Slovenia, Georgia, and Estonia. Clearly, a state is an embodiment of territoriality.

RENEWAL DURING THE 1960s

Political geography has undergone a major revival, if not a renaissance, since the 1960s. As that decade began, political geography's prevailing views of its subject matter and role were largely static and generally accepted the state as the only critical political unit for analysis, giving little attention to the role of political units at other geographic scales. Due to the work of influential geographer Richard Hartshorne, political geographers had adopted a functional approach for their research that focused on the characteristics of successful and unsuccessful states. Well-functioning states, it was argued, were characterized by substantially greater numbers of centripetal forces than of centrifugal forces. Centripetal forces—those binding a state together—might include a common language or ethnicity. If these forces were insufficient to overcome the centrifugal forces pulling a state apart, some political geographers viewed their task as identifying additional centripetal forces for the maintenance of the state. Thus, political geography largely viewed the existing world political system as a given and accepted a role in its preservation.

A number of political, social, and economic world events during the 1960s and 1970s created a climate of thought prompting and facilitating an improved vitality for political geography. Among these was the U.S. civil rights movement, which attracted research attention to issues of race, inequity, and minority political action. Unlike past efforts by political geographers that had focused on the state, the civil rights movement prompted interest in examinations of the political geography of urban areas, among other subnational scales. For example, urban political geographers considered the selection of geographic sites for public facilities in metropolitan areas to determine whether the results of such decisions favored one group over another.

The 1960s and 1970s also witnessed the creation of many new states as former colonies in the Caribbean, Africa, and Asia achieved their formal political independence from colonial powers. These changes challenged the then common assumption that state boundaries and territories were essentially fixed in space and led to a change in perspectives that accepted greater fluidity in the arrangements of national boundaries on the world political map. The period also saw a renewed focus on women's rights and feminist perspectives, concern over environmental issues and policy, and the early

development of supranational organizations such as the European Economic Community, the predecessor to today's European Union. The oil embargo of 1973–1974 and the U.S. withdrawal from Vietnam in 1975 further underscored the interconnectivity of the world's states and the reduced ability of the developed world to easily maintain its vision of the world political map. Collectively, these events and movements led political geographers to undertake studies at geographic scales ranging from the neighborhood, to metropolitan areas, to subnational regions, to worldwide levels. There was also a greater appreciation about the dynamic qualities of potential changes in the division of political space and a growing willingness to critique the political and geographic implications of policies emanating from all scales of government.

Research in political geography during the 1960s and 1970s also reflected changes in methodological perspectives and techniques. During the 1960s, political geographers embraced more systematic and rigorous approaches to their research, frequently using hypothesis testing in conjunction with quantitative modeling. This spatial–analytic approach drew from the other social sciences but added geographic considerations such as distance and location. Research during this period also focused increasingly on societal problems, recognizing that there was a demonstrable geographic pattern to real-world inequities. The radicalization of the era led some political geographers to criticize the spatial–analytic approach as being limited in its ability to facilitate positive societal change. As a result, some instead adopted a political–economic perspective that drew from varied radical literatures, including Marxism.

During the early 1980s, multiple new theoretical perspectives, including world systems analysis, began to inform work in political geography. Developed by sociologist Immanuel Wallerstein, world systems analysis models the world capitalist economy and associated developmental processes inhibiting developing world states in the periphery from becoming developed states in the core. Anthony Giddens's work on structuration theory also allowed political geographers to move away from the comparatively deterministic processes posited in Marxist theory. Giddens's structuration theory provides for a more active role for institutions and individuals in their interactions with deep-seated economic and political systems. Thus, such actors become more than mere puppets in a series of historical transitions over which they have little or no influence. Rather, they are viewed as having the ability to exercise some measure of control in determining outcomes on the political landscape. Finally, during the latter portion of the 1980s, political geography was influenced by postmodernism. Postmodernism questions the objectivity of knowledge, suggesting that the process of knowledge collection is compromised by the limits of language and the conventions of the researcher's discipline. As a result, grand theories and *"isms"* are found lacking and researchers focus on understanding the linguistic conventions guiding the geography of political processes.

THE 1990s AND BEYOND

The changes occurring to the world political map during the late 1980s and early 1990s also led to an expansion of interest in political geography. Among these events was the fall of the Berlin Wall in 1989, subsequently leading to the reunification of East and West Germany, the division of Czechoslovakia into the Czech Republic and the Slovak Republic, the demise of the Soviet Union in 1991, and the subdivision of Yugoslavia, beginning in 1991, with the declarations of independence by both Croatia and Slovenia. These events led not only to dramatic changes on the world political map but also to the discrediting of Marxist ideology. Without a communist Soviet Union, the ideological basis for the cold war and its simplistic geopolitical view of a bipolar world ended, leaving the United States as the sole superpower. A number of important political geography questions have emerged from these events pertaining to changing international boundaries, the political stability and international roles for newly independent states, the diffusion of capitalism to emerging states, and the effect of globalization on the world system, among others. To these changes must be added the genocides in Rwanda and Darfur, concerns over non-state-sponsored terrorism in the wake of the terrorist attacks of September 11, 2001, the dubious justification for a preemptive invasion of Iraq by an Anglo-American coalition in 2003, and the potential for emerging counterbalancing forces for U.S. hegemony over world political processes in light of the Soviet Union's demise. Clearly, there is great potential for substantial additional changes to the world political map in the near future.

CURRENT RESEARCH CLUSTERS AND PERSPECTIVES

These dramatic events have led to continued increases in the vitality and interest in political geography. There are several viable categorizations of ongoing research in political geography. For example, preeminent political geographer John Agnew divided current research in political geography into five overlapping categories informed by three principal research perspectives. The five general categories of current research in political geography are (1) the spatiality of states, (2) geopolitics, (3) geographies of political and social movements, (4) places and the politics of identities, and (5) geographies of nationalism and ethnic conflict. The three research approaches are the spatial–analytic, political–economic, and postmodern perspectives.

The spatial–analytic perspective became popular during the 1960s and posits that geographic variables such as space and distance have a direct influence on political structures and activities. This approach frequently uses existing geographic concepts such as distance decay in its analysis. A distance decay function suggests a progressive decline in a variable's magnitude with increasing distance from a defined point in geographic space. For example, we might hypothesize that the probability that a voter casts a ballot on Election Day is a direct function of the distance between the polling station and the voter's residence. Arguably, it involves less effort and is less costly for a voter to travel to the polling station if he or she lives across the street than if the voter must drive many miles to cast his or her ballot. Research in the spatial–analytic tradition frequently is quantitatively sophisticated and was aided by geography's quantitative revolution during the 1950s and 1960s. It is also notable that research in this tradition continues to be carried out at multiple geographic levels, from urban settings to the global scale.

During the 1970s, some geographers criticized the spatial–analytic perspective for its limited ability to develop theory and its essentially conservative posture. Given the political activism of the period with the ongoing civil rights movement, growing feminist critiques of patriarchal societal structures, and increasing environmental consciousness, this is not surprising. These geographers instead argued in favor of a political–economic perspective for their research that highlighted the role of major institutions in the creation and preservation of demographic and geographic inequities. Drawing from thinkers on the left, such as Karl Marx and Immanuel Wallerstein, this research perspective criticized modern society for these inequities and valued political activism.

During the latter portion of the 1980s, political geographers began to critique the political–economic perspective for varied reasons, including its sometimes grandiose claims of certainty and knowledge. Some geographers began to question whether fully objective research is possible and argued that all research efforts are tainted by the biases of the researcher, the temporal and geographic contexts of their work, and the dominant perspectives of the researcher's discipline. Inspired by influential thinkers such as Jean-François Lyotard, Jean Baudrillard, and Michel Foucault, political geographers adopting the postmodern perspective emphasize the role of political discourse and language in constructing political landscapes. They contend that our understanding of the world is more a function of the language used than of assembled facts of dubious veracity. To some degree, all three of these research perspectives play a role in political geography today, and at times elements of all three can be identified in a single study in one of the five topical research clusters or categories.

Among the five research clusters listed earlier, state spatiality and geopolitics are the oldest. Political geography research classified as falling under studies of state spatiality focuses on the processes of state formation, territoriality, the development of frontiers, boundary creation and disputes, intrastate conflicts, and the internal geographic structures of states. Thus, studies falling under this label may examine federalism, local government organization, the effects of globalization on state structures, and the geography of public finance.

Geopolitics emphasizes competition and conflict between states. It has a long history as an area of emphasis in political geography, dating back at least to Friedrich Ratzel's work during the late 19th century. But clearly the most influential contribution to the development of geopolitics was Sir Halford Mackinder's "heartland theory." Mackinder believed that a geographic heartland, located largely in what today is Russia, would come to be used as a continental fortress to wage war on the rest of the world. Through variable lineages and modifications, this idea and its variations were accepted by many foreign policy specialists and contributed to the development of the cold war, the domino theory, the U.S. foreign aid program, and the policy of containment implemented by the United States during the 1950s. Although geopolitical studies from this perspective are not as plentiful as they once were, they do continue to be conducted by both geographers and nongeographers.

In spite of the reduction in traditional geopolitical studies, this area of focus has experienced a substantial resurgence of interesting and thoughtful research during the past decade. This is particularly the case with critical geopolitics, which rejects its predecessor's state-centric emphasis on global truths in favor an interpretive perspective that emphasizes and critiques the use of language in geopolitical discourse. Critical geopolitics examines current assumptions about power relations among members of the world community of states and sheds the traditional assumption of research neutrality. With the substantial changes in the world political map during the past 15 years or so, the current increase in geopolitical research almost certainly will continue for some time to come.

The third research concentration, the study of the geographies of political and social movements, examines the political geography of collective action. Social and political groups develop and organize on the basis of a number of different issues. Such movements might be focused on racial equality, women's rights, environmental quality, the working conditions of labor, or gay rights. Although historically most social and political movements were localized, during recent decades the scale of their efforts has increased with improved communications. Thus, today there is a worldwide movement emphasizing environmental stewardship and quality.

A substantial emphasis included in this third category of political geography research focuses on the geography of elections. Electoral geography examines the spatial patterns associated with ballots for and against candidates or issues as well as the geographic implications of resulting policy. Thus, an electoral geographer might examine the pattern of votes in congressional elections to determine why different places voted differently and how such decisions affect legislative outputs in Congress. An electoral geographer might also consider the geographic structuring of political behavior in congressional elections through the delineation of congressional districts.

The fourth concentration of current political geography research focuses on places and the politics of identities. This emphasis examines the efforts of population groups to define their differences within a broader society and to have those differences recognized and legitimized. Frequently a group's identity is defined in part by the place in which it is located or a place is employed to buttress the group's identity. Such social groups strive to be accepted as different but respected in the broader society. Typically, groups striving for such recognition emphasize stories, symbols, myths, and rituals to define their contrasting identities. For example, the neo-Confederate movement in the American South is endeavoring to use history, myth, and symbols to reinvigorate a sense of southernness and to argue that the traditions of the Old South are honorable and germane to the region during the 21st century.

The final current area of research emphasis in political geography focuses on geographies of nationalism and ethnic conflict. Simplistically, a nation is an ethnic group that has decided to pursue the creation of its own state. The origins of nationalism date to the emergence of the nation-state during the past few centuries and the associated motivation to delineate state boundaries based on the geographic pattern of nations. The past century clearly has witnessed a number of horrific conflicts growing out of nationalist movements, including World Wars I and II. Nationalism has received substantial attention by social scientists, including political geographers, since the end of the cold war and with the rise in ethnic conflict in the area of the former Soviet Union and Yugoslavia, among others.

Nationalism is commonly employed by ethnic groups for political advantage and to facilitate their desire to exert dominant control over a state. Because most states are multiethnic, such movements frequently involve conflict over efforts to define a nation. The process of creating a nation generally involves the development of an identity emphasizing an ethnic group's history and distinctiveness. The group's goal may then focus on securing territory for a national homeland toward the establishment of a sovereign state. When such efforts are successful, they can create a chain reaction of parallel endeavors by other ethnic groups attempting to secure self-determination in their own states and to insulate themselves from potential violence by their neighbors. Thus, political geographers might focus on how a national identity is created, how conflict with other ethnic groups is used and directed, and how national groups attempt to delineate boundaries around the desired territory. Given the potential for increasing numbers of nationalist movements throughout the world, this category of focus will be active for many years to come.

—*Gerald Webster*

See also Civil Society; Class; Communism; Critical Geopolitics; Decolonization; Democracy; Development Theory; Electoral Geography; Gerrymandering; Globalization; Identity, Geography and; Imaginative Geographies; Justice, Geography of; Labor, Geography of; Law, Geography of; Local State; Marxism, Geography and; Mode of Production; Modernization Theory; Nation-State; Nationalism; Neoliberalism; Orientalism; Political Ecology; Postmodernism; Race and Racism; Radical Geography; Redistricting; Resistance; Underdevelopment; Uneven Development; Urban and Regional Planning; Urban Social Movements; World Economy; World Systems Theory

Suggested Reading

Agnew, J. (Ed.). (1997). *Political geography: A reader.* London: Edward Arnold.

Agnew, J. (2002). *Making political geography.* London: Edward Arnold.

Agnew, J., Mitchell, K., & Toal, G. (Eds.). (2003). *A companion to political geography.* Malden, MA: Blackwell.

Cohen, S. (2003). *Geopolitics of the world system.* Lanham, MD: Rowman & Littlefield.

Cox, K. (2002). *Political geography: Territory, state, and society.* Malden, MA: Blackwell.

Dodds, D. (2000). *Geopolitics in a changing world.* Upper Saddle River, NJ: Prentice Hall.

Flint, C. (1999). Changing times, changing scales: World politics and political geography since 1890. In G. Demko & W. Wood (Eds.), *Reordering the world: Geopolitical perspectives on the 21st century* (2nd ed., pp. 19–39). Boulder, CO: Westview.

Glassner, M., & Fahrer, C. (2004). *Political geography* (3rd ed.). Hoboken, NJ: John Wiley.

Ó Tuathail, G., & Dalby, S. (Eds.). (1998). *Rethinking geopolitics.* London: Routledge.

Ó Tuathail, G., & Shelley, F. (2003). Political geography. In G. Gaile & C. Willcott (Eds.), *Geography in America at the dawn of the 21st century* (pp. 164–184). Oxford, UK: Oxford University Press.

Taylor, P., & Flint, C. (2000). *Political geography: World-economy, nation-state, and locality* (4th ed.). Upper Saddle River, NJ: Prentice Hall.

POPULAR CULTURE, GEOGRAPHY AND

Until the 1980s, geography's engagement with popular culture was relatively thin. While North American cultural geographers described and mapped traditional folk cultures, especially in rural areas, forms of popular culture associated with the mass media were ignored or even disparaged for destroying the local music, crafts, or other practices that made regions and places unique. As geographers have adopted more critical theories of culture and increasingly noted the geographic aspects of many forms of popular culture, however, it has become more common to choose films, music, sports, and various kinds of performances as objects of analysis.

Popular culture often is taken to encompass the material practices and symbolic representations connected to leisure and recreation. Television and radio programs, massive sporting events such as the Olympics, and individual hobbies or activities such as bowling all are examples of popular culture representations and practices that people both consume and produce. Traditionally, popular culture is contrasted with high culture, especially when distinguishing between the arts and literature symbolizing the status of a society's elite classes and the entertainments of the masses. Most current work in human geography, however, uses an anthropological definition of culture and tends to view the practices and symbols associated with popular culture as cultural expressions that may reveal, reproduce, or create aspects of place, space, landscape, and identity.

Critical human geographers have adopted a more political definition of culture that emerged out of the works of Raymond Williams and Stuart Hall, among others. Rather than viewing culture as a fixed attribute of societies, this approach views culture as always emerging out of struggles among groups with diverging interests and unequal access to power. A key term in this definition is *hegemony,* that is, the process by which dominant values and practices are maintained through the cooption and dilution of alternative and often oppositional ways of life. Therefore, critical geographic approaches to popular culture tend to focus on how hegemonic meanings and values are reproduced and/or challenged by particular representations of spaces and identities or by certain spatial practices. The denigration of the Appalachian region and subculture in popular film is an example of this tendency. The perpetuation of the "backward" hillbilly stereotype in *Sergeant York* and *Deliverance* helps urban and suburban Americans to feel better about their own material conditions and lifestyle choices.

Much contemporary geographic research involving popular culture consists of analyses of mass media—especially of the visual media. Photography, film, and television all represent geographies and, therefore, widely disseminate information about places and cultures. Some geographers focus on what they perceive as misrepresentations of places and seek to correct errors found in news stories or even fictional accounts. Other geographers, reacting to critiques of the ability of researchers to objectively know and portray their research subjects that appeared across the social sciences during the 1980s and 1990s, theorize that mass-media images contribute to, or are actually parts of, the production of social spaces and places. Economic development departments in many cities, for example, recruit film production companies to use their streetscapes as movie sets. A successful movie can increase the number of tourists who choose to visit that city. In fact, entire neighborhoods may be altered or preserved for years so that the landscape continues to recall the film in the minds of tourists.

Human geographers also study popular music. Some of this work follows older traditions in cultural geography and maps the hearths and diffusion of particular musical forms. Other research focuses on how music lyrics establish or reproduce particular senses of place. As Arthur Krim noted, the reputation and meaning of Route 66 in American culture, for example, is reproduced every time Nat King Cole's jazz classic is heard on the radio. Finally, John Lovering's study of the global music industry provided insights into how the processes of commodification and globalization interact with local music generation and performance.

There is also an emerging body of work focusing on the bodily practices associated with popular culture. Nigel Thrift uses dance as an example of expressive bodily movements occurring in places that convey thoughts and feelings beyond the reach of spoken or written words. To use an example from tourism geography, places remain tourism destinations only because workers and tourists repeatedly perform certain actions in that space. Such practices include gazing at and photographing a landmark such as the Eiffel Tower, driving a tour bus to designated historic sites, and the seemingly mundane act of donning a bathing suit and lounging on a beach. Thus, play and work, just as much as the representational arts of filming and writing, are pop-cultural performances/practices that constantly reproduce the worlds that people know and experience.

—Stephen Hanna

See also Cultural Geography; Culture; Film, Geography and; Music and Sound, Geography and; Sport, Geography of; Tourism, Geography and/of

Suggested Reading

Aitken, S., & Zonn, L. (Eds.). (1994). *Place, power, situation, and spectacle: A geography of film.* Lanham, MD: Rowman & Littlefield.

Burgess, J., & Gold, J. (1985). *Geography, the media, and popular culture.* New York: St. Martin's.

Crouch, D. (1999). *Leisure/tourism geographies.* London: Routledge.

Krim, A. (1998). "Get your kicks on Route 66!" A song map of postwar migration. *Journal of Cultural Geography, 18,* 46–60.

Lovering, J. (1998). The global music industry: Contradictions in the commodification of the sublime. In A. Leyshon, D. Matless, & G. Revill (Eds.), *The place of music* (pp. 31–56). New York: Guilford.

Thrift, N. (1997). The still point: Resistance, expressive embodiment, and dance. In S. Pile & M. Keith (Eds.), *Geographies of resistance* (pp. 124–151). London: Routledge.

Warren, S. (1993). "This heaven gives me migraines": The problems and promise of landscapes of leisure. In J. Duncan & D. Ley (Eds.), *Place/culture/representation* (pp. 173–186). London: Routledge.

POPULATION, GEOGRAPHY OF

Population geography examines population growth and change and the demographic characteristics of large and small areas on the earth's surface. Although geographers have long been concerned with the study of population and population characteristics, the subfield of population geography traces its origins to the 1953 Association of American Geographers presidential address of Glen Trewartha. Trewartha stated that population never would be adequately covered in the prevailing divisions of physical and cultural geography and that the discipline should be organized around a tripartite structure of population, the physical earth, and cultural landscape. A critical mass of population geographers had emerged by the mid-1960s and closely aligned its interests with spatial demography, logical positivism, and quantitative methods. Early research themes addressed the determinants and consequences of internal migration and residential mobility at the intraurban level using model-oriented and behavioral approaches.

Considerable effort was directed toward understanding the decision to move and toward identifying the forces associated with internal migration flows in the United States and across the world.

Although human movement remains the central theme of population geography, the scale of analysis has shifted from local movements to global flows. This change in scale coincided with the dramatic growth in international migration and a growing recognition that global movement has altered local economies and societies in developing nations, not to mention the social and economic structure of U.S. and European cities. In 2001, more than 140 million people lived outside of their country of birth. These movements resulted from economic globalization, substantial and growing income differences between developed and less developed countries, civil conflicts, natural disasters, and lingering imbalances associated with the fall of the former Soviet Union. Population geographers are concerned with the magnitude and direction of the flows themselves and with their positive and negative consequences for sending and receiving countries. For sending countries, migration tends to relieve pressure from unemployment and to generate substantial remittances. Returning migrants often are agents of modernization. For most receiving countries, immigration provides demographic and economic vitality for aging populations and increases cultural diversity; however, in some places, it strains social, educational, and health services.

Population geographers are particularly interested in refugees, a special class of immigrants defined as those people living outside of their country of nationality and unwilling to return due to a well-founded fear of persecution. Recent research reveals the difficulty of separating refugees from labor force migrants because many refugees are, in fact, motivated by forces similar to those that influence other migrants—regional disparities in income and welfare, the presence of relatives who provide much-needed information and support for new immigrants, and the weakening of traditional values and social ties in the face of modernization. Economic factors often are as important as political violence in understanding the causes of refugee flows. The Eritrean refugee crisis of the 1980s was, for example, a response to agrarian transformation in Sudan.

Transnationalism is a process whereby the everyday lives of international migrants transcend national boundaries. Population geographers are interested in the nature of transnational communities and networks, remittances sent to countries of origin and their effects on the economic and social structures of these countries, and identity formation among people who lead transnational lives. Interest in the latter has led population geographers to supplement the traditional quantitative methods of spatial demography with more qualitative approaches to study human movement. Ethnography deemphasizes the event of migration itself and stresses the larger social world in which it occurs. In this respect, ethnography balances attention to the everyday detail of an individual's life with wider social structures. Allison Mountz and Richard Wright, in a fascinating use of ethnographic techniques to study interconnections between Mexican workers in Poughkeepsie, New York, and family and friends in the rural Mexican community of San Agustin, Oaxaca, found that the mainly male migrants in Poughkeepsie interacted with their wives, children, and families in Mexico on a frequent basis and returned for fiestas, funerals, and other village events; provided remittances; and sent and received messages about their daily lives in the transnational community.

Population geographers, in collaboration with economic, social, and cultural geographers, study immigrant communities, including the nature of their social networks, the success of immigrant businesses, the role of ethnic concentration in the adjustment process of new immigrants, and the way in which ethnic concentration has transformed the cultural landscapes. As mass culture erodes many of the unique features of the North American regional geography, immigrant communities help preserve the color and maintain the distinctiveness of the American cultural landscape. In the suburbs of northern Virginia, Vietnamese place making takes the form of retail clusters while Vietnamese immigrants live in a dispersed manner. In New Orleans, backyard gardens and levee plots perpetuate traditional Vietnamese ways of life, especially among the elderly. The bustling street life in Russian and Ukrainian neighborhoods of Brighton Beach in southern Brooklyn, New York, along with signage, toponyms (place names), and cultural symbols, links these immigrant neighborhoods to their home countries.

Early research on the geography of age involved comparing the age structures of mature and stable populations in more developed societies with those of rapidly growing less developed countries. More recently, attention has focused on the "baby boom" generation and the effects of aging populations in North America, Europe, and Japan. Considerable evidence supports the fact that internal migration in the United States is linked to the aging of generations of varying sizes. The "Easterlin effect" states that individuals from large cohorts (people born at the same point in time) face greater competition for jobs and housing and are less likely to migrate than individuals from smaller cohorts. In addition, people from small cohorts tend to move earlier in their lives than do people from larger cohorts. Sunbelt migration during the 1970s coincided with a period of stagnant job growth in the Midwest and the Northeast. Young adults just entering their high-mobility phase faced a crowded tight labor market and relocated to Sunbelt regions where job opportunities were plentiful. Sunbelt migration was reinforced by large numbers of people 60 to 70 years of age, beginning around 1960 and extending until 1985. Elderly migrants have been attracted to high-amenity regions not only in the United States but also in Europe and Japan.

The spatially focused nature of elderly and labor force migration and regionally varying life expectancies has generated a fair amount of regional demographic diversity in terms of age—with profound implications for healthcare, social security and pensions, the provision of local services, and potential for future growth. In 2000, the percentage of the population over 65 years of age varied from 5.7% in Alaska to 17.6% in Florida. In Canada, local population aging is an economic disadvantage because communities and provinces with limited resources often shoulder a disproportionate burden due to the growth of their elderly populations.

The subject matter and approach of population geography has broadened recently, due in part to the work of British geographers Paul White and Peter Jackson, who criticized traditional population geography for its overemphasis on population events at the expense of the longer biographical history and the wider political economy, preoccupation with data at the expense of wider social theory, blind acceptance of official categories without probing their social meaning, an unhealthy attachment to essentialist categories such as gender and age, and reluctance to delve into the larger social world in which population processes take place. In response, population geography has begun to incorporate social theory and has adopted new approaches to the study of migration. Social theory has led to greater emphasis on gender and race and on the roles of partriarchy and racial discrimination in structuring migration, immigrant adjustment, fertility, aging, and other demographic processes. Ethnographic methods are used to explore human movement as an outgrowth of personal life histories, human aspirations, and future expectations. Humanistic methods use the personal narrative to reveal the struggles of ordinary people and the social relations around which their migration is organized. The new diversity of methods is reflected in the growing number of "gender and migration" studies that view human movement in the context of the household's division of labor, patriarchical decision structures, and socially determined, gender-segregated labor markets. Use of these methods has led to a deeper understanding of the meaning of migration for individuals and for the political and social systems within which they function.

The use of mixed methods and the growing interest in the gendered nature of the migration process has revealed the way in which women's social roles structure their movement behavior. A series of quantitative studies documented that married women most often are affected negatively in their earning power by a family move. The so-called "wife's sacrifice" exists even when the husband and wife have similar levels of education and income-earning potential. Qualitative studies have revealed that social and family responsibilities are more important than job and income considerations in determining the length of stay for Puerto Rican women migrants. The study of marriage migrants in China has shown that peasant women move not only to get married but also to improve their economic and social position by living in a more economically advantaged region. A gendered perspective on migration has unlocked new insights into why women move and what their moves mean to them, to their families, and to the societies in which they live.

There is a long tradition of public policy research in population geography, dating from early work on legislative redistricting. The act of drawing boundaries for voting purposes is, by its very nature, an exercise in applied population geography. The Voting Rights Act of 1965 prevents the drawing of legislative boundaries in a way that weakens the voting strength of a legally protected group. This straightforward mandate is subject to many different interpretations based on the demographic circumstances of a particular community, including the age and citizenship characteristics of minority group members, socioeconomic status and

propensity to vote, and geographic concentration of a particular group. Population geographers have also been active in evaluating whether local school districts have manipulated boundaries to maintain or increase racial segregation and the relative importance of demographic change versus changing boundaries on the racial composition of schools. Much of this work exists in informal consulting reports rather than in scholarly books and journals.

The great challenge for the field of population geography today is to link its core interests in migration and spatial demography to larger trends in the field of human geography. Engagement with social theory has led the field to delve into the deeper social, economic, and political forces that generate migration and regional demographic variability and to articulate the social significance of contemporary migration and demographic variability. In addition, innovations in geographic information sciences offer myriad opportunities for population geographers to represent and visualize demographic processes in new and innovative ways and to integrate them with ongoing environmental and economic change.

—*Pat Gober*

See also Behavioral Geography; Census Tracts; Children, Geography of; Demographic Transition; Fertility Rates; Geodemographics; Labor, Geography of; Migration; Mortality Rates; Population Pyramid; Social Geography; Space, Human Geography and; Urban Geography; Urbanization

Suggested Reading

Bascomb, J. (1993). The peasant economy of refugee resettlement in eastern Sudan. *Annals of the Association of American Geographers, 83,* 320–346.

Fan, C., & Huang, Y. (1998). Waves of rural brides: Female marriage migration in China. *Annals of the Association of American Geographers, 88,* 227–251.

Gober, P., & Tyner, J. (2004). Population geography. In G. L. Gaile & C. J. Willmott (Eds.), *Geography in America at the dawn of the 21st century* (pp. 185–199). New York: Oxford University Press.

Mountz, A., & Wright, R. (1996). Daily life in the transnational migrant community of San Agustin, Oaxaca, and Poughkeepsie, New York. *Diaspora, 5,* 403–425.

Trewartha, G. (1953). A case for population geography. *Annals of the Association of American Geographers, 43,* 71–97.

Wood, J. (1997). Vietnamese place making in northern Virginia. *The Geographical Review, 87,* 58–72.

POPULATION PYRAMID

Except for total size, the most important demographic feature of a population is its age–sex structure. The age–sex structure affects the needs of a population as well as the supply of labor; therefore, it has significant policy implications. A rapidly growing population implies a high proportion of young people under the working age. A youthful population also puts a burden on the education system. When this cohort enters the working ages, a rapid increase in jobs is needed to accommodate it. In contrast, countries with a large proportion of older people must develop retirement systems and medical facilities to serve them. Therefore, as a population ages, its needs change from those of schools to jobs to medical care.

The age–sex structure of a country typically is summarized or described through the use of population pyramids. They are divided into 5-year age groups, with the base representing the youngest group and the apex representing the oldest group. Population pyramids show the distribution of males and females of different age groups as percentages of the total population. The shape of a pyramid reflects long-term trends in fertility and mortality as well as short-term effects of "baby booms," migrations, wars, and epidemics. It also reflects the potential for future population growth or decline.

Two basic representative types of pyramid may be distinguished (Figure 1). One is the squat triangular profile. It has a broad base, concave sides, and a narrow tip. It is characteristic of developing countries having high crude birth rates, a young average age, and relatively few elderly. Natural growth rates in such societies tend to be high.

In contrast, the pyramid for economically developed countries, including the United States, describes a slowly growing population. Its shape is the result of

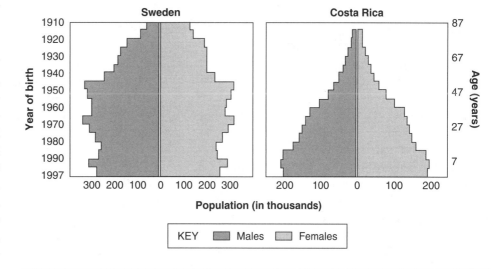

Figure 1   Population Pyramids for Typical Developed and Underdeveloped Countries

SOURCE: Stutz, F. and B. Warf. (2005). *The World Economy: Resources, Location, Trade, and Development*, 4th edition. Upper Saddle River, NJ: Pearson/Prentice Hall. Reprinted with permission of the publisher.

a history of declining fertility and mortality rates, augmented by substantial immigration. With lower fertility, fewer people have entered the base of the pyramid; with lower mortality, a greater percentage of the "births" have survived until old age. In short, the structure of the population pyramid closely reflects the stage of the demographic transition in which a country is positioned.

—*Barney Warf*

See also Demographic Transition; Population Geography

Suggested Reading

Yaukey, D., & Anderton, D. (2001). *Demography: The study of human population* (2nd ed.). Prospect Heights, IL: Waveland.

POSTCOLONIALISM

Postcolonialism, often mistaken as a simple reference to the historical period following the end of European colonialism (i.e., that which happened *after* colonialism), is actually a research tradition, an approach, or even a paradigm in the social sciences and human geography that is concerned with the multiple impacts of colonialism as a cultural, economic, and political practice. First, it examines how, during colonial times, certain types of knowledge were produced to distinguish the colonizers from the colonized, that is, how a body of supposedly objective scientific data was collected to distinguish the colonizers from the colonized. Most important, this distinction always was a hierarchical one, placing the colonizing European powers above their colonized subjects. Second, postcolonialist work studies how this assigning of characteristics and values to entire civilizations meant that their identities, and the representations of their identities in the media (e.g., newspapers, school textbooks, exhibits, popular science literature), were shaped in ways that fit the interests of the colonizing power in establishing long-lasting negative stereotypes of colonized peoples. Third, postcolonial studies show how scientists, writers, and geographers were actively involved in the process of establishing, maintaining, and (later) defending colonial and postcolonial power relations.

As an account of knowledge production, postcolonialism is a critique of scientific, and thus also geographic, practices that first emerged during colonial times. In other words, it serves as a methodological critique that looks at the role of the geographer in the field and his or her relationship to the subjects of research. Postcolonial work especially targets the predominance of social Darwinist research during the colonial period and examines its role in naturalizing and justifying the exploitation of certain people on the basis of their race, gender, or sexuality. For example. at the beginning of the 20th century, the American Museum of Natural History in New York had live exhibits of Inuit from Greenland to depict the racial inferiority of what the museum referred to as a species. In London, a Hottentot woman by the name of Sara Baarthman was displayed and studied (even after her death) due to the large size of her buttocks to illustrate the so-called racial abnormalities of African people. In contrast to such cases where colonizing powers try to reinforce certain stereotypes of others with the help of scientific studies, even the formerly colonized have used stereotyping of former colonizers to support their agenda. In Zimbabwe, for example, President Robert Mugabe has sought to restrict the citizenship rights of gays and lesbians, arguing that such expressions of alternative sexual identity run counter to native traditional cultures in Africa. Thus, a former colony has acted here against what it perceives as the permissive attitudes fostered by colonial powers and

uses the stereotype of the colonizer to reinforce a representation of its own original and native culture. All of these examples show how scientific work by the colonizers and the colonized has been used to establish notions of the ostensible normality of the former versus the ostensible abnormality and inferiority of the latter, and geographers have made significant contributions to uncovering these representations.

Such stereotyping based on scientific work and exploration has been further supported by the media of colonial powers that turned such work into popular accounts of differences between places and regions of the world. Beyond individual groups that were singled out to outline the cultural and racial differences between colonists in power and their colonial subjects, popular media of the colonial period perpetuated stereotypes of entire continents and civilizations and ingrained them in colonizing populations. Perhaps the best work illustrating this process is Edward Said's book *Orientalism,* published in 1978, which analyzed how the simple notion of the Orient was transformed by Western civilization and its colonizers into a system of representations that was abound with raced, gendered, and sexual stereotypes. Said showed how Oriental men were portrayed as effeminate yet dangerous, how Oriental women were portrayed as submissive and exotic, and how the actual diversity of vast Middle Eastern and Asian cultural landscapes was subsumed under such limited stereotypical representations. Most important, Said argued that such stereotypes of the Orient did not cease to exist after the end of British colonial presence in the so-called Orient during the 20th century; rather, the perpetuation of Orientalist thinking continues to this day in the ways in which we talk about and construct the Middle East, its inhabitants, and their cultural traits. His work was distinctly political in that it called for a critical examination of all representations of the Orient or the Middle East and the methods and means that are used today to perpetuate such representations of modern power relations.

Finally, postcolonialism functions as a critique of the role of the sciences and the media in general, and geography in particular, in the manifestations of colonial power relations. It examines how geographers and their work establish and manifest colonial power relations through travel, geographic fieldwork, expeditions, and mapmaking. Literary accounts of colonial relations (Joseph Conrad's *Heart of Darkness*), as well as school textbooks and maps, sought to inscribe the superiority of the colonizers by attributing positive character traits to them. For example, early-20th-century literature on the global distribution of races and their characteristics frequently referred to the British, French, and Germans as muscular, strong, intelligent, dominant, and quick in their responses. In contrast, South American and African tribes might be described as emaciated, weak, slow, and retarded. In addition, colonial maps often used regional attributes that portrayed colonized territories as less advantageous for commerce, unhealthy, or even mysterious, dark, and unexplored. Such attitudes, as portrayed by geographers, were internalized by the general population, whose members continued to support their colonizing government and its practices of exploitation and expansion.

—*Olaf Kuhlke*

See also Colonialism; Cultural Geography; Decolonization; Ethnocentrism; Eurocentrism; Identity, Geography and; Imaginative Geographies; Imperialism; Orientalism; Other/ Otherness; Spaces of Representation; Underdevelopment

Suggested Reading

Blunt, A., & McEwan, C. (Eds.). (2002). *Postcolonial geographies.* New York: Continuum.

Said, E. (1978). *Orientalism.* New York: Pantheon.

Spivak, G. (1999). *A critique of post-colonial reason: Toward a history of the vanishing present.* Cambridge, MA: Harvard University Press.

POSTINDUSTRIAL SOCIETY

First popularized during the 1970s, especially by famed sociologist Daniel Bell and futurist Alvin Toffler, the term *postindustrial* has come to include a loose group of views about the social and spatial structure of advanced capitalism. This perspective became increasingly popular in the wake of the sustained deindustrialization that Europe and North America suffered during the late 20th century. Highly optimistic in nature, it viewed postindustrialism as a natural stage of capitalism in an evolutionary process from agrarian poverty to worldwide cosmopolitanism.

Essentially, the postindustrial society thesis maintained that manufacturing created one society, with a corresponding landscape, and that a services-based society would create a qualitatively different society and corresponding geography. This view largely equated services with information-processing activities, focusing on occupations of skilled, well-educated professionals (producer services) such as clerical activity, executive decision making, telecommunications, and the media. Such a view heralded information processing as a qualitatively new form of economic activity; thus, services were held to represent a historically new form of capitalism. Postindustrialism held that the growth of services signaled a change from a world of work in which people used their bodies to one in which they used their minds. It maintained that the evolution of societies from those dominated by blue-collar forms of work into cleaner, white-collar ones would unleash massive rounds of productivity growth that effectively would put an end to scarcity and hence to poverty and its related social ills. This transformation allegedly would allow for a greater focus on the quality of life, including matters concerning equity rather than efficiency, that is, human needs and social equality rather than simple efficiency and productivity.

Geographically, the postindustrial thesis maintained that the shift from a manufacturing-based economy to a services-based economy entailed a reconfiguration of spatial relations. In particular, this view upheld the central role played by telecommunications that would allow intangibles such as services to be widely distributed via an "electronic cottage." Thus, the postindustrial argument anticipated the Internet by three decades. However, in assuming that all services could be produced, transmitted, and relayed in this manner, postindustrialists exaggerated the argument, holding that the new, dispersed, polynucleated landscapes of electronic cottage workers would obviate the need for commuting, rendering large cities effectively obsolete. Such a view naively assumed that telecommunications only promote the decentralization of activity rather than more complex patterns of simultaneous concentration and deconcentration.

The postindustrial view suffered from several severe analytical flaws. Although many service jobs do involve the collection, processing, and transmission of large quantities of data, many others do not; for example, the trash collector, restaurant chef, security guard, and janitor all work in the service sector, but the degree to which these activities center around information processing is minimal. Indeed, in contrast to early, overly optimistic, postindustrial expectations that a service-based economy would eliminate poverty, a large share of new service jobs pay poorly, offer few benefits, and are part-time or temporary in duration, leading to widespread concerns about the "McDonaldization" or "Kmartization" of the economy. Finally, as the geography of producer services over the past four decades has shown, many advanced services centralize in large cities due to the agglomeration economies available there rather than decentralize to the rural periphery.

—*Barney Warf*

See also Producer Services

Suggested Reading

Bell, D. (1973). *The coming of post-industrial society: A venture in social forecasting.* New York: Basic Books.
Toffler, A. (1970). *Future shock.* New York: Random House.
Webster, F. (2002). *Theories of the information society.* London: Routledge.

POSTMODERNISM

Postmodernism has been defined in diverse and sometimes confusing ways since it first became common in the lexicon of human geographers during the mid to late 1980s. This confusion arose because postmodernism has two closely related definitions: one as *object* and the other as *attitude*. First, postmodernism can be understood as an object or a thing, and in particular this object can be seen as an era. This era is defined by things such as literature, art, and architecture and by processes such as differing forms of capitalist production that result in the context of postmodern thought. Second, postmodernism can be understood as an attitude or a way of understanding the world. In particular, the things and processes that characterize the era of postmodernism both reinforce and remake postmodernism as an attitude. This attitude can perhaps be understood more specifically as an intellectual movement that provides a coherent set of ideas for understanding the world in a particularly postmodern way. In an attempt to capture some of the complexity of the interrelationships between postmodernism as object and postmodernism as attitude, geographers such as Michael Dear have suggested that we further distinguish among postmodernism as epoch, postmodernism as style, and postmodernism as method. Style and epoch help us to understand the notion of postmodernism as an object, whereas method helps us to understand postmodernism as an attitude.

Postmodern style can be understood as a thing to be analyzed, and such analyses suggest that there has been a change in both the style and epoch of contemporary Western societies. Whereas such societies once reflected the somber and univocal face of modernity—especially in the faceless architecture of the international modern style—they now exhibit the more diverse and polyvocal face of postmodern style. In architecture, this has seen the development of more playful buildings such as the sinuous glass, limestone, and titanium walls of the Museo Guggenheim in Bilbao, Spain. Such buildings often are designed specifically as statements of opposition to the utilitarian, monolithic, and (often) ubiquitous skyscrapers of the international modern style. In this sense, postmodern style can be seen as a recognizable literary, artistic, and cultural trend reacting against modernism.

Some researchers consider postmodern style to be part of a larger set of processes that characterize a postmodern epoch. The key set of relationships on which these researchers focus involves social and economic relationships that apparently have undergone a major shift. Capitalist production and accumulation, for example, once were characterized by mass production and scientific management (as exemplified by the Fordist mass production of the automobile), but they are now said to have shifted to a mode of "just-in-time" or flexible production systems.

As a method or an intellectual attitude, postmodernism is characterized by a reaction against the so-called certainty of narratives of progress and Enlightenment that characterize the modernist foundations of Western intellectual activity. In a postmodern method, former boundaries between phenomena such as the economic and the social, art and science, and myth and reality—distinctions so important to modernism—collapse into one another. Postmodern thinkers tend to be wary of what are termed *metanarratives* (or overarching theories), such as secular humanism, historical materialism, and scientific rationalism, which attempt to provide unified and singular explanations for social relations. In fact, one of the key postmodern thinkers, Jean-François Lyotard, defined postmodernism simply as incredulity toward metanarratives.

There has been intense debate about both the existence of postmodernism as an object and the veracity of postmodernism as a method. Moreover, there are those who accept the existence of postmodernism as a condition (style and epoch) but reject it as a method for understanding the world. Interestingly, there has been a gendered geography to such debates, with the most intense struggles over postmodernism occurring between men in places such as the United Kingdom and the United States—places where modernism was perhaps most entrenched and where modernist men may have had much to lose in a shift to postmodernism. While not discounting the strength of feelings that underpinned debates over modernism and postmodernism in other places, and while not wishing to flatten the so-called Third World into a homogeneous mass, there has been a tendency for women and people of color from places outside of the United Kingdom and the United States to adapt and adopt postmodern thinking more readily. Much of this has given us a better understanding of the partiality of knowledge and the situated character of intellectual movements. This has especially been the case when such groups have connected postmodern thinking to feminism, postcolonialism, and poststructuralism and particularly to understanding struggles over identity and place.

Debates over postmodernism today are largely absent in geography because many geographers subsequently have moved on to other intellectual struggles. Nevertheless, it would be fair to suggest that many have—consciously or unconsciously—been affected greatly by postmodernism as a method, an epoch, and a style.

—*Lawrence Berg*

See also Epistemology; Flexible Production; Modernity; Paradigm; Postcolonialism; Poststructuralism; Situated Knowledge

Suggested Reading

Berg, L. (1993). Between modernism and postmodernism. *Progress in Human Geography, 17,* 490–507.

Dear, M. (1988). The postmodern challenge: Reconstructing human geography. *Transactions of the Institute of British Geographers, 13,* 262–274.

Dear, M., & Flusty, S. (1998). Postmodern urbanism. *Annals of the Association of American Geographers, 88,* 50–72.

Graham, J. (1988). Post-modernism and Marxism. *Antipode, 20,* 60–65.

Haraway, D. (1988). Situated knowledges: The science question in feminism and the privilege of partial perspective. *Feminist Studies, 14,* 575–599.

Harvey, D. (1989). *The condition of postmodernity.* Oxford, UK: Blackwell.

Jones, J., III, Natter, W., & Schatzki, T. (1993). *Postmodern contentions: Epochs, politics, space.* New York: Guilford.

Lyotard, J.-F. (1984). *The postmodern condition: A report on knowledge* (G. Bennington & B. Massumi, Trans.). Minneapolis: University of Minnesota Press. (Original work published 1979 [in French])

Marden, P. (1992). The deconstructionist tendencies of postmodern geographies: A compelling logic? *Progress in Human Geography, 16,* 41–57.

Scott, J., & Simpson-Housley, P. (1989). Relativizing the relativizers: On the postmodern challenge to human geography. *Transactions of the Institute of British Geographers, 14,* 231–236.

Warf, B. (1993). Post-modernism and the localities debate: Ontological questions and epistemological implications. *Tijdschrift voor Economische en Sociale Geografie, 84,* 162–168.

POSTSTRUCTURALISM

During the late 1980s and early 1990s, geographers, following the broader trends occurring in social theory, called into question the theoretical suppositions of the major paradigms of spatial science, humanism, structuralism, and realism in geography. Geographers, particularly social and cultural geographers, became concerned with the broader "crisis of representation" in the academy, which suggested that all representations occur within the context of power relations and that those relations artificially construct boundaries around geography's objects of analysis. Thus, poststructuralist geographers questioned the boundaries that separated key concepts in the field, such as objectivity and subjectivity, nature and culture, and authentic and inauthentic, refusing to privilege either side of these binaries. These critiques appealed to geographers interested in a variety of geographic areas of inquiry—both "traditional," such as economic and political geography, and "nontraditional," such as sexuality and body space geographies. In many ways, poststructuralist geographers have destabilized subdisciplinary boundaries and called into question the value of a purely "cultural" or "social" geography.

Poststructuralism emerged during the 1960s and 1970s with the work of Jacques Derrida, Michel Foucault, and others who became increasingly critical of structuralism's assumptions about the underlying mechanisms that structured social life. These scholars argued that the social world, which is mediated through language, has no essential set of characteristics. Instead, social categories are historically contingent and their meanings ebb and flow over time and space. The binary oppositions on which Western thought has also been pinned, such as object/subject and nature/culture, are arbitrary; more important, the privileging of one side of the binary over the other is made possible only through the deployment of social power. This means that Western theorizations of object/subject relationships are based on a principle of either/or; that is, it is either objective or subjective. In this theorization, one side of the binary is also privileged over the other; objectivity becomes the foundational center, and subjectivity becomes the margin. Derrida argued, however, that these binaries are coconstitutive of each other. Instead of thinking of objectivity and subjectivity as an either/or proposition, it is better to think of these as operating in a both/and relationship. Simply put, objectivity cannot exist without its "other," that is, subjectivity. So, the subjective is always part and parcel of any objective inquiry. It is only through the deployment of power that the objective and subjective are torn asunder and assigned separate meanings.

Poststructuralist theorists suggest that any set of binary oppositions always is in tension and can be deconstructed, interrogating how the hierarchies within the binaries came to be constructed as real, natural, and fixed in the first place. Resting at the heart of deconstruction, a preferred methodology for poststructuralists, is an interest not in a singular "truth" but rather in how truths, and the knowledges on which these truths are based, are socially constituted. In deconstruction, the object of analysis is the binary and how one side of the binary becomes the center while the other is maintained in the margin. Thus, excavating the relationship between center and margin opens up the possibility of destabilizing the artificiality of these binaries. Poststructuralists, such as Michel Foucault, have interrogated how particular locations become important sites for the organization and generation of social and spatial meaning. In Foucauldian terms, the clinic, as a location, privileges an objective rational science of medicine and gives hegemonic status to biomedical practitioners who define "health" and "illness." This is possible as societies become interested in socially and spatially isolating mental illness in sanitariums, which is considered marginal to a rational economic actor in 19th-century Europe. This

is done because people who are deemed mentally ill cannot function in an emerging capitalist society and therefore have no use value or exchange value.

Poststructuralism's import into geography, and particularly with the reading of the work of Hélène Cixous, Jacques Derrida, and Michel Foucault in the fields of cultural studies and literary theory, was partially precipitated by a growing concern with postmodern geographies, which appeared to lack theoretical or methodological rigor. In particular, poststructuralist geographers eschewed postmodernism's nihilism and its focus on difference for difference's sake. These geographers offered an antiessentialist reading of geography's objects, but instead of taking this to the extremes of postmodernism, they followed the lead of feminists and postcolonial theorists who had an interest in operationalizing how power relations bind and naturalize particular identity positions and spaces. In this formulation, the focus is not on the essential and authentic characteristics of space and identity; instead, it is on how the plays of power fix and make authentic particular spaces and identities. Place is not real; rather, it is socially constructed through a web of discourses that make it appear to be real. Taking a social constructionist view, geographers began to interrogate how the organization of various spaces produce and reproduce discourses of power such as heterosexism.

In the early theorizations of poststructuralism in geography, scholars also became interested in how various epistemologies (or how we know the world) structure our ontologies (or what the world is like in the first place). This theoretical move inverts the structuralist paradigm, which places its emphasis on ontological structures. No longer desiring to unpack the ontological pregivens of society, such as the underlying mechanisms that create various events, poststructuralist geographers instead focus on how spaces emerge and are structured by how we know and think about them. First, a poststructuralist epistemology provides an arena from which it is possible to critique other geographic modes of inquiry. A poststructuralist critique suggests that the grid epistemology of spatial science structures the world from a distance, presuming the rationality of all subjects and spaces. In "discovering" spatial laws, however, spatial scientists construct the rationalities and spaces they seek to explain. Spatial scientists create the real through the representation of their own authorial vision of how the world should and does operate; they literally construct their own ontological presuppositions.

Second, poststructuralist geography offers more than simply a critique of other geographic paradigms; it also suggests new ways of "doing" geography. Poststructuralist geographers are particularly interested in deconstructing spatial discourses, language, and the textuality of everyday life. With a new focus on textuality (and intertextuality, i.e., the relations between texts), geographers now examine traditional concepts such as the cultural landscape with new epistemological assumptions. Historically, cultural geographers, working from the tradition of Carl Sauer, theorized the landscape as a series of objects that were relics of culture itself. Shifting away from Sauerian approaches to culture, geographers during the 1970s and 1980s offered analyses of the cultural landscape as a place of humanistic meaning and as an object for the sedimentation of the structures of power such as capitalism. During the late 1980s and 1990s, poststructuralists began to retheorize landscapes as texts, that is, composites of multiple objects that are invested with power and meaning but with no essential quality tied to any one particular process such as capitalism, patriarchy, or racism. As such, landscapes are not isolated objects in space but are themselves constitutive of the flows of sociospatial relations emanating from other landscapes and texts; they are overdetermined by the constitutive nature of processes such as capitalism, patriarchy, and racism. Put simply, landscapes are partial representations of meaning and can be "read" only in relation to other texts and contexts, which always are brought to any reading of the landscape as part of the research process. Thus, geographers also need to be self-reflexive and considerate of how their own subject positions mediate their analyses of the world in which they are embedded.

The importance of texts and textuality in geographic inquiry also means that, more broadly, poststructuralist geographers are interested in representations and representational practices. Representations are not ephemeral and immaterial; rather, they are part and parcel of the spaces they claim to represent. Maps, for example, are not neutral objects; they are invested with meaning and are also linked intertextually to other texts and spaces. When someone makes or reads a map, he or she draws from other discourses, practices, and spaces to interpret and use that map. Moving away from a mimetic reading of maps, it is possible to think of maps as spaces themselves, partially representing spaces but also investing spaces with meaning. In this theorization, space is both a representation and a representational process. Geographers interested in the representational aspects of space might interrogate the ways in which meanings are invested in various spaces through the production of other representations such as films, television programs, fiction and nonfiction books, music, and the myriad number of images that are a part of everyday landscapes (e.g., billboards).

Space is not a neutral backdrop to social relations; rather, space is implicated in the production and reproduction of those relations through the ways in which it is both invested with social meaning and constructive of that meaning. Social identities are also spatial ones. Identity is intertextuality linked to various spaces, such as the ghetto and the pub, although the practices of identity—what people do—illustrate that representations and spaces never can fully capture social identities and spaces. So, whereas it might behoove someone in a position of power to invest the ghetto with an identity linked to violence and poverty, others might practice the ghetto as an emancipatory space for developing an alternative identity position and culture that is affirming of difference. Using a deconstructionist methodology, geographers are interested in how particular identities might become sutured to particular spaces and what implications this might have for politics.

At the same time that geographers have focused on constructing an epistemological reading of geography's binaries and on deconstructing the ways in which representations and identities are spatially constituted, some poststructuralist geographers remain interested in theorizing a poststructuralist spatial ontology. Edward Soja's theorization of "thirdspace" and Gillian Rose's conceptualization of "paradoxical space" suggest that the world is structured through a both/and ontology; thus, space always is operating through numerous oppositional moments and tensions that constitute a hybrid set of spaces and identities. In Soja's case, this supposes the operation of a thirdspace that is neither material nor symbolic, functioning beyond binary thinking. Working in particular through the philosophies of Gilles Deleuze and Félix Guattari, Marcus Doel and other geographers have become interested in an ontology of rhizomatic (horizontal) materiality in opposition to a structuralist ontology that digs deep into the arboreal (vertical) relations between mechanisms and events to uncover the true workings of the social and the spatial. In the context of a rhizomatic ontology, the world operates through a series of sociospatial flows. These flows come together temporarily in various spatial folds such as those in which scales, points, and lines temporarily collapse or are, more likely, permanently fractured and in the process of becoming (in contradistinction to a fixed, clearly demarcated bounded space with measurable outcomes of an essential existence as in spatial science).

In conclusion, resting at the heart of most poststructuralist geography is the assumption that spatial discourses operate to construct the world we see and study. Those discourses are constitutive, operating in relation to other discourses and in contexts that demand historical excavation. Paralleling structuralism, poststructuralist geographers remain interested in language, discourse, and representation. Departing from structuralists, however, poststructuralists find interest in identity politics and the ways in which identity and space are constituted intertextually. Methodologically, poststructuralist geographers have been interested in using deconstruction as a way to pry open power relations and understand the operations of those relations in constituting particular spaces, representations, and identities. Finally, poststructuralist geographers recently have become interested in retheorizing spatial ontology and considering how we might better conceptualize how spatiality operates in the construction of knowledge.

—*Vincent Del Casino*

See also Epistemology; Feminisms; Ontology; Postmodernism; Spaces of Representation; Spatiality; Structuralism; Theory

Suggested Reading

Dixon, D., & Jones, J., III. (2004). Poststructuralism. In J. Duncan, N. Johnson, & R. Schein (Eds.), *A companion to cultural geography* (pp. 79–107). Malden, MA, & Oxford, UK: Blackwell.

Doel, M. (1999). *Poststructuralist geographies: The diabolical art of spatial science.* Lanham, MD: Rowman & Littlefield.

Gibson-Graham, J. (1996). *The end of capitalism (as we knew it): A feminist critique of political economy.* Oxford, UK: Blackwell.

Massey, D. (1994). *Space, place, and gender.* Minneapolis: University of Minnesota Press.

Rose, G. (1993). *Feminism and geography.* Cambridge, UK: Polity.

Shurmer-Smith, P. (2002). Poststructuralist cultural geography. In P. Shurmer-Smith (Ed.), *Doing cultural geography* (pp. 41–52). London: Sage.

Soja, E. (1996). *Thirdspace: Journeys to Los Angeles and other real-and-imagined places.* Oxford, UK: Blackwell.

POVERTY

Poverty is a social and economic condition in which access to goods and services lies below a given threshold. Often this threshold is defined according to an individual's or a household's monetary income. Unemployment, low wages, low educational attainment, high housing costs, high healthcare costs, and discrimination all can contribute to poverty. Poverty has an inverse relationship to quality of life, especially when people living in poverty live within close proximity to one another. Many governments establish an official poverty line to determine the income level at which individuals and households can be officially recognized as poor. This gives governments the ability to identify an official poverty rate, measure how much of a country's population is living in poverty at any given time, distribute welfare benefits, and locate the spatial dimensions of poverty.

Geographers use governmental statistics and other data to reveal the spatial characteristics of poverty at various scales. At the global scale, geographers and others note a general divide between the global North and the global South, with wealthier populations living in the former and poorer populations living in the latter. Spatial patterns of poverty also exist within countries. For example, in the United States, poverty is concentrated in the Appalachian Mountains, in the southeastern states, among Native American reservations, and in core sections of large cities. In China, a stark divide between urban and rural residents exposes the spatial distribution of poverty. Geographers have contributed to trends about poverty, including the feminization of poverty and the geographic unevenness of antipoverty legislation.

Poverty has been part of geographic study since the early 1900s. Social scientists research the spatial dynamics of poverty in three main ways. First, they direct attention to the location and experiences of poor people, often within an urban setting. Second, social scientists examine the social and economic conditions that contribute to poverty's spatial form. Third, they focus on the meaning of poverty from a variety of perspectives to critique any singular definition of poverty and its manifestation through space.

During the early part of the 20th century, sociologists at the University of Chicago began studying urban social relations through the lenses of race, ethnicity, and class. Led by Robert Park, the Chicago

School of Sociology developed a set of theories about the city known as human ecology. Using Chicago as the primary case study, these models of the city's human ecology divide urban space into several distinct components that work in conjunction with one another to form the cohesive unit of a city's community and economy. Such studies partition urban space into districts, such as the central business district and ethnic enclaves, based on their perceived primary social and economic function. Each of these parts was held to operate as a relatively autonomous unit in the city's overall social and economic ecology. Such models reveal the social and economic segregation of urban space, whereby poorer residents live in close proximity to one another and wealthier residents live together in different places from where the poor can be found. Although these divisions also point to racial and ethnic segregation within urban space, the racial and ethnic divisions often correlate to the location of poverty because certain racial or ethnic minorities had not been incorporated into the city's economy.

Not only did the Chicago School map poverty within the city, but the school's models of urban society also guided research projects into the lives of poor people. Out of these human ecologies, they produced ethnographies of the urban poor, such as hoboes and immigrants working in factories, to provide detailed descriptions of the social and economic life of urban space and its residents. Contemporary versions of human ecological studies have incorporated the tools of quantitative analysis and spatial science to give greater detail across wider populations of the location and demographic characteristics of poverty.

The Chicago School's human ecology dominated the geographic study of poverty for much of the 20th century and continues to be a prominent field of study (particularly among sociologists interested in urban ghettos). Although the Chicago School made important strides in mapping the location of poverty in urban space, much of the work remained descriptive rather than analytic. Inherent within the school's representations is a concern that urban industrial capitalism is the primary reason for divisions between the rich and the poor and for the expression of this division in the landscape. Even so, few works in the vein of the Chicago School provided analyses of why people end up impoverished or how capitalism contributes to poverty. To fill this gap in knowledge about the geography of poverty, Marxist geography developed a perspective on poverty that incorporated a strong critique of capitalism. Marxist geographers began to understand poverty as a reflection of capitalism's inherent contradictions. Capitalism's geography is visible in various ways—the valuation of land rents, the relationship among multiple places through commodity chains and production processes, and the creation of spaces where socially and economically marginal populations reside. Using the theories of Karl Marx and other political economists, Marxist geographers developed an analysis of poverty in which they posited that poverty is a social and economic condition created by, and maintained through, the dominance of capitalism.

According to this view, certain people live in poverty because of their relationship to capital. Those who do not control the means of production and who sell their labor through a wage system are at the greatest risk of poverty, especially when a crisis in capitalism occurs and wages get reduced or labor markets shrink. Such a negative relationship is reinforced by the geographic processes of capitalism. Poorer people can be found living in places that have been rendered not valuable for capitalism's continued unfolding, have been wholly abandoned by capital, or have a structural relation of dependence that is disadvantageous. The process of deindustrialization in many cities of the global North, such as Birmingham (United Kingdom) and Detroit (United States), gives an empirical basis to the first two outcomes. The widespread poverty of formerly colonized countries in the global South, such as Latin American and sub-Sahara African states, provides examples for the third outcome. Through this turn to political economy, geographers and others analyzed the causes of poverty and also discussed poverty as a dynamic phenomenon that changes in response to changes in the economic base. Therefore, many Marxist analyses also include a political prescription to alter capitalism and reduce poverty.

Marxist geography gained prominence during the 1970s and continues to study the relationship between changes in capitalism and the changing geography of poverty. Although Marxist geography has had a significant influence on how poverty is studied as a phenomenon in cities, in countries, and in an international context, some geographers grew concerned that the focus on political economy curtailed analysis from seeing the multidimensional aspects of poverty. During the 1990s, geographers and others inspired by postmodern social theory turned their attention to the multiple ways in which poverty can be defined depending on the cultural context in which it is to be studied. This notion is particularly geographic in that it focuses on the relationship among locale, place, and meaning. What is poverty is one place might not be poverty in another place. Such a postmodern turn in poverty studies reveals that poverty is not only an economic condition but also a social and cultural one. This insight indicates that alleviating poverty requires more than just increasing income levels. Furthermore, such an argument demonstrates an alternative way in which to understand the relationship between power and poverty. Whereas for Marxist geographers power is manifested through control over capital, for postmodern geographers power is corralled by dominating systems of meaning (e.g., governmental agencies that determine official definitions of poverty through the poverty line).

A postmodern perspective on poverty highlights the subjectivity of poverty; even though someone might have a low monetary income, that person might not consider himself or herself poor. Through the postmodern turn, poverty can be measured through a variety of indexes that may or may not relate to economic standing and connect to the multifarious ways in which people determine a quality of life. According to the protagonists of this postmodern turn, this trend gives greater power to people and places to define themselves along a continuum of impoverishment and allows the so-called poor to avoid some negative connotations and conditions associated with poverty.

—*Jonathan Lepofsky*

See also Central Business District; Chicago School; Class; Core–Periphery Models; Dependency Theory; Developing World; Ghetto; Homelessness; Housing and Housing Markets; Justice, Geography of; Labor, Geography of; Marxism, Geography and; Political Geography; Postmodernism; Poststructuralism; Rustbelt; Segregation; Social Geography; Social Justice; Squatter Settlement; Structural Adjustment; Underdevelopment; Uneven Development; Urban Spatial Structure; World Systems Theory

Suggested Reading

Fan, C. (2003). Rural–urban migration and gender division of labor in transitional China. *International Journal of Urban and Regional Research, 27,* 24–47.

Glasmeier, A. (2005). *An atlas on poverty in America: One nation pulling apart 1960–2003.* University Park: Pennsylvania State University Press.

Harvey, D. (1973). *Social justice and the city.* Baltimore, MD: Johns Hopkins University Press.

Massey, D., & Denton, N. (1993). *American apartheid: Segregation and the making of the underclass.* Cambridge, MA: Harvard University Press.

Wilson, W. (1996). *When work disappears: The world of the new urban poor.* New York: Knopf.

Yapa, L. (1996). What causes poverty? A postmodern view. *Annals of the Association of American Geographers, 86,* 707–728.

POWER

The term *power* refers to practices and processes through which institutions, groups, and individuals arrange the social world and attempt to change it to advance their interests. Traditionally, it has been political geographers who have been concerned with power and power relations. Political geographers have had a strong interest in the territorial aims of groups and nations and in the deployment of power to achieve those aims. Attention to state policies and actions have relied on conceptions of power as force that is gained and wielded by one group over another group. Studies of power within political geography have tended to be descriptive analyses of accumulation and trajectory of force. Cultural geographers tended not to incorporate power centrally in their scholarship because of the conceptualization of landscapes as organic expressions of cultural groups. Marxist geographers made important contributions to theorizations of power by directly addressing structural processes shaping places. In general, however, there has been a pattern of geographers treating power as a realm of formal political institutions at the national and international scales and as separate from everyday life. More recently, scholars across several fields have been developing alternative theorizations of power that problematize traditional conceptualizations of power's structure and its operational scales.

Any one-sentence definition of power can be only a starting point for the understanding of a complex and multidimensional concept. Consequently, this discussion addresses three aspects of the term: (1) power as force, (2) power as a field of social practices, and (3) the issue of scale in the operation of power.

Traditionally, power has been equated with force and been framed as measurable and attributable to one or more groups. There are various types of force such as military, economic, and social. The analysis of power often has centered on the ability to project force or on the capacity to threaten force credibly. National governments gaining territory through military action are prime examples of power as force. There are two significant implications of the concept of power as force. The first is that power is seen as possessed or wielded by institutions. The second is that ascribing power to one social actor implies that other social actors lack power. The ability to use or project force is a stark element of contemporary social relations, as in the case of the ability of the United States to undertake military actions against other nations. Yet although force can be useful in social analysis, it can be understood as a limited perspective on power that ignores complex dynamics.

More recently, geographers have drawn on alternative and more expansive theorizations that problematize the idea of power as a phenomenon that social actors possess. Rather than understanding power as something that is held by individuals or groups, power can be framed as a set of multiple processes in which social actors are located. Actors do not posses power per se; rather, they align themselves in advantageous positions within dynamic processes of power. More expansive theorizations frame power as a multidimensional field in which actors of all types operate (e.g., governments, corporations, special interest groups, community groups, individuals). Actors, such as a mayor and a community organization, can be understood as positioned (either advantageously or disadvantageously) in a system of power. Systems of power have many dimensions such as corporate interests and neighborhood organization agendas. There are also specific constituencies such as public school teachers and both the workers and management of important companies. There are also distinct geographic interests such as downtown businesses and suburban real estate developers. In a system or field of power relations, no one group has a monopoly on power over an extended period of time, and rarely is one group completely powerless. Rather, groups maneuver in a social landscape where power is asserted and contested in numerous ways. Importantly, thinking of power as a process frames it as being produced by a wide range of actors, even those that are not explicitly political. In addition, understanding power as a field incorporates the ability to contest and resist as an important aspect of power. In the analysis of power, the processes can be just as important as, if not more important than, the end results of conflicts and struggles. The ability to project power is often contingent, finding expression and articulation only under specific circumstances. For example, a coalition of pro-growth urban interests groups may have success on a series of initiatives but also may have projects that fail to win acceptance. Similarly, a minority neighborhood may have a history of relative powerlessness in relation to a city government but may find success in stopping one particular project. Understanding power as a field of social relations frames it as structural yet dynamic and as both an outcome and a process. More expansive conceptualizations suggest that people and groups operate within a field of relations in which all groups and organizations have the ability to project and resist power in different ways.

One aspect of power that has not been considered extensively is its operation across multiple scales. Individuals operate simultaneously in an array of power relations across many geographic locations and scales. For example, a married woman with children, a job, and an extensive social network may move among many positions of power. As a parent, she may exert extensive power over her children. She may, however, be in a subordinate position in relation to her husband. If the woman is a manager at her place of employment, she may be positioned advantageously in relation to the employees over whom she has authority, yet within the decision-making hierarchy of her firm she may be marginalized by gender ideologies that limit her opportunities for promotion. In each relation and in each location, power is part and parcel of the woman's life. How different dimensions of power do and do not intersect at specific locations are significant questions that present rich opportunities for geographers to explore.

—*Carlos Tovares*

See also Feminist Geographies; Hegemony; Ideology; Political Geography; Poststructuralism

Suggested Reading

Allen, J. (2003). Power. In J. Agnew, K. Mitchell, & G. Ó Tuathail (Eds.), *A companion to political geography.* Malden, MA: Blackwell.

Dixon, D., & Jones, J., III. (2004). Poststructuralism. In J. Duncan, N. Johnson, & R. Schein (Eds.), *A companion to cultural geography* (pp. 79–107). Malden, MA, & Oxford, UK: Blackwell.

Massey, D. (1993). Power geometry and a progressive sense of place. In J. Bird, B. Curtis, T. Putnam, G. Robertson,

& L. Tickner (Eds.), *Mapping the futures: Local cultures, global change* (pp. 59–69). London: Routledge.

Painter, J. (1995). *Politics, geography, and "political geography": A critical perspective.* London: Edward Arnold.

PRODUCER SERVICES

Producer services are forms of service activity sold primarily to business and government clients. In contrast to retailing and consumer services that have their primary markets with households for final consumption, producer services are sold as inputs to the production process of various industries. They are commonly regarded as "intermediate" services because they are absorbed into the goods and services produced by their clients. In the United States, producer services have long been considered to be a combination of business, legal, and engineering and management services. This might be regarded as a narrow definition because other lines of service activity also have strong intermediate markets. A broader definition of producer services would include a considerable proportion of finance, insurance, and real estate; some categories of membership organizations; the arrangement of transportation services; and certain modes of transportation (e.g., air cargo, pipelines, rail freight, truck freight, waterborne cargo container and bulk commodity movement).

Producer services have grown rapidly in the United States and in other advanced economies. In the United States from 1940 to 1970, employment in narrowly defined producer services expanded from 500,000 to 3.5 million, a gain of nearly 600%. This rapid growth has continued, with employment between 1970 and 2002 expanding to 17.5 million, a gain of 400%. Most of this growth has been in small-business establishments. In 1970 the average producer service establishment employed 11 persons, whereas in 1997 the average establishment employed 12 persons. The number of producer service establishments expanded from 300,000 in 1970 to 1.6 million in 1997. Producer service businesses historically have concentrated in the largest metropolitan areas. However, growth has occurred in this sector in smaller and medium-sized metropolitan areas as well as in nonmetropolitan communities.

The reasons for this rapid growth are multiple. First, the overall economy has grown, thereby requiring more producer service activity. However, producer services employment growth rates have been significantly greater than the growth in overall employment—between 1970 and 2002, U.S. total employment expanded by approximately 75%, while producer services had more than five times this growth rate—implying a greater demand for producer services over time. In part, this growth has been related to the export of services to foreign countries or interregionally in the United States. In part, it has been related to growth in the share of total production costs devoted to the purchase of producer services by industries in all segments of the economy and by governments. New types of producer services, such as computer services and temporary help services, have been accepted in the marketplace. Other producer services have been transformed by information technologies so that they provide a wider array of services to clients, for example, the growth of accountants providing financial advisory services to their clients.

There has been considerable debate over the forces surrounding this rapid growth of producer services. Some have argued that the growth has been fueled largely by downsizing and outsourcing of services previously produced in-house. Studies of the use of specialized producer services by manufacturers, for example, find increases in external purchases and, in some cases, shifts of lines of producer service work from in-house departments to free-standing producer service suppliers. Others have argued that growth has come primarily from expanded purchases of specialized producer services. The lack of expertise on the part of clients, the need for independent third-party advice, government regulations, and the mismatch between the minimum size that would be required of an in-house department and the actual demand for a particular producer service all contribute to the purchase of these services from external suppliers. Complicating the question of the shifting of supply from in-house departments to outside suppliers has been the rapid pace of change in information technologies and the evolution in the nature of producer services offered. Clients of producer service firms often have in-house departments producing certain specialized services (e.g., legal services) but at the same time buy different services from suppliers in the same line of work that complements work done by in-house departments.

The specialized nature of many producer service establishments is associated with the development of niche market positions, intended to convey to clients that the seller of the service differs from its competitors.

This positioning in the marketplace is also related to the development of competitive advantage based on market strategies intended to differentiate the seller from its competitors. In developing market positions of this type, the price of the service is deemed less important as a factor driving the client to the seller than the perceived quality of the service being rendered by the seller.

Historically, the production of such specialized producer services has required face-to-face contact between clients and suppliers of producer services. In some lines of producer services, this contact typically takes place at the office of the supplier (such as with lawyers); in other cases, it typically takes place in the office of the client (such as with management consultants). With the advent of advanced information technologies, it is possible to move complex documents and engage in communications between buyers and sellers at a distance, leading some to speculate that face-to-face communication has become less important in the producer services. However, the preponderance of evidence suggests that face-to-face meetings still are very important, especially in cases where the activity involved is highly specialized, nonroutine, and subject to negotiation, redesign, and high levels of creativity and conceptualization.

Some producer service establishments sell all of their services in their local community, whereas others develop significant regional, interregional, and/or international markets for their services. Although statistics on trade in producer services are not reported systematically, various surveys point to the fact that producer services have become one of the most important components in the economic base of large metropolitan economies. Firms with widespread geographic markets have a degree of locational flexibility, and survey research has found that entrepreneurs in firms of this type often seek locations where the quality of life is high or where they want to live.

—*William Beyers*

See also Agglomeration Economies; Economic Geography

Suggested Reading

Beyers, W. (2000). Cyberspace or human space: Wither cities in the age of telecommunications? In Y. Aoyama, J. Wheeler, & B. Warf (Eds.), *Cities in the telecommunications age: The fracturing of geographies* (pp. 161–180). New York: Routledge.

Beyers, W., & Lindahl, D. (1996). Explaining the demand for producer services: Is cost-driven externalization the major factor? *Papers in Regional Science, 75,* 351–374.

Lindahl, D., & Beyers, W. (1999). The creation of competitive advantage by producer service firms. *Economic Geography, 75,* 1–20.

Ochel, W., & Wegner, M. (1987). *Service economies in Europe: Opportunities for growth.* London: Pinter.

Tschetter, J. (1987, December). Producer services industries: Why are they growing so rapidly? *Monthly Labor Review,* pp. 31–40.

Walker, R. (1985). Is there a service economy? The changing capitalist division of labor. *Science & Society, 49,* 42–83.

PRODUCT CYCLE

One of the major theories in economic geography during the height of location theory was the product cycle, which integrated the supply and demand dimensions of corporate locations as they changed over time. This view posits an abstract sequence in which goods move from being newly introduced innovations to becoming more widely accepted until they finally reach a "mature" status. Different stages in the product cycle are associated with different market conditions, varying organization of production, and geographic locations.

Drawing on perspectives of innovation adoption developed in marketing, this view begins with the changing nature of demand for a firm's product over time, which is typically a bell-shaped curve, with new innovations adopted by a small group of risk takers. As the product becomes better known and more widely accepted, larger numbers of more risk-adverse people will adopt it, sales will increase, the market will expand, and the firm can be confident of future revenues. Finally, when the product is widely known, rates of new purchases will slow down.

As the product proceeds through its cycle, the supply conditions will change accordingly. With new innovations in which the information content is high, firms must expend considerable costs for research and development. Firms with monopolies over the production of products during their early years (e.g., as patent holders) may enjoy superprofits for a limited time. Such companies tend to be relatively small, use higher proportions of skilled labor, and face a great deal of uncertainty in the market. Because they must produce in relatively small quantities, they are not likely to enjoy economies of scale and typically compete more on the basis of quality than on the basis of price. Small firms in the early stages of the product cycle must accept the market price; that is, they are price takers rather than price setters. As the product becomes more widely adopted and the market expands, firms can shift to producing larger quantities, standardizing the production process. As larger firms out-compete smaller ones, the industry becomes more capital intensive in nature and more reliant on economies of scale, and the market becomes increasingly oligopolistic. This transition often is accompanied by a process of vertical integration in which different stages of the production process become incorporated in-house within one firm.

Raymond Vernon offered a series of geographic correlates for different stages of the product cycle. The original analysis was framed at the international scale, although subsequent ones extended this approach to national and regional scales. Thus, during the early stages of the product cycle, firms tend to prefer locations close to large pools of skilled labor, typically in economically developed countries. In subnational modifications of the product cycle, the optimal locations are in the cores of metropolitan regions, where firms can take advantage of agglomeration economies and dense networks of specialized contacts and information. Thus, spatial cores are "seedbeds" of innovation. At the national scale in the United States, such locations were found in the traditional northeastern–midwestern Manufacturing Belt. As the product moves through the cycle and the production process becomes more capital intensive, the need for centralized locations declines and firms can relocate to the metropolitan periphery—or, in the context of the national space of the United States, the Sunbelt. Core regions are left to begin a new cycle of production by innovating in skilled, labor-intensive goods. Thus, the product cycle leads to the hypothesis that the decentralization of manufacturing is accompanied by the process of capital intensification as oligopolistic firms spin off their branch plants to peripheral locations in a search for cheaper and less skilled labor. At the international scale, this process leads to multinational corporations investing in developing countries. This perspective sustains the view that exports from regions with comparative advantages early in the product cycle should be relatively labor intensive, whereas those with advantages later in the cycle should be more capital intensive. As product cycles play out continually over the landscape, markets will reproduce uneven development, in contrast to expectations that the free flow of labor and capital will result in a convergence of economic growth among regions.

Despite its comprehensiveness and appeal, the product cycle has been criticized as being simplistic. Although its description of the empirical nature of industrial change is essentially correct, the approach does not delve deeply enough into the dynamics of the production process, failing to embed it in any wider comprehension of social relations. Thus, it rests on an implicit notion of technological determinism. The model assumes that firms are single-product entities that are always given over to capital intensification and vertical integration in a unilateral trajectory of growth and expansion, although this is not always the case in practice.

—*Barney Warf*

See also Agglomeration Economies; Comparative Advantage; Economic Geography; Flexible Production

Suggested Reading

Norton, R., & Rees, J. (1979). The product cycle and the spatial decentralization of American manufacturing. *Regional Studies, 13,* 141–151.

Storper, M. (1985). Oligopoly and the product cycle: Essentialism in economic geography. *Economic Geography, 61,* 260–282.

Taylor, M. (1986). The product-cycle model: A critique. *Environment and Planning A, 18,* 751–761.

Vernon, R. (1966). International investment and international trade in the product cycle. *Quarterly Journal of Economics, 80,* 190–207.

PRODUCTION OF SPACE

The phrase *production of space* comes from French philosopher Henri Lefebvre, who used it as the title of a famous book published in France in 1974. The book was a sophisticated attempt to bring the analytical rigor and political edge of Marxist theory to bear on questions of space and, with that, to revolutionize Marxist theory itself. In it, Lefebvre challenged the dominant conception of space in social theory that saw it as but an inert container for other more important and, on the whole, historical processes. Instead, Lefebvre argued that society was a fundamentally spatial phenomenon and that capitalism was a spatial process. Lefebvre argued against the view that space could be little more than a passive backdrop of social life and preferred to see it as an active medium.

Lefebvre's emphasis on the production of space reflected the influence of Marxist theories of materialist practice. Social space did not exist for Lefebvre apart from the active practices that created, modified, and sustained it. In part, then, Lefebvre was reacting against an idealist view of space that saw it as but a subjective realm of experience largely divorced from human labor and practice. His ideas on space, however, were equally a reaction against the more structural Marxism of Louis Althusser. Lefebvre rejected the tendency to treat space as if it were only a geometric field without human actions and purpose or simply a product of the subjective mind. Instead, he sought to bring together a rigorous Marxist theory and a humanist strain in Marxism that always sought to ground revolutionary thought in human feeling and desire.

LEFEBVRE'S SPATIAL TRIAD

Philosophically and politically, then, Lefebvre sought to transcend the simplistic dualisms that opposed space and time, subjectivity from the material world, the local from the global. Lefebvre believed that the dominant tendency in Western philosophy toward dualisms had the effect of flattening the richness and complexity of human experience. To break through the philosophical and conceptual impasse of dualistic frameworks, Lefebvre proposed a conceptual triad. Lefebvre's conceptual framework distinguished among three different kinds of space: spaces of representation, representational space, and spatial practices.

Lefebvre defined *spaces of representation* as the rational spaces of planners and engineers. It is, in short, a view of how space ought to be, representing a kind of power over space. In contrast, Lefebvre saw *representational space* as passively experienced, imaginative, and ultimately dominated space. Although both are forms of representation of space, the former is a reflection generally of a hegemonic group and thus has the power to actually build that representation into landscapes. The latter, however, is a space of resistance that challenges those dominant representations of space. The final category, *spatial practices,* involves the activities that collectively serve to produce space and are interwoven with the other two dimensions of Lefebvre's triad.

Lefebvre's production of space framework also contained a historical argument about the development of social space. Here Lefebvre distinguished *abstract space* from what he termed *concrete space*. As he characterized it, concrete space was the space of lived experience. Abstract space, in contrast, was space abstracted from lived experience and embodied in the distant perspective of planners, bureaucrats, and businessmen and crystallized in maps, planning documents, and shopping malls. For Lefebvre, modernity involved the gradual colonialization of concrete space by abstract space. The task of a progressive politics, then, was to reclaim the spaces of everyday life from the homogenizing tendencies of both capital and state power.

IMPACT ON GEOGRAPHY

Lefebvre's theorization of space arguably went significantly beyond that of geographers of the day, who tended to be strongly influenced by either the dominant trend of the blunt positivism of the quantitative revolution or the subjectivist inflection of the emerging humanist geography. His book began to make its mark on Anglophone geography during the 1980s in the work of scholars such as David Harvey, Neil Smith, and Edward Soja, who used the conceptual framework introduced by Lefebvre to significantly advance geographic theories of the complex geographic dynamics of contemporary capitalist development and urbanization.

Interest in Lefebvre exploded in the Anglophone world in 1991 with the publication of an English translation of his book. The notion of the production of space proved to be particularly attractive to geographers interested in bringing together a more traditional analysis of political economy and an emerging concern with language, culture, and the politics of everyday life. More recently, the concept of the production of space has influenced a greater diversity of geographic study. No longer confined to the study of political economy, work that draws on Lefebvrian ideas now includes studies of technology, race, and sexuality.

—Bruce D'Arcus

See also Space, Human Geography and; Spaces of Representation; Structuration Thoery

Suggested Reading

Gregory, D. (1994). *Geographical imaginations*. Oxford, UK: Blackwell.

Kirsch, S. (1995). The incredible shrinking world? Technology and the production of space. *Environment and Planning D, 13,* 529–555.

Lefebvre, H. (1991). *The production of space.* Oxford, UK: Blackwell.

Merrifield, A. (1993). Place and space: A Lefebvrian reconciliation. *Transactions of the Institute of British Geographers, 18,* 516–531.

Smith, N. (1990). *Uneven development: Nature, capital, and the production of space* (2nd ed.). Oxford, UK: Blackwell.

Soja, E. (1989). *Postmodern geographies: The reassertion of space in critical social theory.* London: Verso.

Stewart, L. (1995). Louisiana subjects: Power, space, and the slave body. *Ecumene, 2,* 227–245.

PROFIT

The goal of generating profit is one of the primary attributes of capitalist economic systems. This goal distinguishes capitalist systems from traditional subsistence economies (which rely on the periodic availability of natural resources to maintain livelihoods) and communist economic systems (which are theoretically designed to produce and redistribute resources according to need). Profit is generated when the revenue produced by a business exceeds the cost of production. Profit can take many forms (e.g., as material resources), but in most modern societies the most important form of profit is money or capital. The advent of capital is critically important because it can be used for many different purposes. At the most basic level, the profits generated by production can be reinvested to improve the efficiency, capacity, or overall productivity of a business. In addition, as a medium of exchange, capital can be used to acquire any number of products or services. This may refer to acquisition of resources, purchase of land, or agreements established with laborers. Thus, profit and capital can be used to unify seemingly disparate social interests.

Profit may be generated by a number of different means. In general terms, a business can improve its rate of profit by assessing the impact that the factors of production have on its economic efficiency. Following the insights of neoclassical economics, the key factors of production include material resources, labor, processing costs, transportation costs, the market, and governmental policies. By assessing the relative importance of each of these variables, a business

theoretically can improve the efficiency of its activities and thereby generate a higher rate of profit. These analytical assessments help to identify the sites that are best suited for locating and constructing factories, stores, and/or other facilities necessary for the production and sale of commodities. The ideal geographic distribution of such activities varies according to the specific nature of a business. For some companies, it is preferable to locate facilities near the material resources used during the process of production. The availability of suitable labor (or lack thereof), however, may complicate this process. A culture of unionized labor in certain regions may also inhibit some companies' ability to increase profits. For others, perhaps those dealing in perishable items, it may be necessary to set up economic operations near the marketplace. In all of these situations, transportation costs may greatly influence the relative importance granted to sites of production and sites of consumption. In addition, government regulations regarding taxation, quotas, and/or environmental regulations may substantially improve or reduce some companies' ability to generate profits. Companies may actively negotiate with governments or openly contest specific governmental policies to boost their profitability, but these lobbying activities are not always successful. By weighing these factors and finding an ideal distribution for economic activity, companies can improve, if not maximize, their profitability.

The issue of scale is also critical in this process. From this perspective, the objective of any business is to find the optimal economy of scale to maximize profits. The economy of scale refers to the size of a business (e.g., its infrastructure) or the amount of products and services the company can produce. Theoretically, the optimal economy of scale for any business is reached when the company generates the highest margin of profit per unit of production. Historically, the most common approach of industry has been to increase economies of scale through the concentration of production. The classic example of this approach was the American automotive industry. This involved the construction of large factories that were able to produce large production batches in a short period of time. These factories were vertically integrated, facilitating the efficient transmission of materials and resources through the production process. This was manifest most clearly in the development of assembly lines, which not only helped companies speed up the production process but also regulated and simplified the actions of factory workers. Through this division of labor, companies were able to generate a large output that minimized their cost per unit. Due to its association with the American automotive industry, this approach became known as Fordism.

With the advent of fast communication technologies and more expedient transportation systems, however, the limitations of Fordism became evident during the coming era of globalization. The large production batches and massive infrastructure of many corporations proved to be unable to adapt to quickly changing circumstances. This inflexibility was perhaps seen most clearly in the deskilled labor that Fordism perpetuated and relied on. This placed such companies at a competitive disadvantage that hindered their ability to generate profits. Consequently, during the 1970s, a new business approach emerged that attempted to maximize corporate flexibility. This approach has become known as post-Fordism but also goes by names such as "lean production," "just-in-time production," and "flexible accumulation." The latter phrase refers explicitly to the ability of flexibility to assist in the accumulation of profit.

In general, post-Fordism is a vertically disintegrated form of production. The production of a single product often involves the activities of several different companies. This may involve different types of social and contractual relationships that temporarily unify the interests of many disparate parties. Each company specializes in a particular aspect of the production process. As a result, economies of scale often are reduced as responsibilities are dispersed among a wider number of actors. Contract relationships with "outside" interests permit greater flexibility, outlining the current short-term needs of the participants. In line with such arrangements, small production batches and small inventories are produced just in time to meet consumer demand. Because capital is not locked up in large inventories, products can be quickly modified to meet changing consumer demands without excessive cost to the producers. This allows for greater product diversity. As this suggests, the role of consumers and consumption (as opposed to factors of production) has become increasingly important during recent years. This is symbolized by the increasing importance of name-brand products in the marketplace. Companies must be aware of these changing preferences to maintain profit levels. In keeping with this overall goal, contracts can be adjusted or discontinued when existing

relationships no longer are able to generate sufficient profit.

In conclusion, profit is sustained by a consideration of all of these variables. The factors identified in industrial location theory still prove to be salient. Similarly, whereas Fordism and post-Fordism are theoretically distinct, most large corporations embody a mixture of these distinct principles to generate profit.

—*Christa Stutz*

See also Economic Geography; Economies of Scale; Factors of Production; Flexible Production; Fordism; Location Theory

Suggested Reading

McConnell, C., & Brue, S. (1993). *Macro-economics: Principles, problems, and policies.* New York: McGraw-Hill.

McConnell, J., & Erickson, R. (1986). Geobusiness: An international perspective for geographers. *Journal of Geography,* 85(3), 98–105.

Scott, A., & Storper, M. (Eds.). (1986). *Production, work, territory: The geographical anatomy of industrial capitalism.* Boston: Allen and Unwin.

Stutz, F., & Warf, B. (2005). *The world economy: Resources, location, trade, and development* (4th ed.). Upper Saddle River, NJ: Pearson/Prentice Hall.

PSYCHOANALYSIS, GEOGRAPHY AND

Psychoanalysis was initiated by Sigmund Freud (1856–1939) and concerns the clinical techniques revised and contested between and within its various schools of thought, all of which aim to relieve psychological, physical, and sexual forms of suffering. Psychoanalysis differs from other psychotherapies in that it asserts and relies on the interpretability of unconscious mental processes, the contingencies of human sexuality whereby we are not simply regulated by instinct, the human compulsion to repeat painful acts, and our propensity to transfer emotions among ideas, objects, and language.

Psychoanalysis played a formative role in the development of 20th-century philosophy, critical theory, the "Frankfurt School," poststructural and feminist theories, art criticism, literary theory, and film studies. For the most part, geographers began to engage with psychoanalysis in response to the limitations of the Anglo-American discipline's 1970s and 1980s humanistic and radical theorizations of human subjectivity and agency. Since the mid-1990s, numerous geographers have argued that psychoanalytic categories are thoroughly spatial and used psychoanalysis to explicate the importance of how our attachments, identifications, and dynamic unconscious processes spatially disrupt and maintain sociopolitical antagonism, domination, and division.

There are three main psychoanalytic approaches used in geography. First, geographers have used Freudian concepts to examine how subjectivity, topographies of the body, and unconscious desires can materially and symbolically landscape the urban contexts of racism, sexual politics, and social anxieties. Second, drawing on object relations theory, which deemphasizes the role of the sexual drive by focusing on the embodied psyche's relations to "objects," geographers have examined how public and private spheres are underpinned by intersubjective processes of exclusion, rejection, purification, and transgression to consolidate hegemonic power through amplifying differences between cultural "selves" and "others." Third, Jacques Lacan's psychoanalytic concepts of the "imaginary" (alienation and rivalry), the "symbolic" (language and law), and the "real" (trauma and impossibility), as well as the psychoanalytic work of the French feminists Hélène Cixous, Julia Kristeva, and Luce Irigaray, have prompted geographers to question the patriarchal and masculinist insistence on distinguishing between real and nonreal spaces, subvert binary conceptions of innate and fixed gender differences, and politicize the roles of fantasy and enjoyment in the ideological dimensions of sociospatial practices.

Recently, some geographers have critiqued the "psychoanalytic turn" in geography for gentrifying the more radical elements of psychoanalytic theory to comply with the discipline's culturalist and social constructionist sensibilities. These geographers' calls to critically embrace the more troubling aspects of psychoanalytic theory, such as the emphasis on a dangerous and paradoxical reciprocity between enjoyment and suffering, seem to be more than timely given the recent worldwide intensification of violence, exploitation, domination, and discrimination that make up the increasingly blurred domains of terrorism, democracy, and morality.

—*Paul Kingsbury*

See also Identity, Geography and; Subject and Subjectivity

Suggested Reading

Callard, F. (2003). The taming of psychoanalysis in geography. *Social and Cultural Geography, 4,* 295–312.

Kingsbury, P. (2004). Psychoanalytic approaches. In J. Duncan, N. Johnson, & R. Schein (Eds.), *A companion to cultural geography* (pp. 108–120). Oxford, UK: Blackwell.

Nast, H. (2000). Mapping the "unconscious": Racism and the oedipal family. *Annals of the Association of American Geographers, 90,* 215–255.

Pile, S. (1996). *The body and the city: Psychoanalysis, space, and subjectivity.* New York: Routledge.

PUBLIC SPACE

Public space has many different meanings but usually is thought of as a place that is created and maintained by a government entity for the benefit of the community and that ideally can be used regardless of one's economic or social condition. Typical examples include parks, roads, town squares, sidewalks, and public beaches. Some privately owned spaces, such as shopping malls and sidewalk cafes, give the appearance of public spaces because of community and social interaction that is common in such places. But these semipublic spaces may limit entry to some people based on the owners' discretion or the ability to pay. In the case of shopping malls, the public is invited to experience the space as potential consumers.

When it comes to the built environment in cities, public space has been an essential subject for architects, urban designers, and planners for centuries. The definition of public space in this context, which traditionally has meant streets, squares, and parks in an urban setting, is a topic of lively debate. These spaces are part of the public domain, meaning all places taken together that are perceived as public—streets, squares, and parks as well as privately owned collective spaces that function as public spaces. During the time of the Roman Empire, massive public domains were created for entertaining the urban masses, including chariot races, theatrical and musical performances, wild beast hunts, mock sea battles, public executions, and gladiatorial combat. During the 19th century, Napoleon's prefect of Paris, Georges Haussmann, created large public domains by blasting through the congested spaces of the old medieval city to construct the long, wide boulevards for which Paris is famous today. During the late 20th century, under the guise of modernist visions of urban renewal, planners in many large U.S. cities razed entire neighborhoods to create a world of highways, parks, and other accommodations designed for public use.

Public space can also be coded as masculine space. This division of male space as public and female space as private became especially pronounced during the Industrial Revolution, when gendered divisions of labor began to emerge that saw men entering the public realm to work for salaried wages and women staying home to provide a refuge to which men would return after facing the trials and temptations of the city. Public spaces are where social interaction takes place and where the conduct of business was seen as a cultural norm set aside exclusively for men. Women, generally portrayed as emotional, unstable, and weak, had no place in the public spaces of the city; instead, they were relegated to "feminine" domestic spheres of privacy such as the home. This public–private division reflected a male-dominated hierarchy that saw traditional women's roles as keepers of family and morality. Today, these divisions are not as clear, but they still play a role in contributing to a spatial entrapment of women that defines many public spaces.

Public space often is seen as inclusive of everyone, where interaction between people is spontaneous and nonpolitical. But, in practice, this is rarely the case. Public spaces can be sites of political protest and struggle, where conflict often is inevitable. For example, Tiananmen Square in Beijing, China, has had a long history as a site of political protest, most recently in 1989 when thousands of people took control of the square demanding democratic and cultural reform. In many ways, this public space represents the symbolic heart of the power of the Chinese government, where a loss of control meant a massive government crackdown in 1989 in which hundreds of people were killed and many more were imprisoned. Tiananmen Square is but one example of a long worldwide history of various public spaces playing a role as geographic sites of resistance. These spaces often are used to challenge the powerful by providing a venue to give a voice to the excluded, dominated, and oppressed in society.

Particular cultural meanings often are associated with different public spaces. One way in which to express these meanings is by creating spectacles such as street theaters and parades. These spaces, by their very definition as public, always have been used as sites for performance in which people appropriate a cultural identity and fill it with a particular meaning. For example, the crowds that attend the St. Patrick's Day Parade in New York City become Irish regardless of their ethnicity. Another example is the Civic Center in San Francisco, where one day a year the public space—which is assumed to be a heterosexual space—is appropriated into an exclusively gay and lesbian space during that city's annual Gay Pride celebration. Another way in which to etch cultural meaning into public space is by examining the role that public sculptures, statues, and monuments contribute to people's collective social memory. It is not only the monuments themselves that are designed to tell particular stories but also the geographic locations of these public displays that play an important role in defining cultural meaning. For example, a proposal to locate a statue of African American tennis star and civil rights activist Arthur Ashe along Monument Avenue in Richmond, Virginia, brought about different interpretations of the same public space for black and white occupants of the city. Monument Avenue is the site of the U.S. South's grandest Civil War–era Confederate memorials. These groups defined and redefined the same public space with different meanings, placing Monument Avenue into a Civil War–era versus civil rights–era debate. These are just a few illustrations of the significance of public space in creating meanings in people's lives.

—*Thomas Chapman*

See also Built Environment; Home; Urban and Regional Planning; Urban Geography; Zoning

Suggested Reading

Johnson, N. (2004). Public memory. In J. Duncan, N. Johnson, & R. Schein (Eds.), *A companion to cultural geography* (pp. 316–327). Oxford, UK: Blackwell.

Leib, J. (2002). Separate times, shared spaces: Arthur Ashe, Monument Avenue, and the politics of Richmond, Virginia's symbolic landscape. *Cultural Geographies, 9,* 286–312.

Mitchell, D. (2000). *Cultural geography: A critical introduction.* Oxford, UK: Blackwell.

QUALITATIVE RESEARCH

Qualitative research includes both a series of techniques and a group of approaches to research. Often mistakenly thought of as simply research without numbers, qualitative research lies at the very core of human geography, involving an array of different theoretical, methodological, and philosophical positions to research that together seek to answer questions of meaning.

Research in human geography, especially in the subfields of cultural geography and historical geography, has long had a qualitative base, although it was not until the late 20th century that it explicitly acquired the label *qualitative research*. During the 1920s, Carl Sauer, considered to be the founder of modern cultural geography, was influenced by anthropologist Alfred Kroeber, adapting his anthropological fieldwork techniques (e.g., participant observation) for use in geography. Although Sauer never termed it as such, his methods of landscape interpretation can be seen to be profoundly qualitative, involving archival research, acute observation, interaction with local people, and skilled interpretation—all techniques that continue to flourish, albeit in modified forms, in contemporary human geography. Sauer's focus, however, was not on methods (means of collecting data) or methodology (conceptualizing how research can and should be conducted) but rather on the empirical results his investigations produced. Thus, although generations of students were trained in Sauerian ways, their writings, like Sauer's own, generally left their techniques and the underlying philosophical underpinnings of their research obscure.

At midcentury, geography's quantitative revolution swept the discipline, appearing to allow geography to enter the elite realm of the "hard sciences" (e.g., chemistry, biology, physics) and offering the allure of research results that resembled the "laws of science" with their broad applicability rather than the idiographic results of the Sauerian era—results that, while offering local insight, seldom had been found to be transferable to other areas. Methods that were not quantitative came to be seen as "soft science" and therefore less valuable; much qualitative research was eclipsed. By the 1970s, however, human geographers began to see that, for all the power of the quantitative techniques, such techniques often reduced human geography to a humanless form, where people were represented in aggregate or as averages rather than as thinking and feeling individuals, and qualitative research saw its resurgence in humanism.

For humanistic geographers, research was motivated by precisely the things that quantitative researchers found to be unimportant—the ability to study subjective meanings (e.g., of place, of landscape, of region), meanings held not only by aggregated groups but also by individuals. Humanistic geographers such as Yi-Fu Tuan showed that quantitative research was not able to address complex questions of meaning and that an understanding of the human social world was incomplete without that. By the 1980s, human geography had two traditions, qualitative and quantitative, both of which were vying for supremacy—in publishing, in faculty positions, and in the ability to answer pressing questions of science and society.

Although for some that duel continues today, most human geographers now understand the difference between qualitative research and quantitative research not as one of superiority/inferiority but rather as one that divides the kinds of questions—and answers—that research can address. In fact, the very use of the labels *soft* and *hard* to segregate the two kinds of research is found to be inappropriate because both present significant, albeit different, challenges to the researcher.

Today, qualitative research again is the most frequent approach used across most subfields of human geography. Contemporary geographers using this approach build on the techniques engaged by previous scholars, develop new techniques and theoretical approaches, and speak and write openly and explicitly about both their methods and their methodologies. Qualitative geographers may acknowledge influences from and allegiances to a wide array of philosophical and theoretical positions, including phenomenology, Marxism, feminisms, poststructuralism, postmodernism, social constructionism, and combinations of these (and other) theoretical/philosophical positions.

At the core of contemporary qualitative inquiry is an understanding that objectivity for both natural and human sciences is not possible; each researcher is "situated"—by, among other things, her or his class, gender, ethnicity, and sexual preference as well as her or his individual background, education, beliefs, and theoretical affiliations—and that positionality influences the choice of research topic, the course of the research, and its eventual results. Thus, it is not possible to observe and interpret human interaction from a purely neutral or objective position. This leads many qualitative researchers to an understanding of the socially constructed nature of the human world—the notion that, despite certain irrefutable facts of life (ranging very broadly, e.g., from gravity to racial oppression), the way in which we make sense of the world is both culturally and individually defined, constructed out of experiences and cultural traditions as well as life's irrefutable facts. Thus, subjectivity no longer is feared or seen as less important in human geography; all research and all researchers are necessarily subjective. The techniques of self-reflexivity (attempting to complexly understand one's own situated position in the world and to conduct research and carry out writing in ways that make that position visible rather than obscure), pioneered by feminist geographers such as Heidi Nast and Linda McDowell, help contemporary qualitative researchers in human geography to understand their subjective positions.

Contemporary human geographers have modified ethnographic techniques of participant observation to recognize the researcher not as an impartial and inert observer but rather as an observant participant who both influences and is influenced by the group or community she or he studies. Such a position acknowledges the situated nature of the researcher as well as the action-oriented role many see for research in setting policy and in helping to correct social, political, and economic injustices. Qualitative researchers in human geography may also conduct open-ended in-depth interviews (with individuals or groups) or analyze any of a wide array of texts, including landscapes, maps, visual images, fiction, and archival materials. Most will use a combination of two or more of these (and other) techniques. What links all of them together is what is said to be a "naturalistic" understanding of research, where the perspectives of respondents/informants are validated and placed at the center of the research agenda. Indeed, unlike quantitative researchers who create a testable hypothesis or research question before beginning their research, qualitative researchers may enter the field with only a general notion of what their research topic will be, allowing issues and ideas relevant to the community they study to become evident as the research progresses and seeking complex questions and answers from the individuals and social worlds they work with and embed themselves in.

By the late 1980s, qualitative researchers in human geography also began to write explicitly methodological works, advancing theoretical discussions and attempting to make clear the methods and techniques so long left undisclosed. Today a wide array of texts help to teach qualitative research to new generations of human geographers. This new emphasis on methods and methodology has also led to a deeper understanding of how researchers construct the reality they study in part through their writing. Where previously the omnipotent-appearing researcher remained hidden "behind" the text, today's qualitative researchers are careful not to exclude their own voices, even as they labor to present the voices of those with whom they have conducted the research. As researchers become more visible in their work, qualitative researchers in human geography have come to see writing itself as more than merely summing up activity at the end of the research process; rather, they see it as a formative element of the research itself.

—*Dydia DeLyser*

See also Interviewing; Participant Observation; Writing

Suggested Reading

Hay, I. (2004). *Qualitative research methods in human geography* (2nd ed.). Oxford, UK: Oxford University Press.

Limb, M., & Dwyer, C. (2001). *Qualitative methodologies for geographers: Issues and debates*. London: Edward Arnold.

QUANTITATIVE METHODS

Quantitative methods are a collection of techniques and models used by researchers to assess or measure social phenomena. These methods describe, explain, analyze, or predict observed behaviors or phenomena. Two common ways in which these are used are statistics and models.

STATISTICS

Statistical techniques can be divided in two basic types: descriptive and inferential. Descriptive statistics refer to measures that summarize data. Descriptive statistics assess the distribution, central tendency, and dispersion of data and can be used to identity a simple pattern of observed conditions. To describe the distribution of data, researchers can perform observation frequencies that detail how many observations share the same value. Measures of central tendency include the mean (or average), median, and mode. Dispersion refers to techniques that describe the overall similarity or dissimilarity of empirical observations based on an observed standard deviation, variance, or range. In addition to these measures, descriptive statistics can be expressed graphically as scatterplots, histograms, or boxplots. Finally, statistics can also refer to numerical quantities that describe either a sample or a population.

Inferential statistics are an extension of descriptive statistics and are used to make key inferences about an observed pattern within the data set. The objective of inferential statistics is to determine whether the relationship is statistically significant. To determine whether a pattern is significant, researchers may calculate estimates and/or test hypotheses. One example of hypothesis testing would be to determine whether the wage structures of two regional economies are significantly different. For example, consider that the mean wage of residents in Region X is 30% higher than the mean wage of residents in Region Y. In this case, the null hypothesis is tested using a simple z test to determine whether the difference is statistically significant. Beyond basic hypothesis testing, inferential statistics include a full suite of bivariate and multivariate techniques that determine whether statistically significant relationships can be observed within and between variables and/or explain observed variance within a parameter. These techniques include correlation, regression, and cluster analysis as well as principal components analysis and factor analysis. In addition to parametric techniques, nonparametric statistics, such as chi-square and Spearman's rank order correlation, can be used to determine whether relationships exist within or between nominal variables and within or between ordinal variables, respectively.

MODELS

Mathematical modeling refers to the integration of techniques from other disciplines that can be used to model spatial relationships. Neoclassical models such as central place theory were adapted to chart and explain the hierarchy of places and markets. Neoclassical modeling is closely associated with location theory. Geographers also socialized well-known models from the natural sciences, including the gravity model. These models explained observed spatial interactions between or among two or more locations. Today, geographers are using these models and many others to investigate the full range of spatial and topological relationships. Indeed, the growth and expansion of geographic information science (GIScience) has facilitated the rapid integration of new methods, such as artificial neural networks, into the practice of spatial modeling.

QUANTITATIVE GEOGRAPHY AND THE QUANTITATIVE REVOLUTION

Geographers always have used quantitative data to describe the world around us. In particular, geographers have used statistics as numerical quantities to describe regions or social processes. Yet statistical methods are closely associated with the quantitative revolution of the 1950s and 1960s that sought to embed geographers and the discipline within a new theoretical framework that mirrored trends observed across the academy. Using statistics and models to chart urban change, migration patterns, and other behaviors, theoretical geographers began to explore how methods adopted from other social or natural sciences can or cannot be effectively integrated into the geographic research agenda. Likewise, geographers began to revisit and reconsider the implications of older—more established—explanatory frames with the aid of new statistical tools and models. In some cases, geographers were at the fore in establishing the existence of the difference geography makes or in explaining how geography complicates statistical analysis. One example of this would be the emergence of spatial statistics that examine spatial autocorrelation.

CRITIQUES AND EMERGING APPLICATIONS

By the 1970s, the quantitative revolution slowed and several of the prominent boosters of quantitative methods began to reconsider the relevancy and legitimacy of quantitative techniques. In particular, critics became disillusioned with the necessity to develop normative models that explain human activity and tended to dehumanize geography. Specifically, some individuals believed that the discipline was increasingly about the models and decreasingly about understanding the meaning of everyday life or describing different peoples and regions. In addition, statistics were incapable of describing or unlocking the mechanisms and structures that created empirically observed differences between social groups, genders, or regions. For this reason, three prominent critiques emerged: structuralism, humanism, and behavioralism. Structuralist geographers—many of them Marxists—recognized that some sociospatial relationships and observed conditions were the result of the production process. Humanism sought to reposition humans and their experiences at the center of the discipline in an attempt to unpack the meaning of space, place, and region. Behavioralism was an attempt to insert people into existing models by abandoning rational economizing principles of human behavior and inserting the concept of bounded rationality and the satisficer. In concert, these three dominant critiques—as well as the later emergence of poststructuralist approaches—ushered in a legitimate collection of qualitative research methodologies. In many respects, qualitative methods emerged as a counter to the epistemological hegemony of the quantitative revolution.

During the 1990s, quantitative geography experienced a resurgence. The primary motivation for the resurgence was the expansion of geotechnical applications, such as geographic information systems (GIS), the Global Positioning System (GPS), and remote sensing, and the growing dominance of geotechniques. The new quantitative geography articulated spatially rigorous versions of traditional techniques such as geographically weighted regression. Concomitantly, geographers were arguing for more nuanced applications of established techniques, such as the expansion, by placing quantitative methods within alternative epistemological frameworks. Consequently, the simple quantitative/qualitative dichotomy that initially characterized contemporary geographic research no longer exists. Today, geographers often use a balanced mixture of intensive and extensive methodologies to explore the many facets of everyday life. Indeed, geographers have adopted an approach that emphasizes the complementarity of quantitative and qualitative approaches. In the future, quantitative methods will continue to be used, become more spatially nuanced, and be more fully integrated into the research regime.

—*Jay Gatrell*

See also GIS; Logical Positivism; Model; Qualitative Research; Quantitative Revolution

Suggested Reading

Clark, W., & Hosking, P. (1986). *Statistical methods for geographers.* New York: John Wiley.

Sheppard, E. (2001). Quantitative geography: Representations, practices, and possibilities. *Environment and Planning D, 19,* 535–554.

QUANTITATIVE REVOLUTION

The quantitative revolution is the profound intellectual transformation occurring in Anglo-American geography beginning in the mid-1950s that followed from the use of scientific forms of theorizing and statistical techniques of description and empirical verification. In the process, an older regional geography concerned with describing, cataloguing, and delineating unique places was pushed aside and replaced by the "new geography" directed toward explaining, scientifically proving, and abstractly theorizing spatial phenomena and relations. Geography no longer was rote memorization of regional capitals, major waterways, and principal products; instead, it was now a science, that is, spatial science.

The quantitative revolution's origins are in World War II. Several of the quantitative revolution pioneers were first trained in statistical methods and scientific theory while serving in the military (particularly in the U.S. Air Force). In addition, wartime service convinced a number of geographers conscripted by the U.S. Office of Strategic Services (OSS, an important arm of military intelligence) of the limitations of the older regional geography. Necessary was geographic systematicity, explanatory purchase, and practical focus, none of which was found in the regional geography in which such geographers were trained. This experience proved to be decisive in postwar university classrooms.

It took roughly 10 years for the seeds planted during World War II to germinate. When the quantitative revolution emerged, it initially was highly localized and centered on one or two key individuals. In the United States, pivotal were the geography departments at the University of Washington in Seattle and the University of Iowa. At Washington, it was Edward Ullman (formerly at the OSS) and William Garrison who made the difference. In 1954, Garrison gave the first advanced course in statistical methodology in a U.S. geography department. And in an early advertisement for the department, the chair, Donald Hudson, boasted of the departmental use of an IBM digital computer, another national first. The first cohort of graduate students from that department (the "space cadets") became a "who's who" of geography's quantitative revolution: Brian Berry, Ron Boyce, William Bunge, Michael Dacey, Arthur Getis, Richard Morrill, John Nystuen, and Waldo Tobler. Collectively, this group was critical in diffusing the Washington message and did so by rapidly establishing themselves and their research agenda at several prestigious U.S. universities, including the University of Chicago, Northwestern University, and the University of Michigan. At Iowa, Harold McCarty, the first human geographer to use a regression equation, was decisive. He attracted a number of graduate students who again were vital in spreading the word about numbers at places such as Ohio State University, MacMaster University, and (later) the University of California, Santa Barbara. Outside of North America, Peter Haggett and Richard Chorley in the United Kingdom (the "terrible twins" of British geography) and Torsten Hägerstrand in Sweden were crucial in establishing European beachheads.

The new geography that emerged, and that was solidified by the mid-1960s, was characterized by several features.

• *A thirst for rigorous formal theory and slaked by begging, borrowing, and stealing from at least five sources outside of geography.* First, Newtonian physics provided ideas of gravity and potential as well as the basis for spatial interaction modeling, that is, the analysis of geographic flows of people and things. Second, neoclassical economics gave the rationality postulate used to theorize geographic choice. Third, an older and hitherto forgotten German school of location theory offered mathematically exact models of agricultural land use, industrial location, and urban–economic settlement patterns. Fourth, urban sociology afforded both intra- and intermetropolitan explanatory models of population and their sociological characteristics. Fifth, geometry made available axioms of topology (the mathematical study of spatial forms) used in transportation studies. More generally, there was a belief that rigorous theory would reveal and explain an underlying spatial order and, at the limit, could be couched as a series of geographic laws of the type found in natural science.

• *The use of an increasingly sophisticated set of statistical and mathematical methods.* Initially, statistical hypothesis testing was rudimentary, but it was quickly ratcheted upward. By the mid-1960s, there was widespread use of complex multivariate inferential statistical techniques. In addition, there was pure mathematical modeling in which formal models were logically derived from a set of abstract assumptions expressed precisely. The implicit justification in both cases was that the geographic world and the world of mathematics were fundamentally ordered according to the same rational logic. Mathematics is nature's own language.

• *A reliance on computerization.* The first commercially sold computers were introduced on American campuses during the mid-1950s (the IBM 650 was the first and was introduced at Columbia University in 1954). Initially, there were no formal programming languages, and the capacity of computers to carry out calculations was limited. The pioneers of the quantitative revolution, however, were some of the earliest users of computers in American universities, often not getting their turn until after midnight and improvising programming techniques on the fly. By the mid-1960s, the computer was essential to the new geography. The complex calculations necessitated by multivariate statistical techniques and large-scale data sets could not be undertaken in any other way.

• *A new professional and social structure.* Young, male, very ambitious, very able graduate students and junior faculty primarily forged the quantitative revolution. Initially blocked by a regional old guard, the "young Turks" set up their own dedicated conferences, their own training sessions to teach the rest of the profession the merits of a quantitative sensibility (the Summer Institutes in Quantitative Geography initiated in 1961 were formative), and their own outlets for publication (the various discussion paper series, particularly the Michigan Interuniversity Community of Mathematical Geographers [MICMOG], were crucial, culminating in 1969 with the specialized journal of quantitative geography, *Geographical Analysis*). In this sense, the quantitative revolution was as much a social and institutional transformation as an intellectual one.

• *The emergence of an alternative philosophical justification for geographic research—positivism.* For the most part, early proponents of the quantitative revolution did not understand their work in philosophical terms. By the end of the 1960s, however, considerable philosophical reflection had gone on around the quantitative revolution's larger intellectual justification. David Harvey's *Explanation in Geography,* published in 1969, was the culminating volume arguing that legitimating the quantitative revolution was *logical positivism,* a philosophy averring that true statements were those—and only those—in which logically consistent theory corresponded flawlessly to experientially grounded facts. It was a far cry from the vision of geography as rote memorization.

Only 4 years later, however, Harvey launched a counterrevolution based on Marxism that, within a decade or so, undid the quantifiers. Partly causing the quantitative revolution's fall was its inability to engage with pressing outside social and political issues and writ large during the late 1960s and 1970s around poverty, civil rights, the environment, war, and gender and racial equality. Partly also, there was a new generation of geographers entering the discipline who, like the quantifiers of the mid-1950s, wanted to make a distinctive mark. In their case, however, it was to be through social theory, not scientific theory. The continuity of that noun—theory—was significant. It ensured that human geography remained part of the social sciences, not lapsing back to the netherworld status in which it languished before the

quantitative revolution. This is perhaps where the real revolution lay.

—Trevor Barnes

See also Location Theory; Logical Positivism; Model; Paradigm, Quantitative Methods

Suggested Reading

Barnes, T. (2001). Lives lived, and lives told: Biographies of geography's quantitative revolution. *Environment and Planning D, 19,* 409–429.

Barnes, T. (2004). Placing ideas: Genius loci, heterotopia, and geography's quantitative revolution. *Progress in Human Geography, 29,* 565–595.

Billinge, M., Gregory, D., & Martin, R. (Eds.). (1984). *Recollections of a revolution.* London: Macmillan.

Burton, I. (1963). The quantitative revolution and theoretical geography. *The Canadian Geographer, 7,* 151–162.

Gould, P. (1978). The Augean period. *Annals of the Association of American Geographers, 69,* 139–151.

QUEER THEORY

Queer theory is a deconstructionist movement that interrogates the categorization of identity according to gender and sexuality and also challenges the notion that identities are liberatory, separable, or essential aspects of human subjectivity. The use of the word *queer* is meant to defy the bigotry of homophobia by reinscribing a term traditionally used to derogate those engaging in nonheterosexual activity. The term *queer* suggests that queer theory is anomalous to the norm. Indeed, queer theory is best defined as a movement that is antinormative and antifoundationalist.

Queer theory arose primarily via critique of Western gay and lesbian movements that, during the 1960s and 1970s, used gender and homosexual identities as bases for politics and scholarship. Queer theorists argue that although gay and lesbian identity politics assumes that proclaiming a homosexual identity is a fundamental right of the liberal subject in Western democracies, identity claiming problematically buttresses the heterosexual/homosexual binary. Queer theorists contend that proclaiming gay identities as an effort of inclusion, even in a struggle for civil rights, paradoxically ensnares subjects in a humanistic logic whereby conscious identity choice falsely signals freedom and liberty.

The development of queer theory during the late 1980s and early 1990s also responded to frustrations stemming from identity politics movements that failed to adequately address the intersections of identities and the question of how to prioritize one identity over another in the struggle for gender, sexual, class, or racial liberation. Therefore, queer theory must be placed along a continuum of feminist, gay and lesbian, and racial civil rights movements, although it essentially challenges the bases of these. Queer politics also accompanied the development of queer theory during the 1980s and 1990s, and early works in queer theory considered the political activism of groups such as Queer Nation, Lesbian Avengers, and ACT-UP (AIDS Coalition to Unleash Power). AIDS activism was a central component to these antiestablishment, antiassimilationist politics that sought to recenter the norm away from heterosexuality and straight-acting, assimilating gays.

Queer theorists suggest that identities, particularly sexual identities that are organized around the genders of sexual object choice such as straight (men desiring women, women desiring men), lesbian (women desiring women), and gay (men desiring men), do not reflect inner qualities of people and should not be a basis for any sort of liberatory politics. According to queer theory, sexual identities such as gay, lesbian, and straight are not universal, timeless, or equivalent. Through this argument, queer theorists advocate a genealogical approach to the study of sexuality and argue that sexual categories and the identities that follow them are historically and socially specific and rotate around a normative and dominant heterosexuality. They point to the historical development of terms such as *homosexuality,* which was invented as a primary classification of the human species and was promulgated during the 19th and 20th centuries through sexology, psychology, and medicine. Indeed, homosexuality was classified as a psychological pathology and an aberration of so-called normal (i.e., heterosexual) sexual development. For example, until 1973, homosexuality was cataloged as a mental sickness in the United States by the American Psychiatric Association. It was the multiple effects of this history that gay and lesbian identity politics has sought to dispel by claiming civil rights as sexual minorities.

Queer thinking surely owes a theoretical debt to feminist, poststructuralist, and psychoanalytic philosophies, which demand that social theory reconsider both the categorization of identity and behavior, and to the conceptualization of human subjectivity, which assumes that subjects are fully self-aware,

stable, and self-contained. Queer theory represents a poststructuralist, deconstructionist movement in that it insists on the indeterminacy of identity in the first place, so that identity politics becomes impossible. Perhaps its greatest debt is to French theorist Michel Foucault, whose histories of sexuality have had immeasurable influence. Foucault was one of the first theorists to suggest that subjects are not freed by proclaiming their identities; rather, they are ensnared in powerful discursive requirements to define themselves by normative categories of difference such as sexuality. Such a reading would tie the proclamation of a gay identity with the history of homosexuality's pathology; sexologists invented the term, and in time homosexuality came to stand for an internal essential identity that, ironically, gays and lesbians themselves now reproduce through identity claiming and identity politics that advocate the category.

Queer theorists and activists, in response, encourage disidentification. By refusing to claim a sexual identity at all, including homosexual, queers highlight the historical construction of the word and the identity's inability to define the subject completely. By identifying as neither heterosexual nor homosexual, for instance, queers illustrate how these categories cannot possibly capture the rich and mutable texture of desire, sexual practice, sexual object choice, and changing and multiple identification. Thus, the act of disidentification calls into question the presumed fact that sexual and gender identities are natural biological qualities of men and women and that sexuality should stand as a primary and defining representation of subjectivity.

Queer has also come to be an umbrella term that represents all nonnormative sexual and gender identities and, as such, draws together bisexuals, gay men, lesbians, transsexuals, transgenders, sadomasochists, fetishes, and so forth—even heterosexuals who wish to resist the dominant implications of opposite-sex desire. However, many queer theorists argue that the umbrella use of *queer* neglects the heart of queer theory's critique. Rather, queer theory and politics dismiss the ideal of sexual and gender identity foundations, thereby precluding the possibility of a composite sexuality identity at all, even one that attempts to disrupt heterosexism.

—*Mary Thomas*

See also Feminisms; Gays, Geography and/of; Gender and Geography; Lesbians, Geography of/and; Poststructuralism; Sexuality, Geography and/of

Suggested Reading

Butler, J. (1990). *Gender trouble: Feminism and the subversion of identity.* New York: Routledge.

Foucault, M. (1978). *History of sexuality* (Vol. 1). New York: Vintage Books.

Sedgwick, E. (1990). *Epistemology of the closet.* Berkeley: University of California Press.

Warner, M. (Ed.). (1993). *Fear of a queer planet: Queer politics and social theory.* Minneapolis: University of Minnesota Press.

R

RACE AND RACISM

Race and racism are interlinked concepts. In broad terms, the concept of race tends to be used to identify human differences, often in relation to visible physical differences such as skin color. Racism refers to the practice of enforcing difference and relies on the power relations that inhere in relations between different groups understood as belonging to different races. Early work in geography took race as a biological fact and used this so-called fact to map the geographic extent of racial differences. The mapping of racial difference in turn served to reinforce the idea of race as a biological fact and to reinforce racist beliefs and practices. Although this understanding persists, more recent work in geography has highlighted the ways in which race is a social, cultural, and political construct. This understanding is more cognizant of the ways in which the practice of racism helps construct and maintain particular concepts of race. In all instances, however, geographers interested in race and racism seek to demonstrate and investigate the relationship among race, racism, and space—how race makes space and how space makes race.

MODERN UNDERSTANDINGS OF RACE

During the 19th century, scientists were concerned with the identification and classification of hierarchies of race. Geographers of that period, and of the early 20th century, sought to identify the relationship between these assumed a priori racial categories and place. For many geographers, race and place were intricately connected; particular kinds of places and particular kinds of climate produced particular kinds of racial characteristics. This form of knowledge construction—now described as environmental determinism—played an important role in both colonialism and the eugenics movement and was very influential in early-20th-century geography. It lost favor, however, after concerted attacks from human geographers, most notably Carl Sauer. It was also tainted by its association with Nazism. For a period from the 1940s onward, the issue of race dropped out of geography and geographic research. Geographers, in common with other social scientists, tended to research and write as if race did not exist.

THE PROBLEMS OF RACE

From the mid-1960s onward, American geographers in particular became explicitly interested in issues of race. This shift coincided with the reinvigoration of the U.S. civil rights movement, which drew attention to the ways in which African Americans had been systematically discriminated against by structures of white privilege. This also coincided with a movement within geography to make the discipline more socially relevant. As a consequence, geographers began to engage with social "problems" such as segregation, riots, gerrymandering, and the creation of ghettos. In so doing, geographers sought to use a range of different techniques to map, identify, and analyze the ways in which social problems and race were linked. These included spatial analytic methods with an emphasis on spatial distribution, interactions, and inequalities. Although this started to make race more visible, spatial relations were the primary object of analysis. Race was understood, in this context, as an unproblematic explanatory variable. However, this approach exemplifies one of the three broad ways in which geographers have engaged with issues of racism: through identifying the spatial consequences of racist structures, processes, and practices. This form of engagement continues, most recently with developments in the area of environmental racism and environmental justice.

PROBLEMATIZING RACE

During and after the 1980s, the cultural turn in geography began to have an impact on the ways in which race was understood. Rather than seeing race as a biological fact or as an analytical category, some geographers began to conceptualize race as a social construction. This was particularly the case in Britain, where academics sought to make sense of the changing nature of social relations, in particular the impacts of migration to Britain of residents of its former colonies. Given this changing context, some British geographers began to direct their attention to the discursive construction of race. Drawing on the work of Edward Said and Stuart Hall, among others, they started to interrogate the ways in which representations of race and place were interlinked. Raymond Williams's cultural materialism was an important foundation for this work as well. This resulted in a new focus on racialization—how race (and racism) is socially produced in particular geographic and historical contexts. Within British geography, however, analyses of racialization often were directed to historical contexts. Geographers interested in contemporary events were more likely to address those issues through an analysis of the spatial relations of race and ethnicity. However, geographers based in other contexts, particularly in white settler societies such as Canada, South Africa, Australia, and New Zealand, were more likely to consider the contemporary implications of racialization, especially in relation to the indigenous populations of those societies. Apartheid South Africa was important as a site for these discussions, as were indigenous–white relations in Canada, Australia, and New Zealand.

In addressing the historical geographies of racialization, British geographers paid particular attention to the relationship among geography, race, imperialism, and colonialism. In so doing, they highlighted the ways in which geography served the imperial project through its construction of racial subjects in place and exposed the hidden assumptions about racial superiority that underpinned the work of colonial geographers. These critical historiographies have highlighted the ways in which the history and practice of geography as a discipline rests on a particular form of white privilege, often reinforced by gender and class privileges. Critical geographers have developed this analysis, highlighting the problematic ways in which geographers interested in racialization have focused their attention on explicitly racialized people and places. This emphasis on the social construction of a limited number of racial categories has ignored the ways in which the category of white is constructed. Geographers now are beginning to direct their attention to the ways in which whiteness as a category of privilege is constructed, reinforced, and normalized.

This concern with the process of racialization represents the second broad way in which geographers have engaged with issues of racism. By exploring and exposing the historically and geographically contingent construction of race, geographers highlight the role of human agency in producing and challenging racist practices.

ANTIRACISM

The third broad way in which geographers are engaging with racism is by suggesting ways in which racist structures, processes, and behavior can be reconstituted as antiracist. In some instances, this builds on identifying and attempting to address spatial inequalities connected to race. In other instances, this works by highlighting the ways in which the discursive construction of race reinforces concepts of racial hierarchies. Recently, however, some geographers have been advocating a more radical approach to antiracism in geography that involves explicit activism.

Radical antiracism in geography takes two forms. The first is directed to disciplinary structures and practices. The second is directed to constructing research projects that address issues of racism from a range of different perspectives. In considering the need for antiracism within the discipline, some geographers have highlighted the ways in which white people dominate geography. In so doing, they argue for increased racial diversity in terms of the student population, faculty, and curriculum. More fundamentally, however, they argue for the need to expose and challenge white privilege within the discipline and within the academy. In considering the need for research that is grounded in antiracism, some geographers have argued that it needs to address structural issues of inequality, to address questions of public policy, and to be grounded in individual and community activism. More broadly, advocates of antiracism argue that race interacts with a range of other markers of identity, such as gender, class, ethnicity, nationality, and sexuality, to produce complex relationships of dominance and subordination. An antiracist agenda for geography needs to be cognizant of these complexities and of the ways in which they are embedded in institutions, structures, and the spaces of everyday life. It also needs to be cognizant of the ways in which studies of racialization themselves are racialized.

REPRODUCING RACE AND RACISM

While some geographers are actively involved in identifying racist practices, deconstructing the concept of race, and promulgating antiracist approaches, many others continue to be complicit in the maintenance of racism. This tension, between racist and antiracist practices, has underpinned the development of geography as an academic discipline.

Research and teaching grounded in ideologies of antiracism have, at times, served to reinforce the social practices they aimed to undermine. Although deconstructive approaches to the history of geography have pointed out the discipline's complicity in the articulation of racial hierarchies and racist practices, they help create the impression that the process of racialization is historical rather than ongoing. The identification of race as a "problem," although motivated by a desire to expose racism, often served to reinforce the association of particular races with particular problematic social characteristics and to locate particular races in specific so-called problem places. In some instances, human geographers interested in issues of race and racism focus on the importance of human agency and individual choice, thereby effacing the social relations of power that govern racialized relations in white racist societies. In this way, white privilege often makes it difficult for white geographers to see the power of race in white societies. In addition, attention to the discursive construction of race has also focused attention on groups that are explicitly racialized, such as African Americans in the United States and blacks or Asians in Britain, rather than toward groups whose racial identity is assumed and unquestioned, such as white Americans and white Britons.

However, the majority of geographers continue to work uncritically within frameworks of understanding that are explicitly or implicitly racist. This is particularly the case in relation to the construction of world regional geographies. Descriptions such as *developed* and *developing* countries, or *First* and *Third* Worlds, reinforce the racist epistemologies that underpinned early academic geography by racializing places and by creating hierarchies of these racialized places. Racialized places continue to be constructed as problematic, for example, through discussions of the problems of underdevelopment or through their association with corruption, disorder, and disease. These associations are also apparent in discussions of migration and migrants. Postcolonial geographers are starting to highlight the ways in which geography continues to be shaped by colonial epistemologies, but they have been slow to engage with the implications of this practice for contemporary understandings of race and place. So long as geographers work within these hierarchical categories of race and place, racism will continue to be present within the discipline of geography.

—*Mary Gilmartin*

See also Colonialism; Critical Human Geography; Discourse; Environmental Determinism; Environmental Justice; Ethnicity; Eurocentrism; Ghetto; Human Agency; Identity, Geography and; Imperialism; Justice, Geography of; Migration; Neocolonialism; NIMBY; Orientalism; Other/Otherness; Postcolonialism; Power; Radical Geography; Redistricting; Segregation; Spatial Inequality; Urban Underclass; Whiteness

Suggested Reading

Anderson, K. (1991). *Vancouver's Chinatown: Racial discourse in Canada, 1875–1980*. Montreal: McGill–Queen's University Press.

Bonnett, A. (2000). *Anti-racism*. London: Routledge.

Dwyer, O., & Jones, J. (2000). White socio-spatial epistemology. *Social & Cultural Geography, 1,* 209–221.

Jackson, P., & Penrose, J. (Eds.). (1993). *Constructions of race, place, and nation*. London: UCL Press.

Western, J. (1997). *Outcast Cape Town* (2nd ed.). Berkeley: University of California Press.

Wilson, B. (2000). *Race and place in Birmingham: The civil rights and neighborhood movements*. Lanham, MD: Rowman & Littlefield.

RADICAL GEOGRAPHY

Radical geography began as an explicitly termed area of study in Anglophone geography during the late 1960s amid a context of crisis. Cold war militarism and imperialism had a heavy human cost in Vietnam, extreme race and class stratification of American cities had been accompanied by massive unrest, and the global economy was limping along under inflation, stagnant productivity gains, and a looming international debt crisis. At the same time, some ecologists issued dire warnings of impending doom that accompanied soaring populations. What, some began to ask, did geography have to offer—not just to understanding these deep problems but also to solving them?

The answer for some was a turn to Marxist theory and a radical politics. The formal emergence of the radical turn came with the publication of the first issue of the journal *Antipode* in 1969. *Antipode* billed itself explicitly as a radical journal of geography. It intended to serve as a forum for publishing research with a radical political commitment that its initial founders cast, almost by definition, as one oriented around not only challenging dominant geographic thinking but also—and more important—addressing deep social problems. Early work published in the journal focused on issues such as the connection between inner-city poverty and race, the geographies of imperialism and underdevelopment, and protest politics.

Radical geography interrogated virtually everything about the then-existing dynamics of disciplinary scholarship—everything from the research questions geographers asked to the theoretical, conceptual, and methodological tools they used to answer these questions. In many ways, the radical turn was a turn away from the positivism of the quantitative revolution that had dominated geography during the post–World War II period. For a new generation of geographers, the dominant orientation of the discipline reflected a profound technocratic conservatism, with its practitioners entirely unaware of the degree to which their work perpetuated commonsense understandings of the world that served to naturalize and reinforce existing inequalities. In response, radical geography was shaped by a perspective on scholarship that set it radically apart from positivism. First, for radical geographers, social problems were not technical problems in need of technical solutions; instead, they reflected deep contradictions in the nature of capitalist development. Second, and following, in the same way there could be no value-free technical solutions, there also could be no value-free science. Geography itself was part of the problem to the degree that geographers blindly believed in their own neutrality, which simply had the effect of re-creating the normative worldview that caused the problem in the first place.

Among the more influential of the radical geographers was David Harvey, whose book *Social Justice and the City,* published in 1973, announced his turn away from positivism to historical materialism with breathtaking analytical power and political commitment. The book was an ambitious attempt to take the basic theoretical insights of Karl Marx on the dynamics of capitalism and to recast them as fundamentally spatial dynamics. Marx, for example, discussed the tendency for capitalism toward crisis. Harvey argued that capital circulated not only through time but also through space. The decimation of traditional large-scale manufacturing in the U.S. Northeast was fundamentally tied to a rapid industrialization of places such as East Asia.

Although much of the initial emergence of the radical turn was related to urban geography, it also had a significant impact on other fields, most particularly in development studies and political ecology, which took the basic theoretical principles and political commitments of urban geographers such as Harvey and applied them to studies of environment–society relations.

FROM RADICAL GEOGRAPHY TO CRITICAL GEOGRAPHY

By the early 1980s, radical geography had gone mainstream as its practitioners rose to the vanguard of geographic scholarship. More recently, radical geography arguably has lost its previous influence as broadly left geography has diversified under the banner of *critical geography.* The confidence of the initial development of radical geography has given way to a period of greater uncertainty and internal debate. The shift from radical geography to critical geography is a product of a number of factors. First, the influence of postmodernism and poststructuralism during the 1980s and 1990s severely challenged the theoretical foundations of Marxism at the root of radical geography. At the same time, the changing theoretical winds themselves were rooted in more grounded critiques of the radical geography project. Feminist and antiracist geographers, for example, objected to what they believed was the narrowly class-based commitments of Marxist geography.

Radical geography was perhaps in part a victim of its own success in that its ideas became so well established within geography that they became taken for granted and seen as a kind of new orthodoxy in need of challenge by a newer generation of scholars. Still, the emergence of radical geography began a vibrant period of innovation in geography that came about largely as a result of the forceful quest both to challenge dominant thinking and practice and to make a difference in the geographic world beyond the gates of the academy.

—*Bruce D'Arcus*

See also History of Geography; Justice, Geography of; Marxism, Geography and; Political Geography; Uneven Development

Suggested Reading

Blaut, J. (1970). Geographic models of imperialism. *Antipode, 2,* 65–85.

Hague, E. (2002). Antipode, Inc.? *Antipode, 34,* 655–661.

Harvey, D. (1973). *Social justice and the city.* London: Edward Arnold.

Harvey, D. (1974). Population, resources, and the ideology of science. *Economic Geography, 50,* 256–277.

Peet, R. (Ed.). (1977). *Radical geography.* Chicago: Maaroufa Press.

REALISM

Realism is a term that has a life outside of academia, for example, when someone refers to the need to "be realistic" in the sense of accepting a given situation and dealing with it without fuss or sentiment. Although this everyday use does not indicate the full complexity, and even ambiguity, of the term as it appears within academic debates, it does reflect something quite important about how philosophers and scientists think about the concept. This is because it points, first and foremost, to the need to acknowledge the "reality" of a situation regardless of what the person experiencing it may well wish or have believed and, second, to the desirability of acting in an appropriately "practical" manner. How, then, do academics think through the complexities of reality and practicality?

To take the term *reality* first, it is generally taken as a defining feature of realist thought that matter, in the form of both objects and events, exists *independently* of the observer. What this means is that despite how ideas on the nature of, say, the earth may change over time and across cultures—that is, whether we think of the earth as flat or spherical, as a product of God or the "big bang"—there is no denying the fact that the object itself is everpresent and that this presence would continue if all of humanity were to be terminated. Even objects that are manufactured, such as houses and cars, are considered to have an independent existence in the sense that once they are fashioned they have a physical presence.

A *practical* response to this situation is to attempt to capture *accurately* something of the type and character of these phenomena such that objects can be used more readily and future events can be predicted more readily. For some academics, commonly grouped under the heading of *rational* materialism, this can be accomplished through the rejection of a priori theories and beliefs about the nature of phenomena (including religious, emotional, and subjective observations) and an emphasis on the adoption and careful deployment of the *scientific method*. This consists of, among other things, the quantification of the empirical aspects of phenomena (i.e., that which can be observed and measured through the use of accepted equipment and scales) and the use of those modes of representation that can best approximate the real-world form of those phenomena (e.g., mathematics).

For others, working within *historical* or *dialectical* materialism, it is also taken as a given that an ordered reality of objects and events exists independently of our perception of it and that, in principle, we can gain accurate knowledge of that world. This approach does differ from that just outlined, however, in its insistence on the fact that we cannot rely on empirical observation to accomplish this. This is because, it is argued, the world we observe around us has been fashioned within, or is influenced by, broader-scale social relations of production (e.g., feudalist, capitalist, socialist) that must be deduced to appreciate their nature and efficacy. Moreover, because we are a part of this world, we must acknowledge the manner in which our own and others' bodies, as well as our own and others' thoughts and feelings, are also shaped by these same relations of production. This reality often is hidden from us by the promulgation of false ideologies about the way in which the world works; however, it is argued that we are capable of appreciating and articulating truths concerning society and the individual as well as the interaction between society and nature. On a practical level, the best way in which to proceed is to alert others to the reality of their own situation and to work to produce a society in which misery, exploitation, and degradation are eradicated.

In general, realism is seen to stand in opposition to the notion of idealism, which is the conviction that all of those objects and events we see, hear, touch, and smell, as well as those we learn of from secondary sources, are indeed *dependent* on the observer. This is because there is no actual *proof* one can give of the existence of these phenomena outside of their being observed by oneself or others; we never can know the outside world directly, nor can we gain an external perspective from which to analyze our own ways of knowing. From an idealist perspective, then, realism is akin to a leap of faith in that it relies on a *belief* in the independent existence of phenomena. Moreover, the attempt to provide an accurate or truthful representation of reality is seen to be doomed to failure in that one cannot stand outside of the analytic process and evaluate which theory is indeed the best approximation.

Although this brief outline of realism versus idealism glosses over the complexities of these debates, it does provide a useful insight into why they are often so contentious. This is because, on the one hand, the rejection of the independent existence of phenomena can lead to the charge of *relativism,* where no one theory of how the world works is seen to have the advantage of accuracy, and so surely all theories are just as good as each other. And it has led to the charge of *nihilism* in the sense that if there is no best determination of what is good or evil in the world, surely all ethical frameworks are just as good as each other. On the other hand, the uncritical adoption of the scientific method has led to the charge of *scientism,* where the experimental procedures used within a scientific project are taken to be a guarantee of its practitioners' neutrality and objectivity. And the belief in the ability of materialists to pronounce on the truth concerning the world around them has been labeled a God's eye view, indicative of a transcendent arrogance.

—*Deborah Dixon*

See also Epistemology; Paradigm; Structuration Theory

Suggested Reading

Latour, B. (1999). *Pandora's hope: An essay on the reality of science studies.* Cambridge, MA: Harvard University Press.
Sayer, A. (2000). *Realism and social science.* Thousand Oaks, CA: Sage.

REDISTRICTING

Redistricting is the process of drawing electoral district boundaries for the purpose of electing members to a legislative body. Redistricting is conducted periodically for legislative bodies in countries that elect candidates to office from discrete geographic districts (e.g., Canada, United Kingdom, United States). In the United States, since the 1840s (with rare exceptions), members of the U.S. House of Representatives have been elected from discrete single-member districts. In addition, many members of U.S. legislative bodies at other scales (state, county, and local legislatures) are elected from geographically based districts (whether they be single- or multimember). In the United States, electoral districts generally are redrawn once a decade following the release of the decennial census population data so as to reflect population changes uncovered by the census. In most cases, state legislatures are responsible for redistricting U.S. House and state legislative seats. In some states, however, nonpartisan agencies (e.g., Iowa) or bipartisan commissions (e.g., New Jersey) are given the power to redistrict. Such efforts aim to reduce the scope of partisanship and gerrymandering (the intentional drawing of election district boundaries to the advantage of one group at the expense of another) in the redistricting process.

Since the 1960s, redistricters in the United States have been constrained by federal court decisions in how districts are to be drawn. Starting in the 1960s and continuing to the 1980s, a series of U.S. Supreme Court decisions mandated that U.S. House districts within a state be equal in population and that state legislative districts be nearly equal. The court rulings were aimed at dealing with issues of malapportionment (large differences in the populations of election districts within the same political jurisdiction) that had kept rural interests in control of many state legislatures and had given them disproportionate power in the U.S. House during the early and mid-20th century as the population shifted from rural to urban areas.

During the early 1990s, redistricters greatly increased the number of African American and Latino population–majority congressional and state legislative districts under guidance from the U.S. Department of Justice in its interpretation of amendments to the Voting Rights Act of 1965. Such affirmative districting measures led to large increases in the number of African American and Latino legislators elected. However, in so doing, some of these new minority–majority districts were geographically extensive and cut across well-recognized regional boundaries. During the mid- and late 1990s, federal courts called into question such minority–majority districts with lesser levels of geographic compactness if they were drawn based primarily on racial factors.

During the 2000s, with partisan control of the U.S. House closely contested and determined by a small margin of seats, the incidence of partisan gerrymandering increased as majority party leaders in state legislatures created plans that helped to increase (or at least maintain) their party's share of seats in the U.S. House (e.g., Texas). Although partisan redistricting has occurred since the beginning of political parties in the United States (and has not been declared unconstitutional by the U.S. Supreme Court), such gerrymandering has become easier during recent decades due to advancements in geographic information systems (GIS) technologies that allow district mapmakers to create and analyze a greater multitude of redistricting plans on their computers at much greater speeds than they could previously. At the same time, public interest groups have also invested in GIS technologies to highlight and call into question the partisan nature of such district plans.

—*Jonathan Leib*

See also Electoral Geography; Political Geography

Suggested Reading

Morrill, R. (1999). Electoral geography and gerrymandering: Space and politics. In G. Demko & W. Wood (Eds.), *Reordering the world: Geopolitical perspectives on the 21st century* (pp. 117–138). Boulder, CO: Westview.

Webster, G. (1997). Geography and the decennial task of redistricting. *Journal of Geography, 96,* 61–68.

REGIONAL GEOGRAPHY

In much of the popular conception of geography, the discipline is concerned with the study of regions and little else. Geographers have examined regions at a variety of spatial scales and from a diversity of conceptual perspectives. Although this topic may appear to be relatively free of controversy, in fact the use of regions and their philosophical significance has been the source of considerable debate.

CLASSICAL REGIONAL GEOGRAPHY

Since the classical Greeks, regions have played a central role in geography as a means to collect, organize, and give meaning to spatial distributions. During the 17th century, Bernhard Varenius (1622–1650) wrote the *Geographia Generalis,* a volume that became a major textbook in Europe for the next 150 years and was translated into English by Sir Isaac Newton. Varenius distinguished between specific geography, which was concerned with the unique character of places, and what he called *general geography,* which was concerned with universal laws.

During the 19th century, three figures loomed large in the formalization of regional geography. First, Carl Ritter (1779–1859) wrote the 19-volume *Erdkunde* (published in 1818), a comprehensive world regional geography text that emphasized the comparison and synthesis of facts through a regional approach, largely with religious goals in mind. Geography's purpose was to detect the whole character of places. Comparative local studies were to be the basis through which generalizations could be made. Ritter claimed to see evidence of divine plans in the world's geography, advocating a religious teleological interpretation. Second, Paul Vidal de la Blache (1845–1918), widely seen as the founder of modern French geography, developed the notion of *genres de vie* (or local lifestyles), which celebrated the uniqueness of rural landscapes in French *pays.* Noting the variations across France in the face of a common climate, he maintained that culture—not nature—was primarily responsible, using this theme to bludgeon environmental determinism and introduce possibilism. Third, Alfred Hettner (1859–1941), in the Kantian tradition, defined geography in chorological terms, believing that the discipline's importance was maintained by its comprehensive regional approach rather than by its subject area and that synthesis was its greatest strength.

AREAL DIFFERENTIATION

The American version of regional geography reached its apex between the two world wars with the ascent of the school variously labeled as *areal differentiation, chorology,* or *regional description.* The ascent of this line of thought was to be found in the aftermath of environmental determinism, when the discipline's retreat from theory sharply differentiated it from other social sciences that were making great strides. Its embodiment is Richard Hartshorne (1899–1992) and his definitive landmark text, *The Nature of Geography* (published in 1939). Having studied under Hettner and thus heavily Kantian in outlook, Hartshorne made a variety of claims regarding regional geography as the definition of the discipline's core and claim to uniqueness within the academic division of labor. Geography, like history, was synthetic, integrating the analysis of different phenomena as they were manifested in unique combinations in particular places. Regions allowed the analysis of both human and physical phenomena, transcending the growing schism between two parts of the discipline. Because the complexity of the world is overwhelming, Hartshorne advocated the study of small regions with relatively little internal variation, accreting this into a mosaic that would encompass larger areas. This view subscribed to a crude spatial determinism in which proximity came to stand for causality; where things were enough to ascertain their nature, and closer phenomena were more likely to be related than more distant ones. He well understood that regions are only tools and, in the vein of Kant, maintained that regions are only mental constructs, that is, simplifications of the world that the mind uses to impose order on the world.

Hartshorne essentially regarded regions as static and held little regard for the need to engage in the study of explanatory processes; that is, regional geography was to be only about appearances. This approach represented a spatialized version of the philosophy of empiricism (with its roots in the British Enlightenment), that is, the assumption that facts and data are true or false without appeal to theory. In contrast, more recent approaches argue that all data are theory laden; that is, it is theory that informs us what facts are relevant to an issue at hand. There are no pure facts, given that observations always are couched in terms of a theory; observation requires interpretation, which in turn requires theory. Hartshornian chorology was the essence of inductive empiricism, and its Achilles' heel was the inability to engage in abstraction.

American regional empiricism suffered a devastating set of setbacks during the 1950s as it was pushed aside by an increasingly assertive school of positivists. A famous article by Frederick Schaefer published in 1953 charged that Hartshorne's claim that geography alone studied the unique was naive, arrogant, and immature. All sciences face the problem of uniqueness, and explanation (as opposed to regional description) consisted of embedding the unique within a wider understanding of laws and causal processes. Schaefer opened the door to a heated debate about whether geography should be concerned primarily

with the nomothetic (i.e., law seeking) or be content with the idiographic (i.e., the unique). The rise of positivism saw the triumph of the nomothetic approach and the discipline shift from the study of all aspects of one place to the study of one aspect of many places, a move that entailed a reorientation from the concrete to the abstract, from induction to deduction, from a concern with the specific to a pursuit of the universal. Ultimately, the demise of chorology saw the discipline change from a concern with regions without explanatory laws to one obsessed with laws devoid of regions. Some observers, such as John Hart, maintained that regional geography nonetheless played an important role in minimizing armchair theorizing and in popularizing the discipline at large.

THE NEW REGIONAL GEOGRAPHY

During the 1980s, as postmodernism swept the field, geographers saw a renewed concern for regionalism, often called the localities school. Unlike the atheoretical empiricist chorology of areal differentiation, this approach took theory seriously, embedding regions within wider understandings of capitalism, uneven development, and the spatial division of labor. Originally British in inspiration, it began with observations that ostensibly uniform processes such as deindustrialization occurred very differently in different places; that is, local contexts made a substantial difference in how that process unfolded. Doreen Massey used this line of thought to develop a geological metaphor of regional landscapes in which their position within changing divisions of labor sedimented layer on layer of investments on the landscape. Regions consist of a palimpsest unique to that place and reflecting their historical trajectory.

More broadly, the localities school implied that no social process unfolds in exactly the same way in different places. From the standpoint of the new localities school, *where* processes occur affects *how* they occur; causality and location are intertwined. This argument reflected the broader insertion of space into social theory in general and the mounting concern for a contextual social science. Such a line of thought dovetailed neatly with the growing popularity of views that emphasized the creative capacities of people and the contingency of social life.

Localities were celebrated as places where the rich world of everyday life plays out. Thus, under chorology, the region was fetishized; under logical positivism, the region was dismissed; and under postmodernism, the region is celebrated as a source of difference, in contrast to the smothering homogeneity of globalization.

The recognition of local uniqueness has become widespread throughout human geography today, and the region has, ironically, recaptured some of its time-honored status. Typically, such approaches tend to be multiscalar in focus, invoking social processes at a variety of spatial scales. For example, regional geography today would do more than say that all places are embedded in a global economy. Rather than being seen as a simple process of telescoping from the global to the local, the global and the local should be seen as intertwined, or *glocal,* in nature.

—Barney Warf

See also Chorology; History of Geography; Idiographic; Logical Positivism; Nomothetic; Postmodernism; Quantitative Revolution; Structuration Theory; Tobler's First Law of Geography

Suggested Reading

Berry, B. (1964). Approaches to regional analysis: A synthesis. *Annals of the Association of American Geographers, 54,* 2–11.

Hart, J. (1982). The highest form of the geographer's art. *Annals of the Association of American Geographers, 72,* 1–29.

Hartshorne, R. (1939). *The nature of geography.* Washington, DC: Association of American Geographers.

Lewis, M., & Wigen, K. (1997). *The myth of continents.* Berkeley: University of California Press.

Massey, D. (1984). *Spatial divisions of labor: Social structures and the geography of production.* New York: Methuen.

Sack, R. (1974). Chorology and spatial analysis. *Annals of the Association of American Geographers, 64,* 439–452.

Schaefer, F. (1953). Exceptionalism in geography: A methodological examination. *Annals of the Association of American Geographers, 43,* 226–249.

Warf, B. (1993). Post-modernism and the localities debate: Ontological questions and epistemological implications. *Tijdschrift voor Economische en Sociale Geografie, 84,* 162–168.

RELIGION, GEOGRAPHY AND/OF

The geography of religion is a small subfield of human geography, but one that has engaged geographers since ancient times, when most people held a religious worldview. From the Sumerians' 4,000-year-old maps depicting beliefs about their gods and the cosmos, to 19th-century Christian geographies of the Holy Land, to current concerns with religious conflict in the Near East, religion has had a secure niche within the discipline. Today most introductory human geography textbooks include sections on religion, with distribution maps of the world's major religions, brief descriptions of their religious principles, photographs of religious expression in the landscape such as temples and cemeteries, and discussions of current religious issues.

Much research in the geography of religion may be typified as the geography of *difference* in which religion appears as an aspect of a nationality, an ethnicity, or a political faction. Many religious groups claim a particular religious site or homeland as foundational to their distinctive identities, for example, Roman Catholics' ties to the Vatican and descriptions of Poland as a Catholic country. Many nations designate official state religions, such as the Church of England and Islamic republics, although most respect the rights of secularists and religious minorities.

Of course, the identification of one faith with its proper place does not hold in many parts of the world where two or more groups compete for the right to control religious sites or entire regions. Danièle Hervieu-Léger called such struggles for territory the geopolitics of the religious. Geographers of religion address territorial conflicts such as the Israeli–Palestinian struggle for the West Bank and the neighborhood dynamics between Protestants and Roman Catholics in Northern Ireland. The partitions of present-day India, Bangladesh, and Pakistan in 1947 and 1971 from a formerly extensive single British colony were designed to minimize interfaith conflict through segregating Hindus and Muslims. Similarly, the division of the Balkan Peninsula of Europe into several micro-states during the 1990s was intended to reduce serious friction between populations based on their religion (Roman Catholic, Eastern Orthodox, and Muslim) and ethnicity by giving each its own homeland. Yet such states seldom achieve homogeneity because their complex histories of immigration and diffusion of religious beliefs inevitably result in religious minorities.

Where religious sites are managed as cultural heritage resources and tourism destinations, they often function as *contested space* even without aggressive confrontations because the power of one stakeholder to manage and interpret a religious site may disadvantage a religious minority or a secular population that equally has a stake in the site yet holds to different traditions. For example, several Christian denominations occupy space in the Church of the Nativity in the West Bank city of Bethlehem, a site plagued by management disputes. In England, the prehistoric monument of Stonehenge attracts neopagan worshippers who dispute the government's management of the site for strictly regulated secular tourism.

In treating a religious group as similar to an ethnic group or nationality, some geographers have focused on the society's relationship with its particular locale in showing how faith communities conceptualize, use, and change their landscapes. Depending on the scholar's inclinations, the belief system may appear as central to understanding the group's relationship with the land or may be somewhat peripheral in a focus on the group's economic life, social structure, or human ecology. Historical geographies of Mormon pioneers in the Great Basin and of early California's Catholic missions in the United States, with their development of irrigated agriculture and unique settlement patterns, are two examples of a "people, place, and region" approach. In cities, immigrants' affiliations with their coreligionists, houses of worship, and (sometimes) a visible ethnic neighborhood reaffirm their sense of religious identity and landscape at a smaller scale.

Movement of religious people is another important consideration. Religious minorities in cities may be new immigrants or even refugees from sectarian violence. Roman Catholicism is on the rise in the United States today due to increases in immigration from Latin America, notably Mexico. Orthodox Jews typically live within walking distance of their synagogue because their laws prohibit driving or using mass transit on the Sabbath. Old Order Mennonite and Amish people in North America do not use automobiles at all, so their transportation by horse and buggy limits their maximum range of travel. Some religious faiths promote or even require pilgrimages to holy places, such as the *Haj,* in which devout Muslims travel to and perambulate within sacred sites in Mecca, Saudi Arabia. Religious travelers visit sacred sites and shrines believed to promote healing or transcendent religious experiences such as the Ganges River of India to Hindus. One interesting recent line of research concerns Muslim women's mobility in cities, both in the West and in the Islamic world. Cultural traditions about women's seclusion from public space or women's need to be veiled in public have notable effects on Muslim women's daily activity paths and time–space patterns.

There is a level, however, at which religion is not comparable to social differences or political conflicts. Religious beliefs themselves, and the theologies that interpret them, create alternative ways of understanding the world. How does one view geographic themes such as the environment, a particular region, urban morphology, and personal mobility through religious lenses? People in precommunist China planned their cities and landscapes to replicate their beliefs about the cosmos, for example, through the currently popular design principles of *feng shui.* Traditional Navajo people in the American Southwest understand their territory as bounded by four sacred mountains and shaped by divine beings. Many religious societies distinguish between secular space and sacred space such as temple grounds, shrines, and burial sites of religious leaders. A view of places as "Roman Catholic space" may be inferred from place names such as São Paulo (St. Paul), Brazil, and El Salvador (the savior). The Association of American Geographers has a Bible Geography Specialty Group whose members "explore" the geographic content of the Bible through fieldwork or textual analyses.

Geography has long been concerned with explaining current events, implying that religion will become an increasingly important topic to study. The rise of evangelical Christians in the American South as a powerful voting bloc, neopaganism as a new form of religion, and religiously motivated terrorism are but three examples of issues likely to attract more geographic research in the near future.

—*Jeanne Guelke*

See also Cultural Geography

Suggested Reading

Hervieu-Léger, D. (2002). Space and religion: New approaches to religious spatiality in modernity. *International Journal of Urban and Regional Research, 26,* 99–105.

Kong, L. (2001). Mapping "new" geographies of religion: Politics and poetics in modernity. *Progress in Human Geography, 25,* 211–233.

Park, C. (1994). *Sacred worlds: An introduction to geography and religion.* London: Routledge.

Stump, R. (2000). *Boundaries of faith: Geographical perspectives on religious fundamentalism.* Lanham, MD: Rowman & Littlefield.

RENT GAP

One of the more popular explanations for gentrification, the rent gap hypothesis first advocated by Neil Smith, approaches this issue in terms of the profitability of land in the urban core. In centrally located sites, actualized ground rents generally decline over time as the buildings and infrastructure age. During the post–World War II boom, low rents on the urban periphery were a major attraction to capital. Deindustrialization and the flight of people and capital to suburbia throughout the late 20th century played major roles in lowering the profitability of land, or devalorizing it, in the central business district (CBD). Low-income residents were incapable of generating rents that guaranteed a high rate of profit. Disinvestment from the urban core was manifested in abandoned buildings and lack of repairs.

However, in contrast to actualized rents, the *potential* rents in such locales remained high. As neoclassical economists tend to frame the issue, urban land was not being put to its "highest and best use." In particular, the rapid growth of producer services over the past three decades and their reliance on agglomeration economies elevated the importance of locations in maximally accessible places. Thus, a significant discrepancy arose between potential and actual rents, that is, a rent gap. Real estate developers and others with a vested interest in land sense that higher rates of profit (rental streams) may be earned by attracting corporate capital or higher-income residents. Lured by the potential profitability of these sites, capital—much of it speculative—began to flow into urban cores in vast quantities as corporations reclaimed the spaces of inner cities, stimulating a building boom in luxury residential uses and commercial ones such as offices, waterfront developments, and sports stadia. The emergence of such land uses typically is accompanied by an influx of well-paid professional labor. Such a process typically involved the invasion and displacement of low-income, working-class, often minority communities. Thus, the rent gap was closed, at least temporarily, until future fluctuations in the economy threaten to re-create it.

The rent gap thesis ties waves of building construction in the landscape to the dynamics of capital accumulation, investment, and uneven development. It offers a structuralist, production-oriented explanation of gentrification that does not rely on demographics or the residential preferences of yuppies. Thus, it posits gentrification in terms of the structural dynamics of class, labor, and capital rather than in terms of individual preferences or lifestyles. Critics have alleged that the thesis is economistic and functionalist, leaving little room for human agency.

—*Barney Warf*

See also Central Business District; Gentrification; Urban Geography

Suggested Reading

Smith, N. (1987). Gentrification and the rent-gap. *Annals of the Association of American Geographers, 77,* 462–465.

Smith, N. (1996). *The new urban frontier: Gentrification and the revanchist city.* New York: Routledge.

REPRESENTATIONAL SPACES

SEE SPACES OF REPRESENTATION

RESISTANCE

Resistance is a simple term for a force in opposition. As a theme or concept deployed in human geography, resistance is inseparable from social or cultural analyses of colonialism, imperialism, capitalism, or globalization. Resistance takes shape as the "anti-" preceding each of the latter words. It began to be used as a part of activist radical geography, particularly as a means of identifying with anticolonial resistance movements. The dangers of romancing resistance while rather blindly backing the peasants, workers, and disenfranchised groups came to the fore fairly quickly for many scholars, who then began to see resistance in more nuanced lighting.

The failures of many movements of resistance against colonialism to produce progressive results, even when they seized power, led some scholars to turn to interpretations of resistance found in the works of Italian philosopher Antonio Gramsci. Gramsci's conundrum involved explaining why the Italian working class and peasants had not emerged at the forefront of a socialist revolution but had instead solidly backed Benito Mussolini's fascism. Why did people not resist their oppressors? Why did they instead back them vociferously? In developing his answer, Gramsci articulated a vision that went beyond seeing the acquiescence of subaltern (here meaning subjugated) classes as mere false consciousness but rather saw it as part of a broader web that manufactured their consent. Gramsci suggested ways in which resistance to that manufacturing process might be fostered through an escape from subaltern culture to be led by a vanguard of organic (meaning derived from the proletariat) intellectuals.

Subaltern studies theorists have criticized the Gramscian approach because it does not do enough to articulate the capacity, or incapacity, of subordinated groups to resist. Political anthropologist James Scott developed an influential critique of Gramscian theory that turned scholars, including geographers, toward a different appreciation of the ways in which resistance can be expressed from that which the still heavily Marxist Gramscian line was seen to allow. Scott championed the capacity of the subaltern groups to devise new methods of resisting their oppression that were virtually invisible as such until one looked a second or third time. In Scott's hands, stealing from the boss, showing up late for work, dragging one's feet about easy tasks, and so forth became the *Everyday Forms of Peasant Resistance* of his book's subtitle. Feminists and those angling for a deeper reading of Gramsci's own ideas have taken Scott to task on a number of counts. Another perhaps more geographically oriented anthropologist, Donald Moore, contributed a chapter to the edited volume, *Geographies of Resistance,* in which he grounded the contestations over identity central to resistance in the ever contested politics of place.

—*Garth Myers*

See also Anticolonialism; Other/Otherness; Subaltern Studies

Suggested Reading

Abu-Lughod, L. (1990). The romance of resistance: Transformations of power through Bedouin women. *The American Ethnologist, 17*(1), 41–55.

Moore, D. (1997). Remapping resistance: "Ground for struggle" and the politics of place. In S. Pile & M. Keith (Eds.), *Geographies of resistance* (pp. 87–106). London: Routledge.

Scott, J. (1985). *Weapons of the weak: Everyday forms of peasant resistance.* New Haven, CT: Yale University Press.

RESOURCE

In common parlance, a resource refers to a supply of any living being, inanimate material, service, or information that can be used for a desired outcome. In geography, the term typically refers to a natural resource, any nonhuman resource derived from the earth, including its land, water, and air. Resources do not exist outside of human valuation or use. For example, coal has existed in the earth for millions of years but was not considered a resource until the Industrial Revolution.

Resources frequently are divided into two categories: nonrenewable and renewable. Nonrenewable resources, such as fossil fuels, minerals, and biodiversity, are those for which there is a finite exhaustible supply on human time scales. Renewable resources are those that have the potential to be replenished. Most renewable resources, such as fisheries and groundwater, can be depleted if the rate of use is greater than the rate of replacement. A special class of renewable resources, perpetual resources, is inexhaustible on human time scales; sunlight is a good example.

Warnings about exhaustion of natural resources have long provoked arguments about resource scarcity and sustainability. Thomas Malthus warned in 1798 that population growth inevitably surpasses food supplies. Paul Ehrlich argued in 1968 that population growth soon would outstrip natural resource availability. Others, such as Julian Simon, have argued that technology and human ingenuity always will find new resources to substitute for old ones.

Resource use is sustainable if current use levels do not diminish the potential for future use. Thus, only renewable resources can truly be used sustainably, and only when the rate of harvest is less than or equal to the rate of replenishment. Of greater concern than the sustainable use of any particular resource is whether development as a whole is environmentally sustainable. This condition might involve substituting some resources for others, but in a way that keeps intact ecosystem services necessary to maintain human livelihoods into the indefinite future. Environmental sustainability requires not only the continued availability of renewable resources but also adequate sinks for pollution produced by using those resources.

Concern about sustainability and resource depletion has led to the study of natural resource management, conservation, and preservation. Preservationist views call for limited or no use of certain natural resources, whereas conservationist or utilitarian views suggest that resources should be protected for use that results in the maximum good for the largest number of people. One important factor in determining sustainability is valuation—whether resources are priced or valued to reflect ecological services and intangible qualities such as the beauty of wilderness.

Geographers also study access to resources—which groups have access to what resources and why. The study of access to and control over natural resources is particularly important in political ecology, which analyzes the environmental and social causes and consequences of (often) unequal resource access. Inequitable resource access can be found at all scales of contemporary social organization. At the global scale, the United States uses far more than its fair share of natural resources; with less than 5% of the world's total population, the United States consumes more than one quarter of all oil used.

Finally, access to a resource usually is related to the property rights that govern it. The type of property right—private, public, or common—may or may not correspond to its resource characteristics. Important characteristics include whether or not a resource is subtractable (whether use by one person takes away from use by another) and whether exclusion is easy or difficult. Mismatches between the properties of the natural resource itself and the property regime that governs it often can lead to depletion.

—Emily Yeh

See also Conservation; Nature and Culture; Political Ecology

Suggested Reading

Conca, K., & Dabelko, G. (Eds.). (1998). *Green planet blues: Environmental politics from Stockholm to Kyoto.* Boulder, CO: Westview.

Cutter, S., & Renwick, W. (2004). *Exploitation, conservation, preservation: A geographic perspective on natural resource use.* New York: John Wiley.

RESTRUCTURING

Restructuring is a process that describes broad underlying shifts in prominent economic, social, and cultural systems. The term is invoked most frequently to summarize the periodic changes in the capitalist economy at the global, national, and regional scales. There were restructurings during the 18th, 19th, and 20th centuries, and restructuring is ongoing. Restructuring has a geographic dimension in that the process often leads to a spatial redistribution of human activity. Economic factors, such as the shift by transnational corporations away from Fordist production systems toward flexible production systems, have caused profound spatial adjustments in economic activity. The nature of capitalism itself, with its ceaseless change and expansion, provokes restructurings. David Harvey cited the movement of capital from the primary circuit of production to the secondary circuit of the built environment as an example of a restructuring that resulted in the redevelopment of urban cores along with gentrification that displaced the poor away from the center. Demographic change is also a stimulant for restructurings, with additional spatial dimensions including the change in migratory flow patterns of people. Technological innovations often drive restructurings, with consequent spatial change in the distribution of growth and development of urban systems.

Economic restructurings are correlated with 50-year Kondratiev waves or long waves of economic growth. Each wave is divided into four phases: prosperity, recession, depression, and recovery. Brian Berry observed that Kondratiev waves are characterized by accelerating rates of price increase from deflationary depressions (the 1840s and 1890s) to inflationary peaks (1815, 1865, 1920, and 1980–1981), followed by decade-long plunges from the peaks to primary troughs (1825, 1873, 1929, and 1991), by weak recoveries, and then by sags into the next deflationary depressions. When economic growth declines, firms become reluctant to invest and unemployment rises. Eventually, the trough of the wave will be reached and economic activity will be stirred up again on the basis of key technologies.

John Borchert noted spatial variation in the intensity of American urban development during key economic restructurings and associated them with innovations in transport and energy technologies that produced well-marked epochs. Each restructuring characterized by new technologies gave unique shape to the character of the urban landscape during that particular time. Cities put down physical infrastructure indicative of the state-of-the-art technology of the time. During periods of economic growth, urban transport and infrastructure were laid down and city building increased. Thus, the infrastructure of the present-day built environment is an accumulation derived from these transport technology epochs marked by technology innovations in water power, steam engines, steel rails, automobiles, and jet engines. Borchert's classification scheme, therefore, provides an excellent way in which to demonstrate the spatial aspects of restructuring as each epoch resulted in a different spatial configuration of the American urban hierarchy.

The first epoch, the sail wagon (1790–1830), was characterized by a compact urban system that was economically oriented toward Europe. The 1790 census recognized only 24 urban places, and most were large port cities on the Atlantic coast or river towns that were linked to the Atlantic coast (Figure 1a). No one city dominated at this point in time, as observed by the nearly equal populations of Boston, Philadelphia, and New York. The smaller inland river towns, such as New Haven, Richmond, Hartford, and Albany, all were navigationally linked to the larger coastal cities; thus, their growth depended on that relationship. Many factories and mills of the time used water power for economic production.

The second epoch, the iron horse (1830–1870), saw the replacement of water power with the steam engine in water and land transportation. In the United States, the use of steam in railroad locomotives increased in significance during the 1830s and allowed for integration with the water transit systems, resulting in a national urban system biased toward ports with relatively large harbors. The 1830 census recognized 90 urban places, and although coastal cities remained prominent, river cities located farther in America's interior were beginning to make their mark on the urban settlement system (Figure 1b). The largest urban places now extended all the way west to St. Louis on the upper Mississippi. The growth during this period was attributed to the railroads, which were complementary to the waterways as they were built outward from the major ports. The development of the railroads accelerated the exploitation of coal, and as a result, places such as Pittsburgh were high up on the urban hierarchy.

The third epoch, the steel rail (1870–1920), was marked by low-cost and mass-produced steel that allowed steel rail to replace the iron on both existing and new rail lines. The railroads became the principal transporter of coal, and inland waterway traffic was diminished, resulting in the relative decline of many small river towns. In the 1870 census, the largest 100 urban places reinforced the patterns established during the iron horse epoch; however, the western limits of the American urban settlement system were much more expansive (Figure 1c). Added to the list of top urban places during this epoch were Kansas City, Omaha, San Francisco, and Sacramento as well as the large industrial cities of Chicago, Milwaukee, Minneapolis, Cleveland, and Detroit. An example of spatial change resulting from the restructuring under way during the steel rail epoch was the fall in ranks of St. Louis and Louisville (although still in the top 100) because they were more linked to the steam engine innovation of the iron horse epoch.

The fourth epoch, the auto–air–amenity (1920–1970), was brought on by the introduction of the internal combustion engine in transportation, and the year 1920 was when motor vehicle registrations became significant and petroleum production began its steep climb. The shift to the automobile and truck and a supporting highway network resulted in a more dispersed pattern of urban settlement (Figure 1d). The largest 100 urban places in 1920 showed that since 1870 many new places had been added at the expense of some that had dropped out. The 1920 urban hierarchy reinforced amenities as a major determinant of metropolitan growth; Florida, the states of the Southwest, and southern California all had centers high in the urban hierarchy.

The striking difference between the largest 100 urban places in 1920 and those in 1970, at the end of the epoch, was that the growth pattern driven by amenities and the service economy became more pronounced. The most systemic change was the appearance of cities in the top 100 that had been just on the outskirts of the previous epoch's large cities (Figure 1e). For instance, in 1970 at the end of the auto–air–amenity epoch, Anaheim, Long Beach, Riverside, and Santa Ana—all clustered around Los Angeles—were now in the top 100, and all except Long Beach had joined those ranks for the first time in 1970. Thus, the emergence of edge cities coincided with the restructurings of the auto–air–amenity epoch. A key advantage for a city in this epoch was its linkage to the nation's interstate highway system and airports.

Borchert identified the year 1970 as the end of the auto–air–amenity epoch and the beginning of the electronic and jet propulsion epoch (1970–). A number of trend reversals were used to choose 1970 as the critical

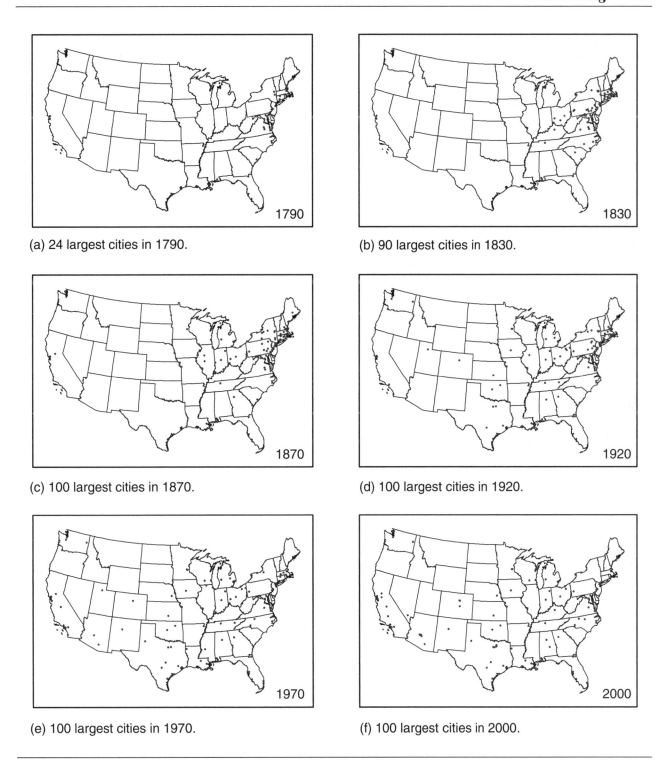

(a) 24 largest cities in 1790.

(b) 90 largest cities in 1830.

(c) 100 largest cities in 1870.

(d) 100 largest cities in 1920.

(e) 100 largest cities in 1970.

(f) 100 largest cities in 2000.

Figure 1   Changing Urban Hierarchy: 1790 to 2000

SOURCE: Data from Gibson, C. (1998). *Population of the 100 largest cities and other urban places in the United States: 1790 to 1990* (Population Division Working Paper No. 27). Washington, DC: U.S. Bureau of the Census.

date, including consumption of paper for recordkeeping, correspondence, magazines, and advertising circulars that accompanied television, computing, word processing, and desktop publishing. The spatial trend of growing edge cities intensified over the 30 years following 1970, with Huntington Beach and Glendale joining ranks in the top 100 cities along with the other Los Angeles suburbs established from the previous epoch (Figure 1f). The phenomenon spread to other metropolitan areas such as Phoenix, with Mesa and Scottsdale now in the top 100 cities list, as well as Dallas and Fort Worth, where Arlington, Plano, Garland, and Irving are, for all intents and purposes, edge cities. Meanwhile, the places that fell out of the top 100 rank between 1970 and 2000 were indicative of the obsolete epochs that gave birth to them, including places such as Evansville, Gary, Providence, Rockford, and Syracuse.

Restructurings that result in broad underlying shifts in economic, social, and cultural systems are associated with Kondratiev waves of economic growth. These restructurings are ongoing and often are stimulated by economic processes as well as technological innovations. The evolution of the American urban system illustrates the role that transportation innovations played in changing dominance patterns within the American urban hierarchy. As new technological innovations were introduced, the map of top cities changed accordingly, reflecting location advantages brought on by these innovations.

—*Richard Greene*

See also Economic Geography; Transportation Geography; Urban Geography

Suggested Reading

Berry, B. (1991). Long waves in American urban evolution. In J. Hart (Ed.), *Our changing cities* (pp. 31–50). Baltimore, MD: Johns Hopkins University Press.

Borchert, J. (1967). American metropolitan evolution. *Geographical Review, 57,* 301–332.

Borchert, J. (1991). Futures of American cities. In J. Hart (Ed.), *Our changing cities* (pp. 218–250). Baltimore, MD: Johns Hopkins University Press.

Harvey, D. (1989). *The condition of postmodernity.* Oxford, UK: Blackwell.

RURAL DEVELOPMENT

The concept of rural development embodies a broad range of ideas and practices that defy easy definition. Loosely, rural development refers to the desire and effort to provide a better quality of life for rural people. But what constitutes a better quality of life and how this may be achieved are hotly debated issues. Indeed, the concept of rural development itself begs two important sets of questions. First, what is rural? Second, what is development? Is rural essentially agricultural? If so, what role, if any, do nonfarm activities play in rural economies? Is development an outcome (e.g., more jobs) or a process (e.g., empowerment)? What aspects of human life are most important? Should development efforts be directed toward expanding economic growth or toward providing basic needs and services to improve human welfare? If poverty alleviation is the goal, how best might this be achieved? Finally, who is development for?

The term *rural* is ambiguous and sometimes is defined as "not urban." Rural areas usually are understood as sparsely populated areas that are dominated by fields, pastures, and forests and where people spend most of their working time on farms or in other resource extraction industries such as fishing, forestry, and mining. As such, they are distinct from more intensively settled urban and suburban areas and from unsettled wilderness or outback areas. Rural areas are highly diverse, and definitions based solely on population density can be problematic given the vast geographic differences in human settlement patterns around the world. Thus, lifestyle factors such as limited access to public services, schools, utilities, and public transport are also used to delineate rural areas.

Since the 1950s, numerous theories, themes, and policy directives have influenced rural development thinking and practice. One of the most predominant of these, according to Ellis and Biggs, is the theory that small-scale subsistence farmers can form the basis of economic development in rural areas. At the time this idea took root, it marked a major departure from the thinking that had prevailed throughout the 1950s, which viewed industrialization and modernization as the path toward development and large-scale agricultural enterprises (e.g., plantations, ranches, commercial farms) as the most viable agents of economic growth. Small-scale subsistence farmers, according to this early perspective, were viewed as inefficient and stagnant; thus, they were best left to provide resources to the modern agricultural sector, which eventually— and inevitably—would replace them.

After the mid-1960s, however, such thinking began to shift as development economists realized that agriculture plays a key role in overall economic growth and that small farmers are just as capable of effectively using new agricultural technologies, such as high-yielding seed varieties and chemical inputs, as are large-scale farmers. It was observed that rising output in the small-farm sector promotes higher growth in nonfarm economic activities in rural areas than do large farms. Furthermore, a focus on small-farm agriculture had potential to generate greater equity in the distribution of the benefits of development than had been achieved by strategies targeting large-scale agriculture. Development practice during this era continued to promote external technology transfers, mechanization, and expansion of output, but its efforts were directed toward small farmers throughout the developing world.

The 1980s and 1990s marked another shift in rural development thinking as top-down "blueprint" approaches gave way to grassroots (or bottom-up) approaches that envisioned rural development as a participatory process empowering rural residents to define and direct their own development needs and priorities. Development objectives were redefined to incorporate redistribution with growth, and strategies shifted emphasis from expansion of output toward employment generation and the provision of basic needs and services. Rural credit services in the form of micro-lending became popular. The unique needs of women in development and issues of environmental sustainability also began receiving attention during this period. Poverty alleviation became the catchphrase of rural development programs throughout the 1990s and into the 2000s.

It is important to note, however, that participatory approaches expanded at the same time that market liberalization and structural adjustment were embraced by the World Bank and other donor agencies. Under pressure from international institutions and donors, governments began withdrawing from the management of rural economies and delivery of rural services, and nongovernmental organizations (NGOs) became key agents of rural development.

During the 2000s, themes such as decentralization, good governance, environment, and sustainability have become important. Social protection in the context of globalization, human rights, and sustainable livelihoods represents a current emerging trend in rural development thought and practice. New thinking on poverty emphasizes the risks of liberalization and the vulnerability of the poor, whereas an emphasis on livelihoods seeks a broader understanding of the multiple and diverse ways in which rural households combine assets and activities to eke out a living.

Rural areas are undergoing a number of changes that challenge rural development practitioners to develop new policies. Rural populations are shrinking relative to urban populations. At the same time, connections of rural areas to roads, power, and communication facilities seem to be improving. Agriculture has declined sharply in relative terms as an employer and as a contributor to exports and gross domestic product, and nonfarm sources of income play an increasingly important role in rural household livelihood strategies. Finally, the number of rural areas affected by conflict and strife around the world has increased.

As a result, rural development policies must grapple with several key issues. First, can agriculture be the engine of rural growth? Second, can small farms survive? Third, can the rural nonfarm economy take up the slack? Fourth, can the state adequately ensure access for the poor to stable markets while also providing necessary safety nets? Fifth, can poverty-reducing democratization be achieved in the absence of strong civil society organizations and local participation? The latest thinking on rural development understands poverty reduction in terms of risk and vulnerability, and it shifts emphasis away from small farms toward diversification and differentiation, with a larger role for the state than what is provided in current conventional wisdom. Promoting the nonfarm sector, finding ways in which to support poor people trapped in conflict, developing better access to markets, and deepening democratic institutions in rural areas are a few of the key areas in which the new thinking now sees potential for rural development.

—*Holly Hapke*

See also Agriculture, Industrialized; Economic Geography; Rural Geography; Rural–Urban Continuum

Suggested Reading

Ashley, C., & Maxwell, S. (2001). Rethinking rural development. *Development Policy Review, 19,* 395–425.

Brohman, J. (1996). *Popular development.* Malden, MA: Blackwell.

Ellis, F., & Biggs, S. (2001). Evolving themes in rural development 1950s–2000. *Development Policy Review, 19,* 437–448.

Flora, C. (2004). *Rural communities: Legacy and change* (2nd ed.). Boulder, CO: Westview.

RURAL GEOGRAPHY

Rural geography as a research focus has evolved as rural areas have changed over the past century. To define rural geography, this entry first addresses the multiple ways in which to define the term *rural*. It then describes the various topics that have been studied under the rural geography rubric.

THE RURAL PROBLEMATIC

For city dwellers, the word *rural* typically conjures up stereotypes of green pastures and quaint villages. However, this image of the rural idyll stands in contrast to the reality of rural areas, where resource extraction, tourism, and other processes create social upheaval and environmental change. Hence, we must address the diverse ways in which we can perceive and define rural areas. There are at least three ways to do this. First, we can define rural dichotomously—city versus country, urban versus rural, or metropolitan versus nonmetropolitan areas. The U.S. Bureau of the Census defines settlements as rural if they have fewer than 2,500 people. An area is defined as metropolitan if it has a central county with at least 50,000 people. Hence, an area is nonmetropolitan if its largest settlement falls below this threshold.

However, these binary classifications mask the diverse geography of rural areas. Hence, geographers have developed indexes of rurality to recognize a rural–urban continuum. The Beale Code used by the U.S. Department of Agriculture places counties into 1 of 10 categories based on their respective population densities. Alternatively, areas may be defined as rural because of their economic base. A region reliant on natural resource extraction (e.g., agriculture, mining) might be viewed as rural. However, some high-amenity tourist regions (e.g., ski resorts) have low population densities but provide predominantly service-sector jobs.

A third way in which to define rural is to see it as being socially constructed. A region is rural if the people who live there perceive themselves as living in the countryside. Others suggest that we categorize communities based on the nature of human relationships. During the late 19th and early 20th centuries, Ferdinand Tönnies suggested that rural areas rely on familial and personal relationships. Urban areas, on the other hand, rely more on impersonal market-based relationships. The implication is that urban areas are more market oriented and therefore more modern—a common view during the 19th century, when a majority of Americans lived in the countryside. However, in the 21st century, nearly 80% of Americans live in urban areas. Therefore, it is prudent to view rurality as a multifaceted concept that has changed over time and is based on particular kinds of economic activity, demographics, and attitudes held by rural residents and visitors alike.

STUDIES OF RURAL GEOGRAPHY

North America became urbanized during the 19th and 20th centuries. Studies within rural geography reflected these demographic changes, and this section describes at least three themes that have emerged. The first theme is agricultural location theory. During the 19th century, Johann Von Thünen suggested that rural land uses changed as a function of the costs to transport agricultural commodities to urban markets. This model presaged central place theory, which explained the rise of urban systems within the context of an agricultural landscape.

A second theme is the focus on rural regions and landscapes. Vidal de la Blache examined communities of the French countryside. In the United States, John Hudson and John Fraser Hart described the Corn Belt and the rise of agricultural landscapes in the American Midwest. These authors relied on rich narratives to describe the unique features of rural communities.

A third theme focuses on rural land use change. The industrialization of agriculture during the 20th century increased productivity, allowing farmers to produce more per acre. However, this suppressed commodity prices, forcing farmers to increase the scale of their operations to survive ever shrinking profit margins. This prompted farm consolidation, rural out-migration, and economic stagnation. Although agriculture remains important at the turn of the 21st century, less than 2% of the American population is directly involved in production agriculture. Consequently, rural regions diversified their economic base—and so did the topics addressed by rural geographers.

Rural geographers continued to study agriculture but adopted new approaches such as focusing on value chains as commodities move from the farm to the retail marketplace. A focus on sustainable forms of agriculture emerged, in part, as a critique of industrial agriculture that while providing low-cost food also caused environmental degradation such as soil erosion and water pollution. Related topics here include the rise of local food systems, organic agriculture, and alternative crops.

Rural regions have become economically more diverse because fewer people are involved in agriculture.

This change prompted some researchers to discuss the emergence of a postagricultural or postproductivist rural landscape. Research from this perspective has focused on social capital and sustainability of rural communities as populations shrink. Other topics include the increasing importance of manufacturing and tourism to rural economic development. As cities sprawl outward onto productive farmland, rural geographers have also studied land use changes at the rural–urban fringe.

Rural regions now are confronting the impacts of globalization and information technology. Many rural communities used lower wages to lure manufacturing from urban areas by marketing their lower wages and property costs. However, many firms have left rural America to relocate in Mexico and other low-cost sites. Many rural communities are trying to compete in the global economy by increasing their information technology infrastructure. This focus on new and alternative technologies can also be seen in the increasing interest in renewable energy generated by wind and biofuels that are abundant in rural America. The continually changing nature of rural landscapes is even evident in the new so-called micropolitan settlement designation introduced by the census bureau for describing towns larger than 10,000 but smaller than 50,000 people.

—*Christopher Merrett*

See also Agriculture, Industrialized; Location Theory; Rural Development; Rural–Urban Continuum

Suggested Reading

Cloke, P. (1997). Country backwater to virtual village? Rural studies and the "cultural turn." *Journal of Rural Studies, 13,* 367–375.

Hart, J. (1998). *The rural landscape.* Baltimore, MD: Johns Hopkins University Press.

Hudson, J. (1994). *Making the Corn Belt.* Bloomington: Indiana University Press.

Woods, M. (2005). *Rural geography.* Thousand Oaks, CA: Sage.

RURAL–URBAN CONTINUUM

The concept of the rural–urban continuum references a diverse continuous landscape that is found across the world. This rural-to-urban transitional landscape,

distinctive in the United States, is characterized by large farms abutting encroaching suburbanization, followed by a gradual transition into a deteriorating urban residential zone and central business district (CBD). Key shaping influences were a host of post–World War II elements—the construction of the interstate highway system, the presence of low-cost automobile fuels, and the presence of low-cost agricultural land available for residential conversion. Inner cities typically have been the most deteriorated sector; critical to this have been the push forces of white flight from the racially integrating cities, underfunded and inadequate schools, increasing urban taxation, and urban crime. Working against inner cities, at the same time, has been the allure of suburban lifestyles associated with perceived desirable places to raise families among other upper-class families of similar attitudes, races, and incomes. In contrast, inner cities in the rest of the world are highly desired residential areas. In the less developed countries, it is the wealthy who live near the city centers, with the poor relegated to and segregated in the suburbs. It is beyond the suburbs where squatter settlements encroach on agricultural land.

Today, more than 60% of U.S. residents live in suburbia on landscapes dotted with shopping malls, big-box retail stores, fast-food restaurants, gasoline stations, dry cleaners, tanning salons, and barber shops. The basic source of income for the suburbs typically has derived from daily commuters who could reach inner-city CBDs quickly and efficiently on the extensive high-speed interstate freeways. More recently, growing numbers of employment nodes have gravitated to the suburbs, and this trend has continued unabated.

The rural landscape is made up of large commercial farms specializing in one to three crops. These farms are mixed with an array of hobby farmers who enjoy working with the land but gain their income from other sources and may make long commutes to the city two or three times per week. Small towns service this rural community but also offer an inviting environment for these periodic commuters.

Except in the Amish community, the days of commercial mixed farming—where farmers raise many crops along with an array of livestock—are gone. Profitable farming is ruled by economies of scale, where extensive areas of the same crop can be planted and harvested efficiently with heavy inputs of chemicals and immense machinery. Here the demands on the farmer's time are intense only a few times during the year, and this frees the farmer to engage in other activities during most of the year. Cropland no longer is fenced to keep animals out. Some farmers engage in small confined livestock production, such as poultry and swine, and operate amid large grain farmers. Large dairy farms capitalize on the availability of food from neighboring grain farmers as well. Profitable beef cattle feed lots have become extremely specialized and relegated to a few areas in the United States where immense economies of scale locate between the feed grain suppliers and the source of calves weaned from the arid ranching lands of the western states. Greeley, Colorado, has become a typical location for beef cattle feed lots.

There is a need to maintain the CBDs for government and for large businesses that require access to large market areas. However, the American CBDs have lost some of their service and retail functions to the suburbs. Some industries are moving out of the highly urbanized areas and locating near undeveloped areas near on-ramps where there is easy access to the interstate highways.

—*F. L. Bein*

See also Agriculture, Industrial; Rural Development; Rural Geography; Suburbs and Suburbanization; Urban Geography; Urban Sprawl

Suggested Reading

Jackson, K. (1985). *Crabgrass frontier.* New York: Oxford University Press.

Knox, P. (1994). *Urbanization.* Englewood Cliffs, NJ: Prentice Hall.

RUSTBELT

Starting in the late 19th century, as the Industrial Revolution unfolded across North America, the region stretching from the northeastern United States across the southern shores of the Great Lakes became the largest complex of manufacturing firms in the world. Scores of industries flourished there, fueled by waves of immigrant workers and a rich infrastructure of rail lines, iron ore, and coal deposits. The textile industry predominated in New England, and garments were widely produced in New York City. Connecticut and

Philadelphia were major centers of shipbuilding. The steel industry gave rise to cities such as Buffalo, Pittsburgh, Youngstown, and Cleveland. Akron was the largest rubber producer on the planet. Chicago was simultaneously a port, a meatpacking center, and a steel producer. Milwaukee became famous as a center of beer production. With the rise of the automobile industry, Detroit became the epicenter of automobile production as well as the model of industrial organization known as Fordism. Numerous smaller cities found niches in automobile parts, machine tools, or agricultural equipment. The Manufacturing Belt was by far the wealthiest part of the United States, with a large middle class generated by industrial jobs. The success and health of this region were evident well past World War II, when it played a major role in ensuring American economic hegemony worldwide.

The restless geographies of capitalism, however, ensure that one region's comparative advantage never lasts for long. Starting in the late 1960s and accelerating rapidly throughout the 1970s and 1980s, the former Manufacturing Belt was besieged by mounting international competition (e.g., from the newly industrializing countries), technological displacement, and a growing crisis of profitability. Deindustrialization of the Manufacturing Belt involved numerous plant and factory closures, mass layoffs, rising unemployment, and the deterioration of standards of living for millions of people. As the multiplier effects from closed manufacturing firms reverberated across local economies, the area witnessed concomitant declines in real estate values and retail sales. What was once the pride of the American economy had now become increasingly stagnant. As the region was abandoned by capital, the Manufacturing Belt became increasingly known as the Rustbelt, often in contrast to the growing population and political power of the Sunbelt. Rustbelt (or even Frostbelt) and Sunbelt, however, are not products of climate but rather reflect the uneven development that is an inherent inescapable part of capitalist landscapes. In this sense, the Rustbelt resembles other sites of deindustrialization such as the British Midlands.

Of course, not all parts of the Rustbelt were equally affected. Some cities, such as Pittsburgh and New York, reinvented themselves around producer services. New England prospered with the growth of high-technology industries. Other places, such as Detroit, sank into a near permanent state of economic collapse. Yet other places hoped for, and sometimes achieved, industrial resurgence with the aid of foreign direct investment. Thus, the Rustbelt serves as a reminder of how rapidly and traumatically capitalism can unmake, as well as create, regional economies.

—*Barney Warf*

See also Deindustrialization; Restructuring; Sunbelt; Uneven Development

Suggested Reading

Bluestone, B., & Harrison, B. (1982). *The deindustrialization of America: Plant closings, community abandonment, and the dismantling of basic industry.* New York: Basic Books.

S

SCALE

Scale is at the heart of all geographic investigations. Yet although it currently is a central concept across the many subfields of geography, 25 years ago few geographers other than cartographers paid much theoretical attention to it. Scale was a concept that interested geographers methodologically, either through questions of cartographic scale or with respect to the influence of an areal unit on statistical analyses and outcomes. *Cartographic scale,* one of the mainstays of the art and science of cartography, is a measure of the relationship between a map's size and the portion of the real world it is intended to represent. The representative fraction (RF), which associates distance on a map to distance on the earth, is the standard way of describing this meaning of scale. In spatial statistics, the predominant concern has been with what is known as *operational scale,* that is, the geographic area (such as the unit size [e.g., a neighborhood vs. a city] or the extent of a watercourse vs. a river system) within which processes are in operation and at which they may be effectively observed. In both of these cases—cartographic and operational—scale is an abstract construct that helps geographers to represent and organize space.

During the past two decades, geographic scale, in contrast to cartographic or operational scale, has become a new focus for human geographers as it is increasingly seen to be key to theorizing and comprehending globalization in its economic, cultural, political, and environmental manifestations. One of the first attempts to explore geographic scale came from Peter Taylor, who produced a "three-scale structure" model consisting of the micro scale of the urban (which he labeled the domain of experience), the meso scale of the nation-state (which he labeled the sphere of ideology), and the macro scale of the world economy (which was understood to be the "scale of reality"). Taylor's model was intended to provide a hierarchy to the horizontal model of world systems theory (core, periphery, and semiperiphery) advanced by Immanuel Wallerstein. The chief strength of Taylor's model was that it helped to show how space could be thought of as vertically organized and how these different levels of space were part of the wider processes of capital accumulation. A significant weakness of Taylor's scale model was that it was static and overly top-down in its emphasis on the macro scale—where capital accumulation predominated—the level that determined what happened in the lower two scales. Building on the work of Taylor, Neil Smith recognized that scale was important to understanding the way in which the world economy worked, but he saw the fixity inherent in Taylor's model to be problematic. Smith suggested instead that cooperation and competition (at all levels of the hierarchy) over capitalist production and accumulation processes were central to understanding scale and that sometimes the macro or global scale might be determinative, whereas at other times it might not be. The central outcome of introducing dynamism into all three levels of Taylor's fixed *geographic scale* hierarchy was to demonstrate that the scales that constitute capitalist space are not a given—the urban, regional, national, and global are not preexisting categories—but rather are *produced* through the complex processes of capitalist expansion. A secondary and especially fertile outcome of Smith's research on scale was the recognition that the production of scale was a *political process,* by which he meant not only that scale was not a fixed hierarchy but also that the production of scale was a contested process by institutions and actors operating locally, regionally, nationally, and internationally within a capitalist framework. Given the different contests in play, and understanding capitalism to be a system that is the product of social forces, it became important to recognize that different scales could be determinative at different times and in different locations. Recognizing that scale production was a political process also enabled Smith to theorize the possibility of *jumping scale,* by which he meant that institutions or actors who had established political claims at one level (e.g., the level of the household) could expand their efforts to another level (e.g., the level of the city) to shape the operations of the capitalist political economy.

There has been a wide range of reactions and elaborations on the early work on scale established through the Marxist theorization of geographic space by Taylor and Smith. The most extensive have attempted to test the theories by applying them to empirical examples (e.g., social movements, immigration, urban planning, governance) to determine how helpful scale theorizing is to understanding the ways in which society and space interact. Neil Brenner, for example, made the case that the social production of scale is a product of *scalar structuration,* that is, the way in which different scales interact to produce a hierarchy of scales that are stable for a time but may lose their fixity under different political and economic conditions. Scalar structuration is a dialectical process such that each scale—with its internal processes and relations—in the hierarchy is constituted through its relationship with other scales. The result is a scale hierarchy that characterizes a particular time and location and that can change as different political and economic conditions emerge and change. An example of scalar structuration can be seen in *glocalization,* the term advanced by Erik Swyngedouw that is meant to convey the determinative scales in the current state of the world economy—where the urban scale and the global predominate in organizing and driving capitalism. When a particular set of relationships between or among scales come to be stable and dominant, scale theorists think of this moment as a *scalar fix.* The current scalar fix is seen to bypass the nation-state (and its restrictive legal and regulatory apparatus) in favor of a linkage between transnational corporations and

governance organizations and cities where capital can be directly invested and profit can be generated.

Using feminist geography as her standpoint, Sallie Marston argued that scale theorizing has tended to ignore social reproduction in favor of economic production in trying to account for the way in which space is produced and transformed. Rather than focusing exclusively on paid labor and national and transnational corporations as the institutions or actors who drive the production of scale, Marston contended that actors in their everyday lives (outside of the paid workplace) and the institutions they create and interact with must also be taken into account when theorizing the production of scale. Like Smith's work, her work is oriented around the contention that there is a politics to scale and that actors beyond the workplace play a central role in that politics and the way in which it shapes the worlds in which we live and work. The focus on the relevance of social reproduction to scale theorizing has also been a means by which questions of difference and identity as they operate on and through the human body could be brought to bear on scale theorizing, exposing them not as generic agents but rather as actors whose social markings and differentiations (e.g., race, ethnicity, sexuality, gender, ability) have significant implications for the ways in which they struggle and produce space.

Widely embraced in social, cultural, political, and economic geography, scale has also come to be significant in the work of environmental geography, particularly through the research and writing of political ecologists who explore the ways in which nature and society interact in the contestation and construction of spatial scales. Swyngedouw called the interaction between nature and society in the production of geographic scale *political–ecological scalar gestalt,* and he regarded these as premised on the ways in which shifting and contested scalar configurations are shaped by local and global processes that shape and are shaped by nature at the same time that they shape and are shaped by society.

Recently, scale has been subjected to a rigorous critique as a handful of human geographers are becoming disenchanted with the hierarchy that structures the conceptualization. Sallie Marston, John Paul Jones, and Keith Woodward suggested that scale should be jettisoned from the geographic lexicon because the levels have no correspondence in real life. Their critique is that place, or more specifically the *social site,* is the only material basis on which to think about contemporary political and economic life and that globalization, if it can be thought of as a material process, is really a set of shifting connections between or among complex social sites that constitute the globe. As a result of this alternative formulation, the vertical hierarchy that constitutes scale is seen as preventing us from recognizing that the material world is horizontally organized through connectivities between people (e.g., government officials, migrant workers), objects (e.g., money, manufactured goods), and ideas or events (e.g., democracy, disasters). There is no consensus yet on the impact this critique will have on our current understanding of scale beyond the fact that it is likely to ignite further debate and disagreement.

For the moment, it is reasonable to posit that scale is a concept that will continue to be scrutinized at the same time that it is employed in a wide range of research endeavors in human geography (as well as in physical geography). Its usefulness as a way in which to comprehend levels of geographic difference from the smallest scale of the body to the largest scale of the globe is likely to persist. As it is currently understood, the concept is premised on the operation of a vertical hierarchy that may be nested (like a Russian doll with smaller dolls/scales contained within larger ones) but also subject to transformation through the ways in which the changing processes of capital expansion and the social reproduction of differently situated actors remake homes, neighborhoods, cities, regions, nation-states, and the globe and all of the levels between or among these.

—*Sallie A. Marston*

See also Cartography; Critical Human Geography; Political Geography; World Systems Theory

Suggested Reading

Brenner, N. (1998). Between fixity and motion: Accumulation, territorial organization, and the historical geography of spatial scales. *Environment and Planning D, 16,* 459–481.

Howitt, R. (2003). Nests, webs, and constructs: Contested concepts of scale in political geography. In J. Agnew, K. Mitchell, & G. Ó Tuathail (Eds.), *A companion to political geography.* Oxford, UK: Blackwell.

Marston, S. (2000). The social construction of scale. *Progress in Human Geography, 24,* 219–242.

Smith, N. (1984). *Uneven development: Nature, capital, and the production of space.* Oxford, UK: Blackwell.

Smith, N. (1992). Contours of a spatialized politics: Homeless vehicles and the production of geographical space. *Social Text, 33,* 54–81.

Swyngedouw, E. (1997). Neither global nor local: "Glocalization" and the politics of scale. In K. Cox (Ed.), *Spaces of globalization: Reasserting the power of the local* (pp. 137–166). New York: Guilford.

Swyngedouw, E. (2000). Authoritarian governance, power, and the politics of rescaling. *Environment and Planning D, 18,* 63–76.

Taylor, P. (1982). A materialist framework for political geography. *Transactions of the Institute of British Geographers, 7,* 15–34.

SEGREGATION

Geographic research on segregation addresses both the processes by which groups of people become spatially isolated from one another and the resulting geographic patterns of isolation. Segregation occurs based on a variety of characteristics (e.g., race, ethnicity, class, gender, sexual orientation, age) and at different spatial scales (e.g., regions, cities, neighborhoods, schools). In practice, researchers have paid the greatest attention to residential segregation in urban environments based on race/ethnicity. A major focus in the United States has been on the sometimes extreme segregation of the black population, but there has also been attention paid to other groups. In Europe, evidence exists for varying degrees of segregation among certain populations, including South Asians and Afro-Caribbeans in Britain, North Africans in France, Turks in Germany, and Roma in Eastern Europe. An example of legally enforced segregation in South Africa was apartheid.

The causes of segregation are complex and vary considerably between contexts. In the United States, segregation of blacks has been attributed to discriminatory real estate and mortgage lending practices, "white flight" (i.e., the out-migration of whites from areas with growing minority populations), the intentional location of public housing in marginal areas, and other factors. Sometimes it is argued that patterns of racial concentration reflect a voluntary choice on the part of group members, but these explanations must be viewed with great caution. For example, recent research in Los Angeles suggests that of four populations interviewed (white, black, Asian, and Hispanic), blacks expressed the strongest preference for living in racially integrated neighborhoods, challenging the notion that the segregation of blacks is voluntary.

Several indexes are used to measure residential segregation, with the most common being the index of dissimilarity. This index, which ranges between 0 and 100, measures the extent to which two populations (e.g., Asians and whites, blacks and non-blacks) are evenly distributed across cities or metropolitan areas. Because most research in the United States relies on census data, the index often uses census tracts (census-defined subdivisions of cities or metropolitan areas) to represent places of residence. For a hypothetical metropolitan area that is 30% black and 70% non-black, an index of 60 means that either 60% of blacks *or* 60% of non-blacks would need to move to another tract for all tracts to achieve a balance equal to the metropolitan area as a whole (30% black and 70% non-black). By this measure, the most segregated U.S. metropolitan areas in 2000 included Detroit (84), Milwaukee (81), Gary (81), Chicago (78), Flint (77), Cleveland (77), Buffalo (76), Cincinnati (74), and Newark (74). Researchers often consider indexes below 30 to be low, between 30 and 60 to be moderate, and above 60 to be high, although this division is rather arbitrary. Studies using indexes of dissimilarity suggest that the segregation of American blacks increased steadily from 1890 until around 1970, with levels declining somewhat in most metropolitan areas since then (although extreme segregation persists in many places).

—*Robert Vanderbeck*

See also Housing and Housing Markets; Neighborhood; Race and Racism; Suburbs and Suburbanization; Urban Geography

Suggested Reading

Bobo, L., Oliver, M., Johnson, J., & Valenzuela, A., Jr. (2001). *Prismatic metropolis: Inequality in Los Angeles.* New York: Russell Sage.

Glaeser, E., & Vigdor, J. (2001). *Racial segregation in the U.S. census: Promising news* (Brookings Institution Survey Series) [Online]. Available: www.brookings.edu/es/urban/census/glaeser.pdf

Sin, C. (2002). The interpretation of segregation indices in context: The case of P* in Singapore. *The Professional Geographer, 4,* 422–437.

SENSE OF PLACE

Sense of place refers to the way in which places are experienced subjectively. It describes a complicated set of emotions and feelings that are evoked by a particular place. Often this is experienced as a sense of attachment and belonging. However, there are also negative emotions such as fear, claustrophobia, and unease that can form a sense of place. The issue of sense of place was not one that geographers engaged with in any depth until the advent of humanistic geography during the 1970s. It was only then that subjectivity, feelings, and emotions were considered worthy of geographic contemplation. Since then, geographers have produced a considerable body of work on the production, maintenance, and transformation of senses of place in a wide array of contexts ranging from the individual and idiosyncratic to the widely shared.

At one level, sense of place refers to a particularly personal set of feelings for some part of the earth's surface. Think, for instance, of the way in which a person feels about his or her childhood home place. All of the memories that are the product of individual biography make such a place particularly evocative. Often these feelings are nostalgic, but they also may be negative—colored by memories of abuse or poverty. Smells, sounds, and tastes all can evoke a particular sense of place that may be completely idiosyncratic.

Beyond the emotions produced through individual experience, there are relatively durable senses of place that are the product of shared lives and shared representations. Many of us, for instance, might have a strong sense of place of New York City or Paris even if we rarely—or never—have been there. This is because these places are constantly being projected into our lives through film, literature, music, and the news media. Indeed, the successful evocation of a sense of place is crucial to the development of fictional works in novels and films. Many people believe that they know the London of Sherlock Holmes or the Los Angeles of Raymond Chandler novels through the successful evocation of place in these novels. There is a long tradition of geographers writing about the senses of place developed in novels.

Despite the power of shared senses of place, they never are completely uncontested. Any place has a myriad of meanings attached to it—some widely shared and some idiosyncratic. For instance, to some people New York City might be a bastion of tolerance and liberalism, whereas to others it might be experienced as a place of oppression and violence. A lot depends on who the person is and how he or she experiences the place. When these different senses of place are both public and (to some degree) shared, there often is conflict over what the appropriate meaning of a place may be. When a Catholic church raised a huge cross adjacent the concentration camp at Auschwitz, for instance, there was an enormous international outcry and debate over the meaning of Auschwitz as a site of memory. Similarly, many travelers, musicians, and artists in Britain during the 1980s saw Stonehenge as a place to be enjoyed through festivals on the site, whereas the people who ran it believed that a heritage site needed to be kept separate from the people so as to be sacred. Indeed, the production and maintenance of senses of place often involve the deliberate exclusion of subversive senses of place that might threaten the image of a cohesive, positive, and uncontroversial place image. This often is the case with the production of heritage sites and places of memory where painful memories of the past (e.g., slavery in the American South) are ignored or glossed over.

Some have suggested that globalization has led to a reduction in the variety of senses of place and that, in fact, places are becoming more or less the same as each other. This creeping "placelessness" is, it is argued, evidenced by the spread of nonplaces such as airports, shopping malls, fast-food outlets, tourist sites, and spaces of transit over an increasing area of the earth's surface. Many of these are spaces of flow where travelers are made to feel "at home" through the production of sensory experiences (e.g., smell, sight, taste, sound) that are uniform throughout the world. Thus, a business traveler now can go to Schiphol Airport in Amsterdam and have a meeting, stay in a Sheraton Hotel that is much like any other Sheraton Hotel, eat a pizza or Thai curry that is like any other pizza or Thai curry, and then return home without ever experiencing anything particularly Dutch or unique to Amsterdam.

Ironically, perhaps because senses of place are so often contested or deemed to be unimportant, a huge industry of place promotion has emerged to create successful images for places around the world that gloss over any conflict around what a place might mean. In a globalized world, it is important for places to create a positive sense of place so as to attract people and investment. It is important for a local government and chamber of commerce to convince people to locate "there" because "there" is special and unique and, indeed, quite unlike anywhere else. Thus, place marketing has become an important business, with even the smallest of towns developing glossy pamphlets and logos to create warm feelings toward them. Former industrial towns in places such as South Wales and Pennsylvania attempt to repackage themselves as heritage sites and tourist attractions. Although a sense of place may appear to be nebulous and of secondary importance to the hard facts of the economy, for instance, it often is sense of place that is used to produce new forms of economic well-being.

—*Tim Cresswell*

See also Humanistic Geography; Literature, Geography and; Phenomenology; Topophilia

Suggested Reading

Charlesworth, A. (1994). Contesting places of memory: The case of Auschwitz. *Environment and Planning D, 12,* 579–593.

Cresswell, T. (2004). *Place: A short introduction.* Oxford, UK: Blackwell.

Relph, E. (1976). *Place and placelessness.* London: Pion.

SEQUENT OCCUPANCE

The concept of sequent occupance can be attributed directly to Derwent Whittlesey (1890–1956). He coined the term and proposed the principle of sequent occupance in his 1929 article of this title. The article opened with a declaration stating that human occupance of area, like other biotic phenomena, carries within itself the seed of its own transformation. He also stated that the view of geography as a succession of stages of human occupance establishes the genetics, or historical development, of each stage in terms of its predecessor. Thus, by implication, he prescribed the study of sequent occupance as integral to historical geography as well as regional geography. With sequent occupance essentially an analogue model, Whittlesey averred that the analogy between sequent occupance in chorology and plant succession in botany would be apparent to all. But he also qualified this, admitting that the botanist's problem is less intricate. The development of plant associations generally involves fewer agents and is more tractable and traceable than the multiplicity of changes, both biophysical and cultural, that combine to create chorologic formations. Having identified sequent occupance as a key to explaining chorological process, or the formation of regions, he also advocated its study in particular areal cases as a method for advancing chorology.

As a demonstration of his model, Whittlesey offered an illustrative case involving a 15-square-mile district in northern New England. Four stages, or occupance sequences, were discernible. The then current occupation, or third stage, consisted of sporadic livestock pasturage amid a landscape of second-growth forest and grassy glades marking former trails and roads. This was preceded by a thoroughgoing subjection of the land to farming. A largely self-sufficient population occupied and exploited all but the most unyielding terrain. The initial occupation was by a few Indians who lived a migrant life. They hunted and gathered, and by implication they had little impact on the virgin mixed forest. The fourth stage, as yet unrealized, was projected to be occupance by forests once more but periodically cut for wood pulp or lumber by nonresident owners. Unremarkably, he closed by affirming that in this New England district each generation of human occupance is linked to its forebear and its offspring. More provocatively, he concluded his short article with the assurance that the life history of each generation discloses the inevitability of the transformation from stage to stage. Notable in this demonstration model was the absence of chronology. In keeping with the generalizing or nomothetic tenor of model building, dating of the transitions or durations are elided in this specific case. Yet Whittlesey intended for his model to be applied to quite specific regional or chorologic units ranging in scale from the microgeographic (e.g., his test New England district) to large regional expanses. He also assumed that once a comparative record of sequent occupancies was established, the number of actual sequence patterns that ever existed would be small.

Whittlesey's article and advocacy must be viewed within the context of the times. During the decade of the 1920s, several geographers besides Whittlesey offered programmatic schemata for advancing human geography, particularly human–environment interactions and the resultant evolution or development of regions and landscapes. In 1923, Harlan Barrows, one of Whittlesey's colleagues at the University of Chicago, delivered his presidential address to the Association of American Geographers. Titled "Geography as Human Ecology," it sought to replace the waning but still influential environmental determinist outlook in North American geography with an emphasis on "human

adjustments" to physical environmental conditions and constraints. Like Whittlesey later, for Barrows the inspiration drawn from biological ecology was more impressionistic than precise. Although Barrows's main affiliation (curiously) was with historical geography, his design for a human ecologized geography was directed mostly toward contemporary cases and questions. The second major statement was published by Carl Sauer in 1925. Titled "Morphology of Landscape," it set forth a program for cultural geography in which cultural landscapes and the historical study of their origins and modifications comprised the organizing principle. Unlike either Barrows or Whittlesey, Sauer eschewed the notion that identifying stages, whether of human occupancy or of landscape change, should be the object of either geographic theory or geographic practice. In part, he was concerned with distancing himself from the Davisian physiographic cycle of erosion with its celebration of staged evolution. He was also skeptical of any system, or accounting of it, that transformed primarily through endogenous or internal change. Sauer soon was to see cultural diffusion, and other forms of exogenous change, as the mainspring of landscape evolution. Although Whittlesey made room for exogenous change in deflecting given trajectories of occupance's sequencings, those geographers who followed his method over the next decade or so were more model bounded rather than less so.

Jan Broek's study of Santa Clara Valley, written in 1932, often is considered to be the genre's exemplar. Preston James's study of Blackstone Valley of southern New England in 1929 actually anticipated Whittlesey's article by a few months. For the most part, however, it was midwestern geographers who put his ideas to test and work. Darrell Davis and Stanley Dodge were early adopters, conducting studies of the Kentucky highlands and the Illinois prairie, respectively. One such practitioner, Alfred Meyer, wrote his dissertation on the Kankakee Marsh of northern Illinois and Indiana using Whittlesey's template. He spent the next two decades publishing amendments to and refinements of this occupancy history. A number of other efforts, particularly midwestern theses and dissertations, followed the logic and format of sequent occupance study over the next decade or so. The heyday of the enterprise, however, really lasted only a decade. Curiously, with the canonization of chorology, or regional studies, in Richard Hartshorne's *The Nature of Geography,* this brief chapter in regional geography's search for sounder groundings was played out. Although echoes of Whittlesey's approach can be found in the subsequent geographic literature, both the frequency and volume have diminished progressively over time.

—*Kent Mathewson*

See also Cultural Geography; History of Geography; Regional Geography; Social Geography; Space, Human Geography and

Suggested Reading

Mikesell, M. (1976). The rise and decline of "sequent occupance": A chapter in the history of American geography. In D. Lowenthal & M. Bowden (Eds.), *Geographies of the mind: Essays in historical geography* (pp. 149–169). Oxford, UK: Oxford University Press.

Whittlesey, D. (1929). Sequent occupance. *Annals of the Association of American Geographers, 19,* 162–166.

SEXUALITY, GEOGRAPHY AND/OF

Sexuality both represents an area of study that interests geographers and is a source of social identity of geographers. Sexuality scholars argue that the two are related. They suggest that because sexual minorities (e.g., lesbians, gays, bisexuals) face discrimination in a dominantly heterosexual discipline, the study of sexual spaces, homosexuality, sexual identity, and dissent, and even the ability to be "out" about sexual orientation, all encounter profound challenges in the discipline. Homophobia, they insist, has curtailed a frank discussion about the centrality of sexuality to space and to geographic inquiry; it also encourages sexuality studies, especially concerning homosexuality, to "stay in the closet." Skeptics seem to think that sexuality is best left in private, perhaps a topic more appropriate for the bedroom than for the classroom, an attitude that has left sexuality to be a social identity severely understudied in geography.

Of course, every geographer has a sexuality, and sex matters are central aspects of every life. However, sexuality, as strongly argued by psychoanalytic theory, is a repressed business, and it is very difficult to have open and productive discussions about sex matters. Although human geographers have written volumes on economic

identities, for example, sexuality still exists on the margins of the discipline's radar screen.

There is room for optimism, however, as those geographers who championed the study of sexuality during the 1970s and 1980s have illustrated the importance of sexuality to the study of human geography and have gone on to influence and encourage other scholars to do the same. There is now, by the early 21st century, a serious contingent of geographers examining a myriad of sexuality topics. Through their work, these scholars are refashioning ideals of academic research and teaching and, thus, are prodding sexuality toward the core debates in political, economic, urban, feminist, and social–cultural geography. Homophobia may remain a serious discrimination, but the combined forces of feminist, critical, and sexuality geography scholars work hard to advance the views of all geographers regardless of their topics of research interest or their sexual practices.

Despite the challenges that prudishness and homophobia present, geographers have made significant inroads to the study of sexuality, sexual spatiality, and the spaces of sex. Studies take quite different conceptual and methodological approaches in their analyses, and certainly they have been influenced over time by debates in feminist, queer, and social theories. One can approach general themes in the geographies of sexuality research, however, by considering their major achievements: reimagining citizenship, critiquing heterosexist spatiality, and theorizing sexual identities.

REIMAGINING CITIZENSHIP

A citizen is someone who receives rights from a state in return for his or her patronage; inclusion is an important component because a citizen must be recognized as belonging to a community to stand as its citizen. Sexuality scholars in geography ask how the concept of citizenship in democracies relies on a public/private spatial distinction to function. For example, in Western democracies, citizens have a right to public speech, that is, to make claims publicly. Certainly, some rights that citizens have publicly claimed have been left unfulfilled such as civil rights to privacy through homemaking, child rearing, and marriage and even (in some places) the right to engage in consensual sexual practices such as sodomy. These examples highlight how civil rights often are extended by states in primarily heterosexual ways given that, obviously, many of these do not extend to homosexuals in most of the world. Homosexual marriage is extremely rare, and even adoption of children by homosexuals is illegal in many places.

Sexuality scholars in geography have sought in their work to challenge the idea of citizenship, and in so doing they have contributed to wider debates in political geography about the changing forms of states, communities, and civil societies. Scholars examine, for example, the demands that different citizens make around issues of sexuality, including sexual activism surrounding acquired immunodeficiency syndrome (AIDS), antidiscrimination, and calls for civil rights. They show that sexual dissidents (including gays and lesbians but also transgendered persons and transsexuals, ill bodies, volunteers in AIDS organizations, etc.) have demanded a public venue for what previously were considered private matters. Sexuality-based activism, they insist, has fractured the confines of private sexuality and brought it to the community, public space, public governance, and public view. As such, sexual citizenship highlights the fragile nature of the public/private sexual/spatial distinction and the need to investigate how spaces come to be produced and experienced through the everyday practices of identity.

CRITIQUING HETEROSEXIST SPATIALITY

Sexuality scholars show that spaces are predominantly heterosexual unless they are marked off as homosexual through the representations, practices, or bodies of homosexuality. For example, imagine a city street. A man and a woman walking hand in hand does not even warrant a second glance, but in most neighborhoods, towns, and cities of the Western world, a man walking hand in hand with another man, or two women walking together intimately, causes a stir. Only when neighborhoods are identifiably gay does such behavior not raise eyebrows. This is just one simple example that illustrates how spaces are coded as heterosexual. And because heterosexuality is enforced by homophobic violence and attacks, or even by the more mundane but equally vicious normative pressure to hide homosexuality or to be straight, we can say that these spaces are heterosexist. They include urban, domestic, work, school, and other spaces.

Geographers have considered what heterosexist space means for all people, whether queer or straight, by examining how gays and queers negotiate straight space and how straight practices enforce the dominant heterosexism that is so widely evident. One aspect of this research is to determine how gays and lesbians have created their own spaces to deflect the limitations placed on them by heterosexist space. Neighborhood-scale studies illustrate that gays have formed their own communities by buying houses in vicinity to each other, often through gentrification processes, and studies also examine how gay neighborhoods form through gay institutions such as bars, newspapers, and coffee shops. These are examples of community building through economic and sexuality practices.

Geographers have also found that sexual minorities strategize in smaller-scale ways as well to avoid the heterosexism of everyday space. They find havens at home, change behavior to hide their sexual identities at work and in the street, or even visit other towns to find sexual partners. Rural sexual queers experience oppression in unique ways compared with urban sexual minorities, who more often may find comfort in numbers. However, scholars say that it would be wrong to think that all things homosexual come from cities given that homosexual perspectives also diffuse from rural to urban.

Sex work, such as prostitution, is a form of dissident sexual practice, but in this case often (but not always) of heterosexuality; there are few, if any, studies on queer sex work in the discipline. Therefore, sex work itself represents a challenge to dominant heterosexual morality. Sex districts, also known as red-light districts, are sexualized spaces and neighborhoods; however, unlike gay neighborhoods, red-light districts do not signify resistant space so easily because the question of sex work presents conceptual difficulties in terms of thinking through issues of sex workers' agencies, trafficking in women, gendered economic and sexual power relations, and (sometimes) age differences between clients and sex workers. Sexuality scholars in geography also consider the regional disparities and gender and racial politics of international sex tourism.

Heterosexist space forms through sexual practices, but these are also informed by other social meanings and representations. The example of international sex work, for example, not only emphasizes gender but also brings region, race, age, and economic class status to the fore. Sexuality studies, therefore, have needed to deal with other identities and social processes, especially as theoretical debates rage about the impossibility of separating social identities in the subject; for example, one could imagine being sexualized but not gendered. Therefore, sexuality studies found their core organizing principle—sexual identity, often either homosexual or heterosexual—under attack.

THEORIZING SEXUAL IDENTITIES

Sexuality studies in human geography began primarily through the examination of gays and lesbians, the spaces they molded and negotiated, and their resistances to heterosexism. This is evidenced, for example, in the research on gay neighborhoods and sexual citizenship. However, theoretical developments that began to question the conceptual usefulness of categories such as homosexual, gay, and lesbian have thrown identity-based research in geography and other social sciences and humanities into turmoil. Queer theory, in particular, has dealt a debilitating blow to the utility of identity-based studies such as gay and lesbian studies. Queer theorists argue that although relying on categories such as homosexual may seem to be liberatory and rights centered, doing so actually reinscribes the centrality of heterosexuality. This is because homosexuality has only come to be defined as the opposite of heterosexuality, thereby confining it through an endless dance around a core dominant heterosexuality.

Feminist theories have also provided fuel for the critique of identity studies because it has been argued, for example, that social identities and categories such as gender and sexuality cannot exist in isolation from other identities such as race, age, and location. Feminists and queer theorists have also insisted that gender and sexuality must be considered in tandem because normative heterosexuality relies on the gender division between men and women to define heterosexuality. They have shown that heterosexuality has meaning only through a corresponding gender division. Certainly, queer theorists agree, and they have taken this argument a step further to call for the dismantling of categories to base politics and theories altogether.

Thus, queer theory presents sexuality studies with a dilemma, arguing paradoxically that sexuality is disastrous for politics and that queers should seek antifoundationalist strategies and concepts to build scholarship and politics. This moves sexuality studies and struggles away from civil rights–based claims to equality and toward the disruption of ideals and stable binaries such as homo/hetero and even gay man/lesbian woman. Spatiality, like sexual identity, must also withstand this critique conceptually, and divisions such as heterosexist space versus homosexual-resistant space have been complicated. Geographers see spaces as more fragile because they cannot be assumed to reflect stable identities. However, this does not make social space or place immaterial or nonexistent. In fact, geographies of sexuality have turned toward examining how material spaces come to have meaning through the practices of sexual subjects. It is just that they have moved away from essentialist assumptions about what those practices reflect.

The effect of queer theory on sexuality studies also is seen in the way in which scholarship goes about the business of exploring resistance. Because identity presents a precarious foundation after queer theory's critique, scholarship also has become more wary of identifying resistance based on nonnormative identities. In addition, the increasingly public gay citizen has presented a conundrum for resistance studies as more gays enjoy the mainstream in the West and as the category shows its dissonance. Psychoanalytic theory also enters to challenge the idea of identities of resistance because it suggests that subjects cannot always fully know themselves, their motivations, their identities, or the complex processes that go into the formation of selves. Geographers of sexuality have also applied psychoanalytic theory to explore psychic (i.e., of the psyche) investments in racial and masculine supremacy.

—*Mary Thomas*

See also AIDS; Gays, Geography and/of; Heterosexism; Homophobia; Identity, Geography and; Lesbians, Geography of/and; Public Space; Queer Theory

Suggested Reading

Bell, D., & Valentine, G. (Eds.). (1995). *Mapping desire: Geographies of sexualities.* New York: Routledge.

Brown, M. (2000). *Closet geographies: Geographies of metaphor from the body to the globe.* New York: Routledge.

Hubbard, P. (2000). Desire/disgust: Mapping the moral contours of heterosexuality. *Progress in Human Geography, 24,* 191–217.

Kempadoo, K., & Doezema, J. (Eds.). (1998). *Global sex workers: Rights, resistance, and redefinition.* New York: Routledge.

Knopp, L. (1992). Sexuality and the spatial dynamics of capitalism. *Environment and Planning D, 10,* 651–669.

Valentine, G. (1993). (Hetero)Sexing Space: Lesbian perceptions and experiences of everyday spaces. *Environment and Planning D, 11,* 395–413.

SITUATED KNOWLEDGE

The term *situated knowledge* is most associated with feminist geographers and their critiques of the process of knowledge production. Drawing inspiration from anthropologist Donna Haraway, who commented critically on the construction of powerful scientific knowledge, feminists have challenged the truth claims of detached disembodied means of knowing the world. Haraway argued for a situated knowledge, referring to the notion that knowledge can be partial, located, and embodied; in other words, and put simply, knowledge always comes from someone somewhere. Conventionally, the Western academy has constructed the most valuable forms of knowledge as ones that are impartial and deeply authoritative because of what Haraway called the "god trick" of seeing everything from nowhere and the refusal to situate claims relative to personal, social, and geographic contexts. Whereas scientific and other "master" knowledges are founded on claims of universality created through supposed objective detachment, feminist researchers argue that situating knowledge enables more critical thinking whereby the transcendent is replaced by partial and politicized knowledge claims. Such feminist approaches must be careful, however, to guard against a romanticized adoption of "subjugated" knowledges (those traditionally denied or ignored) and to avoid relativism (the notion that all knowledges are equally valid).

Such critiques of knowledge production have influenced feminist geographic research in particular, and this has been evident in debates about research methods and power relations in the discipline. These debates in turn have influenced feminist research practice as researchers have tried to employ a range of qualitative research techniques that enable sensitive, nonexploitative, and embodied encounters with a range of actors and agents in a variety of locations. Producing situated knowledge also entails processes of reflexivity—itself a kind of "self-regard"—to avoid the god trick of the objective master gaze and locate oneself relative to others and hegemonic social discourses. However, such processes often are fraught with difficulties because it is challenging to clearly situate oneself in a world that often is conceptualized and experienced as fluid and uncertain. Hence, Haraway's account of situating knowledge challenges feminist and other researchers to destabilize taken-for-granted forms of academic authority, but without problematically "fixing" our positions in a fluid world or simplistically understanding human differences as (just) relative.

—Hester Parr

See also Participant Observation; Qualitative Research

Suggested Reading

England, K. (1994). Getting personal: Reflexivity, positionality, and feminist research. *The Professional Geographer, 46,* 80–89.

Haraway, D. (1991). *Simians, cyborgs, and women: The reinvention of nature.* London: Routledge.

Rose, G. (2002). Conclusion. In L. Bondi, H. Avis, A. Bingley, J. Davidson, R. Duffy, V. Einagel, et al. (Eds.), *Subjectivities, knowledges, and feminist geographies: The subjects and ethics of social research* (pp. 253–258). Lanham, MD: Rowman & Littlefield.

SOCIAL GEOGRAPHY

Social geography is commonly understood in the Anglo-American tradition as a subdiscipline of human geography, partially separate from others such as economic, political, and urban geography. In this way, social geography is recognized as a specialty that concentrates on the social differences, groups, and relations that interact with (and shape) the spaces in which we live. Within some European traditions, social geography has been understood differently as a more overarching and synthetic enterprise, an undertaking that draws together many of the dimensions studied in other subdisciplines (e.g., population, land use, cultural practices) to show the rich sociospatial composite that is produced as people build up layers of social life in a given place, territory, and landscape. In other cases, social geography has not developed as a strong component of human geography. Nonetheless, since the end of World War II, social geography has become increasingly popular and relevant in many Western settings as geographers mapped and planned for postwar progress and then turned to recognize the spatial occurrence of social problems and injustices such as poverty and racism. Geographers have used different theoretical approaches to investigate social issues; thus, as with other subdisciplines of human geography, we can observe the effect of spatial science, humanistic, Marxist, feminist, and postmodern or poststructural approaches. Across this diversity, social geography provides a key way in which to identify and explain the differences that transect a society (e.g., class, gender, sexuality, ethnicity) and the negotiations people must make in their everyday lives as these differences shape, and are shaped by, the spaces in their lives. Future social geographies will continue these traditions and add new considerations of ethics, morals, emotions, posthuman perspectives, and further materialist and action-oriented possibilities.

CONTRASTING APPROACHES TO SOCIAL GEOGRAPHY: THEORETICAL TRADITIONS

Although geographers have long investigated social topics, social geography as a vibrant branch of human geography emerged during the second half of the 20th century alongside wider political and social agendas focused on modernist impulses and social scientific interests in problem solving. Consequently, the practice of geography as a spatial science saw many social geographers recording and modeling the spatial patterns characterizing different populations, their residential patterns, and their social service needs. Then, through the 1970s, humanistic geographers challenged the statistical descriptions and mathematical formulas used to address the social world. They noted the limited use of such techniques for understanding the complex and interwoven factors and issues that defied measurement yet affected individual and group experiences of daily life. Humanistic approaches drew on phenomenological philosophies and ethnographic techniques to create conceptual and empirical accounts of everyday people and their social worlds at a far more intimate and detailed scale than that of most spatial science proponents.

Parallel to humanistic developments, a series of radical geographies were challenging the traditions of mapping and describing sociospatial patterns. Welfare, Marxist, and feminist geographies all highlighted how social conditions and individual experiences should be critiqued rather than accepted as status quo to be mapped and modeled. Different theoretical emphases on basic needs, social justice, modes of production, or gender relations were used, but each approach sought to critically analyze (and support change in) the wider social and economic structures that created unequal power relations and living conditions.

Since the early 1990s, and following wider interest in European social and cultural theory, postmodern and poststructural geographies have practiced considerable skepticism about grand explanatory metanarratives presented by scientific and radical accounts in social geography. Scholars have turned instead to highlight the socially and culturally specific discourses that frame our diverse lives and social worlds and that circulate ideas of social difference, space, place, and power.

CORE FOCI OF SOCIAL GEOGRAPHY

Irrespective of theoretical interests, social geographers have worked on three key subjects or themes that characterize much of the subdiscipline, shaping the focus of investigations and the production of knowledge. First, as part of a problem-solving discipline, social geographers have commonly sought to address *social conditions and problems*. The heritage of regional geography linked social conditions and *genres de vie* to specific places and regions in several European traditions. In the Anglo-American context, the modernist, progress-oriented views circulating society and academia during the 1950s and 1960s saw geographers applying spatial science traditions to map the incidence of contrasting social conditions and analyzing the occurrence and likely spread or containment of social phenomena (e.g., prostitution, the "negro ghetto"). The spatial planning and systematization of geographic information also emerged as a core dimension in European settings. Other geographers have drawn on welfare and Marxist geographies to make a more overtly political move from the description or modeling of social problems to critique the causes of these conditions. David Smith's call for a *welfare geography* approach and David Harvey's conceptualization of *social justice* both sought to highlight the inequalities and uneven power relations underlying spatial variations in living conditions and problems (e.g., housing provision, access to services, unemployment). The social geographer as activist and critic draws from this heritage and continues in the work of many researchers.

A second theme in social geography involves many scholars engaging in sophisticated analyses focusing on categories of social difference. In-depth geographies of class, gender, ethnicity and race, sexuality, and (dis)ability illustrate this trend in Anglo-American geography, although these categories of difference are unevenly (and sometimes rarely) recognized in other contexts. For example, substantial research on gender has introduced a key critical component to Spanish geography, but less work has been published on lesbian and gay issues. Whereas each of these categories of difference is documented in a separate subfield of the English literature, more recent postmodern and poststructural recognition of diversity has encouraged still further geographies of others who face marginalization and social exclusion.

Across these diverse records of social difference, geographers have worked to show the spatiality of difference at many scales and have explored how spaces both reflect and are shaped by these differences. This forms a third overarching theme in social geography concerning geographers' recognition that categories and experiences of social difference do not occur discretely or separately. Indeed, they are interwoven and combined in a *mutually constitutive relation with space*. Consequently, numerous works have recorded the composite power relations and effects of social difference that are negotiated through a range of spaces in people's lives. These works show how combinations of difference are shaped by various spaces and in turn affect how those spaces are maintained and understood. This type of work can be seen in geographies as diverse as those analyzing body spaces, homes, streetscapes, and public spaces. It is also apparent in social geographies of larger arenas such as neighborhoods, communities, workplaces, and whole industries. In these spheres, geographers show how the contrasting spaces reflect dominant social norms (e.g., heterosexuality, masculinity, whiteness) while also including the experiences of a wider heterogeneous set of individuals' and groups' experiences and needs. At times, these groups and needs are expressed strategically in sociopolitical actions that, in turn, illustrate how spaces (from individual work sites to wider community and public spaces) are not neutral phenomena devoid of social character; instead, they are socially scripted and constructed in ways that can open up opportunities for resistance and reconstruction (e.g., gay pride marches, community activism).

THEORIZING SOCIETY AND SPACE

Across the rich fields of social geographic literature, increasingly nuanced ideas of society and space have emerged. Spatial scientific traditions encouraged the observation of space as a socially neutral physical plane across which social patterns and movements could be analyzed systematically. But later approaches increasingly explored the possibility of society–space relations. Humanistic geographers such as Anne Buttimer drew on existential and phenomenological philosophies to argue that space was less of an objective pregiven reality and more of a social phenomenon that would be experienced and given meaning in contrasting social ways as individual "lifeworlds" were created and negotiated. The role of everyday lives and human agency were given special attention in the study of society–space relations, producing geographies of "sense of place" and what Yi-Fu Tuan called "topophilia."

Complementing humanistic rejection of an objective space in positivist science, Marxist and feminist geographers also theorized space as a socially nuanced product—the result of social and material processes. These radical geographers showed how space is produced and controlled via unequal power relations based on contrasting modes of production and gender relations. Marxist geographers drew particularly on Henri Lefebvre's ideas of the production of space to show how a sociospatial dialectic connected material and social relations to specific spatialities reflecting different forms of capitalism and social life.

More recently, postmodern and poststructural theories have underpinned social geographers' further study of space and society as mutually constituted, for instance, the social production of the "postmodern city." Poststructural attention to discursive constitutions of the social world has been particularly invigorating for geographers who have been able to "read" the discursive production of space and place while also treating spaces, sites, and scales as "texts" that can be deconstructed, and even reconstructed, in strategic and political moves. This work has involved cultural and political influences on social geography.

DECONSTRUCTIONS: CULTURAL AND POLITICAL INFLUENCES ON SOCIAL GEOGRAPHY

Social geography has been affected by the cultural turn that swept through human geography during the 1990s. The use of wider cultural theory resulted in mostly poststructural debates and methods being applied to social topics. The resultant overlapping of social and cultural geography saw scholars moving from grand or universal-style analyses of societies and spaces (e.g., the city, the countryside) to make readings of those subjects that emphasized the cultural specificity and socially constructed nature of phenomena (e.g., postmodern cities, diverse ruralities). Geographers also registered the contestable nature of concepts such as space, place, and scale.

Social geographers' continuing interest in difference was joined by the deconstructive potential of poststructural approaches so that the hegemony or normativity of dominant categories of difference could be disturbed and challenged. Categories such as gender and ethnicity no longer were seen as relatively fixed or certain but rather needed to be read alongside multiple other conditions and spatial negotiations. A more detailed conceptualization of identity, power, and action enabled social geography to combine a continuing focus of difference with more critical and complex appreciations of social choices and struggles made in people's everyday lives and spaces. The individual and collective politics of these negotiations was a potent new direction for many social geographers, especially through a focus on the formation and reconstruction of identity—of people and places.

Whereas cultural perspectives produced insights on the meaning and representation of many social phenomena, social geography has retained and enhanced its political dimension throughout this period. Much of the culturally informed work of social geographers has continued to point to the interplay of space and power in the social worlds and specific sites in which we live. In many cases, this involves what Doreen Massey called "spatialized social power" in the everyday places of home, work, and neighborhoods. At other times, it has involved strategic social action via processes that take up and rescript social ideas and uses of space for strategic reasons.

FUTURES FOR SOCIAL GEOGRAPHY

The future shape of social geography will emerge from the rich details and conceptual energies that exist in both ongoing society–space debates and the developments in other complementary subdisciplines. Existing differences between geographic traditions will likely continue, but the hegemony of Anglo-American approaches is now being critiqued vigorously and more attention is being given to what

Lawrence Berg called the context-specific "spatialities of geographic knowledge production."

The position of social geography as a relevant and critical component of human geography will remain and strengthen via the consideration of three key questions that have emerged early in the 21st century.

1. *What is the significance of ethics, morals, and emotions in social geography?*

A series of affective and philosophical developments within and beyond social geography have created new energy and debate concerning the limits of our understandings and investigations of the social world. First, a considerable literature has developed around the ethics of research topics and practices that subsequently have stimulated more reflexive and self-conscious performances of social geography. Second, some scholars have called for more explicit considerations of morality in our design and analysis of social and other human geographic phenomena, reigniting the hopes and aims of some radical geographies and imagining better worlds for future societies. Third, geographers are recognizing that emotions form a significant set of relations shaping the social world and geographers' experiences in researching it. Together, considerations of ethics, morality, and emotions enhance the evaluation of social geography, drawing scholars more closely into the research practices and worlds they wish to understand.

2. *Does the material world hold continuing relevance in a social geography that is so strongly influenced by immaterial and cultural foci?*

Although the cultural turn has had a bracing effect on many fields of social geography, concern has grown about emphasizing cultural specificity, diversity, and discourse at the expense of producing significant social critique or research applications. Calls to recognize continuing sociospatial inequalities and to "rematerialize" social geography serve as two core directions for new, socially relevant, and policy-informing research agendas. Examples of materially attentive endeavors stretch across the socially and politically infused "scales" of bodies and dress; through homes, work, and leisure sites; to national inequalities and social polarization.

3. *What challenges and opportunities do posthuman approaches provide for social geography?*

The deconstruction of human/physical and culture/nature binaries common within the cultural turn has been followed by calls for hybrid and posthuman geographies. These both challenge social geography and support new ways of conceptualizing and researching the social world. Understanding the "human" as a socially and culturally specific construction can threaten some views of the central subject of social geography, or it can refocus the way in which social experience is situated and understood within different settings and worlds. There is no neat separation of human and nonhuman worlds; instead, there is what Sarah Whatmore described as a "messy heterogeneity of being-in-the-world." Studies of human *being* suggest new ways in which to approach social geography, new ways in which to imagine and support people's capacity to act, and new ways in which to understand lived experiences within what Jonathan Murdoch called the "entangled ecologies" of human and more-than-human elements.

Cumulatively, the considerations of these questions provide new ways in which to enhance the core foci of social geography outlined here. Attention to social problems, differences, and sociospatial interrelations can be invigorated and extended by acknowledging affective, ethical, and rematerialized perspectives as well as by recognizing that human subjects always are in the process of being. These themes also suggest new directions for social geography beyond the traditional problems–difference–sociospatial trio. A more socially sensitive, materially embedded, and humanly problematized subdiscipline can develop. For some this will continue a commitment to action-oriented, politically relevant research, whereas for others it will emphasize the multifaceted social geography/earth writing that is yet waiting to be formed.

—*Ruth Panelli*

See also Class; Cultural Geography; Cultural Turn; Ethnicity; Feminist Geographies; Gays, Geography and/of; Gender and Geography; History of Geography; Humanistic Geography; Identity, Geography and; Marxism, Geography and; Political Geography; Postmodernism; Poststructuralism; Production of Space; Queer Theory; Race and Racism; Scale; Sense of Place; Sexuality, Geography and/of; Urban Geography

Suggested Reading

Berg, L. (2004). Scaling knowledge: Towards a critical geography of critical geographies. *Geoforum, 35,* 553–558.

Buttimer, A. (1976). Grasping the dynamism of lifeworld. *Annals of the Association of American Geographers, 66,* 277–292.

Castree, N., & Nash, C. (2004). Introduction: Posthumanism in question. *Environment and Planning A, 36,* 1341–1343.

Garcia-Ramon, M., Albert, A., & Zusman, P. (2003). Recent developments in social and cultural geography in Spain. *Social and Cultural Geography, 4,* 419–431.

Gregson, N. (2003). Reclaiming "the social" in social and cultural geography. In K. Anderson, M. Domosh, S. Pile, & N. Thrift (Eds.), *Handbook of cultural geography* (pp. 43–57). London: Sage.

Jackson, P. (2000). Rematerializing social and cultural geography. *Social and Cultural Geography, 1,* 9–14.

Massey, D. (1999). Spaces of politics. In D. Massey, J. Allen, & P. Sarre (Eds.), *Human geography today* (pp. 279–294). Cambridge, UK: Polity.

Murdoch, J. (2004). Humanising posthumanism. *Environment and Planning A, 36,* 1356–1359.

Pain, R. (2003). Social geography: On action-orientated research. *Progress in Human Geography, 27,* 677–685.

Sibley, D. (1995). *Geographies of exclusion: Society and difference in the West.* London: Routledge.

Timar, J. (2004). More than "Anglo-American," it is "Western": Hegemony in geography from a Hungarian perspective. *Geoforum, 35,* 533–538.

Tuan, Y.-F. (1977). *Space and place: The perspective of experience.* Minneapolis: University of Minnesota Press.

Whatmore, S. (2002). *Hybrid geographies: Natures, cultures, spaces.* London: Sage.

SOCIAL INFORMATICS

As with ecoinformatics or geoinformatics, social informatics lies at the intersection between information technology and a substantive scientific domain, in this case, social science. It has two distinct meanings. In the first, social informatics consists of the study of the impacts of information and communication technology (ICT) on society; in the second, it consists of the use of ICT to advance research in social science. In the second case, social informatics sometimes is contracted to socioinformatics.

THE SOCIAL IMPACTS OF ICT

Over the past few decades, particularly since the early 1990s, there has been a dramatic uptake of information technologies in society; therefore, it is incumbent on the academy to reflect on this process and its obvious and hidden impacts. Early computers were designed for massive processing of numbers in pursuit of the cold war, and it was not until the 1970s that the development of database technology opened the possibility for the creation and use of archives of digital information about people, although the U.S. Bureau of the Census had pioneered the use of information-processing technology several decades earlier. By the 1980s, most large corporations had invested heavily in computing technology, using it to track customers and clients as well as production and distribution, and government agencies were not far behind. With the popularization of the personal computer during the 1980s and of the Internet and World Wide Web during the 1990s, a large proportion of society became computer literate, using the technology for everything from shopping to home entertainment. Massive economies of scale deriving from a uniform digital technology, along with exponential increases in processing and storage power, meant that by the turn of the 21st century, computers with the power of the mainframes of the 1960s were accessible to the average consumer.

All of this has had massive impacts on society and is the focus of the growing field of social informatics. Among the core issues is the digital divide, that is, the sharp and growing difference between those to whom ICT is an inescapable part of life and those to whom it is largely unavailable. These differences exist at all geographic scales—from the international (access to ICT is tightly controlled in many countries and is simply impossible in others), to the regional, and even to the neighborhood level. Another core issue is privacy and related issues of surveillance. ICT has made it possible for extensive monitoring networks to be installed using video cameras capable of recognizing vehicle license plates and even faces. The Global Positioning System (GPS) is now installed in a growing proportion of the U.S. vehicle fleet and is used, for example, by car rental agencies to monitor the travel and speeds of renters. Databases can be readily linked, based on names or street addresses, to assemble frightening amounts of data on individuals. To date, there has been very little legislation in the United States aimed at regulating such databases, particularly in the private sector, whereas legislation in the European Union is somewhat more advanced.

A further set of issues derives from the ability of ICT to copy and disseminate information at electronic speed. Copyright and intellectual property rights are taking on new meaning, and ICT is providing unprecedented opportunities for plagiarism. Traditional publication mechanisms are being replaced by the much more flexible technologies of the World Wide Web, with implications for the business models of university presses, learned journals, and traditional private-sector publishers.

THE USE OF ICT IN SOCIAL SCIENCE

Like all areas of science, the social sciences are making increasing use of ICT. Research databases in digital form emerged during the 1960s, and today a vast amount of data of potential value to researchers is available via the Internet. Software tools for analysis are also widely available, as are communication tools such as electronic mail and video teleconferencing. Both teaching and professional presentation of research results have been strongly influenced by technologies such as Microsoft PowerPoint and computer projection.

However, it is clear that ICT will have an even greater impact on the practice of social science in the future. Terms such as *e-social science* and *cyberinfrastructure* are being adopted to describe a vision of this future world in which social scientists will be able to collaborate and communicate virtually independently of physical separation. In principle, this vision imagines a world in which communication between scholars enabled by ICT is as effective as the face-to-face conversation that historically dominated academic research and instruction. A number of technical and institutional developments will be required, however, to make this vision a reality. Collaborative technologies are still far too invasive, and the more familiar e-mail still imposes an effective filter on many aspects of communication such as gesture and voice inflection. Standards are needed to ensure that information can be shared and understood, particularly across differences of discipline, culture, and language. Many of the functions that today must be installed and run on the researcher's desktop could be more effectively provided from common servers as shared services, but to date very little progress has been made in this direction. Although massive archives of social data exist and are widely used, there still are only a few very limited shareable archives of the tools and processes of social data analysis.

—*Michael F. Goodchild*

See also Communications, Geography of; Cyberspace; Geodemographics; GIS; Humanistic GIScience; Telecommunications, Geography and; Virtual Geographies

Suggested Reading

Kling, R. (1999). What is social informatics and why does it matter? *D-Lib Magazine, 5*(1) [Online]. Available: www.dlib.org/dlib/january99/kling/01kling.html

National Science Foundation. (2003). *Revolutionizing science and engineering through cyber-infrastructure* (Report of the National Science Foundation Blue Ribbon Advisory Panel on Cyberinfrastructure) [Online]. Available: www.communitytechnology.org/nsf_ci_report/

SOCIAL JUSTICE

Social justice is a normative concept concerning the ways in which resources and power should be shared across society. Most conceptions of social justice have something to do with social equality, although they differ in the ways in which they contend society should be equal.

Conceptions of social justice are geographic in a number of ways. First, they vary across space and time. Although attempts usually are made to solidify a particular understanding of social justice as universal, what constitutes social justice for one group of people in one place at one point in history does not necessarily do so for others. Social justice might mean something entirely different for striking union workers in Mexico City than it does for antiwar protesters in London. Second, social justice is concerned with the ways in which resources and power are shared across *space*. According to many conceptions of social justice, an entire city (not just a few neighborhoods) or the whole world (not just a handful of countries) should enjoy social equality. Along this line of thinking, some geographers, as well as social activists, have described the tendency for different places to develop uneven levels of prosperity and influence as socially *unjust*. Instead of having pockets of impoverished neighborhoods next to areas of great affluence, a more socially just geography might involve more spatial equity in the distribution of resources and the exercise of power. Third, notions of social justice are used to address social conditions on particular scales and in fact are integral to the construction of those scales. Measures of social justice have been applied regularly to the urban scale, although the concept is also used to evaluate and change social relations at the body, household, neighborhood, regional, national, and international scales. Consider, for instance, the meanings of social justice that have been used to make international human rights standards or those that have been struggled over when developing reproductive rights for women. Fourth, ideas of social justice often are inscribed right into the landscape, especially in the built environment. The presence, location, and conditions of public housing can say something about a society's conceptions of social justice, as can the presence (or absence) of adjacent high-rise office and apartment buildings. Fifth, ideas of social justice are also applied to inequities in the ways in which different people are forced to endure environmental problems. Environmental justice, as this view sometimes is called, seeks to eradicate dangers such as unsafe drinking water, polluted air, and lead poisoning that disproportionately affect people of color and other low-income communities.

Geographers and philosophers have identified and developed a few particular approaches to social justice that resemble, to greater or lesser degrees, popular attitudes and legislative agendas on the matter. Libertarian conceptions of social justice hold that everyone should have an equal opportunity, or starting point, to attain wealth and power regardless of what may happen thereafter. Unequal levels of income, therefore, would be considered socially just so long as everyone had the same chances to earn money from the beginning. This conception of social justice often is aligned with free market capitalism or neoliberalism. Advocates of this understanding of the concept argue that whatever inequities may emerge from the market are socially just outcomes because the conditions that structured the results ultimately were fair (so long as the market was allowed to operate properly).

Liberal (or social contract) theories of social justice also contend that people should enjoy equal opportunities, but they add that resources should be distributed in a way that benefits the poorest members of society. This idea of social justice has led to, and has derived from, welfare state capitalism and Keynesian economics. Like libertarianism, it emphasizes proportional compensation for individual merit, but it also calls for the intervention of the state to redistribute income in a way that ensures that no one will fall below a certain level of poverty. It acknowledges, furthermore, that such inequities in income are more likely a result of failures in the market economy than of shortcomings of individuals' abilities or efforts.

Some have looked for notions of social justice in the writings of Karl Marx, arguing that the exploitative nature of capitalism is inherently unjust. In particular, Marxists argue that workers are not adequately compensated for their labor. Workers put more value into the products they produce than they receive back from their capitalist employers. The fact that workers do not own the machinery and raw materials necessary to produce these products ensures that capitalists can and will continue to pay them less than they deserve. A socially just society, Marxists argue, cannot be reached simply by redistributing income; it also must include the common ownership of the means of production. Other Marxist thinkers have dismissed the concept of social justice altogether, arguing that it is an ideological tool of the ruling class used to legitimize their position within the broader social order. Social justice, this argument goes, is a term erected by economic leaders to make the social relations of the given society seem fair.

The work of feminists, both inside and outside of the academy, has extended the terrain of social justice from the traditionally assumed arenas of economic and political life to the household. Applications of social justice, they argue, must account for the ways in which gender differences structure how resources are distributed (even in the name of social justice) and power is exercised. Such an argument ultimately leads not only to the household but also to the bedroom and the body because women experience inequities in the distribution of income, especially when performing unpaid domestic work, and face male-dominated legal systems and cultures, which often put their bodies out of their control and in danger.

Poststructuralist approaches, which many feminists have incorporated into their work, have criticized the apparent claims to universalism that mark many libertarian, liberal, and Marxist conceptions of social justice. They maintain that no universally shared meaning of social justice does, or can, effectively exist. They also argue that the bases on which conceptions of social justice have been measured place far too much emphasis on class or economic interests than on other forms of social well-being. This is not to say that poststructuralists necessarily wish to do away with ideas of social justice. Viable conceptions of social justice could exist so long as they incorporate mechanisms to recognize and dismantle the everyday power inequities related to differences in gender, race, ability, and sexuality *in addition to* those associated with class.

Conflicts over which notion of social justice should materialize into landscapes, legislation, the organization of social relations, or even a particular social movement's politics often are just as troubling to interested individuals and groups as are the conditions that give rise to the conceptions in the first place. So, a lot of the work involved with social justice is not only developing fair solutions to social problems but also persuading people that one conception is better than any other. Efforts are being made, however, to bridge the divides between different ideas of social justice.

—*Robert Ross and Clayton Rosati*

See also Environmental Justice; Ethics, Geography and; Justice, Geography of; Law, Geography of; Radical Geography

Suggested Reading

Harvey, D. (1973). *Social justice and the city.* London: Edward Arnold.

Harvey, D. (1996). *Justice, nature, and the geography of difference.* Oxford, UK: Blackwell.

Smith, D. (1994). *Geography and social justice.* Oxford, UK: Blackwell.

Young, I. (1990). *Justice and the politics of difference.* Princeton, NJ: Princeton University Press.

SOCIAL MOVEMENT

Social movements represent the organized endeavors of multiple individuals, communities, or organizations to pursue political objectives within society and are generally seen to act outside formal state or economic spheres. They can be organized around particular groups (e.g., people with disabilities) or particular goals (e.g., environmental conservation), and their demands can be focused on society as a whole, on the state, or on the economy (or any combination of these). Geographers have become increasingly interested in the scope, nature, and orientation of social movements during recent years, although it remains difficult to measure their effects because they often are very dynamic and often are composed of informal or voluntary associations. Social movements also cannot be understood in isolation and out of context because it is necessary to understand and view them within a broader framework of social and geographic analysis.

Many theories of social movements understand them as a phenomenon within civil society, that is, a major arena of political action (along with the state and the economy). Originally, civil societies were seen as a realm of society that was distinct from and complementary to the state and the market, and until the early 20th century little attention was focused on civil societies; therefore, many theories of social movements were understood to be the product of conflict between the state and the economy and derived from theoretical work concerned with these themes (e.g., Marxism). Movements concerned with labor (e.g., trade unions) were understood in these terms as emanating from structural conflicts in the sphere of the state or the economy.

Social movements were popularized in particular by Manuel Castells's work on the city as a space of collective consumption (by which was meant the state's provision of housing, transport, education, etc.). Castells's work, which was rooted in the French political experience, claimed that contradictions and tensions abounded in the state's provision of these goods and services, leading to urban struggles that later transformed into urban social movements aiming to promote social change. Castells argued that crises in state provision led to anticapitalist urban struggles, although his work was criticized for being too economistic and mechanistic as well as silent on the gender composition of such movements.

Since the 1980s, many observers have highlighted how many of these urban social movements are being displaced by new local–global modes of communication, leading to the creation of new social movements (NSMs). New social movement theory developed initially in Europe to help explain a host of new movements that emerged during the 1960s and 1970s and that did not seem to fit the model of Marxian class conflict that had been the predominant model in much European social movement theory. These movements are new because they are more issue specific, use less conventional tactics, cut across class lines, and are less likely to turn to established political parties and channels to achieve their objectives. NSMs are made up of networks of collective actors (both in the North and in the South) with common interests and identities that have the threat of mobilization as the main source of their power. NSMs can be both progressive and regressive and can be very diverse and heterogeneous, including squatter movements, women's associations, human rights organizations, neighborhood groups, indigenous rights groups, youth groups, and self-help organizations of various kinds. For some observers, NSMs may be indicative of wider shifts in postwar civil society and may be a response to the decline in the authority and

legitimacy of the state or a result of the global emergence of neoliberal economic philosophies. Some new social movement theorists emphasize a change in the economic structure of the First World as a structural force shaping the new movements. This is said to involve a shift from an industrial, heavy manufacturing–based Fordist economy (named after Henry Ford's assembly line) to a postindustrial, postmodern, or post-Fordist economy centered more on the service sector and computer-based information industries.

In the global South, social movements often arise as a result of tensions and contradictions in the development process both locally and globally and occasionally have moved beyond issues raised by urban struggles to embrace wider communities and priorities. In many ways, social movements are the visible signs of deep transformations and of the growing emphasis on culture and identity politics but also of wider shifts away from the state-led developmental projects of the past. Many of these groups can involve the formation of multiple identities and solidarities based on class, gender, kinship, or neighborhood. Many are multidimensional, addressing issues of poverty, gender, culture, and ecology in a simultaneous fashion. Some geographers have argued that the discipline can contribute to the understanding of struggles for survival in different cultural contexts and that, given increasing global interdependence, we might also consider ways of attempting to contribute toward these struggles.

Social identities and changing social movements represent a crucial component in any study of power. NSMs also stand in contrast to official political discourse about the global economy, articulating a geopolitics from below—through an evolving international network of groups, organizations, and social movements. In terms of the economy, social movements in the South often articulate conflicts around the productive resources in a society such as forest or water resources, calling for new services, new forms of access, and more equitable distributions. Many of these movements seem to express political struggles for power and resources on one level but also articulate cultural struggles over identity and attachment to place. The social and geographic diversity of these struggles makes it difficult to generalize; however, exclusion does form a common ground, and this has even been claimed as the basis of the autonomy of some movements. Over the past decade or so, there has been an increasing recognition of the value of organizing transnationally through organizations such as the World Social Forum (WSF). Furthermore, many older social movements (principally labor and trade unions) are also being drawn into these processes of mobilization. Thus, a variety of social movements have demanded meaningful participation in their societies. The ability of these movements to create linkages, disrupt boundaries and hierarchies, and produce new meanings and identities is likely to have an important bearing on the shape of politics (both locally and globally) in the 21st century.

—*Marcus Power*

See also Political Geography; Resistance; State; Urban Social Movements

Suggested Reading

Castells, M. (1983). *The city and the grassroots: A crosscultural theory of urban social movements.* Berkeley: University of California Press.

Della Porta, D., & Diani, M. (1999). *Social movements: An introduction.* London: Blackwell.

Miller, B. (2000). *Geography and social movements.* Minneapolis: University of Minnesota Press.

Power, M. (2003). Theorising back: Views from the South and the globalisation of resistance. In M. Power, *Rethinking development geographies* (pp. 194–218). London: Routledge.

Routledge, P. (1998). Anti-geopolitics: Introduction. In G. Ó Tuathail, S. Dalby, & P. Routledge (Eds.), *The geopolitics reader* (pp. 245–255). London: Routledge.

SOCIALISM

Socialism is defined by state ownership, rather than private ownership, of the means of production (natural resources and infrastructure). In addition, the state must be controlled democratically by its citizens. Control over the means of production as well as the distribution of goods and services may be placed into cooperative ownership and may involve a degree of centralized economic planning. A socialist ideology emphasizes that the economy is to be managed for the benefit of the majority of people instead of for the benefit of a few groups of people. Economic cooperation, rather than economic competition, is promoted by socialist views. The term *socialist* sometimes is used specifically in reference not to an actual state (country) but rather to a type of economy or society.

Many types of socialism have emerged based on differing views related to how socialist systems are to be achieved (e.g., revolution, reform, some combination of the two); how a state is to manage the means of production; whether or not the state system should be a direct democracy or a representative democracy; whether the economy should be a totally socialized one, a mixed economy that includes some features of a capitalist economy, or something in between; and whether injustice in current systems is caused by problems in the distribution of goods and services or by ownership of the means of production being in the hands of a few.

Karl Marx argued that a socialist society, in which industrial workers (rather than an elite capitalist bourgeoisie) would be dominant, was the logical and revolutionary follow-up to capitalist systems and a precursor to communism. In his view, the socialist society would enable the state to wither away and be replaced by a classless communist society.

The former Soviet Union and its sphere of influence in Eastern Europe had socialist economies insofar as the means of production and most of the economy were controlled by the state. However, because the Communist party monopolized control of the state and people in those places had little or no democratic influence on the government, some would argue that the Soviet Union was not structured around a truly socialist economy. In addition, some might argue that the Soviet elite who controlled the economy only continued the trend of the pre-Soviet, tsarist-era capitalist class to manage the economy predominantly for their own benefit.

Some capitalist countries implement socialist policies in sectors of strategic interest to the populace. Examples of such policies include the nationalization (or state control) of defense, the postal service, and even critical industries such as airlines and steel production. Other socialist policies involve worker representation in corporate decision making and in profit-sharing opportunities. Services such as social welfare and insurance for unemployed citizens are other socialist policies.

—*Shannon O'Lear*

See also Communism; Marxism, Geography and; Nation-State; Political Geography; World Systems Theory

Suggested Reading

Schumpeter, J. (1987). *Capitalism, socialism, and democracy* (6th ed.). London; Boston: Unwin Paperbacks.

SOVEREIGNTY

Although it is a much-debated concept, sovereignty traditionally has been defined as a condition of final and absolute authority in a political community. Essentially, sovereignty includes the recognized independent right and inherent power of a state (or country) to stand alone from all other states and for that state to lawfully and independently make and follow its own laws. Since 1648 and the Treaty of Westphalia (which codified the modern system of international politics), sovereignty has been invested in states that have authority over the land and people within their territories. It is not possible to become sovereign just through self-declaration, and so sovereignty never is a matter for a single state but rather is part of an interstate arrangement that requires reciprocal recognition. This reciprocated sovereignty is very much a feature of the capitalist world economy, and this emergence of territorially based sovereignty has created the modern interstate system that is at the heart of contemporary political geography.

Sovereignty operates both internally (outside powers cannot intervene in the affairs of a state unless they are invited to do so) and externally (this depends on reciprocity and a mutual recognition of sovereignty by other states). Thus, sovereignty provides one of the most fundamental ground rules of international politics today in that it defines who is and who is not part of the international interstate system. Some states that have been created have not been recognized by other states and so have not been considered a part of the interstate system (e.g., the republic set up in the northern half of Cyprus following the Turkish invasion of 1974). Since 1945, many new states have been created (particularly since the end of imperialism), and the main way in which they have had their sovereignty recognized is by joining the United Nations. This process was again repeated following the breakups of the Soviet Union and Yugoslavia when the new states these breakups created applied to join the United Nations to prove and mark their entry onto the stage of international politics. Previously, the Soviet Union regularly interfered in the internal affairs of socialist countries in breach of those countries' sovereignty, as in the case of the Soviet interventions in Czechoslovakia (in 1968) and Afghanistan (in 1979).

These twin principles of territory and sovereignty as the basis of international law mean that states are the political units around which laws are framed and that the rights of states have priority over the interests of other institutions. Far from always providing order and stability in the international system, sovereignty often has been a source of conflict, especially when its territorial basis has been contested. If sovereign states do not have precisely delimited boundaries and borders, disputes and wars often result, as in the case of the disputed border between Ethiopia and Eritrea during the 1980s and 1990s. Territorial claims by one state against another state can have historical and cultural roots or can derive from geographic proximity and pressures for integration and/or national self-determination. In addition, conflict may occur where nations and ethnic communities are subjected to the sovereign authority of a particular state and do not recognize themselves as citizens and subjects of that state.

Sovereignty extends upward (to include air space) as well as downward (to include subterranean resources such as oil). The question of territorial sovereignty is very much relevant today in debates about the impacts of globalization. Given that globalization is seen as a transnational process that transcends the boundaries of states, some commentators have argued that this marks the end of state sovereignty. Thus, they now see the sovereignty of states as driven by new processes of flexible accumulation or by transnational corporations. Global financial institutions and markets arguably prevent states from regulating their own currencies, whereas media agencies and crime syndicates can also challenge the authority of states by disseminating information and goods across borders in ways that elude state regulation and control. It should be remembered, however, that this process (like globalization itself) is not some new or novel phenomenon. Occasionally, globalization is (mistakenly) seen to refer to some new recent process in which states no longer are said to be the primary units of decision making and only now are finding themselves located in something called the world market.

It is important, however, to ask what fundamental changes in the state, and in the analysis of the state, have been stimulated by economic globalization. In the course of interactions with global markets and regulatory agencies, so-called Asian tiger countries such as Malaysia and Indonesia have created new economic possibilities, social spaces, and political constellations. Because sovereignty typically refers to the existence of a highest or supreme power over a set of people, things, or places, commentators increasingly have questioned whether sovereignty can be legitimately located in an agent such as a state. Aihwa Ong suggested that the concept of social sovereignty offers us a way in which to understand not only how both state and nonstate actors together can be central to the governance of social domains but also how communities can fashion sovereign "societies" as opposed to states. Ong noted how the shifting relations among market, state, and society in Asia have resulted in the state's flexible experimentations with sovereignty. Ong proposed the concept of graduated sovereignty, in which the state subjects individuals to *different* regimes of control and valuation, thereby creating different *zones* of law internally. The Asian financial crisis of the late 1990s demonstrates the concept of graduation in that the market-oriented agenda can mean different things, strengthening state power and protections in certain areas but not in others. Taking Malaysia as a primary example, with its three ethnicities (Malays, Chinese, and Indians), Ong noted the presence of six zones of graduated sovereignty: the low-wage manufacturing sector, the illegal labor market, the aboriginal periphery, the refugee camp, the cyber corridor, and the growth triangle (the last of which is made up of three border-straddling economic development areas). Thus, it is important to detail the different modes of law and state intervention that discipline each zone. Therefore, sovereignty today is defined at a variety of spatial scales (local, regional, and global) and by a range of actors (both state and nonstate), and different forms of sovereignty can exist within the same political territory.

—*Marcus Power*

See also Civil Society; Critical Geopolitics; Geopolitics; Globalization; Nation-State; Political Geography; World Systems Theory

Suggested Reading

Braden, K., & Shelley, F. (2000). *Engaging geopolitics.* Englewood Cliffs, NJ: Prentice Hall.
Ong, A. (1999). *Flexible citizenship: The cultural logics of transnationality.* Durham, NC: Duke University Press.
Painter, J. (1995). *Politics, geography, and "political geography": A critical perspective.* London: Edward Arnold.
Taylor, P., & Flint, C. (2000). *Political geography: World economy, nation-state, and locality.* Englewood Cliffs, NJ: Prentice Hall.

SPACE, HUMAN GEOGRAPHY AND

Space is without doubt the most important concept within the discipline of geography. As ideas of what space is and how it is defined have changed, so too has the identity of the discipline. To understand the significance of the definition of the concept of space, it needs to be examined both chronologically and conceptually. What emerges is the realization that the ways in which geographers have identified this concept often are tied into larger scientific paradigms and social contexts.

A brief survey of the different concepts and types of space that have been identified over the past 60 years alone highlights the multitude of spatial concepts discussed in the discipline. There are absolute, abstract, architectural, concrete, discursive, material, performative, produced, relational, relative, representational, social, and many other concepts of space. Among these, the development of absolute, relative, and relational concepts of space is examined here.

THE EMERGENCE AND DEFEAT OF ABSOLUTE SPACE

Although distinct theoretical work on the concepts of space and place was largely absent in Anglo-American geography before the 1930s, this began to change with the growing popularity and influence of Richard Hartshorne. His major contribution to geography, a landmark publication titled *The Nature of Geography* (published in 1939), was the first to spark a significant debate about the goals of geography as a discipline and the definition of a concept of space. Although Hartshorne's work enjoyed tremendous popularity and remained virtually unchallenged for more than a decade, it received extensive criticism in 1953 with the publication of Fred Schaefer's article "Exceptionalism in Geography: A Methodological Examination." The resulting scholarly exchange not only redefined geography as a discipline and ushered in the quantitative revolution but also rewrote the concept of space.

Hartshorne regarded geography as a discipline whose primary goal is the understanding of the functional integration of phenomena over space. For him, geography was chorology, a concept he borrowed from German geographer Alfred Hettner. As chorology, Hartshorne saw it as geography's goal to study the causal relationships between geographic phenomena occurring within particular regions. Geography's exceptional character as a scholarly discipline was to be both systematic and regional. Accordingly, the focus of regional geography was to study the arrangement of phenomena insomuch that the phenomena themselves—not the logic behind their spatial arrangement—are at the center of the study. Emphasis was put on the uniqueness of spaces and regions—not on what science normally does—to understand reality in terms of governing laws and regularities. Subsequently, Hartshorne's concept of space was one based on absolute fixed entities, that is, on the arrangement of objects anchored in space in a specific manner.

The first to significantly challenge both Hartshorne's definition of geography as a discipline and his concepts of space was Schaefer, a young economist at the University of Iowa. Schaefer wanted to make geography more scientific by following the scientific method. He argued that Hartshorne's idea of geography as chorology is based on the German model of science (*Wissenschaft*), which is considerably more inclusive when it comes to the methodological approaches that can be classified as scientific. In German *Wissenschaft*, science is any organized body of knowledge, not a distinct, agreed-on, and singular scientific method that produces a uniform body of knowledge based on hypothesis testing and a rigorous validation and verification process.

For Schaefer, geography must pay attention to the spatial arrangement of the phenomena in an area and not so much to the phenomena themselves. Spatial relations are the only ones that matter in geography; therefore, the discipline ought to study the laws concerning spatial arrangements. Geography, as regional science like Hartshorne envisioned it, does not go beyond mere description of spatial structures. Schaefer wanted geography to be purely analytical. He criticized geography as being phenomena oriented. For him, geography focused too much on the absoluteness of phenomena and their change over time and space, and geography pointed too much toward the uniqueness of spatial structures. This dualistic focus on regional and systematic analysis made it different from other sciences in terms of methodological approaches, and Schaefer resisted this exceptionalism that Hartshorne supported so vehemently. In contrast, Schaefer wanted to explore and emphasize purely geographic laws that stand apart from process laws. An example of purely geographic universal laws that Schaefer sought to promote was the idea of distance decay, which states that the intensity and frequency of interaction decreases with distance. Two places, people, companies, or similar entities are much more likely to interact if they are close together,

whereas two more distant objects or subjects will be much less likely to interact. In other words, proximity and distance guide any form of spatial interaction. Such generalized laws can be applied to economic, social, political, and even cultural systems; they are universally applicable and strictly geographic—not historical or social—in nature.

Ultimately, what Schaefer sought to accomplish was a paradigm shift in spatial concept thinking. He attempted to get away from the analysis of absolute unique spaces and spatial arrangements of Hartshorne's regional geography. Instead, a relative concept of space, where the pure spatial relations were at the core of geographic inquiry, was developed. For Schaefer, the universal laws of spatial relations were at the heart of geography's scientific mission.

FROM RELATIVE CONCEPTS TO THE PRODUCTION OF SPACE

Following Schaefer's initial philosophical argument, a group of geographers, primarily located at the University of Washington in Seattle, started what is widely known as the quantitative revolution in geography by elaborating on the variety of core geographic themes that were based distinctly on relative concepts of space. Emphasizing the replicability of research and methodological clarity, these geographers outlined distance, pattern, position, and site (among others) as the core concepts of geography. The examination of space now was merely about the laws, as well as geometrical and mathematical patterns, governing the relations between events or objects, and the social and historical placement of these entities disappeared from the analysis entirely. In fact, as many critiques have noted, much of the work produced by the quantitative revolution reduced geography to geometry and eliminated the subjective stories, biographies, narratives, emotions, and motivations of human life from geographic analysis. Space became an empty concept insofar as it described, in a very abstract way, only what was *between* objects in space. At the same time, this relative concept of space neglected the forces driving the objects and events in space and their spatial arrangement.

It was again the arrival of critical geography during the 1970s that significantly challenged such an abstract, numbers-driven concept of space and sought to replace it with distinctly *relational* ideas. The relational view of space conflates both absolute and relative concepts of space and argues that these are in fact inseparable. Structures and processes—the tangible and the abstract—are interrelated and cannot be separated. Although we can measure space in relational terms and establish rules and laws of spatial relations in terms of distance, direction, and other concepts, we must also always examine the social systems that produce distinct, real, and material spatial structures. Social and economic systems play out in absolute spaces that are material and tangible, but these are the product of abstract socioeconomic processes that are also distinctly spatial. For example, capitalism, like any other economic system, not only is based on economic laws and rules of profit generation but also depends on and produces spatial inequalities that support such a system indefinitely.

The best and most comprehensive analysis of the interplay of absolute and relative spaces can be found in Henri Lefebvre's seminal work *The Production of Space.* Originally published in 1974 and examined by only a few Francophile geographers, this work received tremendous attention in the Anglo-American geography following its English-language release in 1991. The major contribution of this work lies in its detailed analysis of the connections among modes of production (i.e., ways of producing and consuming goods and services), the visual logic that accompanies each mode of production, and the distinct material forms of social organization that result from these. Lefebvre argued that each economic, social, and cultural system, be it primitive, feudal, mercantile, or modern capitalism, has its own requirements for how space needs to be produced and organized for the system to function best. He often illustrated these connections by referring specifically to cities and their form, that is, their morphology.

In primitive societies with little social organization, the spatial organization of settlement landscapes is analogical, based on local interpretations of the environment. Small settlements are organized according to local conventions that best reflect adaptation to the environment and ensure the survival of the community. For example, small Amazonian villages might be characterized by a division of male, female, and family spaces and have very little distinction between private and public spaces. The spatial logic that organizes this community is practical as well as spiritual and is a reflection of a specific adaptation to an environment that manifests itself in localized productions of meaning. In contrast, the ancient Greek city of Athens is characterized by a distinct mode of production that shapes the city into a coherent whole and makes it into a material manifestation of a cosmological model. For example, temples in ancient Athens were arranged according to stellar patterns, and the political system reflected such a search for universality.

Feudal (late medieval and early modern) cities no longer were based on cosmological models, but their mode of production was distinctly symbolic. In France, the royal palace at Versailles under Louis XIV reflects such ideals. The radial design of Versailles, where all roads lead to and from the central palace, reflects the ideas of absolute rulership. Thus, landscape design had become symbolic and representative of the political system in which it was placed.

In contrast, the modern capitalist city is not ruled by cosmological or politically symbolic design; rather, it simply reflects the logical outcome of capitalism. The arrangement of industrial and residential areas, for example, follows a distinctly capitalist logic. As class conflict theorist David Gordon argued, modern urban design, as can be found in the United States today, might not be the result of a natural growth and expansion process; rather, it may be the result of a process that was distinctly driven by the decision makers of a capitalist economy. Changes in urban design and planning, as they happened throughout the 20th century, are reflections of changes in capitalist thinking and strategizing that are driven by adjustments in profit maximization strategies. For example, at the beginning of the 20th century, most American cities still were relatively compact and characterized by a close proximity of industrial sites and laborer residences. As the labor movement grew stronger and organized major strikes during the first two decades of the 20th century, capitalists noticed that this was the result of the close proximity of work sites and residences. Union activists could easily walk from door to door to organize unrest and spread their agendas. As a response to such threats that interrupted the production process and thus the generation of profit, many factory owners began to move their operations to less densely populated areas on the outskirts of cities. Laborers who worked at these new sites now were coming from more dispersed residential areas and subsequently were less likely to organize or participate in disputes. This process of moving production sites led to the first decentralization of American cities and their gradual expansion over larger and larger territories. As this example illustrates, it is particularly the logic of capitalist thinking that had a profound effect on urban form. Thus, capitalism produced a distinct form of spatial organization.

Recent geographic work on the production of space not only argued that capitalism has had profound effects on the spatial organization of urban environments but also concluded that such capitalist spaces are distinctly gendered, racialized, and sexed, meaning that they influenced not only where work sites and residences were placed but also where men, women, ethnic and racial groups, and sexual minorities are placed. In summary, the concept of the production of space has provided geographers with a comprehensive tool that not only regards space as absolute and relational but also argues that space must be considered as one of the elements of the productive forces of society.

—*Olaf Kuhlke*

See also Berkeley School; Chicago School; Chorology; Circuits of Capital; Communications, Geography of; Core–Periphery Models; Critical Human Geography; Cultural Geography; Cyberspace; Development Theory; Division of Labor; Economic Geography; Environmental Determinism; Eurocentrism; Feminist Geographies; Historical Geography; History of Geography; Human Agency; Humanistic Geography; Justice, Geography of; Labor, Geography of; Logical Positivism; Marxism, Geography and; Orientalism; Phenomenology; Political Ecology; Production of Space; Quantitative Revolution; Radical Geography; Social Geography; Spaces of Representation; Time Geography; Transportation Geography; Underdevelopment; Uneven Development; Urban Geography; Urbanization

Suggested Reading

Gottdiener, M. (1985). *The social production of urban space.* Austin: University of Texas Press.

Hartshorne, R. (1939). *The nature of geography: A critical survey of current thought in light of the past.* Washington, DC: Association of American Geographers.

Harvey, D. (1985). *The urbanization of capital.* Oxford, UK: Blackwell.

Lefebvre, H. (1991). *The production of space.* Oxford, UK: Blackwell.

Schaeffer, F. (1953). Exceptionalism in geography: A methodological examination. *Annals of the Association of American Geographers, 43,* 226–249.

SPACES OF REPRESENTATION

In its most literal sense, geography is the science or art of Earth writing or describing. Thus, how spaces on the earth's surface are described or made visible through language is an important concern. This issue involves an account of (a) the nature of signs, which are expressions of objects or ideas; (b) the process of abstraction, which is the process by which one thing stands for another; and (c) the symbolic mediation of experience, that is, the ways in which we bring reality into consciousness. The roots of this line of thought may be traced to medieval questions of hermeneutics (from Hermes, messenger of the gods) and attempts to find the meaning of the Bible.

Traditional positivist views of representations held to a strict separation between observer and observed, between sign and referent, between facts and values, and between epistemology and ontology. In this reading, claims to truth are mirrors of the objective world, representation is only a technical problem, and interpretation is unproblematic. Because there is only one objective reality, there is only one true meaning of a sign.

The positivist notion of signs and representations was steadily undermined during the 20th century as language and questions of meaning became a central issue of debate. The growth of structuralist linguistics, such as the work of Ferdinand Saussure, argued that signs have no meanings in and of themselves and acquire significance only through their position relative to other signs. Philosopher Ludwig Wittgenstein took this line of thought further, arguing that all linguistic structures were essentially arbitrary and that language was not a neutral avenue for the construction and sharing of meanings but rather an opaque medium in its own right that limited its users' views of reality. Thus, language is not only a mirror of the world but also the means by which we live in the world. By the 1980s, textual and literary analysis had moved to the highly relativistic conception of deconstruction in which texts (and by extension all social objects) have no essential meanings of their own (the original goal of hermeneutics) but instead have multiple, shifting, or even contradictory meanings. Thus, every representation of the world is a simplification—too simple to capture its diversity and complexity—and analysts of language must focus on its silences, limitations, and contradictions.

Social philosophers focused on representations as social constructions. To have meaning, representations must be shared intersubjectively; that is, they are understood only within the context of other representations or intertextually. Discourses are socially produced sets of representations that simultaneously enable and constrain our understanding of the world. Representations can be Frankenstein-like, escaping the intentions of their creators, given that the ways in which representations are interpreted or consumed is not necessarily how they are intended or produced. Michel Foucault argued that representations always are linked to power; that is, they inevitably serve someone's interest. Hegemonic discourses (i.e., representations that serve the dominant powers in a society) tend to legitimize social practices and naturalize the status quo, in the process constituting subjects in everyday life. Thus, the political understanding of representations maintains that they always have social consequences, albeit not always intended ones (i.e., as dominant and subversive discourses), and that knowledge is less a mirror of the world than a contested battleground of views linked to different social interests. Thus, representations are part of the reality they help to construct; word making is also world making (i.e., discourses do not simply mirror the world, they constitute it). In this way, epistemology and ontology always are intertwined. By politicizing the production and consumption of meaning, poststructuralists moved from the postmodern politics of difference to a concern for discursive regimes of truth.

The most significant pioneer in making explicit the social nature of spatial representations was French philosopher Henri Lefebvre (1901–1991), whose book *The Production of Space,* which appeared in English in 1991, had an enormous influence over the discipline's view of this topic. Lefebvre played a critical role in portraying geographic landscapes as part and parcel of capitalism; capitalist societies reproduce themselves only by constructing spaces ranging from the individual body to the global economy. Lefebvre's work centered on a famous triad consisting of (a) spatial practice, that is, the material reality in which the body is situated and nature is transformed; (b) representational space, that is, the scientific diagrams used to understand space such as engineering charts and blueprints; and (c) spaces of representation, that is, the artistic views of geographic reality that allow room for the imagination and unrealized possibilities. In this triad, all three elements are intertwined, producing a

fusion of consciousness and space so that the line between the real and the imagined becomes blurred in creative ways.

If positivist and empiricist geographies of the past implicitly held that interpretation was an unproblematic act of immaculate perception, poststructuralists denaturalized how space is encoded and brought into the domain of consciousness. A tradition of critical cartography inaugurated by Brian Harley and Denis Wood portrayed maps as social constructions linked to power interests; maps may hide the interests that bring them into being, but the deconstructionist critique has increasingly made their social significance clear. Like all ideologies, the power of maps is at its greatest when we take them for granted. Maps enable the past to be part of the present and are embodiments of others' experiences. They create an aura of detached objectivity that is powerful to the extent that their authors disappear. The historical role of cartography, for example, both explicitly served colonial interests (e.g., in the geographing of boundaries) and implicitly legitimated the Eurocentric panopticonic view. In the same vein, John Pickles offered a powerful critical assessment of the social implications of geographic information systems (GIS), discussion of which typically is framed in technocratic terms as if it were devoid of social roots and consequences. Most analyses in and of GIS are framed in a highly technocratic language of engineering, efficiency, control, and manipulation, a discourse in which method substitutes for theory. In the process of rendering representation as only a technical problem and not a social one, GIS asserts a Cartesian view of space, implicitly condoning a positivist or empiricist view of geography. More recently, however, some GIS theorists have turned to social constructivism in an attempt to socialize the pixel while pixelizing the social.

Other geographers have focused on the nature of representations as they reflected and sustain broader historical configurations of power and space. Derek Gregory's *Geographical Imaginations* demonstrated how seemingly different modernist paradigms, such as positivism and Marxism, reflect a common underlying scopic regime that presumes the existence of a detached, all-knowing, objective observer, an assumption that has become increasingly questionable in an age of mounting relativism. However, rather than comprising some objective truth independent of historical experience, what Gregory called the modernist world-as-exhibition is but one scopic regime among many, a masculinist way of framing the world that rose concomitant with the historical process of commodifying social life and space. Because the legitimacy of representational systems is derived from their connections to institutionalized commodity production and consumption, the periodic restructuring of capitalism inescapably initiates concomitant changes in symbolic systems. Thus, the late-20th-century crisis of representation was spawned by the massive rounds of time–space compression unleashed by contemporary globalization.

—Barney Warf

See also Cartography; Critical Human Geography; Discourse; Empiricism; Eurocentrism; Geodemographics; Ideology; Imaginative Geographies; Literature, Geography and; Logical Positivism; Orientalism; Other/Otherness; Popular Culture, Geography and; Postmodernism; Poststructuralism; Production of Space; Space, Human Geography and; Time, Representation of; Travel Writing, Geography and; Vision; Writing

Suggested Reading

Barnes, T., & Duncan, J. (Eds.). (1992). *Writing worlds: Discourse, text, and metaphor in the representation of landscape.* London: Routledge.

Gregory, D. (1994). *Geographical imaginations.* Cambridge, MA: Blackwell.

Gregory, D. (1995). Imaginative geographies. *Progress in Human Geography, 19,* 447–485.

Harley, J. (1989). Deconstructing the map. *Cartographica, 26,* 1–20.

Lefebvre, H. (1991). *The production of space.* Oxford, UK: Blackwell.

Pickles, J. (Ed.). (1995). *Ground truth: The social implications of geographic information systems.* New York: Guilford.

Pickles, J. (2003). *A history of spaces: Cartographic reason, mapping, and the geo-coded world.* London: Routledge.

Wood, D. (1992). *The power of maps.* New York: Guilford.

SPATIAL ANALYSIS

Analysis can be defined as spatial if the locations of the objects in space matter by affecting the results of the analysis. A simple regression analysis of the relationship between the number of golf courses in a country and its gross national product would not be spatial, for example, because countries could be relocated without affecting the results; the analysis would not require the locations of the countries. To a geographer, spatial analysis almost always implies that the space of interest is geographic space—the surface and near surface of the earth—but methods of spatial analysis might also, in principle, be applied to phenomena distributed within the space of the human brain, along the linear space of the human genome, or on the surface of another planet. In all such cases, methods of spatial analysis might be used to reveal what is otherwise hidden or only partially visible to the user in the form of patterns, anomalies, clusters, or correlations. For example, a classic problem in spatial analysis is to determine whether or not clusters exist in the point patterns created by instances of disease and the implications of those clusters for epidemiological processes.

Many methods of spatial analysis have been devised to answer a wide range of questions. One way in which to organize them is by the type of data they address. Thus, there are methods to analyze the following:

- Patterns of points such as records of the instances of a disease, the locations of crimes, or the locations of nesting birds

- Patterns of lines such as the tracks of hurricanes or migrating birds

- Patterns of areas such as voting districts, sales districts, or patches of habitat

- Surfaces of continuous variation such as images collected from Earth-observing satellites or the earth's topographic surface

- Patterns of interactions between places such as those created by migration, airline travel, or e-mail messages

Today, most methods of spatial analysis are available in the form of computer software. In many cases, software incorporates functions to create, edit, and display data as well as functions for analysis, and such packages are termed geographic information systems (GIS) if they include facilities for handling data referring to locations on the surface of the earth. In other cases, routines for spatial analysis may have been added to standard statistical packages, such as S and SAS, or to mathematical packages, such as MatLab. However, these generally lack the elaborate supporting functionality for dealing with spatial data.

Several distinct ways of organizing the numerous methods of spatial analysis have been proposed to make it easier for investigators to find appropriate methods and appropriate tools. The five data types listed earlier form the foundation for one method, and several textbooks have been organized in this way, discussing in turn methods available for analyzing each major data type. The theory of geographic information suggests, however, that a more fundamental distinction between spatial data types exists and that the point/line/area/surface/interaction classification masks this distinction and so can be confusing. Geographic information scientists first distinguish between continuous fields and discrete objects as two alternate and mutually exclusive conceptualizations of the geographic world.

In the continuous field conceptualization, any point on the earth's surface can be characterized by the values of a finite number of variables, each of which is measurable at every location. For example, atmospheric temperature, soil pH, and surface elevation all are measurable at points and normally are conceptualized as functions of location. Other field variables, such as soil class, land ownership, and county name, are nominal. In the discrete object conceptualization, the earth's surface is empty except where it is occupied by one or more objects. Objects may be differentiated by one or more variables, commonly termed *attributes*. Objects are countable, and their identities typically persist through time. Humans, vehicles, trees, and buildings all are better conceptualized as discrete objects than as continuous fields. Many naturally occurring geographic features, however, are problematic; mountains, weather disturbances, and clouds are examples of phenomena that sometimes are treated as countable discrete objects and other times are better treated as continuous fields (of elevation, atmospheric pressure, and sky opacity, respectively).

Methods of spatial analysis can be characterized as either exploratory or confirmatory. In the first case, spatial analysis is used to discover patterns, anomalies, or relationships and subsequently to support new hypotheses, generalizations, or abstractions about the world; spatial analysis in time leads to new theory. This is the classic inductive style of research. On the other hand, spatial analysis can also be used to test previously formulated theories by comparison with real data in an approach that implements the classic deductive style of research. Exploratory spatial analysis requires a flexible visual approach to the design of tools, exemplified in packages such as GeoDa (www

.csiss.org/clearinghouse/GeoDa/), whereas confirmatory spatial analysis tends to rely on formal testing of hypotheses.

All methods of spatial analysis rely on the investigator to choose and execute appropriate forms of analysis and to interpret the results. One of the major advantages of spatial analysis as distinct from other forms of analysis is that it provides context to observations and encourages the investigator to interpret results within his or her own knowledge of the study area and the phenomena of interest. Thus, for example, a simple regression analysis might yield residuals measuring the differences between the observed value of a variable at a sample point and the value predicted from the regression equation and the value of the independent variable at that point. A simple map of residuals often will suggest additional independent variables, inviting the formulation of new, more elaborate models of higher predictive power. In what follows, it is implicit that spatial analysis involves collaboration between the analytic tool, in the form of software, and the investigator's own personal knowledge and reasoning abilities.

The following sections follow a classification of spatial analysis based on the conceptual complexity of the method.

SIMPLE QUERY

The most rudimentary form of spatial analysis consists of the execution of simple queries. For example, an investigator might request a list of all discrete objects having values on a specific attribute variable above a defined threshold or within a defined range. Because this query involves only the attributes and results in a list, the locations of objects are irrelevant and the query fails to meet the definition of spatial analysis given earlier. However, a query such as "list all of the discrete objects within 5 kilometers of this point" clearly does satisfy the definition. Queries in GIS frequently are formulated in Standard Query Language (SQL), which recently has been extended to include specifically spatial queries.

MEASUREMENT

One of the very earliest motivations for the development of GIS stemmed from the difficulty of making simple measurements from paper maps by hand.

Lengths of curved lines on a map can be measured using wheels, and areas can be measured using planimeters or by counting dots, but all of these methods are tedious, labor intensive, and inaccurate. If a geographic feature can be represented digitally, however, simple algorithms are available for all of these measurement functions. The ratio of the perimeter to the square root of area provides a simple measure of the shape of an area feature, and several methods are available for estimating properties such as slope and aspect from representations of terrain. In short, GIS can be a powerful tool for extracting geometric properties of simple line and area features and for measuring the distance between features.

TRANSFORMATIONS

Fundamental to many forms of spatial analysis is the ability to transform features into new features. For example, a point, a line, or an area can be dilated (or buffered) to create a new feature that includes all of the area within an investigator-defined distance of the original feature. This is useful in many planning applications, in identifying areas likely to be affected by pollution or noise from a new highway, or in identifying areas within the market of a new shopping center. The concept of a spatial join is another example of a transformation; in this operation, attributes of one set of features are concatenated with attributes of another set based on simple geometric relationships such as containment and adjacency. For example, attributes of a set of points could be aggregated (summed or averaged) and allocated to the appropriate one of a set of areas such as counties. This so-called point in polygon operation is very useful in the analysis of accidents, incidences of disease, or crimes in relation to the attributes of neighborhoods or administrative areas. Another common form of spatial join occurs in polygon overlay, that is, when two disjoint sets of areas are combined by computing the intersections between them and concatenating the respective attributes.

DESCRIPTIVE SUMMARIES

Many techniques of spatial analysis attempt to summarize the nature of a collection of points, lines, areas, or interactions, or of a surface, in a few appropriately designed statistics. The simplest of these are analogs of the familiar mean, median, and mode, each of which is designed to capture the central value of a set

of observations of a variable—but extend them to the two or three spatial dimensions. For example, one might want to summarize the location of a collection of points by calculating their mean center, centroid, or center of gravity (all synonymous terms) and tracking how the centroid changes through time or across different subsets of the points. The center of population computed by the U.S. Bureau of the Census is of this type, presenting a summary of the decennial redistribution of the U.S. population and showing how the center of gravity has moved steadily west since the country declared its independence in 1776. The spatial analog of the standard deviation or variance is the dispersion, defined as the average distance of a set of points from their mean center. Centers are also useful as sites that minimize travel distance and thus form logical locations for central facilities to serve spatially dispersed demand.

Other descriptive summaries include measures of spatial autocorrelation, which quantify how similarity of value is related to similarity of position. Patterns in which like values tend to cluster, with high values surrounded by high values and low values surrounded by low values, are said to exhibit positive spatial autocorrelation, a property that is widely observed for geographic phenomena. Patterns in which neighboring values are less similar than distant values are said to exhibit negative spatial autocorrelation. Patterns in which there is no relationship between proximity in space and similarity of value show no spatial autocorrelation or spatial independence.

Landscape ecologists are particularly interested in a class of measures known as fragmentation statistics, which summarize the degree to which a landscape is fragmented into distinct habitat types. They include the number of patches, the variance of patch size, the fractional dimension of patch boundaries, and many others. There is interest in the degree to which the decisions of landowners over time lead to greater or lesser fragmentation and thus to variation in the ability of species to find adequate habitats.

OPTIMIZATION

Methods of spatial analysis can be applied both to the search for pattern and to the design of idealized patterns that meet criteria specified by the user. For example, one might ask for the shortest path through a road network between an origin and a destination, for the optimal path for a new highway across a landscape, for the optimal locations for new fire stations to achieve the fastest possible response time, or for the optimal locations for new retail stores to maximize profit. Well-defined methods are available for the solutions of all these problems and are implemented in many GIS packages. Spatial analysis can also be used to assess plans such as when a public health specialist assesses a planned program to be offered through a series of primary health clinics, perhaps in terms of the average distance to be traveled by patients. Systems that are designed to deliver such forms of analysis often are termed spatial decision support systems.

HYPOTHESIS TESTING

Mention was made earlier of the confirmatory style of analysis, in which models or theories are tested against data, and measures that assess the degree to which the data fit the models or theories are determined. A common strategy is to pose a null theory or hypothesis that states the absence of an effect of interest and then to show that the null theory can be rejected, by implication confirming the effect of interest. This is none other than the classic strategy of statistical hypothesis testing or inference adapted to the special case of spatial data.

For example, a pattern of points representing an outbreak of a disease might be analyzed to see whether it shows evidence of clustering. Clusters could arise through two distinct mechanisms. First-order effects are defined as variations in point density due to variation in the strength of causal factors and would be exhibited, for example, when rates of influenza infection show an inverse correlation with atmospheric temperature. Second-order effects produce clusters through interactions between points such as through contagious processes. Although clusters can be produced by first-order effects, second-order effects, or both, it is generally impossible to determine which type dominates through examination of a point pattern; however, if each case can be dated, contagion will leave a distinct pattern in the spread of the disease through time.

Several null hypotheses are of general value in spatial analysis. Complete spatial randomness (CSR) occurs when discrete objects are equally likely to occur anywhere in space and when the location of one object makes other objects neither more nor less likely in the object's vicinity. A common strategy, therefore, is to confirm clustering through the rejection of CSR. Two null hypotheses are used to test for spatial autocorrelation. The randomization null proposes that the observed values are randomly distributed over the objects of study, whereas the resampling null randomly distributes new values from a distribution estimated to be that from which the observed values were drawn. Both can be used to establish the presence of either positive or negative spatial autocorrelation through rejection of the null.

In general, however, spatial data present a problem for traditional hypothesis testing because of the widespread occurrence of spatial autocorrelation. Many of the null hypotheses used in statistical inference, such as the two-sample null ("both samples were obtained randomly and independently from a single distribution," commonly a Gaussian distribution) and the relationship null ("the values of two variables were obtained randomly and independently from their parent distributions," commonly Gaussian distributions), assert that values were sampled independently, an assertion that is clearly false for many spatial data types, allowing the null to be rejected out of hand. Another widely observed property of geographic data, the tendency for conditions to vary over the earth's surface in what a statistician would call nonstationarity (the common spatial analytic term is *spatial heterogeneity*), is also problematic for spatial analysis and during recent years has led to the development of so-called local statistics. For example, the local version of Moran's *I* statistic of spatial autocorrelation compares each value with those of its neighbors, allowing the investigator to identify "hot spots" where anomalously high values are surrounded by similarly high values.

MODELING AND SIMULATION

The inferential approach described in the previous section is inherently negative; effects are identified through the rejection of their absence. In contrast, modeling and simulation compare explicit theories about the landscape directly with data. For example, rather than rejecting a null hypothesis of CSR, an epidemiologist might hypothesize that disease is directly related to proximity to known pollution sources and simulate the patterns that would result under this hypothesis, using explicitly hypothesized values for the relevant parameters such as the rate of decay of disease incidence with distance and direction from the pollution source. With luck, the analysis would lead to improved models through a process of fitting the parameters to the data.

During recent years, there has been an explosion of interest in simulation models of this type. Cellular models consist of sets of rules about the values assigned to cells in a raster and have been widely used to simulate processes of urban growth and sprawl. For example, a cell's probability of changing from a state of "undeveloped" to a state of "developed" might depend on the states of its neighboring cells, on the slope of the land in the cell, on its status in the area's official plan, and on proximity to transportation arteries. Agent-based models ascribe patterns of behavior to agents, which then influence the landscape through the decisions they make, and have been used to model processes of residential segregation and landscape fragmentation.

—*Michael Goodchild*

See also GIS; Logical Positivism; Model; Quantitative Methods

Suggested Reading

Bailey, T., & Gatrell, A. (1995). *Interactive spatial data analysis.* Harlow, UK: Longman.

Goodchild, M., & Janelle, D. (2004). *Spatially integrated social science.* New York: Oxford University Press.

Haining, R. (2003). *Spatial data analysis: Theory and practice.* Cambridge, UK: Cambridge University Press.

Longley, P., Goodchild, M., Maguire, D., & Rhind, D. (2001). *Geographic information systems and science.* New York: John Wiley.

SPATIAL AUTOCORRELATION

A time series is said to be autocorrelated if it is possible to predict the value of the series at some time from recent values of the series. For example, yesterday's temperature at noon often is a good predictor of today's temperature at noon, and the value of stock market indexes similarly bears stronger resemblance to very recent values than to historic values. Underlying these observations is the notion that some phenomena vary relatively slowly through time. Spatial autocorrelation refers to similar behavior in space; however, unlike the temporal case, space may be two- or even three-dimensional. A general statement by Waldo Tobler, often termed Tobler's first law of geography, asserts that spatial autocorrelation is positive for nearly all geographic phenomena.

Numerous indexes of spatial autocorrelation are in common use. Many are based on a simple extension of the Pearson coefficient of bivariate correlation, which is defined as the covariance between the two variables divided by the product of the standard deviations. In the case of autocorrelation, there is only one variable, so the denominator is the variable's variance and the covariance is the mean product of each value with neighboring values rather than the mean product of each value with the corresponding value of the other variable.

The definition of neighboring depends on the nature of the sampling scheme. When the variable is sampled over a raster, two cells can be regarded as neighbors if they share a common edge or if they share either an edge or a corner. When the variable is sampled over an irregular tessellation, as with summary statistics from the census, it is common to define two cases as neighbors if they share a common edge. More generally, when w_{ij} is defined as the weight used in comparing the value cases i and j of the variable, these schemes can be seen as providing ways of defining w as a binary indicator of adjacency. Continuous scaled definitions of w are available based on decreasing functions of distance, for example, negative exponential functions.

Spatial autocorrelation is of interest in numerous disciplines, and the precise ways in which it is commonly measured vary substantially. In human geography, where data often are encountered in the form of summary statistics for irregularly shaped reporting zones, the common measures are the indexes defined by Moran and Geary, notated I and c, respectively. I is essentially the Pearson correlation coefficient defined as in the previous paragraph using a user-defined matrix of weights. Thus, its fixed points are 0 when there is no tendency for neighboring values to be more similar than distant values (the precise expected value of the index is $-1 / [n - 1]$, where n is the number of observations), positive when neighboring values tend to be more similar than distant values, and negative when neighboring values tend to be less similar than distant values. The Geary index's numerator is the mean weighted sum of differences between values and has a confusingly different set of fixed points: between 0 and 1 when spatial autocorrelation is positive, 1 when it is absent, and greater than 1 when it is negative.

In physical geography, on the other hand, it is more likely that observations will have been made at points. Whereas the reporting zones that dominate human geography provide an implicit definition of scale, in physical geography the variation of spatial autocorrelation with scale is more likely to be of interest. As a result, measurement of spatial autocorrelation usually occurs within the conceptual and theoretical framework of geostatistics or the theory of regionalized variables. By comparing observations at pairs of points at increasing distances apart, it is possible to construct either a correlogram (based on the covariances between paired values) or a variogram (based on the squared differences between paired values). By showing how spatial autocorrelation varies with scale, these diagrams provide a much richer description than the scalar Moran and Geary indexes.

The form of the variogram often is the subject of interpretation, and may also be used as the basis for interpolation of values at points where no samples were taken, in a process commonly termed *spatial interpolation* and known in this specific case as *kriging* (named after South African mining engineer Danie Krige). Variograms are commonly found to rise monotonically to a distance known as the range, at which they reach an asymptotic value known as the sill. Variograms may also exhibit a "nugget"—a nonzero intercept with the *y*-axis—if repeated measurements at or near a point fail to yield identical values.

After the mean and variance, spatial autocorrelation is perhaps the most important property of any geographic variable and, unlike them, is explicitly concerned with spatial pattern. It can be used to measure the spatial extent over which a process appears to persist, as in the case of statistics on the prevalence of a disease; for example, strong positive spatial autocorrelation in cancer rates between counties would indicate that the causal factors responsible for varying rates persist over areas larger than counties, whereas zero spatial autocorrelation would indicate that they vary much more rapidly in space. Negative spatial autocorrelation often is interpreted in terms of competition for space, and the ability of high values to drive away other high values, in applications such as the analysis of retail distributions. At the same time, spatial autocorrelation often is perceived as a particularly problematic aspect of working with spatial data because many statistical methods assume that samples have been drawn independently from a parent distribution and the presence of spatial autocorrelation

clearly violates that assumption. In practice, investigators are forced to adopt one of three strategies: (1) to discard samples closer together than the range exhibited by the data (and no investigator is happy about discarding data), (2) to abandon inferential statistics entirely and limit the interpretation to the description of the sample, or (3) to incorporate spatial effects explicitly in any model.

—Michael Goodchild

See also GIS; Humanistic GIScience; Quantitative Methods; Spatial Dependence; Tessellation; Tobler's First Law of Geography

Suggested Reading

Isaaks, E., & Srivastava, R. (1990). *Applied geostatistics.* New York: Oxford University Press.

Sui, D. (2004). Tobler's first law of geography: A big idea for a small world? *Annals of the Association of American Geographers, 94,* 269–277.

Tobler, W. (1970). A computer movie simulating urban growth in the Detroit region. *Economic Geography, 46,* 234–240.

SPATIAL DEPENDENCE

Spatial dependence is a characteristic of the distribution of geographic data, that is, data where the location of observations is explicitly taken into account. It combines the notion of attribute similarity with that of locational similarity. Not only is there dependence (correlation) between observations for a given variable, but also this dependence shows a spatial structure such as closer locations being more similar than locations that are farther apart. This is a formal expression of Tobler's first law of geography. Spatial dependence stands in contrast to spatial randomness, that is, the absence of any spatial structure in the data. The concept of spatial randomness is the reference or null hypothesis against which potential patterns are compared.

In practice, spatial dependence is quantified by means of a measure of spatial autocorrelation, that is, a summary of the data that formalizes closeness (as distance or contiguity) and similarity (as cross-product correlation or squared difference). The specific measures used depend on the nature of the spatial observations, that is, whether they are considered to be locations of events (e.g., the locations of accidents), discrete observations (e.g., variables observed in counties), or samples from a continuous surface (e.g., measurements obtained at weather stations). Each of these three settings requires a different statistical framework, referred to as point pattern analysis, lattice data analysis, or geostatistics, respectively. For each of these frameworks, a number of test statistics have been developed to assess whether the null hypothesis of spatial randomness can be rejected. Detecting spatial autocorrelation is important because the standard assumptions of statistical inference (independent random sample) no longer hold. In addition, formal models of spatial dependence are required for the estimation and prediction of theories of spatial interaction and other spatially explicit phenomena.

Regression analysis applied to observations that are cross sections in space can be extended to incorporate the notion of spatial correlation (i.e., in spatial regression analysis) or spatial econometrics. Not only does this allow for the modeling of spatial interaction, but also the correction for the presence of spatial correlation may result in better parameter estimates. In addition, explicitly accounting for spatial autocorrelation yields improved spatial prediction in geostatistical models (i.e., kriging).

Spatial autocorrelation statistics can be classified either as indicators of global spatial autocorrelation (clustering) or as indicators of local spatial autocorrelation (clusters). Global spatial autocorrelation is a characteristic of the spatial distribution of the data as a whole and does not suggest *where* the patterns are located. There are two basic results: positive spatial autocorrelation and negative spatial autocorrelation. Positive spatial autocorrelation indicates a clustering of like observations, although this may be high values, low values, or a combination of both. Negative spatial autocorrelation suggests a checkerboard pattern, where high values tend to be surrounded by low values and vice versa.

Local indicators of spatial autocorrelation (LISA) are specific to each observation. They suggest whether locations are part of a local cluster or are local spatial outliers. Local clusters are collections of observations that are similar, showing either high values or low values. They often are characterized as "hot spots." Local spatial outliers are locations that have values that are much higher or much lower than neighboring locations.

—*Luc Anselin*

See also Quantitative Methods; Spatial Analysis; Tobler's First Law of Geography

Suggested Reading

Haining, R. (2003). *Spatial data analysis: Theory and practice.* Cambridge, UK: Cambridge University Press.

Waller, L., & Gotway, C. (2004). *Applied spatial statistics for public health data.* Hoboken, NJ: John Wiley.

SPATIAL HETEROGENEITY

Spatial heterogeneity is a characteristic of the distribution of geographic data, that is, data where the location of observations is explicitly taken into account. It is a special form of structural instability, that is, the situation where a property of a distribution is not constant across the sample. The lack of constancy becomes *spatial* when it conforms to an underlying spatial structure such as the presence of subregions or the existence of a spatial trend. The spatial heterogeneity may pertain to moments of a distribution such as the mean, variance, or covariance (spatial autocorrelation); to the distribution itself; or to the parameters of a model such as the slope coefficients in a regression model. Detecting and estimating spatial heterogeneity is important because most spatial statistical methods assume homogeneity (i.e., stability or, more formally, spatial stationarity). The lack of such homogeneity not only may invalidate standard techniques but also, more important, may point to the failure of the maintained hypothesis or model to hold uniformly throughout the landscape. Proper treatment of spatial heterogeneity allows for the identification of locations or subregions that do not conform to the overall model. A focus on spatial heterogeneity also corresponds with a growing interest in *local* phenomena.

The spatial structure of the heterogeneity can be categorized as discrete variation or continuous variation. In the discrete case, the sample can be organized into a number of discrete and spatially contiguous subsets or subregions that often are referred to as *spatial regimes*. In the continuous case, the variability corresponds to a spatial surface, for example, describing the spatial change (or spatial drift) in a regression coefficient. Spatial heterogeneity is easily visualized by constructing a map of the values of parameters in subsets of the data or by showing a predicted surface that illustrates the spatial drift.

In dealing with discrete spatial heterogeneity, interest typically focuses on the degree to which the mean of the distribution is the same across subregions or on whether regression slope coefficients are constant. The former is investigated by means of spatial analysis of variance, that is, a special case of the familiar ANOVA technique where the "control" consists of a number of spatial subregions. Spatial analysis of variance often is complicated by the presence of spatial autocorrelation that requires the application of specialized techniques. A test for regional homogeneity assesses the degree to which the coefficients in a regression model are constant across spatial regimes. This is a special case of a test on structural stability (e.g., the familiar Chow test in econometrics), but its application is again complicated by the possible presence of spatial autocorrelation (resulting in a spatial Chow test).

Continuous spatial heterogeneity most often is encountered in regression models with spatially varying coefficients. The variation can be modeled in functions of auxiliary variables, as in the so-called spatial expansion method. This is similar to the specification of individual coefficient variability in multilevel (hierarchical) regression models. A recent approach is the geographically weighted regression (GWR), a special case of locally weighted regression, where a kernel estimate is obtained reflecting the variability of regression coefficients around each location in the sample.

—Luc Anselin

See also Spatial Analysis; Spatial Dependence; Quantitative Methods

Suggested Reading

Anselin, L. (1988). *Spatial econometrics, methods, and models.* Boston: Kluwer Academic.

Fotheringham, A., Brunsdon, C., & Charlton, M. (2002). *Geographically weighted regression: The analysis of spatially varying relationships.* Chichester, UK: Wiley.

SPATIAL INEQUALITY

The term *spatial,* when applied in the context of human geography, refers to the geographic dimensions of human relations and practices. The term *inequality* refers to the concept that relationships are uneven or unjust, reflecting differences in power, representation, mobility, and/or access to resources. The two terms combined, therefore, refer to the existence of an unevenness in the relationships between different places in a way that is qualitatively different, suggesting contrasts that may be viewed as morally unfair (e.g., selected neighborhoods within a city may have extensive access to municipal resources, whereas neighboring communities do not) and that also reflect discrepancies in power.

Developing out of research during the late 1970s in the area of welfare geography, which was particularly concerned with differential access to social and governmental services, the concept of spatial inequality has diversified substantially to include not only difference between places or countries but also difference within social groups in the same place. Drawing on Marxist theories, some geographers used the concept of uneven development to highlight the ways in which practices such as colonialism, capitalism, racism, and immigration policies have facilitated the ongoing economic and political differences between wealthy and low-income countries, for example, distinct sociospatial patterns of a global South and a global North.

Spatial inequalities exist at a variety of scales. This is illustrated through the work of feminist geographers who have examined the ways in which the geography of discriminatory—or unequal—practices, such as patriarchy, homophobia, and violence, are present at the levels of the body, street, town, region, government, and international relations. Assumptions about child care, such as when it is presumed that a nuclear family is the norm and that child care will be undertaken primarily by a mother at home, reflect a patriarchal ideology (i.e., a culture in which male dominance is asserted). This may be institutionalized through government policies (e.g., limited maternity leave, no paternity leave, no recognition of same-sex parenting policies), used in dominant images of the nation (e.g., those depicted in popular art or national crests), enacted at a local level in particular towns or cities (e.g., by having limited child care available at places of employment, by having public transportation that is difficult to access with a child's stroller), and embodied individually by parents (e.g., females being criticized for continuing with careers, males being critiqued for becoming full-time caregivers, parents with physical disabilities being discouraged from becoming parents, guilt being internalized due to peer pressure to perform certain "motherly" duties or activities in particular circumscribed places such as breastfeeding in public spaces and dressing in "appropriate" maternity clothing). All of these illustrate specific contexts in which gender inequalities become spatially manifested as work/home divisions, mobility limitations, narrowly defined performances of parents, and policy exclusions for female and male parents. These different scales of inequality may coexist, and may even contradict each other, while also working toward the ongoing exclusion and discriminatory treatment of specific social groups.

Human geographers have also studied the construction and maintenance of spatial inequalities in the context of economic investment and social infrastructures. One recent example can be seen in relation to the differential accessibility of healthcare in varying regional, rural, and urban settings for varying social groups with a range of physical abilities, particularly in relation to studies of human immunodeficiency virus/acquired immunodeficiency syndrome (HIV/AIDS) and preventive healthcare and outreach service. This research has illustrated high costs, limited resources, and overstretched health centers in low-income countries (with rapidly increasing HIV infection rates), whereas relatively large-scale investment and availability of healthcare can be found in wealthier countries. Again, within countries, regions, and cities, geographers have shown that there appears to be inconsistent availability of educational programs and healthcare resources by district and region as well as by race, class, physical ability, and gender. These spatial (and social) inequalities are also seen as being demonstrative of wider economic, political, and historic power differentials.

Spatial inequalities exist not only in terms of material conditions that people manage and negotiate but also in relation to the way in which they produce and negotiate images of place and culture. Representation is a key process that is interwoven with our understandings of space and identity and, as such, is a significant component in the process of depicting who belongs where, understanding what difference or inequalities are viewed as "natural," and being aware of challenges to inequalities that may exist but have been largely ignored. Media images (via television, newspapers, cinema, and radio) are examples of systems of representation that have helped to frame our understandings and negotiations of the world. For example, geographers have noted that if we explore mainstream images of a spatial process such as migration combined with representations of migrants, nativist sentiments stating that the destination country is being placed under unmanageable pressures or that migrants pose a "threat" to the cultural identity of the

host nation become an important part of our spatial imaginary of international movements of people—of how we view the receiving and sending locations and of how we imagine (and respond to) migrants. Geographers have argued that these systems of representation often rely on stereotypes of places and people as a shorthand that depicts the host nations as being unfairly pressured, whereas the larger economic and political inequalities that may be a part of this movement or conflicts within the home countries are largely hidden. In addition, only limited voice may be given to individual migrants in media images of migration, and so official authorities or organizations may come to be seen as symbolic of the destination country's response as a whole, whereas the populations of the other countries are, by their exclusion, viewed largely as inarticulate or voiceless (particularly in the case of refugees and undocumented migrants). Therefore, representation can reflect, reproduce, influence, and challenge spatial inequalities that are grounded in specific localities and events. Of course, these representations and inequalities can also be resisted and challenged by new social movements and activist organizations, and human geographers increasingly are examining the ways in which activist media, organizations, and networks have worked to create greater social inclusion and to address issues of social justice, ethics, and representation.

—*Susan P. Mains*

See also AIDS; Ethics, Geography and; Feminist Geographies; Justice, Geography of; Postcolonialism; Uneven Development

Suggested Reading

Del Casino, V., Jr. (2001). Enabling geographies? NGOs and the empowerment of people living with HIV and AIDS. *Disability Studies Quarterly, 21*(4), 19–29.

Gleeson, B. (1999). *Geographies of disability.* New York: Routledge.

Harvey, D. (1982). *The limits to capital.* Oxford, UK: Blackwell.

Massey, D. (1994). *Space, place, and gender.* Cambridge, UK: Polity.

Massey, D., & Allen, J. (Eds.). (1988). *Uneven re-development: Cities and regions in transition.* London: Hodder & Stoughton.

Naficy, H. (Ed.). (1999). *Home, exile, homeland: Film, media, and the politics of place.* New York: Routledge.

Pile, S., & Keith, M. (Eds.). (1997). *Geographies of resistance.* New York: Routledge.

Rose, G. (1993). *Feminism and geography: The limits of geographical knowledge.* Cambridge, UK: Polity.

SPATIALITY

In current academic parlance, particularly in the humanities and cultural studies, spatiality is an increasingly used but ill-defined foundational term. Spatiality can be defined as *any* property relating to or occupying space. The flexibility of the concept, as both a noun and an adjective, can be seen in its various uses—to refer to a quality of material space, to call on the power of spatial metaphors, and (at times) to imply that material space and spatial metaphors increasingly are the same. The term *spatiality,* then, can refer to actual material space and jurisdiction, virtual space, assumptions about the nature of space, and the ways in which everyday experiences of space undermine these same assumptions. The term refers to material spaces and spatial metaphors, often at the same time. Therefore, spatiality, like globalization or hegemony, is an analytical term increasingly made to bear the weight of a substantive term such as space itself. Of particular importance is the rise of the term's popularity at the very moment that networked information technologies, such as the Web and other new media (e.g., virtual reality), give rise to a belief in a parallel or virtual space produced by and dependent on these technologies and the experiences of their users. Authors writing about these technologies have claimed that electronic networks increasingly serve as the organizing concept for the spaces of everyday life.

Applied to networks and their ability to link people globally, spatiality connotes the materiality of space in terms of virtual networks. For example, cyberspace has been argued to possess a *quality* of spatiality by virtue of its ability to convey to users experiences of immersion and interaction. When used to imply quality, spatiality conveys a sense of its own agency or even jurisdiction, an agency that operates to place an individual within a metaphorical space. The term, therefore, can also act as a metonymic bridge between metaphor and personal experience. Metaphors themselves are inherently spatial forms of written and verbal intellection. Spatiality, used metaphorically to suggest the spatial nature of electronic networks, also implies a spatial *quantity* that works to support claims that the virtual spaces such networks support are not much different from the material spaces of embodied everyday life.

In its meanings as both a quality of space and yet also of space itself, the term, with its allegorical ability to blur material and linguistic practices, may lead to an indistinction or blurring of meanings among metaphors of space, actual spaces, and their different but increasingly interpenetrating jurisdictions unless the writer or speaker is careful to note the specific contexts of use (i.e., the lived, the material, and the conceptual or any combination of these three) each time the term is employed. Although space and experience rely on one another for their meanings, an underconsidered use of the term *spatiality* introduces the potential to blur meaningful distinctions between the two. For example, spatiality, when used to promote the materiality of virtual space, relies on a widespread understanding that experiences of virtual space are real even though such experiences must necessarily rely on spatial metaphors to assist users in orienting themselves within networks. As in the case with any neologism, the wide array of assumed but potentially conflicting definitions that underlie the contemporary use of the term may lead to an ironic outcome where readers are at a loss to understand what is being argued or the limits to the argument.

—*Ken J. Hillis*

See also Cyberspace; Space, Human Geography and

Suggested Reading

Lefebvre, H. (1991). *The production of space.* Oxford, UK: Blackwell.
Smith, N., & Katz, C. (1993). Grounding metaphor: Towards a spatialized politics. In M. Keith & S. Pile (Eds.), *Place and the politics of identity* (pp. 67–80). London: Routledge.

SPATIALLY INTEGRATED SOCIAL SCIENCE

Spatially integrated social science (SISS) derives its principles and practices from the integration of spatial analytical methods with the theories and thematic problems of the social sciences. SISS is based on the premise that a wide variety of social processes and problems are understood more clearly through the mapping of phenomena and the analysis of spatial patterns. The locational properties of information often are obscured in traditional tabular formats, but maps permit the visualization of this information to reveal patterns and trends not easily seen in tables. Spatial association, regional differentiation, diffusion, spatial interaction, and pattern detection are key concepts of spatial thinking. Through applications of analytical cartography, spatial statistics, spatial econometrics, and geographic information systems (GIS), these concepts facilitate the integration of theory with empirical analyses and aid in both the interpretation of research findings and the presentation of research results.

CONCEPTS

Spatial association links information sets, social processes, and problems to geographic coordinates and regions. For example, maps of environmental quality and human health can be overlaid to examine correlations that may suggest clues for further research. Spatial associations among diverse phenomena provide bases for hypothesis testing and for inferential analysis in the social sciences. The mapping of residuals from the expected values from regression models is a standard approach for identifying explanatory factors to enhance the understanding of social patterns.

Regional differentiation characterizes the outcomes of processes that distinguish regions and that allocate space to different populations and activities. For instance, the territorial division of cities, based on ethnicity, demographic processes, and social class, plays a significant role in determining social changes and social needs. Similarly, the projection of trends in land use patterns is an important factor in implementing policy measures regarding metropolitan structure and human activity patterns (e.g., commuting and shopping behavior). Geographic information systems play a significant role in monitoring changes in the uses and regional differentiation of space.

Diffusion and spread effects underlie a range of problems at a variety of spatial and temporal scales. Public health researchers are concerned with contagion effects in the spread of diseases, and changes in public opinion may reflect social diffusion processes that underlie political movements, shifts in value systems, and changing norms of human behavior. Cartographic visualization of these processes through animated maps represents one method of depicting temporal patterns in the geographic spread of such phenomena.

Spatial interactions refer to movements among places—migrations among nations, traffic flows within cities, commodity flows among regions, and so

forth. The analysis and modeling of flows is an important focus for resolving problems in transportation studies, in explaining trade patterns in relationship to regional development issues, and in understanding demographic changes that alter the demand for social services.

Physical arrangement and clustering of phenomena are keys to pattern detection, for example, in identifying the patterns of crime occurrences in cities and in being able to discern whether such patterns arise by chance or through some underlying associations of social and economic conditions that occur within regions and their surrounding areas.

PRINCIPLES

The primary principles of SISS are integration, spatiotemporal context, spatially explicit modeling, and place-based analysis and interpretation.

The *integration* perspective of SISS focuses on location as a natural basis for ordering and combining diverse information sources and for seeing the resolution of social science problems as fundamentally multidiscipline in character. For example, GIS and other spatial tools can facilitate an integration of perspectives from several disciplines (e.g., anthropology, economics, geography, political science, sociology) to help understand social processes such as economic globalization and gentrification. Confining investigations of such issues to the realm of one discipline fails to capture the complexity of processes and interactions across geographic scales.

The principle of *spatiotemporal context* maintains that the understanding of social processes must consider all dimensions of possible variation. As a corollary, data collection and analysis at multiple scales provide advantages for comprehension and modeling of complex processes. Information about the areas surrounding observations in time and space can assist in the interpretation of social processes. Thus, instances of crime may be better understood when mapped to reflect the order and timing of occurrences in relationship to surrounding neighborhoods.

Spatially explicit modeling treats space and distance as explicit facilitators or hindrances to human interactions. Parameters and weightings of distance frequently are incorporated into descriptive and predictive models of interaction and diffusion processes and often are embedded in the behavioral decision rules of agent-based models.

Place-based analysis makes use of data and methods to integrate diverse information about specific places. One of the interesting attributes of a place (e.g., a neighborhood, a city, a county) is that multiple social processes occur simultaneously and span perspectives that capture the conceptual strengths and tools of a broad range of disciplines. Place and location are also valuable for organizing and searching for information. Using geographic location as a primary key, new Internet search technologies provide a basis for linking diverse data sources.

Geographic information systems provide the primary tools for integrating diverse information sources to keep track of and map spatial patterns, spatial changes, and spatial associations. In combination with spatial econometrics and modeling tools such as agent-based simulation, geographic information systems have emerged as a fundamental underpinning of spatially integrated social science. This is reflected in an expanding body of literature across the social sciences and in the introduction of courses on spatial analysis and GIS in several social science disciplines (www.csiss.org).

—*Donald Janelle and Michael Goodchild*

See also Digital Earth; GIS; Humanistic GIScience

Suggested Reading

Goodchild, M., Anselin, L., Appelbaum, R., & Harthorn, B. (2000). Toward spatially integrated social science. *International Regional Science Review, 23,* 139–159.

Goodchild, M., & Janelle, D. (2004). Thinking spatially in the social sciences. In M. Goodchild & D. Janelle (Eds.), *Spatially integrated social science* (pp. 3–21). New York: Oxford University Press.

SPORT, GEOGRAPHY OF

The field of sports geography examines the intersection of sports with human geography and, to a lesser extent, physical geography. The field began during the 1960s with the work of American geographer John Rooney and continues with the work of British geographer John Bale. However, sports have been understudied by geographers, and few geography departments in the United States offer courses in sports geography. That geographers have by and large ignored calls for a critical engagement with sports is surprising given both the important role that sports play in modern American culture and the more serious treatment of sports in disciplines such as economics, history, and sociology.

At the most basic level, geography and sports intersect in terms of the impact of physical and human geography on athletic performance and on the outcome of sporting competitions. Physical geography has had dramatic impacts in terms of sports performances in more extreme physical locations (e.g., the 1968 Mexico City Summer Olympics, Major League Baseball in Denver), whereas students of sports have long noted social geography phenomena such as home field advantage, where teams tend to win more games when playing at home than when playing away from home (as was the case for all 29 teams in the National Basketball Association during the 2003–2004 season).

The most studied area in sports geography has been the recruitment of players for college and professional sports teams. Answering the question of where players come from gives hints as to the areas and regions where specific sports are most popular. Recent decades have seen a growing trend toward globalized recruiting in North American professional sports leagues, especially in professional basketball (with players coming to play from all parts of the world), professional hockey (with players coming from Northern and Eastern Europe and the former Soviet Union), and professional baseball (with players coming from Latin America and East Asia). At the same time as players have been recruited from various parts of the world to play professional sports in North America, team owners and leagues have been marketing their teams and sports in other parts of the world through broadcasting and licensing agreements in other countries and by scheduling regular season and exhibition games in Europe, Latin America, and Asia.

The globalization of sports is also seen in recent cross-national ownership of teams such as the Nintendo Corporation's ownership of the Seattle Mariners baseball team and the owner of the Tampa Bay Buccaneers football team's recent purchase of England's most famous soccer team, Manchester United. At the same time as North American team sports are being exported to the rest of the world, American fans still show their longtime resistance to the importation of the world's most popular sport, soccer, despite recent success by the men's and women's U.S. national soccer teams at the world level. Many of the best American soccer players play professionally in other countries to enhance their skills and achieve higher pay and professional status.

However, the intersection of sports and geography goes well beyond that of the recruitment of players. The importance of sports is seen in its high-profile role in nationalism and geopolitics, including the rise of pride in patriotism following important national sporting success (as hockey victories by Canada's all-star team over the Soviet Union in the 1972 Summit series and by the U.S. Olympic team over the Soviet Union in 1980 demonstrated) and sports' role in geopolitics both in helping to reduce tensions among enemies (as in U.S.–China "ping pong" diplomacy in 1971 and recent cricket test matches between India and Pakistan) and in using sports as a geopolitical weapon (as in the cold war boycotts by the United States at the 1980 Moscow Winter Olympics and by the Soviet Union at the 1984 Los Angeles Summer Olympics). In modern Western society, sports also play a role in identity formation and forging attachments between people and the places they live, from the role of high school football in west Texas, to stock car racing in parts of the American South, to hockey in Canada.

Geographers have also studied the locations of sports franchises at both the regional and local levels. The four major North American sports leagues have had different spatial patterns over time but have generally diffused from being predominantly regional leagues to nationwide leagues. Although much of this change is attributable to the growth and diffusion of population, other factors also play a role in franchise relocation. Such spatial redistribution of franchises is also apparent at the level of minor league sports and training sites for major league sports teams, although the reasons for such shifts may differ.

One issue in the retention and relocation of professional sports franchises is the building of new sports facilities and their impact on the urban political economy. Whereas prior to World War II most professional sports facilities were privately financed, the post–World War II period saw a trend toward publicly financed facilities. The past 15 years or so have seen numerous political battles over the issue of using taxpayer money to build new facilities for professional sports teams. Proponents argue that such new facilities have a positive economic impact by spurring growth and (re)development, especially when built in a downtown area. Proponents argue that spending public money on a sports facility has positive public benefits when the facility is the centerpiece of a downtown redevelopment strategy. However, most academic studies of the impact of public spending on professional sports facilities (primarily by economists and public policy analysts) have been highly critical of this argument, questioning the overall economic benefit of publicly financed facilities.

Sports also leave an indelible imprint on the physical and cultural landscapes. Examples include the building of golf courses, with recent trends toward these courses being the centerpieces of real estate golf developments, and changes over time in the style and architecture of baseball stadia, as seen in recent attempts to invoke earlier eras of baseball through stadium construction (as in the "retro" stadium building boom of the 1990s). Where sports facilities are built within the urban landscape has also changed over time, from the building of professional sports stadia and arenas on the built-up edges of urban cores prior to World War II, to the movement of facilities to the suburbs after World War II, to the trend toward building facilities in or near downtowns over the past 15 years.

In these and numerous other ways, geography and sports intersect. Given the importance of sports in modern American culture, geographers can add much to our understanding of the topic.

—*Jonathan Leib*

See also Cultural Geography; Popular Culture, Geography and

Suggested Reading

Bale, J. (2003). *Sports geography* (2nd ed.). London: Routledge.

Markovitz, A., & Hellerman, S. (2001). *Offside: Soccer and American exceptionalism.* Princeton, NJ: Princeton University Press.

Raitz, K. (Ed.). (1995). *The theater of sport.* Baltimore, MD: Johns Hopkins University Press.

Rooney, J. (1974). *A geography of American sport: From Cabin Creek to Anaheim.* Reading, MA: Addison-Wesley.

Rosentraub, M. (1999). *Major League losers: The real cost of sports and who's paying for it* (rev. ed.). New York: Basic Books.

SQUATTER SETTLEMENT

The meaning of squatter settlement varies widely from country to country. In general, it is recognized as a residential area in an urban locality inhabited by the very poor who have no access to tenured land of their own and who therefore squat on vacant land that is either privately or publicly owned. The following terms are also used interchangeably with the term *squatter settlements:* informal settlements, shack settlements, low-income settlements, semipermanent settlements, shantytowns, spontaneous settlements, unauthorized settlements, unplanned settlements, and uncontrolled settlements.

Squatter settlements are dense settlements comprising communities housed in self-constructed shelters under conditions of informal or traditional land tenure. They are common features of developing countries and typically are the product of an urgent need for shelter by the urban poor. As such, they are characterized by a dense proliferation of small makeshift shelters built from diverse materials, by degradation of the local ecosystem, and by severe social problems.

Informal settlements occur when the current land administration and planning fail to address the needs of the whole community. At a global scale, squatter settlements are a significant problem, especially in parts of the world housing the global poor.

Squatter settlements tend to be viewed by governments, local authorities, and even academics as aberrations on the urban landscape. However, in 2003 an estimated 85% of urban residents in the developing world occupied property illegally.

Geographers have documented and drawn attention to the highly differentiated character of squatter settlements around the world. Geographers have also noted that despite the high level of variability among squatter settlements, these settlement types are found on marginal land around rapidly growing cities and usually in the developing world. Notably, geographers and others have shown how squatter settlements tend to be located on hazardous, marginal, unstable, sloping, or allegedly worthless stretches of urban land. The marginal character of land on which squatter settlements develop augments the cycle of poverty within which inhabitants are trapped. This process sometimes is referred to as *spatial entrapment.*

The concentration of poverty, and with it a shared alienation from urban resources and infrastructure, can also create a sense of community in squatter settlements that sometimes can lead to political mobilization, unrest, instability, and (in extreme cases) violence. An examination of studies by geographers

tends to draw attention to the physical, social, and legal character of these settlement types.

PHYSICAL CHARACTERISTICS

A squatter settlement usually lacks traditional urban services and infrastructure. If these services are present, they tend to be below minimum standards. Such services are both physical and social—water supply, sanitation, electricity, roads and drainage; schools, health centers, and marketplaces.

SOCIAL CHARACTERISTICS

Squatter settlement households tend to belong to lower-income groups. Their residents either work as wage laborers or in various informal-sector enterprises. On average, most earn wages at or near the minimum-wage level. But household income levels can also be high due to many income earners and part-time jobs. Squatters are predominantly migrants—either rural–urban or urban–urban. In some parts of the world, however, it is not uncommon to find second- or third-generation squatters because of the spatial entrapment process mentioned earlier.

LEGAL CHARACTERISTICS

The key defining characteristic of squatter settlements is their lack of ownership of the land parcels on which residents have built their houses. These could be vacant government or public land or marginal land parcels.

So long as urban areas offer economies of scale and agglomeration economies, large cities will continue to grow, attracting migrants from rural and smaller urban areas faster than the cities can provide housing and services, and so squatting will continue to occur. Considering the inevitability of squatting, the need is primarily for a change in attitude toward squatting, squatters, and squatter settlements. One such approach that has been receiving considerable attention from various government and public authorities has been the "enabling" approach, whereby governments, instead of taking a confrontationist attitude, work to create an enabling environment in which people, using and generating their own resources, can find unique local solutions to their housing and shelter problems. There is considerable controversy surrounding the enabling approach. Critics argue that such an approach does little to address the deeper structural issues that lead to housing shortages and create informal housing settlements in the first place.

—*Glen Elder*

See also Developing World; Housing and Housing Markets; Urban Geography; Urbanization

Suggested Reading

Davis, M. (2004, February 3). Mega slums. *New Left Review,* pp. 5–34.

Mangin, W. (1967). Squatter settlements. *Scientific American, 217,* 21–29.

Rakodi, C. (1995). Poverty lines or household strategies? A review of the conceptual issues in the study of urban poverty. *Habitat International, 19,* 407–426.

Saff, G. (2001). Exclusionary discourse towards squatters in suburban Cape Town. *Ecumene, 8*(1), 87–107.

STATE

The state refers to a geographic area delineated by national borders, the population that inhabits this territory, and the political unit and institutions that govern the social and economic relationships among people. This term can describe three levels of government: federal, state/provincial, and local. At each level, the state is an assemblage of institutions, networks, administrative functions, and people organized by bureaucratic organizations. When used alone, the term most often references the nation-state or country. Whereas a nation refers to a group of people with some common identification, a nation-state is a political expression of this commonality. The boundaries of nation and state can be at odds.

In addition to functioning as a set of institutions on the ground, the state has a lively conceptual existence in the social sciences, generally referring to theories of the state. There likely exist more theories of the state than kinds of states. Geography and geographers have engaged unevenly with theories of the state ever since the inception of the discipline. The state has largely been the domain of political geographers but has seen a recent surge of interest across subdisciplinary fields.

HISTORY OF THE STATE

The division of the globe into nation-states is a relatively recent phenomenon, and the increasing commonality among kinds of contemporary states is also relatively recent. Until the emergence of the modern nation-state in 16th-century Europe, societies were grouped and governed in more varied ways, some of which could be seen as precursors to the state. Although nomadic tribes, clans, and feudal groups respected boundaries and territories, their boundaries shifted in daily practice and these populations tended not to operate under an independent governance structure. However, agrarian societies traded surplus, amassed wealth, divided by social class, and documented population, albeit in localized fashion. Eventually, feudal systems gave way to absolute monarchs who assembled military power. City-states and empires alike carved territories on the map in a more fluid fashion. They shifted frequently in form and geography. With industrialization, class differentiation became more complex and the need for a powerful central governance structure became more acute.

States tended to evolve in relation to natural landscapes where conquest and/or defense were possible as well as through the diffusion of ideas and ways of life and the migration of people. Eventually, out of differing modes of war, historical contingencies, and geographically uneven forms of power, there emerged what is known as the modern nation-state. The modern nation-state developed administrative institutions that surrounded a monarch. Whereas political geographers have tended to privilege Eurocentric understandings of states and their development, ancient states of the Middle East, China, South America, and West Africa made key contributions to the evolution of the state. The modern nation-state differed from earlier forms of political organization by institutionalizing the relationship between territoriality and membership, thereby clearly demarcating an inside and an outside to territorial belonging. The operation of power also changed. The modern nation-state also had more geometrically drawn borders and smaller territories. These often were the product of colonization, whereby colonial powers would carve colonies into administrative territories without respect for local practices such as the boundaries of tribes and clans in Africa. Whereas monarchs were interested primarily in collecting taxes and building military power, the administrative function of the modern state involved the governance of the daily activities of citizens.

States came to resemble one another more closely during the 20th century. The latter half of that century was also characterized by the proliferation of nation-states and the emergence of suprastate institutions, such as the United Nations, to govern this community of states. Decolonization, the dissolution of empires, and social movements organized around ethnic separatism all contributed to this process. This proliferation is predicted to continue and presents an interesting contrast to arguments made during the late 20th century that globalization would signal the decline of the nation-state.

CONTEMPORARY TYPOLOGIES

All states exercise power through sovereignty, but they do so with various methods of governance. Joe Painter drew a rough typology of three kinds of contemporary states: transitional economies, developers, and welfare states. Transitional economies are in transition from centrally planned socialist governments to market-based economic systems. These include many of the states of Eastern Europe and the former Soviet Union and China. Developers promote social and economic development. These include newly industrializing countries, oil-exporting countries, and early and late postcolonial states. Welfare states are the liberal democracies in North America, Western Europe, Japan, Australia, and New Zealand. These are complex capitalist societies with substantial wealth and stable governments that invest heavily in social services.

There exist other spectra along which states are divided. Federal states institutionalize the embodiment of distinct belief systems, languages, and cultural practices within their boundaries, whereas unitary states are organized around one system. Democratic states empower the involvement of the citizenry in decision making and institutionalize checks and balances so that power does not rest with one individual office. Authoritarian states, on the other hand, centralize power with one figure or dictator.

Political geographers' studies of state formation have dwelled on European models that have influenced political organization throughout the world through colonialism. However, non-Western states also contributed to global processes of state formation and differ in contemporary form.

EPISTEMOLOGIES

Geographers have relied on diverse epistemological approaches to the state, with all of these approaches holding in common the project of spatializing theories of the state. Friedrich Ratzel conceptualized states as organic entities that grow as plants grow. He drew on social Darwinism to argue that the fittest states would survive. This idea of the organic state was influential in geography for many years but ultimately derided as environmental determinism. Years later, Richard Hartshorne advocated a functionalist approach that considered the structures and functions of states as well as their success according to centripetal and centrifugal forces. Political geographers then expanded the field to study the functional integration of states into political or world systems.

During recent decades, geographers have been influenced by ideas of Marxian political economy, embodied particularly in regulation theory or ways in which to regulate modes of production. These more structural understandings of state practices have been countered by scholars who see the state as a manifestation of ideas made powerful by the belief in its existence. Such critiques have called on an understanding of the cultural contexts in which political systems arise and are perpetuated. Feminist and poststructural scholars have studied the power of the state to alternatively subordinate and empower various people. They counter the notion that states are monolithic institutions with common agendas with the notion that states are sets of embodied practices. These institutions embody networks of people and practices, function as nodes of power, and have diverse effects that include the simultaneous empowerment and alienation of various peoples.

GEOGRAPHY AND THE STATE

The state always has been among the topics central to the work of political geographers. Early political geographers conceptualized the state in terms that mapped territoriality and power in relation to practices of war and conquest. For example, Halford Mackinder's imperialist heartland theory predicted victory through conquest of states with the best geographic possibilities for expansion. Over time, political geographers tended to conceptualize states less in terms of expansion and more as containers of people, economic systems, cultural practices, and military power. The idea that states are containers permeated many of the traditional topics of political geography during the 20th century such as territoriality and state formation, sovereignty, war, and geopolitics.

Marxist and neo-Marxist geographers drove interest in the role of the state in facilitating production and accumulation within a capitalist system. These include studies of institutions of the state such as the state apparatus. Not coincidentally, studies of the more insidious effects of the state emerged as globalization and neoliberal policies took hold. Geographers became interested in the internationalization of states and in devolution or the role of the shadow state. Some scholars argued that the liberal welfare state was being hollowed out as a result of global flows, whereas others looked at its more scattered reconfiguration as diasporic populations and global networks facilitated the reterritorialization of communities and the structures that govern them. During recent years, geographers have become less interested in states as containers and more interested in states as spaces of global flows and networks where new configurations of power materialize.

As geographers explored the work of Henri Lefebvre, they became more interested in theorizing scale. Political geographers studied how states produced space at various scales and, conversely, how states themselves were produced at various scales. This change in epistemology accompanied a radical economic shift to a postindustrial global economy where power is amassed with the trafficking of knowledge and ideas rather than the trafficking of raw resources and production. The role of the state changed, as did theories of the state and empirical practices of sovereignty and citizenship.

Following the cultural turn and influenced by Michel Foucault and other poststructural and postmodern scholars, geographers grew less interested in the state and more invested in diverse ways of thinking about power. A healthy literature on governmentality has arisen. Whereas states once asserted power through repression, this literature took into account the power of the state to operate in productive ways, producing identities and subgroups.

The start of the 21st century has seen a resurgence of interest in the state across the social sciences, driven in part by the hegemony of the United States and the reemergence and reconfiguration of the concept of empire. Political geographers are studying the new roles and governance of international borders, new configurations and practices of security and sovereignty, and new ways of conceptualizing identity in relation to the state. Postcolonial and feminist perspectives have brought histories of violence to bear on understandings of the state. Geographers are jumping scale in their understandings of states to understand at once the roles of states in the global economy and the highly localized daily practices at work inside the states. They are studying the reconfiguration of spatialities of the state and discourses of sovereignty. Through critical geopolitics, they are simultaneously critiquing and displacing the predominance of state-centric views of international relations. Political geographers are exploring alternative forms of organization that exist outside of or transcend the state such as transnational social movements, alternative forms of political organization, multiple and nontraditional practices of citizenship, and stateless zones and individuals.

—*Alison Mountz*

See also Critical Geopolitics; Geopolitics; Local State; Nation-State; Nationalism; Political Geography; Sovereignty; World Systems Theory

Suggested Reading

Agnew, J. (2002). *Making political geography*. London: Edward Arnold.

Brenner, N., Jessop, B., Jones, M., & MacLeod, G. (Eds.). (2003). *State/Space: A reader*. Oxford, UK: Blackwell.

Clark, G., & Dear, M. (1984). *State apparatus: Structures and language of legitimacy*. London: Allen & Unwin.

Dodds, K., & Atkinson, D. (Eds.). (2000). *Geopolitical traditions: A century of geopolitical thought*. London: Routledge.

Ó Tuathail, G. (1996). *Critical geopolitics*. Minneapolis: University of Minnesota Press.

Painter, J. (1995). *Politics, geography, and "political geography": A critical perspective*. London: Edward Arnold.

STRUCTURAL ADJUSTMENT

During the 1970s and 1980s, many poorer nonpetroleum-producing countries took out loans to help them pay their hard currency bills for imported goods (including oil) and services. With lower-than-expected economic growth rates in the debtor countries and rising interest rates, the amount of debt and debt service payments (interest on the loans and fees) grew rapidly. With this situation at first viewed as a temporary "liquidity" problem, additional loans were made to indebted economies. However, with the default of Mexico in August 1982, when it failed to service its enormous debt, the problem became redefined as the debt crisis. Heavily indebted countries found themselves in a tight situation—unable to pay off or even service their growing debt burdens.

The International Monetary Fund (IMF, set up at the 1944 Bretton Woods Conference), based in Washington, D.C., began negotiations with debtor countries. The IMF typically recommended a package of policies aimed at stabilizing the national currency and controlling inflation. Further loans from the IMF were conditional on the prescribed stabilization measures being put into effect. The World Bank (also set up at the Bretton Woods Conference) worked on broader economic stabilization agreements with similar conditionalities.

Although they are not homogeneous and have changed over time, the policies favored by the IMF and the World Bank in their dealings with indebted countries, along with those advocated by other multilateral development banks and bilateral aid organizations, have enough common elements to be characterized broadly as structural adjustment policies (SAPs). SAPs, sometimes also known as austerity programs, are based on neoliberal economic ideas—a set of ideas about the best way in which to manage an economy that was increasingly in vogue during the 1980s and that was reflected in the policies of the major economies at that time. The overall trajectory of SAPs was toward the free market and entailed the deliberate withdrawal and shrinkage of state or government involvement in the economy.

There are six major elements typically found in any SAP. The first is a strict tight monetary policy. This is meant to stabilize a country's currency, but often this is achieved only through one or more major devaluations of the currency. Controlling the money supply is also aimed at controlling inflation and preventing high rates of inflation. The second element is a decrease in government spending, especially social spending. Government spending is not supposed to be deficit spending. Reducing social expenditures is done characteristically through the weakening or abolition of subsidies for basics such as food, cooking and heating fuel, public transport, healthcare, and education while at the same time not establishing policies that would tend to increase wages for ordinary working people. User fee systems, whereby payments need to be made to access formerly public services or facilities, often are instituted. Government spending on prisons, the military, and police often is exempted from these policies. The third element is a reduction in public ownership of the economy. It involves the privatization of

state-owned companies and social assets, including land, communications, utilities, and water. The fourth element is a liberalization of trade regime. This is done particularly through the reduction or elimination of barriers to imports of goods and services and the lessening of restrictions on foreign direct investment (FDI) and foreign ownership of businesses. The fifth element is the promotion of, and provision of incentives for, exports. Exports of commodities, manufactured goods, and services are encouraged through the liberalization of FDI and the devaluation of the currency (making the exports relatively cheap on the world market). The sixth element is the general deregulation of the economy (including of the banking system) and reform of the tax structure to lower the tax burden on corporations and high earners, to broadly encourage business and business classes, and to expand the market (the price system) into previously nonmarketized domains.

Clearly, SAPs are not just economic given that they can deeply affect social relations in countries. For this reason, SAPs have, from their very beginnings, been criticized, challenged, and resisted. Although street protests against the IMF and the World Bank have occurred recently in the United States and other so-called developed countries, major protests and riots against SAPs have been occurring all over the world since their beginnings. Some analysts have recorded more than 100 major social protests over SAPs in countries from Algeria to Zambia. In addition to their redistributive consequences for adopting countries, SAPs have been criticized for reinforcing global geographies of inequality rather than being programs that have led to a narrowing of the development gap. Even sympathetic insider critics such as William Easterly and Joseph Stiglitz have noted the unequal power that undergirds any pretense at negotiation between the leaders of indebted countries and IMF officials. Stiglitz, like many critics of SAPs, pointed out the parallels between the operation of SAPs and old colonial relations of power. As expressions of a metaframing of both development and geopolitics, SAPs reinforce a relation in which the richer countries insist on the poorer countries following rules made up by the rich countries. A basic problem of SAPs is that they are aimed first at ensuring repayment of debt—not at poverty reduction or development—despite official rhetoric that assumes the latter will follow from the former.

In part as a response to the critics, the IMF and the World Bank have promoted policies in lieu of SAPs.

These policies and programs have had new names and varied emphases, including poverty reduction, the inclusion of civil society (usually in the form of formal nongovernmental organizations), anticorruption and transparency, and so forth. Many critics have noted that these new policies, in the main, are just variations on the familiar precepts of SAPs examined earlier. For example, under the Highly Indebted Poor Countries (HIPC) initiative, initiated in 1996, SAPs continued in all but name.

There have been many calls to cancel and/or reduce the debt burdens of the world's poorest countries. These campaigns have raised public awareness in the rich countries of the extent of debt and the effects of SAPs. Among several recent developments, in 2005 the G8 (a leading forum for the world's richest countries to formulate economic policy) moved to cancel some debt.

Despite these events, there remains a stubborn and deeply significant geography of indebtedness and SAPs. Of the 38 current countries classified as HIPCs, 32 are in sub-Saharan Africa. Many countries are not classified as HIPCs but still are paying large amounts in fees and interest on old loans annually. SAPs have become a more or less permanent fact of life for the majority of the world's population.

—*Susan M. Roberts*

See also Debt and Debt Crisis; Developing World; Globalization; Neoliberalism

Suggested Reading

Easterly, W. (2001). *The elusive quest for growth: Economists' adventures and misadventures in the tropics.* Cambridge: MIT Press.

Mohan, G., Brown, E., Milward, B., & Zack-Williams, A. (Eds.). (2000). *Structural adjustment: Theory, practice, and impacts.* New York: Routledge.

Porter, P., & Sheppard, E. (1998). *A world of difference: Society, nature, development.* New York: Guilford.

Stiglitz, J. (2002). *Globalization and its discontents.* New York: Norton.

STRUCTURALISM

French social theorists working in the fields of linguistics, anthropology, and (later) psychoanalysis throughout the early to mid-20th century theorized that language, culture, and the psyche were guided by a set of unseen underlying societal structures. Thus, structuralism, as a philosophy and methodology, was guided by an interest in uncovering the logic of those structures. By nature of their theories of the world, structuralists are antihumanistic and do not believe that human agency exists outside of broader structures such as capitalism. Instead, structuralism suggests that the social (and spatial) world is mediated by structures that manifest themselves in particular events "out there" in the world (e.g., capitalism structures relations between workers and factory owners).

Early work in the area of structural linguistics, particularly through the work of anthropologist Claude Lévi-Strauss, suggested that social scientists should focus on the interrelationships between words rather than treating each word or phrase as an independent object. In such a conceptualization, it is necessary to investigate the relationships between signs and symbols as signifying larger social structures, such as capitalism, patriarchy, and racism, through the emergence of hierarchized binaries of words and objects such as entrepreneur and worker, men and women, and white and black. This suggests not only that social scientists should be interested in the signs present in everyday life (e.g., magazine representations, the built environment) but also that it is necessary to interrogate the ideological presuppositions that mediate those signs and the relationships between signifiers (e.g., words, images) and the signified (e.g., abstract mental constructs). In so doing, it is possible to construct a set of generalizable laws about how those ideologies operate in structuring everyday social interactions and relations.

Structuralism first entered geography through the work of Marxist geographers and as part of a larger critique of spatial science that began to emerge during the 1970s. This critique included the work of humanistic geographers who were interested in challenging spatial science's limited understanding of human experience. Marxist geographers, working through structuralism, rejected this naive reading of human agency and instead focused their attention on unpacking the depths of the spatial economy or the social landscape. The work of David Harvey and Manuel Castells used differing readings of structuralism in their analysis of urbanization and cities more generally. Harvey argued that the structures of capitalism, and its contradictions, produced space and spatial relations in ways intended to maximize capitalist accumulation, whereas Castells drew from French Marxist Louis Althusser in suggesting that space and spatial relations were inherently social, that is, organized as part of particular modes of production. In both readings of structuralism, it is important to remember that structures are not natural phenomena; rather, they are made up of a complex of constitutive practices and events. Thus, the structures of capitalism are made up of the events of capitalist relations such as accumulation and tensions between capitalists and proletariats.

The import of structuralism, particularly structural Marxism, into geography provided an opportunity for geographers to further critique the empiricist project of spatial scientists. In that critique, some structuralist geographers suggested that ideology and language, and the organization of these social formations at different conjunctions of space–time, were critical components in the construction of the cultural landscape. This notion is epitomized in the work of Denis Cosgrove, whose interest in signs suggested important ways for geographers to study how the embeddedness of social formations in symbolic landscapes ensured that culture was reproduced in ways that benefited dominant ideologies (and social classes). Cosgrove's analysis tied into the work of other geographers interested in examining the role of symbols and language in the study of geographic landscapes. Indeed, it could be argued that these particular readings of structuralism in geography suggested alternative ways of conducting geographic studies and that this opened up the possibility that a plethora of images, such as landscape paintings, photography, and art, were legitimate areas of geographic inquiry, something in which feminist geographers became interested during the 1980s and 1990s. Furthermore, structuralist readings of the landscape suggest the important ways in which dominant ideologies function to naturalize power relations through the production of landscapes that hide as much as they reveal.

The lasting legacy of structuralism in geography, however, has been the interest it has generated in language and discourse. Although structuralism in geography itself has been largely left behind in the wake of the emergence of structuration theory, realism, and poststructuralism, the importance of examining discourse, language, power, and ideology remains. Thus, the challenges of structuralism emboldened the position of Marxists, and later feminists and other radical geographers, interested in interrogating the underlying structures that produce and reproduce domination, uneven development, and inequitable social relations.

—*Vincent Del Casino*

See also Marxism, Geography and; Poststructuralism; Realism; Structuration Theory; Symbols and Symbolism

Suggested Reading

Castells, M. (1997). *The urban question: A Marxist approach* (A. Sheridan, Trans.). Cambridge: MIT Press.

Cosgrove, D. (1984). *Social formation and the symbolic landscape.* London: Croom Helm.

Harvey, D. (1973). *Social justice and the city.* Baltimore, MD: Johns Hopkins University Press.

Peet, R. (1998). *Modern geographical thought.* Oxford, UK: Blackwell.

STRUCTURATION THEORY

Structuration theory is an important part of contemporary social and geographic theory that effectively overcame the long-standing division between micro approaches that focused on individual humans (e.g., phenomenology, other humanistic views) and macro approaches that began and ended with social structures but ignored the dynamics of individual behavior (e.g., structural Marxism). Structuration theory began with and formed an important part of the career of Anthony Giddens, one of the 20th century's leading sociologists.

Giddens argued that orthodox theory lacked an adequate theory of the subject as a conscious actor who possesses the capacity to choose and to exert power. Rather, actors often are portrayed as unwitting dupes, and social change is erroneously held to occur "above their heads" or "behind their backs." Thus, he argued that the behavioral and structural dimensions of human life, and their corresponding theoretical perspectives, must be seen as mutually complementary. Giddens favored a *duality* between structure and agency, in which they are simultaneously determinant and mutually recursive, rather than a simplistic dualism of opposing forces.

Structuration begins with the phenomenological recognition that only humans are sentient knowledgeable agents; that is, they have consciousness about themselves and their world, however limited. Everyone, in this sense, is a sociologist. Giddens drew on the rich humanistic and behavioral traditions concerned with perception, cognition, and language. Moving beyond the usual definitions of culture as the sum total of learned behavior or a "way of life," structuration theory portrays culture as what people take for granted, that is, common sense or the matrix of ideologies that allow actors to negotiate their way through their everyday worlds. Culture defines what is normal and what is not, what is important and what is not, and what is acceptable and what is not within each social context. Culture is acquired through a lifelong process of socialization; individuals never live in a social vacuum but rather are socially produced from cradle to grave.

Structure in this view is seen to consist only of the rules and resources that are instantiated in social systems. In their daily lives, actors draw on these rules and resources, which in turn structure their actions; hence, the structural qualities that generate social action are continually reproduced through these very same actions. The socialization of the individual and the reproduction of society and place are two sides of the same coin; that is, the macrostructures of social relations are interlaced with the microstructures of everyday life. People reproduce the world, largely unintentionally, in their everyday lives, and in turn the world reproduces them through socialization. In forming their biographies every day, people re-create and transform their social worlds primarily without meaning to do so; individuals are both produced by and producers of history and geography. Hence, everyday thought and behavior not only mirror the world but also constitute it. History and geography are produced through the dynamics of everyday life, the routine interactions and transient encounters through which social formations are reproduced. Thus, "time" is not some abstract independent process but rather is synonymous with historical change (but not progress) and the capacity of people to make and transform their worlds. Social change in this light is not lawless in the sense of being anarchical and unpredictable, nor is it so completely subject to laws that the outcomes of action are predictable with confidence; in short, social organization and change are simultaneously structured and contingent.

Power, in all of its complex multiple forms, has long played a central role in this analysis. As transformative capacity, power is intrinsically tied to human agency, the capability of actors to secure outcomes. Giddens maintained that there are two primary structures of domination: allocative resources (material wealth and technology) and authoritative resources (the social organization of time, space, and the body).

The stretching of human activities and relations across greater and lesser amounts of time and space not only is fundamental to the organization and exertion of power in all of its forms but also is itself historically specific and contingent. Giddens called

this process time–space distanciation, a notion very similar to what geographers had called time–space convergence or time–space compression. The very malleability of time and space revealed them to be not "natural" or external to social relations but rather a product (and a producer) of those relations, that is, as human constructions. Thus, time and space are as plastic and mutable as the social structures of which they are a part. Time–space distanciation understandably was of great utility to geographers concerned with flows of people, goods, and information as well as the social construction of scale.

In the theory of structuration, Giddens offered a compelling analysis of how capitalism radically changed the fundamental contours of class relations and culture, extending commodity relations into various spheres of life, dramatically accelerating the tempo of production and reproduction, and marking a decisive break from the past. In precapitalist (class-divided) societies, he noted, power was exerted primarily through the state as a mechanism to extract surplus value and implement social control. Under capitalism and the private appropriation of wealth (class societies), in contrast, these relations occur through the market, that is, with an apolitical patina in which power lay in the private domain rather than in the public domain. Simultaneously, control over time and space, which is central to the exertion of power, shifted from the city-states of Italy to the incipient nation-states of northwestern Europe. Structuration theory, therefore, emphasizes the contingent nature of social life to confront functionalist and evolutionary interpretations of historical change.

—*Barney Warf*

See also Body, Geography of; Critical Human Geography; Existentialism; Humanistic Geography; Identity, Geography and; Marxism, Geography and; Phenomenology; Production of Space; Realism; Social Geography; Space, Human Geography and; Structuralism; Time Geography; Time–Space Compression

Suggested Reading

Giddens, A. (1981). *A contemporary critique of historical materialism.* Berkeley: University of California Press.

Giddens, A. (1984). *The constitution of society: Outline of the theory of structuration.* Berkeley: University of California Press.

Giddens, A. (1987). *The nation-state and violence.* Berkeley: University of California Press.

SUBALTERN STUDIES

Through the subaltern studies group, Ranijat Guha and a collective of South Asian historians sought to explore the question of the possibilities of resistance to British imperialism that had been missing from the historiography of the Raj. The term *subaltern* has two rather different places of origin that intersect in their work. First, *subaltern* was the term used for minor functionaries of the British colonial regime in India, including Indian, Anglo-Indian, and Anglo officials of a low rank and having origins as a term for lower ranks in the British military forces. Second, and more tellingly, it is a term often associated with the work of Italian Marxist political philosopher Antonio Gramsci, who used it to refer to the working class, peasants, and proletariat who were excluded from, and who stood in opposition to, the dominant historical bloc of a given social formation. At first, Guha and this collective built explicitly on Gramsci's approach to Italian history in an attempt to rewrite Indian history from below. Gradually, other influences, such as those of Michel Foucault and Jacques Derrida, entered the picture. Besides Guha, Gayatri Chakrovarty Spivak is the scholar who geographers are most likely to associate with the use of the term *subaltern*. Essentially, this is due to her essay titled "Can the Subaltern Speak?", a staple of postcolonial studies and cultural studies readers for nearly two decades. Spivak cautioned against the easy acceptance of any possibility for rewriting history from below, asking whether subaltern voices can indeed be recognized given the degree of their marginalization. Her particular examples of illiterate rural Indian subaltern women complicated the task to which the subaltern studies group had assigned itself. Geographers analyzing imperialism and colonialism, or those concerned with questions of the power dynamics of highly unequal societies, have been captivated by the work of the subaltern studies group.

—Garth Myers

See also Anticolonialism; Colonialism; Other/Otherness; Postcolonialism; Resistance

Suggested Reading

Crehan, K. (2002). *Gramsci, culture, and anthropology.* Berkeley: University of California Press.

Gramsci, A. (1971). *Selections from the prison notebooks of Antonio Gramsci* (Q. Hoare & G. Smith, Eds. & Trans.). New York: International Publishers.

Guha, R., & Spivak, G. (Eds.). (1988). *Selected subaltern studies.* Oxford, UK: Oxford University Press.

SUBJECT AND SUBJECTIVITY

The couplet *subject/subjectivity* connotes a multifaceted set of issues addressed in distinctive ways by different schools of human geographic research. Before outlining different approaches to the topic, two broad contexts in which issues around subject/subjectivity become pertinent to human geography can be identified:

• How human subjects, through their actions and values, actually produce and make meaningful the geographies that we study—and, conversely, how particular subject positions (particular identities, attitudes, and practices) are shaped by wider social, political, cultural, and economic processes, discourses, and structures. This involves issues such as the determination of individual or personal geographies by social–cultural–economic circumstances; questions of agency, creativity, free will, and resistance; and gendered, racialized, and sexualized identities and their concomitant geographies.

• The role and status of subjectivity in the production of geographic knowledge. How does the subjectivity of researchers (their opinions and background as well as their gender, ethnicity, and sexuality) feed into the research they do (the topics they choose to study, the methods they employ, and how they interpret results)? Should subjective values inform research, or should researchers strive to remain neutral and objective? What role might biographical or autoethnographic writing play in human geographic research?

Debates around the human subject and subjectivity have come to the fore in human geography via the range of postpositivist approaches that have influenced the discipline over the past 30 years. Marxist/Radical approaches, for example, position the individual human subject as the product of economic conditions under capitalism. Thus, the individual is understood primarily in terms of social class role and position, and individual beliefs, desires, and motivations—indeed, the very concept of the individual as a free autonomous agent—are held to be produced within the cultural and political ideologies of capitalism. In marked contrast, the varieties of humanistic geography that also emerged during the 1970s focused on the human subject as a locus and fountainhead of creative and imaginative abilities and explicitly positioned subjective beliefs and practices as central topics of geographic study. Structuration theory, associated with the work of Anthony Giddens and influential in human geography during the 1980s, may be understood as an attempt to find a middle ground between these conceptions of the subject as either wholly determined or wholly autonomous by arguing that knowledgeable and capable human subjects operate within broader social, political, and economic structures that both enable and constrain subjective actions.

The advent of geographic feminisms throughout the 1980s and 1990s has had a major impact on the ways in which human geographers conceptualize and engage with issues of subjectivity. This impact has been substantive insofar as feminist geographers have conducted numerous studies highlighting the structured gendered nature of human experience and the gendering of material and symbolic spaces. It has also been methodological and epistemological in that feminism is in part a critique of traditional modes of social science knowledge production emphasizing objectivity, detachment, and neutrality as virtues on the part of the researcher. Feminist geographers have been instrumental in the development of qualitative research methods within human geography, methods that involve acknowledgment of the human subjectivity of both the researcher and those being researched via principles of empathy and dialogue and that place subjective knowledges, emotions, and beliefs at the center of geographic study.

Parallel to and often intertwined with feminist geographies, work under the aegis of the *cultural turn* has added considerable depth and complexity to human geographic studies of processes of subject formation. Social, cultural, historical, and political geographies throughout the 1990s focused on the manifold ways in which spaces and identities are socially and culturally constructed, contested, and practiced. In tandem with the rise of cultural and postcolonial studies, geographers have paid particular attention to the role of gender, race, ethnicity, sexuality, and (dis)ability within patternings of subjectivity and spatiality. Such geographies have drawn a further large measure of

inspiration from continental poststructuralism, particularly in the work of Michel Foucault and Jacques Derrida. That this inspiration has led to a plethora of geographic work on the human subject is somewhat paradoxical given that both Foucault and Derrida may be described as posthumanistic writers. Their work details how subjectivity, and the entire Western conception of individuality, may be understood as an effect of systems of meaning and constellations of discursive power (rather than being understood as that on which power acts).

Nonetheless, there has occurred a concordance of sorts within human geographies of the cultural turn between a focus on quasi-structural explanations of spatiality and identity in terms of power, language, discourse, and so forth and a valorization of subjectivity in terms of individual agency, expression, and ability to flexibly inhabit, resist, and subvert regimes of cultural meaning.

Three additional, still-emerging strands of human geographic work engaging with issues of subjectivity may be identified. First, there is the growing interest in issues of embodiment and performativity. Initial, chiefly feminist and culturalist engagements stressed the body (and thus the subject) as a site where cultural and social meanings were scripted and contested. More recent work has moved toward an active conception of the body as a lived milieu of practice, performance, and expression. Drawing inspiration from the work of Judith Butler, for example, feminist geographers have worked toward a concept of gender not in terms of norms and constructs but rather in terms of citational bodily performances that constitute gendered subjectivities. Equally, taking cue from performance studies, from the phenomenology of Maurice Merleau-Ponty, and from ethnomethodologies of everyday life, human geographers have begun to interrogate the body–subject through studies of practices such as walking, driving, and eating.

A second strand of recent work concerns the application of psychoanalytic and therapeutic theories and principles to studies of spatiality and subjectivity. Early interventions in this subfield focused on the connections between psychoanalytic concepts of identity and geographic processes of exclusion in which certain identities and subjectivities were culturally and physically positioned as marginal. More recently, attention has turned to the relations between subjectivity and spatiality encountered within states such as agoraphobia and depression and the use of therapeutic tools and procedures as means of engaging, or reengaging, self and place.

Work on actor–network theory and hybrid geographies constitutes a third and final area of current geographic engagement with issues of subjectivity. Here, in a sense further developing the antihumanistic emphases of poststructuralism, the accent of research is on extending the traditional attributes of subjectivity (e.g., agency, will, expression) beyond the specifically human subject in which these customarily have been located. In addition to understanding subjectivity as a relational effect rather than a unique or inherent quality, this work draws attention to nonhuman subjects, particularly technological objects and animals, as assemblages of agency, intention, and effectivity in their own right.

—John Wylie

See also Body, Geography of; Feminist Geographies; Humanistic Geography; Identity, Geography and; Phenomenology; Psychoanalysis, Geography and; Structuration Theory

Suggested Reading

Cresswell, T. (1996). *In place/Out of place: Geography, ideology, and transgression.* Minneapolis: University of Minnesota Press.

Gregson, N., & Rose, G. (2000). Taking Butler elsewhere: Performativities, spatialities, and subjectivities. *Environment and Planning D, 18,* 433–452.

McDowell, L. (1999). *Gender, identity, and place: Understanding feminist geographies.* Cambridge, UK: Polity.

Pile, S., & Thrift, N. (1995). Mapping the subject. In S. Pile & N. Thrift (Eds.), *Mapping the subject: Geographies of cultural transformation* (pp. 199–225). London: Routledge.

Rose, G. (1993). *Feminism and geography: The limits of geographical knowledge.* Cambridge, UK: Polity.

Whatmore, S. (2002). *Hybrid geographies: Natures, cultures, spaces.* London: Sage.

Woman and Geography Study Group. (1997). *Feminist geographies: Explorations in diversity and difference.* Essex, UK: Addison-Wesley Longman.

SUBURBS AND SUBURBANIZATION

Suburbs usually are defined as politically independent jurisdictions located outside of a larger central city but still sharing social or economic ties with the city. Suburbs, together with the central city, comprise the larger metropolitan area.

Suburbs have existed in some form since antiquity, but during ancient times only the less well-off tended to live outside of the city walls. Later, in Europe, the wealthy developed country estates outside of the city's boundaries. Prior to the mid-19th century, the growth and size of cities were constrained by the lack of rapid mass-transportation networks. This meant that workers and owners lived in close proximity both to each other and to their places of employment. The result was compact cities whose size was determined by the ability of people to walk between various parts of the city (the pedestrian city). During the 19th century, industrialization together with waves of migration and immigration into British and U.S. cities made the central city a less desirable location for the affluent. From the 1840s onward, the development of new public transit systems, such as horse cars, streetcars, trains, and subway systems, enabled the affluent in British cities, and later in American cities, to begin moving out of the central city and into the open spaces on the city's outskirts. This enabled greater locational separation between both different classes in society and different types of land use and, thus, also between residence and employment. This process was rooted in the particular social and cultural norms of British bourgeois society, whose members desired a clear class separation from the working class for their nuclear families. The design of private detached housing away from the central city fulfilled this need.

Although the roots of the modern suburb are English, the greatest expression of the modern suburb is found in the United States (together with countries such as Canada and Australia). The development of the Anglo-American suburb always has had two important components: a desire for homogeneity (e.g., class, race, ethnicity) and a desire to control one's revenues (through local government). An additional crucial element for the development of suburbs was the availability of cheap land on the outskirts of cities and property speculators willing to buy and develop this land for profit. Thus, Anglo-American suburban development cannot be separated from a speculative private property–driven land market. As transportation moved from fixed lines to the flexible routes made possible by the automobile, the suburbs spread farther from the central city and suburban infill occurred between the tracks. This resulted in the vast low-density sprawl characteristic of modern American cities (Los Angeles is the paramount example). The cheap availability of the automobile, and then the wider availability of mortgages (through the establishment of the Federal Housing Authority in 1934), made suburbanization more accessible and affordable to the middle class. This process accelerated rapidly after World War II with the mass production of new suburban tract housing. These new suburban developments (e.g., Levittown, Pennsylvania) were marketed as places where young families could enjoy relatively low-cost homeownership, have ample space, have access to new amenities and appliances, have a say in controlling their taxes and schools, and (importantly) leave the problems and ethnic diversity of the cities behind.

American suburbanization was unique because suburbs were places where the affluent or middle classes resided in large lots, as homeowners, and often had considerable commutes to their places of work. Upward of two thirds of Americans own their own homes, a proportion greatly exceeding the proportions in other developed countries. A consequence of this is that it gives homeowners a vested interest in upholding property values, and this often is manifested in the exclusionary nature of suburbia. Levittown, for example, did not allow any African American purchasers until federal law made this illegal during the 1960s. Thus, it is not surprising that although 73% of American whites live in suburbs, only roughly 40% of Hispanics and African Americans do—and much of this suburbanization has occurred in older housing close to central cities or in segregated sections of existing suburbs.

Two further criticisms of suburbs are that they are cultural wastelands and that they have isolated women from communities and the workplace. The former criticism was attacked as early as the 1960s, and suburbs arguably have produced their own cultures, often as varied as those of cities. Although the early suburban isolation of women has been well documented, the rise of the dual-income family has meant that fewer suburban women remain isolated in their tract homes.

Since the 1970s, American suburbs have undergone a significant shift as most new commercial construction has occurred within suburbs, resulting in the majority of suburban residents being employed within suburbs rather than the central city. This has brought many of the familiar urban problems of cities to the suburbs (e.g., traffic jams, pollution, noise). Some maintain that this commercialization has resulted in the end of suburbia and that instead we now have a multicentered metropolis where suburbs and the

city are virtually indistinguishable from one another. Although this view may be extreme, it is clear that American suburbs continue to evolve as more of the affluent turn to closed or gated communities and others escape the problems of suburbia by moving to the ever expanding exurban metropolitan fringe.

—*Grant Saff*

See also Exurbs; Housing and Housing Markets; Rural–Urban Continuum; Urban Geography; Urban Sprawl

Suggested Reading

Fishman, R. (1987). *Bourgeois utopias*. New York: Basic Books.
Ganz, H. (1967). *The Levittowners*. New York: Pantheon.
Hayden, D. (2002). *Redesigning the American dream: Gender, housing, and family life*. New York: Norton.
Jackson, K. (1985). *Crabgrass frontier*. New York: Oxford University Press.
Knox, P. (1994). *Urbanization*. Englewood Cliffs, NJ: Prentice Hall.
Massey, D., & Denton, N. (1994). *American apartheid*. Cambridge, MA: Harvard University Press.

SUNBELT

The Sunbelt is an imprecise popular region, most often characterized as the 13 states of the southern tier of the conterminous United States that lie south of roughly 37 degrees north latitude (Figure 1). Also included is the part of California that extends southward from the San Francisco Bay Area to the Mexican border as well as southernmost Nevada. Geographers would never choose to define such an amorphous area as a single region because it does not exhibit the unifying criteria specified by the discipline's regional concept. Nonetheless, wide popular use by the media has led social scientists to at least consider the merits of the Sunbelt idea. Overwhelmingly, they were unimpressed but came to realize that more attention needed to be paid to the relentless migration of people and activities to the fast-developing South and West during the final decades of the 20th century. By 2000, human geographers had begun to focus more of their studies on this part of the country, but such studies have produced no evidence showing that the southern tier states are developing collectively as a coherent

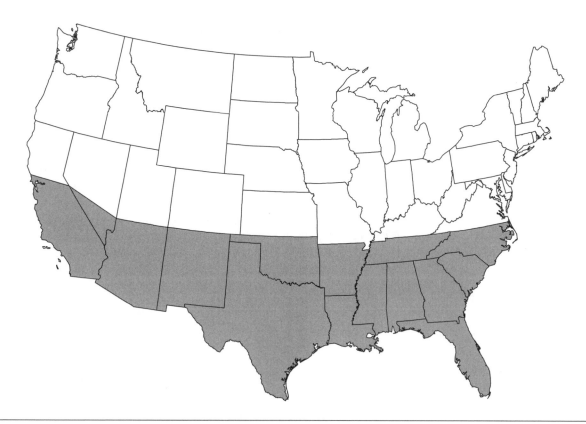

Figure 1   The Sunbelt Region of the United States

unit of spatial organization. In fact, the term *Sunbelt* has almost disappeared from the popular media, whose dominant view of the broad partitioning of the United States in 2005 centered on the "red states/blue states" politico–geographic split in the aftermath of the two most recent presidential elections.

The term originated during World War II with military planners who pursued the strategic advantages of dispersing defense industries from the Northeast to the "Sunshine Belt" of the South and West. A quarter of a century later, political commentator Kevin Phillips first popularized the Sunbelt notion in his writings on the South's emerging Republican majority, but its broader implications did not take hold in the national imagination until the publication of Kirkpatrick Sale's *Power Shift* in 1975. Although scholars soon began to explore the consequences of the intensifying southwest-ward shift, most of their work has concentrated on local-ized impacts, particularly the transformations at the metropolitan scale that have shaped key urban com-plexes such as southern California, Phoenix, Houston, the Dallas–Fort Worth metroplex, Atlanta, and Florida's Interstate 4 corridor and southeastern Gold Coast.

Several forces underlie the recent growth and rising prosperity of the southern tier states. The increasing footlooseness of manufacturing and higher-order ser-vices allowed employers to take advantage of cheaper land, labor, energy supplies, and other costs of doing business. Globalization accelerated the relocation trend, especially the emergence of the West Coast as a major player in the burgeoning Pacific Rim. Amenities, notably the warm winters and recreational opportuni-ties, played a leading role as well. And the technological innovations of the past half century also reinforced the break with the nation's historic northeastern core, par-ticularly the revolutions in air and highway travel, telecommunications and information processing, per-sonal computing, and universal air-conditioning.

After four decades of unremitting growth, the increasingly besieged southern tier states exhibit trends that are converging with those in the rest of the United States, especially problems of income inequality, social conflict, congestion and sprawl, and environmental degradation. Internally, most of these states resemble an economic–geographic checkerboard marked by stark contrasts between spaces of poverty and affluence, with the latter comprising a constellation of "sunspots" that prosper amid vast overwhelmingly rural areas still untouched by Sunbelt development.

—*Peter O. Muller*

See also Rustbelt; Urban Geography

Suggested Reading

Abbott, C. (1987). *The new urban America: Growth and politics in Sunbelt cities* (2nd ed.). Chapel Hill: University of North Carolina Press.

Bernard, R., & Rice, B. (Eds.). (1983). *Sunbelt cities: Politics and growth since World War II.* Austin: University of Texas Press.

Mohl, R. (Ed.). (1990). *Searching for the Sunbelt: Historical perspectives on a region.* Knoxville: University of Tennessee Press.

SUSTAINABLE DEVELOPMENT

The term *sustainable development* is intended to convey the idea that economic development and environmental sustainability should and can be made compatible. It has become one of the dominant ideas in development thinking today and one of the most prominent phrases in development discourse. The most common definition of sustainable development is based on the World Commission on Environment and Development (WCED) report of 1987 (published as *Our Common Future*), which stated that development should meet the needs of the current generation without compromising the ability of future generations to meet their own needs. With that definition, sustainable development encompassed inter- and intragenerational equity and promoted the idea that, using appropriate technology, economic growth could be achieved without exceeding environmental limits.

Sustainable development is best understood within the context of its institutional history. The United Nations Conference on the Human Environment held in Stockholm in 1972 usually is identified as having laid the groundwork for the formulation of the idea at the international level. It was here that the term *eco-development* was introduced and that the United Nations Environment Program (UNEP) was established. The conference drew explicit links between environment and development. For example, participating nations agreed that development need not be impaired by

environmental protection, that development was needed to improve the environment, and that this would require assistance such as funding for environmental safeguards.

In 1980, UNEP, along with the International Union for the Conservation of Nature and the World Wildlife Fund, published the *World Conservation Strategy* (*WCS*), the document in which sustainable development was first codified. The *WCS* identified three categories of conservation objectives: (1) the maintenance of essential ecological processes and life support systems, (2) the preservation of genetic diversity, and (3) the sustainable use of species and ecosystems. The *WCS* had strong roots in wildlife conservation and was intended to bring these concerns to bear on questions of development by providing a framework for resource management.

In contrast, the WCED's 1987 report focused on economic development as the primary mechanism for achieving an equitable and environmentally sound world economy. Published during a decade of declining (or, in some countries, negative) economic growth rates, *Our Common Future* (also known as the Brundtland Report) identified poverty as the primary cause of environmental degradation and prescribed economic growth as the cure. The rising burden of debt servicing and declining terms of trade and austerity programs were identified as inhibiting sustainable development in the Third World. *Our Common Future,* which was presented to the United Nations (UN) General Assembly, served to link discussion of development and the environment at the international level. Widely read, it also popularized the idea of sustainable development.

As a follow-up to *Our Common Future,* the UN scheduled a conference on environment and development to be held 5 years after its publication. The United Nations Conference on Environment and Development in Rio de Janeiro (the "Earth Summit") in 1992 was the largest environmental conference ever held, being attended by 172 states, more than 100 heads of state, and 9,000 media representatives, with an additional 28,000 participants at the coinciding International Nongovernmental Organization Forum. The primary output of the official conference was *Agenda 21,* a large document (600-plus pages) so named because it was intended to demonstrate how to make the planet sustainable by the start of the 21st century.

Thus, we can see that sustainable development emerged as a concept articulated in a series of documents produced under the auspices of multilateral institutions, primarily the UN. This has been labeled *mainstream sustainable development* (MSD), for which the following characteristics have been identified. First, MSD is modernist and industrialist, holding the view that developed countries represent the model for developing countries, that poverty is a primary cause of environmental degradation, and that the market and economic growth hold the solution to poverty. Second, MSD adheres to the principle of market environmentalism, that is, that problems can be addressed by setting prices for environmental "goods" and "services." Environmental economics and ecological economics are important approaches for developing MSD policies. Third, MSD sees the intertwined issue of development and the environment from an internationalist perspective, defining it as a global problem requiring multilateral action. Fourth, MSD is reformist, that is, working for change *within* the global capitalist system.

The most general criticisms of MSD are that it is vague and contradictory. First, in the attempt to achieve consensus in multilateral negotiations, the meaning of sustainable development has been diluted. Second, although MSD has been credited with bringing environmental issues to the fore in international fora, critics say that it has done so by linking incompatible goals—of economic growth and environmental protection—in a single phrase. Third, because the focus of MSD has shifted from conservation to development, it cannot be distinguished from development discourse in general. Fourth, the rhetoric of sustainable development has been used to "greenwash" ecologically destructive neoliberal economic policies. Some scholars argue that since their participation at the Earth Summit, transnational corporations have set the agenda for sustainable development at the global scale, focusing on more easily managed problems instead of pressing ecological problems that would require systemic change.

With human–environment relations as one of its primary disciplinary foci, geography has contributed substantially to the literature on sustainability. Geographers have responded with theoretical and empirical research that critically evaluates the precepts of sustainable development, for example, considering the problems associated with the privatization of common resources and the expansion of markets for previously subsistence goods. They have undertaken case studies to assess the sustainability of particular land use practices and of resource extractive activities and to evaluate the causal links between poverty and environmental degradation. Geographers have employed the concept of scale to develop applied models for implementing and evaluating sustainable policies and also to analyze the discourse of sustainable development as articulated at multiple levels. Geographers working in economic geography, rural geography, critical geopolitics, and resource management have addressed various aspects of sustainability, whereas the approach of political ecology is most closely identified with studying the relation between environment and development to address the question of what is being sustained and for whom.

—*Gail Hollander*

See also Developing World; Development Theory; Environmental Justice; Nature and Culture; Resource; Wilderness

Suggested Reading

Adams, W. (2001). *Green development: Environment and sustainability in the Third World* (2nd ed.). London: Routledge.

Bruntland, H. (1987). *Our common future.* Oxford, UK: Oxford University Press.

Castro, C. (2004). Sustainable development: Mainstream and critical perspectives. *Organization and Environment, 17,* 195–225.

Ilbery, B., Chiotti, Q., & Rickard, T. (Eds.). (1997). *Agricultural restructuring and sustainability: A geographical perspective.* New York: CAB International.

McManus, P. (1996). Contested terrains: Politics, stories, and discourses of sustainability. *Environmental Politics, 5,* 48–71.

Peet, R., & Watts, M. (Eds.). (2004). *Liberation ecologies: Environment, development, social movements* (2nd ed.). London: Routledge.

Robinson, J. (2004). Squaring the circle? Some thoughts on the idea of sustainable development. *Ecological Economics, 48,* 369–384.

Williams, C., & Millington, A. (2004). The diverse and contested meanings of sustainable development. *Geographical Journal, 170,* 99–104.

SYMBOLS AND SYMBOLISM

Symbols are important objects of analysis for geographic inquiry. Historically, the study of symbols, symbolism, and symbolization has been a key aspect of cartographic design, particularly in discussions of geographic visualization. Cartographers have long been interested in the use of symbols and their role in representing points and patterns across geographic space. In this context, symbols might be thought of as literal stand-ins for real-world objects. Over time, critical cartographers have challenged this reading of cartographic symbols in their interrogation of maps as cultural and political products. As such, the analysis of symbols has broader theoretical application in human geography, where investigating cultural symbolism is of interest to cultural and social geographers. In this expanding analysis, symbols have been examined as part of geography's interest in symbolic interactionism, a social theoretical tradition that has long been important in anthropology and sociology. Put simply, scholars interested in symbolic interactionism presume that the world of social categories and practices—and the symbols that constitute those categories and practices—is constructed through social (and spatial) interactions among people.

The study of symbols and symbolism in the social sciences (e.g., anthropology, sociology) during the 20th century focused on the investigation of culture and society as symbolic systems constituted through spoken language, exchanges, gestures, writing, photography, and so forth. A cultural symbol, in a naive anthropological sense, is anything that is put in place to represent something else. Complicating this naive definition, it is theorized that through cultural exchange and interaction, symbols are pushed through various interpretive cultural frameworks. Because symbols are part of broader culture systems, it is impossible to disentangle the individual use of a particular symbol from the cultural framework or society in which a person chooses to deploy that symbol. Therefore, methodologically, to comprehend a symbol and symbolism (the representation of the symbol), it is necessary to investigate the sociocultural context in which the symbol has been assigned meaning. In anthropology, these investigations often are qualitative and include ethnography, life history interviews, and participant observation. In these studies, symbols of the economy, political systems, and religion, for example, all are important aspects of symbolism (i.e., sociocultural practice); thus, symbols are thought to play important roles in all aspects of society. Put simply, social scientists interested in symbols and symbolism have theoretically extended how we think about the complex cultural construction and production of these objects and processes.

In concert with anthropology's and sociology's broad concern with representation and representational practices, particularly in the realm of the symbolic, geographers have maintained an interest in critically investigating their own symbols and symbolic processes. Geographers have, for example, critically interrogated cartographic symbols and symbolism. J. B. Harley and others have analyzed the map and its symbols as cultural texts. This view pushes the tradition of analyzing maps away from the emphasis on their mimetic capabilities and accuracy. In so doing, critical cartographers have argued that map symbols do not escape the broader social context in which they have been chosen and that geographers must examine both the position of the mapmaker and the sociocultural and spatial context of the map's use when considering the symbolic meaning of a particular map or map image. The analysis of maps and mapping as cultural products has been extended to other representational systems in geography such as geographic information systems (GIS).

Geographers with an interest in other nonpositivistic forms of inquiry have also used symbolic interactionism, traces of which can be seen in humanistic geography, urban social geography, and cultural geography. In all of these cases, geographers have been interested in the subjective experience of both place and community (read: identity) and the interactions between the two. More recently, cultural geographers arguing that culture is political have made use of symbols and symbolism in their discussions of how the cultural landscape is infused with power and meaning. Cultural symbols never are benign and therefore are subject to critical investigation for how the process of symbolism (i.e., how symbols emerge and are assigned meaning) operates within webs of sociospatial relations constituting particular meanings.

Denis Cosgrove made a number of significant contributions to the study of the symbolic landscape in his cultural geographic research. Cosgrove analyzed the ways in which symbolic representations in objects such as landscape paintings represent more than banal images of an idyllic past. Rather, landscape paintings are infused with a bourgeois ideology that privileges elite representations of landscape and community. Work and labor remain marginal in such symbolic formations—an invisible "other" without voice. Thus, those on the margins lack the ability to constitute their own symbolic landscapes. The cultural landscape of a privileged bourgeoisie naturalizes capitalistic accumulation and wealth.

Gillian Rose also interrogated the symbolic landscapes and ocular-centric (visual-centric) perspectivalism of geography's project in her feminist critique of the discipline. Through an analysis of landscape painting, Rose demonstrated how the symbolism of landscape images naturalizes more than class relations; it also naturalizes gender relations and silences the masculine pleasures integral to Western representations of the landscape. Landscapes are feminized as nature and thus are subject to the masculine gaze of geographers who speak and view the landscape from a position of cultural hegemony over nature. What this suggests is that symbolism and symbolic interactions more generally do not take place on a neutral field. Rather, symbols are subject to various deployments of power.

Cultural geographers, particularly those working within a framework of "new cultural geography," have rethought geography's own engagement with symbols, symbolism, and symbolization in interrogating various cultural practices and spaces. James Duncan first launched a critique of cultural geography's "superorganicism" from within a context of symbolic interactionism, arguing that symbolically culture is socially constructed and not an a priori pregiven ontological fact. Don Mitchell also invoked the importance of the symbolic is his retheorization of culture as politics. Finally, the realm of the symbolic has not escaped geographers in other subfields of the discipline, including economic and political geography, where the appeal of postpositivistic cultural geography has significant import. These geographers examine the economy and politics as symbolic systems rather than as fixed categories or as backdrops to sociospatial relations. Thus, the economy and politics are fully implicated in the deployment of cultural systems and power.

—*Vincent Del Casino*

See also Cartography; Cultural Geography; Feminisms; Humanistic Geography; Poststructuralism; Spaces of Representation; Text and Textuality

Suggested Reading

Cosgrove, D. (1984). *Social formation and the symbolic landscape.* London: Croom Helm.

Duncan, J. (1980). The superorganic in American cultural geography. *Annals of the Association of American Geographers, 70,* 181–198.

Harley, J. (1989). Deconstructing the map. *Cartographica, 26,* 1–20.

Mitchell, D. (2000). *Cultural geography: A critical introduction.* Oxford, UK: Blackwell.

Pickles, J. (Ed.). (1995). *Ground truth: The social implications of geographic information systems.* New York: Guilford.

Pickles, J. (2003). *A history of spaces: Cartographic reason, mapping, and the geo-coded world.* London: Routledge.

Rose, G. (1993). *Feminism and geography.* Minneapolis: University of Minnesota Press.

Sparke, M. (1995). Between demythologizing and deconstructing the map: Shawnadithit's New-Found-Land and the alienation of Canada. *Cartographica, 32,* 1–21.

Wood, D. (1992). *The power of maps.* New York: Guilford.

TELECOMMUNICATIONS, GEOGRAPHY AND

As the production and marketing of goods and services have steadily become more information intensive, technological changes have accelerated and product cycles have shortened. And as a deregulated worldwide market has increased uncertainty and increased the competition among places for investment and jobs, economic activities have become stretched over ever larger distances and the need to transmit information has grown accordingly. As a result, contemporary economic landscapes are closely tied to the deployment and use of telecommunications systems. Telecommunications has been critical to this process for more than 150 years, accelerating the flow of information across distances and bringing places closer to one another in relative space through time–space compression.

Telecommunications is not a new phenomenon. With the invention of the telegraph in 1844, the transmission of information over long distances was made possible and communications became detached from transportation. For decades after the invention of the telephone in 1876, telecommunications was synonymous with simple telephone service. Just as the telegraph was instrumental to the colonization of the American West during the late 19th century, the telephone became critical to the growth of the American city system, allowing firms to centralize their headquarters functions while they "spun off" branch plants to smaller towns. Even today, despite the proliferation of several new technologies, the telephone remains by far the most commonly used form of telecommunications for businesses and households.

During the late 20th century, as the cost of computing capacity dropped and the power increased rapidly with the microelectronics revolution, new technologies, particularly fiber optics and satellites, drastically increased the capacity of telecommunications. With the digitization of information during the late 20th century, telecommunications steadily merged with computers to form integrated networks, most spectacularly through the Internet. New technologies such as fiber optics have complemented, and at times substituted for, telephone lines.

MISCONCEPTIONS ABOUT TELECOMMUNICATIONS AND GEOGRAPHY

There is considerable confusion about the real and potential impacts of telecommunications, in part due to the long history of exaggerated claims that have been made, particularly by those subscribing to postindustrial theory. Often such views, which are widespread among academics and planners, hinge on a simplistic, utopian technological determinism that ignores the complex, and often contradictory, relations between telecommunications and local economic, social, and political circumstances. For example, proclamations that telecommunications would allow everyone to work at home via telecommuting, dispersing all functions and spelling the obsolescence of cities, have fallen flat in the face of the persistence of growth in dense urbanized places. In fact, telecommunications is generally a poor substitute for face-to-face meetings, the medium through which most sensitive corporate interactions occur, particularly when the information involved is irregular, proprietary, and/or unstandardized in nature. Most managers spend the bulk of their working time engaged in face-to-face contact, and no electronic technology can yet allow for the subtlety and nuances critical to such encounters. For this reason, a century of technological change has left most high-wage, white-collar, administrative command-and-control functions clustered in downtown areas. In contrast, telecommunications is ideally suited for the transmission of routinized standardized forms of data, facilitating the dispersal of functions involved with their processing to low-wage regions. Popular notions that telecommunications will render geography meaningless are simply naive. Although the costs of communications have decreased, other factors have risen in importance, including local regulations, the cost and skills of the local labor force, and infrastructural investments. Economic space, in short, will not evaporate because of the telecommunications revolution.

TELECOMMUNICATIONS AND FINANCE

Telecommunications probably has had its most important economic impacts in financial markets. Banks, insurance companies, and securities firms, all of which engage in very information-intensive activities, have been at the forefront of the construction of an extensive worldwide network of leased and private communication networks. Electronic funds transfer systems form the nervous system of the international financial economy, allowing banks to move capital around at a moment's notice, arbitraging interest rate differentials, taking advantage of favorable exchange rates, and avoiding political unrest. With the breakdown of the Bretton Woods agreement in 1971 and the collapse of fixed currency exchange rates, electronic trade in national currencies skyrocketed. Telecommunications networks give banks an ability to move money around the globe at stupendous rates. In securities markets, global telecommunications systems have facilitated the linking of stock and bond dealers through computerized trading programs. Subject to digitization, traveling at the speed of light as nothing but assemblages of zeros and ones, information and capital have become two sides of the same coin. As large quantities of funds cross borders with mounting ease, national financial policies have become increasingly

questionable in their effectiveness, making monetary controls over exchange, interest, and inflation rates ever harder to sustain.

One of the most significant repercussions of this trend has been the growth of global cities, particularly London, New York, and Tokyo, the command-and-control centers of the world system. Through vast tentacles of investment, trade, migration, and telecommunications, each city is tied to clients, markets, suppliers, and competitors scattered around the world. All global metropolises are endowed with enormous telecommunications infrastructures that allow corporate headquarters to stay in touch with global networks of branch plants, back offices, customers, subcontractors, subsidiaries, and competitors. However, telecommunications simultaneously threatens the agglomerative advantages of urban areas, particularly those obtained through face-to-face communications.

Telecommunications has also affected the financial industries through the growth of offshore banking. Usually in response to highly favorable tax laws implemented to attract foreign firms, offshore banking has become important to many microstates in the Caribbean, in Europe, in the Middle East, and in the South Pacific. As the technological barriers to moving money around internationally have fallen, legal and regulatory barriers have increased in importance and financial firms have found the topography of regulation to be of the utmost significance in choosing locations.

TELECOMMUNICATIONS AND BACK-OFFICE RELOCATIONS

Another economic activity that has been greatly affected by telecommunications is routinized back-office functions. Back offices essentially perform data entry of office records, payroll and billing, bank checks, insurance claims, and magazine subscriptions. These tasks involve unskilled or semiskilled labor (primarily women) and frequently operate on a 24-hour-per-day basis. Back offices have few of the interfirm linkages associated with headquarters activities and require extensive data processing facilities, reliable sources of electricity, and sophisticated telecommunications networks.

Historically, back offices have been located next to headquarters activities in downtown areas to ensure close management supervision and rapid turnaround of information. However, given the increasing locational flexibility afforded by satellites and interurban fiber-optics lines, back offices have also begun to relocate on a much broader continental scale, making them increasingly footloose. Many financial and insurance firms and airlines have moved their back offices from New York, San Francisco, and Los Angeles to low-wage communities in the Midwest and South. Internationally, this trend has taken the form of the offshore office. Offshore back offices are established not to serve foreign markets but rather to generate cost savings for U.S. firms by tapping low-wage Third World labor pools. Several New York–based life insurance companies, for example, have relocated back office facilities to Ireland. Likewise, the Caribbean, particularly Jamaica and Barbados, has become a particularly important locus for American back offices. Such trends indicate that telecommunications may accelerate the "offshoring" of many low-wage, low-value-added jobs from the United States, with dire consequences for unskilled workers.

TELECOMMUNICATIONS AND URBAN SPACE

Telecommunications has had important effects on the urban organization of space. One increasingly important effect of new information systems is telework or telecommuting, in which workers substitute some or all of their working days at a remote location (almost always their homes) for time usually spent at their offices. Telework is most appropriate for jobs involving mobile activities or routine information handling such as data entry and directory assistance. Proponents of telework claim that it enhances productivity and morale, reduces employee turnover and office space, and leads to reductions in traffic congestion (especially at peak hours), air pollution, energy use, and accidents. Further growth of telecommuting will encourage more decentralization of economic activity in suburban areas. However, there are countervailing reasons why telecommunications may actually *increase* the demand for transportation rather than decrease it. First, although telecommuters spend fewer days at their workplaces, it is not at all clear that they have shorter *weekly* commutes overall; indeed, because telecommuting may allow them to live farther from their workplaces, the total distances traveled might actually rise. Second, time freed from commuting may be spent traveling for other purposes such as shopping and recreation; telecommuting may alter the reasons for travel but not necessarily the frequency or volume of travel. Third, by reducing congestion,

telecommuting may lead to significant induced effects, whereby others formerly inhibited from driving may be induced to do so. In short, whether or not an actual trade-off between telecommunications and commuting occurs (their substitutability rather than complementarity) is not clear.

Another potential, and growing, impact of information systems on urban form concerns transportation informatics, including a variety of improvements in surface transportation such as smart metering, electronic road pricing, synchronized traffic lights, automated toll payments on turnpikes, automated road maps, information for trip planning and navigation, travel advisory systems, electronic tourist guides, remote traffic monitoring and displays, and computerized traffic management and control systems, all of which are designed to minimize congestion and optimize traffic flow (particularly at peak hours), enhancing the efficiency, reliability, and attractiveness of travel. Wireless technologies such as cellular phones allow more productive use of time otherwise lost to congestion. Such systems do not comprise new technologies so much as they enhance existing technologies.

The widespread use of telecommunications has eased the locational mobility of firms and reinforced changes in the nature of urban planning. Desperate for jobs, many localities vie with one another by making ever greater concessions to attract firms, forming a kind of auction that resembles a zero-sum game. The effects of such a competition are hardly beneficial to those with the least purchasing power and political clout. Left to sell themselves to the highest corporate bidder, localities frequently find themselves in a race to the bottom in which entrepreneurial governments promote growth—but do not regulate its aftermath—via tax breaks, subsidies, training programs, looser regulations, low-interest loans, infrastructure grants, and zoning exemptions. As a result, local planners have become increasingly less concerned with issues of social redistribution, compensation for negative externalities, provision of public services, and so forth and more enthralled with questions of economic competitiveness, investment capital, and the production of a favorable business climate.

THE INTERNET

Among the various networks that comprise the world's telecommunications infrastructure, the largest and most famous is the Internet, an unregulated electronic network that in 2005 connected an estimated 900 million people in more than 180 countries—roughly 15% of the world's population. From its military origins in the United States during the 1960s, the Internet emerged at a global scale through the integration of existing telephone, fiberoptic, and satellite systems that was made possible by the technological innovation of packet switching and the Integrated Services Digital Network (ISDN), in which individual messages may be decomposed, with the constituent parts transmitted by various channels and then reassembled—virtually instantaneously—at the destination. Spurred by declining prices of services and equipment, the Internet has grown worldwide at rapid rates.

Systems such as the Internet and the World Wide Web have had, and will continue to have, substantial (if unanticipated) effects on the social fabric and geographic space. Information systems may reinforce existing disparities in wealth, connecting elites in different nations who may be increasingly disconnected from the local environments of their own cities and countries. Indeed, in a social–psychological sense, cyberspace may allow for the reconstruction of communities without propinquity, that is, groups of users who share common interests but not physical proximity. In an age where social life is increasingly mediated through computer networks, the reconstruction of interpersonal relations around the digitized spaces of cyberspace is of the utmost significance.

Significant discrepancies exist in terms of access to the Internet, largely along the lines of wealth, gender, and race. Worldwide, access to computers linked to the Internet, either at home or at work, is highly correlated with income, education level, and employment in professional occupations. Social and spatial differentials in access to the skills, equipment, and software necessary to access the Internet threaten to create a large, predominantly minority underclass deprived of the benefits of cyberspace. Modern economies are increasingly divided between those who are comfortable and proficient with digital technology and those who neither understand nor trust it, disenfranchising the latter group from the possibility of citizenship in cyberspace. At the international level, inequalities in access to the Internet reflect the long-standing bifurcation between the First World and the Third World. Outside of North America and Europe, the vast bulk of the world's people, particularly the Third World, have little to no access to the Internet. The global

move toward deregulation in telecommunications likely will lead to more use-based pricing (the so-called "pay-per" revolution), in which users must bear the full costs of their calls, and fewer cross-subsidies among different groups of users (e.g., between commercial and residential ones), a trend that likely will make access to cyberspace even less affordable to low-income users.

—Barney Warf

See also Communications, Geography of; Cyberspace; Economic Geography; Global Cities; Postindustrial Society; Producer Services; Time–Space Compression; Virtual Geographies

Suggested Reading

Akwule, R. (1992). *Global telecommunications: The technology, administration, and policies.* Boston: Focal Press.

Brooks, J. (1975). *Telephone: The first hundred years.* New York: Harper & Row.

Brunn, S., & Leinbach, T. (Eds.). (1991). *Collapsing space and time: Geographic aspects of communications and information.* London: HarperCollins.

De Sola Pool, I. (Ed.). (1977). *The social impact of the telephone.* Cambridge: MIT Press.

Graham, S., & Marvin, S. (1996). *Telecommunications and the city: Electronic spaces, urban places.* London: Routledge.

Marvin, C. (1988). *When old technologies were new: Thinking about electric communication in the late nineteenth century.* Oxford, UK: Oxford University Press.

Solomon, E. (1997). *Virtual money: Understanding the power and risks of money's high-speed journey into electronic space.* Oxford, UK: Oxford University Press.

Warf, B. (1995). Telecommunications and the changing geographies of knowledge transmission in the late 20th century. *Urban Studies, 32,* 361–378.

TERMS OF TRADE

A country's terms of trade refer to the relative values of its exports and imports. When the terms of trade are good, a country sells, on average, relatively high-valued commodities (e.g., manufactured goods) and imports relatively low-valued ones (e.g., agricultural products). This situation generates foreign revenues that allow the country to earn revenues with which to pay off debt or reinvest in infrastructure or new technologies.

Unfortunately, many less developed countries (LDCs), with economies distorted by centuries of colonialism in which they became suppliers of minerals and foodstuffs, have poor terms of trade; that is, they export relatively low-valued goods and must import expensive, high-valued ones. Primary commodities account for roughly 70% of the total exports of developing countries. Many LDCs rely heavily on exports of goods such as petroleum, copper and iron ore, timber, bananas, coffee, and fruits. Lacking domestic production capacities, they must import items such as automobiles, pharmaceuticals, and machine tools. The worldwide glut in raw materials, including petroleum, wheat, and many mineral ores, has depressed the prices for many LDC exports. Without indigenous manufacturing, many are forced to sell low-valued goods such as bananas for high-valued ones such as computers.

The proceeds from these exports are needed to pay for imports of manufactures, which are vital for continuing industrialization and technological progress. Shifts in the relative prices of commodities and manufactures, therefore, can change the purchasing power of the exports of LDCs dramatically. The situation is exacerbated because many of these low- and middle-income countries are vulnerable, single-commodity-dependent ones.

The economies of many LDCs are characterized by structural rigidity. They cannot alter the composition of exports rapidly in response to changing relative prices. Thus, if their commodity export prices decrease, they have no alternative but to accept declines in their terms of trade. This situation makes it difficult to generate foreign revenues and exacerbates these countries' debt problems. Furthermore, poor terms of trade perpetuate these LDCs' cycle of poverty; low foreign revenues yield little to reinvest, helping to create a shortage of capital and resulting in low rates of productivity.

Another factor that may lead to worsening terms of trade is technological advances in developed countries. Advanced technology enables industrial economies to (a) reduce the primary content of final products, (b) produce high-quality finished products from less valuable or lower-quality primary products, and (c) produce substitutes for existing primary products (e.g., synthetic rubber for naturally grown rubber). These developments are irreversible. The demand for many primary products may be inelastic for price decreases, but in the long run it may be very elastic for price increases. A rise in the price of a raw material provides an incentive for industrial research geared to economizing on the commodity, substituting

something else for it, or producing it in the importing country.

—*Barney Warf*

See also Comparative Advantage; Trade

Suggested Reading

Stutz, F., & Warf, B. (2005). *The world economy: Resources, location, trade, and development* (4th ed.). Upper Saddle River, NJ: Pearson/Prentice Hall.

TESSELLATION

A tessellation is a subdivision of a space into nonoverlapping regions that fill the space completely. Tessellations are considered regular if the regions are of the same size and shape (e.g., a remotely sensed image composed of square pixels) and are considered irregular otherwise (e.g., the provinces of Sri Lanka). Tessellations are said to nest if they are organized hierarchically, with the individual regions of a tessellation at one level being subdivided at a lower level; for example, each of the states forming a tessellation of the continental United States may be subdivided into counties. When a tessellation is used to display data collected for its regions by means of shading symbols, it is called a choropleth map.

The most frequently encountered tessellations in human geography are various types of administrative units. However, some phenomena, such as the trade areas of supermarkets in a city, sometimes are modeled as tessellations even though the individual regions are not mutually exclusive. For some tessellations the individual regions might have one or more associated internal entities (e.g., counties and county seats), whereas for others they might not (e.g., census tracts). Even when the phenomenon under consideration does not take the form of a tessellation, it still may be possible to generate a tessellation from the original objects. One example, in which the regions are known as Voronoi (or Thiessen or proximal) polygons, often is used in locational decision making involving service facilities. Here a region is created for each facility by assigning to it all locations in space that are closer to that facility than to any other facility so that the regions collectively form a tessellation.

Because irregular tessellations of administrative units are created by humans, they involve a degree of arbitrariness and are modifiable. For example, gerrymandering is said to occur when boundaries of electoral districts are drawn in a manner that is most advantageous for a particular political party. Often more than one tessellation may exist for a given space. For example, a metropolitan area may be divided into school districts, health districts, police precincts, and so forth. Furthermore, any one of these tessellations may change over time. Reconciling different data collected from more than one tessellation usually is problematic and requires some form of areal interpolation. When choropleth maps are constructed for a variable collected at an individual level using different tessellations, different representations can occur because the tessellations may have different relationships to the underlying spatial pattern of the variable. Similarly, the relationships between variables may also change with different tessellations. The same thing may also be observed when a variable is mapped using tessellations at different levels. These two situations are referred to as the zoning problem and the aggregation problem, respectively; collectively, they are referred to as the modifiable areal unit problem (MAUP).

—*Barry Boots*

See also GIS; Humanistic GIScience; Spaces of Representation

Suggested Reading

O'Sullivan, D., & Unwin, D. (2003). *Geographic information analysis.* Hoboken, NJ: John Wiley.

TEXT AND TEXTUALITY

Text and textuality, a metaphor commonly used in cultural geography that likens landscape interpretation to reading a written document, refer to a system of signs that are unstable and open to multiple understandings and repeated revisions. Such an expansive notion of text leading well beyond its traditional association with the written page to include a vast array of cultural productions—including landscapes but also maps, pictures, social institutions, political regimes, and even the world itself—is characteristic of the poststructuralist perspective at the heart of this concept. Arguably, the single feature most distinctive of poststructuralism is the linguistic turn toward text,

discourse, reading, and interpretation and the accompanying close connection among language, power, and knowledge. In human geography, that turn is reflected most directly in the metaphor of landscape-as-text.

Some of geography's earliest advocates of cultural landscape study were interested in how the earth's surface was, in some sense, authored. Carl Sauer did not use this metaphor explicitly, but his intention was to read the landscape for signs of the culture that created it. Peirce Lewis, a geographer not formally associated with the Berkeley School but associated with theoretical and methodological sympathies for it, applied this metaphor more overtly in his 1979 essay, "Axioms for Reading the Landscape." Lewis proposed that reading the cultural landscape was, in important ways, like reading a book but that, in two specific ways, it was more difficult. First, Lewis argued that cultural landscapes seem to be messy and disorganized, sort of like a book with missing, torn, and smudged pages; like books, landscapes *can* be read, but unlike books, landscapes were not *meant* to be read. Second, Lewis was convinced that Americans were not used to reading landscapes. Thus, driven by a concern to counter such geographic illiteracy and by the belief that such reading brought both pleasure and a deeper understanding of American culture, he offered guidelines—what he called self-evident axioms—designed to help students envision the landscape as a text to be read.

Lewis's essay proved to be quite influential and sparked the interest in several generations of human geographers to see the cultural landscape as our unwitting autobiography—as a cultural text that reflected our tastes, values, aspirations, and fears in a concrete and visible form. Curiously, given the text metaphor's obvious debt to the literary arts, Lewis derived his own inspiration not from literary theory but rather from what he called traditional geomorphology and traditional plant ecology, fields that encouraged scientists to use their eyes and think about what they saw. Perhaps as a result, some critics charge, Lewis was uninterested in how landscapes were written and read in actuality as well as in practice. By failing to investigate the specific relationship between authoring or writing the landscape and reading it, Lewis left unanswered many of the most interesting questions about how such texts were created and what it means to read them.

Over the past two decades, cultural geographer James Duncan has sought to address such questions by problematizing the very concept of reading. Working with several coauthors, including Nancy Duncan and Trevor Barnes, Duncan sought to show that both experts (geographers like himself and Lewis) and ordinary people read the landscape all of the time as part of everyday life. But rather than conceptualizing landscapes as mere reflections of culture—our unwitting autobiographies—he argued that such texts are constitutive of the world and its contents. Such an expanded view of text derives from a broadly postmodern view that sees texts as constituting reality rather than merely mimicking it. It also relies on the concept of intertextuality, that is, the idea that the context of any text is other texts. Such other texts include the sorts of things that we traditionally call texts—literary works, legal documents, letters, and so forth—as well as cultural productions such as botanical gardens and a country's legal system. Indeed, from this perspective, everything is a text—a signifying system ready to be read for meanings, decoded and deciphered for symbolic messages, and continually rewritten. Thus, we all are cultural readers whose very act of reading alters the text itself.

Deriving his central theoretical positions from diverse literary theorists such as Roland Barthes, Jacques Derrida, and Paul Ricoeur as well as from cultural anthropologists such as Clifford Geertz and James Clifford, Duncan pointed to four reasons to understand the world (or landscape) as a text. First, meaning in written discourse is made concrete when inscribed or textualized, just as aspects of social life take on a similar fixity. Second, texts escape their authors, just as social actions carry unintended consequences. Third, both written texts and social occurrences are subject to continual reinterpretation. Fourth, the meaning of a text is inherently unstable and, like social institutions, is dependent on a wide range of understandings from specific textual communities. This last point is especially important and one that positions this interpretive framework at the center of contemporary cultural studies and poststructuralist theory. By arguing that it is impossible to assign permanent and universal meanings to a text, Duncan and those influenced by his method insisted that the inherent instability of texts makes them ripe for countering hegemonic power. With all texts open for interpretation and reinterpretation, alternative individual and collective readings of different cultural productions—like landscapes—quite frequently unmask system-supporting ideologies.

Human geographers and critical theorists invoking the landscape-as-text metaphor often deploy it for the specific task of interrogating power. In his study of the precapitalist Kandyan Kingdom of Sri Lanka, Duncan focused on how landscapes signify power relations. He found that narratives of the glorious past were incorporated into landscape designs, thereby showing how the power of the Kandyan king was spatially and temporally contiguous with the power of the gods and hero-kings of long ago; such texts legitimize a ruthless king's rule. Richard Schein, with nods to both Duncan and Lewis, set forth a framework for interpreting a residential suburb that envisions landscape as the tangible visible articulation of competing texts. Schein drew from examples of specific texts—what he called discourse materialized in the landscape (landscape architecture, historic preservation, neighborhood associations, and zoning)—to demonstrate that such texts often serve to naturalize social relations as embodied in the landscape.

But it is not only landscape that has occupied geographers interested in the text metaphor. Critical historical geographers, most notably J. Brian Harley, have shown how maps might be fruitfully understood as texts that are no more stable than those deconstructed by literary theorists. Cartography, Harley argued, belongs to the terrain of the social world in which it is produced; maps are, in other words, intertextual and constitutive of power relations. Because maps operate behind a mask of seemingly neutral science—what Harley called maps' sly rhetoric of neutrality—they are especially powerful texts that hide and deny their social dimensions at the same time that they legitimate those very social relations (Figure 1).

As important as the text metaphor remains in cultural geography, it is not without its critics. Some have argued that an exclusive focus on the textual and discursive qualities of landscape has led to its dematerialization, that is, to a weakening of the substantive meaning of landscape as a place of human habitation and environmental concern. Others have objected to it on more political grounds, arguing that envisioning landscape as a text diminishes the nontextual, nonlinguistic practices at work in making a landscape or that, as Henri Lefebvre put it, an overestimation of texts leads to an underestimation of material circumstances. Don Mitchell shares this view, arguing for a greater focus on the material production of landscape and less on the politics of reading, language, and text. Advocates of the textual method argue that such

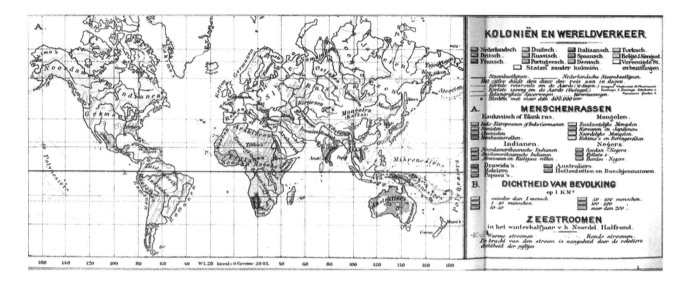

Figure 1   *Menschenrassen* (Human Races), an inset map of a larger map titled *Koloniën en Wereldverkeer* (Colonies and World Transportation). When read critically as a text, even seemingly straightforward thematic maps can be shown to carry messages of power. This school atlas map classifies human "races" into discrete categories that simplified dramatically the complex world of cultural difference.

SOURCE: Reprinted from *Bos' Schoolatlas der Geheele Aarde* (Groningen, Netherlands: J. B. Wolters, 1908).

criticisms are beside the point because texts cannot be separated from other aspects of reality, including the material world. Either way, during the past several years, text and textuality have been supplemented by other interpretive frameworks, including spectacle, stage, and theatricality—metaphors that are based less on representation than on performance.

—*Steve Hoelscher*

See also Berkeley School; Cartography; Cultural Geography; Cultural Landscape; Cultural Turn; Hegemony; Postmodernism; Poststructuralism; Power; Spaces of Representation; Symbols and Symbolism; Writing

Suggested Reading

Barnes, T., & Duncan, J. (Eds.). (1992). *Writing worlds: Discourse, text, and metaphor in the representation of landscape.* London: Routledge.

Duncan, J. (1990). *The city as text: The politics of landscape interpretation in the Kandyan Kingdom.* Cambridge, UK: Cambridge University Press.

Duncan, J., & Duncan, N. (1988). (Re)Reading the landscape. *Environment and Planning D, 6,* 117–126.

Harley, J. (2002). *The new nature of maps: Essays in the history of cartography.* Baltimore, MD: Johns Hopkins University Press.

Lewis, P. (1979). Axioms for reading the landscape: Some guides to the American scene. In D. Meinig (Ed.), *The interpretation of ordinary landscapes: Geographical essays* (pp. 11–32). New York: Oxford University Press.

Schein, R. (1997). The place of landscape: A conceptual framework for interpreting an American scene. *Annals of the Association of American Geographers, 87,* 660–680.

THEORY

Theory is a set of assumptions and propositions that offer explanation. There is no one unified theory that is applied by all social and physical scientists to explain the world. Instead, there are myriad explanatory frameworks that invoke different theoretical paradigms and approaches, some of which are more dominant at different points in time and space. In human geography in particular, there are four major metatheoretical paradigms that have developed over the late 20th and early 21st centuries and that inform the majority of geographic inquiry: spatial science, humanism, critical realism, and poststructuralism. Metatheories are constituted by epistemological assumptions (how we know the world) and ontological assumptions (how the world is structured to produce knowledge). These epistemological and ontological assumptions vary depending on one's paradigmatic perspective. Moreover, these broad paradigms have multiple theoretical modifications that privilege key objects of analysis. It is possible to argue that feminism, which focuses its attention on the power dynamics that are part and parcel of gender, cuts across all four paradigms, although many feminists today favor rethinking the epistemological assumptions of critical realism and poststructuralism in defining a feminist theoretical approach more so than those of spatial science and humanism. Needless to say, it is possible to argue for a distinct theoretical approach called feminist poststructuralism.

As John Paul Jones explained, epistemology and ontology in Western thought historically are organized around key sets of binaries. In geography, some of the most common epistemological binaries include objectivity and subjectivity, nomothetic and ideographic, general and particular, and explanation and interpretation. Ontologically speaking, geographers are concerned with the relationships between order and chaos, nature and culture, materiality and discourse, and society and space. These are but a few examples. Ontology and epistemology, taken in conjunction, frame our theories about how the world might work and how best to study that world. A spatial scientist understands the world as ontologically ordered and rational. Within that context of order, the spatial scientist would view the world as objectively quantifiable. From a different epistemological framework, a poststructuralist geographer may view the world as chaotic. Ontologically, the poststructuralist might privilege discourse over the material in studies of the webs of chaotic social relations that are part of the relationships between, for example, space and social identity.

Spatial science historically is based in the philosophical tradition of logical positivism. Epistemologically, positivism favors objectivity; ontologically, it favors order. Positivism operates within the assumption that through the generation of hypotheses and the empirically testing of those hypotheses, it is possible to generate laws about how the world works. Also underpinning positivism's assumptions about the objective world is that once laws are proven, they may be generalizable for universal application. The overarching goal of positivism is not to overturn the epistemological and ontological assumptions of this paradigm but rather to refine, through empirical investigation and

methodological nuance, the theories that emerge about the objective and orderly world. Thus, human geographers seek to advance spatial laws about how the world operates, working across an ontologically flat and absolute space in their study of what John Nystuen called geography's primitives, for example, distance and location. The spatial scientist, operating from the position of objectivity, treats the field of study as something "out there"; the subject position of the researcher is distinct and separate from the people the researcher claims to study. Despite the fact that spatial scientists operate within a framework of objectivity, they use both deductive (top-down) and inductive (bottom-up) analyses. Inductively, it is possible to generate new theories of the world beginning with empirical investigation, although the presumption is that those theories still operate within a world that is based in order and rationality.

Humanism in geography, unlike spatial science and positivism, favors subjective experience over objective law. Humanism is an idealistic philosophy concerned with how individuals think about the world around them. Epistemologically, whereas spatial science favors generalizability, idealism favors individuality. Humanists, through a focus on subjective experience, eschew the notion that it is possible to test hypotheses of the social world empirically; thus, they shy away from deductive theories. Humanism in geography pays particular attention to the importance of place and place meaning or a "sense of place." Theoretically, humanism's interests lie in the area of emotions, individual control and agency, and embodied experience. At the same time, humanists theoretically believe the world to be orderly, although that ordered world is situated and experienced at the individual level. Because humanism centers its theories about the world on subjective experience, it also has a strong interest in intersubjectivity, that is, relations between subjects and relations between subjects and their place in the world.

Critical realism, which is based in part in structuralism, provides another important theoretical framework in geographic inquiry. Within the confines of this approach, oppositional tendencies to bifurcate objectivity and subjectivity, as in spatial science and humanism, are collapsed. Ontologically, critical realists still assert that there is a real set of processes that structure social life but argue that those processes are mediated by human knowledge and experience. However, the ontological ordering of the binary oppositions, such as chaos and order, are theorized as contextually related. Critical realists theoretically argue that the world is ordered through a series of necessary and contingent relations. As an example, Marxist critical realists would argue that the necessary structures that undergird social life are capitalistic. Contingent on those necessary structures, however, are processes such as patriarchy, and that is why there is a gendered division of labor under capitalism. Critical realists, despite their ontological and epistemological flexibility, do not discharge the notion that they can uncover and unpack the core truths and operations of larger structures and thus tend to side with spatial science's privileging of objectivity. Theoretically, critical realists understand the world through empirical investigations of it, although critical realists have become increasingly interested in the role of discourse and language as objects of analysis in investigations of contingent relations.

Poststructuralism relies, importantly, on an array of disciplinary and political engagements with the limitations of structuralist thought, including concerns from feminism, cultural studies, and queer theory, to name a few. Epistemologically, poststructuralism operates under the assumptions that the world is understood through subjective experience; ontologically, the world is chaotic. Unlike humanism, poststructuralism retains the important qualities of structuralism, particularly its focus on language and discourse. Also, poststructuralism reasserts the importance of individual position and the situatedness of experience. Thus, it is possible to understand how various discourses can come to temporarily fix certain spaces and identities. The assumptions that experience is subjective but that subjectivity is mediated by the deployment of power through representations and representational practices, therefore, allow poststructuralists to turn their attention not only to the social construction of space but also to the ways in which theorists themselves construct the world through the use of their own epistemological lenses.

Theories and theoretical thinking are important because they inform the overall research process. Sandra Harding argued, in fact, that theories are more than simply our epistemological and ontological assumptions; they are also part and parcel of our methodology. A methodology, in Harding's terms, is a meso-level theoretical construct that allows researchers to translate their epistemological and ontological assumptions into data. Different theories of the world focus attention on different objects of analysis. So, a humanistic geographer theoretically considers

subjective experience to be more central to understanding the social world than is objective rationality. Methodologically, then, a humanistic geographer focuses his or her attention on "place" as the key object of analysis. Those same methodological theories of the world also guide conceptualizations of key research terms such as validity, reliability, and research ethics. Spatial science, like positivism, presumes that for a study to be valid, one must be able to replicate findings over and over again. This suggests that the universal application of generalizable law is central to spatial scientific thinking. That being the case, spatial scientific data will be limited methodologically to what can be tested and retested, marginalizing subjective experience. In the context of research ethics, the spatial scientist seeks to maintain objectivity so as not to taint the research findings and results. This is a radical departure from a poststructuralist view of methodology, where the data always are messier than our theories of them. As such, poststructuralists do not presume to be marginal from the experiences they study. Instead, poststructuralism, drawing strongly from feminism, considers researchers to always be part of the process they claim to be studying. Instead of seeking objectivity, researchers need to be reflexive and critical of their own subject position within the confines of the study.

Thus, methodology is an important part of theory and the theoretical process. It presupposes method, which is the actual "nuts and bolts" of the research process. So, as Harding suggested, it might be more valuable to think of a feminist methodology and epistemology than to think specifically about a feminist method. Methodology, and its application, might suggest that a life history interview method is a more valuable research tool when considering subjective experiences of place, as a humanistic geographer would want to do. In "doing research" or in reading the research of other geographers, therefore, it is valuable to begin by critically reflecting on how epistemological and ontological assumptions inform the questions being asked, the types of data being collected, and the relationship between the researcher and the researched. No matter what "paradigm" or metatheory is being applied, theories are only partial explanations of how the world operates. More important, it is important to remember that although it is possible to consider metatheoretically how the relationships between objectivity and subjectivity operate, it is more likely that the boundaries between theories and their objects of analysis remain blurry. What this suggests is that it is possible to create theories that are multiparadigmatic and multimethodological and that it might be best to consider what various theories have in common rather than drawing the boundaries around what might be different.

—Vincent Del Casino

See also Epistemology; Feminisms; History of Geography; Humanistic Geography; Logical Positivism; Marxism, Grography and; Ontology; Phenomenology; Postmodernism; Poststructuralism; Realism; Spatially Integrated Social Science; Structuralism

Suggested Reading

Del Casino, V., Jr., Grimes, A., Hanna, S., & Jones, J., III. (2001). Methodological frameworks for the geographic study of organizations. *Geoforum, 31,* 523–539.

Duncan, J., Johnson, N., & Schein, R. (Eds.). (2004). *A companion to cultural geography.* Oxford, UK: Blackwell.

Harding, S. (1987). *Feminism and methodology.* Bloomington: University of Indiana Press.

Johnston, R., & Sidaway, J. (2004). *Geography and geographers: Anglo-American human geography since 1945* (6th ed.). London: Edward Arnold.

Jones, J., III. (2004). Introduction: Reading geography through binary oppositions. In K. Anderson, M. Domosh, S. Pile, & N. Thrift (Eds.), *Handbook of cultural geography* (pp. 511–519). Thousand Oaks, CA: Sage.

Nystuen, J. (1968). Identification of some fundamental spatial concepts. In B. Berry & D. Marble (Eds.), *Spatial analysis* (pp. 35–41). Englewood Cliffs, NJ: Prentice Hall.

TIME, REPRESENTATION OF

Our world is a dynamic place. Everything in our environment is changing constantly—from land use and traffic patterns to climate and wildlife habitat. Understanding spatial phenomena requires examining not only patterns over space but also how those patterns change over time as a process. Earth-related processes tend to be highly complex, involving many aspects that interact at various space–time scales. This tendency is further complicated by the fact that space–time lags are ubiquitous in cause–effect associations. Examining how things change over time allows us to discover temporal patterns (cycles and rhythms of occurrence). From this, we gain an understanding of cause and effect so that we can predict and plan effectively.

Although geographic information systems (GIS) have become an essential software tool for examining and analyzing just about any Earth-related phenomenon, representing temporal dynamics within GIS has proven to be a significant challenge. Geographic information systems maintain a static view of the world inherited from cartographic tradition. Although not currently possible in any straightforward manner because the data models and query languages used in current GIS do not explicitly incorporate the temporal dimension, some simple questions that might be asked using GIS in an example application of land use change could include the following:

- What did the distribution of single-family housing look like 15 years ago?

- What residential areas were added between 1980 and 1998?

- What areas changed from a predominance of agricultural to residential land use since the enactment of agricultural preservation regulations?

- What agricultural areas are most likely to be converted to other uses within the next 10 years?

BASIC APPROACHES FOR DIGITAL REPRESENTATION OF SPACE–TIME DATA

There have been recent introductions of GIS products for analyzing the occurrence of discrete point data over space and time (e.g., incidence of disease or crime) and for tracking point objects as they travel through space and time (e.g., delivery vans). The Tracking Analyst from the Environmental Systems Research Institute (ESRI) is one such GIS product. Dealing with areal data, however, is more challenging. Areal objects, such as residential districts, wetlands, and areas of poor air quality, not only move but also can grow, shrink, and change shape. Implementations for dealing with this type of data currently involve improvising with the (static) raster (i.e., grid) and vector data models currently inherent in GIS.

The primary conceptual approach currently used for representation of areal space–time data in GIS is known as the snapshot approach—sequential images or snapshots of values for a given variable over a given area for a known point in time (Figure 1a). These snapshots are conceptually equivalent to slices extracted from the continuous space–time cube. Another conceptual approach for representing space–time data is what Gail Langran called "base state with amendments" (Figure 1b), where only the changes known at specific times are recorded. A third conceptual scheme is the space–time composite, which can be thought of as the result of

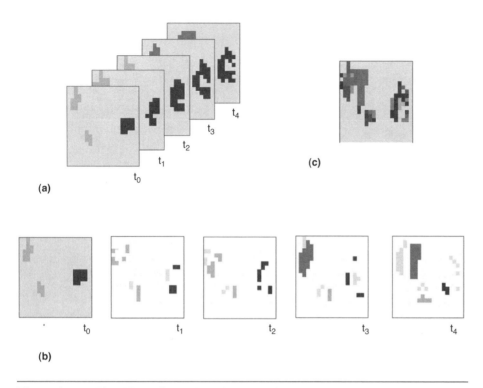

(a)

(c)

(b)

t_0 t_1 t_2 t_3 t_4

Figure 1   The Three Space-Based Approaches to Space–Time Representation: (a) Snapshots, (b) Base State With Amendments, and (c) Space–Time Composites

stacking all successive changes on top of one another as virtual transparencies, flattening out the space–time cube into a single cumulative depiction (Figure 1c).

Sequent snapshots usually are implemented as gridded raster images, although vector implementation can also be done. This approach is easy to implement using the standard raster and vector data models available in existing GIS by "tricking" the software. Instead of individual grids representing a set of thematic layers (e.g., elevation, vegetation, soil type), each spatially registered layer represents the state at a given moment in time for one of these themes (e.g., a vegetation sequence). Because modern computer graphics use rasters, the gridded implementation of sequent snapshots allows these to be displayed as a movie similar to the radar map on the Weather Channel. Under software control, the display sequence can be speeded up, slowed down, or reversed. Temporal change can also be calculated by finding the differences between two snapshots.

There are, however, inherent drawbacks to the sequent snapshot approach. What happens between snapshots is not known because this representation records information at discrete times. It is possible that

a whole series of changes occurred between the two times for which data are recorded. Another drawback is that values for the entire area (the state of all locations or entities at the time represented by the snapshot) must be recorded whether or not a particular location or entity had changed. This redundancy greatly inflates the total volume of stored data.

Implementing the base state with amendments can also be done in either grid or vector form. As expected, a grid-based implementation records a complete snapshot as the first in the sequence, and then grids representing subsequent times record only changes, noting the new values in the appropriate coordinate (row, column) location. The vector implementation is similar except that it is the changes to entities, instead of locations, that are stored in each successive time layer. The base state consists of all objects, or entities, as they are configured at the starting time (i.e., the original state) as points, lines, and polygons. Changes to point entities are noted as single x and y coordinates, and changes to line and polygon entities are noted as strings of x and y coordinates representing change vectors. At each time layer, new and obsolete locations are noted for specific entities as they appear, disappear, grow, shrink, and move. The base state with amendments is a more efficient approach from an implementation standpoint because it eliminates the redundant data storage problem of sequent snapshots. However, the other characteristics, both pro and con, are shared with the sequent snapshot implementation.

Each of these representational approaches attempts to fold time into space in some way. The space–time cube portrays time as a spatial dimension. Sequent snapshots imply the temporal dimension through sequences of static world states in a time series. With the base state with amendments approach, changes rather than complete states are represented in what

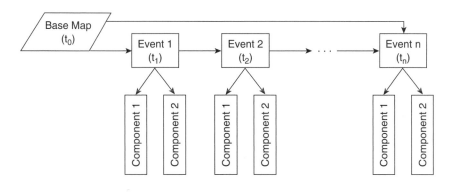

Figure 2   The Timeline Representation

amounts to abbreviated snapshots. Time is discrete, whereas space is continuous, with both the snapshot and base state with amendments approaches. With the space–time composite, time is completely collapsed into space. There is no organization of the data on the basis of time—no time sequencing. Rather, time simply becomes an attribute value.

All of these approaches, both implicitly and explicitly, are based on a space-based view of the world that time can be represented within the context of space. Although time-based questions such as those mentioned earlier can be answered using the four conceptual approaches described above, a time-based representation would be more effective. The timeline representation, as shown in Figure 2, allows information to be retrieved directly on the basis of events, or changes, ordered along a timeline.

Implementation of these conceptual schemes can be achieved, with difficulty, using the database management systems (DBMS) facilities available within current GIS. Even with successful implementation, however, temporal query operators (e.g., before, after, during) to allow effective temporal data retrieval and analysis are not yet available. Development of temporal GIS is an active area of research.

—Donna Peuquet

See also GIS; Humanistic GIScience; Time Geography

Suggested Reading

Langran, G. (1992). *Time in geographic information systems.* London: Taylor & Francis.

Langran, G., & Chrisman, N. (1988). A framework for temporal geographic information. *Cartographica, 25*(3), 1–14.

Peuquet, D., & Duan, N. (1995). An event-based spatiotemporal data model (ESTDM) for temporal analysis of geographical data. *International Journal of Geographical Information Systems, 9*(1), 7–24.

TIME GEOGRAPHY

Time geography posits that individuals, groups, and institutions navigate through time–space and, therefore, that everyday life is defined by the availability of these two interrelated resources. Time and space serve as a shared social canvas on which people have the opportunity to interact with other individuals and institutions. At the same time, time–space interactions are inherently limited and highly dependent on the daily geographies of people.

BACKGROUND

The notion of a human geography contextualized by time was first forwarded by Torsten Hägerstrand during the 1960s. Drawing on influences from human ecology, Hägerstrand's concept evolved from earlier work on diffusion and his observations concerning the highly contextualized nature of innovation and the diffusion process. By the 1970s, time geography had become an established conceptual framework for understanding the interconnectedness of everyday life and the many (often taken-for-granted) events that shape and contextualize geography "on the ground." Beyond geography, the model has been adapted for use in archaeology, anthropology, and sociology. In this sense, time geography establishes human geography as the investigation of localized structures and individual agency.

THE WEB MODEL OF TIME GEOGRAPHY

Modeling the localization of structures and human agency in time–space requires that the following assumptions be made:

- Time and space are limited.
- Constraints are placed on individual action.
- Constraints promote structured interaction.
- Constraints limit free time–space activity.

In each of these assumptions, time geography posits that the everyday activity space of individuals—known as paths—is structured by various constraints. Structural constraints on individual time–space paths include the spatial limits of everyday life (i.e., travel time and locations), modes of production (i.e., work spaces), and limited access to a particular resource in time–space (i.e., business hours). Based on these structural limitations, maximum activity spaces—also known as prisms or domains—are created, and it is within these spaces that everyday life occurs. Within prisms, the activities and time–space paths of individuals are bundled with the paths of others at key locations or "stations." In concert, the combination of constraints and opportunities determines the time–space path of individuals and effectively weaves a "web" of interconnected and shared social experiences (Figure 1). It is the mapping of this web and the network social relations of individuals that can be used to elucidate key sociospatial bundles, or bound interactions in space–time, that produce the material and symbolic landscape.

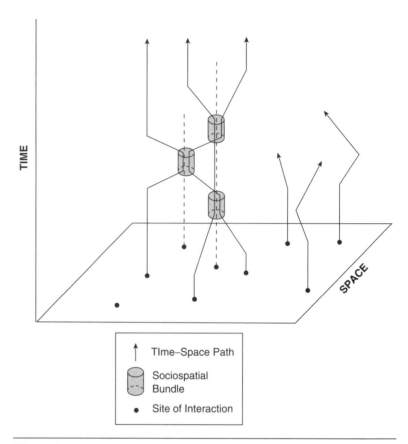

Figure 1   Hägerstrand's Time–Space Model

CRITIQUES

Three basic critiques have been leveled. First, time geography is inherently reductionist. That is, humans are reduced to actors traveling paths to and from stations (or locations) determined primarily by capability, coupling, and authority constraints. As such, critics have suggested that time geography can be reduced to little more than a graphic exercise. Second, the method is intensive and requires a great deal of resources to effectively establish the time geography of an individual, place, or local culture. Because of the intensity of the field-work necessary to map individual paths, the scale of the analysis is highly localized. As a result, time geography has been regarded as overly ambitious and problematic beyond the local scale. Finally, time geography became embedded within a larger structure and agency debate that consumed the discipline from the late 1970s through the mid-1980s. As might be expected, time geography was critiqued by humanists and structuralists alike as either too dependent on structure or too dependent on agency, respectively.

CONTEMPORARY TIME GEOGRAPHY

The contemporary relevance of time geography is the model's implicit observation that people are purposeful

and that individual geographies and histories are meaningful. In concert, these themes marry spatial modeling and "explanation" with more fuzzy concepts such as meaning. One example of time geography that provides both an explanation of a place and economy and a contextualized understanding of everyday life is the collected works of Allan Pred. In 1984, Pred grafted the concept onto that of biography formation and theory of structuration. Pred believed that by charting the histories of individuals and roughly mapping their time geographies, time geography could be used to operationalize the principles outlined in Anthony Giddens's theory.

Today, time geography has seen a renewal of sorts as themes of place, the region, and social network theory have emerged to explain unique geographies. In many cases, time geography is used to explain extraordinary and allegedly counterfactual cases by charting the biographies of key individuals, the interconnectedness of local (and global) social networks, and the place-specific constraints that structured sociospatial interactions. Time geography has been used to reinforce, albeit perhaps not explicitly, a deeper understanding of humans, communities, and attachment to place. Gillian Rose's article, "Imagining Poplar in the 1920s," charted the time geography of a neighborhood and how daily life paths are reimagined, reengineered, or romanticized to create and contest spatial communities.

Beyond the adapted works of structuralists and humanists, time geography also informs the works of regional science and geographic information science (GIScience). Indeed, time geography is being used to unlock and model the everyday time–space paths of individuals. Today, time–space geographies are at the forefront of geospatial modeling in many subfields, including urban, economic, medical, and transportation geographies. As the most recent revisionings of time geography suggest, answering Hägerstrand's question of where the people are in regional science, or Harvey Miller's updated question of where the people are in GIScience, continues to be central to the work of geographers, regional scientists, and many other social scientists striving to understand place, region, and community.

—*Jay Gatrell*

See also Structuration Theory; Time–Space Compression

Suggested Reading

Giddens, A. (1984). *The constitution of society.* Berkeley: University of California Press.

Hägerstrand, T. (1967). *Innovation diffusion as a spatial process.* Chicago: University of Chicago Press.

Hägerstrand, T. (1970). What about people in regional science? *Papers of the Regional Science Association, 24,* 7–21.

Kwan, M., Janelle, D., & Goodchild, M. (Eds.). (2003). Accessibility in time and space [special issue]. *Journal of Geographical Systems, 5*(1).

Miller, H. (2003). What about people in geographic information science? *Computers, Environment, and Urban Systems, 27,* 447–453.

Pred, A. (1984). Place as historically contingent process: Structuration and the time-geography of becoming places. *Annals of the Association of American Geographers, 74,* 279–297.

Rose, G. (1990). Imagining Poplar in the 1920s: Contested concepts of community. *Journal of Historical Geography, 16,* 425–437.

TIME–SPACE COMPRESSION

If geography is the study of how humans are stretched over the earth's surface, a vital part of that process is how we know and feel about space and time. Although space and time appear as "natural" and outside of society, they are in fact social constructions; every society develops different ways of dealing with and perceiving them. For example, how people experience time and space in a nomadic herding society is very different from how they do so in an advanced producer services economy. Thus, time and space are socially created, plastic, mutable institutions that profoundly shape individual perceptions and social relations. Sociologist Anthony Giddens used the term *distanciation* to describe how societies are stretched over time and space. Because the economy cannot be detached from other realms of social life, time–space compression is more than simply an economic phenomenon. By changing the time–space prisms of daily life—how people use their time and space, the constraints they face, and the meanings they attach to them—time–space compression is simultaneously cultural, social, political, and psychological in nature. This issue elevates the analytical significance of relative space, in which distances are measured through changing metrics of time and cost, above that of absolute space, the traditional Cartesian form that characterized most Enlightenment forms of geography.

Time–space compression is as old as human civilization itself. Early changes in transportation technology such as the wheel and keel, for example, were instrumental in the development of centralized empires. Many cultures (e.g., the Romans) developed road or canal networks to shuttle people and goods between places. The Mongols and Incas created messenger systems using riders on horseback and runners, respectively. The rise of capitalism, and the Industrial Revolution in particular, generated enormous improvements in transportation and communications that significantly reduced travel times between places. For example, the travel time between Edinburgh and London, a distance of 640 kilometers, was roughly 20,000 minutes by stagecoach in 1658. By the 1850s, with the arrival of the steam locomotive, travel time had been reduced by 96% to 800 minutes. By 1988, the rail journey between Edinburgh and London took 275 minutes. When the line was electrified in 1995, travel time was reduced to less than 180 minutes. By airplane, it takes less than 60 minutes today (Figure 1). Thus, although the absolute distance between the two cities remained the same, the relative distance was reduced by 1,940 minutes over 350 years—or roughly 5 minutes per year.

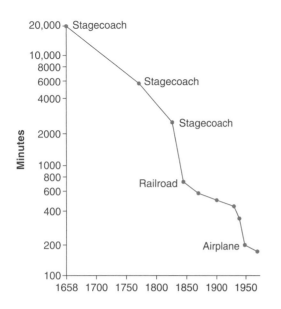

Figure 1   Falling Travel Times Between Edinburgh and London, 1658–1950

SOURCE: Redrawn from Janelle, D. (1969). Spatial reorganization: A model and concept. *Annals of the Association of American Geographers, 59,* 348–365. Oxford, UK: Blackwell. Reprinted with permission of Blackwell Publishing.

Air transportation provides spectacular examples of time–space convergence. During the 1930s, it took a plane between 15 and 17 hours to fly the United States from coast to coast, whereas modern jets now cross the country in roughly 5 hours. In 1934, planes took 12 days to fly between London and Brisbane, Australia, whereas today the Boeing 747 is capable of flying any commercially practicable route nonstop. The result is that any place on the earth is within 24 hours of any other place using the most direct route.

Closely related to time–space compression is cost–space compression, that is, the steady reduction in the cost of moving goods, people, or information between places. For example, the opening of the Erie Canal in 1825 reduced the cost of transport between Buffalo and Albany from $100 to $10 per ton and, ultimately, to $3 per ton. Railroad freight rates in the United States dropped 41% between 1882 and 1900. Between the 1870s and 1950s, improvements in the efficiency of ships reduced the real cost of ocean transport by roughly 60%.

Thus, time–space compression is intimately associated with the ways in which capitalism generates new geographies over time. In David Harvey's famous term, capitalism regularly produces a *spatial fix,* that is, a set of landscapes conducive to capital accumulation at given historical junctures. Falling transport times have created a steadily "shrinking world" (Figure 2) that changes the suite of opportunities and constraints faced by individual places. Globally, cheaper, more efficient modes of transport widened the range over which goods could be shipped profitably, allowing regions and countries to realize a comparative advantage and stimulating large-scale production and an increasingly internationalized geographic division of labor.

Within countries, transportation improvements contributed to changing patterns of urban accessibility. Cities in the industrialized world grew from compact walking and horse-car cities (pre-1800–1890), to electric streetcar cities (1890–1920), and finally to dispersed automobile cities in the recreational automobile era (1920–1945), the freeway era (1945–1970), the edge city era (1970–1990), and the exurban era (1990–present).

Telecommunications represents the latest, and perhaps the most profound, in the long series of chapters of time–space compression, effectively reducing the communications time between places to zero. The telegraph of the 19th century, for example, connected the global economy's expanding networks almost instantaneously, allowing news to travel at the speed of light, reducing uncertainty for producers, allowing multiestablishment firms to connect headquarters and branch plants, and allowing markets to expand over ever larger spatial domains. The telephone had similar effects throughout the 20th century, and declines in the cost of telephone service generated considerable cost–space compression with multiple economic, social, and psychological impacts.

Starting in the 1980s and the microelectronics revolution, a worldwide skein of fiber-optics lines came to form the nervous system of the international financial economy, creating the time–space compression of the flexible, post-Fordist, digitized global economy. Electronic funds transfer systems allow financial institutions to switch vast sums of capital from one place to another instantaneously, in the process arbitraging interest rate differentials, speculating in foreign exchange, and investing or disinvesting in a wide variety of instruments in a diverse array of markets. The Internet represents another dimension of this process. Connecting roughly 15% of the world's population in 2005, the interlinked series of computers that comprise the Internet and World Wide Web allow almost instantaneous time–space compression.

Time–space compression does not affect all places and peoples equally. Spatially, this process endlessly generates new geographies of centrality and peripherality, bringing some—but not all—places closer together. In general, elites tend to enjoy the earliest, and often the greatest, advantages from reduced costs and transmission times of transportation and communications. Even within the best connected of global cities, there are substantial pockets of "off-line" poverty.

A common, but erroneous, perspective of this issue is that telecommunications entails the end of geography. Often such a view hinges on a simplistic, utopian technological determinism that ignoresthe complex relations between telecommunications and local economic, social, and political circumstances. For example, repeated predictions that telecommunications would allow everyone to work at home via telecommuting, dispersing all functions and spelling the obsolescence of cities, have fallen flat in the face of the persistent growth in densely inhabited urbanized places and global cities, where face-to-face contact is essential. Thus, popular notions that telecommunications will render geography meaningless are naive. Although the costs of communications have decreased, as they did with transportation, other factors have risen in importance, including local regulations,

1500–1850

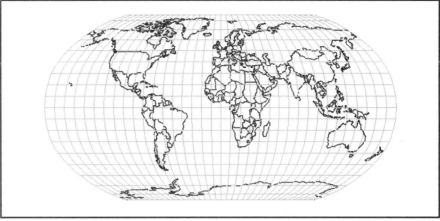

Best average speed of horse-drawn coaches or sailing ships was 10 m.p.h.

1850–1930

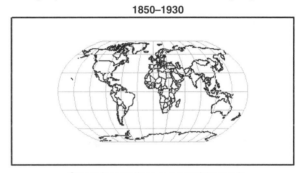

Steam locomotives averaged 65 m.p.h.
Steamships averaged 38 m.p.h.

1950s

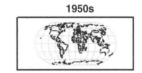

Propellered aircraft 300–400 m.p.h.

1960s

Jet passenger aircraft 600–700 m.p.h.

Figure 2   A Shrinking World

the cost and skills of the local labor force, government policies, and infrastructural investments.

Cyberspace, for example, has changed the everyday lives of hundreds of millions (if not billions) of people, altering not only what we know but also how we know it, creating a prosthetic extension that fuses people and machines. Globally, in a shrinking world, distant strangers become less and less distant and strange. Harvey attributed the growth of postmodern culture—the fragmented set of discourses emphasizing difference and heterogeneity that flourished during the late 20th century—to the compression unleashed by digitized, globalized time–space compression.

—Barney Warf

See also Cyberspace; Structuration Theory; Telecommunications, Geography and; Time Geography; Virtual Geographies

Suggested Reading

Giddens, A. (1984). *The constitution of society: Outline of the theory of structuration.* Berkeley: University of California Press.

Harvey, D. (1989). *The condition of postmodernity.* Oxford, UK: Blackwell.

Janelle, D. (1969). Spatial reorganization: A model and concept. *Annals of the Association of American Geographers, 59,* 348–365.

Kern, S. (1983). *The culture of time and space 1880–1918.* Cambridge, MA: Harvard University Press.

Thrift, N. (1995). A hyperactive world. In R. Johnston, P. Taylor, & M. Watts (Eds.), *Geographies of global change* (pp. 18–35). Oxford, UK: Blackwell.

TOBLER'S FIRST LAW OF GEOGRAPHY

Tobler's first law of geography (TFL) refers to the statement made by Waldo Tobler in an article published in 1970: "Everything is related to everything else, but near things are more related than distant things." Embedded in TFL are two interwoven theses: the pervasive interrelatedness among all things and how they vary spatially. TFL is also conceptually consistent with the notion of distance decay (also known as the inverse distance effects or distance lapse rate) that geographers developed a long time ago.

TFL captures the characteristics of spatial dependence—a defining feature of spatial structures. TFL normally is interpreted as a gradual attenuating effect of distance as we traverse across space while considering that the effect of distance is constant in all directions. The acceptance of TFL implies either a continuous, smooth, decreasing effect of distance on the attributes of adjacent or contiguous spatial objects or an incremental variation in values of attributes as we traverse space. TFL now is widely accepted as an elementary general rule for spatial structures, and it also serves as a starting point for the measurement and simulation of spatially autocorrelated structures.

Although often deployed only implicitly in social physics (e.g., the gravity model) and in some quantitative methods (e.g., the inverse distance weighting method for spatial interpolation, regionalized variable theory for *kriging*), TFL is central to the core of geographic conceptions of space as well as spatial analytical techniques. With continuing progress in spatial analysis and advances in geographic information systems (GIS) and geographic information science (GIScience), new life will continue to breathe into TFL as we become better equipped to conduct detailed analyses of the "near" and the "related." New measures for spatial autocorrelation (e.g., local indicators of spatial autocorrelation [LISA]) have been developed to empirically test TFL in physical, socioeconomic, and cultural domains.

New developments in telecommunications technologies have altered spatial relationships in society in many fundamental ways, and the universality of TFL has been questioned during recent years. Critics of TFL, often grounded in the poststructural or social construction of scientific literature, reject TFL as a *law*—much less as the first law of geography. Instead, they argue that all universal laws are necessarily local knowledge in disguise. The complexity and diversity of the real world render lawlike statements impossible, especially in the social arena. Instead of calling it the first law of geography, critics consider that TFL should better be regarded as local lore. Obviously, whether TFL should be treated as the *first* law of geography or local knowledge will have profound implications at the ontological, epistemological, methodological, and even ethical levels.

—Daniel Sui

See also Gravity Model; Spatial Autocorrelation; Spatial Dependence; Spatial Heterogeneity

Suggested Reading

Sui, D. (2004). Tobler's first law of geography: A big idea for a small world? *Annals of the Association of American Geographers, 94,* 269–277.

Tobler, W. (1970). A computer movie simulating urban growth in the Detroit region. *Economic Geography, 46,* 234–240.

TOPOPHILIA

As defined by the geographer Yi-Fu Tuan, topophilia is the affective bond between people and place. His 1974 book set forth a wide-ranging exploration of how the emotive ties with the material environment vary greatly from person to person and in intensity, subtlety, and mode of expression. Factors influencing one's depth of response to the environment include cultural background, gender, race, and historical circumstance, and Tuan also argued that there is a biological and sensory element. Topophilia might not be the strongest of human emotions—indeed, many people feel utterly indifferent toward the environments that shape their lives—but when activated it has the power to elevate a place to become the carrier of emotionally charged events or to be perceived as a symbol.

Aesthetic appreciation is one way in which people respond to the environment. A brilliantly colored rainbow after gloomy afternoon showers, a busy city street alive with human interaction—one might experience the beauty of such landscapes that had seemed quite ordinary only moments before or that are being newly discovered. This is quite the opposite of a second topophilic bond, namely that of the acquired taste for certain landscapes and places that one knows well. When a place is home, or when a space has become the locus of memories or the means of gaining a livelihood, it frequently evokes a deeper set of attachments than those predicated purely on the visual. A third response to the environment also depends on the human senses but may be tactile and olfactory, namely a delight in the feel and smell of air, water, and the earth.

Topophilia—and its very close conceptual twin, sense of place—is an experience that, however elusive, has inspired recent architects and planners. Most notably, new urbanism seeks to counter the perceived placelessness of modern suburbs and the decline of central cities through neotraditional design motifs. Although motivated by good intentions, such attempts to create places rich in meaning are perhaps bound to disappoint. As Tuan noted, purely aesthetic responses often are suddenly revealed, but their intensity rarely is long-lasting. Topophilia is difficult to design for and impossible to quantify, and its most articulate interpreters have been self-reflective philosophers such as Henry David Thoreau, evoking a marvelously intricate sense of place at Walden Pond, and Tuan, describing his deep affinity for the desert.

Topophilia connotes a positive relationship, but it often is useful to explore the darker affiliations between people and place. Patriotism, literally meaning the love of one's *terra patria* or homeland, has long been cultivated by governing elites for a range of nationalist projects, including war preparation and ethnic cleansing. Residents of upscale residential developments have disclosed how important it is to maintain their community's distinct identity, often by casting themselves in a superior social position and by reinforcing class and racial differences. And just as a beloved landscape is suddenly revealed, so too may landscapes of fear cast a dark shadow over a place that makes one feel a sense of dread or anxiety—or *topophobia.*

—*Steve Hoelscher*

See also Humanistic Geography; Nationalism; New Urbanism; Place; Sense of Place

Suggested Reading

Tuan, Y.-F. (1990). *Topophilia: A study of environmental perception, attitudes, and values* (2nd ed.). New York: Columbia University Press.

Tuan, Y.-F. (2001). The desert and I: A study in affinity. *Michigan Quarterly Review, 40*(1), 7–16.

TOURISM, GEOGRAPHY AND/OF

Tourism is routinely acknowledged as the world's largest industry. According to the World Tourism Organization (WTO), in 2003 tourism involved 694 million international visitors, generated $514 billion (more than 10% of global gross domestic product), and employed at least 200 million people (approximately 10% of the global labor force). Although most international tourism traffic is between North America and Europe, the portion accruing to the global South is steadily increasing and now accounts for approximately 25% of all international arrivals.

As a consequence of its scale and complexity, tourism is studied by numerous disciplines, including economics, psychology, anthropology, sociology, and geography. Although definitional or conceptual consensus on tourism has not been achieved, researchers usually agree on one empirical point, namely that exponential increases in international tourism arrivals since the early 1980s have fundamentally altered the tourist industry. Widely recognized are three main forms of tourism. First, conventional *mass tourism* is commercial, seasonal, and typically located along coastal areas; it involves a high volume of tourists who generally adhere to their own cultural norms. Second is the post-1980s growth in *alternative tourism,* a term that encompasses a range of sustainable, "low-impact," and holistically planned strategies that purport to offer benign alternatives to the economic, cultural, and environmental problems associated with mass tourism. Finally, the post-1990s emergence of *mass ecotourism* or *resort ecotourism* consists of the adoption and adaptation of sustainable and/or alternative tourism planning policies by conventional mass-tourism corporations.

Geographers have studied tourism since the early 1930s. These Anglo-American geographic inquiries quantified the economic impacts of recreation by mapping the changing land use patterns of rural areas predominantly in the U.S. Northeast and Upper Midwest. Geographic scholarship on tourism from the 1940s to the 1970s typically is characterized as an era dogged by a lack of theoretical rigor, disciplinary coherence, and empirical case studies other than at the intranational scale. Stephen Britton's Marxian analyses of Fiji during the early 1980s were the first major geographic inquiries to use economic dependency theory and to specifically identify tourism as an exploitative form of capitalistic development. Moreover, Britton's studies prompted numerous geographers to document how monopolistic, foreign-owned mass-tourism corporations in the global South created the following problems: loss of control of local resources, low multiplier and spread effects outside of tourism enclaves, lack of articulation with other domestic sectors, and high foreign exchange "leakages." Following Britton's call for greater attention to the theorization of tourism, since the early 1990s geographers have eschewed a tradition of empiricism exemplified by impact studies and the modeling of tourism flows.

Influenced by the critical aspirations of new cultural geography and the seminal interdisciplinary works on tourism by Dean MacCannell and John Urry, contemporary geographic study of tourism is defined by considerable theoretical debate and methodological innovation. The recent appearance of two interdisciplinary geography journals on tourism, *Tourist Studies* and *Tourism Geographies,* evinces the growing importance of tourism as a subfield in geography. Although geographers have shown an enduring interest in critically documenting the environmental

problems associated with tourism such as soil erosion, water and air pollution, loss of biodiversity, detrimental visual impacts, and overloading of key infrastructure, for the most part geographers have sketched a research agenda that focuses on the socioeconomic and cultural problems of tourism.

Today, critical tourism geographers are united by the assertion that tourism never can be understood as just innocent enjoyment or a peripheral component of modern life. Embracing the theories of Marxism, feminism, poststructuralism, phenomenology, performativity, and psychoanalysis, many geographers examine tourism phenomena through grounded analyses to document geographies of power, knowledge, practices, technologies, identities, representations, and social memories. Such geographies of tourism consist of uneven, relational, discursive, and material spaces that form a key component of people's everyday practices and landscapes at local and global scales.

For geographers, not only is tourism the world's largest industry, it also is profoundly unequal and normative in terms of the lived, landscaped, and contested dimensions of class, race, gender, sexuality, and so forth. Geographers, for example, have argued that tourism is profoundly gendered, with women concentrated in the low-paid service or informal sectors. Elsewhere, geographers have insisted that tourism is profoundly normative, with African American travel agents struggling to survive economically because of racist perceptions and gay cruise ships struggling to acquire docking permission from morally conservative yet bankrupt governments in the Caribbean. Tourism's power relations, then, are spatially complex and exclusionary; tourism is the international activity par excellence that spatially concentrates and divides people (tourists, workers, managers, and locals) with extremely diverse economic resources and cultural identities from various local and international locations. Today, the urgency and vitality of tourism as an established field of study in geography is such that any claims that accord with the discipline's past tendency to regard tourism as a "trivial" topic now are not only antiquated but also laughably irrelevant.

—*Paul Kingsbury*

See also Cultural Geography; Globalization

Suggested Reading

Britton, S. (1991). Tourism, capital, and place: Towards a critical geography of tourism. *Environment and Planning D, 9,* 451–478.

Coleman, S., & Crang, M. (Eds.). (2003). *Tourism: Between place and performance*. New York: Berghan Books.

Hanna, S., & Del Casino, V. (Eds.). (2003). *Mapping tourism: Representation, identity, and intertextuality*. Minneapolis: University of Minnesota Press.

Shaw, G., & Williams, A. (2002). *Critical issues in tourism: A geographical perspective* (2nd ed.). Malden, MA: Blackwell.

TRADE

International trade is defined as the movement of goods and services between countries. As such, international trade is an example of spatial interaction. International trade occurs when one country purchases goods or services from another country. Why do countries engage in international trade? Why does each country not simply produce all of the goods and services that it requires internally within its own borders? One answer to this question can be found in the writings of David Ricardo, a 19th-century English economist. Ricardo developed the principle of comparative advantage to explain the benefits to be gained by countries from engaging in international trade. According to this principle, a country should produce those products that it is best (most efficient) at producing and export them to other countries. Conversely, a country should import those products that it is less efficient at producing. By specializing in the production of those products that it can produce most efficiently, a country puts its resources (e.g., land, labor) to their optimal use. In a free market economy, the principle of comparative advantage should ensure a global production system whereby goods and services are produced at their optimal locations. The world economic system, however, is not a purely free market, and there are a number of factors (e.g., trade barriers) that distort the pattern of production and trade that might occur under free market conditions.

PATTERNS OF INTERNATIONAL TRADE

In 2000, the total value of goods and services exchanged between countries as a result of international trade was roughly $6.9 trillion. Approximately 80% of this trade involved the exchange of goods, with services accounting for the remaining 20%. Nearly half of merchandise (goods) trade involves office and telecommunications equipment, automotive products, chemicals, food, and fuels. A relatively small number of countries account for the majority of international trade. Indeed, 10 countries—including the United States, Japan, and Western Europe—account for nearly 60% of trade in goods. Most international trade takes place between countries whose economies are highly developed. The United States is the world leader in merchandise trade, accounting for 12% of exports and 18% of imports. The same countries that dominate merchandise trade also dominate international trade in services, with the United States again being the lead exporter and importer.

When a country engages in international trade, it prefers that its exports exceed its imports. When this occurs, international trade generates income for a country. For many countries, however, the value of their imports exceeds the value of their exports. These countries experience what is termed a trade deficit. For example, the 2005 trade deficit of the United States (the excess of imports over exports) exceeded $700 billion. The United States has trade deficits with every one of its major trading partners. During recent years, the size of the U.S. trade deficit with China has grown significantly as the result of China's growing importance as a manufacturing location. China's large population provides it with the world's largest labor force. Chinese manufacturing wages are approximately 2% of what they are in the United States. Today, Americans and Europeans purchase a wide range of products (everything from toys to television sets) that are manufactured in China. Increasing imports into a country are an indication that consumers prefer to purchase goods and services produced in other countries rather than goods and services produced in their own country. This generally occurs when goods and services produced in other countries are cheaper and/or of better quality than domestically produced goods and services. This, however, has serious implications for the economy of the importing country. Falling sales for domestic businesses can result in the need to lay off workers, resulting in higher unemployment levels. When a company (or sometimes a whole industry) sees that its sales are being negatively affected by imports, it often turns to its government for assistance.

FREE TRADE AND BARRIERS TO TRADE

Free trade is the idea that goods and services should be traded between countries without any interference from the governments of those countries. Truly free trade, however, rarely exists. The governments of countries erect both tariffs and quotas that affect the free flow of goods and services between countries. A tariff is a tax that must be paid when a particular good or service is imported into a country. A quota is a numerical limitation placed on the volume of a particular good or service that can be imported into a country. Governments establish tariffs and quotas to protect industries (and jobs) within their own countries.

For example, during the early 1980s the United States placed a quota on the number of automobiles that Japanese automakers could export to the United States. In 1970, only 3.7% of new automobiles sold in the United States were imported from Japan. During the 1970s, however, the popularity of Japanese automobiles among American consumers increased significantly. Japanese cars were generally less expensive, more fuel efficient, and more reliable than automobiles produced in the United States by American automakers. As a result, imports of automobiles from Japan increased, and by 1980 nearly 20% of all new cars sold in the United States were imported from Japan.

The increased popularity of Japanese automobiles meant decreased sales for American automakers. Decreased sales meant that some American autoworkers were laid off and forced to join the ranks of the unemployed. In 1981, one U.S. automaker, Chrysler Corporation, nearly went bankrupt. Faced with declining sales, U.S. automakers asked the government to take action on their behalf. The government responded, and in 1981 a quota of 1.68 million vehicles per year was placed on Japanese automobile exports to the United States. In 1985, the quota was increased to 2.3 million vehicles per year. By establishing quotas, the U.S. government intended to limit the supply of Japanese automobiles available to American consumers. Limiting supply in an environment of increasing demand should have the effect of raising the price of imported Japanese automakers. Rising prices for Japanese automobiles should make them relatively less attractive to American consumers, who (the government hopes) would be more likely to purchase American-made automobiles instead. If Americans bought more American-made automobiles, more American jobs would be saved. Estimates suggest that the quotas had the impact of increasing the average price of an imported Japanese automobile by approximately $1,000. At the same time, it was estimated that the quotas saved approximately 26,000 American jobs in the automobile industry.

Another factor that influenced the price of imported Japanese automobiles during the 1980s was the increasing value of the Japanese yen relative to the U.S. dollar. This had the impact of making imported Japanese automobiles more expensive to American consumers. To circumvent the impact of the quotas (and the rising yen), Japanese automakers took two distinct measures. First, they built their own automobile assembly plants in the United States. Japanese automobiles that were assembled in the United States were not subject to quotas. Today, every Japanese automaker assembles automobiles in the United States. This has reduced the import of automobiles from Japan and increased the number of automobiles made inside the United States by American workers. By the year 2000, fewer than 10% of the new automobiles sold in the United States were assembled in Japan. The second tactic employed by Japanese automakers to circumvent the negative impact of quotas was to start exporting more expensive luxury automobiles (e.g., models such as Acura, Infiniti, and Lexus). The profit margins on luxury automobiles are higher than those on nonluxury models.

Quotas, similar to those placed on the export of Japanese automobiles to the United States, abound in the world economy. In some cases, quotas represent a complex web of quota agreements among large numbers of countries. A good example is the Multi-Fiber Agreement (MFA), which went into effect in 1971. The MFA established quotas on the trade of textile products from all of the major exporting countries to all of the major importing countries.

THE WORLD TRADE ORGANIZATION

The World Trade Organization (WTO) was founded in 1995. The stated objective of the WTO is to help trade flow smoothly, freely, fairly, and predictably. To achieve its objective, the WTO performs a number of functions, including administering trade agreements between countries, acting as a forum for trade negotiations, and settling trade disputes. Prior to the establishment of the WTO, world trade was regulated by a series of negotiated agreements that were made within the framework of the General Agreement on Tariffs and Trade (GATT). The first GATT was signed in 1947 by only 23 countries. Today, nearly 150 countries belong to the WTO.

TRADING BLOCS

Recognizing the benefits to be gained from free trade, a number of countries have entered into agreements with other countries that have manifested themselves in the form of what are termed *trading blocs.* A trading bloc generally consists of countries that are located in the same geographic region. There are more than two dozen trading blocs functioning in the world today, including the European Union, the North American Free Trade Agreement, Mercosur, the Association of South East Asian Nations (ASEAN), the Arab Common Market, and the Andean Community. Recognizing the benefits to be derived from free trade, trading blocs are established to break down barriers to trade and to encourage and facilitate trade among bloc members. The members of a trading bloc reduce or remove trade barriers among themselves. At the same time, trade barriers are maintained between the trading bloc and countries outside the trading bloc. Trading bloc membership provides unhindered access to a larger market for companies located inside the trading bloc.

The European Union, the world's largest and most advanced trading bloc, provides a good example of the benefits of membership. The European Union abolished trade barriers among member counties starting as far back as 1968. Since then, it has made progress on other integrative measures such as the introduction of a single currency (the Euro) in 2002. The largest beneficiaries of trading blocs often are small countries whose internal markets can be very limited in size. For example, Luxembourg, with a population of only a half million, companies located in the country normally would have access to a very small internal market. However, because Luxembourg is a member of the European Union, companies with business facilities located in that country have access to a market of 456 million.

—*Neil Reid*

See also Comparative Advantage; Economic Geography; Globalization; Terms of Trade; Transnational Corporations; World Economy

Suggested Reading

Krugman, P. (1994). *Rethinking international trade.* Cambridge: MIT Press.

TRANSNATIONAL CORPORATIONS

Also called multinational corporations, transnational corporations (TNCs) are among the most powerful agents in the world economy today. They play a dominant role in international trade, investment, finance, development, technology transfer, and geopolitics.

The origins of TNCs may be traced back to the chartered monopolies that arose with the global expansion of capitalism during the 16th century. Many played key roles in colonial conquests and administration such as the British Hudson Bay Company in Canada, the British East India Company in India, the British West India Company, and the Dutch East Indies Company in South Africa and Indonesia. Such institutions were important in the establishment of plantations and the commodification of labor in many regions. However, the decades after World War II mark the period when TNCs reached new heights of wealth and power. Given the unsurpassed size and importance of the United States during the *Pax Americana,* American TNCs have been disproportionately represented among the world's largest firms. In 2001, of the world's 100 largest TNCs, 30 were American, 18 were Japanese, 11 were British, 11 were French, and 9 were German. More recently, some TNCs have emerged from newly industrializing countries. Today, some 30,000 TNCs control two thirds of world trade and employ 100 million people. Many of these are household names across the globe such as Gulf, Mitsubishi, Hyundai, Fujitsu, AT&T, Citicorp, and Hilton Hotels. Some TNCs have annual revenues larger than the gross national products (GNPs) of countries; for example, Exxon–Mobil, with sales of $235 billion annually, exceeds the GNP of each of Saudi Arabia, Chile, and Venezuela.

TNCs make a variety of investments overseas. The best known of these is foreign direct investment (FDI), which involves tangible assets either in newly constructed plants and factories or via acquisitions. In 2001, global FDI surpassed $1.4 trillion. However, other forms of foreign investment include intangibles such as purchases of foreign stocks and bonds (both public and private). Often, TNCs engage in joint ventures with other firms, including domestic ones, in the forms of strategic alliances or purchases of subsidiaries. TNCs are represented in many sectors of the economy, including the primary sector (e.g., petroleum, mining, lumber, food processing), the secondary sector (e.g., automobiles, electronics, pharmaceuticals), and the tertiary sector (e.g., banking, hotels, airlines, telecommunications).

Because they operate in more than one country, TNCs typically create an intracorporate division of labor that allows them to minimize costs and maximize efficiency. The most common form of this phenomenon is to locate their headquarters in their country of origin, generally in a large metropolitan area in which high-wage, white-collar administrative, managerial, and research functions have access to the agglomeration economies that such regions offer. In contrast, TNCs often relocate less-skilled, low-wage, blue-collar assembly functions in lower-income countries via networks of branch plants. However, despite the stereotype that all TNCs are global behemoths with operations in every continent, the reality is that most such institutions invest in only two or three countries. Firms such as Ford Motor Company, with plants throughout the world, are the exceptions rather than the norm. Moreover, contrary to the stereotype that TNCs always seek out low-wage pools of Third World labor, the reality is that most TNCs, which originate in capital-rich developed countries, invest in other capital-rich developed countries. For example, most U.S. TNCs invest in Canada and Europe, and British and German TNCs invest most heavily in Europe and North America. Japanese TNCs tend to invest in East Asia, indicating that national origins do matter and that something of a distance decay function away from the home country exists. Among developing countries, China is by far the largest recipient of TNCs' FDI, with $50 billion in 2003, followed by Mexico, Brazil, Singapore, Russia, Argentina, and Chile.

TNCs have a variety of motivations to invest abroad. Obviously, lower labor costs figure prominently; wages in much of the developing world may be as low as 5% of those in the West. Moreover, workers in many developing countries often are likely to be relatively docile and compliant, with few powerful unions, although this varies over space and time. However, low labor costs alone do not explain the geography of FDI; labor productivity is also essential. A second motivation to invest abroad is that FDI allows TNCs to avoid protectionist tariffs and quotas and allows them to minimize the impacts of exchange rate fluctuations. Finally, many governments, particularly in the age of neoliberalism and hypermobile capital, eagerly offer incentives to lure TNCs, including tax breaks and holidays, infrastructure subsidies, and labor training programs. Often such programs are linked to export processing platforms.

The costs and benefits of TNCs' presence, especially in developing countries, have been hotly debated. Advocates of free trade, typically working from the perspective of neoclassical economics, tend to stress the benefits of TNCs, including the generation of jobs, technology transfer, rising skills, higher productivity, and improved exports, foreign revenues, and terms of trade. Critics of TNCs, often invoking perspectives such as dependency theory, generally hold that such firms often are capital intensive, generating relatively few jobs; that skilled managerial, technical, and executive positions are filled by residents of the home country; that TNCs may drive local producers into bankruptcy (i.e., job displacement is greater than job creation); that the profits of foreign operations often are repatriated to the TNCs' home country; that the technologies transferred may be inappropriate to the conditions of the host country (e.g., grasslands agricultural equipment in a tropical ecosystem); that nonlocal subcontracts yield very low local employment or output multipliers, minimizing the potential for long-term development; that government subsidies and concessions to TNCs deprive them of the ability to provide public services; and that TNCs have a long history of political interference, including bribery, corruption, and the invoking of foreign powers, particularly the United States, in coups d'état against what they perceive as unfriendly regimes. Such criticisms amount to an argument that TNCs perpetuate a state of neocolonialism.

The impacts of TNCs are likely to be contingent on temporally and geographically specific mixtures of both costs and benefits that reflect the bargaining power of both TNCs and the host country involved. Acknowledging this complexity means situating them conceptually between the extremes of stereotypes that portray them, as harmless apolitical investors facilitating a comparative advantage, on the one hand, and as brutally exploitative parasites, on the other.

—Barney Warf

See also Dependency Theory; Developing World; Economic Geography; Export Processing Zones; Factors of Production; Global Cities; Globalization; Neocolonialism; Neoliberalism; Trade; World Economy; World Systems Theory

Suggested Reading

Barnet, R., & Muller, R. (1974). *Global reach: The power of the multinational corporations.* New York: Simon & Schuster.

Chandler, A., & Mazlish, B. (Eds.). (2005). *Leviathans: Multinational corporations and the new global history.* Cambridge, UK: Cambridge University Press.

Dicken, P. (2003). *Global shift: The internationalization of economic activity* (4th ed.). New York: Guilford.

Tolentino, P. (2001). *Multinational corporations: Emergence and evolution.* London: Routledge.

TRANSPORTATION GEOGRAPHY

Long an important subfield of human geography, transportation geography seeks to understand the movement of people and goods between locations and the spatial organization of the systems that facilitate these exchanges. Traditionally, transportation geography emphasized topics such as the analysis of transportation networks (e.g., highways, airline route structures), the development of models for estimating the amount of spatial interaction between locations (e.g., the gravity model), and the effects of policy on the spatial provision of transport services (e.g., airline deregulation). The work of transport geographers routinely overlaps with other areas of human geography, most notably urban and economic geography. Moreover, transport geography is a highly interdisciplinary subfield of human geography in that its subject matter frequently brings its practitioners into contact with civil engineers, computer scientists, environmental specialists, and urban and regional planners. Furthermore, the use of geographic information systems (GIS) to tackle questions in transportation geography is widespread, and so-called geographic information systems for transportation (GIS-T) have become a fruitful area of research and application. Contemporary transportation geographers investigate a broad range of social problems, ranging from ameliorating traffic congestion and related automobile fuel consumption to understanding the global implications of transportation systems.

HISTORICAL EVENTS SHAPING CONTEMPORARY TRANSPORT GEOGRAPHY

Research in transportation geography ranges from the theoretical to the applied, addressing issues at multiple scales in both developing and developed nations. Several streams of contemporary research in transportation geography can be traced to important developments that occurred decades ago. For example, much of the current urban-oriented transport geography can be traced to events of the 1950s and 1960s. One of the most critical of these events was the Chicago Area Transportation Study (CATS) conducted during the 1950s. This project was the first large-scale transportation study, and among its goals was developing predictive models of traffic flows in the Chicago metropolitan region. The methods established in the CATS have been refined over the years and applied in many other places. In fact, predictive traffic modeling generally has persisted to the present as a major transportation research focus. As a second related example, research on urban housing and labor markets during the 1960s laid a theoretical foundation for a generation of urban transportation geographers to address basic issues of residential and employment location choice and their implications for people's travel behavior.

Long a popular topic, the analysis of transportation networks, approached by abstracting highway, airline, rail, and other forms of transportation infrastructure to their essential geometric objects (i.e., representing them as points and lines), grew out of earlier work during the late 1960s and early 1970s. Many quantitative tools rooted in principles of analytic geometry and linear algebra were created to measure the spatial properties of networks such as their internal connectivity, their redundancy, and the accessibility they provided. Some of these tools have been incorporated into the analytic capabilities of modern-day GIS, thereby making these systems available to users of the software for implementation in various problem-solving situations.

Transportation geographers continue to experiment extensively with the spatial interaction model (or gravity model). A series of books and articles during the 1950s and 1960s described how the gravity model of planetary physics could be adapted and applied to problems of a geographic nature. The range of geographic questions that have been addressed with gravity models since those innovations is vast, although a commonality of these applications is the notion that interaction between locations can be estimated based on a proxy for the size and types of activities at locations and the geographic distances between them. This approach has had wide appeal for addressing spatial transportation problems. It has provided geographers with a means of estimating vehicle traffic flows between parts of a city, predicting trade volumes of goods between nations and air traveler exchanges between major cities, and determining interactions in many other types of scenarios.

TRENDS AND RESEARCH QUESTIONS IN CONTEMPORARY TRANSPORT GEOGRAPHY

Many of the questions that have intrigued transport geographers for decades still attract great attention today. Improving and applying techniques of network analysis, for instance, continues to be an active research area. It has been fueled by recent interest in measuring the spatial structure of commercial telecommunications and Internet infrastructure as well as by governmental need to identify and protect critical network infrastructure (e.g., bridges, pipelines) in light of terrorism concerns post-2001.

The movement of freight flows around the globe continues to draw attention from transport geographers, particularly as manufacturing activities increase in size and scope in developing countries. The characteristics and spatial configuration of intermodal facilities that handle the interchange of goods from one transportation mode to another (e.g., air to rail) are of great interest to transportation geographers. Transport geographers have examined in detail some of the firms involved in global logistics (e.g., Federal Express, United Parcel Service).

Several new research foci have emerged over the past decade. For example, there is interest in understanding the extent to which modern-day transport systems, and the institutions that shape them, are sustainable. One finds sustainability questions at the forefront of many current research efforts and underpinning the aims of numerous others. The sustainability movement, as it pertains to transportation, picked up momentum during the late 1990s and stresses improving transportation systems to minimize their environmental, social, and economic impacts. For example, issues such as developed nations' heavy reliance on automobiles for personal travel present a challenge to sustainability given that the consumed fuel resources are limited and automobiles contribute to congestion and pollution.

The idea that some people in society are unable to effectively access the goods and services they need to fulfill their life needs (e.g., food, healthcare) also calls into question the sustainability of modern transport systems. Recently, transport geographers have explored variations on this issue with respect to the concept of accessibility and have sought to identify the factors that make it difficult for people to reach the places they need to reach, including analyzing how people's personal constraints (e.g., whether they are employed, whether they have children, whether they have fixed appointments) affect their abilities to access locations in space. In this equation, transportation systems must be considered because if automobiles are not available, individuals must rely on other means to access what they need such as public bus transport. Accordingly, transportation geographers' work has sought to improve the performance and operations of public transportation systems to better serve individuals.

Very recently, transportation geographers have suggested that if needed goods and services are systematically inaccessible to individuals and groups within society, research must probe whether these people are excluded and not participating in society to the fullest extent possible. Thus, a substantial body of new research in transportation geography that seeks to define and understand the nature of social exclusion in urban areas is being produced. Similar to other topics in transportation geography, recent research on social exclusion is highly interdisciplinary, drawing contributions from engineers, planners, sociologists, and urban specialists.

Technology itself has greatly influenced the way in which transportation geographers conduct research and also has presented new issues for them to study. Today, geographic information systems are widely used to address problems in transport geography. This is because the functionality of GIS technology has vastly improved over the past decade, allowing a broader range of transport research questions to be considered. At the same time, computing technology has become more affordable, and the proliferation of the commercial Internet has provided GIS users with a convenient and inexpensive way in which to exchange information. Finally, the current environment can be characterized as data rich because a substantial number of government and private firms use GIS in their work and create spatial databases that are shared regularly and freely among transport researchers throughout the world.

Very recently, geographers and many others have begun asking questions about the collective effects of information and communications technologies (ICTs) on transportation. Some of these technologies include GIS and the Global Positioning System (GPS), but also germane to this discussion is the commercial Internet and the routing and navigating technologies it allows people to access as well as personal devices such as cellular telephones. An important question is whether these new technologies ultimately will help to make transportation systems more accessible and easier to use or whether they will simply reinforce existing problems with accessing transportation services.

—*Mark W. Horner*

See also GIS; Gravity Model

Suggested Reading

Black, W. (2003). *Transportation: A geographical analysis.* New York: Guilford.

Hanson, S., & Giuliano, G. (2004). *The geography of urban geography* (3rd ed.). New York: Guilford.

Taaffe, E., Gauthier, H., & O'Kelly, M. (1996). *Geography of transportation* (2nd ed.). Englewood Cliffs, NJ: Prentice Hall.

TRAVEL WRITING, GEOGRAPHY AND

Human geographers have a long-standing and rather basic interest in "travel" insofar as travel of a sort is assumed within a wide range of people's movements, from large-scale migration flows and settlement patterns to a host of everyday activities such as the journey to work. However, studies of travel *writing*—written representations of journeys and experiences away from home—form a distinct but eclectic body of critical scholarship that helps us to understand relationships between places traveled *from* and places traveled *to*. Descriptions of people, places, and experiences in travel writing also provide a window to the effect that travel has on the personal subjectivity of the writer. Popular mainstream interest in travel writing underwent a renaissance of sorts during the late 20th century, as display counter after display counter at bookstores attested to the newly commodified forms of travel and tourism. Scholars of travel writing, working in a range of disciplines such as literary criticism, anthropology, geography, social history, and postcolonial and cultural studies, have been concerned with travel writing as both cultural practice and cultural product. Importantly, they recognize that travel texts do not merely reflect some taken-for-granted reality about places; they also help to constitute that reality.

A heterogeneous array of texts might be categorized as travel writing, including exploration accounts, settler diaries and letters, war correspondence, reports of religious pilgrimages, wilderness and nature writing, and the travel sections of newspapers and magazines. Nonetheless, many scholars recognize travel writing or the "travel book" as a historical genre that emerged during the 18th- and 19th-century bourgeois era. Thus, travel writing was distinguished as "autobiography in motion," that is, a type of text that combines a narrative figure as protagonist (whether foregrounded or impersonal) with descriptions of places, people, events, and objects. The emergence of the travel book or essay coincided with new forms of mass transportation (especially the steamship and railroad during the 19th century), new wealth available to a leisure class of travelers concerned with their own self-improvement and education, and new global relationships and interactions that provided contexts for travel. Examples included working in colonial administration, visiting relatives who had emigrated, and visiting fashionable tourist destinations, such as Yosemite and Yellowstone National Parks, accessible through the transcontinental "grand tour" of North America by railroad.

Mode of transportation affects the travel experience and thus can produce new types of travel writing and perhaps even new "identities." Modes of transportation determine the types and duration of social encounters; affect the organization and passage of space and time; determine itineraries, destinations, rhythms of travel; and also affect perception and knowledge—how and what the traveler comes to know and write about. The completion of the first U.S. transcontinental highway during the 1920s (Route 30, the "Lincoln Highway"), for example, inaugurated a new genre of travel literature about the United States—the automotive or road narrative. Such narratives highlight the experiences of mostly male protagonists "discovering themselves" on their journeys, emphasizing the independence of road travel and the value of rural folk traditions.

Travel writing's relationship to empire building—as a type of "colonialist discourse"—has drawn the most attention from academicians. Close connections have been observed between European (and American) political, economic, and administrative goals for the colonies and their manifestations in the cultural practice of writing travel books. Travel writers' descriptions of foreign places have been analyzed as attempts to validate, promote, or challenge the ideologies and practices of colonial or imperial domination and expansion. Mary Louise Pratt's study of the genres and conventions of 18th- and 19th-century exploration narratives about South America and Africa (e.g., the "monarch of all I survey" trope) offered ways of thinking about travel writing as embedded within relations of power between metropole and periphery, as did Edward Said's theories of representation and cultural imperialism. Particularly Said's book, *Orientalism,* helped scholars understand ways in which representations of people in travel texts were intimately bound up with notions of self, in this case, that the Occident defined itself through essentialist, ethnocentric, and racist representations of the Orient. Said's work became a model for demonstrating cultural forms of imperialism in travel texts, showing how the political, economic, or administrative fact of dominance relies on legitimating discourses such as those articulated through travel writing.

Feminists have taken the study of travel writing as colonialist discourse one step further by exposing it as a deeply gendered practice. Feminist geographers' studies of travel writing challenge the masculinist history of geography by questioning who and what are relevant subjects of geographic study and, indeed, what counts as geographic knowledge itself. Such questions are worked through ideological constructs that posit men as explorers and women as travelers—or, conversely, men as travelers and women as tied to the home. Studies of Victorian women who were professional travel writers, tourists, wives of colonial administrators, and other (mostly) elite women who wrote narratives about their experiences abroad during the 19th century have been particularly revealing. From a "liberal" feminist perspective, travel presented one means toward female liberation for middle- and upper-class Victorian women. Many studies from the 1970s onward demonstrated the ways in which women's gendered identities were negotiated differently "at home" than they were "away," thereby showing women's self-development through travel. The more recent poststructural turn in studies of Victorian travel writing has focused attention on women's diverse and fragmented identities as they narrated their travel experiences, emphasizing women's sense of themselves as women in new locations, but only as they worked through their ties to nation, class, whiteness, and colonial and imperial power structures. Critics have demonstrated the heterogeneous, ambivalent, and (sometimes) counter-hegemonic positions such women occupied with respect to colonized people. Their work in turn helped to further postcolonial studies of travel writing by shifting focus to the experiences, voices, and resistances of the colonized subjects of travel writing.

In general, geographers are uniquely positioned to sensitize other scholars to the need to register the materiality of spaces through which travel writers move. Geographers are attuned to the importance of the *spatiality of representation*—the multiple sites at which travel writing takes place and the means by which travelers engage them.

—*Karen Morin*

See also Cultural Geography; Spaces of Representation; Text and Textuality; Tourism, Geography and/of; Writing

Suggested Reading

Duncan, J., & Gregory, D. (Eds.). (1999). *Writes of passage: Reading travel writing.* London: Routledge.

Pratt, M. (1992). *Imperial eyes: Travel writing and transculturation.* London: Routledge.

Said, E. (1978). *Orientalism.* London: Routledge & Kegan Paul.

Smith, S. (2001). *Moving lives: 20th-century women's travel writing.* Minneapolis: University of Minnesota Press.

Speake, J. (Ed.). (2003). *The literature of travel and exploration* (3 vols.). London: Fitzroy Dearborn.

U

UNCERTAINTY

Geography refers literally to representations of phenomena on the earth's surface and near surface. Traditionally, geographic information is communicated with paper maps that depict various spatial entities and phenomena. Using computers, geographic information systems (GIS) have revolutionized the way in which maps are produced and used, thereby permitting great efficiency in processing a large quantity of spatial data for numerous applications.

However, maps in both analog and digital formats may disguise unseen generalization, selection, and approximation typical in the process of mapmaking. Moreover, due to measurement error, the input data sources themselves are subject to inaccuracy, as exemplified in photogrammetric plotting and classification of remotely sensed images. In a GIS environment where map overlaying and predictive modeling are performed, errors in input map layers will be propagated to the resultant maps, usually in complex ways. Thus, spatial data and their derivatives are subject to error, raising questions as to whether map users are willing to accept map data at their face value and what confidence limits should be placed on information products derived from imperfect data. As GIS applications are expanding at an unprecedented scale, answers to such questions are needed more urgently than ever before, and decades of research into error issues have produced some constructive results.

It is important to consider how the real world is conceptualized and, for the purpose of digital geoprocessing, made discrete. Conceptually, the real world can be considered as populated by discrete entities or as consisting of single-valued functions defined everywhere in space, known as objects and fields, respectively, in geographic information science (GIScience). The object-based models clearly are suited to represent well-defined features, such as human-made structures, human settlements, and transportation infrastructure, and have played an important role in cartography. On the other hand, the field-based models are better geared to represent spatially varied phenomena such as land cover, where class labels can be assigned at pixels or patches of land, and elevation, which can be defined for every location on the terrain surface. Clearly, fields are important data models in physical geography, and many phenomena about humans, human activities, and the space and/or structures used to conduct human activities are making greater use of fields as well as objects for representation and geoprocessing.

In object models, spatial entities are represented by position and attributes. Often it is the positional data that are examined with respect to error in objects, whereas attribution is assumed to be the responsibility of specialists such as census workers. There are three types of objects in terms of geometry: points, lines, and areas. Positional errors in points and lines usually are described by error ellipses and epsilon error bands, respectively, whereas positional errors in area objects are described by their boundary inaccuracy, which is effectively a combination of errors in lines constituting areas' boundaries. In field models, errors in variables measured at a continuous scale are described by standard deviations, and on the assumption of Gaussian distribution it is possible to state that the true value is within the interval of measured or estimated values plus or minus twice the standard deviation with a probability of 95%. For field variables that represent discrete labels, probability can be used to describe the accuracy of labeling at certain locations. For continuous data, including positional data and discrete data, accuracy assessment is commonly performed on the basis of independent reference data, which usually are of higher accuracy, to act as ground truth to benchmark the discrepancy of a test data set from it. Root mean squared errors (RMSEs) usually are calculated to measure the closeness of a data set to the reference, whereas percentage correctly classified (PCC) pixels that result from summarizing a confusion matrix measure the likelihood that a randomly chosen pixel classified as a certain class actually belongs to that class according to reference classification.

Although conventional statistics provides some of the useful tools for describing error in spatial data, it is widely recognized that spatial data and their errors should be handled in a spatially explicit way. Due to the presence of spatial dependence in both spatial data and their errors, conventional statistics might not be suitable for quantifying variance in spatial variables or their derivatives because confidence limits assuming spatial independence tend to be unduly narrower than in actuality. Geostatistics is specifically developed to deal with spatially correlated phenomena. A spatial prediction method commonly known as *kriging* can provide the best linear unbiased estimation for a variable at unsampled locations on the basis of sampled locations, assuming stationarity in means and covariance. An advantageous aspect of kriging is that it provides location-specific estimation of kriging variance in addition to its estimation of the underlying variable, reflecting the dispersion of prediction from the true values.

Beyond spatial prediction, geostatistics can be applied for quantifying spatial uncertainty. Consider the case of error modeling in slope calculated from digital elevation data. Knowledge of location-specific estimation about elevation and variance is not adequate for predicting the variance in the resultant slope estimate because that would require information on the spatial covariance between the elevation values at the locations concerned. Another example is about estimating variance in areal extents of a certain class, say, Class A, as predicted from classification of remote sensor images. Again, knowledge of pixel-specific PCCs does not permit calculation of variance of the joint occurrence of Class A in the pixels labeled as

"A" over the problem domain. To facilitate error modeling in complex geoprocessing, stochastic simulation is needed to characterize error propagation, especially when it implies operation in a spatial neighborhood and involves multiple variables. This is because stochastic simulation can generate a large number of equally probable realizations for each input map that are subjected to specific geoprocessing, such as map overlaying, resulting in many versions of output maps, each representing a possible version of the incompletely known world and, as a whole, carrying information about uncertainty in the given data set or maps. That way, spatial uncertainty in resultant information products can be quantified flexibly.

The concept of uncertainty has evolved from conventional notions of error and precision to a broader one implying both error of a random nature and vagueness due to inherently poorly defined spatial and semantic extents, with the former rightly approached by probability theory and the latter by fuzzy sets describing membership values in the interval of 0 and 1. Both probability theory and fuzzy sets are useful in dealing with multifaceted uncertainty. There are many geographic examples that demonstrate the relevance of both in dealing with imperfect data and information, including the following: An error ellipse for a measured location implies probabilistic interpretation about the positional accuracy, whereas the concept of urbanization and its boundaries certainly have fuzzy implications.

It is anticipated that uncertainty as an important dimension of geographic information will become more and more easily implemented in routine spatial analysis, especially in GIS environments, as many of the techniques that formerly were scattered in some specialist groups around the world have come along, and will continue to come along, to various applications of geographic information. The vision of seeing the maps not only as single representations but also as realizations drawn from many possible versions is central to a new era of geography and information society.

—Jingxiong Zhang

See also GIS; Quantitative Methods

Suggested Reading

Zhang, J., & Goodchild, M. (2002). *Uncertainty in geographical information.* London: Taylor & Francis.

UNDERDEVELOPMENT

As a compound word, *underdevelopment* conveys that something is lacking, suggesting development that is of a lesser degree or somehow insufficient. Within the body of development thought, underdevelopment generally has been conceptualized in one of two ways. On the one hand, some have used the term to suggest the state of a country or region prior to economic development. This view was associated with modernization theories of development, which posit a linear form of progress in which developed countries show the way for underdeveloped countries. In this case, underdevelopment is a condition that development can redress. On the other hand, a more widespread understanding of the term refers to the theories of underdevelopment that were first postulated during the 1950s, albeit with longer roots in 19th-century discussions of regional economic difference. These theories argued that the condition of underdevelopment would not be alleviated by the *diffusion* of capitalist development but instead had been caused by processes of *uneven* capitalist development. From this perspective, underdevelopment in one place is part of the process and is a destructive result of development elsewhere.

Although the term *underdevelopment* is associated most closely with 20th-century dependency theory, its conceptualization finds close parallels in 19th- and early-20th-century discussions that took place in Russia over economic reform in the context of Russia's comparative economic backwardness. There the debate hinged on whether the country, as a latecomer in the capitalist world economy, was at such a disadvantage in securing export markets and in developing domestic markets that the prospect of capitalist industrialization was foreclosed. That question led to the critical debate—prompting Vladimir Lenin's theory of imperialism—as to whether Russia could advance directly from feudalism to socialism without first passing through a stage of capitalist industrialization.

Theorization of underdevelopment proceeded apace following World War II in the writings of scholars such as Paul Baran and Andre Gunder Frank. In theorizing the roots of backwardness, Baran stressed the importance of transfers of wealth from non-European countries to European ones, thereby placing resources in the hands of European capitalists and providing an exogenous contribution that boosted Western Europe's development. On the other side of the ledger were the underdeveloped countries, which suffered a serious setback to their primary accumulation of capital. Once underdeveloped, these countries then found indigenous industrial expansion to be inhibited because with limited internal demand, manufactured goods were easily supplied from abroad and so there were few opportunities for profitable investment. Postcolonial societies' development was constrained because foreign investors and domestic elite appropriated the economic surplus without investing in local production for domestic needs. Baran drew primarily from the historical experiences of India and the Soviet Union, with Japan providing a counterexample as the only country in Asia, Africa, and Latin America that escaped being turned into a colony or dependency.

Frank was one member of a group of Latin American scholars whose work contributed to the development of dependency theory. Building on the work of Baran, among others, Frank argued that the historical development of the capitalist system generated underdevelopment in what he termed the *peripheral satellites,* whose economic surplus was appropriated while generating economic development in the metropolitan centers in which that surplus was accumulated. His metropolitan–satellite model stood for relations between developed and underdeveloped countries as well as, within underdeveloped countries, between town and country, with attention paid to agrarian structure. Frank's historical arguments were based primarily on the examples of Chile and Brazil, cases he used to argue that the problem was not that capitalism had failed to reach remote areas but rather that these regions had been linked to global capitalism for centuries, with the result being the development of underdevelopment and the underdevelopment of development. Because Baran and Frank were concerned with interdependencies between developed and underdeveloped countries, their work was foundational to the genesis of world systems theory.

Criticisms of this theorization of underdevelopment are numerous. First, critics point out that dependency theory merely inverts Adam Smith's idea of trade as progress by substituting the opposite, that is, trade as the means of underdevelopment. This is in part because exchange is given analytical priority over production relations. Second, underdevelopment seems to be as inevitable to dependency theorists as development is to neoclassical and neoliberal modernization theorists.

Third, the focus on the national scale neglected class relations both within and between countries. Finally, although dependency theory used historical examples, it focused on what did not happen—development—and generalized to the point of obscuring important differences among countries.

Another line of criticism targets both the dependency and modernization conceptualizations of underdevelopment, arguing that these two theorizations share the assumption that economic development is a desirable objective. That is, they simply differ on whether it is attainable within the global capitalist system. From an anti- or postdevelopment perspective, this economically centered view is challenged. These critics suggest that the very terms of the debate have marginalized other cultures, ideas, and values by discursively reducing the diversity of much of the world by labeling its condition as one of underdevelopment and, in so doing, preparing the way for potentially destructive or homogenizing development projects.

Questions of underdevelopment as a manifestation of processes of uneven development have been central to geographic research. Geographers are especially concerned with the ways in which capital moves geographically to exploit the conditions of underdevelopment such as low wages without incurring the cost of development, thereby using mobility as a spatial fix. Geographers have pointed out that theories of underdevelopment are inherently geographic, emphasizing space over class to highlight interregional relations and surplus transfer. They have also argued that underdevelopment theory has a normative component concerned with issues of social, spatial, and (increasingly) environmental justice. Moreover, as the wealth and income gap widens at all spatial scales, geographers now are critically engaged with analyzing uneven globalization, not only its causes but also its impacts such as marginalization, polarization, and underdevelopment.

—*Gail Hollander*

See also Dependency Theory; Development Theory; Economic Geography; Globalization; Marxism, Geography and; Uneven Development; World Systems Theory

Suggested Reading

Baran, P. (1957). *The political economy of growth.* New York: Monthly Review Press.

Cowen, M., & Shenton, R. (1996). *Doctrines of development.* London: Routledge.

Frank, A. (1967). *Capitalism and underdevelopment in Latin America.* New York: Monthly Review Press.

Peet, R. (1991). *Global capitalism: Theories of societal development.* London: Routledge.

Porter, P., & Sheppard, E. (1998). *A world of difference: Society, nature, development.* New York: Guilford.

UNEVEN DEVELOPMENT

Uneven development refers to the unequal distribution of people, resources, and wealth that is a fundamental characteristic of human geography. Uneven development is evident at the global, regional, national, and urban scales. Concern with uneven development stems from the fact that vast differences in human well-being exist at each of these scales.

Human geographers seek to describe and explain the processes by which uneven development is created. Furthermore, many human geographers are concerned with addressing the social injustices that are part and parcel of uneven development.

Early attempts to explain differential patterns of human geography relied on environmentally based explanations. Here the argument is that uneven development reflects the widely varying resource endowments of different locations. Thus, the rapid development of European industrialization was said to be the result of the availability of resources such as coal.

When taken to their extremes, such explanations were unabashedly determinist. Among their most egregious claims, environmental determinists contended that climatic variations were responsible for the allegedly pronounced spatial variation in human intelligence and behavior. According to Ellsworth Huntington, the climatic variability typical of mid-latitude climates produced vigorous and entrepreneurial people, whereas the unchanging tropical climate caused the lassitude and lack of ambition claimed to be characteristic of humans who lived in tropical locations.

Environmental determinism was not supported by empirical evidence and was soundly rejected by human geographers early in the 20th century. In its stead came theories such as possibilism. Possibilists argued that spatial patterns of uneven development were caused by differences in the ways in which humans used the resource endowments available to them. Spatial variation in human activity was not determined by natural resources; rather, levels of development depended on choices made in how to use available resources. However, possibilists also argued that some regions were better endowed with natural resources than were others.

Uneven development has also been studied through the lens of regional geography. Regional geographers contended that uneven development was the product of distinctive regional ensembles of culture and nature. Space was viewed as a set of containers, with each one holding a distinctive combination of natural resources and human culture. With this in mind, regional geographers set themselves with the task of identifying and explaining the evident (to them) patterns of regional differentiation. Uneven development was simply taken for granted as the natural outcome of the many unique combinations of culture and resource endowment.

The emphasis on regional geography largely disappeared during the quantitative revolution of the 1960s. Statistically oriented human geographers rejected the notion of unique regions. Instead, quantitative economic geographers argued that spatial variation in factors of production such as labor and transportation cost explained why some places specialized in the production of manufactured goods while other locations produced agricultural products. Therefore, uneven development is reflective of spatial variation in cost of production. Location choices are made according to economically rational criteria; whatever can be produced for the least cost (and thereby the most profit) is what will be produced in a particular location. According to this model, locational specialization occurs as each place concentrates on what it can produce most efficiently (i.e., at the least cost). Because it is most efficient for each place to specialize, patterns of trade develop and enable each location to secure what it needs while focusing on what it can produce most efficiently. It is argued that as economically efficient specialization increases over time, regional convergence will occur, thereby reducing—if not eliminating—uneven development. Uneven development is viewed as a temporary situation that will, in time, disappear.

During the 1970s, human geographers began to focus more on spatial variation in human well-being. Recognizing that uneven development meant that levels of human well-being varied considerably across the globe, some human geographers began asking whether a more just distribution of resources could be accomplished. Interest in issues of spatial justice was stimulated by the 1973 publication of David Harvey's *Social Justice and the City*. Explicitly rejecting his earlier reliance on spatial analysis, Harvey argued that uneven development (in this case at the urban scale) was an inherent part of the capitalist space economy. Harvey's concern with social justice, and his recognition that uneven development was produced by the capitalist economy, was a landmark event in human geography. Most noteworthy was the explicit recognition that there is a direct and intimate relationship between places that are well-off and places where the inhabitants suffer from the burdens of poverty, racism, and economic exploitation. The concern for social justice implications of uneven development spread (unevenly) throughout human geography, and many began to document patterns of spatial inequality. Particularly noteworthy was David Smith's book, *Where the Grass Is Greener*, which examined uneven development at a variety of spatial scales.

Interest in uneven development was further bolstered with the 1984 publication of Neil Smith's *Uneven Development*. Taking an explicitly Marxian approach, Smith argued that uneven development was a product of the contradictions inherent in capitalism. Thus, some places are sites of rapid and damaging disinvestment, whereas others receive flows of new capital that enable their rapid development.

Uneven development continues to be the spatial expression of social injustice. Global, national, and local conflicts are fueled by the growing spatial inequality between wealthy and poor areas as well as between wealthy and poor people. Perhaps when a more even development of human well-being is achieved, the world will become a more peaceful place.

—Jeff Crump

See also Dependency Theory; Social Justice; Underdevelopment

Suggested Reading

Harvey, D. (1973). *Social justice and the city*. London: Edward Arnold.

Smith, D. (1979). *Where the grass is greener: Geographical perspectives on inequality*. London: Croom Helm.

Smith, N. (1984). *Uneven development: Nature, capital, and the production of space*. Oxford, UK: Blackwell.

URBAN AND REGIONAL PLANNING

Urban and regional planning, in its essence, is concerned with deliberate efforts by societies to influence the geography of human activities. In modern societies, this typically—but not exclusively—has involved laws and regulations that seek to control the geography of land uses and the built environment as patterns and processes of human–environment relations have changed. Thus, planning, in its simplest formulation, is about managing and responding to change—what types of development will occur, where they will occur, when will they occur, and so forth. Most often, planning is described in terms of its role in protecting the health, welfare, and safety of a community. These catchall terms, however, can encompass both progressive and regressive policies. Of course, urban and regional planning is but a single facet in the complex web of both micro- and macro-scale processes that create and re-create places in the sociospatial dialectic. The specific dimensions of planning have reflected, and continue to reflect, the problems and potential solutions identified by society. At the core of planning practice are planners themselves. Planners, as a diverse community of professionals, illustrate many of society's tensions. Are planners simply technocrats implementing and managing the development of plans and land use regulations? Are they facilitators of the status quo or agents of regressive or progressive change? How concerned should planners be with issues of professional ethics and social and economic justice? Such questions have formed part and parcel of the evolution of urban and regional planning. They are also the core ideas in the codes of ethics adopted by the planning profession.

THE HISTORY OF PLANNING

The origins of city design and planning can be detected in many ancient societies and cultural traditions. In a more contemporary sense, the legal and jurisdictional evolution of urban and regional planning is highly variable—both between and within countries. Nevertheless, a collection of guiding principles and practices can be observed in the United States, Canada, and the United Kingdom. Planning, particularly at the municipal scale, appears to have evolved most significantly during the late 19th and early 20th centuries. The formalization and professionalization of planning, as one facet of societal response to the impacts of rapid industrialization and conflict over the external effects of incompatible land uses, can be easily traced. The instruments and practices of planning that emerge can be seen as one

component of the evolving sociospatial dialectic at various spatial scales.

In its early forms, the legal environment of planning was strongly influenced by the common law of nuisance. At a societal level, this involved the notion that there were certain economic activities that were incompatible with housing and other urban land uses; some of these were considered dangerous or unhealthy, whereas others simply were undesirable. Ordinances and regulations on these bases typically prohibited the activities inside of city boundaries in a series of collective NIMBY ("not in my backyard")-like actions.

As industrialization increased, broader concerns over health and safety were conjoined with efforts to respond to social, cultural, and economic conflict. In one dimension, planning became part of private responses to rapid social and economic change in the conceptualization and design of new suburban and exurban private developments and also in the utopian visions of thinkers such as Ebenezer Howard and his ideas of new garden cities. In a second public dimension, a concern for quality of life—whether conceived in idealistic utopian terms or in the practical designs for street patterns and transportation, water mains and sewers, public parks and open spaces, and the form and function of housing—brought architects, landscape architects, engineers, and an array of social and political reformers into the planning mix. In its physical manifestation, planning for infrastructure and public investment also had the potential for graft and corruption as it created positive and negative external effects. At the scale of cities, in their guise as agents of social and economic change, planners could be placed at the center of rapidly evolving capitalist social relations. The excesses of rapid economic growth and industrialization and the vicissitudes of capitalism in the urban context also drew the attention of national governments. Calls for regulation of laissez-faire economic practices included concerns over the central tenets of what evolved into urban and regional planning. It is important to note, however, that capitalists who might have opposed regulations and state oversight in general could support, and have continued to support, certain facets of planning that rendered investments in land and development more predictable and that might moderate the negative impacts of sudden land use changes.

Tangible facets of this stage of the maturation of planning were the development of professional associations, a system of national and state/provincial laws, and the proliferation of municipal (and some regional) planning processes. These typically have involved the development and implementation of plans for public and private actions. In the public dimension, this typically has involved planning for the location, timing, and financing of investments in infrastructure. In the regulation of the private sphere, ordinances and/or regulations have influenced the broad, and sometimes specific, patterns and timing of changes in land use and the built environment. In different countries, and even within particular countries, the extent to which cities and regions could exercise the controls over public and private actions has varied. For example, in the United States during the 1920s, standard legislation, generally called model legislation, was prepared by the federal government to enable states to enact planning laws. These would, in turn, allow local-scale planning processes to emerge in a consistent manner. Many states declined to enact such laws, and planning lagged in those geographic areas. In other states, where municipalities and rural communities were empowered to act, many did not act then and have not acted to this day. In a very real sense, there is a geography of planning; some states and jurisdictions exhibit innovation, whereas others do not. In the United Kingdom, national planning legislation following World War II changed the geographic scale of planning to make counties a more important component of land use regulation. Also in the United Kingdom, national-scale planning policies promoted the industrial development of some regions over others. Regional planning initiatives for economic development are also evident in the United States. Human geographers' interest in the complexities and interdependencies of geographic scale has drawn them to examinations of urban and regional planning.

THE TOOLS OF LOCAL-SCALE LAND USE PLANNING: ANGLO-AMERICAN EXAMPLES

Three basic tools of land use planning merit specific discussion: zoning, subdivision control, and eminent domain. The term *zoning* collectively characterizes a set of regulatory practices in which a community exerts control over what individual landowners (or groups of landowners) can and cannot do with their land. Typically, a long-term vision of human activities is captured in the tangible form of a comprehensive plan and a map of the allowed land uses. These support the system of ordinances and regulations designed to implement this long-term view. Initially, zoning systems were designed to separate activities that were deemed to be incompatible. In addition, particularly after the development of long-term amortized mortgages, zoning played a role in efforts to protect long-term investments. (These goals could also be accomplished to some degree by provisions in the deeds of individual properties through the use of restrictive covenants; zoning was a collective instrument to accomplish some of the same goals.) In its purest and earliest forms, zoning formalized the separation of industrial, commercial, and residential activities from each other. Residential activities were further subdivided into single-family homes, duplexes, and apartments. In the United States, for example, it was deemed a legitimate and constitutional action to separate single-family homes from duplexes and apartments and other businesses. Although some of the earliest advocates of zoning were directly concerned with the health and safety of lower-income groups in inner cities, zoning became the hallmark of class and racial division—albeit via indirect regulation. These facets of zoning would later become the basis for legal challenges to its exclusionary dimensions. In the United States, the availability of affordable and accessible housing has been a repeated theme in these challenges to exclusionary or "snob" zoning.

Zoning is not confined to the regulation of land use. Ordinances could also regulate the minimum size of individual parcels and the area, location, and physical characteristics of a building (or buildings) within a particular property parcel. An emphasis on uniformity within particular zones was typical.

In terms of the sociospatial dialectic, the separation of activities, regulated uniformity, and economic and racial exclusivity produced the types of neighborhoods and communities that many now decry as sterile, boring, and socially dysfunctional.

Contemporary zoning has moved away from homogeneity in land uses to some degree, with more flexibility on the mixing of land uses and building types within those land uses. This is seen most frequently in larger mixed-use developments rather than in zoning districts per se. However, it can also be seen in more flexibility in the locations of individual houses within a broader development. In some jurisdictions, exclusionary practices have been moderated by so-called inclusionary zoning, which, among other things, can offer density bonuses in return for the inclusion of affordable housing units.

Subdivision control is the second major tool of municipal land use planning. The premise of this facet of planning is that before a landowner can divide an existing parcel of land into separate saleable lots, a set of conditions must be satisfied. In most jurisdictions, a permit to subdivide is allocated in a two-stage process. First, the landowner proposing to subdivide the land provides a plan and information about such actions. This plan is evaluated in procedural terms (e.g., compliance with information requirements, design requirements, financial requirements), and then, in a second stage, the proposal is evaluated on its merits.

The content of this more substantive review can include myriad items such as street patterns and designs, alterations of topography, management of water courses, and soil erosion. Subdivision control sometimes is discussed in terms of public responsibility to protect naive land purchasers from unscrupulous sellers. This represents a clear exception to the caveat emptor ("let the buyer beware") doctrine and predates consumer protection laws more generally. The extent to which governments have exercised this responsibility, however, has varied considerably, with some jurisdictions exerting minimal control and allowing land speculators to turn quick profits with little or no protection for land buyers (the sale of swampland in Florida often is cited here) and with other jurisdictions requiring proof of the availability of water, sewage disposal, and so forth. During the past few decades in the United States, some states have enacted state-level review of significant residential subdivisions or commercial/industrial developments. In some jurisdictions, planning authorities have also regulated the development of individual buildings by using procedures similar to subdivision review through the tool of site plan review.

The third planning tool, most often linked with the provision of public utilities, transportation routes, and other public land uses, is *eminent domain* (known as *expropriation* in Canada and as *compulsory purchase* in the United Kingdom). Eminent domain allows a legally empowered entity (municipal, regional, or state/provincial agency) to take privately owned land for public use without the consent of the landowner. Reasonable compensation must be provided. In some cases, private or quasi-government agencies may be empowered to use eminent domain. This tool for planning can be quite controversial because it can create both positive and negative externalities, and the resulting public uses can alter historic patterns of spatial interaction. In addition, where used for large-scale redevelopment or economic development projects, the definition of the public interest can be seen to benefit particular sets of interests. In the United States, the era of urban renewal probably is the best example of large-scale use of eminent domain. During more recent years, the practice has been used in redevelopment projects as larger cities have engaged in the entrepreneurial developments of hotels, convention centers, and sports stadia.

CONTEMPORARY PLANNING ISSUES

Contemporary planning continues to focus on the development and implementation of the myriad dimensions of community change. More recent emphases include explicit consideration of the principles of sustainable development/environmental protection and more aggressive management of the geography of growth ("urban sprawl"). In the United Kingdom, for example, the Agenda 21 principles that emerged from the United Nations Conference on Environment and Development in Rio de Janeiro in 1992 have influenced the development of local Agenda 21 plans and yielded a national commission on sustainable development. Individual cities in the United Kingdom have become models for urban and environmental planning in other countries. In the United States, sustainable development themes have, to varying degrees, become incorporated into many dimensions of local and regional planning in spite of a stronger neoliberal bent at the national level. Discussions of the problem of urban sprawl have permeated urban and regional planning and led to some changes in the nature of land use regulation. In methodological terms, technological developments—especially those related to geographic information science (GIScience)—have become a central component of urban and regional planning.

CONCLUSION

Urban and regional planning plays an important role in the sociospatial dialectic creating and re-creating places. Planning is a component of societal processes that illustrates the complexity and interdependency of geographic scale. It can be part of an accepted and routine set of practices responding to economic and environmental changes, or it can be the focus of intense conflict over the processes of planning and distributional outcomes. Although many commonalities between countries, and among regions within countries, can be seen, there is a clear geography of planning with significant variation in legal and jurisdictional systems. In the era of globalization, planning will continue to get the attention of human geographers interested in the processes of community change.

—*Richard Kujawa*

See also Central Business District; Gentrification; Infrastructure; New Urbanism; NIMBY; State; Suburbs and Suburbanization; Urban Entrepreneurialism; Urban Fringe; Urban Geography; Urban Managerialism; Urban Social Movements; Urban Sprawl; Zoning

Suggested Reading

American Institute for Certified Planners. (2005). *Ethics for the certified planner* [Online]. Available: www.planning.org/ethics/

Cullingworth, J. (1993). *The political culture of planning: American land use planning in comparative perspective.* New York: Routledge.

Dalton, L., Hoch, C., & So, F. (Eds.). (2000). *The practice of local government planning* (3rd ed.). Washington, DC: International City/County Management Association.

Fainstein, S. (1991). Promoting economic development: Urban planning in the United States and Great Britain. *Journal of the American Planning Association, 57,* 22–33.

Florida, R., & Jonas, A. (1991). U.S. urban policy: The postwar state and capitalist regulation. *Antipode, 23,* 349–384.

Gaffikin, F., & Warf, B. (1993). Urban policy and the post-Keynesian state in the United Kingdom and the United States. *International Journal of Urban and Regional Research, 17,* 67–84.

Leitner, H. (1990). Cities in pursuit of economic growth: The local state as entrepreneur. *Political Geography Quarterly, 9,* 146–170.

Sanderock, L. (Ed.). (1998). *Making the invisible visible: A multicultural planning history.* Berkeley: University of California Press.

URBAN ECOLOGY

Ecology is a branch of biology that studies relationships between and among organisms and their environment. When the environment under study is located in a city, the subfield is called urban ecology. A city is larger than a small town. However, definitions vary as to what minimum population is required to constitute an urban environment. The U.S. Census Bureau requires a population of at least 2,500 to be considered an urban location.

The urban ecology scale of analysis can be a small patch of land within a city, an urban area determined by city boundaries, an urban area determined by settlement boundaries, or a region containing several cities. Sometimes the urban extent of settlement is under- or overbounded by the city limits, making it difficult to use political boundaries to study naturally bounded ecosystems. To solve this problem, some scholars suggest using watershed or alluvial basin as the appropriate naturally bounded unit of study, including one or more urban areas and all of the adjoining agricultural and natural landscapes within them.

Urban ecology is important and useful because it provides a bridge between urban environmental research and the more practical aspects of urban land use planning. Even though cities constitute only a small percentage of the surface of the earth, the anthropogenic impacts are profound—pollution, fragmentation or elimination of habitat, and climate modification, to name a few. With the goal of solving some of the problems associated with urbanization, urban ecology practitioners hope to achieve an efficient and equitable management of resources to improve the quality of life of city residents.

The meaning and methodology of urban ecology among scholars, practitioners, and the general public have evolved significantly over time. Ecology has its roots among naturalists of the 19th century and was not applied to cities until the 1920s and 1930s. Scholars at the University of Nebraska and the University of Chicago sought to use concepts and methods of plant and animal ecology for the explanation of human spatial patterns and changes over time in rapidly growing cities of early-20th-century America. For example, concepts of distribution and growth of animal communities were applied to the distribution of crime and disease and the subsequent spread of these blights in city neighborhoods. Popular among sociologists, economists, and geographers, this emerging subfield was first called human ecology and was not called urban ecology until several decades later. Social scientists of the 1940s criticized this biological approach to studying human populations and began to emphasize the importance of culture and social context. Mapping of urban socioeconomic characteristics was used to develop the earliest urban models that described the ideal city in terms of zones of wealth and poverty as well as areas of concentration and segregation of several generalized socioeconomic groups.

Environmental social movements and resultant legislation of the 1970s propelled research, grassroots involvement, and government mandates to build environmentally healthy and sustainable cities. Rapidly growing cities, increasing urban sociological problems, and ever worsening urban environmental pollution drew the attention of ecologists. Social scientists shifted from past methods of looking only at the socioeconomic variables of city life to considering the environmental concerns of urban sustainability. Issues such as clean water, clean air, waste management, infrastructure development impacts, and interactions with the nonhuman natural environment were of urgent interest. In response, many towns, cities, and states formed governmental and nongovernmental organizations (NGOs) to better understand and promote urban ecology.

The approaches to urban ecology continue to differ among life scientists and social scientists. A life science urban ecologist studies the way in which urban environments affect nonhuman biological resources such as the flora and fauna, whereas a social science urban ecologist examines the way in which urban social structure adapts to and interacts with natural resources. However, new frontiers in ecology are being advanced through interdisciplinary studies supported by large funding agencies such as the National Science Foundation (NSF). During the mid-1990s, the NSF augmented funding of its Long-Term Ecological Research project (a project extending over a period of 30 years and still in progress) to include social science and, for the first time, urban study sites. Two cities were selected for study: Phoenix and Baltimore.

Other large funding agencies supporting interdisciplinary urban ecological research include the U.S. Environmental Protection Agency, the National Aeronautical and Space Administration, the U.S. Geological Survey, and the U.S. Department of Agriculture. In addition to Phoenix and Baltimore, other areas being studied include Detroit, Seattle, and Willamette Valley (Oregon). Interaction with earth scientists, environmental engineers, and social and economic scientists is essential for the field of urban ecology. Some scholars refer to this convergence as a *biosocial* approach, that is, an approach that studies the flows of biological and social resources and the mechanisms of allocation of these resources. Scientists

hope that, through interdisciplinary collaboration and integrated frameworks, it will be possible to advance appropriate methods and theories to better understand urban systems, to better serve the needs of urban dwellers, and to truly create livable natural cities.

The practical application of urban ecology is tied closely to ideas associated with sustainable development of cities. Again, there is a difference between those practitioners who focus on the biological aspects and those who focus on the sociological aspects. For example, the National Park Service of the U.S. Department of the Interior has a Center for Urban Ecology that focuses on natural habitat and biological resources, engaging in studies such as habitat fragmentation resulting from urbanization and urban impacts on biodiversity. In contrast, the stated goal of Urban Ecology, a San Francisco Bay Area NGO, is to help develop ecologically friendly, socially just, and economically fair communities. This cultural and sociological focus of Urban Ecology emphasizes urban morphology, architecture, transportation, health, justice, and policy. In existence since 1975, this organization hosted the first international ecocity conference in 1990, drawing participants from 13 countries.

Ecologists of the past traveled to remote exotic locations to conduct research in seemingly pristine locations. Now they are increasingly focusing their attention on the ecology of cities, studying urban habitats of native and introduced species and researching the profound impact of urban human intervention on the natural landscape. Urban ecology centers, institutes, courses, and NGOs exist not only in the United States but also around the world. Examples of topics of study and application include urban watershed and habitat, creating and sustaining healthy urban environments, brownfield redevelopment, lead poisoning prevention, improved water and air quality, environmentally sound urban waterfront development, biological surveys, reestablishment of native vegetation, and educational programs. An example of an innovative urban ecology project is one that joins isolated areas of urban vegetation to increase the connectedness of wildlife habitat so as to create a more robust food web and thereby improve biodiversity in cities.

—*Betty Smith*

See also Chicago School; Nature and Culture; Urban Geography

Suggested Reading

Collins, J., Kinzing, A., Grimm, N., Fagan, W., Hope, D., Wu, J., et al. (2000). A new urban ecology. *American Scientist, 88,* 416–425.

Grimm, N., & Redman, C. (2004). Approaches to the study of urban ecosystems: The case of Central Arizona–Phoenix. *Urban Ecosystems, 7,* 199–213.

Grove, J., & Burch, W., Jr. (1997). A social ecology approach and applications of urban ecosystem and landscape analyses: A case study of Baltimore, Maryland. *Urban Ecosystems, 1,* 259–275.

Lord, C., Strauss, E., & Toffler, A. (2003). Natural cities: Urban ecology and the restoration of urban ecosystems. *Virginia Environmental Law Journal, 21,* 317–381.

Theodorson, G. (1982). *Urban patterns: Studies in human ecology.* University Park: Pennsylvania State University Press.

URBAN ENTREPRENEURIALISM

Geographer David Harvey coined the term *urban entrepreneurialism* during the late 1980s. He sought to characterize what he saw as a definite shift in urban governance from processes concerned mostly with the provision of services to those concerned with economic development. Harvey saw this evolution as quite different from existing forms of urban "boosterism" in that it involved an explicit form of public–private partnership. These partnerships were more speculative in nature and frequently involved risk-bearing scenarios for the deployment of local-scale fiscal resources. Moreover, the development of urban entrepreneurialism was seen to be both a response to and constitutive of broader changes in urbanization processes connected to the demise of Fordism and the emergence of flexible accumulation strategies. One key to Harvey's conceptualization was his insistence that urban governance involved a broader array of actors and stakeholders in government, quasi-government, and the private sector. Urban entrepreneurialism, then, can be characterized as ephemeral public–private partnerships focused on a single project, or it can involve the construction of more significant and lasting partnerships among a wide array of actors at the municipal scale. In the United States, early signs of urban entrepreneurialism can be see in the relatively short-lived Urban Development Action Grant federal program. Designers and advocates of this program exhorted cities to act like businesses, take risks, and develop public–private partnerships for redevelopment. Similarly, large-scale development projects such as festival spaces, aquaria, museums, financial centers, and sports stadia began to feature public risk taking as cities competed for capital in an urban-scale version of neoliberalism.

Urban entrepreneurialism, in its myriad forms, has become a dominant feature of urban development and redevelopment in contemporary capitalism as cities of widely varying sizes and positions in their national urban hierarchies have become involved in development strategies that reach beyond the urban scale per se. This has been the context for both empirical analyses of and prescriptions for competitive economic development policy. For example, the U.S. Department of Housing and Urban Development has examined the contribution of public–private partnerships in an urban entrepreneurialist vein to national economic growth. Here local responses to the challenges faced in economic restructuring are the focus.

However, the evolution of urban entrepreneurialism has also formed the focus for continued analyses of the form and function of state and quasi-state apparatuses in global capitalism in all of its complexity and unevenness. Some have seen the processes of urban entrepreneurialism as one component in a complex rescaling of global economic competition that is not only global but also local and regional, hence its appeal for geographers and others interested in the interdependence of geographic scale in globalization.

In addition, urban entrepreneurialism has been the focus of research that seeks to specify the processes of postindustrial place making, replete with its language of entrepreneurialism and new forms of social and cultural control. This postindustrial place making has varied from the competition for set-piece development projects for waterfronts or older industrial districts to place-specific museums and other manifestations of cultural symbolism. Complex place-based relationships, in many cases, appear to be reduced to the language of interurban competition as populations are "sold" the ideology of the public–private partnership for urban regeneration and the importance of image and perception, even as the public financial commitments marshaled to support the process further marginalize and impoverish neighborhoods and populations. Again, some have seen this as part of a neoliberal agenda and a repressive and punitive attitude toward marginalized and vulnerable populations.

Concerns for social and economic justice appear to be undermined as the nearly mythlike narrative of urban entrepreneurialism and the trickling down of its benefits maintain and reproduce its orthodoxy.

Popular and academic literature is swollen with case study–based discussions of urban entrepreneurialism that reveal its tensions and contradictions. From the development of the Guggenheim in Bilbao, Spain; to the shifts in planning processes in Cape Town, South Africa; to the competition for (and retention of) professional sports franchises and their requisite stadia in the United States; to the spatial injustice of the urban renaissance of Glasgow, Scotland, it is clear that the concept of urban entrepreneurialism will form an important facet of research and teaching in human geography.

—Richard Kujawa

See also Neoliberalism; Urban and Regional Planning; Urban Geography; Urban Managerialism

Suggested Reading

Brenner, N. (2003). "Glocalization" as a state spatial strategy: Urban entrepreneurialism and the new politics of uneven development in Western Europe. In J. Peck & H. Yeung (Eds.), *Remaking the global economy* (pp. 197–215). Thousand Oaks, CA: Sage.

Gonzáles Ceballos, S. (2004). The role of the Guggenheim Museum in the development of urban entrepreneurial practices in Bilbao. *International Journal of Iberian Studies, 16,* 177–186.

Harvey, D. (1989). From managerialism to entrepreneurialism: The transformation in urban governance in late capitalism. *Geografiska Annaler B, 71*(1), 3–17.

Leitner, H. (1990). Cities in pursuit of economic growth: The local state as entrepreneur. *Political Geography Quarterly, 9,* 146–170.

MacLeod, G. (2002). From urban entrepreneurialism to a "revanchist city"? On the spatial injustices of Glasgow's renaissance. *Antipode, 34,* 602–624.

Wood, A. (1998, August). Making sense of urban entrepreneurialism. *Scottish Geographical Magazine,* pp. 120–123.

URBAN FRINGE

The urban fringe is an area bordering a metropolitan location within the rural–urban land continuum. This area has a mixture of urban, suburban, and rural land uses and socioeconomic characteristics. The fringe is in a constant state of flux as rural land is overcome and becomes a more urban area as people and businesses seek open land and proximity to the market. When the new area is completely urbanized, a new fringe emerges—expanding farther out from the center. The area's size and specific characteristics vary from locale to locale due to many factors such as the urban center's size, institutional controls (e.g., zoning), and socioeconomic conditions. Found in both Western and non-Western countries, the fringe has a mixture of farmers and nonfarmers (usually former city dwellers). The nonfarmers consist mostly of families with children looking for houses with big yards, new employment opportunities, and so forth. This eclectic mix of people causes conflicts about the damage of crops by nonfarming recreation and the increase of land values due to speculation that drive farmers off their land.

The closeness of the urban environment increases the demand for the fringe (especially if the land will quickly convert to become classified as truly urban), which in turn raises the price of the land and forces many farmers—particularly small farmers—to sell their land, or to partially sell their land, reducing their farm plots. Developers become the new owners of the land and eventually exploit the land themselves or resell it to other interested developers. Infrastructures in the fringe are considered to be somewhat deficient (i.e., public transportation might not be as widespread in fringe areas as it is in the urbanized area). This fringe also becomes the location of certain uses that are kept out by zoning laws in the urbanized areas such as landfills and waste dumps. Some farming is successful in the fringe because the closeness to the metropolitan area helps farmers sell goods that have a short shelf life.

This fringe has also been generally modeled into two components: an inner fringe and an outer fringe. The *inner fringe* is formerly rural land that is being converted increasingly into urban land. The *outer fringe* is differentiated by rural land uses in which segments of the urban land use have penetrated the area (e.g., single-family houses, major transportation routes). Beyond this area are two additional zones: the urban shadow and the rural hinterland. The *urban shadow* is the area where the physical urban landscape has made a minor imprint, especially in reference to land ownership being controlled by nonfarmers. The *rural hinterland* is the location of urban dwellers' second homes found mixed in with large amounts of agricultural land. One difficulty in identifying this area is especially acute when attempts are made to map its spatial extent. The U.S. Bureau of the Census uses the urban fringe concept in its data sets; however, these data are difficult to use in specific locations due to the census aggregation levels. Researchers have been able to use remote sensing and geographic information systems (GIS) in the examination of development within the fringe. The remote sensing images enable users to identify the mixture of urban and rural land uses.

To combat expansion of urbanized areas, governments have attempted to use different programs to protect agriculture land from urban development. Some government programs provide tax incentives, whereas others set up agricultural districts. There are many types of tax incentive programs, including simply reducing taxes on agricultural land and assessing the land at an agricultural level rather than true market value. Another possibility of controlling this conversion is through purchase or transfer of development right programs. In these programs, governments buy the developmental rights to the land and hold on to them. Finally, there are also some nongovernmental programs that attempt to halt the development of agricultural land, including land trusts, through donations of land or money to buy developmental rights.

—*Thomas Mueller*

See also Edge Cities; Exurbs; Rural–Urban Continuum; Suburbs and Suburbanization; Urban Sprawl; Zoning

Suggested Reading

Bryan, R., & McLelland, J. (1982). *The city's countryside: Land and its management in the rural–urban fringe.* London: Longman.

Furuseth, J., & Lapping, M. (Eds.). (1999). *Contested countryside: The rural–urban fringe in North America.* Hampshire, UK: Ashgate.

Pryor, R. (1968). Defining the rural–urban fringe. *Social Forces, 47,* 202–215.

URBAN GEOGRAPHY

Like geography itself, urban geography is defined more by a way of thinking than by its content. A wide range of topics and issues are viewed through the lenses of space, place, and environment as well as by the categories *city* and *urban*. What exactly city and urban mean has been the subject of much theoretical and philosophical debate. That dissonance has led different people to very different notions of what urban geography is. This entry summarizes some of the chief theoretical and philosophical approaches toward urban geography that are represented in the field and includes brief discussions of topics typically addressed in the field.

Some of the earliest approaches to urban geography developed out of an attempt by early- to mid-20th-century human geographers to spatialize 19th- and early 20th-century urban sociology. They married an emerging morphological human geography to a more process-oriented urban sociology, such that various social patterns and behaviors could be identified and understood as responses to a specifically urban environment. The logics of Émile Durkheim, Georg Simmel, and Ferdinand Tönnies, and later those of Louis Wirth, Herbert Gans, and Claude Fischer, were foundational here. They are perhaps best represented by the geographic models of urban structure made famous by Ernest Burgess, Homer Hoyt, Chauncy Harris, and Edward Ullman.

Also inspired by 19th- and early-20th-century social theory are a number of approaches that emphasize social structures and hierarchies of power. Karl Marx, Friedrich Engels, and Max Weber, among others, provided a continuing intellectual basis for various political–economic approaches to urban geography, where cities and urban space are both products of and inputs to capitalist social relations. Less materialist variants of these approaches are represented by, among other things, analyses of urban landscapes and neighborhood change in which the roles of culture and meaning systems are seen as crucial parts of the social structures and hierarchies of power in question. Recently, these have highlighted not only capitalist social relations but also patriarchal, racist, and heterosexist social relations.

The quantitative revolution of the mid-20th century led to an array of positivist approaches. On the one hand, there emerged a body of work focused on the morphology of cities and city systems, including location models (e.g., central place, bid–rent) and systems theories and models (e.g., rank–size rule). These spatial science approaches to the city presumed that human social life and its outcomes mimic the laws of geometry, physics, and/or biology. On the other hand, behavioral approaches purported to account for the vicissitudes of human behavior and the human condition by applying quantitative techniques and the scientific method to human behavior. For example, urban geography from these perspectives often traced out the time–space routines of individuals in an urban environment and their outcomes.

During the late 20th and early 21st centuries, the apparent failures of modernism (e.g., science, technology, reason) to deliver on their promises of a better world led to devastating critiques of positivism and structuralism. So-called postmodern and poststructuralist geographies have focused on fluid and unstable realities, multiple truths, and the politics of representation (including particularly the political power of semiotics and rhetoric). Often fueled by the anger and frustration of social movements whose participants felt excluded by more conventional approaches (including Marxism), these new approaches tend to approach the urban as texts in need of deconstruction (both analytically and politically). They focus on the often colliding and fragmented meaning systems reflected and reproduced in urban landscapes and designs as well as their connections to often elusive and highly contingent—but nonetheless real—exercises of power.

Although the actual intellectual and empirical agendas of urban geography can be quite complex and hybrid, it is possible to identify at least six topical subfields to which these approaches have been applied: the built environment, its construction, and its morphology; human–environment relations in an urban context; social geography and social patterns in an urban context; macro-scale perspectives on city systems and functions; micro-scale perspectives on city systems and functions; and issues of urban planning, policy, design, and architecture. Summaries of these themes follow.

THE BUILT ENVIRONMENT

During its early phases, urban geography considered the pattern and morphology of buildings in cities in concert with land use categories. Housing, commercial activities, and industrial activities, for example, were seen to exhibit particular patterns at particular historical junctures. In many formulations, these were connected with technological change or market equilibrium models. For each land use category, specific urban locations were deemed optimal under different economic and political conditions. This descriptive morphology was much less contested than its explanation. For instance, more radical geographers

emphasized class struggle as the driving force for the suburbanization of industry rather than behavioral decisions or transportation innovations. Similarly, the logics for the development of downtowns and central business districts, specific geographies of housing types, and the nature and location of retail and commercial activities all have been contested by researchers with different approaches.

Analyses of the built environment have also focused on the development process itself, with examinations of the role of private-sector and state actors, the availability of development finance (and its geography), and the groups and individuals directly involved. Specific examples here include studies of urban redevelopment, gentrification, sprawl, and edge cities. Most recently, examinations of construction projects spawned by public–private partnerships connected with place marketing, urban entrepreneurialism, and interurban competition have formed a core element in urban geographers' analyses. The simplified nexus between the built environment and land use has been disaggregated as geographies of production and consumption have become more complex.

HUMAN–ENVIRONMENT RELATIONS IN AN URBAN CONTEXT

Studies of human–environment relations have undergone considerable transformation during the past few decades. Early studies of urban location emphasized the opportunities and constraints posed by physical conditions. In some cases, these would relate to materialist and economic assessments with an emphasis on propitious locations or required human-made changes for trade or industry. In other cases, more attention was paid to health concerns, especially in the context of disease and human health. In spatial terms, marginalized groups appeared to be disproportionately subjected to natural and human-produced environmental and health hazards. Studies of the geography of locally unwanted land uses (LULUs) and the NIMBY ("not in my backyard") phenomenon have been part of the environmental dimension of urban geography that has considered environmental justice in an urban setting. This has led to a consideration of the political and economic processes involved in the production of urban environmental infrastructures (e.g., drainage, sewage treatment, water supply, solid waste disposal) and in the specification of the role of the state in addressing environmental problems at the urban scale.

In broader terms, studies of the urban physical environment traditionally have been an extension of the systems thinking of physical geography. Topics here include studies of urban runoff and its impact on the ecology of streams, rivers, and so forth; atmospheric, water, and noise pollution resulting from increased population densities, industrialization, transportation, and waste disposal; and local atmospheric/climate changes in the form of "heat islands."

A more recent trend has involved examination of the viability of specific urban ecosystems and the development of urban policies in this regard. Some examples include studies of urban forestry practices, the vegetation ecology of public parks, and the ecologies of other public spaces. Restoration and remediation of industrial-era pollution has been part of these processes. Again, issues of social and environmental justice are important.

Finally, urban geographers have studied the impacts of a broader societal concern with sustainable development and the global impact of human activities. The urban dimensions of these concerns have revealed themselves in local-scale initiatives for reduction in the production of carbon dioxide and other greenhouse gases and in other environmentally friendly changes in transportation, housing, and other activities.

URBAN SOCIAL GEOGRAPHY

Urban *social* geography has tended to concern itself with social groups and processes and their various spatialities such as residential segregation, (sub)culturally meaningful spaces, territorial practices, and (recently) how difference is constructed in and through urban space as part of larger systems of power and control. The theoretical perspectives informing these run the gamut noted previously.

Some of the most influential early work in the field was produced by researchers in sociology at the University of Chicago during the early to mid-20th century. Burgess's famous "concentric zone" model held that social groups (especially immigrants and the assimilated) were like species competing for scarce resources and that the city was their ecosystem. In one sweeping and very parsimonious theoretical model, changing patterns of segregated land use and residences were understood as part of a single coherent process in which spatially defined inequality was naturalized and upward social mobility and assimilation were seen as inevitable. Later variants of this approach, emphasizing slightly different economic and methodological logics, included Hoyt's sectoral model and Harris and Ullman's multiple-nuclei models. Other, more substantial critiques operating from within rather different, but still broadly ecological, paradigms ranged from highly quantitative "social area" analyses to much more humanistic formulations focused on community.

Later, more critical formulations were informed by a variety of both materialist and meaning-oriented social-theoretic frameworks that emphasized issues of power. Locational conflict geographers linked grassroots community empowerment and political resistances to the territorial practices of various social groups in cities as well as those of their powerful adversaries. Such work helped to rigorously spatialize Marxism. Eventually, feminist, lesbian/gay, critical race, and disability scholars (among others)—many of whom initially were trained (and in some cases still work) within quite structuralist approaches to capitalism, patriarchy, homophobia/heterosexism, and racism—became influenced by postmodernist and poststructuralist deconstructions of Marxian categories and processes and developed new spatial theories of the urban. Some theorized connections among gender relations, the domestic economy, home life in general, and the physical design of capitalist cities. Later, others developed explicitly antiessentialist ways of conceptualizing race and gender (and indeed capitalism) such that the categories themselves are seen as time–space-specific sociospatial constructions. Still others, working with queer theories, developed approaches that detail highly contingent connections among desire, sexuality, gender, and other axes of difference in the constitution of urban places and processes. Issues pertaining to the human body (including ability/disability, movement, and health issues in general) have also emerged as important themes.

CITY SYSTEMS AND FUNCTIONS: MACRO SCALE

At the interurban level, urban geography has evolved dramatically from descriptions and interpretations of the influence of physical conditions on the viability of urban patterns or spatial interaction. In these macro-scale analyses, the pathways of individual cities could be "personalized" and imbued with the characteristics of individuals. These narratives, especially those drawn from urban historians, often were replete with elitist notions of city leaders and prescient decision making regarding economic development or transportation innovations as well as the importance of propitious site characteristics.

The central concepts subsequently deployed often reflected locational analyses of the factors of production and the impact of shifts in transportation technology. For example, in North America, the importance of a system of gateways and hinterlands has been seen as the basis of the modern urban system. Conceptualizations of natural, initial, and comparative advantage were important to interpretations of the evolution of a hierarchical urban system. Descriptive models of urban primacy, such as the rank–size rule, later were conjoined with more positivist models of central place theory. Quantitative analysis of the functional specializations of cities and the processes of regional industrial development represented the heyday of urban systems analysis. These were reflected in discussions of the evolution of manufacturing belts, regional shifts in manufacturing, deindustrialization, and the development of metropolitan regions.

Phases or periods were, and still are, key heuristic hallmarks of macro-scale urban geography. Time periods were characterized by transportation innovations or investor and consumer choice. Marxist urban geographers, in contrast, deployed broader analyses based on evolving relationships among labor, capital, and the state through the lexicon of Fordism, post-Fordism, and flexible production that still infuse most contemporary analyses. Economic restructuring as both a cause and an effect of political change, for example, was featured in discussions of the expansion of the North American Sunbelt. More recently, analyses of urban hierarchies have included the concept of *world cities* (although the phrase had been coined decades earlier). In this context, urban geographers have considered the consequences of further time/cost–space convergence, neoliberal policies, and the new international division of labor.

CITY SYSTEMS AND FUNCTIONS: MICRO SCALE

At the more regional and local scale, urban geography has tended to focus on issues of, or changes in, land use, neighborhoods, place-based social movements, and public versus private space. All of these are recognized as being heavily influenced by larger-scale (including global) formations and processes such as neoliberal political economies. But they are also seen as formative and significant in their own right as "the global" is increasingly seen as constituted in and through its other, "the local."

In terms of land use and land use change, a number of Marxist urban geographies, for example, have examined the land use structure of North American cities from a perspective that emphasizes the increasingly complicated and locally specific form that capitalist class (and other) relations take. They emphasize various nexuses between local and nonlocal interests (and environments) that produce the specificities of "place" (including individual districts in cities) as resources (or, more narrowly, as commodities).

Ironically, at the same time that there has been an increase in interest in marginalized and oppressed populations in urban geography, there has been a decrease in emphasis on the neighborhood spaces that such groups traditionally occupy. For instance, studies of ghettos, barrios, and favelas have gone somewhat out of vogue, whereas gentrified neighborhoods, export processing zones, privatized consumer spaces, upscale suburbs, gated communities, and parade/festival spaces have become more studied. There has been a substantial amount of attention paid to issues of public space and its significance as well, especially in the context of the recent global trend toward its privatization. These studies explore everything from the implications of urban public space for democracy and citizenship to its role in producing/reproducing collective identities.

This shift to a focus on upscale urban spaces probably is more characteristic of work on (and in) so-called Western societies than of work on non-Western ones. Urban geography in a development context has retained more of a focus on marginalized peoples and spaces than on its more developed world counterparts. Still, even in a Third World context, an increased focus on spaces of affluence and consumption is evident, as illustrated by a rapidly growing (but still small) body of work on spaces of Western-oriented tourism (and especially sex tourism) in Southeast Asia and the Middle East.

For those committed to breaking down binaries such as materialist/discursive, these developments in urban geography may be welcomed; for others, such as those who lament the passing of an unambiguous politics of class, race, or gender, they may very well represent a triumph of divide-and-conquer tactics.

Nevertheless, they clearly have broadened the range of topics considered by urban geographers.

URBAN PLANNING, POLICY, AND DESIGN

Urban and regional planning and policy have varied in their relevance to research and discussion in urban geography. In general, they have reflected macroeconomic policies (e.g., setting of interest rates, design of the tax code, national-scale budget allocations for infrastructure). Such exogenous forces have a direct and significant impact on urban development. When national-scale budget allocations become regional and local, uneven development processes may be exacerbated or ameliorated. There also may be international differences in the development and use of these macroeconomic instruments. A good example in North America is the difference between the tax treatment of mortgage interest in the United States and that in Canada.

In the United States, the development of specific urban-scale policies for urban renewal and public housing goes well beyond macro-scale economic policy. In addition to the allocation of fiscal resources, changes in the regulatory environment have been important. National-scale policy in this dimension, however, is refracted both by differences in governmental structures and by local and regional interests. These refractions have drawn the attention of urban geographers. In some analytical frameworks, policy changes are captured as distortions in the calibration of spatial economic modeling or modifiers of behavioral choice models. In others, especially those influenced by political economy perspectives, the creation and re-creation of places is a direct product of the evolution of conflict. Analyses of urban growth machines or regimes are good examples.

Urban and regional planning has also been an object of geographers' research. Municipal and regional planning is highly variable and spatially uneven. However, planning generally can exert an influence on the timing, density, and type of land use. Here the specific geographies of allowable land uses and types of development, as well as the geographies of public facilities (e.g., parks, schools) and infrastructure, can be specified. It is in the interstices of legal and regulatory frameworks and their implementation that the impact of planning on the development of urban landscapes and communities is examined. Planning has drawn particular attention in the development of suburban and exurban landscapes, where it has formed a context for analyses of the sociospatial dialectic and the complex role of state and quasi-state apparatuses in land use and community change.

Urban geographers have also studied architecture and landscape design. One tack has focused on how planning attempts to solve urban problems spatially, for example, through the design of working-class housing or the design of parks. More functional or determinist perspectives have linked changes in the built environment to changes in construction technologies, changes in the overall economy, or the amelioration of social problems. Examples here include the development of the elevator and structural steel in the evolution of the built environment of central business districts and the need for single-story factory buildings to house assembly lines and hence suburbanization of industry. Geographers have also studied the specific work of architects and designers. Here they critique those urban artifacts semiotically both as reflections of class, gender, and cultural relations and as constitutive of them in specific urban and suburban contexts. Examples of this include discussions of the design and geography of skyscrapers in American cities, the role of architecture and design in dystopic postmodern cities, and the retro designs of new urbanist suburbs.

—Larry Knopp and Richard Kujawa

See also Built Environment; Central Business District; Chicago School; Cultural Landscape; Exurbs; Flexible Production; Fordism; Gated Community; Gentrification; Geodemographics; Ghetto; Global Cities; Growth Machine; Historic Preservation; Home; Homelessness; Industrial Districts; Locally Unwanted Land Uses; Location Theory; Neighborhood; Neighborhood Change; NIMBY; Postmodernism; Poststructuralism; Rent Gap; Rural–Urban Continuum; Segregation; Space, Human Geography and; Squatter Settlement; Suburbs and Suburbanization; Sunbelt; Time Geography; Transportation Geography; Uneven Development; Urban and Regional Planning; Urban Ecology; Urban Entrepreneurialism; Urban Fringe; Urban Managerialism; Urban Social Movements; Urban Sprawl; Urban Underclass; Urbanization; Zoning

Suggested Reading

Beaverstock, J., Smith, R., & Taylor, P. (2000). World–city network: A new metageography? *Annals of the Association of American Geographers, 90,* 123–134.

Castells, M. (1983). *The city and the grassroots: A cross-cultural theory of urban social movements.* Berkeley: University of California Press.

Davis, M. (1990). *City of quartz.* London: Verso.

Harvey, D. (1973). *Social justice and the city.* Baltimore, MD: Johns Hopkins University Press.

Harvey, D. (1985). *The urbanization of capital.* Baltimore, MD: Johns Hopkins University Press.

Jackson, K. (1985). *Crabgrass frontier.* New York: Oxford University Press.

Knox, P., & Pinch, S. (2000). *Urban social geography* (4th ed.). Upper Saddle River, NJ: Prentice Hall.

Kunstler, J. (1993). *The geography of nowhere.* New York: Touchstone.

Logan, J., & Molotch, H. (1987). *Urban fortunes.* Berkeley: University of California Press.

Sassen, S. (2002). *Global networks: Linked cities.* New York: Routledge.

Wilson, W. (1987). *The truly disadvantaged: The inner city, the underclass, and public policy.* Chicago: University of Chicago Press.

URBAN MANAGERIALISM

Urban managerialism originally was conceived as an analytical framework to consider the role of urban managers in the sociospatial processes of capitalist urbanization. It has also been used to denote a period in urban governance (usually seen as the 1970s and early 1980s) when the focus of urban governments was seen to be concerned predominantly with the provision of services or collective consumption. This entry focuses on the first of these meanings.

Urban managerialism emerged from housing research conducted in urban sociology in the United Kingdom. It rapidly became the focus of broader theoretical and empirical discussions in urban geography and urban studies. The term originally coalesced in the writings of British urban sociologist Ray Pahl. In distributional analyses of access to scarce urban resources and facilities, Pahl isolated the role of what he termed *urban managers* in both the public and private sectors. In this earliest formulation, the scarce resources and facilities often were in the housing sector with urban managers—or gatekeepers, as they sometimes were called—directly concerned with the development, marketing, management, and financing of both the rental and ownership sectors of the market. In Pahl's formulation, inequality in this consumption sector also influenced the life chances of individuals and groups who already were subjected to inequality in the production sector. It directly reflected a Weberian-style analysis of this particular facet of society. The focus on urban managers involved careful consideration of their individual value systems, the role of political and class ideology, and other facets of both individuals and organizations.

Research driven by the ideas of urban managerialism made its mark on research in urban geography during the late 1970s and early 1980s. A key facet in the evolution of urban managerialist ideas in urban geography, and for that matter in urban sociology itself, was the tension between so-called managerial autonomy in these distributional processes and the overarching power of broader political and economic structures. To be fair, this tension had been highlighted from the very inception of the managerialist thesis and had drawn critical words from the likes of David Harvey and Manuel Castells. One can detect all of the elements of the so-called structure–agency debates in such tension. Pahl himself had struggled with this broader tension and with more focused critiques that questioned which managers were worthy of study and for what reasons. In fact, he had reformulated his original thesis to move away from a direct causal role for managers. For many, these adaptations further weakened or marginalized the relevance of the urban managerialist thesis.

In urban geography, however, further attention was given to ways in which the core ideas of urban managerialism and other facets of organizational analysis could be explored to meet the structure–agency contradiction head-on and hence to synthesize the insights of urban political economy with institutional analyses. Key contributors to this resuscitated urban managerialism were Peter Williams and (later) David Wilson. In this milieu, Wilson and others explored the meaning of human agency in specific organizational contexts as they were created and re-created. Wilson's empirical work, which examined both urban revitalization and the implementation of the Community Development Block Grant program, revealed the potential of an examination of the constrained decisions of urban managers in the evolving processes of urbanization. Crucial to Wilson's analyses of the decision-making processes of managers, and their impact, was the recognition that although specific historic and material circumstances (including broader political and economic changes as well as programmatic structures and local conditions) constrain the behavior of "managers," they do not determine them. In this decision-making space, the applicability of some of the original tenets of the managerialist thesis reemerged with a focus on the active and creative behavior of individuals. This analysis, however, was not restricted to distributional aspects of housing and other resources; rather, it was also applied to urban revitalization and commercial redevelopment.

While the attention of urban geographers moved more toward the analyses of urban entrepreneurialism and public–private partnerships for economic growth, the reformulated urban managerialism, both as an analytical approach and as a vehicle for the examination of the interplay of structure and agency, has had considerable value for urban geographers in their analyses of institutional and organizational change. Insights from urban managerialism have also been deployed in analyses of the housing and service economies of the transitional economies of Eastern Europe and China.

—*Richard Kujawa*

See also Urban and Regional Planning; Urban Entrepreneurialism; Urban Geography; Urban Social Movements

Suggested Reading

Domanski, B. (1991). Gatekeepers and administrative allocation of goods under socialism: An alternative perspective. *Environment and Planning C, 9,* 281–293.

Pahl, R. (1970). *Whose city?* Harmondsworth, UK: Penguin.

Williams, P. (1982). Restructuring urban managerialism: Toward a political economy of urban allocation. *Environment and Planning A, 14,* 95–105.

Wilson, D. (1987). Urban revitalization on the upper west side of Manhattan: An urban managerialist assessment. *Economic Geography, 63,* 35–47.

Wilson, D. (1988). Toward a revised urban managerialism: Local managers and community development block grants. *Political Geography Quarterly, 8,* 21–41.

URBAN SOCIAL MOVEMENTS

Urban social movements (USMs) consist of individuals who share real or perceived inequality within a larger society and who engage in strategies and activities to achieve collective goals. The inequality may be social, cultural, political, or economic. The common objective of the movement is to achieve or prevent change. Members of USMs seek to elucidate and solve problems of housing, environment, municipal services, cost of living, transportation, jobs, security, redevelopment, and other issues of urban life.

At some point, those who participate in USMs have become impatient with existing institutions and mobilize with the aim of transforming their cities into better places to live and work. Participants come from all classes of society and are most active in political environments of participatory democracies. Less frequently, USMs are found in nondemocratic areas of the world such as the Middle East and China. The geographic scale of membership in USMs varies and may simultaneously be local, state, regional, national, international, and transnational. There is a nested organizational hierarchy in which local grassroots movements affiliate with larger regional or national organizations that are further affiliated with international or global USMs. For example, an individual may identify with several local USMs while simultaneously identifying with a national and transnational USM.

USMs have a long history, dating back at least two centuries to when citizens identified the need for improved health and safety conditions in cities. Although some USMs are long-lasting, others may be of shorter duration. The life cycle of a USM usually begins with mobilization around a common urban problem, followed by local actions and coordination with other movements. If the USM is long-lasting, it may formalize into associations, societies, or nongovernmental organizations. If the goals of the USM become institutionalized and meaningful change is achieved, the movement may cease to exist. Likewise, if USM participants become overly discouraged or violently repressed, the movement may cease to exist. Clearly, democratic civil societies are more conducive to USM development and sustainability than are oppressive authoritarian states.

Scholars debate whether USMs must be reactive or simply provide an innovative alternative way of doing things. Analyses of USMs cross disciplinary boundaries and occur in the private and public sectors, at universities, and at government agencies. Geographers provide a spatial perspective by mapping the uneven distributions of socioeconomic variables associated with USMs. They conduct fieldwork, analyze the changing morphology of cities, use census data to identify areas of potential USM development, consider the role of urban space in constructing social processes, and contribute to modern thinking and modeling of USMs. Research and insight into the development of USMs occurs in sociology, political science, anthropology, history, geography, and other fields such as American studies, ethnic studies, and gender studies. Some examples of themes around which USMs organize in the United States include environmental, nuclear, civil rights, disability rights, workers' rights, housing, redevelopment, globalization, free trade, black nationalism, race and ethnicity, gender, socialism, acquired immunodeficiency syndrome (AIDS), and abortion. Some movements are citywide, others are national in scale, and still others cross borders. For example, USMs may form solidarity with workers in other countries, form alliances for social justice among persons damaged by environmental contaminants in cities, forge collective pan-Amerindian identity to demand historical political rights, protest child labor and transnational "sweat shops," and protest international sex trafficking among major cities. In each case, members of a movement share the experience and common identity of being denied rights, opportunities, or respect because of who they are.

With the advent of globalization comes a new geography of USMs. Increasingly, participant identities cross established political, social, economic, and cultural boundaries. The inequalities that drive USMs in the cities of the world derive from an uneven distribution of political power. The inequitable distribution of wealth and power has been institutionalized by the global market, dominated by multinational corporations, and compounded by illegal commerce and corruption. Transnational USMs today often support the national and local election of anti-neoliberal populist leaders in developing countries, hoping for economic growth with increased social, political, and cultural equity.

The global distribution of USMs captures the geographic essence of the North–South dichotomy of dependency theory. This is the core–periphery view of the world in which large cities in the economically developed world gain wealth and power by exploiting resources and people in less developed countries. For example, the export of primary products from developing countries of the Southern Hemisphere is supplemented today by the export of human resources, whereby massive out-migration to cities in the wealthiest countries of the global North results in remittance income to developing countries of the migrants' origin. While living abroad as guest workers or adjusting to permanent residency or new citizenship, migrants find solidarity in mobilization with established labor or social service USMs. A

mobilizing source of likely participants in USMs is poor urban neighborhoods. Typically, the urban poor of Europe and North America reside in the inner cities, whereas the majority of urban poor in developing countries of Africa, Latin America, and the Caribbean live in the urban periphery. Thus, grassroots development of USMs occurs in different geographic urban spaces depending on the region of the world.

The need for USMs to address social, political, and economic injustice continues in areas such as the role of women in society, working conditions, housing, and the environment. Among representative democratic governments are trends toward weaker labor unions and greater inequality of income distribution. Today, transnational USMs are most active in the contentious arena of globalization, addressing the problems of those members of society who are marginalized by free trade agreements, focusing on the conflicts between economic development and environmental protection, demanding the equitable provision of basic public municipal services such as clean water and waste disposal, and bringing attention to urban discrimination based on gender and race. Growing urbanization in most democratic countries suggests an increasing importance of cities and USMs.

—*Betty Smith*

See also Core–Periphery Models; Dependency Theory; Globalization; Neoliberalism; NIMBY; Power; Resistance; Urban Geography

Suggested Reading

Ayres, J. (2002). Transnational political processes and contention against the global economy. In J. Smith & H. Johnston (Eds.), *Globalization and resistance: Transnational dimensions of social movements* (pp. 191–205). Lanham, MD: Rowman & Littlefield.

Castells, M. (1983). *The city and the grassroots: A cross-cultural theory of urban social movements.* London: Edward Arnold.

Castells, M. (2004). *Power of identity: The information age— Economy, society, and culture* (Vol. 2). Oxford, UK: Blackwell.

Hamel, P., Lustiger-Thaler, H., Nederveen Pieterse, J., & Roseneil, S. (Eds.). (2001). *Globalization and social movements.* New York: Palgrave.

Tarrow, S. (1998). *Power in movement: Social movements and contentious politics.* Cambridge, UK: Cambridge University Press.

URBAN SPATIAL STRUCTURE

The study of urban spatial structure is concerned with the examination of urban form (or morphology) and the cultural, economic, social, and political processes behind the production of these forms. It emerged as a research tradition in urban geography and sociology and has been heavily involved with the development of schematic models of urban form that specifically emphasize the idealized flow of goods, services, and people within a city.

In the United States, the study of urban spatial structure was influenced strongly by the so-called Chicago School of urban ecology, starting in the 1920s in the Department of Sociology at the University of Chicago. Whereas this early examination of urban spatial structure used ecological models to describe the genesis of urban form, later challenges to the Chicago School provided a series of neoclassical economic, Marxist, and postmodern approaches to the development of distinct urban patterns that are discussed in what follows.

URBAN ECOLOGY AND THE CHICAGO SCHOOL

The Chicago School of urban ecology comprises a number of schematic models of urban spatial structure, all of which start with one preconception—that cities are embodying and representing the characteristics of human nature and thus must be understood as such. Humans are essentially territorial beings who struggle with others over the control of space and resources in the game of survival. In this ecological conceptualization of human competition, the spatial arrangements of urban settlements represent the different levels and opportunities of adaptation of humans and their social groups to the physical environment. In other words, the ecological urban models of the Chicago School draw heavily on establishing correlations between ecological patterns and social processes. This tradition goes back to Herbert Spencer, who was the first to use Charles Darwin's evolutionary theory to explain social processes. Spencer interpreted the processes of social organization by comparing human interactions and their resulting spatial patterns to the principles of species competition, where in personal competition for resources the individuals selected are those who are best suited for the task. This view became predominant in

the early theoretical work of the so-called Chicago School of sociology, to which Robert Park, Ernest Burgess, Roderick McKenzie, Chauncy Harris, Edward Ullman, and Homer Hoyt became the most significant contributors. In their work, they documented how the city became the playground for human competition where social groups compete for jobs, opportunities, and land. The outcome of this work was schematic models of idealized urban environment that portrayed the spatial arrangement of urban structures as a process in which certain social groups, at certain points in urban development, occupy certain parts of the city. In Park's concentric zone model, for example, the city of Chicago is portrayed as an idealized flat surface with concentric rings of different functions and land uses organized around it. According to Park's theory of urban development, zones are centered around a central business district (CBD). Immediately around the CBD could be found a transitional zone that was characterized by older, smaller, less valuable and attractive houses mixed in with light industry and office space. This zone generally attracted newcomers to the city, especially immigrants, due to the fact that housing there was generally cheap, available, and temporary. Succeeding this transitional zone could be found another low-income residential area that was more stable in terms of the in- and out-migration of residents and that was established over longer periods of time, with a culturally and socially stable community. Farther outward in another ring could be found the increasingly larger dwelling units of the middle and upper classes whose members were willing and able to commute considerable distances to their places of work.

Although this idealized model of urban structure was not to be taken literally, it provided a pattern of social differentiation in cities and also sought to explain the process behind it. For Park and Burgess, the city was a place of human competition for resources in the struggle to survive and succeed. As newcomers entered the city, they typically were poor and in search for work. The transitional zone provided both residence and work opportunities for these new arrivals. Once established, these workers would save resources, make more money, and consequently look for more permanent and attractive housing in more stable communities. As individuals climbed up the social ladder, they relocated farther away from the center of the city in more attractive neighborhoods. Others who were not so lucky or successful might stay within an individual zone or move outward only slightly, reflecting their lack of success in the struggle for resources and prestige in urban settings.

Other urban researchers subsequently refined Park and Burgess's model and portrayed the city as an organically grown entity resulting from human competition in starlike, sectoral, or zonelike patterns. Also, these researchers would take into account more factors that explained the process creating these patterns by including in their analyses not only a historical progression of upward mobility but also international influences and a variety of broad economic forces. Yet what all of these models had in common was that they assumed the development of urban spatial structures to be the natural outcome of economic competition among rationally behaving humans.

MARXIST POLITICAL ECONOMY, URBAN FORM, AND BEYOND

In contrast to such macro- and microeconomic models of human competition in capitalist societies stands a whole body of literature that approaches urban spatial structures from a Marxist, socialist, or critical geographic perspective. Although the literature on Marxist analysis of cities is wide-ranging and complex, two general approaches to understanding urban environments from this ideological perspective are highlighted here: class conflict theory and capital accumulation theory. These two groups of models do not necessarily provide new models and outlines of urban form, but they criticize the explanations of the process that led to contemporary urban forms as the Chicago School attempted to sketch them.

Class conflict theory postulates that urban spatial structure can be understood as the outcome of class struggle. In general, it assumes that in a city, urban forms are changed and manipulated by a ruling and landowning class whose members rearrange their production facilities according to their need for profit maximization and security. Thus, the working class becomes a passive responder to the needs of those in control of power and resources, and the working class is forced to move according to the changing realities of production in a capitalist system. As technologies for production improve, new facilities are built and residential neighborhoods develop around them. Although early industrial cities show a close proximity of workplaces and residences, this began to change at the turn of the 20th century with the increasing influence of labor unions and the social organization of labor. With strong labor organization, the initial

ideal of profit maximization by having workers close to the places of capitalist production no longer fit the needs of industrialists. Proximity became a hindrance due to the fact that labor organizations could quickly organize unrest or strikes and block access to the production sites. As such, it became necessary for the ruling class to isolate the workers from labor organizations. This was accomplished by the collective efforts of capitalists to move their production sites out of the densely populated inner-city settings to less crowded adjacent areas. This shift toward the outskirts of the city started a process of decentralization and sprawl that was aided by the arrival of effective transportation technologies such as the railroad, the streetcar, and the automobile. In contrast to such early, simplistic Marxist models of class conflict that focus on the mechanisms behind urban growth and change stands the literature on capital accumulation theory.

Capital accumulation theory assumes that urban spatial structures are the result of the need of capital owners (those possessing land, money, and resources) to accumulate and continuously reinvest their financial resources to ensure the continued growth of their corporate operations. Thus, the surplus production of capital and its reinvestment, circulation, and organization directly determine urban form. The ways in which the location of production facilities, residences, and the flow of goods and services in a city are set up spatially reflect the profit maximization strategies of those in control of the resources in a capitalist society. For example, three ways in which individuals and corporations in a capitalist society can make money is by rent, interest, and profit. Urban development and expansion is driven by the need of corporations to grow, even during times of low demand, and to invest capital to gain more surplus resources by renting out spaces or by acquiring land to build new structures that can be sold at a profit and have to be financed by new owners with interest-bearing loans. Urban growth and change becomes a necessity to keep the process of making more and more profit running. Owners of capital and property intervene in urban structures whenever the prospects of cheap land and structures cross their way, only to return these structures to the market at a profit. Although these observations barely scratch the surface of the complex analysis of class conflicts and capital accumulation, it is necessary to reiterate that Marxist political economic analysis of urban spatial structures focuses on uncovering the multiple mechanisms that guide the development of our constantly changing urban landscapes. These theories explain how the logic of capitalism, particularly attempts to avoid class conflict and maximize profits, drive the design of urban spaces that fit the needs of a limited number of people—those in control of land, financial resources, and political resources.

During recent years, a distinct school of urban thought that actively challenges many of these metatheoretical large-scale approaches has emerged. During the 1990s, a group of scholars whose work focused on the city of Los Angeles argued for new alternative models of spatial structure. The work of Michael Dear, Steve Flusty, and Mike Davis has been instrumental in this context. Using the examples of southern California's urban landscapes, they argued that the empirical analysis of urban forms ought to account for the multiple shapes and arrangements of urban settlements instead of attempting to generalize the processes and patterns that might explain the city to us. Late-20th-century cities no longer fit into traditional urban ecological or Marxist models; rather, they often are the result of arbitrary, contradictory, and chaotic development that needs to be analyzed in the context of local and regional economies, cultures, and political systems.

—*Olaf Kuhlke*

See also Central Business District; Chicago School; Exurbs; Global Cities; Housing and Housing Markets; Marxism, Geography and; Production of Space; Space, Human Geography and; Suburbs and Suburbanization; Urban and Regional Planning; Urban Ecology; Urban Entrepreneurialism; Urban Fringe; Urban Geography; Urban Managerialism; Urban Social Movements; Urban Sprawl; Urban Underclass; Urbanization; Zoning

Suggested Reading

Gottdiener, M. (1994). *The social production of urban space* (2nd ed.). Austin: University of Texas Press.

Harvey, D. (1973). *Social justice and the city*. Baltimore, MD: Johns Hopkins University Press.

URBAN SPRAWL

The term *urban sprawl* is associated with the growth, form, and composition of urban areas and has several commonly used meanings. First, existing development within the generally accepted bounds of an urban or metropolitan area is referred to as urban sprawl if characterized by low-density/intensity uses that are mostly segregated from one another and

spread out over the landscape. Second, new development that occurs at the urban fringe or in surrounding rural areas is referred to as urban sprawl particularly if it is scattered (i.e., interspersed with undeveloped lands), leaps over undeveloped lands, or radiates out from the existing urban area (typically along roadways). Third, the process of urban growth generally characterized by the outward expansion and deconcentration of urban activities and land uses into the surrounding countryside is referred to as urban sprawl.

Although chiefly described as a land use pattern, urban sprawl is necessarily linked to the transportation system, other public infrastructure and services, and economic activities (both local and global). For example, systems of high-capacity roadways designed to collect commuters from distant locales and transport them efficiently to workplaces have developed concurrently with a sprawling development pattern.

Most urban planners view urban sprawl as an undesirable development pattern with multiple externalities. It is commonly described as inefficient, costly, unattractive, indirectly linked with the decline of the existing urban core, and excessively consumptive of resources such as agricultural lands, natural areas, and rural landscapes. This contrasts with an idealized urban form characterized by high-density residential and high-intensity office, commercial, and industrial uses intermixed in a compact pattern that is economically and fiscally efficient as well as socially equitable and vibrant.

Yet urban sprawl continues mostly unabated as the primary residential choice of the middle class as suburbs and exurbs continue to proliferate. This paradoxical result is significant to the discourse on urban sprawl that reflects the contested nature of growth and development. Urban sprawl is the pejorative term used for a particular pattern of development that may otherwise be viewed positively. Arguably, the continued proliferation of sprawl is the unintended result of actions of state, capital, and civil society that create a dispersed, low-density settlement pattern while seeking to accomplish other desirable objectives.

Households frequently view suburban and exurban developments as safer, cleaner, and less congested, and as having superior facilities and services (particularly schools), compared with older urban areas. Developers take advantage of lower land prices at the urban fringe to increase profits. Numerous government policies and programs serve to facilitate sprawl, including federal home mortgage programs, interstate highways, and tax breaks.

Thus, urban sprawl can be viewed as describing a complex web of land uses and infrastructure. It is the product of capital, state, and civil society, and as a socially defined spatial structure it interacts back on those systems. Numerous theories attempt to explain the production of urban sprawl and the more generalized expansion and deconcentration of urban areas. These theories vary in their emphases on the primary causes of sprawl, including economics, technology, social factors, and state policies that subsidize or otherwise facilitate sprawl.

Any analysis of the causes and effects of, and potential solutions to, urban sprawl requires a precise and (often) locally specific operationalization of the concept. Major variables include the level of density and intensity of development, amount and mix of uses of development, spatial relationship to other development (including accessibility among land uses), quality and/or use of the land prior to development (e.g., environmental or agricultural), provision of infrastructure, and the equitable distribution of initial and future costs of the development. These variables must be considered individually and collectively. The numerous urban sprawl studies conducted since the 1970s have failed to address these variables consistently, resulting in disagreements as to the impacts of urban sprawl.

Over the past several decades, urban planning has sought to discourage urban sprawl, with limited success. Current methods include new urbanism, smart growth, urban service areas, urban growth boundaries, and various efforts at central-city revitalization. The success of these initiatives is mixed, however, because the desire to promote economic development along with capital profits and an increased tax base often results in permissive land use planning and regulation.

—*Robert Pennock*

See also Edge Cities; Exurbs; Rural–Urban Continuum; Suburbs and Suburbanization; Sunbelt; Transportation Geography; Urban Fringe; Urban Geography; Urbanization; Zoning

Suggested Reading

Gillham, O., & Maclean, A. (2002). *The limitless city: A primer on the urban sprawl debate.* New York: Island Press.

Squires, G. (Ed.). (2002). *Urban sprawl: Causes, consequences, and policy responses.* Washington, DC: Urban Institute.

URBAN UNDERCLASS

Within the cores of American cities, and to a lesser extent in other industrialized countries (although most have better national safety nets than does the United States), lives a sizable group of impoverished people often known as the *urban underclass.* The term has been the object of much debate because some argue that it dehumanizes those to which it refers; however, the concept is widely acknowledged to have considerable validity. In the American context, the urban underclass refers to the poor, overwhelmingly African American population whose members populate the ghettos and low-income neighborhoods in the centers of most metropolitan areas. The story of how the urban underclass came to be formed reveals much about the race-specific dynamics of urban growth in the United States.

Some observers explain the origins of the underclass in terms of the three centuries of slavery and violent suppression practiced against African Americans, the only immigrant group whose members did not come voluntarily to the New World. White racism and institutionalized discrimination were largely responsible for the low rates of intergenerational mobility found among this population. Although the traditional core region of black Americans was in the South, by the early 20th century large numbers had begun to migrate north toward the ample jobs found in the then booming Manufacturing Belt. By the 1920s, large African American working-class communities had emerged in cities such as Boston, New York, Philadelphia, Pittsburgh, Cleveland, Detroit, Chicago, and Milwaukee, where many found employment in a wide range of semiskilled or unskilled occupations in industries such as the railroads, meatpacking, coal mining, shipyards, warehousing, steel, machine tools, and automobile production. Labor shortages accentuated by congressional quotas limiting immigration and the demand for labor during World War II contributed to these opportunities. Many African Americans developed vibrant healthy communities with low rates of crime, stable families, relatively high rates of homeownership, and incremental improvements in their standards of living. Harlem, in northern Manhattan, New York, emerged as the center of black intellectual and cultural life.

However, following World War II, and particularly toward the end of the postwar boom during the 1960s, structural changes in the American economy and society began to erode the foundations of these communities. Steady sustained suburbanization saw the evacuation of much of the white middle class to the urban periphery, taking with it much of the tax base necessary to sustain public services in the inner city, particularly schools. Because suburbanization was an escape that was not generally open to most African Americans, who lacked the incomes to buy suburban houses and often were the targets of discriminatory zoning ordinances, the proportions of African American populations in inner cities rose steadily even as their economic prospects began to dwindle.

Deindustrialization, brought on by mounting international competition and technological displacement of workers, saw numerous factories close in the Northeast and Midwest. Many factories shut down in the national core, only to open up in the burgeoning Sunbelt or overseas. As manufacturing opportunities declined, African Americans—generally the least skilled and educated and with the fewest alternatives—bore the brunt of the costs in terms of declining employment and incomes. In short, deindustrialization affected blacks considerably more than it did whites, who generally made the transition into services more smoothly. As a result, black unemployment, particularly among males, rose steadily during the 1960s and 1970s, and black poverty rates—already higher than those for whites—reached new highs. Today, African American family incomes are generally around 70% of those of white families, although there has been some increase in the size of the black middle class.

The growing crisis of African Americans in labor markets was compounded by changes in housing markets. Blacks had long been the target of discriminatory practices by landlords and by financial institutions in the form of redlining, that is, refusal to extend mortgages to low-income, minority-inhabited areas. During the 1950s, federal government slum clearance programs annihilated many working-class black (and some white) neighborhoods to clear urban space for freeways and office buildings. Housing shortages, accentuated by an overall disinvestment in housing (including abandonment) and lack of repairs by landlords, created a growing problem of affordable housing in the inner city.

The result of these conjoined trends—deindustrialization, suburbanization and "white flight," a declining municipal tax base, and deteriorating housing affordability—was to concentrate impoverished African Americans in the inner-city urban underclass. Although there is considerable diversity within such communities in terms of standard of living and life chances, for the most part such areas are typified by decrepit building stocks and relatively few employment opportunities. Often in such circumstances, the informal economy becomes significant, including a variety of semilegal or illegal practices that include sales of illegal drugs, prostitution, casual day labor, and crime. With low expectations of achieving steady employment, many African American youth make the decision to drop out of high school.

Conservative views of the black ghetto often maintain that it was produced by federal government reforms introduced during the 1960s War on Poverty, which ostensibly removed the incentive to find employment. However, it is revealing to note that this position ignores the macroeconomic context that underpins the changes in the inner-city economy. Moreover, welfare payments have decreased steadily in real (postinflationary) purchasing power over time, and employment opportunities for the urban underclass rarely extend beyond minimum-wage retail trade jobs.

Sociologist William Julius Wilson introduced another dimension to this argument, noting that the black family, historically a strong institution, collapsed under the weight of economic and social stress during the mid-20th century. Specifically, rising unemployment generated a shortage of marriageable young men, leading many young women to forgo marriage. Prior to the 1950s, black women were more likely than white women to marry; however, by the late 1960s, this situation had been reversed. The steep rise in black out-of-wedlock births (to roughly 70% of black births today) created a different family unit, one in which children and mothers were deprived of the benefits of two wage-earning parents (in contrast to the situation with whites) and the role models of adults who worked on a daily basis. Wilson's critique is not a moral argument but rather an economic one, namely that families headed by single women are almost invariably poorer than those with two working adults present.

Today, in an increasingly globalized, service-oriented economy, the urban underclass remains a major social predicament for American society. In general, inner-city neighborhoods populated by African Americans and the working poor witness a lack of well-paying jobs. Low rates of high school graduation and college attendance deprive many residents of opportunities in the growing producer services economy. Inner-city schools in the United States are notoriously underfunded and often are crowded and violent with dilapidated buildings. Gentrification in many such places has contributed to a growing problem of housing affordability and has helped to swell the ranks of the homeless, who are disproportionately minorities. Moreover, neoliberal political programs that cut aid to the poor (e.g., housing subsidies, winter preparedness, employment assistance) have accentuated inner-city poverty. Public healthcare and medical care in such communities often is a national disgrace; lacking health insurance, many families are forced to use emergency rooms as a source of primary medical care. Rates of drug addiction and human immunodeficiency virus (HIV) infection are far higher than the national average. Moreover, crime rates, particularly those for violent crimes, among inner-city residents tend to be much higher than those among residents in suburbs.

In sum, the formation of the inner-city ghetto and urban underclass reveals the intersecting dynamics of race and class in the United States, that is, the ways in which geographies of advantage and disadvantage play out in particular ways and in particular places.

—*Barney Warf*

See also Deindustrialization; Ghetto; Poverty; Race and Racism; Rustbelt; Social Justice; Urban Spatial Structure

Suggested Reading

Wilson, W. (1987). *The truly disadvantaged: The inner city, the underclass, and public policy.* Chicago: University of Chicago Press.

URBANIZATION

Urbanization is the general process by which essentially rural societies and the regions they occupy transform into predominantly urban ones, usually occurring over long periods of time, involving substantial redistributions of people in space, and concentrating proportionally more and more of them in towns and cities. Urban places and their communities differ principally from rural ones by exceeding certain thresholds of population size, physical density, territorial extent, administrative status, functional complexity,

and morphological diversity, beyond which they are considered unquestionably urban in character.

A host of interrelated transforming processes drive urbanization. Foremost among them is economic change, but others include demographic, political, cultural, social, and technological changes as well as shifts in natural resource use. All of these are shaped fundamentally by geographically and historically contingent circumstances.

In theory, the process has a beginning, when towns and cities appear for the first time within a previously wholly rural settlement area, and an end, when virtually all people live concentrated in urban places. Most regions fall somewhere between these extremes of urbanized degree (at the continental scale between 37% and 80% in 2000), depending on the type and extent of their economic development and the interplay of these and other factors. Urbanization is seen as a permanently transforming societal process that is diffusing historically across the globe. Consequently, the earth's regions consistently display immense geographic variation in the extent and form of their urbanization. The organization of urban systems (i.e., networks of cities), land use, social geography, infrastructure and built environment, and townscape character reflects such variation.

Urbanization can be observed, measured, analyzed, and understood at many geographic scales—from localities of a few square kilometers or miles, to regions such as metropolitan areas, to regions of a larger extent such as states, to nations, to continents.

PERIODS IN URBANIZATION HISTORY

Human society can be said to have passed through a series of fundamental technological developments in exploiting Earth resources—chiefly the invention of agriculture, the discovery of trade, and the perfection of industrial manufacture—with urban centers playing an increasing role at each stage. This view led some theorists to posit that the world has passed through three periods of urban experience: (1) an essentially preurban world in which pre- or protourban centers were few, were scattered, and existed largely for ceremonial purposes; (2) an urban world in which cities arose as distinct trading and manufacturing nodes spatially distributed across vast agricultural regions to coordinate complementary flows of commodities and services within and between them; and (3) a possible posturban world in which widespread population decentralization has blurred the distinctions between urban and rural living and in which most people occupy a ubiquitous web of urbanlike settings strewn across and mingling with vast rural-like zones used for producing consumer essentials and providing various lifestyle environments. Even so, the historic urban centers have continued to be crucial in providing services, knowledge and innovation, and connectivity.

It is generally recognized that the spread of the Industrial Revolution, which began in Britain during the 18th century and was based on the harnessing of inanimate sources of power, proved to be the biggest catalyst in the widespread acceleration of urbanization at a global scale. Prior to 1900, the world's population as a whole was little urbanized. By 1950, fully 29% of the world's population was considered urban, and by 2000 the proportion had reached 47%. It is estimated to reach 60% by 2030.

TRANSFORMING PROCESSES IN URBANIZATION

Cities, as inherently non-self-sufficient settlements, depend on a wide array of food, raw materials, manufactured products, and services produced elsewhere to survive. As cities have grown in number and size, their collective draw on the earth's resources has increased substantially. With the improvement of transport and the increased demand for low-density urban development, the proportion of land consumed by urban places has risen steadily. Cities are considered to be largely antithetical to nature. Urbanization creates specialized natural ecologies in which many vegetative and animal species have learned to adapt and thrive, but many forms of urban development have created natural instabilities, including urban-exacerbated flooding and ozone-depleting air pollution.

The most enduring economic force driving urbanization is the seemingly relentless historical process of progressive specialization in and division of labor, a trend that various political–economic systems, from capitalism to socialism, have encouraged and exploited. Industrial capitalism has encouraged the extensive enlargement of manufacturing capacity, which has brought aggressive urbanization. In advanced economies, this has created several phases of industrial renewal and relocation, each time spurring a new round of regional urbanization and reurbanization. The decline of traditional imperialism and the intensifying globalization of trade and finance

during the 20th century have spread economic development pressures to regions previously little affected and have produced significant urbanization as a result.

The mechanization of most forms of production triggers business relocations to more efficient production sites, often in "greenfield" areas peripheral to established urban centers, thereby leading to the urbanization of formerly rural places. Mechanization also diffuses across and between regions, stimulating new urbanization. Technological change brings new modes of physical transport into being, each with its own spatial attributes and influences. Railroads once encouraged urban concentration; superhighways encourage decentralization. Air freight services permit regional deconcentration and long-distance impulses toward new rounds of urbanization in dispersed localities. Electronic communication supports both centralization and decentralization.

All major demographic forces—birth rates, death rates, and migration patterns, and so forth—can fuel urbanization (both singly and in combination). Natural increases through higher birth rates in rural areas than in urban areas stimulates cityward migration that enlarges cities, especially when rural areas cannot absorb the increased population. In large, economically developed urban regions, urban areas can achieve their own net natural growth. In more local settings, the reclassification of rapidly developing rural areas to urban status alters the basis for calculating urbanization levels and rates.

Urbanization is both a reflection and a driver of cultural change. The acceptance in a society of radical economic change, for example, results in the greater importance of cities. This produces greater cityward migration to fill new jobs and take advantage of better living conditions available in towns. Urbanization, in turn, acts as a diffusion mechanism for cultural change through the wider regional society, further altering the broader society's openness to cultural change. The development of a mass consumer culture, for instance, combines with transport mobility and improved communications to spread urban appetites and lifestyles in rural areas, stimulating cityward migration and the greater receptivity of rural populations to urban commodities and values.

Urbanization tends to promote social heterogeneity and interaction. As dynamic centers of economic exchange, cities always have attracted migrants from diverse regions with less opportunity to offer, migrants whose connections were needed for interregional trade, or migrants whose labor and skills were needed in new industrial ventures. This has promoted far greater diversity of regional, ethnic, and religious backgrounds among residents of urban areas than among residents of rural areas. Urban migration often is seen as draining rural areas of their forward-looking home-grown talent, to the benefit of already socially diverse and (often) more socially tolerant cities.

Cities have long been the loci of political power within their surrounding regions; therefore, urbanization is associated with the concentration of such power. Governments of practically all types, whether monarchies, dictatorships, or democracies, benefit from urbanization and in turn seek to influence the degree and location of urban growth. Changes in political organization often resulted in the multiplication of capital cities, as well as the advance of border urbanization, because of differences in political–economic conditions on either side of political boundaries. In modern times, growing dependence across intergovernmental frontiers, represented by transnational agreements and global corporate activity, has strengthened supranational bonds while partially weakening national states.

GEOGRAPHIC PRODUCTS OF URBANIZATION

Urbanization creates a distinct set of geographic consequences, six of which deserve primary mention. First, cities are so numerous across the face of the earth that at all scales, and in all regions, they form urban systems and regional subsystems, that is, hierarchically structured networks of cities and towns in space. Thus, one can speak of a world urban system in which global cities occupy the top ranks of the hierarchy. Second, urban complexity generates within urban areas differentiated patterns of land (and building) use, generally classified as commercial, residential, institutional, industrial, transport, and recreational uses, all of which display their own specific spatial structures. Third, this results in a built environment composed not only of buildings and circulation spaces but also of an infrastructure of varied utilities such as water and sewer systems, power lines, and telecommunications networks. Fourth, the diversity of work patterns, earning capacity, sociocultural heterogeneity, and psychological factors produces a complex social ecology within cities in which work and residence often are sharply differentiated in space. Fifth, the concentration of activities and disciplines unique

to urban places creates variable *urban* "ways of life" that differ greatly not only from those of rural areas but also from one urban region to another. Sixth, all of these geographic features of urbanization are embodied in townscape and cityscape character, visually as well as symbolically, both reflecting and influencing perception and behavior.

GEOGRAPHIC SCALE AND PLACE

There is great variation worldwide in defining types of urban places that influence the scale, mode, and rate of urbanization. Although villages and hamlets are considered to belong to the rural realm, nearly all other terms for agglomerated settlements denote places that reflect some sort of urbanization. This usually is considered in relation to thresholds of size and importance above which there are discrete size classes in an ascending scale of urban-ness. Urban growth may advance individual places from one class to another, but the size range of common terms such as *town, commune, city,* and *metropolis* are culturally defined and vary widely from country to country and even within regions. Criteria are either socially based (number of inhabitants or specially defined forms of community), place based (geographic character of locality), or a combination of both.

Among the socially based measures, the statistical threshold for considering urbanization to have begun ranges from 200 inhabitants (Norway) to 10,000 inhabitants (Malaysia). Interestingly, Syria recognizes only places with more than 20,000 inhabitants as urban unless they are Mohafaza and Mantika centers or official cities, which may be of any size. The threshold in the United States is 2,500 inhabitants. Elsewhere, administrative or legal status as towns or cities and a plurality of nonagricultural families ensure recognition as urban. Alternatively, place-based criteria may apply, usually defining some sort of urbanized area according to extent or density, for example, at least 100 dwellings (Peru), at least 400 persons per square kilometer (Canada), or houses less than 200 meters apart (France). Sometimes the measures are vague, for example, possession of "urban characteristics" (Indonesia) or simply "localities" (Finland). In Nicaragua, these are defined as urban if they have "streets and electric light." Such variations make the calculation of comparative urbanization indexes problematic.

Intense urbanization on a large scale produces metropolitanization in which urban areas often spread beyond municipal boundaries, producing suburbanization and conurbation, often across provincial, state, and national boundaries. In the United States, metropolitan areas are defined as clusters of counties that meet certain size, density, contiguity, and commuting criteria.

Counterurbanization may be considered a special type of urbanization in which complex societal urbanization pressures lead to strong urban-to-rural migration and the consequent growth of rural areas and small towns at the partial expense of large centers. This is particularly characteristic of developed regions at an advanced stage of their evolution and does not necessarily mean the deurbanization of metropolitan areas. Deurbanization occurs under sharp reversals of established urbanization trends in particular circumstances. Wars can produce periodic and localized deurbanization, and Mao Zedong's governmental policy in China for a generation created a strong forced shift in population and production away from cities.

GEOGRAPHIC VARIATION IN THE PACE AND SPREAD OF URBANIZATION

Historically, strong urbanization has been associated with developed countries, especially in Europe and North America. During the late 20th century, the developing world saw the most dramatic rise in urbanization, although rates varied widely among countries. Excluding formerly colonial city-states (e.g., Singapore), as recently as 1975, Belgium was alone in the world in being more than 90% urbanized. By 2000, 15 countries had exceeded this level, and it is estimated that 27 states will surpass this benchmark by 2030. Many of the new cases will be in Africa, South America, and the Middle East. Despite Asia containing some of the world's largest urban agglomerations,

it is predicted that the region's general urbanization will not occur until after 2030. Urbanization has not necessarily paralleled rates of economic development, and this has placed uneven pressure on natural resources and further widened differences in global migration patterns. Urbanization may represent a homogenizing force in human society, but its changing geography will long continue to be one of its most salient features.

—*Michael Conzen and Nicholas Dahmann*

See also Built Environment; Circuits of Capital; Diffusion; Division of Labor; Economic Geography; Exurbs; Global Cities; Globalization; Industrial Revolution; Rural–Urban Continuum; Suburbs and Suburbanization; Urban Ecology; Urban Geography; Urban Spatial Structure

Suggested Reading

Champion, T. (2001). Urbanization, suburbanization, counterurbanization, and reurbanization. In R. Paddison (Ed.), *Handbook of urban studies* (pp. 143–161). London: Sage.

Knox, P., & McCarthy, L. (2005). *Urbanization: An introduction to urban geography* (2nd ed.). Upper Saddle River, NJ: Prentice Hall.

Lampard, E. (1965). Historical aspects of urbanization. In P. Hauser & L. Schnore (Eds.), *The study of urbanization* (pp. 519–554). New York: John Wiley.

United Nations. (2003). *Demographic yearbook 2001*. New York: Author.

United Nations. (2004). *World urbanization prospects: The 2003 revision*. New York: Author.

U.S. DEPARTMENT OF HOUSING AND URBAN DEVELOPMENT

SEE HUD

VIRTUAL GEOGRAPHIES

The term *virtual geography* originated during the 1990s. Broadly, it refers to geographic aspects of information and communication technologies (ICTs), especially the Internet, as well as related geographies of the social, cultural, and political spheres. It is an example of a heterogeneous assemblage of material and symbolic relations.

Early commentators on virtuality tended to differentiate it from physical spaces. Not only was it new or unprecedented, but unlike physical space it also was globalized, decentered, immediate, and placeless. Although writers were divided on whether virtual spaces were liberatory or surveillant, democratizing or repressive, and enabling or unequal, many of them agreed that virtual spaces were fostering a digital era of globalization. Whereas some emphasized an inherently negative logic (citing increased corporatization and surveillant characteristics), others celebrated the ability of virtual spaces to foster community or found links between democratic achievement and Internet access. Manuel Castells epitomized this work by arguing that virtual geographies were made up of "spaces of flows" rather than places. This led other writers to speculate that an old idea in geography, time–space compression or convergence, signaled the "death of distance."

These approaches were criticized on two grounds. First, they tend to homogenize and essentialize virtual spaces. Rather than understanding virtual spaces as separate from physical spaces, critics emphasized the mutual relations between virtual spaces and material spaces. Virtual spaces (or "cyberspace") were produced by sociopolitical conditions of material spaces, and it

was also realized that there was a simultaneous impact on our material geographies by the virtual. This "co-construction" of virtual space and physical space is now a dominant understanding of cyberspace. Second, it was argued that virtual spaces were not placeless or decentered; rather, they had a distinct geography.

Virtual geographies include not only the Internet but also cellular and wireless technologies, video monitoring systems, and the associated social practices (e.g., discussion groups, blogging, online commerce, voting, online mapping, and geographic information systems [GIS] services). These complex assemblages of material-discursive objects have precluded commentators from easily defining their field of study. Nevertheless, those interested in virtuality and geography have pursued several themes of study.

MAPPING CYBERSPACES

Some of the earliest and most evocative work on virtuality addressed the following questions. Where is cyberspace? Can the flows of information be mapped? Who is connected to whom? Many researchers captured the geography of the Internet, as measured by its networks, flows, and nodes, in a variety of maps. From its origins in the United States during the 1960s, the Internet (and other ICTs) has diffused to nearly every country in the world, although with marked disparities.

DIGITAL DIVIDE

Maps of the Internet soon revealed what many commentators already suspected, namely that virtual spaces are not placeless or equally accessible. This "digital divide" refers not only to unequal access to

ICTs but also to the skills and services of the knowledge economy. As such, it has profound implications for the spatial nature of economic systems. A number of geographers have traced inequalities in the Internet industry and argued that it is characterized by investment in existing economic and intellectual infrastructures (e.g., Silicon Valley, research universities in the United States). Although the Internet is global, there remain profound differences not only in access itself but also in content and control. These disparities have been identified at multiple geographic scales below that of the state (e.g., at the local or county level in the United States). The lack of data on connectivity (and its rapidly changing status) has led some commentators to model the divide using economic indicators such as the Gini coefficient and the Lorenz curve.

LOCATION-BASED SERVICES

Exploiting the ability of ICTs to provide real-time locational updates, a variety of products and services that provide geographic information have emerged. Although many of these are commercial or governmental in nature (e.g., asset tracking, emergency vehicle tracking), individuals have developed a number of them. These "map hacks" and "geoblogs" have incorporated the spatial into people's everyday lives. Locationally aware devices and services proliferated during the early 2000s.

LEGAL, ETHICAL, AND POLITICAL IMPLICATIONS

As a result of these differences, it has become increasingly clear that cyberspace will be the site of political intervention, either to alleviate the divide through an appeal to cyber rights or by seeking to protect centers of investment and to maintain control over content. Although the globalizing effects of virtual geographies challenge the primacy of the nation-state, a number of companies based in one country have complied with laws on content in another country. Other geographers have pointed out that the impact of the virtual on people's everyday lives has been profound. Geographic tracking tools (e.g., the Global Positioning System [GPS], cellular technology) have prompted warnings of geoslavery or locational surveillance. In the context of America's wars in Afghanistan and Iraq during the early 2000s, the United States passed laws that not only criminalized cyberattacks but also equated them with terrorism and threats to homeland security. Thus, the virtual had come to be identified as a real territorial extension of the nation-state.

Today, most geographers understand that there is a dialectical relation between physical spaces and virtual spaces. The fact that in many cases the same geolocational tools are used with very different purposes indicates that technology is not neutral; rather, it is constructed within the messy situatedness of spatial relations. Although pronouncements that are dependent on rapidly changing technology are risky, geographers seem justified in concluding that rather than deterritorialization, virtual geographies have engendered a whole range of *reterritorializations*.

—*Jeremy W. Crampton*

See also Cyberspace; Geoslavery; GIS; Location-Based Services; Telecommunications, Geography and; Time–Space Compression

Suggested Reading

Cairncross, F. (1997). *The death of distance: How the communications revolution will change our lives.* Boston: Harvard Business School Press.

Chakraborty, J., & Bosman, M. (2005). Measuring the digital divide in the United States: Race, income, and personal computer ownership. *The Professional Geographer, 57,* 395–410.

Crampton, J. (2003). *The political mapping of cyberspace.* Chicago: University of Chicago Press.

Crang, M., Crang, P., & May, J. (Eds.). (1999). *Virtual geographies: Bodies, space, and relations.* London: Routledge.

Dodge, M., & Kitchin, R. (2001). *Atlas of cyberspace.* Reading, MA: Addison-Wesley.

Dodge, M., & Kitchin, R. (2001). *Mapping cyberspace.* London: Routledge.

Graham, S. (1998). Spaces of surveillant simulation: New technologies, digital representations, and material geographies. *Environment and Planning D, 16,* 483–504.

Kitchin, R. (1998). *Cyberspace: The world in the wires.* Chichester, UK: Wiley.

Misa, T., Brey, P., & Feenberg, A. (Eds.). (2003). *Modernity and technology.* Cambridge: MIT Press.

Zook, M. (2005). *The geography of the Internet industry: Venture capital, dot-coms, and local knowledge.* New York: Blackwell.

VISION

Geography has a deep tradition as a visual discipline. In late-19th- to mid-20th-century treatises defining the field as a descriptive science, geographers were called to describe what they saw thoroughly and objectively. Whether the objects of this gaze were landscapes, regions, or cultures, few questioned the ability of geographers to see such objects in their entirety and to represent them to others objectively via the map and the word. Although the definition of the discipline and its methods of inquiry have changed and broadened over the past 30 years or so, the dominance of vision and the visual continues in contemporary geographic inquiry. Cultural geographers read landscapes as visual texts and note that tourists consume places through the gaze. Cartographers and geographic information scientists use high-resolution imagery and three-dimensional graphics to create geographic visualizations. And an increasing number of geographers examine the reproduction of spaces and places in a variety of visual media.

It is common to make a distinction between vision and visuality. Vision can be defined as what the human eye is physiologically capable of seeing. The ability of the eye to perceive certain wavelengths of radiation, for example, defines the visible portion of the electromagnetic spectrum. Not surprisingly, most remote sensing systems used by geographers are designed to detect, record, and image this type of radiation. Visuality refers to the cultural construction of what and how people see. Children must be taught to identify the objects they see and to interpret the meanings of the objects in their own cultural contexts. In a more political realm, groups and individuals suffering some form of oppression often work with members of the media to make their struggles visible to a larger population.

The primacy of the visual among the human senses often is presented as a characteristic of modernity in Western society. As both Martin Jay and Chris Jenks noted, the rise of printing, the reliance on the written word for communication, and the use of the telescope and microscope to bring the distant and the invisibly small into view all contributed to the tendency to equate seeing with knowing. To a large extent, the development of geography as a modern discipline rested on this assumption. Derek Gregory described how 19th-century geographers followed careful procedures to describe, classify, and map the resources and "native" peoples they saw as they worked to enlarge the West's "scientifically known" and politically controlled world.

More recent work in geography has been heavily influenced by the increasing prevalence and importance of visual images in Western society. Beginning in the 1980s, geographers have examined how place, space, landscape, and identity are represented and/or reproduced in visual media such as films, television programs, and advertisements. Much early work in this field tried to identify misrepresentations of particular places or identities by comparing the images with reality as seen/known by the geographers.

Other contemporary geographic research seeks to retheorize the relationships between vision/visuality and space, place, and landscape. Some of this work relies on postmodern critiques of the modernist relationship between sight and knowledge described earlier. If images of bodies, places, landscapes, and the entire planet are increasingly central to everyday experiences, it becomes impossible to see/know a reality outside of its representation. An individual's experience in Los Angeles, for example, may be heavily influenced by knowledge gained through previous consumption of visual images of that city on television or in films. Therefore, many cultural geographers view all visual phenomena, including material landscapes, as cultural texts that must be decoded. Gillian Rose's *Visual Methodologies* reviewed several approaches to such research, including content analysis, semiology, psychoanalysis, and discourse analysis.

Such approaches may be used to reveal the uneven social relations of power involved in the production and consumption of images. Feminist geographers note that some visuals appear to be authoritative and objective both because the position of the image producer is hidden from image consumers and because the resulting representation appears to show objects in their entirety. Such visualizations tend to marginalize the seemingly partial perspectives of oppressed groups, including those based on sexuality, class, gender, and race. To counter this kind of absence, feminists and other critical human geographers argue that views of spaces, places, landscapes, and identities always should be represented as partial, multiple, and/or fragmented.

The link between sight and power is also central in Michel Foucault's work on surveillance in modern social institutions. Using the metaphor of Jeremy Bentham's Panopticon, a prison plan that would make every inmate constantly visible to prison guards, Foucault argued that individuals' actions/behaviors in space are regulated, or self-regulated, by their visibility. J. Brian Harley, for example, noted that maps make things subject to surveillance and control. Similar critiques have been leveled at geographic information systems (GIS) and the Global Positioning System (GPS) that allow governments or businesses to create spatial units that permit systematic observation.

Finally, some geographers point out that continuing dominance of the visual in the discipline's approach to space, place, and landscape limits the attention paid to how people apprehend their worlds through the other senses and marginalizes the role of the body in producing both geographies and geographic knowledge. For example, a place is reproduced as a tourism destination through the constant repetition of certain bodily practices such as strolling through a historic district and lounging on a beach. The challenge, according to Tim Cresswell, is for geographers to understand that landscapes, for example, are made and understood through both visual and bodily practices.

—*Stephen Hanna*

See also Feminisms; Modernity; Photography, Geography and; Postmodernism; Spaces of Representation

Suggested Reading

Cresswell, T. (2003). Landscape and the obliteration of practice. In K. Anderson, M. Domosh, S. Pile, & N. Thrift (Eds.), *Handbook of cultural geography* (pp. 269–281). London: Sage.

Deutsch, R. (1991). Boys Town. *Environment and Planning D, 9,* 5–30.

Gregory, D. (1994). *Geographical imaginations.* Oxford, UK: Blackwell.

Harley, J. (2002). *The new nature of maps: Essays in the history of cartography.* Baltimore, MD: Johns Hopkins University Press.

Jay, M. (1994). *Downcast eyes: The denigration of vision in twentieth-century French thought.* Berkeley: University of California Press.

Jenks, C. (1995). *Visual culture.* London: Routledge.

Jones, J., III. (1995). Making geography objectively: Ocularity, representation, and *The Nature of Geography.* In W. Natter, T. Schatzki, & J. Jones (Eds.), *Objectivity and its other* (pp. 67–92). New York: Guilford.

Rose, G. (2001). *Visual methodologies.* London: Sage.

WHITENESS

Whiteness, according to Ruth Frankenberg, can be understood to have three interrelated components. First, whiteness can be seen as a location of structural advantage that white people occupy in society. Second, whiteness is a standpoint from which white people understand the world and their position in it. Third, whiteness is a set of cultural practices that—in white settler societies such as the United States and Canada—usually are dominant but also unmarked and unnamed. Accordingly, in places such as the United States and Canada, whiteness is hidden as the normative "way of life" by which all other cultural ways of being are measured; it forms the taken-for-granted and hidden framework that gives meaning to events, social actions, and phenomena, and it privileges white people over all others in such spaces.

As a location of structural advantage, whiteness can be seen as a form of unacknowledged social privilege that confers advantage on white people. For social scientists generally and for critical geographers more specifically, then, it is the power to define the terms under which social reality is understood without ever needing to acknowledge its own cultural specificity that makes whiteness so problematic. In this sense, whiteness becomes a set of practices that confers advantage—or "white privilege"—on particular members of society, usually without their knowledge. Thus, it is through the banal practices of everyday life that whiteness is constituted and reconstituted, and in so doing it can be seen as a key component of what has been called the "racial formation" of places such as the United States and Canada.

Until very recently, whiteness had not been studied much by geographers in the United States for the very simple reason that academic geography in the United States is itself dominantly white and that whiteness has been, until very recently, a hidden category for white academics. Thus, white geographers have been unaware of the white privilege that structures their relations with colleagues of color. Studies of whiteness in geography developed as a result of a wider interest in place and the politics of identity that arose during the early 1990s as a part of the cultural turn in geography, where geographers began to critically analyze socially constructed categories of identity such as race, class, sexuality, gender, age, and ability. Within this context, critical human geographers began to investigate the social construction of dominant identity categories such as masculinity and heterosexuality. Studies of the dominant category of whiteness have developed only within the past 10 years in geography.

Now that geographers are moving beyond programmatic arguments (i.e., arguments about the need to study whiteness), we are starting to see substantive empirical studies of whiteness that help us to understand the complexities, contradictions, and nuances of white identities and their spatial configurations. Thus, geographers have begun to study phenomena such as the historical–geographic antecedents of some of the structural advantages conferred on white people in the context of dominantly (or hegemonically) white everyday life, historical geographies of the banal practices and taken-for-granted sociospatial relations that serve to constitute specific spaces of whiteness, and the complexities—and especially the *contradictions*—of whiteness as they are played out in particular places at specific times. Overall, such studies provide us with a better understanding of the social geographies of racialization that constitute society and space in the dominant white spaces of the United States and Canada.

—*Lawrence Berg*

See also Ethnicity; Race and Racism; Segregation

Suggested Reading

Bonnett, A. (1997). Geography, race, and whiteness: Invisible traditions and current challenges. *Area, 29,* 193–199.

Bonnett, A. (2000). *White identities: Historical and international perspectives.* London: Routledge.

Dyer, R. (1997). *White: Essays on race and culture.* London: Routledge.

Frankenberg, R. (1997). *Displacing whiteness.* Durham, NC: Duke University Press.

Gilmore, R. (2002). Fatal couplings of power and difference: Notes on racism and geography. *The Professional Geographer, 54,* 15–24.

Jackson, P. (1998). Constructions of whiteness in the geographical imagination. *Area, 30,* 99–106.

Kobayashi, A., & Peake, L. (2000). Racism out of place: Thoughts on whiteness and an anti-racist geography in the new millennium. *Annals of the Association of American Geographers, 90,* 392–403.

Pulido, L. (2000). Rethinking environmental racism: White privilege and urban development in Southern California. *Annals of the Association of American Geographers, 90,* 12–40.

WILDERNESS

Wilderness is a wild and uncultivated area marked by minimal human influence on the natural environment and its processes. The word is derived from the Anglo-Saxon term *Wil(d)deor,* meaning "wild deer" or "wild beast." Wilderness has a long history of use in Western culture. In medieval Bibles, it referred to an arid and unsettled wasteland that was a sign of God's displeasure. Yet it also was a place that drew hermits seeking to escape human temptation and cleanse their souls. Hence, a foundation existed for the later positive transformation of the term. Wilderness continued to mean a disordered—even dangerous—place until the end of the 18th century. Settlers along the eastern North American coast used the term to describe the expansive forest to the west and vigorously sought to erase and thereby civilize it.

One of the earliest signs of a changing concept of wilderness appeared in the work of landscape architects. Orderly gardens dominated by human structures were replaced by natural-looking scenes that exhibited "sublimity," that is, a vaguely disturbing wild character. Artists and writers also began celebrating wild places, as did social organizations such as the American Civic Association, partly in response to industrialization. Authors John Muir and Henry David Thoreau, poet William Wordsworth, and artist Thomas Moran brought positive views of wilderness to the settled populations, especially in the eastern United States.

The transformation of wilderness to mean a positive, physically, and emotionally necessary place spawned a powerful preservation movement whose followers sought protection for such areas. In the United States, preservationists Aldo Leopold, Bob Marshall, and Howard Zahniser convinced Congress to create tangible wilderness areas with explicit legal restrictions. The Wilderness Act of 1964 defined wilderness as a place where the earth and its community of life are "untrammeled" by humans who visit but do not remain.

The exportation of this idea of wilderness proved to be controversial when other countries tried to establish such reserves in long-settled areas. Less developed countries, in particular, rejected it as a Western concept that denies use of large tracts by indigenous peoples. Scholars such as William Cronon and Max Oelschlaeger contended that wilderness is a cultural idea or a quality rather than a real place. The weight of this latest transformation of the meaning of wilderness has forced preservation groups to adapt their defense of legislated wilderness areas, especially in North America. They now define wilderness as a place without mechanization and where visual evidence of modern civilization is absent.

—*Lary Dilsaver*

See also Nature and Culture

Suggested Reading

Cronon, W. (1995). The trouble with wilderness. In W. Cronon (Ed.), *Uncommon ground: Toward reinventing nature* (pp. 69–90). New York: Norton.

Nash, R. (1982). *Wilderness and the American mind* (3rd ed.). New Haven, CT: Yale University Press.

Oelschlaeger, M. (1991). *The idea of wilderness.* New Haven, CT: Yale University Press.

Plumwood, V. (1998). Wilderness skepticism and wilderness dualism. In J. Callicott & M. Nelson (Eds.), *The great new wilderness debate* (pp. 652–690). Athens: University of Georgia Press.

WORLD ECONOMY

The world economy refers to the aggregate sum of all economic activities going on all over the world and usually implies an emphasis on international economic interaction. The world economy includes the production of primary commodities (e.g., minerals, lumber), agricultural crops, and manufacturing. It also includes service activities, from car washing services to accounting and financial services. The movements of commodities (trade) and money (financial flows) are additional integrative elements of the world economy.

DIFFICULTIES IN DEFINING THE TERM

Before elaborating further on the nature of world economy, it is worth noting a few serious difficulties in defining the term. First, as it is often invoked, the term includes only the formal economy. Such a restricted conceptualization misses the so-called informal economy—all of the work and objects that get exchanged that, although often monetized, operate under the radar of state authorities and therefore are not counted in official depictions or measurements of the national economy. Also missing from most accounts of the world economy are the massive amounts of labor, resources, and money connected with the world's three largest illicit international trading markets: those in drugs, arms, and art and artifacts. Less technically and more politically and conceptually, the labor of those who are unpaid also is hardly ever counted as part of the economy at any scale. Household labor, child care, communal production, subsistence farming, and other livelihood strategies often fall outside of the market and thus outside of the economy as it is framed. This is despite the fact that these are very significant and productive activities, even though they often are not directly income-generating ones.

A second major difficulty associated with the world economy pertains to the way in which the economy often is treated as if it were a separate and distinct sphere of human activity, often called the market, operating according to its own rules. This is particularly a problem when the world economy is taken as globalizing of its own accord. Such accounts of the world economy tend to naturalize it, that is, make it seem like something natural instead of something that humans are creating and recreating, however haphazardly, every day. They can also make it difficult to think of alternative arrangements. How can we hope for and work for different sorts of economic arrangements if we accept that the way in which the world economy functions is the outcome of some extra-social dynamics beyond human control?

These conceptual difficulties with the world economy do not make the term worthless. Indeed, perhaps the most amazing transformations of the earth have occurred over the past five centuries or so as something we call a world economy, more or less organized as capitalism, has been created. The term *world economy* conveys a sense of a dynamic and highly uneven set of activities and relations in which people and places are embedded differentially, although we should bear in mind that it is not comprehensive and that it can be used as a naturalizing description. Although some people and some places appear to be much more embedded in the world economy than are others, few are outside of it completely.

FIVE CENTURIES IN THE MAKING

Scholars argue about whether the emergence and consolidation of the world economy is a new or recent phenomenon. Geographers have been greatly influenced by the work of historians and world systems theorists who documented the importance of enduring networks of long-distance trade. Writers such as Immanuel Wallerstein argued that it was during the so-called long 15th century that the world economy was born. The growth of modern empires—first centered on Spain and Portugal and later on Great Britain, France, the Netherlands, and others—and the rise of the world economy went hand in hand as a European world economic core developed through the exploitation of a newly incorporated periphery elsewhere, including stolen and mined bullion and enslaved people, patterning and connecting an emerging world economy. During the 19th century, industrially manufactured goods and their associated technologies also traveled widely and knitted together spaces and their populations—again with the resultant income and wealth being concentrated in the core. Today, the cores of the world economy are also the world's financial centers, commanding the flows of diverse and complex financial products. The world economy has an important history and historical geography, and what we see and experience today as the world economy is built on at least five centuries of production and exchange of everything from silver to seeds to people. The historical geography of the world economy also teaches us that it is a highly dynamic phenomenon. Even though accounts of China and India as the new powerhouses of the world economy may be exaggerated, it is certain that the order of things in the world economy will not stay the same and that there will be new types of cores and peripheries patterning the geography of the world economy to come.

THE CONTEMPORARY WORLD ECONOMY

The world economy is in part the sum of all the national economies in the world. Using official estimates of countries' gross domestic products (GDPs) as a measure of the size of national economies, we can see which countries' economies account for the greatest shares of gross world product. The U.S. economy accounts for roughly one fifth of the world economy. The Euro Area (countries that had adopted the Euro as their currency) plus Great Britain accounts for slightly less (18.4%), whereas China alone has 13.2% of the world economy and Japan has 6.9%. Together, these countries or groupings (in the case of Europe) make up more than half of the world economy. Despite having some significant national economies within them, whole regions such as Africa (3.3% of the world economy), Central and South America (7.5%), and the republics of the former Soviet Union (3.8%) have economies that are dwarfed by the world's largest ones. This basic and highly uneven pattern of distribution of formal economic activity is, as noted, a legacy of the historical geography of the world economy, but it is dynamic as well. It also has enormous consequences for who gets to call the shots in the world economy.

The geography of economic activity is also a geography of power. What happens (whether intended or not) in the U.S. economy, by dint of its sheer size, has a ripple effect on other economies—no matter how far away. More directly political, however, because the United States has the world's largest economy, it has taken—or is accorded—certain agenda-setting privileges when it comes to the governance of the world economy. For example, the United States is allocated more than 17% of the total votes of the International Monetary Fund's (IMF) decision-making body, the board of governors, and a similar percentage of the World Bank's component institutions' votes. The World Bank and the IMF, together with the World Trade Organization (WTO), are the key institutions set up to stabilize and regulate the world economy. The WTO sets the rules for international trade and has as its priority ensuring that trade flows as smoothly, predictably, and freely as possible. Trade, in everything from steel to bananas to ideas (i.e., intellectual property), is a particularly contentious but crucial dimension of the world economy.

The United States accounts for 10.4% of world trade (as measured in terms of exports), the Euro Area plus Britain accounts for 35.8%, and China accounts for 5.9%. Simply by comparing the shares of world trade with those of world production, it becomes apparent that different national economies are positioned differentially in the world economy. Some are much more involved in trade than are others. What is not apparent from these data, however, are the major asymmetries that characterize world trading relations in the current world economy. Today, for example, the United States imports far more than it exports and has a staggering annual trade deficit of more than $700 billion (in U.S. dollars). Because of the U.S. economy's sheer size, it can run such a large deficit by issuing official debt (Treasury bills) that the trading surplus countries (notably China, Japan, and Taiwan) purchase. How long this particular asymmetry will endure, and what exactly its consequences are or will be, is much debated. Overall, even with such imbalances and with cycles of rapid and slow expansion, since the 1950s the volume of world trade has been growing faster than world production, meaning that parts of the world economy are becoming increasingly interconnected.

Complicating any understanding of the geography of world trade is the fact that much international trade is conducted within transnational or multinational corporations. Most multinational corporations are not huge, but there has emerged a set of very large companies that have impressive global capacities. The top 10 multinational corporations in the world economy today include banking and financial companies and oil companies. It has become somewhat common to point out that the sales of these corporations are bigger than the GDPs of many countries; for example, there are only 26 countries in the world whose GDPs are bigger than Exxon's sales.

However, it is not just the size of multinational corporations that can make them very powerful shapers of the geography of the world economy; it is also that they can move investments around the world, and this mobility or its threat can give them leverage in negotiations with any less mobile organization such as a government or a union.

DEVELOPMENT AND THE WORLD ECONOMY

Although the spatial patterns of production and trade are the building blocks of a geographic understanding of the world economy, economic activity is about strategies to ensure and enhance human lives. Given this, it is appropriate to consider how the world economy is distributed not only among nationally mapped spaces but also among people. As a first step, we can take the national level as a unit and briefly examine gross national product (GNP) per capita or what happens when the sheer volume of economic activity in a certain national space (GNP) is divided by the number of people living within that space. Although this is a crude measure and has come under internal and external critique, it nevertheless provides a useful beginning to critically considering the highly uneven geographies of the world economy. In terms of GNP per capita, the United States is approximately $40,000, the Euro Area plus Britain is roughly $29,400, and China is roughly $3,400. The GNP per capita in sub-Saharan Africa is substantially less—$500 if measured as GNP per capita or $1,750 if measured by adjusting for relative purchasing power. The corresponding figures for South Asia (with more than a billion people) are $500 and $2,640, respectively. Of course, GNP per capita is a crude measure and is not at all holistic as an indicator of human well-being. Because it is an average, it also reveals nothing about social or spatial differences within a country. However, as a beginning, these enormous differences in GNP per capita among countries do give a picture of inequality in the world economy given that countries with low GNP per capita figures also tend to have high infant mortality rates and short life expectancies (to take just two significant, albeit insufficient, indicators of well-being), whereas countries with high GNP per capita

data tend to have low infant mortality rates and long life expectancies. There are many differing analyses of how this uneven geography has emerged and how best to tackle it. The world economy seems to have enabled development for only some people around the globe, and the question of whether this will change remains a pressing one for billions of people.

GEOGRAPHIC APPROACHES TO THE WORLD ECONOMY

When it comes to explaining the geography of the world economy, monocausal explanations—although they may tap into generally accepted ideas—are unsatisfactory. Thus, an explanation for the world economy's spatial patterns and processes that rests on the geography of, say, natural resources only cannot explain Luxembourg even though it helps to explain Bahrain. However, even in this example, the fact that contemporary capitalism is so reliant on petroleum and not some other resource is a historically contingent fact. Geographers have a range of ways in which to understand the contingent nature of the world economy. Some focus on the historical development of a world system, increasingly integrated by flows of goods, services, money, ideas, and people. The uneven development of the world economy is explained as the result of tendencies inherent in capitalism and/or in empires. Other geographers focus more on development and might stress the ways in which national-level policies can affect the relative position of an economy and a population in the world economy (e.g., as they appear to have done in East Asia). Still others approach the world economy as the net result of billions of decisions made every day by different sorts of actors. Chief among them is the multinational corporation, and by examining the locational decisions and spatial strategies of firms, such analysts offer insight into the role of corporations in shaping the world economy. Corporations, however footloose they might appear to be, are embedded in different regions and cities. Indeed, many geographers consider regions and cities to be the most significant elements of the world economy and so study how regions and cities work as space economies. Whatever their approach, and whatever the serious differences among them, geographers of all sorts might agree that it is a mistake to explain the world economy in only its own terms. That is, to explain a complicated and differentiated set of relations such as is signaled by the term *world economy* takes analyses that are multifaceted and can relate the nitty-gritty of people's attempts to secure their livelihoods to the more abstracted logics of the so-called market.

—*Susan Roberts*

See also Colonialism; Core–Periphery Models; Debt and Debt Crisis; Dependency Theory; Developing World; Development Theory; Economic Geography; Globalization; Gross Domestic Product; Modernization Theory; Terms of Trade; Trade; Transnational Corporations; Uneven Development; World Systems Theory

Suggested Reading

Dicken, P. (2003). *Global shift: Reshaping the global economic map in the 21st century* (4th ed.). New York: Guilford.

Gibson-Graham, J. (1996). *The end of capitalism (as we knew it): A feminist critique of political economy.* Oxford, UK: Blackwell.

Knox, P., & Agnew, J. (1994). *The geography of the world economy* (2nd ed.). London: Edward Arnold.

WORLD SYSTEMS THEORY

Established during the 1970s through the work of sociologist Immanuel Wallerstein, world systems theory has been a fundamental challenge to the way in which social scientists think. The underlying premise of world systems theory is that society and social change should be defined at the broadest scale of social interaction, namely the historical social system. Three types of historical social systems were identified: mini-systems (e.g., small tribes), world empires (e.g., the Roman Empire, feudal Europe), and the capitalist world economy (the contemporary and sole existing social system). Beginning in Europe in the mid-1400s, the capitalist world economy has diffused across the whole globe, incorporating mini-systems (e.g., the tribes of Africa) and world empires (e.g., Chinese dynasties, feudal Japan) in the process. Wallerstein detailed the creation and diffusion of the capitalist world economy in a three-volume study, but it is the conceptual ingredients of the capitalist world economy that have captured the attention of human geographers.

Five main concepts from world systems theory have been incorporated into human geography: core–periphery, the challenge to developmentalism, geographic scale, hegemony, and world cities.

Wallerstein identified two sets of processes in the capitalist world economy: *core* and *periphery*. Core processes are associated with high-end economic activities, high salaries, and high levels of consumption. Periphery processes are the opposite. Areas of the world where the former predominate may be called core areas, and those where the latter predominate may be called periphery areas. For example, in the United States, core processes predominate, although it is also possible to find areas of low-paying jobs and poverty. The opposite may be said for, say, Sudan. In some parts of the world, say, Brazil and Russia, there exists a relatively even balance of core and periphery processes and these areas are identified as the semiperiphery. The areas are connected in that the core exploits the semiperiphery and periphery, whereas the periphery is exploited by both the core and semiperiphery. It is important to remember that it is not the geographic areas that are inherently core or periphery; instead, the geographic zones are defined by the processes within them. Hence, a particular part of the globe may change its status over time. The United States initially was incorporated into the capitalist world economy as a colonial periphery. Over time, it has become home to some of the most important core processes such as the international finance center of Wall Street. Geographers have been able to map different manifestations of the spatial pattern of core and periphery processes and their implications—poverty, demographics, political systems, and types of warfare.

The mapping of core and periphery processes relates to world systems theory's challenge to *developmentalism*. During the 1950s and 1960s, geographers and other social scientists were using models that contended that each and every country could follow a singular path to development if it so desired. The economic, population, and political characteristics of the "undeveloped" countries were given set stages that they could pass through until they gained the appearance of "developed" countries. The stages and final outcome were taken from the historical paths of Great Britain and the United States. World systems theory challenged developmentalism. Each country was not a separate unit that could willingly follow the path of, for example, the United States. Peripheral countries possessed many burdens. Especially their position within the periphery of the world economy tied them into the exploitive interests of powerful countries and businesses. The choices that developmentalism posited did not exist in reality. Instead, the undeveloped countries were the poorest areas of the world economy, a system requiring a core–periphery hierarchy. Although some countries have been able to improve the livelihood of their populations relative to other countries, on the whole, peripheral countries have remained peripheral.

World systems theory's identification of states as the political units of a larger system provided the opportunity for geographers to discuss the political role of geographic scale. Peter Taylor's seminal contributions identified three geographic scales: the world economy (the scale of reality), the nation-state (the scale of ideology), and the locality (the scale of experience). An individual experiences unemployment or ethnic cleansing, for example, within particular places or localities. The causes of these experiences ultimately are traced to the structures and processes of the capitalist world economy. Nation-states are the scale of ideology for two reasons, namely that (1) governments try to ameliorate the impacts of the world economy and (2) political activity is targeted toward national policies rather than the scale that *really* matters—the world economy. Although these scales have been criticized for being too structurally defined and not paying enough attention to how they are constructed by people and groups, Taylor's framework paved the way for the central role of scale in contemporary human geography.

The concept of *hegemony* has also been adopted in human geography. The hegemonic power is a country that dominants the globe economically and politically. Currently, the United States is the hegemonic power, as it was for much of the 20th century. Geographers have studied how hegemonic countries extend their power into the sovereign spaces of other countries through the diffusion of political, economic, military, and cultural practices.

Despite the attempt of world systems theory to direct attention away from states, most research has still looked at countries within the broader systemic context. One important development is the *world cities* project. Instead of focusing on the territorial geography of nation-states, world cities are seen as important nodes in global economic (especially finance services), political, and cultural networks. Research has emphasized how cities are connected horizontally across space and vertically in a hierarchy of cities, with places such as New York and London at the top. The network of world cities manages flows of information, money, and goods that establish and reinforce the core–periphery structure of the world economy.

—*Colin Flint*

See also Core–Periphery Models; Dependency Theory; Development Theory; Political Geography; Scale; World Economy

Suggested Reading

Chase-Dunn, C. (1998). *Global formation: Structures of the world-economy* (2nd ed.). Lanham, MD: Rowman & Littlefield.

Dunaway, W. (Ed.). (2003). *Emerging issues in the 21st century world-system.* Westport, CT: Praeger.

Shelley, F., & Flint, C. (2000). Geography, place, and world-systems analysis. In T. Hall (Ed.), *A world-systems reader* (pp. 69–82). Lanham, MD: Rowman & Littlefield.

So, A. (1990). *Social change and development: Modernization, dependency, and world-system theories.* Newbury Park, CA: Sage.

Taylor, P. (1994). The state as container: Territoriality in the modern world-system. *Progress in Human Geography, 18,* 151–162.

Taylor, P., & Flint, C. (2000). *Political geography: World-economy, nation-state, and locality* (4th ed.). Upper Saddle River, NJ: Prentice Hall.

Wallerstein, I. (1974). *The modern world-system: Capitalist agriculture and the origins of the European world-economy in the sixteenth century.* New York: Academic Press.

WRITING

At one time, geographers and others thought about writing as a straightforward mode of communication, one that could be used to directly communicate the results of their research. During recent years, however, geographers and others have examined writing more closely, and we now understand it as a complex form of expression that may indeed be used to convey the results of our research, but never in simple, direct, or unmediated ways, for each person writes from somewhere—from a physical place, as well as from a variety of theoretical and philosophical positions, and based on a personal background—and that influences our writing in every way. Furthermore, the very act of writing, far from simply a summing-up activity related to the presentation of research, has come to be understood as a powerful act of persuasion as well as a formative part of the research process itself; through our writing, we convince others of our arguments, but we also think, discover, and come to understand what those very arguments are, in part, as we write. Although human geographers and others see these issues as complexly intertwined, we can separate them into three groups: how writers present their data and arguments, how writers represent themselves and others in their work, and how writing shapes the research itself.

Since at least the early 20th century, geographers have debated the ways in which geographic research should be written up. For much of that time, most have agreed that geographic writing should be clear and, for that reason, should be written in "plain English"—the language of everyday speech—avoiding the use of jargon as much as possible. Although such advice seems simple, it has been contentious, for even the term *plain English* here actually refers to that version of the language spoken by those with a university education; for many others, it might not be "plain" at all. Furthermore, although jargon has been derided by some who associate it with pretentious terms used by a small in-group to deliberately obfuscate otherwise obvious meanings (and thereby speak only to members of the in-group), others see it as a necessary shorthand (for its ability to describe, say, a complex concept in one word), and still others see it as an essential part of the development of geography as a field and as a science (arguing that no one would question specialized language in, say, biology or chemistry, and so they should not do so in geography either). What is more, it has also become clear that even apparently plain language can convey unintended or loaded messages, for example, through its use of modifiers, especially superlatives. For example, when I wrote "most" earlier, did I really know for certain that it was the majority or did I just guess? Also, when I wrote "many," how many did I have in mind and would others agree that this number constitutes "many" in this case? Because simple language can mask meanings and messages, some have argued that purposely prolix prose can, ironically, make those more apparent; by forcing the reader to move slowly through a text, contemplating the multiple meanings of every word, complex writing may actually seek to disrupt the (deviously?) persuasive power of narratives, exposing the writing and research, as well as the researcher, for what they are.

Closely connected to this debate is the role of the writer/researcher in his or her text. For decades, scholars in many fields found it inappropriate for the writers to make themselves noticed, for that was seen to distract from the perceived purity of the science and the clarity of the data (a model inherited from laboratory testing where the researchers themselves were once thought to play no role in experiment outcomes). Writers, even those who played very active and personal roles in their research, would write themselves out of their own research by using the passive voice, describing how "interviews were conducted," and then would write their participants out as well, generalizing the people they worked with into seemingly homogeneous groups with statements such as "respondents agreed." Then some writers, at first particularly feminists in a variety of fields, noticed that such language use not only masked the active role of each writer in her or his work (in both research and writing) but also made the world seem as if it were made up of only white, middle-class, Western, heterosexual men—that it silenced the voices of the majority of the world's citizens—and then academic writing entered a new era. Today, most human geographers write about their research in the active voice, use the first person ("I conducted interviews"), and no longer assume that "he" stands for everybody. Beyond that, many of us also seek to understand and convey some fundamental aspects of ourselves to readers because we understand that who we are matters in terms of how we decide which topic to research, how we approach the topic, how we carry out the research, how others respond to us during the research process (e.g., in interviews), and how we communicate the results of our research. Understanding the importance of these issues—of reflexivity and positionality (attempting to complexly understand one's own situated position in the world and to conduct research and carry out writing in ways that make that position visible rather than obscure)—has led geographers and others to understand writing as a complexly situated activity, far from the neutral means of presenting findings that some once simplistically understood.

Finally, writers in geography and other fields now also understand what "creative" writers have long known, namely that writing itself is more than a mechanical act; it is a process through which we engage with ourselves and our ideas in creative, challenging, and insightful ways. We do not simply "write up" the results of our research; rather, we partly create them as we write—in the language we use as well as in the energized process of thinking/writing. Partly in recognition of the formative role of writing itself, some academics now present their work through "alternative" ways of writing using poetry, journaling, nonlinear narratives, and other expressive techniques, whereas others elect to work within the structures of academic writing conventions to harness their creativity, for all writing, whether in poetry or prose, in academic journals or literary fiction, is creative—in its form, in its expression, and in its content.

—Dydia DeLyser

See also Qualitative Research; Text and Textuality; Travel Writing, Geography and

Suggested Reading

Berg, L., & Mansvelt, J. (2005). Writing qualitative geographies, constructing geographical knowledges. In I. Hay (Ed.), *Qualitative research methods in human geography* (2nd ed., pp. 248–265). Melbourne, Australia: Oxford University Press.

DeLyser, D., & Pawson, E. (2005). From personal to public: Communicating qualitative research for public consumption. In I. Hay (Ed.), *Qualitative research methods in human geography* (2nd ed., pp. 266–274). Melbourne, Australia: Oxford University Press.

Z

ZONING

In the United States, zoning refers to the practice of dividing a municipality or county into districts to control land use. It is the most widely used technique for achieving community planning goals. A comprehensive zoning ordinance specifies permitted uses, density, height, bulk, lot size, setbacks, yards, off-street parking, signage, accessory uses, and other regulations designed to protect neighborhood character. The ordinance text defines the specific meanings of the zoning terminology, and a map shows zone location. Some ordinances include special exceptions or conditional uses, subject to municipal approval, provided additional requirements are fulfilled. Zoning boards of appeal decide whether area or use variances to the regulations due to hardship are permitted in specific instances and locations. If a zoning change is made, new construction must follow the amended regulation, but existing nonconforming uses may be able to continue or might need to be eliminated after a specified time period. Zoning ordinances are local laws, and the power to zone is delegated by the states to municipalities through the Standard State Enabling Act of 1922. Zoning laws are enacted to safeguard the public health, morals, safety, and welfare of a community—a government responsibility termed *police power*.

ZONING HISTORY

At the beginning of the 20th century, only a few American cities had controls regulating building heights. Rapid urban growth had spurred some cities to adopt regulations controlling tenements to protect public health. However, it was not until 1916 that New York City adopted the first comprehensive zoning ordinance that controlled building height, density, and land use. The impetus for this law was the northward spread of garment manufacturing and immigrant workers from lower Manhattan that threatened to disrupt the ambiance of retail trade for influential Fifth Avenue merchants. By limiting land uses, New York's ordinance threatened the Fifth Amendment to the U.S. Constitution, which prohibits the taking of private property without compensation. Lawyer Edward Bassett, the creator of the New York ordinance, had anticipated a legal challenge and so promoted the benefits of zoning in stabilizing neighborhoods and protecting real estate investments across the country.

The U.S. Supreme Court case that eventually upheld zoning as a valid use of the police power was the 1926 case of *Village of Euclid v. Ambler Realty Company.* Property owned by Ambler Realty had been zoned by Euclid for residential use only, and Ambler claimed that the land had lost value as a result. But the Supreme Court finally decided that this use of the police power served a legitimate public purpose aimed at solving community development problems. Subsequently, zoning ordinances were widely adopted by municipalities throughout the United States as a method of land use control and became the major planning mechanism. In the original enabling statutes, zoning regulations were meant to be adopted in accordance with a comprehensive or master plan, but many communities failed to develop or legally adopt such long-range plans. Regulations that are based on a comprehensive plan are more likely to withstand legal challenges.

INNOVATIVE ZONING

Zoning has proved to be a crude tool for planning development and future land use, so a variety of techniques have been developed to make it more flexible. *Incentive zoning* offers a bonus to the developer for providing community amenities. The bonus might be additional units or extra building stories in exchange for the developer providing a park or affordable housing. *Inclusionary zoning* is not optional and requires developers who build more than a certain number of units to include a percentage of low- or moderate-income (or other types of) housing units. *Performance zoning* involves assessing the potential impact of a proposed development and requiring that certain standards be met. These may involve limiting levels of noise, limiting smoke or pollution emissions, mitigating environmental impacts such as stormwater runoff, or protecting scenic views. *Clustering* allows a developer to redesign a site plan to group homes more densely than the ordinance permits in specific areas, provided that the allowable total site density is not exceeded. This popular approach can preserve open space and lower infrastructure costs. *Overlay zones,* such as floodplain, waterfront, or historic districts, impose additional restrictions to the underlying zoning district to protect property. *Floating zones* give municipalities flexibility in locating desired uses while examining significant projects for future impacts. These zones are not located until a proposal is made, the project meets municipal standards, and the project is approved by the local legislative body. The *transfer of development rights* (TDR) allows the right to develop to be separated from an owner's "bundle of rights" so that the owner may retain the land but sell the development rights for use on other properties. TDR can be used to preserve farmland or historic structures in a sending area while transferring the increased development density to an appropriate receiving area.

LEGAL BATTLES

The practice of zoning has been shaped by legislation, litigation, and judicial decisions. After *Euclid v. Ambler,* the U.S. Supreme Court did not hear another zoning case until 1987, and its recent decisions have focused on narrowing the interpretation of takings. This has meant a refining of how much reduction in economic value is allowed, and for how long, before a land use regulation is judged to be unconstitutional. However, there have been many zoning cases heard in state courts based on a wide variety of constitutional challenges. The legal record is ambiguous because the pattern of court decisions differs considerably within and between states across the country. Exclusionary zoning, or the use of regulations to exclude minorities or housing types from a municipality, has been found to be unconstitutional. However, there are no effective mechanisms to routinely enforce fair treatment through land use regulation other than the threat of legal action.

—*Jo Margaret Mano*

See also Urban and Regional Planning; Urban Geography

Suggested Reading

Babcock, R. (1979). Zoning. In F. So (Ed.), *The practice of local government planning* (pp. 416–443). Washington, DC: International City Management Association.

Meshenberg, M. (1976). *The language of zoning.* Chicago: American Society of Planning Officials.

Smith, H. (1983). *The citizen's guide to zoning.* Chicago: APA Planners Press.

Master Bibliography

Abbott, C. (1987). *The new urban America: Growth and politics in Sunbelt cities* (2nd ed.). Chapel Hill: University of North Carolina Press.

Abeyie, D., & Harries, K. (Eds.). (1980). *Crime: A spatial perspective.* New York: Columbia University Press.

Abu-Lughod, J. (1989). *Before European hegemony: The world system A.D. 1250–1350.* New York: Oxford University Press.

Abu-Lughod, L. (1990). The romance of resistance: Transformations of power through Bedouin women. *The American Ethnologist, 17*(1), 41–55.

Acs, Z. (2000). *Regional innovation, knowledge, and global change.* London: Pinter.

Acs, Z., & Varga, A. (2002). Geography, endogenous growth, and innovation. *International Regional Science Review, 25,* 132–148.

Adams, M., Blumenfeld, W., Castaneda, R., Hackman, H., Peters, M., & Zuniga, X. (Eds.). (2000). *Readings for diversity and social justice: An anthology on racism, anti-Semitism, sexism, heterosexism, ableism, and classism.* New York: Routledge.

Adams, P. (1995). A reconsideration of personal boundaries in space–time. *Annals of the Association of American Geographers, 85,* 267–285.

Adams, P., Hoelscher, S., & Till, K. (Eds.). (2001). *Textures of place: Exploring humanist geographies.* Minneapolis: University of Minnesota Press.

Adams, W. (2001). *Green development: Environment and sustainability in the Third World* (2nd ed.). London: Routledge.

Agnew, J. (Ed.). (1997). *Political geography: A reader.* London: Edward Arnold.

Agnew, J. (2002). *Making political geography.* London: Edward Arnold.

Agnew, J. (2004). Nationalism. In J. Duncan, N. Johnson, & R. Schein (Eds.), *A companion to cultural geography* (pp. 223–237). Oxford, UK: Blackwell.

Agnew, J., Mitchell, K., & Toal, G. (Eds.). (2003). *A companion to political geography.* Malden, MA: Blackwell.

Ahmad, A. (1992). *In theory: Classes, nations, literatures.* London: Verso.

Aiken, C. (1977). Faulkner's Yoknapatawpha County: Geographical fact into fiction. *Geographical Journal, 67,* 1–21.

Aitken, S. (2001). *Geographies of young people: The morally contested spaces of identity.* New York: Routledge.

Aitken, S., Cutter, S., Foote, K., & Sell, J. (1989). Environmental perception and behavioral geography. In G. Gaile & C. Willmott (Eds.), *Geography in America* (pp. 218–238). Columbus, OH: Merrill.

Aitken, S., & Zonn, L. (Eds.). (1994). *Place, power, situation, and spectacle: A geography of film.* Lanham, MD: Rowman & Littlefield.

Akwule, R. (1992). *Global telecommunications: The technology, administration, and policies.* Boston: Focal Press.

Alderman, D. (2002). School names as cultural arenas: The naming of U.S. public schools after Martin Luther King, Jr. *Urban Geography, 23,* 601–626.

Allen, B. (2002). *The Faber book of exploration: An anthology of worlds revealed by explorers through the ages.* London: Faber & Faber.

Allen, J. (2003). Power. In J. Agnew, K. Mitchell, & G. Ó Tuathail (Eds.), *A companion to political geography.* Malden, MA: Blackwell.

American Institute for Certified Planners. (2005). *Ethics for the certified planner* [Online]. Available: www.planning.org/ethics/

Amin, A. (1994). Post-Fordism: Models, fantasies, and phantoms of transition. In A. Amin (Ed.), *Post-Fordism: A reader* (pp. 1–39). Oxford, UK: Blackwell.

Amin, A. (Ed.). (1994). *Post-Fordism: A reader.* Oxford, UK: Blackwell.

Amin, A., & Thrift, N. (Eds.). (2004). *The Blackwell cultural economy reader.* Oxford, UK: Blackwell.

Amin, S. (1976). *Imperialism and unequal development.* Hassocks, UK: Harvester.

Anderson, B. (1991). *Imagined communities: Reflections on the origin and spread of nationalism.* London: Verso.

Anderson, K. (1991). *Vancouver's Chinatown: Racial discourse in Canada, 1875–1980.* Montreal: McGill–Queen's University Press.

Anderson, K., Domosh, M., Pile, S., & Thrift, N. (Eds.). (2003). *Handbook of cultural geography.* London: Sage.

Anderson, P. (1974). *Lineages of the absolutist state.* London: Verso.

Anselin, L. (1988). *Spatial econometrics, methods, and models.* Boston: Kluwer Academic.

Appadurai, A. (1991). Global ethnoscapes: Notes and queries for a transnational anthropology. In R. Fox (Ed.), *Recapturing anthropology* (pp. 191–210). Santa Fe, NM: School of American Research Press.

Appadurai, A. (1996). *Modernity at large: Cultural dimensions of globalization.* Minneapolis: University of Minnesota Press.

Archer, C., & Shelly, F. (1986). *American electoral mosaics.* Washington, DC: Association of American Geographers.

Archer, J., Lavin, S., Martis, K., & Shelley, F. (2002). *Atlas of American politics, 1960–2000.* Washington, DC: Congressional Quarterly Press.

Archer, J., Shelley, F., Taylor, P., & White, E. (1988). The changing geography of America's presidential elections. *Scientific American, 268,* 44–51.

Ashenhurst, R. (1996). Ontological aspects of information modeling. *Minds and Machines, 6,* 287–394.

Ashley, C., & Maxwell, S. (2001). Rethinking rural development. *Development Policy Review, 19,* 395–425.

Atkinson, P., & Hammersley, M. (1998). Ethnography and participant observation. In N. Denzin & Y. Lincoln (Eds.), *Strategies of qualitative inquiry* (pp. 110–136). Thousand Oaks, CA: Sage.

Audretsch, D., & Feldman, M. (1996). R&D spillovers and the geography of innovation and production. *American Economic Review, 86,* 630–640.

Auge, M. (1995). *Non-places: Introduction to an anthropology of supermodernity.* London: Verso.

Autio, E., & Kloftsen, M. (1998). A comparative study of two European business incubators. *Journal of Small Business Management, 36,* 30–43.

Ayres, J. (2002). Transnational political processes and contention against the global economy. In J. Smith & H. Johnston (Eds.), *Globalization and resistance: Transnational dimensions of social movements* (pp. 191–205). Lanham, MD: Rowman & Littlefield.

Azaryahu, M. (1997). German reunification and the politics of street names: The case of East Berlin. *Political Geography, 16,* 479–493.

Babcock, R. (1979). Zoning. In F. So (Ed.), *The practice of local government planning* (pp. 416–443). Washington, DC: International City Management Association.

Bailey, R. (1988). Problems with using overlay mapping for planning and their implications for geographic information systems. *Environmental Management, 12*(1), 11–17.

Bailey, T., & Gatrell, A. (1995). *Interactive spatial data analysis.* Harlow, UK: Longman.

Baker, A. (2003). *Geography and history: Bridging the divide.* Cambridge, UK: Cambridge University Press.

Baker, J. (1931). *A history of geographical discovery and exploration.* London: Harrap.

Bale, J. (2003). *Sports geography* (2nd ed.). London: Routledge.

Baran, P. (1957). *The political economy of growth.* New York: Monthly Review Press.

Barber, B. (1995). *Jihad vs. McWorld: How globalism and tribalism are reshaping the world.* New York: Ballantine.

Barnes, T. (2001). Lives lived, and lives told: Biographies of geography's quantitative revolution. *Environment and Planning D, 19,* 409–429.

Barnes, T. (2004). Placing ideas: Genius loci, heterotopia, and geography's quantitative revolution. *Progress in Human Geography, 29,* 565–595.

Barnes, T., & Duncan, J. (Eds.). (1992). *Writing worlds: Discourse, text, and metaphor in the representation of landscape.* London: Routledge.

Barnet, R., & Muller, R. (1974). *Global reach: The power of the multinational corporations.* New York: Simon & Schuster.

Bascomb, J. (1993). The peasant economy of refugee resettlement in eastern Sudan. *Annals of the Association of American Geographers, 83,* 320–346.

Basgen, B., & Blunden, A. (Eds.). (2004). *Encyclopedia of Marxism: Glossary of terms.* Available: www.marxists .org

Bassett, T., & Zimmerer, K. (2003). Cultural ecology. In G. Gaile & C. Willmott (Eds.), *Geography in America at the dawn of the 21st century* (pp. 97–112). Oxford, UK: Oxford University Press.

Batty, M. (2005). *Cities and complexity: Understanding cities with cellular automata, agent-based models, and fractals.* Cambridge: MIT Press.

Batty, M., Couclelis, H., & Eichen, M. (1997). Urban systems as cellular automata. *Environment and Planning B, 24,* 159–305.

Bauman, Z. (1991). *Modernity and ambivalence*. Ithaca, NY: Cornell University Press.

Beaverstock J., Smith, R., & Taylor, P. (2000). World–city network: A new metageography? *Annals of the Association of American Geographers, 90,* 123–134.

Becker, K. (2004). *The informal economy: Fact finding study*. Stockholm: Swedish Agency for International Development Cooperation.

Bednarz, R., & Bednarz, S. (2004). Geography education: The glass is half full and it's getting fuller. *The Professional Geographer, 56,* 22–27.

Bell, D. (1973). *The coming of post-industrial society: A venture in social forecasting*. New York: Basic Books.

Bell, D., & Valentine, G. (Eds.). (1995). *Mapping desire: Geographies of sexualities*. New York: Routledge.

Bell, D., & Valentine, G. (1997). *Consuming geographies: We are where we eat*. London: Routledge.

Benenson, I., & Torrens, P. (2004). *Geosimulation: Automata-based modeling of urban phenomena*. London: Wiley.

Beneria, L. (1985). *Women and development: The sexual division of labor in rural societies*. New York: Praeger.

Berg, L. (1993). Between modernism and postmodernism. *Progress in Human Geography, 17,* 490–507.

Berg, L. (2004). Scaling knowledge: Towards a critical geography of critical geographies. *Geoforum, 35,* 553–558.

Berg, L., & Longhurst, R. (2003). Placing masculinities and geography. *Gender, Place, and Culture, 10,* 351–360.

Berg, L., & Mansvelt, J. (2005). Writing qualitative geographies, constructing geographical knowledges. In I. Hay (Ed.), *Qualitative research methods in human geography* (2nd ed., pp. 248–265). Melbourne, Australia: Oxford University Press.

Berman, M. (1982). *All that is solid melts into air: The experience of modernity*. New York: Penguin Books.

Bernard, R., & Rice, B. (Eds.). (1983). *Sunbelt cities: Politics and growth since World War II*. Austin: University of Texas Press.

Berry, B. (1964). Approaches to regional analysis: A synthesis. *Annals of the Association of American Geographers, 54,* 2–11.

Berry, B. (1991). Long waves in American urban evolution. In J. Hart (Ed.), *Our changing cities* (pp. 31–50). Baltimore, MD: Johns Hopkins University Press.

Betts, R. (1998). *Decolonization (The making of the contemporary world)*. London: Routledge.

Beyers, W. (2000). Cyberspace or human space: Wither cities in the age of telecommunications? In Y. Aoyama, J. Wheeler, & B. Warf (Eds.), *Cities in the telecommunications age: The fracturing of geographies* (pp. 161–180). New York: Routledge.

Beyers, W., & Lindahl, D. (1996). Explaining the demand for producer services: Is cost-driven externalization the major factor? *Papers in Regional Science, 75,* 351–374.

Billig, M. (1995). *Banal nationalism*. Thousand Oaks, CA: Sage.

Billinge, M., Gregory, D., & Martin, R. (Eds.). (1984). *Recollections of a revolution*. London: Macmillan.

Black, W. (2003). *Transportation: A geographical analysis*. New York: Guilford.

Blaikie, P. (1985). *Political economy of soil erosion in developing countries*. New York: John Wiley.

Blaut, J. (1970). Geographic models of imperialism. *Antipode, 2,* 65–85.

Blaut, J. (1993). *The colonizer's model of the world: Geographical diffusionism and Eurocentric history*. New York: Guilford.

Blaut, J. (2000). *Eight Eurocentric historians*. New York: Guilford.

Blomley, N. (1994). *Law, space, and the geographies of power*. New York: Guilford.

Blomley, N., Delaney, D., & Ford, R. (Eds.). (2001). *The legal geographies reader: Law, power, and space*. Oxford, UK: Blackwell.

Bluestone, B., & Harrison, B. (1982). *The deindustrialization of America: Plant closings, community abandonment, and the dismantling of basic industry*. New York: Basic Books.

Blunt, A., & McEwan, C. (Eds.). (2002). *Postcolonial geographies*. New York: Continuum.

Boal, F., & Douglas, N. (Eds.). (1982). *Integration and division: Geographical perspectives on the Northern Ireland problem*. London: Academic Press.

Bobo, L., Oliver, M., Johnson, J., & Valenzuela, A., Jr. (2001). *Prismatic metropolis: Inequality in Los Angeles*. New York: Russell Sage.

Bonanno, A., Busch, L., Friedland, W., Gouveia, L., & Mingione, E. (Eds.). (1994). *From Columbus to ConAgra: The globalization of agriculture and food*. Lawrence: University Press of Kansas.

Bondi, L. (1991). Gender divisions and gentrification: A critique. *Transactions of the Institute of British Geographers, 16,* 190–198.

Bondi, L., Davidson, J., & Smith, M. (2006). *Emotional geographies*. London: Ashgate.

Bonnett, A. (1997). Geography, race, and whiteness: Invisible traditions and current challenges. *Area, 29,* 193–199.

Bonnett, A. (2000). *Anti-racism.* London: Routledge.

Bonnett, A. (2000). *White identities: Historical and international perspectives.* London: Routledge.

Boots, B., Okabe, A., & Thomas, R. (Eds.). (2002). *Modeling geographic systems: Statistical and computational applications.* Boston: Kluwer Academic.

Borchert, J. (1967). American metropolitan evolution. *Geographical Review, 57,* 301–332.

Borchert, J. (1991). Futures of American cities. In J. Hart (Ed.), *Our changing cities* (pp. 218–250). Baltimore, MD: Johns Hopkins University Press.

Boserup, E. (1965). *The conditions of agricultural growth: The economics of agrarian change under population pressure.* Chicago: Aldine.

Boudeville, J. (1966). *Problems of regional economic planning.* Edinburgh, UK: Edinburgh University Press.

Boughton, J. (2001). *Silent revolution: The International Monetary Fund, 1979–1989.* Washington, DC: International Monetary Fund.

Bounds, M. (2003). *Urban social theory: City, self, and society.* Oxford, UK: Oxford University Press.

Bourguet, M.-N. (1999). The explorer. In M. Vovelle (Ed.), *Enlightenment portraits* (pp. 257–315). Chicago: University of Chicago Press.

Bourne, L. (1981). *The geography of housing.* London: Edward Arnold.

Braden, K., & Shelley, F. (2000). *Engaging geopolitics.* Englewood Cliffs, NJ: Prentice Hall.

Brantingham, P., & Brantingham, P. (1981). *Patterns in crime.* New York: Macmillan.

Brenner, N. (1998). Between fixity and motion: Accumulation, territorial organization, and the historical geography of spatial scales. *Environment and Planning D, 16,* 459–481.

Brenner, N. (2003). "Glocalization" as a state spatial strategy: Urban entrepreneurialism and the new politics of uneven development in Western Europe. In J. Peck & H. Yeung (Eds.), *Remaking the global economy* (pp. 197–215). Thousand Oaks, CA: Sage.

Brenner, N. (Ed.). (2003). *State/Space: A reader.* Boston: Blackwell.

Brenner, N., Jessop, B., Jones, M., & MacLeod, G. (Eds.). (2003). *State/space: A reader.* Oxford, UK: Blackwell.

Brenner, N., & Theodore, N. (2002). *Spaces of neoliberalism: Urban restructuring in North America and Western Europe.* Oxford, UK: Blackwell.

Bresnahan, T., & Gambardella, A. (Eds.). (2004). *Building high-tech clusters: Silicon Valley and beyond.* New York: Cambridge University Press.

Britton, S. (1991). Tourism, capital, and place: Towards a critical geography of tourism. *Environment and Planning D, 9,* 451–478.

Brohman, J. (1996). *Popular development.* Malden, MA: Blackwell.

Brookfield, H. (1969). On the environment as perceived. *Progress in Geography, 1,* 51–80.

Brooks, J. (1975). *Telephone: The first hundred years.* New York: Harper & Row.

Brown, M. (1995). Ironies of distance: An ongoing critique of the geographies of AIDS. *Environment and Planning D: Society and Space, 13,* 1391–1396.

Brown, M. (2000). *Closet geographies: Geographies of metaphor from the body to the globe.* New York: Routledge.

Brunn, S., & Leinbach, T. (Eds.). (1991). *Collapsing space and time: Geographic aspects of communications and information.* London: HarperCollins.

Bruntland, H. (1987). *Our common future.* Oxford, UK: Oxford University Press.

Bryan, R., & McLelland, J. (1982). *The city's countryside: Land and its management in the rural–urban fringe.* London: Longman.

Bryson, J., Daniels, P., & Warf, B. (2004). *Service worlds: People, organizations, and technologies.* London: Routledge.

Buisseret, D. (Ed.). (2005). *The Oxford companion to exploration.* New York: Oxford University Press.

Bulmer, M. (1984). *The Chicago School of sociology: Institutionalization, diversity, and the rise of sociological research.* Chicago: University of Chicago Press.

Burgess, J., & Gold, J. (1985). *Geography, the media, and popular culture.* New York: St. Martin's.

Burnham, D. (1983). *The rise of the computer state.* New York: Random House.

Burton, I. (1963). The quantitative revolution and theoretical geography. *The Canadian Geographer, 7,* 151–162.

Butler, J. (1990). *Gender trouble: Feminism and the subversion of identity.* New York: Routledge.

Butler, J. (1993). *Bodies that matter: On the discursive limits of "sex."* New York: Routledge.

Butler, R., & Parr, H. (Eds.). (1999). *Mind and body spaces: Geographies of illness, impairment, and disability.* London: Routledge.

Buttimer, A. (1976). Grasping the dynamism of lifeworld. *Annals of the Association of American Geographers, 66,* 277–292.

Butzer, K. (1989). Culture ecology. In G. Gaile & C. Willmott (Eds.), *Geography in America* (pp. 192–208). Columbus, OH: Merrill.

Cairncross, F. (1997). *The death of distance: How the communications revolution will change our lives.* Boston: Harvard Business School Press.

Callard, F. (2003). The taming of psychoanalysis in geography. *Social and Cultural Geography, 4,* 295–312.

Calthorpe, P., Corbett, M., Duany, A., Moule, E., Plater-Zyberk, E., & Polyzoides, S. (n.d.). *The Ahwahnee Principles* [Online]. Available: www.lgc.org/ahwahnee/principles.html

Campbell, D. (1992). *Writing security: United States foreign policy and the politics of identity.* Minneapolis: University of Minnesota Press.

Castells, M. (1983). *The city and the grassroots: A cross-cultural theory of urban social movements.* London: Edward Arnold; Berkeley: University of California Press.

Castells, M. (1996). *The rise of the network society.* Oxford, UK: Blackwell.

Castells, M. (1997). *The urban question: A Marxist approach* (A. Sheridan, Trans.). Cambridge: MIT Press.

Castells, M. (2004). *Power of identity: The information age—Economy, society, and culture* (Vol. 2). Oxford, UK: Blackwell.

Castree, N. (2001). Socializing nature: Theory, practice, and politics. In N. Castree & B. Braun (Eds.), *Social nature: Theory, practice, and politics* (pp. 1–21). Malden, MA: Blackwell.

Castree, N., & Braun, B. (2001). *Social nature: Theory, practice, and politics.* Oxford, UK: Blackwell.

Castree, N., & Nash, C. (2004). Introduction: Posthumanism in question. *Environment and Planning A, 36,* 1341–1343.

Castro, C. (2004). Sustainable development: Mainstream and critical perspectives. *Organization and Environment, 17,* 195–225.

Césaire, A. (1972). *Discourse on colonialism* (J. Pinkham, Trans.). New York: Monthly Review. (Original work published 1955)

Chakraborty, J., & Bosman, M. (2005). Measuring the digital divide in the United States: Race, income, and personal computer ownership. *The Professional Geographer, 57,* 395–410.

Chamberlain, M. (1999). *Decolonization: The fall of European empires* (2nd ed.). Malden, MA: Blackwell.

Champion, T. (2001). Urbanization, suburbanization, counterurbanization, and reurbanization. In R. Paddison (Ed.), *Handbook of urban studies* (pp. 143–161). London: Sage.

Chandler, A., & Mazlish, B. (Eds.). (2005). *Leviathans: Multinational corporations and the new global history.* Cambridge, UK: Cambridge University Press.

Charlesworth, A. (1994). Contesting places of memory: The case of Auschwitz. *Environment and Planning D, 12,* 579–593.

Chase-Dunn, C. (1989). *Global formation: Structures in the world economy.* Oxford, UK: Blackwell.

Chase-Dunn, C. (1998). *Global formation: Structures of the world-economy* (2nd ed.). Lanham, MD: Rowman & Littlefield.

Chauncey, G. (1995). *Gay New York.* New York: Basic Books.

Chen, X. (1994). Substitution of information for energy: Conceptual background, realities, and limits. *Energy Policy, 22,* 13–23.

Chomsky, N. (2003). *Hegemony or survival: America's quest for global dominance.* New York: Metropolitan Books.

Chorley, R., & Haggett, P. (Eds.). (1967). *Models in geography.* London: Methuen.

Claritas. (2005). *A corporate history* [Online]. Available: www.claritas.com/claritas/default.jsp?ci=6&si=2

Clark, A. (1954). Historical geography. In P. James & C. F. Jones (Eds.), *American geography: Inventory and prospect* (pp. 70–105). Syracuse, NY: Syracuse University Press.

Clark, G., & Dear, M. (1984). *State apparatus: Structures and language of legitimacy.* London: Allen & Unwin.

Clark, S., & Gaile, G. (1998). *The work of cities.* Minneapolis: University of Minnesota Press.

Clark, W., & Hosking, P. (1986). *Statistical methods for geographers.* New York: John Wiley.

Clayton, D. (2003). Critical imperial and colonial geographies. In K. Anderson, M. Domosh, S. Pile, & N. Thrift (Eds.), *Handbook of cultural geography* (pp. 354–368). London: Sage.

Clifford, N., & Valentine, G. (Eds.). (2003). *Key methods in geography.* London: Sage.

Cloke, P. (1997). Country backwater to virtual village? Rural studies and the "cultural turn." *Journal of Rural Studies, 13,* 367–375.

Cloke, P., Cook, I., Crang, P., Goodwin, M., Painter, J., & Philo, C. (2004). *Practising human geography.* London: Sage.

Cloke, P., Milbourne, P., & Widdowfield, R. (2002). *Rural homelessness: Issues, experiences, and policy responses.* Bristol, UK: Policy Press.

Cloke, P., Philo, C., & Sadler, D. (1991). *Approaching human geography.* New York: Guilford.

Cochrane, W. (1993). *The development of American agriculture* (2nd ed.). Minneapolis: University of Minnesota Press.

Cohen, B. (1973). *The question of imperialism: The political economy of dominance and dependence.* New York: Basic Books.

Cohen, B. (1998). *The geography of money.* Ithaca, NY: Cornell University Press.

Cohen, R. (1997). *Global diasporas: An introduction.* London: UCL Press.

Cohen, S. (2003). *Geopolitics of the world system.* Lanham, MD: Rowman & Littlefield.

Coleman, S., & Crang, M. (Eds.). (2003). *Tourism: Between place and performance.* New York: Berghan Books.

Collins, J., Kinzing, A., Grimm, N., Fagan, W., Hope, D., Wu, J., et al. (2000). A new urban ecology. *American Scientist, 88,* 416–425.

Comrie, B., Matthews, S., & Polinsky, M. (Eds.). (1996). *The atlas of languages.* New York: Quarto.

Conca, K., & Dabelko, G. (Eds.). (1998). *Green planet blues: Environmental politics from Stockholm to Kyoto.* Boulder, CO: Westview.

Congress for New Urbanism. (n.d.). [Online]. Available: www.cnu.org

Connell, J., & Gibson, C. (2003). *Sound tracks: Popular music, identity, and place.* London: Routledge.

Connell, R. (1995). *Masculinities.* Berkeley: University of California Press.

Conzen, M. (1993). The historical impulse in geographical writing about the United States, 1850–1990. In M. Conzen, T. Rumney, & G. Wynn (Eds.), *A scholar's guide to geographical writing on the American and Canadian past* (pp. 3–90). Chicago: University of Chicago Press.

Cook, I. (1997). Participant observation. In R. Flowerdew & D. Martin (Eds.), *Methods in human geography* (pp. 127–149). Harlow, UK: Longman.

Cook, I., & Crange, P. (1996). The world on a plate: Culinary culture, displacement, and geographical knowledges. *Journal of Material Culture, 1,* 131–153.

Cooke, P. (1996). The contested terrain of locality studies. In J. Agnew, D. Livingston, & A. Rodgers (Eds.), *Human geography: An essential anthology* (pp. 476–491). London: Blackwell.

Corbridge, S. (1986). *Capitalist world development.* Lanham, MD: Rowman & Littlefield.

Corbridge, S. (1990). Post-Marxism and development studies: Beyond the impasse. *World Development, 18,* 623–640.

Corbridge, S. (1992). *Debt and development.* Cambridge, MA: Blackwell.

Corbridge, S., Martin, R., & Thrift, N. (1994). *Money, power, and space.* Oxford, UK: Blackwell.

Cornes, R., & Sandler, T. (1996). *The theory of externalities, public goods, and club goods.* Cambridge, UK: Cambridge University Press.

Coser, L., & Durkheim, É. (1997). *The division of labor in society.* New York: Free Press.

Cosgrove, D. (1984). *Social formation and the symbolic landscape.* London: Croom Helm.

Cosgrove, D. (1994). Contested global visions: One-world, whole-earth, and the Apollo space photographs. *Annals of the Association of American Geographers, 84,* 270–294.

Cosgrove, D. (1998). *Social formation and symbolic landscape* (2nd ed.). Madison: University of Wisconsin Press.

Cosgrove, D., & Daniels, S. (Eds.). (1988). *The iconography of landscape: Essays on the symbolic representation, design, and use of past environments.* Cambridge, UK: Cambridge University Press.

Couclelis, H. (1988). Of mice and men: What rodent populations can teach us about complex spatial dynamics. *Environment and Planning A, 20,* 99–109.

Cowen, M., & Shenton, R. (1996). *Doctrines of development.* London: Routledge.

Cox, K. (2002). *Political geography: Territory, state, and society.* Malden, MA: Blackwell.

Cox, K., & Golledge, R. (Eds.). (1981). *Behavioral problems in geography revisited.* London: Methuen.

Crampton, J. (2003). *The political mapping of cyberspace.* Chicago: University of Chicago Press.

Crang, M., Crang, P., & May, J. (Eds.). (1999). *Virtual geographies: Bodies, space, and relations.* London: Routledge.

Crehan, K. (2002). *Gramsci, culture, and anthropology.* Berkeley: University of California Press.

Cresswell, T. (1996). *In place/Out of place: Geography, ideology, and transgression.* Minneapolis: University of Minnesota Press.

Cresswell, T. (2003). Landscape and the obliteration of practice. In K. Anderson, M. Domosh, S. Pile, & N. Thrift (Eds.), *Handbook of cultural geography* (pp. 269–281). London: Sage.

Cresswell, T. (2004). *Place: A short introduction.* Oxford, UK: Blackwell.

Cresswell, T., & Dixon, D. (Eds.). (2002). *Engaging film: Geographies of mobility and identity.* Lanham, MD: Rowman & Littlefield.

Crewe, L., & Lowe, M. (1995). Gap on the map? Towards a geography of consumption and identity. *Environment and Planning A, 27,* 1877–1885.

Cromley, E., & McLafferty, S. (2002). *GIS and public health.* New York: Guilford.

Cronon, W. (1983). *Changes in the land: Indians, colonists, and the ecology of New England.* New York: Hill & Wang.

Cronon, W. (1995). The trouble with wilderness. In W. Cronon (Ed.), *Uncommon ground: Toward reinventing nature* (pp. 69–90). New York: Norton.

Crouch, D. (1999). *Leisure/tourism geographies.* London: Routledge.

Crow, D. (Ed.). (1996). *Geography and identity.* Washington, DC: Maisonneuve Press.

Cullingworth, J. (1993). *The political culture of planning: American land use planning in comparative perspective.* New York: Routledge.

Curry, M. (1997). The digital individual and the private realm. *Annals of the Association of American Geographers, 87,* 681–699.

Curtis, S. (2004). *Health and inequality: Geographical perspectives.* Thousand Oaks, CA: Sage.

Cutler, D., Glaeser, E., & Vigdor, J. (1999). The rise and decline of the American ghetto. *Journal of Political Economy, 107,* 455–506.

Cutter, S., & Renwick, W. (2004). *Exploitation, conservation, preservation: A geographic perspective on natural resource use.* New York: John Wiley.

Cutter, S., Holm, D., & Clark, L. (1996). The role of scale in monitoring environmental justice. *Risk Analysis, 16,* 517–526.

D'Emilio, J. (1983). *Sexual politics, sexual communities: The making of a homosexual minority in the United States, 1940–1970.* Chicago: University of Chicago Press.

Dalby, A. (2003). *Language in danger: The loss of linguistic diversity and the threat to our future.* New York: Columbia University Press.

Dalby, S. (1990). American security discourse: The persistence of geopolitics. *Political Geography Quarterly, 9,* 171–188.

Dalton, L., Hoch, C., & So, F. (Eds.). (2000). *The practice of local government planning* (3rd ed.). Washington, DC: International City/County Management Association.

Daniels, S. (1993). *Fields of vision: Landscape imagery and national identity in England and the United States.* Cambridge, UK: Verso.

Daniels, T. (1999). *When city and country collide: Managing growth in the metropolitan fringe.* Washington, DC: Island Press.

Davidson, J., & Milligan, C. (2004). Embodying emotion, sensing space: Introducing emotional geographies. *Social and Cultural Geography, 5,* 523–532.

Davis, M. (1990). *City of quartz.* London: Verso.

Davis, M. (2004, February 3). Mega slums. *New Left Review,* pp. 5–34.

De Graaf, J., Wann, D., & Naylor, T. (2001). *Affluenza: The all-consuming epidemic.* San Francisco: Berrett-Koehler.

De Sola Pool, I. (Ed.). (1977). *The social impact of the telephone.* Cambridge: MIT Press.

De Tocqueville, A. (2001). *Democracy in America.* New York: Signet. (Original work published 1835)

De Waal, A. (1998). *Famine crimes: Politics and the disaster relief industry in Africa.* Bloomington: Indiana University Press.

Dear, M. (1988). The postmodern challenge: Reconstructing human geography. *Transactions of the Institute of British Geographers, 13,* 262–274.

Dear, M. (1994). Postmodern human geography. *Erdkunde, 48,* 2–12.

Dear, M. (Ed.). (2002). *From Chicago to L.A.: Making sense of urban theory.* Thousand Oaks, CA: Sage.

Dear, M., & Flusty, S. (1998). Postmodern urbanism. *Annals of the Association of American Geographers, 88,* 50–72.

Del Casino, V., Jr. (2001). Enabling geographies? NGOs and the empowerment of people living with HIV and AIDS. *Disability Studies Quarterly, 21*(4), 19–29.

Del Casino, V., Jr., Grimes, A., Hanna, S., & Jones, J., III. (2001). Methodological frameworks for the geographic study of organizations. *Geoforum, 31,* 523–539.

Della Porta, D., & Diani, M. (1999). *Social movements: An introduction.* London: Blackwell.

DeLyser, D. (2003). Ramona memories: Fiction, tourist practices, and placing the past in Southern California. *Annals of the Association of American Geographers, 93,* 886–908.

DeLyser, D., & Pawson, E. (2005). From personal to public: Communicating qualitative research for public consumption. In I. Hay (Ed.), *Qualitative research methods in human geography* (2nd ed., pp. 266–274). Melbourne, Australia: Oxford University Press.

Deutsch, R. (1991). Boys Town. *Environment and Planning D, 9,* 5–30.

Dicken, P. (2003). *Global shift: Reshaping the global economic map in the 21st century* (4th ed.). New York: Guilford.

Dicken, P. (2003). *Global shift: Transforming the world economy* (4th ed.). New York: Guilford.

Dickinson, R. (1969). *The makers of modern geography.* New York: Praeger.

Dixon, D., & Jones, J., III. (2004). Poststructuralism. In J. Duncan, N. Johnson, & R. Schein (Eds.), *A companion to cultural geography* (pp. 79–107). Malden, MA, and Oxford, UK: Blackwell.

Dobson, J. (1993). A conceptual framework for integrating remote sensing, GIS, and geography. *Photogrammetric Engineering and Remote Sensing, 59,* 1491–1496.

Dobson, J. (1993). The geographic revolution: A retrospective on the age of automated geography. *The Professional Geographer, 45,* 431–439.

Dobson, J. (2000, May). What are the ethical limits of GIS? *GeoWorld,* pp. 24–25.

Dobson, J. (2003, May). Think twice about kid-tracking. *GeoWorld,* pp. 22–23.

Dobson, J., & Fisher, P. (2003). Geoslavery. *IEEE Technology and Society Magazine, 22*(1), 47–52.

Dodds, D. (2000). *Geopolitics in a changing world.* Upper Saddle River, NJ: Prentice Hall.

Dodds, K., & Atkinson, D. (Eds.). (2000). *Geopolitical traditions: A century of geopolitical thought.* London: Routledge.

Dodge, M., & Kitchin, R. (2001). *Atlas of cyberspace.* Reading, MA: Addison-Wesley.

Dodge, M., & Kitchin, R. (2001). *Mapping cyberspace.* London: Routledge.

Doel, M. (1999). *Poststructuralist geographies: The diabolical art of spatial science.* Lanham, MD: Rowman & Littlefield.

Domanski, B. (1991). Gatekeepers and administrative allocation of goods under socialism: An alternative perspective. *Environment and Planning C, 9,* 281–293.

Domosh, M. (1996). *Invented cities: The creation of landscape in nineteenth-century New York and Boston.* New Haven, CT: Yale University Press.

Domosh, M. (1998). Those "gorgeous incongruities": Polite politics and public space on the streets of nineteenth-century New York City. *Annals of the Association of American Geographers, 88,* 209–226.

Domosh, M., & Seager, J. (2001). *Putting women in place: Feminist geographers make sense of the world.* New York: Guilford.

Dorling, D. (1995). *A new social atlas of Britain.* Chichester, UK: Wiley.

Dreyfus, H. (1992). *What computers still can't do.* Cambridge: MIT Press.

Driver, F. (1992). Geography's empire: Histories of geographical knowledge. *Environment and Planning D, 10,* 23–40.

Driver, F. (2001). *Geography militant: Cultures of exploration and empire.* Oxford, UK: Blackwell.

Duara, P. (Ed.). (2004). *Decolonization (Rewriting histories).* London: Routledge.

Dunaway, W. (Ed.). (2003). *Emerging issues in the 21st century world-system.* Westport, CT: Praeger.

Duncan, J. (1980). The superorganic in American cultural geography. *Annals of the Association of American Geographers, 70,* 181–198.

Duncan, J. (1990). *The city as text: The politics of landscape interpretation in the Kandyan Kingdom.* Cambridge, UK: Cambridge University Press.

Duncan, J., & Duncan, N. (1988). (Re)Reading the landscape. *Environment and Planning D, 6,* 117–126.

Duncan, J., & Duncan, N. (2003). *Landscapes of privilege: The politics of the aesthetic in an American suburb.* London: Routledge.

Duncan, J., & Gregory, D. (Eds.). (1999). *Writes of passage: Reading travel writing.* London: Routledge.

Duncan, J., Johnson, N., & Schein, R. (Eds.). (2004). *A companion to cultural geography.* Oxford, UK: Blackwell.

Duncan, J., & Ley, D. (1982). Structural Marxism and human geography. *Annals of the Association of American Geographers, 72,* 30–59.

Duncan, J., & Ley, D. (Eds.). (1993). *Place/culture/representation.* London: Routledge.

Duncan, S., & Goodwin, M. (1987). *The local state and uneven development.* New York: St. Martin's.

Dunn, K. (2005). Interviewing. In I. Hay (Ed.), *Qualitative research methods in human geography* (2nd ed., pp. 79–105). Oxford, UK: Oxford University Press.

Dwyer, C. (2002). "Where are you from?" Young British Muslim women and the making of home. In A. Blunt & C. McEwan (Eds.), *Postcolonial geographies* (pp. 184–199). New York: Continuum.

Dwyer, O., & Jones, J. (2000). White socio-spatial epistemology. *Social & Cultural Geography, 1,* 209–221.

Dyer, R. (1997). *White: Essays on race and culture.* London: Routledge.

Earle, C. (1992). *Geographic inquiry and American historical problems.* Stanford, CA: Stanford University Press.

Easterly, W. (2001). *The elusive quest for growth: Economists' adventures and misadventures in the tropics.* Cambridge: MIT Press.

Edney, M. (1997). *Mapping an empire: The geographical construction of British India, 1765–1843.* Chicago: University of Chicago Press.

Egenhofer, M., & Mark, D. (1995). Naive geography. In A. U. Frank & W. Kuhn (Eds.), *Spatial information theory: A theoretical basis for GIS* (pp. 1–15). Berlin: Springer-Verlag.

Elder, G., Knopp, L., & Nast, L. (2003). Sexuality and space. In G. Gaile & C. Willmott (Eds.), *Geography in America at the dawn of the 21st century* (pp. 200–208). Oxford, UK: Oxford University Press.

Ellin, N. (Ed.). (1996). *Postmodern urbanism.* Cambridge, MA: Blackwell.

Ellis, F., & Biggs, S. (2001). Evolving themes in rural development 1950s–2000. *Development Policy Review, 19,* 437–448.

Ellwood, W. (2002). *No-nonsense guide to globalization.* London: Verso.

Elson, D. (1994). Micro, meso, macro: Gender and economic analysis in the context of policy reform. In I. Bakker (Ed.), *The strategic silence: Gender and economic policy* (pp. 33–45). London: Zed Books.

Elson, D. (2003). Gender justice, human rights, and neo-liberal economic policies. In M. Molyneux & S. Razavi (Eds.), *Gender justice, development, and rights* (pp. 78–114). Oxford, UK: Oxford University Press.

England, K. (1994). Getting personal: Reflexivity, positionality, and feminist research. *The Professional Geographer, 46,* 80–89.

Enloe, C. (1989). *Bananas, beaches, and bases: Making feminist sense of international relations.* Berkeley: University of California Press.

Entrikin, J. (1976). Contemporary humanism in geography. *Annals of the Association of American Geographers, 66,* 615–632.

Escobar, A. (1995). *Encountering development: The making and unmaking of the Third World.* Princeton, NJ: Princeton University Press.

Fabian, J. (1983). *Time and the other: How anthropology makes its object.* New York: Columbia University Press.

Fainstein, S. (1991). Promoting economic development: Urban planning in the United States and Great Britain. *Journal of the American Planning Association, 57,* 22–33.

Fainstein, S., Gordon, I., & Harloe, M. (1992). *Divided cities: New York and London in the contemporary world.* Oxford, UK: Blackwell.

Falconer Al-Hindi, K., & Till, K. (Eds.). (2001). The new urbanism and neotraditional town planning [special issue]. *Urban Geography, 22*(3).

Falzon, M.-A. (2004). Paragons of lifestyle: Gated communities and the politics of space in Bombay. *City and Society, 16*(2), 145–167.

Fan, C. (2003). Rural–urban migration and gender division of labor in transitional China. *International Journal of Urban and Regional Research, 27,* 24–47.

Fan, C., & Huang, Y. (1998). Waves of rural brides: Female marriage migration in China. *Annals of the Association of American Geographers, 88,* 227–251.

Fanon, F. (1963). *The wretched of the earth* (C. Farrington, Trans.). New York: Grove. (Original work published 1961)

Farber, D. (Ed.). (1998). *The struggle for ecological democracy: Environmental justice movements in the United States.* New York: Guilford.

Featherstone, M. (Ed.). (1990). *Global culture: Nationalism, globalization, and modernity.* London: Sage.

Feldman, M. (1994). *The geography of innovation.* Dordrecht, Netherlands: Kluwer.

Feldman, M., & Florida, R. (1994). The geographic sources of innovation: Technological infrastructure and product innovation in the United States. *Annals of the Association of American Geographers, 84,* 210–229.

Fellman, J., Getis, A., & Getis, J. (1992). *Human geography: Landscapes of human activities.* Dubuque, IA: W. C. Brown.

Finch, V. (1939). Geographical science and social philosophy. *Annals of the Association of American Geographers, 29*(1), 1–28.

Fisher, P., & Dobson, J. (2003). Who knows where you are, and who should, in the era of mobile geography? *Geography, 88,* 331–337.

Fishman, R. (1987). *Bourgeois utopias.* New York: Basic Books.

Flake, G. (1998). *The computational beauty of nature: Computer explorations of fractals, chaos, complex systems, and adaptation.* Cambridge: MIT Press.

Flint, C. (1999). Changing times, changing scales: World politics and political geography since 1890. In G. Demko & W. Wood (Eds.), *Reordering the world: Geopolitical perspectives on the 21st century* (2nd ed., pp. 19–39). Boulder, CO: Westview.

Flint, C. (2001). A timespace for electoral geography: Economic restructuring, political agency, and the rise of the Nazi party. *Political Geography, 20,* 301–329.

Flora, C. (2004). *Rural communities: Legacy and change* (2nd ed.). Boulder, CO: Westview.

Florida, R., & Jonas, A. (1991). U.S. urban policy: The postwar state and capitalist regulation. *Antipode, 23,* 349–384.

Fone, B. (2000). *Homophobia: A history.* New York: Picador.

Foner, N., & Frederickson, G. (Eds.). (2004). *Not just black and white.* New York: Russell Sage.

Fontana, A., & Frey, J. (2000). The interview: From structured questions to negotiated text. In N. Denzin & Y. Lincoln (Eds.), *Handbook of qualitative research* (2nd ed., pp. 645–672). Thousand Oaks, CA: Sage.

Food and Agriculture Organization of the United Nations. (1984). *Changes in shifting cultivation in Africa.* Rome: Author.

Food and Agriculture Organization of the United Nations. (2003). *The state of food insecurity in the world 2003.* Rome: Author. Available: www.fao.org/sof/sofi/index_ en.htm

Foote, K., Hugill, P., Mathewson, K., & Smith, J. (Eds.). (1994). *Re-reading cultural geography.* Austin: University of Texas Press.

Ford, L. (1994). *Cities and buildings: Skyscrapers, skid rows, and suburbs.* Baltimore, MD: Johns Hopkins University Press.

Foster-Carter, A. (1978, January–February). The modes of production controversy. *New Left Review,* pp. 47–77.

Fotheringham, A. (2000). A bluffer's guide to a solution to the ecological inference problem. *Annals of the Association of American Geographers, 90,* 582–586.

Fotheringham, A., Brunsdon, C., & Charlton, M. (2002). *Geographically weighted regression: The analysis of spatially varying relationships.* Chichester, UK: Wiley.

Foucault, M. (1977). *Discipline and punish.* London: Tavistock.

Foucault, M. (1978). *History of sexuality* (Vol. 1). New York: Vintage Books.

Frank, A. (1967). *Capitalism and underdevelopment in Latin America.* New York: Monthly Review.

Frank, A. (1996). The development of underdevelopment. In C. Wilber & K. Jameson (Eds.), *The political economy of development and underdevelopment* (pp. 105–115). New York: McGraw-Hill. (Original work published 1967)

Frank, A. (1998). *ReOrient: Global economy in the Asian age.* Berkeley: University of California Press.

Frankenberg, R. (1997). *Displacing whiteness.* Durham, NC: Duke University Press.

Franzen, T. (2005). *Gödel's theorem: An incomplete guide to its use and abuse.* Wellesley, MA: A. K. Peters.

Friedman, T. (1999). *The lexus and the olive tree.* New York: Farrar, Straus and Giroux.

Friedman, T. (2005). *The world is flat: A brief history of the 21st century.* New York: Picador USA.

Friedmann, H. (1993). The political economy of food. *New Left Review, 197,* 29–57.

Fröbel, F., Heinrichs, J., & Kreye, O. (1979). *The new international division of labour.* Cambridge, UK: Cambridge University Press.

Fuller, D., & Kitchin, R. (Eds.). (2004). *Radical theory/ Critical praxis: Making a difference beyond the academy?* [Online]. Available: www.praxis-epress.org/ availablebooks/radicaltheorycriticalpraxis.html

Fulton, W. (1997). *The reluctant metropolis: The politics of urban growth in Los Angeles.* Baltimore, MD: Johns Hopkins University Press.

Furuseth, J., & Lapping, M. (Eds.). (1999). *Contested countryside: The rural–urban fringe in North America.* Hampshire, UK: Ashgate.

Gaffikin, F., & Warf, B. (1993). Urban policy and the post-Keynesian state in the United Kingdom and the United States. *International Journal of Urban and Regional Research, 17,* 67–84.

Ganz, H. (1967). *The Levittowners.* New York: Pantheon.

Garcia-Ramon, M., Albert, A., & Zusman, P. (2003). Recent developments in social and cultural geography in Spain. *Social and Cultural Geography, 4,* 419–431.

Garey, M. R., & Johnson, D. S. (1979). *Computers and intractability: A guide to the theory of NP-completeness.* New York: Freeman.

Garreau, J. (1991). *Edge city: Life on the new frontier.* New York: Anchor.

Garvin, A., & Berens, G. (1997). *Urban parks and open space.* Washington, DC: Urban Land Institute and Trust for Public Land.

Geertz, C. (1973). *The interpretation of cultures.* New York: Basic Books.

Gelernter, D. (1997). *Machine beauty: Elegance and the heart of technology.* New York: Basic Books.

Geography Education Standards Project. (1994). *Geography for life: National Geography Standards 1994.* Washington, DC: National Geographic Research and Exploration.

George, S. (1988). *A fate worse than debt: A radical analysis of the Third World debt crisis.* Harmondsworth, UK: Penguin Books.

Gereffi, G. (1990). International trade and the industrial upgrading in the apparel commodity chain. *Journal of International Economics, 48,* 37–70.

Gereffi, G. (1990). *Manufacturing miracles: Paths of industrialization in Latin America and East Asia.* Princeton, NJ: Princeton University Press.

Gereffi, G., & Korzeniewicz, M. (Eds.). (1994). *Commodity chains and global capitalism.* Westport, CT: Greenwood.

Gertler, M. (1992). Flexibility revisited: Districts, nation-states, and the forces of production. *Transactions of the Institute of British Geographers, 17,* 259–278.

Gesler, W., & Kearns, R. (2002). *Culture/place/health.* London: Routledge.

Gewin, V. (2004). Mapping opportunities. *Nature, 427,* 376–377.

Gibb, R. (1994). Regionalism in the world economy. In R. Gibb & W. Michalak (Eds.), *Continental trading blocs: The growth of regionalism in the world economy.* New York: John Wiley.

Gibson-Graham, J. (1996). *The end of capitalism (as we knew it): A feminist critique of political economy.* Cambridge, MA; Oxford, UK: Blackwell.

Giddens, A. (1981). *A contemporary critique of historical materialism.* Berkeley: University of California Press.

Giddens, A. (1981). Introduction. In M. Weber (Ed.), *The Protestant ethic and the spirit of capitalism* (pp. 1–12). London: Routledge.

Giddens, A. (1984). *The constitution of society: Outline of the theory of structuration.* Berkeley: University of California Press.

Giddens, A. (1987). *The nation-state and violence.* Berkeley: University of California Press.

Giddens, A. (1990). *The consequences of modernity.* Stanford, CA: Stanford University Press.

Giddens, A. (1991). *Modernity and self-identity: Self and society in the late modern age.* Stanford, CA: Stanford University Press.

Gillham, O., & Maclean, A. (2002). *The limitless city: A primer on the urban sprawl debate.* New York: Island Press.

Gilmore, R. (2002). Fatal couplings of power and difference: Notes on racism and geography. *The Professional Geographer, 54,* 15–24.

Glacken, C. (1967). *Traces on the Rhodian shore: Nature and culture in Western thought from ancient times to the end of the eighteenth century.* Berkeley: University of California Press.

Glacken, C. (1973). Environment and culture. In P. Wiener (Ed.), *Dictionary of the history of ideas* (Vol. 2, pp. 127–134). New York: Scribner.

Glaeser, E., & Vigdor, J. (2001). *Racial segregation in the U.S. census: Promising news* (Brookings Institution Survey Series) [Online]. Available: www.brookings .edu/es/urban/census/glaeser.pdf

Glasmeier, A. (2005). *An atlas on poverty in America: One nation pulling apart 1960–2003.* University Park: Pennsylvania State University Press.

Glassner, M. (1993). *Political geography.* New York: John Wiley.

Glassner, M., & Fahrer, C. (2004). *Political geography* (3rd ed.). Hoboken, NJ: John Wiley.

Gleeson, B. (1999). *Geographies of disability.* New York: Routledge.

Gober, P., & Tyner, J. (2004). Population geography. In G. Gaile & C. Willmott (Eds.), *Geography in America at the dawn of the 21st century* (pp. 185–199). New York: Oxford University Press.

Gold, M. (1984). A history of nature. In D. Massey & J. Allen (Eds.), *Geography matters! A reader* (pp. 12–33). New York: Cambridge University Press.

Golledge, R. (2002). The nature of geographic knowledge. *Annals of the Association of American Geographers, 92,* 1–14.

Golledge, R., & Stimson, R. (1997). *Spatial behavior: A geographic perspective.* New York: Guilford.

Gonzáles Ceballos, S. (2004). The role of the Guggenheim Museum in the development of urban entrepreneurial practices in Bilbao. *International Journal of Iberian Studies, 16,* 177–186.

Goodchild, M. (1992). Geographical information science. *International Journal of Geographical Information Systems, 6,* 31–45.

Goodchild, M., Anselin, L., Appelbaum, R., & Harthorn, B. (2000). Toward spatially integrated social science. *International Regional Science Review, 23,* 139–159.

Goodchild, M., & Janelle, D. (2004). *Spatially integrated social science.* New York: Oxford University Press.

Goodchild, M., & Janelle, D. (2004). Thinking spatially in the social sciences. In M. Goodchild & D. Janelle (Eds.), *Spatially integrated social science* (pp. 3–21). New York: Oxford University Press.

Goodchild, M., & Longley, P. (1999). The future of GIS and spatial analysis. In P. Longley, M. Goodchild, D. Maguire, & D. Rhind (Eds.), *Geographical information systems: Principles, techniques, management, and applications* (pp. 567–580). New York: John Wiley.

Goodman, D., & Redclift, M. (1990). *Refashioning nature: Food, ecology, and culture.* London: Routledge.

Gore, A. (1992). *Earth in the balance: Ecology and the human spirit.* Boston: Houghton Mifflin.

Gore, A. (1994). *Department of Housing and Urban Development: Accompanying report of the National Performance Review.* Washington, DC: Office of the Vice President.

Goss, J. (1993). "The magic of the mall": An analysis of form, function, and meaning in the contemporary retail built environment. *Annals of the Association of American Geographers, 83,* 18–47.

Goss, J. (1995). "We know where you are and we know where you live": The instrumental rationality of geodemographic information systems. *Economic Geography, 71,* 171–198.

Goss, J. (1999). Once upon a time in the commodity world: An unofficial guide to the Mall of America. *Annals of the Association of American Geographers, 98,* 45–75.

Gottdiener, M. (1985). *The social production of urban space.* Austin: University of Texas Press.

Gottdiener, M. (1994). *The social production of urban space* (2nd ed.). Austin: University of Texas Press.

Gould, P. (1978). The Augean period. *Annals of the Association of American Geographers, 69,* 139–151.

Gould, P. (1991). Dynamic structures of geographic space. In S. Brunn & T. Leinbach (Eds.), *Collapsing space and time: Geographic aspects of communication and information* (pp. 3–30). London: HarperCollins Academic.

Gould, P., & White, R. (1974). *Mental maps.* London: Penguin.

Graham, B., & Nash, C. (Eds.). (1999). *Modern historical geographies.* Harlow, UK: Longman.

Graham, J. (1988). Post-modernism and Marxism. *Antipode, 20,* 60–65.

Graham, S. (1998). Spaces of surveillant simulation: New technologies, digital representations, and material geographies. *Environment and Planning D, 16,* 483–504.

Graham, S. (Ed.). (2004). *The cybercities reader* (Urban Reader series). London: Routledge.

Graham, S., & Marvin, S. (1996). *Telecommunications and the city: Electronic spaces, urban places.* London: Routledge.

Graham, S., & Marvin, S. (2001). *Splintering urbanism: Networked infrastructures, technological mobilities, and the urban condition.* London: Routledge.

Gramsci, A. (1971). *Selections from the prison notebooks of Antonio Gramsci* (Q. Hoare & G. N. Smith, Eds. & Trans.). New York: International Publishers.

Grant, J., & Mittelsteadt, L. (2004). Types of gated communities. *Environment and Planning B, 31,* 913–930.

Greenberg, D. (1990). *The construction of homosexuality.* Chicago: University of Chicago Press.

Gregory, D. (1978). *Ideology, science, and human geography.* London: Hutchinson.

Gregory, D. (1981). Human agency and human geography. *Transactions of the Institute of British Geographers, 6,* 1–18.

Gregory, D. (1994). *Geographical imaginations.* Cambridge, MA; Oxford, UK: Blackwell.

Gregory, D. (1995). Between the book and the lamp: Imaginative geographies of Egypt, 1849–50. *Transactions, Institute of British Geographers, 20,* 29–57.

Gregory, D. (1995). Imaginative geographies. *Progress in Human Geography, 19,* 447–485.

Gregory, D. (2000). Ideology. In R. Johnston, D. Gregory, G. Pratt, & M. Watts (Eds.), *The dictionary of human geography* (4th ed., pp. 369–370). Oxford, UK: Blackwell.

Gregory, D. (2004). *The colonial present: Afghanistan, Palestine, and Iraq.* Malden, MA; Oxford, UK: Blackwell.

Gregson, N. (2003). Reclaiming "the social" in social and cultural geography. In K. Anderson, M. Domosh, S. Pile, & N. Thrift (Eds.), *Handbook of cultural geography* (pp. 43–57). London: Sage.

Gregson, N., Crewe, L., & Brooks, K. (2002). Shopping, space, and practice. *Environment and Planning D, 20,* 597–617.

Gregson, N., & Rose, G. (2000). Taking Butler elsewhere: Performativities, spatialities, and subjectivities. *Environment and Planning D, 18,* 433–452.

Grimbly, S. (Ed.). (2001). *Atlas of exploration.* Chicago: Fitzroy Dearborn.

Grimm, N., & Redman, C. (2004). Approaches to the study of urban ecosystems: The case of Central Arizona–Phoenix. *Urban Ecosystems, 7,* 199–213.

Grove, J., & Burch, W., Jr. (1997). A social ecology approach and applications of urban ecosystem and landscape analyses: A case study of Baltimore, Maryland. *Urban Ecosystems, 1,* 259–275.

Gruffudd, P. (1999). Nationalism. In P. Cloke, P. Crang, & M. Goodwin (Eds.), *Introducing human geographies* (pp. 199–207). London: Edward Arnold.

Guarino, N. (1997). Understanding, building, and using ontologies. *International Journal of Human–Computer Studies, 46,* 293–310.

Guarino, N., & Giaretta, P. (1995). Ontologies and knowledge bases: Towards a terminological clarification. In N. Mars (Ed.), *Towards very large knowledge bases* (pp. 25–32). Amsterdam: IOS Press.

Guelke, L. (1978). Geography and logical positivism. In D. Herbert & R. Johnston (Eds.), *Geography and the urban environment: Progress in research and applications* (Vol. 1, pp. 35–61). Chichester, UK: Wiley.

Guha, R., & Spivak, G. (Eds.). (1988). *Selected subaltern studies.* Oxford, UK: Oxford University Press.

Guinness, P. (2002). *Migration.* London: Hodder Murray.

Gwynne, R. (1986). *Industrialization and urbanization in Latin America.* Baltimore, MD: Johns Hopkins University Press.

Habermas, J. (1975). *Legitimation crisis.* Boston: Beacon.

Hackett, S., & Dilts, D. (2004). A systematic review of business incubation research. *Journal of Technology Transfer, 29,* 55–82.

Hägerstrand, T. (1965). A Monte Carlo approach to diffusion. *European Journal of Sociology, 6,* 43–67.

Hägerstrand, T. (1967). *Innovation diffusion as a spatial process.* Chicago: University of Chicago Press.

Hägerstrand, T. (1970). What about people in regional science? *Papers of the Regional Science Association, 24,* 7–21.

Haggett, P. (1965). *Locational analysis in human geography.* London: Edward Arnold.

Hague, E. (2002). Antipode, Inc.? *Antipode, 34,* 655–661.

Haining, R. (2003). *Spatial data analysis: Theory and practice.* Cambridge, UK: Cambridge University Press.

Hall, S. (Ed.). (1997). *Representation: Cultural representations and signifying practices.* London: Sage.

Hall, T., & Hubbard, P. (1998). *The entrepreneurial city: Geographies of politics, regime, and representation.* Chichester, UK: Wiley.

Hamel, P., Lustiger-Thaler, H., Nederveen Pieterse, J., & Roseneil, S. (Eds.). (2001). *Globalization and social movements.* New York: Palgrave.

Hamnett, C. (1994). Social polarization in global cities: Theory and evidence. *Urban Studies, 31,* 401–424.

Hanna, S., & Del Casino, V. (Eds.). (2003). *Mapping tourism: Representation, identity, and intertextuality.* Minneapolis: University of Minnesota Press.

Hanson, S., & Giuliano, G. (2004). *The geography of urban geography* (3rd ed.). New York: Guilford.

Haraway, D. (1988). Situated knowledges: The science question in feminism and the privilege of partial perspective. *Feminist Studies, 14,* 575–599.

Haraway, D. (1991). *Simians, cyborgs, and women: The reinvention of nature.* New York; London: Routledge.

Harding, S. (1986). *The science question in feminism.* Ithaca, NY: Cornell University Press.

Harding, S. (1987). *Feminism and methodology.* Bloomington: University of Indiana Press.

Harley, J. (1989). Deconstructing the map. *Cartographica, 26,* 1–20.

Harley, J. (2002). *The new nature of maps: Essays in the history of cartography.* Baltimore, MD: Johns Hopkins University Press.

Harrington, J., & Warf, B. (1994). *Industrial location: Principles, practice, and policy.* London: Routledge.

Harris, C. (2002). *Making native space: Colonialism, resistance, and reserves in British Columbia.* Vancouver: University of British Columbia Press.

Harris, R., Sleight, P., & Webber, R. (2005). *Geodemographics, GIS, and neighborhood targeting.* New York: John Wiley.

Hart, J. (1982). The highest form of the geographer's art. *Annals of the Association of American Geographers, 72,* 1–29.

Hart, J. (1998). *The rural landscape.* Baltimore, MD: Johns Hopkins University Press.

Hart, J. (2003). *The changing scale of American agriculture.* Charlottesville: University of Virginia Press.

Hartshorne, R. (1939). *The nature of geography: A critical survey of current thought in light of the past.* Washington, DC: Association of American Geographers.

Hartwick, E. (1998). Geographies of consumption: A commodity-chain approach. *Environment and Planning D, 16,* 423–437.

Hartwick, E. (2000). Towards a geographical politics of consumption. *Environment and Planning A, 32,* 1177–1192.

Harvey, D. (1969). *Explanation in geography.* London: Edward Arnold.

Harvey, D. (1973). *Social justice and the city.* Baltimore, MD: Johns Hopkins University Press; London: Edward Arnold.

Harvey, D. (1974). Population, resources, and the ideology of science. *Economic Geography, 50,* 256–277.

Harvey, D. (1976). Labor, capital, and class struggle around the built environment in advanced capitalist societies. *Politics & Society, 6,* 265–295.

Harvey, D. (1978). The urban process under capitalism: A framework for analysis. *International Journal of Urban and Regional Research, 2,* 101–131.

Harvey, D. (1982). *The limits to capital.* Chicago: University of Chicago Press; Oxford, UK: Blackwell.

Harvey, D. (1984). On the history and present condition of geography: An historical materialist manifesto. *The Professional Geographer, 36,* 1–10.

Harvey, D. (1985). The geopolitics of capitalism. In D. Gregory & J. Urry (Eds.), *Social relations and spatial structures* (pp. 128–163). New York: St. Martin's.

Harvey, D. (1985). *The urbanization of capital.* Baltimore, MD: Johns Hopkins University Press; Oxford, UK: Blackwell.

Harvey, D. (1989). *The condition of postmodernity: An enquiry into the origins of cultural change.* Cambridge, MA; Oxford, UK: Blackwell.

Harvey, D. (1989). From managerialism to entrepreneurialism: The transformation in urban governance in late capitalism. *Geografiska Annaler B, 71*(1), 3–17.

Harvey, D. (1996). *Justice, nature, and the geography of difference.* Oxford, UK: Blackwell.

Harvey, D. (1997, Winter–Spring). The new urbanism and the communitarian trap. *Harvard Design Magazine,* pp. 66–69.

Harvey, D. (1999). *The limits to capital* (2nd ed.). London: Verso.

Harvey, D. (2000). *Spaces of hope.* Berkeley: University of California Press.

Harvey, D. (2001). *Spaces of capital: Towards a critical geography.* New York; London: Routledge.

Harvey, D. (2003). *The new imperialism.* Oxford, UK: Oxford University Press.

Harvey, D. (2003). *Paris, capital of modernity.* New York: Routledge.

Harvey, F. (2003). Knowledge and geography's technology: Politics, ontologies, representations in the changing ways we know. In K. Anderson, M. Domosh, S. Pile, & N. Thrift (Eds.), *Handbook of cultural geography* (pp. 532–543). London: Sage.

Hay, I. (2004). *Qualitative research methods in human geography* (2nd ed.). Oxford, UK: Oxford University Press.

Hayden, D. (2002). *Redesigning the American dream: Gender, housing, and family life.* New York: Norton.

Hayden, D. (2003). *Building suburbia: Green fields and urban growth, 1820–2000.* New York: Vintage Books.

Heiman, M. (1966). Waste, race, and class: New perspectives on environmental justice. *Antipode, 28,* 111–121.

Heiman, M. (1990). From "not in my backyard!" to "not in anybody's backyard!" *Journal of the American Planning Association, 56,* 359–362.

Henderson, G. (2002). *California and the fictions of capital.* Philadelphia: Temple University Press.

Hensher, D., Button, K., Haynes, K., & Stopher, R. (2004). *Handbook of transport geography and spatial systems.* London: Elsevier.

Herb, G., & Kaplan, D. (Eds.). (1999). *Nested identities: Nationalism, territory, and scale.* Lanham, MD: Rowman & Littlefield.

Herbert, D. (2001). Literary places and the tourism experience. *Annals of Tourism Research, 28,* 312–333.

Herod, A. (1997). From a geography of labor to a labor geography: Labor's spatial fix and the geography of capitalism. *Antipode, 29,* 1–31.

Herod, A. (2001). *Labor geographies: Workers and the landscapes of capitalism.* New York: Guilford.

Herod, A. (Ed.). (1998). *Organizing the landscape: Geographical perspectives on labor unionism.* Minneapolis: University of Minnesota Press.

Herod, A., Ó Tuathail, G., & Roberts, S. (Eds.). (1998). *An unruly world? Globalization, governance, and geography.* London: Routledge.

Hervieu-Léger, D. (2002). Space and religion: New approaches to religious spatiality in modernity. *International Journal of Urban and Regional Research, 26,* 99–105.

Hewitson, B., & Crane, R. (Eds.). (1994). *Neural networks: Applications in geography.* Dordrecht, Netherlands: Kluwer Academic.

Higgs, D. (Ed.). (1999). *Queer sites: Gay urban histories since 1600.* London: Routledge.

Hobsbawm, E., & Ranger, T. (Eds.). (1983). *The invention of tradition.* Cambridge, UK: Cambridge University Press.

Hoelscher, S. (1998). The photographic construction of tourist space in Victorian America. *The Geographical Review, 88,* 548–570.

Hoelscher, S. (2003). Making place, making race: Performances of whiteness in the Jim Crow South. *Annals of the Association of American Geographers, 93,* 657–686.

Hoggart, K., Lees, L., & Davies, A. (2002). *Researching human geography.* London: Edward Arnold.

Holder, J., & Harrison, C. (Eds.). (2003). *Law and geography.* Oxford, UK: Oxford University Press.

Holloway, S., & Valentine, G. (2000). *Children's geographies: Playing, living, learning.* New York: Routledge.

Holloway, S., & Valentine, G. (2000). Corked hats and Coronation Street: British and New Zealand children's imaginative geographies of the other. *Childhood, 7,* 335–357.

Holt-Jensen, A. (2003). *Geography: History and concepts* (3rd ed.). London: Sage.

Hooson, D. (Ed.). (1994). *Geography and national identity.* Oxford, UK: Blackwell.

Hopcroft, J. E., Motwani, R., & Ullman, J. D. (2001). *Introduction to automata theory, languages, and computation* (2nd ed.). Reading, MA: Addison-Wesley.

Howitt, R. (2003). Nests, webs, and constructs: Contested concepts of scale in political geography. In J. Agnew, K. Mitchell, & G. Ó Tuathail (Eds.), *A companion to political geography.* Oxford, UK: Blackwell.

Hubbard, P. (2000). Desire/disgust: Mapping the moral contours of heterosexuality. *Progress in Human Geography, 24,* 191–217.

Hubbard, P., Kitchin, R., Bartley, B., & Fuller, D. (2002). *Thinking geographically: Space, theory, and contemporary human geography.* London: Continuum.

Hubbard, P., Kitchin, R., & Valentine, G. (Eds.). (2004). *Key thinkers on space and place.* London: Sage.

Hudson, J. (1994). *Making the Corn Belt: A geographical history of middle-western agriculture.* Bloomington: Indiana University Press.

Huntington, S. (1968). *Political order in changing societies.* New Haven, CT: Yale University Press.

Hyndman, J. (2004). Mind the gap: Bridging feminist and political geography through geopolitics. *Political Geography, 23,* 307–322.

Ilbery, B., Chiotti, Q., & Rickard, T. (Eds.). (1997). *Agricultural restructuring and sustainability: A geographical perspective.* New York: CAB International.

Imrie, R. (1996). *Disability and the city.* London: Paul Chapman.

International Labour Organization. (2002). *Globalization and the informal economy: How global trade and investment impact on the working poor.* Geneva, Switzerland: Author.

International Labour Organization. (2002). *Women and men in the informal economy: A statistical picture.* Geneva, Switzerland: Author.

Isaaks, E., & Srivastava, R. (1990). *Applied geostatistics.* New York: Oxford University Press.

Isard, W. (1956). *Location and space economy.* Cambridge: MIT Press.

Ittelson, W. (Ed.). (1973). *Environment and cognition.* New York: Seminar Press.

Jackson, J. (1997). *Landscape in sight: Looking at America* (H. L. Horowitz, Ed.). New Haven, CT: Yale University Press.

Jackson, K. (1985). *Crabgrass frontier.* New York: Oxford University Press.

Jackson, P. (1991). The cultural politics of masculinity: Towards a social geography. *Transactions of the Institute of British Geographers, 16,* 199–213.

Jackson, P. (1992). *Maps of meaning: An introduction to cultural geography.* London: Routledge.

Jackson, P. (1998). Constructions of whiteness in the geographical imagination. *Area, 30,* 99–106.

Jackson, P. (2000). Rematerializing social and cultural geography. *Social and Cultural Geography, 1,* 9–14.

Jackson, P., & Penrose, J. (Eds.). (1993). *Constructions of race, place, and nation.* London: UCL Press.

Jackson, P., & Smith, S. (Eds.). (1981). *Social interaction and ethnic segregation.* London: Academic Press.

Jakle, J., & Wilson, D. (1992). *Derelict landscapes: The wasting of America's built environment.* Lanham, MD: Rowman & Littlefield.

James, P. (1972). *All possible worlds: A history of geographical ideas.* Indianapolis, IN: Bobbs-Merrill.

Janelle, D. (1969). Spatial reorganization: A model and concept. *Annals of the Association of American Geographers, 59,* 348–365.

Janelle, D., & Hodge, D. (Eds.). (2000). *Information, place, and cyberspace: Issues in accessibility.* Berlin: Springer-Verlag.

Jay, M. (1994). *Downcast eyes: The denigration of vision in twentieth-century French thought.* Berkeley: University of California Press.

Jenks, C. (1995). *Visual culture.* London: Routledge.

Joad, C. (1950). *A critique of logical positivism.* Chicago: University of Chicago Press.

Johnson, C. (2004). *The sorrows of empire: Militarism, secrecy, and the end of the republic.* New York: Metropolitan Books.

Johnson, N. (1995). Cast in stone: Monuments, geography, and nationalism. *Environment and Planning D, 13,* 51–65.

Johnson, N. (2002). The renaissance of nationalism. In R. Johnston, P. Taylor, & M. Watts (Eds.), *Geographies of global change* (2nd ed., pp. 130–142). Oxford, UK: Blackwell.

Johnson, N. (2004). Public memory. In J. Duncan, N. Johnson, & R. Schein (Eds.), *A companion to cultural geography* (pp. 316–327). Oxford, UK: Blackwell.

Johnston, R., Gregory, D., Pratt, G., & Watts, M. (Eds.). (2001). *Dictionary of human geography* (4th ed.). Oxford, UK: Blackwell.

Johnston, R., & Sidaway, J. (2004). *Geography and geographers: Anglo-American human geography since 1945* (6th ed.). London: Edward Arnold.

Johnston, R., Taylor, P., & Shelley, F. (Eds.). (1990). *Developments in electoral geography.* London: Routledge.

Jones, G., Caldwell, J., D'Souza, R., & Douglas, R. (Eds.). (1998). *The continuing demographic transition.* Oxford, UK: Clarendon.

Jones, J., III. (1995). Making geography objectively: Ocularity, representation, and *The Nature of Geography.* In W. Natter, T. Schatzki, & J. Jones (Eds.), *Objectivity and its other* (pp. 67–92). New York: Guilford.

Jones, J., III. (2004). Introduction: Reading geography through binary oppositions. In K. Anderson, M. Domosh, S. Pile, & N. Thrift (Eds.), *Handbook of cultural geography* (pp. 511–519). Thousand Oaks, CA: Sage.

Jones, J., III, Nast, H., & Roberts, S. (Eds.). (1997). *Thresholds in feminist geography.* Lanham, MD: Rowman & Littlefield.

Jones, J., III, & Natter, W. (1999). Space and representation. In A. Buttimer, S. Brunn, & U. Wardenga (Eds.), *Text and image: Social construction of regional knowledges* (pp. 239–247). Leipzig, Germany: Institut für Landerkunde Leipzig.

Jones, J., III, Natter, W., & Schatzki, T. (1993). *Postmodern contentions: Epochs, politics, space.* New York: Guilford.

Jones, M. (1998). Restructuring the local state: Economic governance or social regulation? *Political Geography, 17,* 959–988.

Jumper, S., Bell, T., & Ralston, B. (1980). *Economic growth and disparities: A world view.* Englewood Cliffs, NJ: Prentice Hall.

Jürgens, U., & Gnad, M. (2002). Gated communities in South Africa: Experiences from Johannesburg. *Environment and Planning B, 29,* 337–353.

Kahn, H. (1976). *The next 200 years: A scenario for America and the world.* New York: William Morrow.

Kalipeni, E., Craddock, S., Oppong, J., & Ghosh, J. (Eds.). (2004). *HIV and AIDS in Africa: Beyond epidemiology.* Boston: Blackwell.

Kaplan, D., & Holloway, S. (1998). *Segregation in cities.* Washington, DC: Association of American Geographers.

Katz, C. (2004). *Growing up global: Economic restructuring and children's everyday lives.* Minneapolis: University of Minnesota Press.

Kay, C. (1989). *Latin American theories of development and underdevelopment.* London: Routledge.

Kay, J. (1991). Landscapes of women and men: Rethinking the regional historical geography of the United States. *Journal of Historical Geography, 17,* 435–452.

Keating, W., Krumholz, N., & Star, P. (Eds.). (1996). *Revitalizing urban neighborhoods.* Lawrence: University Press of Kansas.

Keith, M., & Pile, S. (Eds.). (1993) *Place and the politics of identity.* London: Routledge.

Kempadoo, K., & Doezema, J. (Eds.). (1998). *Global sex workers: Rights, resistance, and redefinition.* New York: Routledge.

Kenzer, M. (Ed.). (1987). *Carl O. Sauer: A tribute.* Corvallis: Oregon State University Press.

Kern, S. (1983). *The culture of time and space 1880–1918.* Cambridge, MA: Harvard University Press.

Kertzer, D., & Arel, D. (Eds.). (2002). *Census and identity: The politics of race, ethnicity, and language in national censuses.* Cambridge, UK: Cambridge University Press.

Kincaid, D. (1993). Peasants into rebels: Community and class in rural El Salvador. In D. Levine (Ed.), *Constructing culture and power in Latin America* (pp. 119–154). Ann Arbor: University of Michigan Press.

King, G. (1997). *A solution to the ecological inference problem.* Princeton, NJ: Princeton University Press.

Kingsbury, P. (2004). Psychoanalytic approaches. In J. Duncan, N. Johnson, & R. Schein (Eds.), *A companion to cultural geography* (pp. 108–120). Oxford, UK: Blackwell.

Kinsman, P. (1995). Landscape, race, and national identity: The photography of Ingrid Pollard. *Area, 27,* 300–310.

Kirby, K. (1996). *Indifferent boundaries: Spatial concepts of human subjectivity.* New York: Guilford.

Kirk, D. (1996). Demographic transition theory. *Population Studies, 50,* 361–388.

Kirsch, S. (1995). The incredible shrinking world? Technology and the production of space. *Environment and Planning D, 13,* 529–555.

Kitchin, R. (1994). Cognitive maps: What are they and why study them? *Journal of Environmental Psychology, 14,* 1–19.

Kitchin, R. (1998). *Cyberspace: The world in the wires.* Chichester, UK: Wiley.

Kitchin, R., & Blades, M. (2002). *The cognition of geographic space.* New York: I. B. Tauris.

Kitching, G. (1989). *Development and underdevelopment in historical perspective.* London: Methuen.

Klak, T. (Ed.). (1997). *Globalization and neoliberalism.* Lanham, MD: Rowman & Littlefield.

Kling, R. (1999). What is social informatics and why does it matter? *D-Lib Magazine, 5*(1) [Online]. Available: www.dlib.org/dlib/january99/kling/01kling.html

Kniffen, F. (1965). Folk housing: Key to diffusion. *Annals of the Association of American Geographers, 55,* 549–577.

Knopp, L. (1986). Social theory, social movements, and public policy: Recent accomplishments of the gay and lesbian movements in Minneapolis. *International Journal of Urban and Regional Research, 11,* 243–261.

Knopp, L. (1992). Sexuality and the spatial dynamics of capitalism. *Environment and Planning D, 10,* 651–669.

Knox, P. (1994). *Urbanization.* Englewood Cliffs, NJ: Prentice Hall.

Knox, P. (1995). World cities and the organization of global space. In R. J. Johnston, P. Taylor, & M. Watts (Eds.), *Geographies of global change* (pp. 232–247). Oxford, UK: Blackwell.

Knox, P., & Agnew, J. (1994). *The geography of the world economy* (2nd ed.). London: Edward Arnold.

Knox, P., Agnew, J., & McCarthy, L. (2003). *The geography of the world economy* (4th ed.). London: Edward Arnold; New York: Oxford University Press.

Knox, P., & McCarthy, L. (2005). *Urbanization: An introduction to urban geography* (2nd ed.). Upper Saddle River, NJ: Prentice Hall.

Knox, P., & Pinch, S. (2000). *Urban social geography* (4th ed.). Upper Saddle River, NJ: Prentice Hall.

Kobayashi, A., & Peake, L. (2000). Racism out of place: Thoughts on whiteness and an anti-racist geography in the new millennium. *Annals of the Association of American Geographers, 90,* 392–403.

Kofman, E. (1996). Feminism, gender relations, and geopolitics: Problematic closures and opening strategies. In E. Kofman & G. Youngs (Eds.), *Globalization: Theory and practice* (pp. 209–224). New York: Pinter.

Kong, L. (2001). Mapping "new" geographies of religion: Politics and poetics in modernity. *Progress in Human Geography, 25,* 211–233.

Kors, A. (Ed.). (2003). *Encyclopedia of the Enlightenment* (4 vols.). New York: Oxford University Press.

Krim, A. (1998). "Get your kicks on Route 66!" A song map of postwar migration. *Journal of Cultural Geography, 18,* 46–60.

Krugman, P. (1991). *Geography and trade.* Cambridge: MIT Press.

Krugman, P. (1994). *Rethinking international trade.* Cambridge: MIT Press.

Kuhn, T. (1962). *The structure of scientific revolutions.* Chicago: University of Chicago Press.

Kuhn, W. (2001). Ontologies in support of activities in geographical space. *International Journal of Geographic Information Science, 15,* 613–631.

Kunstler, J. (1993). *The geography of nowhere.* New York: Touchstone.

Kvale, S. (1996). *Interviews: An introduction to qualitative research interviewing.* Thousand Oaks, CA: Sage.

Kwan, M.-P. (2002). Feminist visualization: Re-envisioning GIS as a method in feminist geographic research. *Annals of the Association of American Geographers, 92,* 645–661.

Kwan, M.-P., Janelle, D., & Goodchild, M. (Eds.). (2003). Accessibility in time and space [special issue]. *Journal of Geographical Systems, 5*(1).

Laclau, E. (1971, May–June). Feudalism and capitalism in Latin America. *New Left Review,* pp. 19–55.

Lake, R. (Ed.). (1987). *Resolving locational conflict.* New Brunswick, NJ: Center for Urban Policy Research.

Lake, R. (1993). Rethinking NIMBY. *Journal of the American Planning Association, 59,* 87–93.

Lal, D. (1983). *The poverty of development economics.* London: Institute of Foreign Affairs.

Lampard, E. (1965). Historical aspects of urbanization. In P. Hauser & L. Schnore (Eds.), *The study of urbanization* (pp. 519–554). New York: John Wiley.

Landes, D. (1969). *The unbound Prometheus: Technological change and industrial development in Western Europe from 1750 to the present.* Cambridge, UK: Cambridge University Press.

Lang, R. (2003). *Edgeless cities: Exploring the elusive metropolis.* Washington, DC: Brookings Institution.

Langran, G. (1992). *Time in geographic information systems.* London: Taylor & Francis.

Langran, G., & Chrisman, N. (1988). A framework for temporal geographic information. *Cartographica, 25*(3), 1–14.

Latour, B. (1999). *Pandora's hope: An essay on the reality of science studies.* Cambridge, MA: Harvard University Press.

Lauria, M. (Ed.). (1997). *Reconstructing urban regime theory: Regulating urban politics in a global economy.* Thousand Oaks, CA: Sage.

Lauria, M., & Knopp, L. (1985). Towards an analysis of the role of gay communities in the urban Renaissance. *Urban Geography, 6,* 152–169.

Laurie, N., Dwyer, C., Holloway, S., & Smith, F. (1999). *Geographies of new femininities.* Essex, UK: Longman.

Laurier, E. (2003). Participant observation. In N. Clifford & G. Valentine (Eds.), *Key methods in geography* (pp. 133–159). Thousand Oaks, CA: Sage.

Le Goix, R. (2005). Gated communities: Sprawl and social segregation in Southern California. *Housing Studies, 20,* 323–343.

Le Sueur, J. (Ed.). (2003). *The decolonization reader.* London: Routledge.

Lee, M. (Ed.). (2000). *The consumer society reader.* Oxford, UK: Blackwell.

Lees, L. (1994). Rethinking gentrification: Beyond the positions of economics or culture. *Progress in Human Geography, 18,* 137–150.

Lefebvre, H. (1991). *The production of space.* Oxford, UK: Blackwell.

Leib, J. (2002). Separate times, shared spaces: Arthur Ashe, Monument Avenue, and the politics of Richmond, Virginia's symbolic landscape. *Cultural Geographies, 9,* 286–312.

Leighley, J. (Ed.). (1963). *Land and life: A selection from the writings of Carl Ortwin Sauer.* Berkeley: University of California Press.

Leisinger, K., Schmitt, K., & Pandya-Lorch, R. (2002). *Six billion and counting: Population growth and food security in the 21st century.* Washington, DC: International Food Policy Research Institute.

Leitner, H. (1990). Cities in pursuit of economic growth: The local state as entrepreneur. *Political Geography Quarterly, 9,* 146–170.

Leonard, W., & Crawford, M. (2002). *Human biology of pastoral populations.* Cambridge, UK: Cambridge University Press.

Leontief, W. (1936). Quantitative input–output relations in the economic system of the United States. *American Economic Review, 64,* 823–834.

Levison, M., & Haddon, W. (1965). The area adjusted map: An epidemiological device. *Public Health Reports, 80*(1), 55–59.

Lewis, D. (2002). *Does technology incubation work? A critical review of the evidence.* Athens, OH: National Business Incubation Association.

Lewis, M., & Wigen, K. (1997). *The myth of continents.* Berkeley: University of California Press.

Lewis, P. (1979). Axioms for reading the landscape: Some guides to the American scene. In D. Meinig (Ed.), *The interpretation of ordinary landscapes: Geographical essays* (pp. 11–32). New York: Oxford University Press.

Lewis, W. (1954). Economic development with unlimited supplies of labour. *Manchester School of Economics and Social Studies, 22*(2), 139–191.

Ley, D., & Mercer, J. (1980). Locational conflict and the politics of consumption. *Economic Geography, 56,* 89–109.

Ley, D., & Samuels, M. (1978). *Humanistic geography: Prospects and problems.* Chicago: Maaroufa Press.

Leys, C. (1996). *The rise and fall of development theory.* London: James Currey.

Leyshon, A., Matless, D., & Revill, G. (Eds.). (1998). *The place of music.* New York: Guilford.

Leyshon, A., & Thrift, N. (1992). Liberalisation and consolidation: The single European market and the remaking of European financial capital. *Environment and Planning A, 24,* 49–81.

Leyshon, A., & Thrift, N. (1997). *Money/space: Geographies of monetary transformation.* London: Routledge.

Limb, M., & Dwyer, C. (2001). *Qualitative methodologies for geographers: Issues and debates.* London: Edward Arnold.

Lindahl, D., & Beyers, W. (1999). The creation of competitive advantage by producer service firms. *Economic Geography, 75,* 1–20.

Linge, G. (1991). Just-in-time: More or less flexible? *Economic Geography, 67,* 316–332.

Linkon, S., & Russo, J. (2002). *Steel Town USA: Work and memory in Youngstown.* Lawrence: University Press of Kansas.

Lipietz, A. (1987). *Mirages and miracles: The crisis of global Fordism.* London: Verso.

Liverman, D. (1999). Geography and the global environment. *Annals of the American Association of Geographers, 89,* 107–120.

Livingstone, D. (1992). *The geographical tradition.* Oxford, UK: Blackwell.

Livingstone, D., & Withers, C. (Eds.). (1999). *Geography and Enlightenment.* Chicago: University of Chicago Press.

Locke, J. (1980). *Second treatise of government.* Indianapolis, IN: Hackett.

Logan, J., & Molotch, H. (1987). *Urban fortunes: The political economy of place.* Berkeley: University of California Press.

Longhurst, R. (2001). *Bodies: Exploring fluid boundaries.* London: Routledge.

Longley, P., & Barnsley, M. (2004). The potential of remote sensing and geographical information systems. In J. Matthews & D. Herbert (Eds.), *Common heritage, shared future: Perspectives on the unity of geography* (pp. 62–80). London: Routledge.

Longley, P., Goodchild, M., Maguire, D., & Rhind, D. (2001). *Geographic information systems and science.* New York: John Wiley.

Longley, P., Goodchild, M., Maguire, D., & Rhind, D. (2005). *Geographic information systems and science* (2nd ed.). Chichester, UK: Wiley.

Loomba, A. (1998). *Colonialism/Postcolonialism.* New York: Routledge.

Lord, C., Strauss, E., & Toffler, A. (2003). Natural cities: Urban ecology and the restoration of urban ecosystems. *Virginia Environmental Law Journal, 21,* 317–381.

Lovering, J. (1998). The global music industry: Contradictions in the commodification of the sublime. In A. Leyshon, D. Matless, & G. Revill (Eds.), *The place of music* (pp. 31–56). New York: Guilford.

Low, M., Cox, K., & Robinson, J. (Eds.). (2003). *Handbook of political geography.* London: Sage.

Lowenthal, D., & Prince, H. (1965). English landscape tastes. *Geographical Review, 55,* 186–222.

Lowman, J. (1986). Conceptual issues in the geography of crime: Toward a geography of social control. *Annals of the Association of American Geographers, 76,* 81–94.

Lynch, K. (1960). *The image of the city.* Cambridge: MIT Press.

Lynn, W. (1998). Contested moralities: Animals and moral value in the Dear/Symanski debate. *Ethics, Place, and Environment, 1,* 223–242.

Lyon, D. (2003). *Surveillance as social sorting: Privacy, risk, and automated discrimination.* New York: Routledge.

Lyotard, J.-F. (1984). *The postmodern condition: A report on knowledge* (G. Bennington & B. Massumi, Trans.). Minneapolis: University of Minnesota Press. (Original work published 1979 [in French])

MacEachren, A. (1995). *How maps work: Representation, visualization, and design.* New York: Guilford.

MacIntyre, A. (1984). *After virtue* (2nd ed.). Notre Dame, IN: University of Notre Dame Press.

MacLeod, G. (2002). From urban entrepreneurialism to a "revanchist city"? On the spatial injustices of Glasgow's renaissance. *Antipode, 34,* 602–624.

Maharidge, D., & Williamson, M. (1996). *Journey to nowhere: The saga of the new underclass.* New York: Hyperion. (Original work published 1985)

Malczewski, J. (1999). *GIS and multicriteria decision analysis.* New York: John Wiley.

Mandelbrot, B. B. (1983). *The fractal geometry of nature.* San Francisco: W. H. Freeman.

Mangin, W. (1967). Squatter settlements. *Scientific American, 217,* 21–29.

Manning, W. (1913). The Billerica town plan. *Landscape Architecture, 3*(3), 108–118.

Manzi, T., & Smith-Bowers, B. (2005). Gated communities as club goods: Segregation or social cohesion? *Housing Studies, 20,* 345–359.

Marcus, M. (1979). Coming full circle: Physical geography in the twentieth century. *Annals of the Association of American Geographers, 69,* 521–532.

Marden, P. (1992). The deconstructionist tendencies of postmodern geographies: A compelling logic? *Progress in Human Geography, 16,* 41–57.

Mark, D., Freksa, C., Hirtle, S., Lloyd, R., & Tversky, B. (1999). Cognitive models of geographic space. *International Journal of Geographical Information Science, 13,* 747–774.

Markovitz, A., & Hellerman, S. (2001). *Offside: Soccer and American exceptionalism.* Princeton, NJ: Princeton University Press.

Markusen, A., & Gwiasda, V. (1994). Multipolarity and the layering of functions in world cities: New York City's struggle to stay on top. *International Journal of Urban and Regional Research, 18,* 167–193.

Marsden, T., & Wrigley, N. (1999). Regulation, retailing, and consumption. *Environment and Planning A, 27,* 1899–1912.

Marshall, A. (1890). *Principles of economics.* London: Macmillan.

Marston, S. (2000). The social construction of scale. *Progress in Human Geography, 24,* 219–242.

Martin, R. (Ed.). (1999). *Money and the space economy.* New York: John Wiley.

Martin, R., Sunley, P., & Wills, J. (1996). *Union retreat and the regions: The shrinking landscape of organised labour.* London: Jessica Kingsley.

Marvin, C. (1988). *When old technologies were new: Thinking about electric communication in the late nineteenth century.* Oxford, UK: Oxford University Press.

Marx, K. (1987–1992). *Capital* (3 vols.). New York: International Publishers.

Marx, K., & Engels, F. (1998). *The communist manifesto: A modern edition.* London: Verso. Available: www.anu.edu.au/polsci/marx/classics/manifesto.html

Massey, D. (1984). *Spatial divisions of labor: Social structures and the geography of production.* New York: Methuen.

Massey, D. (1993). Power geometry and a progressive sense of place. In J. Bird, B. Curtis, T. Putnam, G. Robertson, & L. Tickner (Eds.), *Mapping the futures: Local cultures, global change* (pp. 59–69). London: Routledge.

Massey, D. (1994). The political place of locality studies. In D. Massey, *Space, place, and gender* (pp. 125–145). Cambridge, UK: Polity.

Massey, D. (1994). *Space, place, and gender.* Cambridge, UK: Polity; Minneapolis: University of Minnesota Press.

Massey, D. (1995). *Spatial divisions of labor: Social structures and the geography of production* (2nd ed.). New York: Routledge.

Massey, D. (1999). Spaces of politics. In D. Massey, J. Allen, & P. Sarre (Eds.), *Human geography today* (pp. 279–294). Cambridge, UK: Polity.

Massey, D., & Allen, J. (Eds.). (1988). *Uneven re-development: Cities and regions in transition.* London: Hodder & Stoughton.

Massey, D., & Denton, N. (1993). *American apartheid: Segregation and the making of the underclass.* Cambridge, MA: Harvard University Press.

Mathewson, K., & Kenzer, M. (Eds.). (2003). *Culture, land, and legacy: Perspectives on Carl O. Sauer and Berkeley School geography* (Geoscience and Man, Vol. 37). Baton Rouge, LA: Geoscience Publications.

Matless, D. (1998). *Landscape and Englishness.* London: Reaktion.

Matthews, G. (1990). *Historical atlas of Canada* (Vol. 3). Toronto: University of Toronto Press.

Mattingly, D., & Falconer-Al-Hindi, K. (1995). Should women count? A context for the debate. *The Professional Geographer, 47,* 427–436.

McCarthy, J. (2002). First World political ecology: Lessons from the wise use movement. *Environment and Planning A, 34,* 1281–1302.

McClintock, A. (1995). *Imperial leather: Race, gender, and sexuality in the colonial contest.* New York: Routledge.

McConnell, C., & Brue, S. (1993). *Macro-economics: Principles, problems, and policies.* New York: McGraw-Hill.

McConnell, J., & Erickson, R. (1986). Geobusiness: An international perspective for geographers. *Journal of Geography, 85*(3), 98–105.

McDowell, L. (1992). Doing gender: Feminism, feminists, and research methods in human geography. *Transactions of the Institute of British Geographers, 17,* 399–416.

McDowell, L. (1999). *Gender, identity, and place: Understanding feminist geographies.* Cambridge, UK: Polity; Minneapolis: University of Minnesota Press.

McHarg, I. (1969). *Design with nature.* New York: Natural History Press.

McManus, P. (1996). Contested terrains: Politics, stories, and discourses of sustainability. *Environmental Politics, 5,* 48–71.

Meade, M., & Earickson, R. (2000). *Medical geography* (2nd ed.). New York: Guilford.

Meadows, D., Randers, J., & Meadows, D. (2004). *Limits to growth: The 30-year update.* White River Junction, VT: Chelsea Green.

Meffe, G., Nielsen, L., Knight, R., & Schenborn, D. (2002). *Ecosystem management: Adaptive, community-based conservation.* Washington, DC: Island Press.

Meinig, D. (Ed.). (1979). *The interpretation of ordinary landscapes: Geographical essays.* New York: Oxford University Press.

Meinig, D. (1986–2004). *The shaping of America: A geographical perspective on 500 years of history* (Vols. 1–4). New Haven, CT: Yale University Press.

Merrett, C. (2004). Social justice: What is it? Why teach it? *Journal of Geography, 103,* 93–101.

Merrifield, A. (1993). Place and space: A Lefebvrian reconciliation. *Transactions of the Institute of British Geographers, 18,* 516–531.

Meshenberg, M. (1976). *The language of zoning.* Chicago: American Society of Planning Officials.

Meyer, W., & Brown, M. (1989). Locational conflict in a nineteenth century city. *Political Geography Quarterly, 8,* 107–122.

Midgley, M. (1983). *Animals and why they matter.* Athens: University of Georgia Press.

Mikesell, M. (1976). The rise and decline of "sequent occupance": A chapter in the history of American geography. In D. Lowenthal & M. Bowden (Eds.), *Geographies of the mind: Essays in historical geography* (pp. 149–169). Oxford, UK: Oxford University Press.

Miller, B. (2000). *Geography and social movements.* Minneapolis: University of Minnesota Press.

Miller, D. (Ed.). (1995). *Acknowledging consumption.* New York: Routledge.

Miller, H. (2003). What about people in geographic information science? *Computers, Environment, and Urban Systems, 27,* 447–453.

Miller, R., & Blair, P. (1985). *Input–output analysis: Foundations and extensions.* Englewood Cliffs, NJ: Prentice Hall.

Mills, C. (1993). Myths and meanings of gentrification. In J. Duncan & D. Ley (Eds.), *Place/culture/representation* (pp. 149–170). London: Routledge.

Mintz, S. (1973). A note on the definition of peasantries. *Journal of Peasant Studies, 1,* 91–106.

Mintz, S. (1985). *Sweetness and power: The place of sugar in modern history.* New York: Penguin Books.

Misa, T., Brey, P., & Feenberg, A. (Eds.). (2003). *Modernity and technology.* Cambridge: MIT Press.

Mitchell, D. (1995). There is no such thing as culture: Towards a reconceptualization of the idea of culture in geography. *Transactions of the Institute of British Geographers, 20*(1), 102–116.

Mitchell, D. (1996). *The lie of the land: Migrant workers and the California landscape.* Minneapolis: University of Minnesota Press.

Mitchell, D. (2000). *Cultural geography: A critical introduction.* Oxford, UK: Blackwell.

Mitchell, D. (2003). *The right to the city: Social justice and the fight for public space.* New York: Guilford.

Mitchell, T. (1988). *Colonizing Egypt.* Cambridge, UK: Cambridge University Press.

Mohan, G., Brown, E., Milward, B., & Zack-Williams, A. (Eds.). (2000). *Structural adjustment: Theory, practice, and impacts.* New York: Routledge.

Mohanty, C., Russo, A., & Torres, L. (Eds.). (1991). *Third World women and the politics of feminism.* Bloomington: Indiana University Press.

Mohl, R. (Ed.). (1990). *Searching for the Sunbelt: Historical perspectives on a region.* Knoxville: University of Tennessee Press.

Mollenkopf, J., & Castells, M. (Eds.). (1991). *Dual city: Restructuring New York.* New York: Russell Sage.

Monmonier, M. (2002). *Spying with maps: Surveillance technologies and the future of privacy.* Chicago: University of Chicago Press.

Moore, D. (1997). Remapping resistance: "Ground for struggle" and the politics of place. In S. Pile & M. Keith (Eds.), *Geographies of resistance* (pp. 87–106). London: Routledge.

Moraga, C., & Anzaldúa, G. (Eds.). (1983). *This bridge called my back: Writings by radical women of color.* Brooklyn, NY: Kitchen Table Press.

Morgenthau, H. (1949). *Politics among nations: The struggle for power and peace.* New York: Knopf.

Morrill, R. (1999). Electoral geography and gerrymandering: Space and politics. In G. Demko & W. Wood (Eds.),

Reordering the world: Geopolitical perspectives on the 21st century (pp. 117–138). Boulder, CO: Westview.

Moss, P. (Ed.). (2002). *Feminist geography in practice.* Oxford, UK: Blackwell.

Mouffe, C. (Ed.). (1979). *Gramsci and Marxist theory.* London: Routledge.

Mountz, A., & Wright, R. (1996). Daily life in the transnational migrant community of San Agustin, Oaxaca, and Poughkeepsie, New York. *Diaspora, 5,* 403–425.

Muller, J.-C. (1984). Canada's elastic space: A portrayal of route and cost distances. *The Canadian Geographer, 18,* 46–62.

Murdoch, J. (2004). Humanising posthumanism. *Environment and Planning A, 36,* 1356–1359.

Murtagh, W. (1993). *Keeping time: The history and theory of preservation in America.* New York: Sterling.

Naficy, H. (Ed.). (1999). *Home, exile, homeland: Film, media, and the politics of place.* New York: Routledge.

Nagar, R., Lawson, V., McDowell, L., & Hanson, S. (2002). Locating globalization: Feminist (re)readings of the subjects and spaces of globalization. *Economic Geography, 78,* 257–284.

Nash, P., & Carney, G. (1996). The seven themes of music geography. *The Canadian Geographer, 40,* 79–74.

Nash, R. (1982). *Wilderness and the American mind* (3rd ed.). New Haven, CT: Yale University Press.

Nast, H. (1994). Women in the field: Critical feminist methodologies and theoretical perspectives. *The Professional Geographer, 46,* 54–102.

Nast, H. (1996). Unsexy geographies. *Gender, Place, and Culture, 5,* 191–206.

Nast, H. (2000). Mapping the "unconscious": Racism and the oedipal family. *Annals of the Association of American Geographers, 90,* 215–255.

Nast, H., & Pile, S. (Eds.). (1998). *Places through the body.* London: Routledge.

National Science Foundation. (2003). *Revolutionizing science and engineering through cyber-infrastructure* (Report of the National Science Foundation Blue Ribbon Advisory Panel on Cyberinfrastructure) [Online]. Available: www.communitytechnology.org/nsf_ci_report/

Nelson, R. (2005). *Private neighborhoods and the transformation of local government.* Washington, DC: Urban Institute Press.

Newman, D. (Ed.). (1999). *Boundaries, territory, and postmodernity.* London: Frank Cass.

Noble, A., & Dhussa, R. (1990). Image and substance: A review of literary geography. *Journal of Cultural Geography, 10,* 49–65.

North, D. (1989). Institutions and economic growth: An historical introduction. *World Development, 17,* 1319–1332.

Norton, R., & Rees, J. (1979). The product cycle and the spatial decentralization of American manufacturing. *Regional Studies, 13,* 141–151.

Nystuen, J. (1968). Identification of some fundamental spatial concepts. In B. Berry & D. Marble (Eds.), *Spatial analysis* (pp. 35–41). Englewood Cliffs, NJ: Prentice Hall.

Ó Tuathail, G. (1996). *Critical geopolitics.* Minneapolis: University of Minnesota Press.

Ó Tuathail, G., & Dalby, S. (Eds.). (1998). *Rethinking geopolitics.* London: Routledge.

Ó Tuathail, G., & Shelley, F. (2003). Political geography. In G. Gaile & C. Willcott (Eds.), *Geography in America at the dawn of the 21st century* (pp. 164–184). Oxford, UK: Oxford University Press.

Ochel, W., & Wegner, M. (1987). *Service economies in Europe: Opportunities for growth.* London: Pinter.

O'Connor, J. (1984). *Accumulation crisis.* Oxford, UK: Blackwell.

Oelschlaeger, M. (1991). *The idea of wilderness.* New Haven, CT: Yale University Press.

Ogborn, M. (1998). *Spaces of modernity: London's geographies, 1680–1780.* New York: Guilford.

O'Loughlin, J. (1982). The identification and evaluation of racial gerrymandering. *Annals of the Association of American Geographers, 72,* 165–184.

Olson, J. (1976). Noncontiguous area cartograms. *The Professional Geographer, 28,* 371–380.

Ong, A. (1999). *Flexible citizenship: The cultural logics of transnationality.* Durham, NC: Duke University Press.

Openshaw, S., & Openshaw, C. (1997). *Artificial intelligence in geography.* New York: John Wiley.

O'Sullivan, D., & Unwin, D. (2003). *Geographic information analysis.* Hoboken, NJ: John Wiley.

Owen, W. (2004). Dulce et decorum est Pro patria mori. In W. Owen, *The poems of Wilfred Owen* [Online]. Available: www.pitt.edu/~pugachev/greatwar/owen.html

Pacione, M. (2002). *Applied geography.* London: Routledge.

Pahl, R. (1970). *Whose city?* Harmondsworth, UK: Penguin.

Pain, R. (2003). Social geography: On action-orientated research. *Progress in Human Geography, 27,* 677–685.

Painter, J. (1995). *Politics, geography, and "political geography": A critical perspective.* London: Edward Arnold.

Palen, J., & London, B. (Eds.). (1984). *Gentrification, displacement and neighborhood revitalization.* Albany: State University of New York Press.

Pandit, K., & Casetti, E. (1989). The shifting pattern of sectoral labor allocation during development: Developed versus developing countries. *Annals of the Association of American Geographers, 97,* 329–344.

Park, C. (1994). *Sacred worlds: An introduction to geography and religion.* London: Routledge.

Park, R. (1926). The urban community as a spatial pattern and a moral order. In C. Peach (Ed.), *Urban social segregation* (pp. 21–31). London: Longman.

Park, R., Burgess, E., & McKenzie, R. (Eds.). (1925). *The city.* Chicago: University of Chicago Press.

Parr, H. (2005). Emotional geographies. In P. Cloke, P. Crang, & M. Goodwin (Eds.), *Introducing human geographies.* London: Edward Arnold.

Patel, D. (1980). *Exurbs: Residential development in the countryside.* Washington, DC: University Press of America.

Payne, R., & Nassar, J. (2005). *Politics and culture in the developing world: The impact of globalization.* New York: Longman.

Peach, C. (2000). Discovering white ethnicity and parachuted plurality. *Progress in Human Geography, 24,* 620–626.

Peck, J. (1992). Labor and agglomeration: Control and flexibility in local labor markets. *Economic Geography, 68,* 325–347.

Peck, J. (1996). *Work–place: The social regulation of labor markets.* New York: Guilford.

Peck, J. (2002). Labor, zapped/growth, restored? The moments of neoliberal restructuring in the American labor market. *Journal of Economic Geography, 2,* 179–220.

Peet, R. (Ed.). (1977). *Radical geography.* Chicago: Maaroufa Press.

Peet, R. (1985). The social origins of environmental determinism. *Annals of the Association of American Geographers, 75,* 309–333.

Peet, R. (1987). *International capitalism and industrial restructuring.* Boston: Allen and Unwin.

Peet, R. (1991). *Global capitalism: Theories of societal development.* London: Routledge.

Peet, R. (1998). *Modern geographical thought.* Oxford, UK: Blackwell.

Peet, R. (1999). *Theories of development.* New York: Guilford.

Peet, R., & Watts, M. (Eds.). (2004). *Liberation ecologies: Environment, development, social movements* (2nd ed.). London: Routledge.

Peters, G., & Larkin, R. (2002). *Population geography: Problems, concepts, and prospects.* New York: Kendall/Hunt.

Peters, W. (1988). *Slash and burn: Farming in the Third World forest.* Moscow: University of Idaho Press.

Petulla, J. (1977). *American environmental history: The exploitation and conservation of natural resources.* San Francisco: Boyd & Fraser.

Peuquet, D., & Duan, N. (1995). An event-based spatiotemporal data model (ESTDM) for temporal analysis of geographical data. *International Journal of Geographical Information Systems, 9*(1), 7–24.

Phillips, R. (1997). *Mapping men and empire: A geography of adventure.* London: Routledge.

Philo, C., & Wilbert, C. (Eds.). (2000). *Animal spaces, beastly places: Critical geographies.* London: Routledge.

Pickles, J. (1985). *Phenomenology, science, and geography: Spatiality and the human sciences.* Cambridge, UK: Cambridge University Press.

Pickles, J. (Ed.). (1995). *Ground truth: The social implications of geographic information systems.* New York: Guilford.

Pickles, J. (2003). *A history of spaces: Cartographic reason, mapping, and the geo-coded world.* London: Routledge.

Pile, S. (1993). Human agency and human geography revisited: A critique of "new models" of the self. *Transactions of the Institute of British Geographers, 18,* 122–139.

Pile, S. (1996). *The body and the city: Psychoanalysis, space, and subjectivity.* London; New York: Routledge.

Pile, S., & Keith, M. (Eds.). (1997). *Geographies of resistance.* New York: Routledge.

Pile, S., & Thrift, N. (1995). Mapping the subject. In S. Pile & N. Thrift (Eds.), *Mapping the subject: Geographies of cultural transformation* (pp. 199–225). London: Routledge.

Pile, S., & Thrift, N. (Eds.). (1995). *Mapping the subject: Geographies of cultural transformation.* London: Routledge.

Plumwood, V. (1998). Wilderness skepticism and wilderness dualism. In J. Callicott & M. Nelson (Eds.), *The great new wilderness debate* (pp. 652–690). Athens: University of Georgia Press.

Podmore, J. (2001). Lesbians in the crowd: Gender, sexuality, and visibility along Montreal's Boul. St-Laurent. *Gender, Place, and Culture, 8,* 333–355.

Popke, E. (2003). Poststructuralist ethics: Subjectivity, responsibility, and the space of community. *Progress in Human Geography, 27,* 298–316.

Popper, D., Lang, R., & Popper, F. (2000). From maps to myth: The census, Turner, and the frontier. *Journal of American and Comparative Culture, 23,* 91–102.

Popper, F. (1985). The environmentalist and the LULU. *Environment, 27*(2), 7–11, 37–40.

Porter, M. (1990). *The competitive advantage of nations.* New York: Free Press.

Porter, M. (1998). *The competitive advantage of nations* (2nd ed.). New York: Free Press.

Porter, P., & Sheppard, E. (1998). *A world of difference: Society, nature, development.* New York: Guilford.

Porter, R., & Teich, M. (Eds.). (1981). *The Enlightenment in national context.* Cambridge, UK: Cambridge University Press.

Power, M. (2003). Development thinking and the mystical "kingdom of abundance." In M. Power, *Rethinking development geographies* (pp. 71–84). London: Routledge.

Power, M. (2003). Theorising back: Views from the South and the globalisation of resistance. In M. Power, *Rethinking development geographies* (pp. 194–218). London: Routledge.

Pratt, M. (1992). *Imperial eyes: Travel writing and transculturation.* London: Routledge.

Prebisch, P. (1962). The economic development of Latin America and its principal problems. *Economic Review of Latin America, 7*(1), 1–22.

Pred, A. (1984). Place as historically contingent process: Structuration and the time-geography of becoming places. *Annals of the Association of American Geographers, 74,* 279–297.

Prewitt, K. (2003). *The American people: Census 2000.* New York: Russell Sage.

Price, M., & Lewis, M. (1993). The reinvention of cultural geography. *Annals of the Association of American Geography, 83,* 1–17.

Pringle, R. (1999). Emotions. In L. McDowell & J. Sharp (Eds.), *A feminist glossary of human geography* (pp. 68–69). London: Arnold.

Proctor, J., & Smith, D. (Eds.). (1999). *Geography and ethics: Journeys in a moral terrain.* London: Routledge.

Pryor, R. (1968). Defining the rural–urban fringe. *Social Forces, 47,* 202–215.

Pulido, L. (2000). Rethinking environmental racism: White privilege and urban development in Southern California. *Annals of the Association of American Geographers, 90,* 12–40.

Purcell, M. (1997). Ruling Los Angeles: Neighborhood movements, urban regimes, and the production of space in Southern California. *Urban Geography, 18,* 684–704.

Raisz, E. (1934). The rectangular statistical cartogram. *Geographical Review, 24,* 292–296.

Raitz, K. (Ed.). (1995). *The theater of sport.* Baltimore, MD: Johns Hopkins University Press.

Rakodi, C. (1995). Poverty lines or household strategies? A review of the conceptual issues in the study of urban poverty. *Habitat International, 19,* 407–426.

Randall, R. (2001). *Place names: How they define the world—and more.* Lanham, MD: Scarecrow Press.

Ravenstein, E. (1885). The laws of migration. *Journal of the Statistical Society of London, 48,* 167–235.

Ravenstein, E. (1889). The laws of migration. *Journal of the Royal Statistical Society, 52,* 241–305.

Reilly, W. (1931). *The law of retail gravitation.* New York: Knickerbocker.

Relph, E. (1976). *Place and placelessness.* London: Pion.

Reynolds, D. (1999). *There goes the neighborhood: Rural school consolidation at the grass roots in early twentieth-century Iowa.* Iowa City: University of Iowa Press.

Riches, G. (1997). *First World hunger: Food security and welfare politics.* New York: St. Martin's.

Ricketts, T., Savitz, L., Gesler, W., & Osborne, D. (Eds.). (1994). *Geographic methods for health services research: A focus on the rural–urban continuum.* Lanham, MD: University Press of America.

Riffenburgh, B. (1994). *The myth of the explorer.* Oxford, UK: Oxford University Press.

Robinson, A., Morrison, J., Muehrcke, P., Kimerling, A., & Guptill, S. (1995). *Elements of cartography.* New York: John Wiley.

Robinson, J. (2004). Squaring the circle? Some thoughts on the idea of sustainable development. *Ecological Economics, 48,* 369–384.

Roddick, J. (1988). *The dance of the millions: Latin America and the debt crisis.* London: Latin America Bureau.

Rodger, E. (1995). *Diffusion of innovations.* New York: Free Press.

Roediger, D. (1991). *The wages of whiteness: Race and the making of the American working class.* London: Verso.

Rooney, J. (1974). *A geography of American sport: From Cabin Creek to Anaheim.* Reading, MA: Addison-Wesley.

Rose, D. (1993). On feminism, method, and methods in human geography: An idiosyncratic overview. *The Canadian Geographer, 37,* 57–61.

Rose, G. (1990). Imagining Poplar in the 1920s: Contested concepts of community. *Journal of Historical Geography, 16,* 425–437.

Rose, G. (1993). *Feminism and geography: The limits of geographical knowledge.* Cambridge, UK: Polity; Minneapolis: University of Minnesota Press.

Rose, G. (2001). *Visual methodologies: An introduction to the interpretations of visual materials.* London: Sage.

Rose, G. (2002). Conclusion. In L. Bondi, H. Avis, A. Bingley, J. Davidson, R. Duffy, V. Einagel, et al. (Eds.), *Subjectivities, knowledges, and feminist geographies: The subjects and ethics of social research* (pp. 253–258). Lanham, MD: Rowman & Littlefield.

Rosentraub, M. (1999). *Major League losers: The real cost of sports and who's paying for it* (rev. ed.). New York: Basic Books.

Rostow, W. (1960). *The stages of economic growth: A non-communist manifesto.* Cambridge, UK: Cambridge University Press.

Routledge, P. (1998). Anti-geopolitics: Introduction. In G. Ó Tuathail, S. Dalby, & P. Routledge (Eds.), *The geopolitics reader* (pp. 245–255). London: Routledge.

Rowles, G. (1978). *Prisoners of space? Exploring the geographical experience of older people.* Boulder, CO: Westview.

Rubenstein, J. M. (2005). *The cultural landscape: An introduction to human geography* (8th ed.). New York: Macmillan.

Ruddick, S. (1996). *Young and homeless in Hollywood.* New York: Routledge.

Rushton, J. (2000). *Race, evolution, and behavior.* Port Huron, MI: Charles Darwin Research Institute.

Ruth, M. (1995). Information, order, and knowledge in economic and ecological systems: Implications for material and energy use. *Ecological Economics, 13,* 99–114.

Ryan, J. (1997). *Picturing empire: Photography and the visualization of the British Empire.* London: Reaktion Books.

Saarinen, T. (1999). The Euro-centric nature of mental maps of the world. *Research in Geographic Education, 1*(2), 136–178.

Sack, R. (1974). Chorology and spatial analysis. *Annals of the Association of American Geographers, 64,* 439–452.

Saff, G. (2001). Exclusionary discourse towards squatters in suburban Cape Town. *Ecumene, 8*(1), 87–107.

Said, E. (1978). *Orientalism.* New York: Pantheon; London: Routledge & Kegan Paul.

Said, E. (1979). *Orientalism.* New York: Vintage Books.

Said, E. (1993). *Culture and imperialism.* New York: Knopf.

Said, E. (1994). *Orientalism* (2nd ed.). New York: Vintage Books.

Samuels, M. (1978). Existentialism and human geography. In M. Samuels & D. Ley (Eds.), *Humanistic geography: Prospects and problems* (pp. 22–40). London: Croon Helm.

Sandberg, L., & Marsh, J. (1988). Literary landscapes: Geography and literature. *The Canadian Geographer, 32,* 266–276.

Sanderock, L. (Ed.). (1998). *Making the invisible visible: A multicultural planning history.* Berkeley: University of California Press.

Sarat, A., Douglas, L., & Umphrey, M. (Eds.). (2003). *The place of law.* Ann Arbor: University of Michigan Press.

Sassen, S. (1991). *The global city: New York, London, Tokyo.* Princeton, NJ: Princeton University Press.

Sassen, S. (2002). *Global networks: Linked cities.* New York: Routledge.

Sauer, C. (1925). The morphology of landscape. *University of California Publications in Geography, 2*(2), 19–53.

Sauer, C. (1941). Foreword to historical geography. *Annals of the Association of American Geographers, 31,* 1–24.

Sauer, C. (1952). *Agricultural origins and dispersals.* New York: American Geographical Society.

Sauer, C. (1963). The morphology of landscape. In J. Leighly (Ed.), *Land and life: A selection of writings of Carl Ortwin Sauer* (pp. 315–350). Berkeley: University of California Press. (Original work published 1925)

Saxenian, A. (1996). *Regional advantage: Culture and competition in Silicon Valley and Route 128.* Cambridge, MA: Harvard University Press.

Sayer, A. (2000). *Realism and social science.* Thousand Oaks, CA: Sage.

Scaramuzzi, E. (2002). *Incubators in developing countries: Status and development perspectives.* Washington, DC: World Bank.

Schaeffer, F. (1953). Exceptionalism in geography: A methodological examination. *Annals of the Association of American Geographers, 43,* 226–249.

Schaffer, R., & Smith, N. (1986). The gentrification of Harlem? *Annals of the Association of American Geographers, 76,* 347–365.

Schecht, S., & Haggis, J. (2000). *Culture and development: A critical introduction.* Oxford, UK: Blackwell.

Schein, R. (1997). The place of landscape: A conceptual framework for interpreting an American scene. *Annals of the Association of American Geographers, 87,* 660–680.

Scheyvens, R., & Storey, D. (Eds.). (2003). *Development fieldwork: A practical guide.* London: Sage.

Schiller, J., & Voisard, A. (Eds.). (2004). *Location-based services.* New York: Morgan Kaufmann.

Schlosser, E. (2001). *Fast food nation: The dark side of the all-American meal.* Boston: Houghton Mifflin.

Schmidt, J. (Ed.). (1996). *What is Enlightenment? Eighteenth-century answers and twentieth-century questions.* Berkeley: University of California Press.

Schulten, S. (2001). *The geographical imagination in America, 1880–1950.* Chicago: University of Chicago Press.

Schumpeter, J. (1987). *Capitalism, socialism, and democracy* (6th ed.). London; Boston: Unwin Paperbacks.

Schwartz, J. (1996). The geography lesson: Photographs and the construction of imaginative geographies. *Journal of Historical Geography, 22,* 16–45.

Schwartz, J., & Ryan, J. (Eds.). (2003). *Picturing place: Photography and the geographical imagination.* London: I. B. Tauris.

Scott, A. (1988). *New industrial spaces.* London: Pion.

Scott, A. (1997). *The limits of globalization.* London: Routledge.

Scott, A., & Storper, M. (Eds.). (1986). *Production, work, territory: The geographical anatomy of industrial capitalism.* Boston: Allen and Unwin.

Scott, J. (1985). *Weapons of the weak: Everyday forms of peasant resistance.* New Haven, CT: Yale University Press.

Scott, J., & Simpson-Housley, P. (1989). Relativizing the relativizers: On the postmodern challenge to human geography. *Transactions of the Institute of British Geographers, 14,* 231–236.

Sedgwick, E. (1990). *Epistemology of the closet.* Berkeley: University of California Press.

Selections on "Women in the Field." (1994). *The Professional Geographer, 46*(1).

Sen, A. (2000). *Development as freedom.* New York: Anchor Books.

Shain, Y. (1999). *Marketing the American creed abroad: Diasporas in the U.S. and their homelands.* New York: Cambridge University Press.

Shanin, T. (Ed.). (1988). *Peasants and peasant societies: Selected readings* (2nd ed.). Oxford, UK: Blackwell.

Shannon, G., Pyle, G., & Bashshur, R. (1991). *The geography of AIDS: Origins and course of an epidemic.* New York: Guilford.

Sharp, J. (2000). *Condensing the cold war:* Reader's Digest *and American identity, 1922–1994.* Minneapolis: University of Minnesota Press.

Shaw, G., & Williams, A. (2002). *Critical issues in tourism: A geographical perspective* (2nd ed.). Malden, MA: Blackwell.

Shelley, F., Archer, J., Davidson, F., & Brunn, S. (1996). *Discovering America's political geography.* New York: Guilford.

Shelley, F., & Flint, C. (2000). Geography, place, and world-systems analysis. In T. Hall (Ed.), *A world-systems reader* (pp. 69–82). Lanham, MD: Rowman & Littlefield.

Sheppard, E. (2001). Quantitative geography: Representations, practices, and possibilities. *Environment and Planning D, 19,* 535–554.

Sheppard, E., & Barnes, T. (Eds.). (2000). *A companion to economic geography.* Oxford, UK: Blackwell.

Shotts, K. (2002). Gerrymandering, legislative composition, and national policy outcomes. *American Journal of Political Science, 46,* 398–414.

Shurmer-Smith, P. (2002). Poststructuralist cultural geography. In P. Shurmer-Smith (Ed.), *Doing cultural geography* (pp. 41–52). London: Sage.

Sibley, D. (1995). *Geographies of exclusion: Society and difference in the West.* London: Routledge.

Sin, C. (2002). The interpretation of segregation indices in context: The case of P* in Singapore. *The Professional Geographer, 4,* 422–437.

Sir Halford Mackinder and "The geographical pivot of history" [special issue]. (2004). *The Geographical Journal, 170,* 291–383.

Skocpol, T. (1981). Political response to capitalist crisis: Neo-Marxist theories of the state and the case of the New Deal. *Politics & Society, 10,* 155–201.

Slocum, T., McMaster, R., Kessler, F., & Howard, H. (2005). *Thematic cartography and geographic visualization* (2nd ed.). Upper Saddle River, NJ: Prentice Hall.

Smith, A. (1995). *Nations and nationalism in a global era.* Cambridge, UK: Polity.

Smith, A. (2003). *The wealth of nations.* New York: Bantam Classics. (Original work published 1776)

Smith, B. (1996). *Understanding Third World politics: Theories of political change and development.* Bloomington: Indiana University Press.

Smith, D. (1979). *Where the grass is greener: Geographical perspectives on inequality.* London: Croom Helm.

Smith, D. (1994). *Geography and social justice.* Oxford, UK: Blackwell.

Smith, D. (2000). *Moral geographies: Ethics in a world of difference.* Edinburgh, UK: Edinburgh University Press.

Smith, H. (1983). *The citizen's guide to zoning.* Chicago: APA Planners Press.

Smith, M., McLoughlin, J., Large, P., & Chapman, R. (1985). *Asia's new industrial world.* London: Methuen.

Smith, N. (1979). Toward a theory of gentrification: A back to the city movement by capital, not people. *Journal of the American Planners Association, 45,* 538–548.

Smith, N. (1984). *Uneven development: Nature, capital, and the production of space.* Oxford, UK: Blackwell.

Smith, N. (1987). Gentrification and the rent-gap. *Annals of the Association of American Geographers, 77,* 462–465.

Smith, N. (1990). *Uneven development: Nature, capital, and the production of space* (2nd ed.). Oxford, UK: Blackwell.

Smith, N. (1992). Contours of a spatialized politics: Homeless vehicles and the production of geographical space. *Social Text, 33,* 54–81.

Smith, N. (1996). *The new urban frontier: Gentrification and the revanchist city.* New York: Routledge.

Smith, N., & Katz, C. (1993). Grounding metaphor: Towards a spatialized politics. In M. Keith & S. Pile (Eds.), *Place and the politics of identity* (pp. 67–80). London: Routledge.

Smith, S. (1986). *Crime, space, and society.* Cambridge, UK: Cambridge University Press.

Smith, S. (2001). *Moving lives: 20th-century women's travel writing.* Minneapolis: University of Minnesota Press.

Smyth, C. (2001). Mining mobile trajectories. In H. Miller & J. Han (Eds.), *Geographic data mining and knowledge discovery* (pp. 337–367). London: Taylor & Francis.

So, A. (1990). *Social change and development: Modernization, dependency, and world-system theories.* Newbury Park, CA: Sage.

Soja, E. (1980). The socio-spatial dialectic. *Annals of the Association of American Geographers, 70,* 207–225.

Soja, E. (1989). *Postmodern geographies: The reassertion of space in critical social theory.* London: Verso.

Soja, E. (1996). *Thirdspace: Journeys to Los Angeles and other real-and-imagined places.* Oxford, UK: Blackwell.

Solomon, E. (1997). *Virtual money: Understanding the power and risks of money's high-speed journey into electronic space.* Oxford, UK: Oxford University Press.

Solomon, R. (1999). *Money on the move: The revolution in international finance since 1980.* Princeton, NJ: Princeton University Press.

Sorensen, G. (1997). *Democracy and democratization: Processes and prospects in a changing world.* Boulder, CO: Westview.

Sparke, M. (1995). Between demythologizing and deconstructing the map: Shawnadithit's New-Found-Land and the alienation of Canada. *Cartographica, 32,* 1–21.

Speake, J. (Ed.). (2003). *The literature of travel and exploration* (3 vols.). London: Fitzroy Dearborn.

Spectorsky, A. (1955). *The exurbanites.* Philadelphia: J. B. Lippincott.

Spencer, C., & Blades, M. (1986). Pattern and process: A review essay on the relationship between behavioral geography and environmental psychology. *Progress in Human Geography, 10,* 230–248.

Speth, W. (1999). *How it came to be: Carl O. Sauer, Franz Boas, and the meanings of anthropogeography.* Ellensburg, WA: Ephemera.

Spiekermann, M., & Wegener, M. (1994). The shrinking continent. *Environment and Planning B, 21,* 651–673.

Spivak, G. (1999). *A critique of post-colonial reason: Toward a history of the vanishing present.* Cambridge, MA: Harvard University Press.

Spooner, B. (1998). *The cultural ecology of pastoral nomads.* Reading, MA: Addison-Wesley.

Spreng, D. (1993). Possibilities for substitution between energy, time, and information. *Energy Policy, 21,* 15–27.

Squires, G. (Ed.). (2002). *Urban sprawl: Causes, consequences, and policy responses.* Washington, DC: Urban Institute.

Staeheli, L., & Lawson, V. (1994). A discussion of "women in the field": The politics of feminist fieldwork. *The Professional Geographer, 46,* 96–102.

Starrs, P., & DeLyser, D. (Eds.). (2001). Doing fieldwork [special issue]. *Geographical Review, 91*(1–2).

Stearns, P. (2001). *Consumerism in world history: The global transformation of desire.* London: Routledge.

Steinitz, C., Parker, P., & Jordan, L. (1976). Hand-drawn overlays: Their history and prospective uses. *Landscape Architecture, 66,* 444–455.

Steward, J. (1955). *The theory of culture change: The methodology of multilinear evolution.* Urbana: University of Illinois Press.

Stewart, J. (1947). Empirical mathematical rules concerning the distribution and equilibrium of population. *Geographical Review, 37,* 461–485.

Stewart, L. (1995). Louisiana subjects: Power, space, and the slave body. *Ecumene, 2,* 227–245.

Stiglitz, J. (2002). *Globalization and its discontents.* New York: Norton.

Stilgoe, J. (1982). *Common landscape of America, 1580–1845.* New Haven, CT: Yale University Press.

Storper, M. (1985). Oligopoly and the product cycle: Essentialism in economic geography. *Economic Geography, 61,* 260–282.

Storper, M., & Walker, R. (1989). *The capitalist imperative.* Oxford, UK: Blackwell.

Strange, S. (1998). *Mad money: When markets outgrow governments.* Ann Arbor: University of Michigan Press.

Stump, R. (2000). *Boundaries of faith: Geographical perspectives on religious fundamentalism.* Lanham, MD: Rowman & Littlefield.

Stutz, F., & Warf, B. (2005). *The world economy: Resources, location, trade, and development* (4th ed.). Upper Saddle River, NJ: Pearson/Prentice Hall.

Sui, D. (2004). GIS, cartography, and the third culture: Geographical imaginations in the computer age. *The Professional Geographer, 56,* 62–72.

Sui, D. (2004). Tobler's first law of geography: A big idea for a small world? *Annals of the Association of American Geographers, 94,* 269–277.

Sui, D., & Rejeski, D. (2001). Environmental impacts of the emerging digital economy: The E-for-environment E-commerce? *Environmental Management, 29,* 155–163.

Swope, C. (2002, December). HUD the unlovable. *Governing,* pp. 26–30.

Swyngedouw, E. (1997). Neither global nor local: "Glocalization" and the politics of scale. In K. Cox (Ed.), *Spaces of globalization: Reasserting the power of the local* (pp. 137–166). New York: Guilford.

Swyngedouw, E. (2000). Authoritarian governance, power, and the politics of rescaling. *Environment and Planning D, 18,* 63–76.

Taaffe, E., Gauthier, H., & O'Kelly, M. (1996). *Geography of transportation* (2nd ed.). Englewood Cliffs, NJ: Prentice Hall.

Takahashi, L. (1996). A decade of understanding homelessness in the USA: From characterization to representation. *Progress in Human Geography, 20,* 291–310.

Takahashi, L., Wiebe, D., & Rodriguez, R. (2001). Navigating the time–space context of HIV and AIDS: Daily routines and access to care. *Social Science and Medicine, 53,* 845–863.

Tarrow, S. (1998). *Power in movement: Social movements and contentious politics.* Cambridge, UK: Cambridge University Press.

Tatham, G. (1951). Environmentalism and possibilism. In G. Taylor (Ed.), *Geography in the twentieth century* (pp. 128–164). London: Methuen.

Taylor, M. (1986). The product-cycle model: A critique. *Environment and Planning A, 18,* 751–761.

Taylor, P. (1982). A materialist framework for political geography. *Transactions of the Institute of British Geographers, 7,* 15–34.

Taylor, P. (1994). The state as container: Territoriality in the modern world-system. *Progress in Human Geography, 18,* 151–162.

Taylor, P. (2000). World cities and territorial states under conditions of contemporary globalization. *Political Geography, 19,* 5–32.

Taylor, P., & Flint, C. (2000). *Political geography: World-economy, nation-state, and locality.* Englewood Cliffs, NJ: Prentice Hall; Harlow, UK: Pearson Education.

Taylor, R. (1974). *The word in stone: The role of architecture in national socialist ideology.* Berkeley: University of California Press.

Theodorson, G. (1982). *Urban patterns: Studies in human ecology.* University Park: Pennsylvania State University Press.

Thill, J.-C. (Ed.). (1999). *Multicriteria decision-making and analysis: A geographic information sciences approach.* Brookfield, VT: Ashgate.

This godless communism. (1961). *Treasure Chest* (Vol. 17, Nos. 2–20). Available: www.authentichistory.com/images/1960s/treasure_chest/godless_communism.html

Thrift, N. (1995). A hyperactive world. In R. Johnston, P. Taylor, & M. Watts (Eds.), *Geographies of global change* (pp. 18–35). Oxford, UK: Blackwell.

Thrift, N. (1997). The still point: Resistance, expressive embodiment, and dance. In S. Pile & M. Keith (Eds.), *Geographies of resistance* (pp. 124–151). London: Routledge.

Thrift, N., & Leyshon, A. (1994). A phantom state? The de-traditionalization of money, the international financial system, and international financial centres. *Political Geography, 13,* 299–327.

Thrift, N., & Olds, K. (1996). Reconfiguring the economic in economic geography. *Progress in Human Geography, 20,* 311–337.

Tikunov, V. (1988). Anamorphated cartographic images: Historical outline and construction techniques. *Cartography, 17*(1), 1–8.

Timar, J. (2004). More than "Anglo-American," it is "Western": Hegemony in geography from a Hungarian perspective. *Geoforum, 35,* 533–538.

Tobler, W. (1970). A computer movie simulating urban growth in the Detroit region. *Economic Geography, 46,* 234–240.

Tobler, W. (2004). Thirty-five years of computer cartograms. *Annals of the Association of American Geographers, 94*(1), 58–73.

Toffler, A. (1970). *Future shock.* New York: Random House.

Tolentino, P. (2001). *Multinational corporations: Emergence and evolution.* London: Routledge.

Tololyan, K. (1996). Rethinking diaspora(s): Stateless power in the transnational moment. *Diaspora, 5*(1), 3–36.

Torrens, P., & Benenson, I. (2005). Geographic automata systems. *International Journal of Geographic Information Science, 19,* 385–412.

Torrieri, N., & Ratcliffe, M. (2003). Applied geography. In G. L. Gaile & C. J. Willmott (Eds.), *Geography in*

America at the dawn of the 21st century (pp. 543–551). New York: Oxford University Press.

Trewartha, G. (1953). A case for population geography. *Annals of the Association of American Geographers, 43,* 71–97.

Tschetter, J. (1987, December). Producer services industries: Why are they growing so rapidly? *Monthly Labor Review,* pp. 31–40.

Tuan, Y.-F. (1971). Geography, phenomenology, and the study of human nature. *The Canadian Geographer, 15,* 181–191.

Tuan, Y.-F. (1974). *Topophilia.* Englewood Cliffs, NJ: Prentice Hall.

Tuan, Y.-F. (1977). *Space and place: The perspective of experience.* Minneapolis: University of Minnesota Press.

Tuan, Y.-F. (1990). *Topophilia: A study of environmental perception, attitudes, and values* (2nd ed.). New York: Columbia University Press.

Tuan, Y.-F. (2001). The desert and I: A study in affinity. *Michigan Quarterly Review, 40*(1), 7–16.

Turkle, S. (1997). *Life on the screen: Identity in the age of the Internet.* New York: Touchstone.

Turner, B., II. (1989). The specialist–synthesis approach to the revival of geography: The case of cultural ecology. *Annals of the Association of American Geographers, 79,* 88–100.

Tyler, N. (2000). *Historic preservation: An introduction to its history, principles, and practice.* New York: Norton.

United Church of Christ Commission for Racial Justice. (1987). *Toxic waste and race in the United States.* New York: United Church of Christ.

United Nations. (2003). *Demographic yearbook 2001.* New York: Author.

United Nations. (2004). *World urbanization prospects: The 2003 revision.* New York: Author.

U.S. Department of Agriculture. (2004). *Quick Stats: Agricultural statistics data base* [computer database]. Available: www.nass.usda.gov/QuickStats/

Valentine, G. (1993). (Hetero)Sexing Space: Lesbian perceptions and experiences of everyday spaces. *Environment and Planning D, 11,* 395–413.

Valentine, G. (1998). Sticks and stones may break my bones: A personal geography of harassment. *Antipode, 30,* 305–332.

Valentine, G. (1999). A corporeal geography of consumption. *Environment and Planning D, 17,* 329–341.

Valentine, G. (2001). *Social geographies: Space and society.* Harlow, UK: Prentice Hall.

Vasishth, A., & Sloane, D. (2002). Returning to ecology: An ecosystem approach to understanding the city. In

M. Dear (Ed.), *From Chicago to L.A.: Making sense of urban theory* (pp. 343–366). Thousand Oaks, CA: Sage.

Vernon, R. (1966). International investment and international trade in the product cycle. *Quarterly Journal of Economics, 80,* 190–207.

Von Braun, J., Webb, P., & Tesfaye, T. (1999). *Famine in Africa: Causes, responses, and prevention.* Washington, DC: International Food Policy Research Institute.

Wagner, P., & Mikesell, M. (Eds.). (1962). *Readings in cultural geography.* Chicago: University of Chicago Press.

Walker, R. (1985). Is there a service economy? The changing capitalist division of labor. *Science & Society, 49,* 42–83.

Waller, L., & Gotway, C. (2004). *Applied spatial statistics for public health data.* Hoboken, NJ: John Wiley.

Wallerstein, I. (1974). *The modern world-system: Capitalist agriculture and the origins of the European world-economy in the sixteenth century.* London: Academic Press.

Wallerstein, I. (1979). *The capitalist world economy.* Cambridge, UK: Cambridge University Press.

Warf, B. (1993). Post-modernism and the localities debate: Ontological questions and epistemological implications. *Tijdschrift voor Economische en Sociale Geografie, 84,* 162–168.

Warf, B. (1995). Telecommunications and the changing geographies of knowledge transmission in the late 20th century. *Urban Studies, 32,* 361–378.

Warner, M. (Ed.). (1993). *Fear of a queer planet: Queer politics and social theory.* Minneapolis: University of Minnesota Press.

Warren, S. (1993). "This heaven gives me migraines": The problems and promise of landscapes of leisure. In J. Duncan & D. Ley (Eds.), *Place/culture/representation* (pp. 173–186). London: Routledge.

Waters, M. (1995). *Globalization.* London: Routledge.

Watts, M. (1995). A new deal in emotions: Theory and practice and the crisis of development. In J. Crush (Ed.), *Power of development* (pp. 44–62). New York: Routledge.

Watts, M. (2000). Political ecology. In T. Barnes & E. Sheppard (Eds.), *A companion to economic geography* (pp. 257–274). Oxford, UK: Blackwell.

Weatherby, J., Evans, E., Gooden, R., Long, D., Reed, I., & Carter, O. (2004). *The other world: Issues and politics of the developing world.* New York: Longman.

Weber, M. (1978). *Economy and society.* Berkeley: University of California Press.

Webster, F. (2002). *Theories of the information society.* London: Routledge.

Webster, G. (1997). Geography and the decennial task of redistricting. *Journal of Geography, 96,* 61–68.

Webster, G. (1997). The potential impact of recent Supreme Court decisions on the use of race and ethnicity in the redistricting process. *Cities, 14*(1), 13–19.

Weeks, J. (1985). *Sexuality and its discontents: Myths, meanings, and modern sexualities.* London: Routledge and Kegan Paul.

Weeks, W. (1997). *Beyond the ark: Tools for an ecosystem approach to conservation.* Washington, DC: Island Press.

Weinberg, A. (1982). On the relation between information and energy systems: A family of Maxwell's demons. *Interdisciplinary Science Review, 7,* 47–52.

Weiss, A., Lutkus, A., Hildebrant, B., & Johnson, M. (2002). *The nation's report card: Geography 2001* (Office of Educational Research and Improvement, National Center for Education Statistics, NCES 2002–484). Washington, DC: U.S. Department of Education.

Western, J. (1997). *Outcast Cape Town* (2nd ed.). Berkeley: University of California Press.

Whatmore, S. (2002). *Hybrid geographies: Natures, cultures, spaces.* London: Sage.

Whittlesey, D. (1929). Sequent occupance. *Annals of the Association of American Geographers, 19,* 162–166.

Wilk, R. (2002). Consumption, human needs, and global environmental change. *Global Environmental Change, 12,* 5–13.

Williams, C. (1997). *Consumer services and economic development.* London: Routledge.

Williams, C. (2003). Nationalism in a democratic context. In J. Agnew, K. Mitchell, & G. Toal (Eds.), *A companion to political geography* (pp. 356–377). Oxford, UK: Blackwell.

Williams, C., & Millington, A. (2004). The diverse and contested meanings of sustainable development. *Geographical Journal, 170,* 99–104.

Williams, P. (1982). Restructuring urban managerialism: Toward a political economy of urban allocation. *Environment and Planning A, 14,* 95–105.

Williams, R. (1976). *Keywords: A vocabulary of culture and society.* New York: Oxford University Press.

Williams, R. (1977). *Marxism and literature.* Oxford, UK: Oxford University Press.

Wills, J. (1996). Geographies of trade unionism: Translating traditions across space and time. *Antipode, 28,* 352–378.

Wilson, B. (2000). *Race and place in Birmingham: The civil rights and neighborhood movements.* Lanham, MD: Rowman & Littlefield.

Wilson, D. (1987). Urban revitalization on the upper west side of Manhattan: An urban managerialist assessment. *Economic Geography, 63,* 35–47.

Wilson, D. (1988). Toward a revised urban managerialism: Local managers and community development block grants. *Political Geography Quarterly, 8,* 21–41.

Wilson, W. (1987). *The truly disadvantaged: The inner city, the underclass, and public policy.* Chicago: University of Chicago Press.

Wilson, W. (1996). *When work disappears: The world of the new urban poor.* New York: Knopf.

Winner, L. (1980). Do artifacts have politics? *Daedalus, 109*(1), 121–136.

Wirth, L. (1938). Urbanism as a way of life. *American Journal of Sociology, 44,* 1–24.

Wolch, J., & Emel, J. (Eds.). (1998). *Animal geographies: Place, politics, and identity in the nature–culture borderlands.* London: Verso.

Wolf, E. (1966). *Peasants.* Englewood Cliffs, NJ: Prentice Hall.

Wolf, E. (1982). *Europe and the people without history.* Berkeley: University of California Press.

Wolff, R., & Resnick, S. (1987). *Economics: Marxian vs. neoclassical.* Baltimore, MD: Johns Hopkins University Press.

Women and Geography Study Group. (1997). *Feminist geographies: Explorations in diversity and difference.* Harlow, UK: Addison-Wesley Longman.

Women and Geography Study Group. (2004). *Gender and geography reconsidered.* Glasgow, UK: Author.

Wong, D. (2003). The modifiable areal unit problem (MAUP). In D. Janelle, B. Warf, & K. Hansen (Eds.), *WorldMinds: Geographical perspectives on 100 problems* (pp. 571–575). Dordrecht, Netherlands: Kluwer Academic.

Wood, A. (1998, August). Making sense of urban entrepreneurialism. *Scottish Geographical Magazine,* pp. 120–123.

Wood, D. (1992). *The power of maps.* New York: Guilford.

Wood, J. (1997). Vietnamese place making in northern Virginia. *The Geographical Review, 87,* 58–72.

Woods, M. (2005). *Rural geography.* Thousand Oaks, CA: Sage.

World Bank. (1991). *Export processing zones.* Washington, DC: Author.

World Bank. (2002). *World Development Report 2002: Building institutions for markets.* Washington, DC: Author.

Wu, F., & Webber, K. (2004). The rise of "foreign gated communities" in Beijing: Between economic globalization and local institutions. *Cities, 21,* 203–213.

Yapa, L. (1996). What causes poverty? A postmodern view. *Annals of the Association of American Geographers, 86,* 707–728.

Yaukey, D., & Anderton, D. (2001). *Demography: The study of human population* (2nd ed.). Prospect Heights, IL: Waveland.

Young, I. (1990). *Justice and the politics of difference.* Princeton, NJ: Princeton University Press.

Young, R. (1995). *Colonial desire: Hybridity in theory, culture, and race.* London: Routledge.

Zelinsky, W. (1983). Nationalism in the American place–name cover. *Names, 31,* 1–28.

Zelinsky, W. (2001). *The enigma of ethnicity: Another American dilemma.* Iowa City: University of Iowa Press.

Zhang, J., & Goodchild, M. (2002). *Uncertainty in geographical information.* London: Taylor & Francis.

Zimmerer, K., & Bassett, T. (Eds.). (2003). *Political ecology: An integrative approach to geography and environment–development studies.* New York: Guilford.

Zook, M. A. (2005). *The geography of the Internet industry: Venture capital, dot-coms, and local knowledge.* New York: Blackwell.

Index